Some Physical Constants

Quantity	Symbol	Value[a]
Atomic mass unit	u	$1.660\ 538\ 782\ (83) \times 10^{-27}$ kg
		$931.494\ 028\ (23)$ MeV/c^2
Avogadro's number	N_A	$6.022\ 141\ 79\ (30) \times 10^{23}$ particles/mol
Bohr magneton	$\mu_B = \dfrac{e\hbar}{2m_e}$	$9.274\ 009\ 15\ (23) \times 10^{-24}$ J/T
Bohr radius	$a_0 = \dfrac{\hbar^2}{m_e e^2 k_e}$	$5.291\ 772\ 085\ 9\ (36) \times 10^{-11}$ m
Boltzmann's constant	$k_B = \dfrac{R}{N_A}$	$1.380\ 650\ 4\ (24) \times 10^{-23}$ J/K
Compton wavelength	$\lambda_C = \dfrac{h}{m_e c}$	$2.426\ 310\ 217\ 5\ (33) \times 10^{-12}$ m
Coulomb constant	$k_e = \dfrac{1}{4\pi\epsilon_0}$	$8.987\ 551\ 788\ldots \times 10^9$ N·m^2/C^2 (exact)
Deuteron mass	m_d	$3.343\ 583\ 20\ (17) \times 10^{-27}$ kg
		$2.013\ 553\ 212\ 724\ (78)$ u
Electron mass	m_e	$9.109\ 382\ 15\ (45) \times 10^{-31}$ kg
		$5.485\ 799\ 094\ 3\ (23) \times 10^{-4}$ u
		$0.510\ 998\ 910\ (13)$ MeV/c^2
Electron volt	eV	$1.602\ 176\ 487\ (40) \times 10^{-19}$ J
Elementary charge	e	$1.602\ 176\ 487\ (40) \times 10^{-19}$ C
Gas constant	R	$8.314\ 472\ (15)$ J/mol·K
Gravitational constant	G	$6.674\ 28\ (67) \times 10^{-11}$ N·m^2/kg^2
Neutron mass	m_n	$1.674\ 927\ 211\ (84) \times 10^{-27}$ kg
		$1.008\ 664\ 915\ 97\ (43)$ u
		$939.565\ 346\ (23)$ MeV/c^2
Nuclear magneton	$\mu_n = \dfrac{e\hbar}{2m_p}$	$5.050\ 783\ 24\ (13) \times 10^{-27}$ J/T
Permeability of free space	μ_0	$4\pi \times 10^{-7}$ T·m/A (exact)
Permittivity of free space	$\epsilon_0 = \dfrac{1}{\mu_0 c^2}$	$8.854\ 187\ 817\ldots \times 10^{-12}$ C^2/N·m^2 (exact)
Planck's constant	h	$6.626\ 068\ 96\ (33) \times 10^{-34}$ J·s
	$\hbar = \dfrac{h}{2\pi}$	$1.054\ 571\ 628\ (53) \times 10^{-34}$ J·s
Proton mass	m_p	$1.672\ 621\ 637\ (83) \times 10^{-27}$ kg
		$1.007\ 276\ 466\ 77\ (10)$ u
		$938.272\ 013\ (23)$ MeV/c^2
Rydberg constant	R_H	$1.097\ 373\ 156\ 852\ 7\ (73) \times 10^7$ m^{-1}
Speed of light in vacuum	c	$2.997\ 924\ 58 \times 10^8$ m/s (exact)

Note: These constants are the values recommended in 2006 by CODATA, based on a least-squares adjustment of data from different measurements. For a more complete list, see P. J. Mohr, B. N. Taylor, and D. B. Newell, "CODATA Recommended Values of the Fundamental Physical Constants: 2006." *Rev. Mod. Phys.* **80**:2, 633–730, 2008.

[a]The numbers in parentheses for the values represent the uncertainties of the last two digits.

Solar System Data

Body	Mass (kg)	Mean Radius (m)	Period (s)	Mean Distance from the Sun (m)
Mercury	3.30×10^{23}	2.44×10^6	7.60×10^6	5.79×10^{10}
Venus	4.87×10^{24}	6.05×10^6	1.94×10^7	1.08×10^{11}
Earth	5.97×10^{24}	6.37×10^6	3.156×10^7	1.496×10^{11}
Mars	6.42×10^{23}	3.39×10^6	5.94×10^7	2.28×10^{11}
Jupiter	1.90×10^{27}	6.99×10^7	3.74×10^8	7.78×10^{11}
Saturn	5.68×10^{26}	5.82×10^7	9.29×10^8	1.43×10^{12}
Uranus	8.68×10^{25}	2.54×10^7	2.65×10^9	2.87×10^{12}
Neptune	1.02×10^{26}	2.46×10^7	5.18×10^9	4.50×10^{12}
Pluto[a]	1.25×10^{22}	1.20×10^6	7.82×10^9	5.91×10^{12}
Moon	7.35×10^{22}	1.74×10^6	—	—
Sun	1.989×10^{30}	6.96×10^8	—	—

[a]In August 2006, the International Astronomical Union adopted a definition of a planet that separates Pluto from the other eight planets. Pluto is now defined as a "dwarf planet" (like the asteroid Ceres).

Physical Data Often Used

Average Earth–Moon distance	3.84×10^8 m
Average Earth–Sun distance	1.496×10^{11} m
Average radius of the Earth	6.37×10^6 m
Density of air (20°C and 1 atm)	1.20 kg/m^3
Density of air (0°C and 1 atm)	1.29 kg/m^3
Density of water (20°C and 1 atm)	1.00×10^3 kg/m^3
Free-fall acceleration	9.80 m/s^2
Mass of the Earth	5.97×10^{24} kg
Mass of the Moon	7.35×10^{22} kg
Mass of the Sun	1.99×10^{30} kg
Standard atmospheric pressure	1.013×10^5 Pa

Note: These values are the ones used in the text.

Some Prefixes for Powers of Ten

Power	Prefix	Abbreviation	Power	Prefix	Abbreviation
10^{-24}	yocto	y	10^1	deka	da
10^{-21}	zepto	z	10^2	hecto	h
10^{-18}	atto	a	10^3	kilo	k
10^{-15}	femto	f	10^6	mega	M
10^{-12}	pico	p	10^9	giga	G
10^{-9}	nano	n	10^{12}	tera	T
10^{-6}	micro	μ	10^{15}	peta	P
10^{-3}	milli	m	10^{18}	exa	E
10^{-2}	centi	c	10^{21}	zetta	Z
10^{-1}	deci	d	10^{24}	yotta	Y

FIFTH EDITION

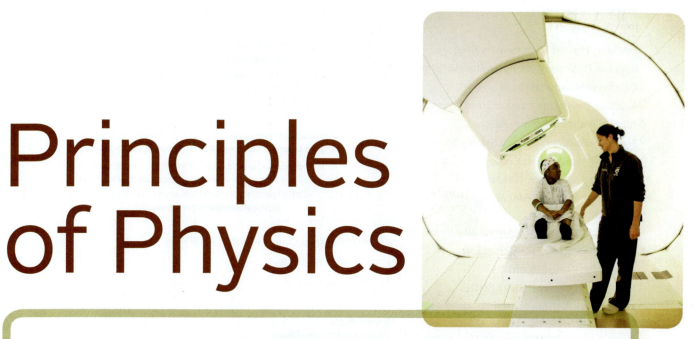

Principles of Physics

A CALCULUS-BASED TEXT

Volume 2

Raymond A. Serway

Emeritus, James Madison University

John W. Jewett, Jr.

Emeritus, California State Polytechnic University, Pomona

BROOKS/COLE
CENGAGE Learning

Australia • Brazil • Japan • Korea • Mexico • Singapore • Spain • United Kingdom • United States

BROOKS/COLE
CENGAGE Learning

Principles of Physics, Fifth Edition, Volume 2
Raymond A. Serway
John W. Jewett, Jr.

Publisher, Physical Sciences: Mary Finch

Publisher, Physics and Astronomy: Charles Hartford

Development Editor: Ed Dodd

Associate Development Editor: Brandi Kirksey

Editorial Assistant: Brendan Killion

Senior Media Editor: Rebecca Berardy Schwartz

Marketing Manager: Jack Cooney

Marketing Coordinator: Julie Stefani

Marketing Communications Manager:
Darlene Macanan

Senior Content Project Manager: Cathy Brooks

Senior Art Director: Cate Rickard Barr

Print Buyer: Diane Gibbons

Manufacturing Planner: Doug Bertke

Rights Acquisition Specialist:
Shalice Shah-Caldwell

Production Service: MPS Limited, a Macmillan
Company

Compositor: MPS Limited, a Macmillan Company

Text Designer: Brian Salisbury

Cover Designer: Brian Salisbury

Cover Images:

• Child prepares for proton therapy at the Roberts
Proton Therapy Center in Philadelphia, PA:
Copyright © 2011 Ed Cunicelli

• A full moon rises through an iridescent len-
ticular cloud at Mt. Rainier: © Caren Brinkema/
Science Faction/Getty Images

• Supernova remnant: NASA

• Infinity Bridge on the River Tees: © Paul Downing/
Flickr Select/Getty Images

For product information and technology assistance, contact us at
Cengage Learning Customer & Sales Support, 1-800-354-9706

For permission to use material from this text or product,
submit all requests online at **www.cengage.com/permissions**.
Further permissions questions can be emailed to
permissionrequest@cengage.com

Library of Congress Control Number: 2011937185

ISBN-13: 978-1-133-11028-6

ISBN-10: 1-133-11028-2

Brooks/Cole
20 Channel Center Street
Boston, MA 02210
USA

Cengage Learning is a leading provider of customized learning solutions with
office locations around the globe, including Singapore, the United Kingdom,
Australia, Mexico, Brazil and Japan. Locate your local office at
international.cengage.com/region

Cengage Learning products are represented in Canada by Nelson Education, Ltd.

For your course and learning solutions, visit **www.cengage.com**.

Purchase any of our products at your local college store or at our preferred
online store **www.cengagebrain.com**.

Instructors: Please visit **login.cengage.com** and log in to access
instructor-specific resources.

We dedicate this book to our wives Elizabeth and
Lisa and all our children and grandchildren for
their loving understanding when we spent time on
writing instead of being with them.

Printed in the United States of America
1 2 3 4 5 6 7 15 14 13 12 11

Welcome to your MCAT Test Preparation Guide

The **MCAT Test Preparation Guide** makes your copy of *Principles of Physics,* **Fifth Edition**, the most comprehensive MCAT study tool and classroom resource in introductory physics. The grid, which begins below and continues on the next two pages, outlines twelve concept-based **study courses** for the physics part of your MCAT exam. Use it to prepare for the MCAT, class tests, and your homework assignments.

Vectors

Skill Objectives: To calculate distance, calculate angles between vectors, calculate magnitudes, and to understand vectors.

Review Plan:

Distance and Angles: Chapter 1
- Section 1.6
- Active Figure 1.4
- Chapter Problem 33

Using Vectors: Chapter 1
- Sections 1.7–1.9
- Quick Quizzes 1.4–1.8
- Examples 1.6–1.8
- Active Figures 1.9, 1.16
- Chapter Problems 34, 35, 43, 44, 47, 51

Motion

Skill Objectives: To understand motion in two dimensions, to calculate speed and velocity, to calculate centripetal acceleration, and acceleration in free fall problems.

Review Plan:

Motion in 1 Dimension: Chapter 2
- Sections 2.1, 2.2, 2.4, 2.6, 2.7
- Quick Quizzes 2.3–2.6
- Examples 2.1, 2.2, 2.4–2.9
- Active Figure 2.12
- Chapter Problems 3, 5,13, 19, 21, 29, 31, 33

Motion in 2 Dimensions: Chapter 3
- Sections 3.1–3.3
- Quick Quizzes 3.2, 3.3
- Examples 3.1–3.4
- Active Figures 3.5, 3.7, 3.10
- Chapter Problems 1, 11, 13

Centripetal Acceleration: Chapter 3
- Sections 3.4, 3.5
- Quick Quizzes 3.4, 3.5
- Example 3.5
- Active Figure 3.14
- Chapter Problems 23, 31

Force

Skill Objectives: To know and understand Newton's laws, to calculate resultant forces and weight.

Review Plan:

Newton's Laws: Chapter 4
- Sections 4.1–4.6
- Quick Quizzes 4.1–4.6
- Example 4.1
- Chapter Problem 7

Resultant Forces: Chapter 4
- Section 4.7
- Quick Quiz 4.7
- Example 4.6
- Chapter Problems 29, 37

Gravity: Chapter 11
- Section 11.1
- Quick Quiz 11.1
- Chapter Problem 5

Equilibrium

Skill Objectives: To calculate momentum and impulse, center of gravity, and torque.

Review Plan:

Momentum: Chapter 8
- Section 8.1
- Quick Quiz 8.2
- Examples 8.2, 8.3

Impulse: Chapter 8
- Sections 8.2–8.4
- Quick Quizzes 8.3, 8.4
- Examples 8.4, 8.6
- Active Figures 8.8, 8.9
- Chapter Problems 5, 9, 11, 17, 21

Torque: Chapter 10
- Sections 10.5, 10.6
- Quick Quiz 10.7
- Example 10.8
- Chapter Problems 23, 30

Work

Skill Objectives: To calculate friction, work, kinetic energy, power, and potential energy.

Review Plan:

Friction: Chapter 5
- Section 5.1
- Quick Quizzes 5.1, 5.2

Work: Chapter 6
- Section 6.2
- Chapter Problems 3, 5

Kinetic Energy: Chapter 6
- Section 6.5
- Example 6.6

Power: Chapter 7
- Section 7.6
- Chapter Problem 29

Potential Energy: Chapters 6, 7
- Sections 6.6, 7.2
- Quick Quiz 6.6
Chapter 7
- Chapter Problem 3

Waves

Skill Objectives: To understand interference of waves, to calculate basic properties of waves, properties of springs, and properties of pendulums.

Review Plan:

Wave Properties: Chapters 12, 13
- Sections 12.1, 12.2, 13.1, 13.2
- Quick Quiz 13.1
- Examples 12.1, 13.2
- Active Figures 12.1, 12.2, 12.6, 12.8, 12.11
Chapter 13
- Problem 7

Pendulum: Chapter 12
- Sections 12.4, 12.5
- Quick Quizzes 12.5, 12.6
- Example 12.5
- Active Figure 12.13
- Chapter Problem 35

Interference: Chapter 14
- Sections 14.1, 14.2
- Quick Quiz 14.1
- Active Figures 14.1–14.3

Matter

Skill Objectives: To calculate density, pressure, specific gravity, and flow rates.

Review Plan:

Density: Chapters 1, 15
- Sections 1.1, 15.2

Pressure: Chapter 15
- Sections 15.1–15.4
- Quick Quizzes 15.1–15.4
- Examples 15.1, 15.3
- Chapter Problems 2, 11, 23, 27, 31

Flow Rates: Chapter 15
- Section 15.6
- Quick Quiz 15.5

Sound

Skill Objectives: To understand interference of waves, calculate properties of waves, the speed of sound, Doppler shifts, and intensity.

Review Plan:

Sound Properties: Chapters 13, 14
- Sections 13.2, 13.3, 13.6, 13.7, 14.3
- Quick Quizzes 13.2, 13.4, 13.7
- Example 14.3
- Active Figures 13.6, 13.7, 13.9, 13.22, 13.23
Chapter 13
- Problems 11, 15, 28, 34, 41, 45
Chapter 14
- Problem 27

Interference/Beats: Chapter 14
- Sections 14.1, 14.5
- Quick Quiz 14.6
- Active Figures 14.1–14.3, 14.12
- Chapter Problems 9, 45, 46

Light

Skill Objectives: To understand mirrors and lenses, to calculate the angles of reflection, to use the index of refraction, and to find focal lengths.

Review Plan:

Reflection: Chapter 25
- Sections 25.1–25.3
- Example 25.1
- Active Figure 25.5

Refraction: Chapter 25
- Sections 25.4, 25.5
- Quick Quizzes 25.2–25.5
- Example 25.2
- Chapter Problems 8, 16

Mirrors and Lenses: Chapter 26
- Sections 26.1–26.4
- Quick Quizzes 26.1–26.6
- Thinking Physics 26.2
- Examples 26.1–26.5
- Active Figures 26.2, 26.25
- Chapter Problems 27, 30, 33, 37

Electrostatics

Skill Objectives: To understand and calculate the electric field, the electrostatic force, and the electric potential.

Review Plan:

Coulomb's Law: Chapter 19
- Sections 19.2–19.4
- Quick Quiz 19.1–19.3
- Examples 19.1, 19.2
- Active Figure 19.7
- Chapter Problems 3, 9

Electric Field: Chapter 19
- Sections 19.5, 19.6
- Quick Quizzes 19.4, 19.5
- Active Figures 19.11, 19.20, 19.22

Potential: Chapter 20
- Sections 20.1–20.3
- Examples 20.1, 20.2
- Active Figure 20.8
- Chapter Problems 3, 5, 8, 11

Circuits

Skill Objectives: To understand and calculate current, resistance, voltage, and power, and to use circuit analysis.

Review Plan:

Ohm's Law: Chapter 21
- Sections 21.1, 21.2
- Quick Quizzes 21.1, 21.2
- Examples 21.1, 21.2
- Chapter Problem 8

Power and Energy: Chapter 21
- Section 21.5
- Quick Quiz 21.4
- Examples 21.4
- Active Figure 21.11
- Chapter Problems 21, 25, 31

Circuits: Chapter 21
- Sections 21.6–21.8
- Quick Quizzes 21.5–21.7
- Examples 21.6–21.8
- Active Figures 21.14, 21.15, 21.17
- Chapter Problems 31, 39, 47

Atoms

Skill Objectives: To understand decay processes and nuclear reactions and to calculate half-life.

Review Plan:

Atoms: Chapters, 11, 29
- Section 11.5
- Sections 29.1–29.6
Chapter 11
- Problems 37–43, 61

Decays: Chapter 30
- Sections 30.3, 30.4
- Quick Quizzes 30.3–30.6
- Examples 30.3–30.6
- Active Figures 30.8–30.11, 30.13, 30.14
- Chapter Problems 18, 23, 25

Nuclear Reactions: Chapter 30
- Section 30.5
- Active Figure 30.18
- Chapter Problems 32, 35

Contents

About the Authors

Raymond A. Serway received his doctorate at Illinois Institute of Technology and is Professor Emeritus at James Madison University. In 2011, he was awarded with an honorary doctorate degree from his alma mater, Utica College. He received the 1990 Madison Scholar Award at James Madison University, where he taught for 17 years. Dr. Serway began his teaching career at Clarkson University, where he conducted research and taught from 1967 to 1980. He was the recipient of the Distinguished Teaching Award at Clarkson University in 1977 and the Alumni Achievement Award from Utica College in 1985. As Guest Scientist at the IBM Research Laboratory in Zurich, Switzerland, he worked with K. Alex Müller, 1987 Nobel Prize recipient. Dr. Serway also was a visiting scientist at Argonne National Laboratory, where he collaborated with his mentor and friend, the late Dr. Sam Marshall. Dr. Serway is the coauthor of *College Physics,* ninth edition; *Physics for Scientists and Engineers,* eighth edition; *Essentials of College Physics; Modern Physics,* third edition; and the high school textbook *Physics,* published by Holt McDougal. In addition, Dr. Serway has published more than 40 research papers in the field of condensed matter physics and has given more than 60 presentations at professional meetings. Dr. Serway and his wife Elizabeth enjoy traveling, playing golf, fishing, gardening, singing in the church choir, and especially spending quality time with their four children, nine grandchildren, and a recent great grandson.

John W. Jewett, Jr. earned his undergraduate degree in physics at Drexel University and his doctorate at Ohio State University, specializing in optical and magnetic properties of condensed matter. Dr. Jewett began his academic career at Richard Stockton College of New Jersey, where he taught from 1974 to 1984. He is currently Emeritus Professor of Physics at California State Polytechnic University, Pomona. Throughout his teaching career, Dr. Jewett has been active in promoting effective physics education. In addition to receiving four National Science Foundation grants, he helped found and direct the Southern California Area Modern Physics Institute (SCAMPI) and Science IMPACT (Institute for Modern Pedagogy and Creative Teaching). Dr. Jewett's honors include the Stockton Merit Award at Richard Stockton College in 1980, selection as Outstanding Professor at California State Polytechnic University for 1991–1992, and the Excellence in Undergraduate Physics Teaching Award from the American Association of Physics Teachers (AAPT) in 1998. In 2010, he received an Alumni Lifetime Achievement Award from Drexel University in recognition of his contributions in physics education. He has given more than 100 presentations both domestically and abroad, including multiple presentations at national meetings of the AAPT. Dr. Jewett is the author of *The World of Physics: Mysteries, Magic, and Myth,* which provides many connections between physics and everyday experiences. In addition to his work as the coauthor for *Principles of Physics* he is also the coauthor on *Physics for Scientists and Engineers,* eighth edition, as well as *Global Issues,* a four-volume set of instruction manuals in integrated science for high school. Dr. Jewett enjoys playing keyboard with his all-physicist band, traveling, underwater photography, learning foreign languages, and collecting antique quack medical devices that can be used as demonstration apparatus in physics lectures. Most importantly, he relishes spending time with his wife Lisa and their children and grandchildren.

Preface

Principles of Physics is designed for a one-year introductory calculus-based physics course for engineering and science students and for premed students taking a rigorous physics course. This fifth edition contains many new pedagogical features—most notably, an integrated Web-based learning system and a structured problem-solving strategy that uses a modeling approach. Based on comments from users of the fourth edition and reviewers' suggestions, a major effort was made to improve organization, clarity of presentation, precision of language, and accuracy throughout.

This textbook was initially conceived because of well-known problems in teaching the introductory calculus-based physics course. The course content (and hence the size of textbooks) continues to grow, while the number of contact hours with students has either dropped or remained unchanged. Furthermore, traditional one-year courses cover little if any physics beyond the 19th century.

In preparing this textbook, we were motivated by the spreading interest in reforming the teaching and learning of physics through physics education research (PER). One effort in this direction was the Introductory University Physics Project (IUPP), sponsored by the American Association of Physics Teachers and the American Institute of Physics. The primary goals and guidelines of this project are to:

- Reduce course content following the "less may be more" theme;
- Incorporate contemporary physics naturally into the course;
- Organize the course in the context of one or more "story lines";
- Treat all students equitably.

Recognizing a need for a textbook that could meet these guidelines several years ago, we studied the various proposed IUPP models and the many reports from IUPP committees. Eventually, one of us (RAS) became actively involved in the review and planning of one specific model, initially developed at the U.S. Air Force Academy, entitled "A Particles Approach to Introductory Physics." An extended visit at the Academy was spent working with Colonel James Head and Lt. Col. Rolf Enger, the primary authors of the Particles model, and other members of that department. This most useful collaboration was the starting point of this project.

The other author (JWJ) became involved with the IUPP model called "Physics in Context," developed by John Rigden (American Institute of Physics), David Griffiths (Oregon State University), and Lawrence Coleman (University of Arkansas at Little Rock). This involvement led to National Science Foundation (NSF) grant support for the development of new contextual approaches and eventually to the contextual overlay that is used in this book and described in detail later in the Preface.

The combined IUPP approach in this book has the following features:

- It is an evolutionary approach (rather than a revolutionary approach), which should meet the current demands of the physics community.
- It deletes many topics in classical physics (such as alternating current circuits and optical instruments) and places less emphasis on rigid object motion, optics, and thermodynamics.
- Some topics in contemporary physics, such as fundamental forces, special relativity, energy quantization, and the Bohr model of the hydrogen atom, are introduced early in the textbook.
- A deliberate attempt is made to show the unity of physics and the global nature of physics principles.
- As a motivational tool, the textbook connects applications of physics principles to interesting biomedical situations, social issues, natural phenomena, and technological advances.

Other efforts to incorporate the results of physics education research have led to several of the features in this textbook described below. These include Quick Quizzes, Objective Questions, Pitfall Preventions, **What If?** features in worked examples, the use of energy bar charts, the modeling approach to problem solving, and the global energy approach introduced in Chapter 7.

Objectives

This introductory physics textbook has two main objectives: to provide the student with a clear and logical presentation of the basic concepts and principles of physics, and to strengthen an understanding of the concepts and principles through a broad range of interesting applications to the real world. To meet these objectives, we have emphasized sound physical arguments and problem-solving methodology. At the same time, we have attempted to motivate the student through practical examples that demonstrate the role of physics in other disciplines, including engineering, chemistry, and medicine.

Changes in the Fifth Edition

A number of changes and improvements have been made in the fifth edition of this text. Many of these are in response to recent findings in physics education research and to comments and suggestions provided by the reviewers of the manuscript and instructors using the first four editions. The following represent the major changes in the fifth edition:

New Contexts. The context overlay approach is described below under "Organization." The fifth edition introduces two new Contexts: for Chapter 15, "Heart Attacks," and for Chapters 22–23, "Magnetism in Medicine." Both of these new Contexts are aimed at applying the principles of physics to the biomedical field.

In the "Heart Attacks" Context, we study the flow of fluids through a pipe, as an analogy to the flow of blood through blood vessels in the human body. Various details of the blood flow are related to the dangers of cardiovascular disease. In addition, we discuss new developments in the study of blood flow and heart attacks using nanoparticles and computer imaging.

The "Magnetism in Medicine" Context explores the application of the principles of electromagnetism to diagnostic and therapeutic procedures in medicine. We begin by looking at historical uses of magnetism, including several "quack" medical devices. More modern applications include remote magnetic navigation in cardiac catheter ablation procedures for atrial fibrillation, transcranial magnetic stimulation for the treatment of depression, and magnetic resonance imaging as a diagnostic tool.

Worked Examples. All in-text worked examples have been recast and are now presented in a two-column format to better reinforce physical concepts. The left column shows textual information that describes the steps for solving the problem. The right column shows the mathematical manipulations and results of taking these steps. This layout facilitates matching the concept with its mathematical execution and helps students organize their work. The examples closely follow the General Problem-Solving Strategy introduced in Chapter 1 to reinforce effective problem-solving habits. In almost all cases, examples are solved symbolically until the end, when numerical values are substituted into the final symbolic result. This procedure allows students to analyze the symbolic result to see how the result depends on the parameters in the problem, or to take limits to test the final result for correctness. Most worked examples in the text may be assigned for homework in Enhanced WebAssign. A sample of a worked example can be found on the next page.

ENHANCED **WebAssign** Most worked examples are also available to be assigned as interactive examples in the Enhanced WebAssign homework management system.

Each solution has been written to closely follow the General Problem-Solving Strategy as outlined on pages 25–26 in Chapter 1, so as to reinforce good problem-solving habits.

Example 6.6 | A Block Pulled on a Frictionless Surface

A 6.0-kg block initially at rest is pulled to the right along a frictionless, horizontal surface by a constant horizontal force of 12 N. Find the block's speed after it has moved 3.0 m.

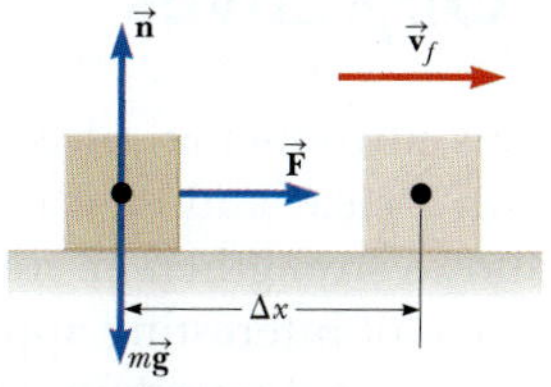

Figure 6.14 (Example 6.6) A block pulled to the right on a frictionless surface by a constant horizontal force.

SOLUTION

Conceptualize Figure 6.14 illustrates this situation. Imagine pulling a toy car across a table with a horizontal rubber band attached to the front of the car. The force is maintained constant by ensuring that the stretched rubber band always has the same length.

Categorize We could apply the equations of kinematics to determine the answer, but let us practice the energy approach. The block is the system, and three external forces act on the system. The normal force balances the gravitational force on the block, and neither of these vertically acting forces does work on the block because their points of application are horizontally displaced.

Analyze The net external force acting on the block is the horizontal 12-N force.

Use the work–kinetic energy theorem for the block, noting that its initial kinetic energy is zero:

$$W_{\text{ext}} = K_f - K_i = \tfrac{1}{2}mv_f^2 - 0 = \tfrac{1}{2}mv_f^2$$

Each step of the solution is detailed in a two-column format. The left column provides an explanation for each mathematical step in the right column, to better reinforce the physical concepts.

Solve for v_f and use Equation 6.1 for the work done on the block by $\vec{\mathbf{F}}$:

$$v_f = \sqrt{\frac{2W_{\text{ext}}}{m}} = \sqrt{\frac{2F\Delta x}{m}}$$

Substitute numerical values:

$$v_f = \sqrt{\frac{2(12\text{ N})(3.0\text{ m})}{6.0\text{ kg}}} = 3.5\text{ m/s}$$

Finalize It would be useful for you to solve this problem again by modeling the block as a particle under a net force to find its acceleration and then as a particle under constant acceleration to find its final velocity.

What If? Suppose the magnitude of the force in this example is doubled to $F' = 2F$. The 6.0-kg block accelerates to 3.5 m/s due to this applied force while moving through a displacement $\Delta x'$. How does the displacement $\Delta x'$ compare with the original displacement Δx?

Answer If we pull harder, the block should accelerate to a given speed in a shorter distance, so we expect that $\Delta x' < \Delta x$. In both cases, the block experiences the same change in kinetic energy ΔK. Mathematically, from the work–kinetic energy theorem, we find that

$$W_{\text{ext}} = F'\Delta x' = \Delta K = F\Delta x$$

$$\Delta x' = \frac{F}{F'}\Delta x = \frac{F}{2F}\Delta x = \tfrac{1}{2}\Delta x$$

and the distance is shorter as suggested by our conceptual argument.

What If? statements appear in about 1/3 of the worked examples and offer a variation on the situation posed in the text of the example. For instance, this feature might explore the effects of changing the conditions of the situation, determine what happens when a quantity is taken to a particular limiting value, or question whether additional information can be determined about the problem situation. This feature encourages students to think about the results of the example and assists in conceptual understanding of the principles.

The final result is symbolic; numerical values are substituted into the final result.

Line-by-Line Revision of the Questions and Problems Set. For the Fifth Edition, the authors reviewed each question and problem and incorporated revisions designed to improve both readability and assignability. To make problems clearer to both students and instructors, this extensive process involved editing problems for clarity, editing for length, adding figures where appropriate, and introducing better problem architecture by breaking up problems into clearly defined parts.

Data from Enhanced WebAssign Used to Improve Questions and Problems. As part of the full-scale analysis and revision of the questions and problems sets, the authors utilized extensive user data gathered by WebAssign, from both instructors who assigned

and students who worked on problems from previous editions of *Principles of Physics*. These data helped tremendously, indicating when the phrasing in problems could be clearer, thus providing guidance on how to revise problems so that they are more easily understandable for students and more easily assignable by instructors in Enhanced WebAssign. Finally, the data were used to ensure that the problems most often assigned were retained for this new edition. In each chapter's problems set, the top quartile of problems assigned in Enhanced WebAssign have blue-shaded problem numbers for easy identification, allowing professors to quickly and easily find the most popular problems assigned in Enhanced WebAssign.

To provide an idea of the types of improvements that were made to the problems, here is a problem from the fourth edition, followed by the problem as it now appears in the fifth edition, with explanations of how the problems were improved.

Problem from the Fourth Edition . . .

35. (a) Consider an extended object whose different portions have different elevations. Assume the free-fall acceleration is uniform over the object. Prove that the gravitational potential energy of the object–Earth system is given by $U_g = Mgy_{CM}$, where M is the total mass of the object and y_{CM} is the elevation of its center of mass above the chosen reference level. (b) Calculate the gravitational potential energy associated with a ramp constructed on level ground with stone with density 3 800 kg/m³ and every-where 3.60 m wide (Fig. P8.35). In a side view, the ramp appears as a right triangle with height 15.7 m at the top end and base 64.8 m.

Figure P8.35

. . . As revised for the Fifth Edition:

37. Explorers in the jungle find an ancient monument in the shape of a large isosceles triangle as shown in Figure P8.37. The monument is made from tens of thousands of small stone blocks of density 3 800 kg/m³. The monument is 15.7 m high and 64.8 m wide at its base and is everywhere 3.60 m thick from front to back. Before the monument was built many years ago, all the stone blocks lay on the ground. How much work did laborers do on the blocks to put them in position while building the entire monument? *Note*: The gravitational potential energy of an object–Earth system is given by $U_g = Mgy_{CM}$, where M is the total mass of the object and y_{CM} is the elevation of its center of mass above the chosen reference level.

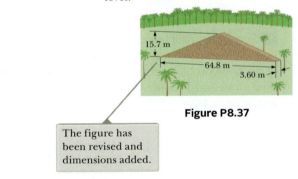

15.7 m
64.8 m
3.60 m

Figure P8.37

> A storyline for the problem is provided.

> The requested quantity is made more personal by asking for work done by humans rather than asking for the gravitational potential energy.

> The expression for the gravitational potential energy is provided, whereas it was requested to be proven in the original. This allows the problem to work better in Enhanced WebAssign.

> The figure has been revised and dimensions added.

Revised Questions Organization. We reorganized the end-of-chapter questions for this new edition. The previous edition's Questions section is now divided into two sections: *Objective Questions* and *Conceptual Questions*.

Objective Questions are multiple-choice, true/false, ranking, or other multiple guess-type questions. Some require calculations designed to facilitate students' familiarity with the equations, the variables used, the concepts the variables represent, and the relationships between the concepts. Others are more conceptual in nature and are designed to encourage conceptual thinking. Objective Questions are also written with the personal response system user in mind, and most of the questions could easily be used in these systems.

Conceptual Questions are more traditional short-answer and essay-type questions that require students to think conceptually about a physical situation.

Problems. The end-of-chapter problems are more numerous in this edition and more varied (in all, over 2 200 problems are given throughout the text). For the convenience of both the student and the instructor, about two-thirds of the problems are keyed to specific sections of the chapter, including Context Connection sections. The remaining problems, labeled "Additional Problems," are not keyed to specific sections. The

BIO icon identifies problems dealing with applications to the life sciences and medicine. Answers to odd-numbered problems are provided at the end of the book. For ease of identification, the problem numbers for straightforward problems are printed in **black**; intermediate-level problem numbers are printed in **blue**; and those of challenging problems are printed in **red**.

New Types of Problems. We have introduced four new problem types for this edition:

Q|C Quantitative/Conceptual problems contain parts that ask students to think both quantitatively and conceptually. An example of a Quantitative/Conceptual problem appears here:

The problem is identified with a **Q|C** icon.

55. Q|C A horizontal spring attached to a wall has a force constant of $k = 850$ N/m. A block of mass $m = 1.00$ kg is attached to the spring and rests on a frictionless, horizontal surface as in Figure P7.55. (a) The block is pulled to a position $x_i = 6.00$ cm from equilibrium and released. Find the elastic potential energy stored in the spring when the block is 6.00 cm from equilibrium and when the block passes through equilibrium. (b) Find the speed of the block as it passes through the equilibrium point. (c) What is the speed of the block when it is at a position $x_i/2 = 3.00$ cm? (d) Why isn't the answer to part (c) half the answer to part (b)?

Parts (a)–(c) of the problem ask for quantitative calculations.

Part (d) asks a conceptual question about the situation.

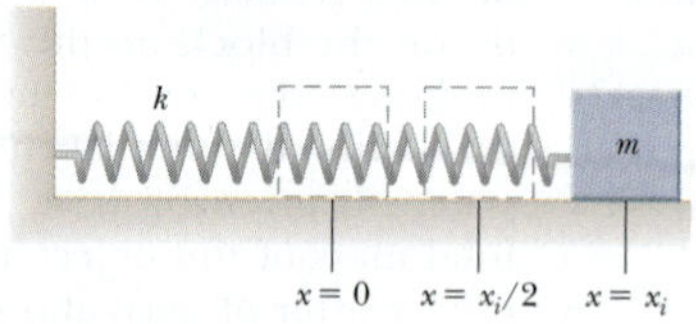

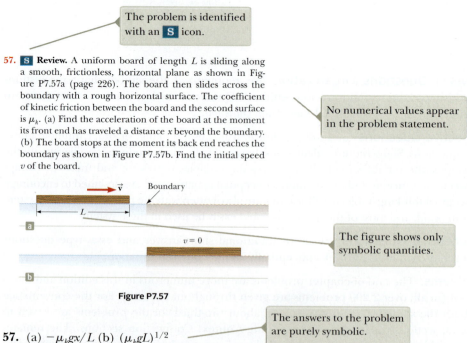

Figure P7.55

S Symbolic problems ask students to solve a problem using only symbolic manipulation. A majority of survey respondents asked specifically for an increase in the number of symbolic problems found in the text because it better reflects the way instructors want their students to think when solving physics problems. An example of a Symbolic problem appears here:

The problem is identified with an **S** icon.

57. S Review. A uniform board of length L is sliding along a smooth, frictionless, horizontal plane as shown in Figure P7.57a (page 226). The board then slides across the boundary with a rough horizontal surface. The coefficient of kinetic friction between the board and the second surface is μ_k. (a) Find the acceleration of the board at the moment its front end has traveled a distance x beyond the boundary. (b) The board stops at the moment its back end reaches the boundary as shown in Figure P7.57b. Find the initial speed v of the board.

No numerical values appear in the problem statement.

The figure shows only symbolic quantities.

Figure P7.57

57. (a) $-\mu_k gx/L$ (b) $(\mu_k gL)^{1/2}$

The answers to the problem are purely symbolic.

GP Guided Problems help students break problems into steps. A physics problem typically asks for one physical quantity in a given context. Often, however, several concepts must be used and a number of calculations are required to obtain that final answer. Many students are not accustomed to this level of complexity and often don't know where to start. A Guided Problem breaks a standard problem into smaller steps, enabling students to grasp all the concepts and strategies required to arrive at a correct solution. Unlike standard physics problems, guidance is often built into the problem statement. Guided Problems are reminiscent of how a student might interact with a professor in an office visit. These problems (there is one in every chapter of the text) help train students to break down complex problems into a series of simpler problems, an essential problem-solving skill. An example of a Guided Problem appears here:

The problem is identified with a **GP** icon.

The goal of the problem is identified.

Analysis begins by identifying the appropriate analysis model.

Students are provided with suggestions for steps to solve the problem.

The calculation associated with the goal is requested.

28. **GP** A uniform beam resting on two pivots has a length $L = 6.00$ m and mass $M = 90.0$ kg. The pivot under the left end exerts a normal force n_1 on the beam, and the second pivot located a distance $\ell = 4.00$ m from the left end exerts a normal force n_2. A woman of mass $m = 55.0$ kg steps onto the left end of the beam and begins walking to the right as in Figure P10.28. The goal is to find the woman's position when the beam begins to tip. (a) What is the appropriate analysis model for the beam before it begins to tip? (b) Sketch a force diagram for the beam, labeling the gravitational and normal forces acting on the beam and placing the woman a distance x to the right of the first pivot, which is the origin. (c) Where is the woman when the normal force n_1 is the greatest? (d) What is n_1 when the beam is about to tip? (e) Use Equation 10.27 to find the value of n_2 when the beam is about to tip. (f) Using the result of part (d) and Equation 10.28, with torques computed around the second pivot, find the woman's position x when the beam is about to tip. (g) Check the answer to part (e) by computing torques around the first pivot point.

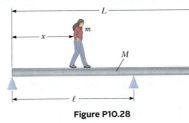

Figure P10.28

Impossibility Problems. Physics education research has focused heavily on the problem-solving skills of students. Although most problems in this text are structured in the form of providing data and asking for a result of computation, two problems in each chapter, on average, are structured as impossibility problems. They begin with the phrase *Why is the following situation impossible?* That is followed by the description of a situation. The striking aspect of these problems is that no question is asked of the students, other than that in the initial italics. The student must determine what questions need to be asked and what calculations need to be performed. Based on the results of these calculations, the student must determine why the situation described is not possible. This determination may require information from personal experience, common sense, Internet or print research, measurement, mathematical skills, knowledge of human norms, or scientific thinking.

These problems can be assigned to build critical thinking skills in students. They are also fun, having the aspect of physics "mysteries" to be solved by students individually or in groups. An example of an impossibility problem appears here:

The initial phrase in italics signals an impossibility problem.

A situation is described.

No question is asked. The student must determine what needs to be calculated and why the situation is impossible.

51. *Why is the following situation impossible?* Albert Pujols hits a home run so that the baseball just clears the top row of bleachers, 24.0 m high, located 130 m from home plate. The ball is hit at 41.7 m/s at an angle of 35.0° to the horizontal, and air resistance is negligible.

Increased Number of Paired Problems. Based on the positive feedback we received in a survey of the market, we have increased the number of paired problems in this edition. These problems are otherwise identical, one asking for a numerical solution and one asking for a symbolic derivation. There are now three pairs of these problems in most chapters, indicated by tan shading in the end-of-chapter problems set.

Thorough Revision of Artwork. Every piece of artwork in the Fifth Edition was revised in a new and modern style that helps express the physics principles at work in a clear and precise fashion. Every piece of art was also revised to make certain that the physical situations presented correspond exactly to the text discussion at hand.

Also added for this edition is a new feature: "focus pointers" that either point out important aspects of a figure or guide students through a process illustrated by the artwork or photo. This format helps those students who are more visual learners. Examples of figures with focus pointers appear below:

Figure 10.28 Two points on a rolling object take different paths through space.

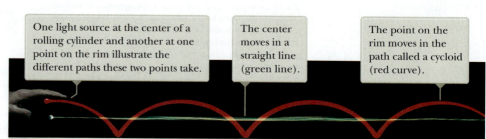

One light source at the center of a rolling cylinder and another at one point on the rim illustrate the different paths these two points take.

The center moves in a straight line (green line).

The point on the rim moves in the path called a cycloid (red curve).

Henry Leap and Jim Lehman

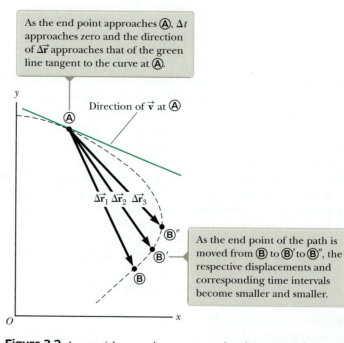

As the end point approaches Ⓐ, Δt approaches zero and the direction of $\Delta \vec{r}$ approaches that of the green line tangent to the curve at Ⓐ.

Direction of $\vec{v}$ at Ⓐ

$\Delta \vec{r}_1$ $\Delta \vec{r}_2$ $\Delta \vec{r}_3$

As the end point of the path is moved from Ⓑ to Ⓑ′ to Ⓑ″, the respective displacements and corresponding time intervals become smaller and smaller.

Figure 3.2 As a particle moves between two points, its average velocity is in the direction of the displacement vector $\Delta \vec{r}$. By definition, the instantaneous velocity at Ⓐ is directed along the line tangent to the curve at Ⓐ.

Expansion of the Analysis Model Approach. Students are faced with hundreds of problems during their physics courses. Instructors realize that a relatively small number of fundamental principles form the basis of these problems. When faced with a new problem, a physicist forms a *model* of the problem that can be solved in a simple way by identifying the fundamental principle that is applicable in the problem. For example, many problems involve conservation of energy, Newton's second law, or kinematic equations. Because the physicist has studied these principles extensively and understands the associated applications, he or she can apply this knowledge as a model for solving a new problem.

Although it would be ideal for students to follow this same process, most students have difficulty becoming familiar with the entire palette of fundamental principles that are available. It is easier for students to identify a *situation* rather than a fundamental principle. The *Analysis Model* approach we focus on in this revision lays out a standard set of situations that appear in most physics problems. These situations are based on an entity in one of four simplification models: particle, system, rigid object, and wave.

Once the simplification model is identified, the student thinks about what the entity is doing or how it interacts with its environment, which leads the student to identify a particular analysis model for the problem. For example, if an object is falling, the object is modeled as a particle. What it is doing is undergoing a constant acceleration due to gravity. The student has learned that this situation is described by the analysis model of a particle under constant acceleration. Furthermore, this model has a small number of equations associated with it for use in solving problems, the kinematic equations in Chapter 2. Therefore, an understanding of the situation has led to an analysis model, which then identifies a very small number of equations to solve the problem, rather than the myriad equations that students see in the chapter. In this way, the use of analysis models leads the student to the fundamental principle the physicist would identify. As the student gains more experience, he or she will lean less on the analysis model approach and begin to identify fundamental principles directly, more like the physicist does. This approach is further reinforced in the end-of-chapter summary under the heading *Analysis Models for Problem Solving*.

Content Changes. The content and organization of the textbook are essentially the same as in the fourth edition. Several sections in various chapters have been streamlined, deleted, or combined with other sections to allow for a more balanced presentation. Chapters 6 and 7 have been completely reorganized to prepare students for a unified approach to energy that is used throughout the text. Updates have been added to reflect the current status of several areas of research and application of physics, including information on discoveries of new Kuiper belt objects (Chapter 11), comparisons of competing theories of pitch perception in humans (Chapter 14), progress in using grating light valves for optical applications (Chapter 27), new experiments to search for the cosmic background radiation (Chapter 28), developments in the search for evidence of a quark–gluon plasma (Chapter 31), and the status of the Large Hadron Collider (Chapter 31).

Organization

We have incorporated a "context overlay" scheme into the textbook, in response to the "Physics in Context" approach in the IUPP. This feature adds interesting applications of the material to real issues. We have developed this feature to be flexible; it is an "overlay" in the sense that the instructor who does not wish to follow the contextual approach can simply ignore the additional contextual features without sacrificing complete coverage of the existing material. We believe, though, that the benefits students will gain from this approach will be many.

The context overlay organization divides the text into nine sections, or "Contexts," after Chapter 1, as follows:

Context Number	Context	Physics Topics	Chapters
1	Alternative-Fuel Vehicles	Classical mechanics	2–7
2	Mission to Mars	Classical mechanics	8–11
3	Earthquakes	Vibrations and waves	12–14
4	Heart Attacks	Fluids	15
5	Global Warming	Thermodynamics	16–18
6	Lightning	Electricity	19–21
7	Magnetism in Medicine	Magnetism	22–23
8	Lasers	Optics	24–27
9	The Cosmic Connection	Modern physics	28–31

Each Context begins with an introductory section that provides some historical background or makes a connection between the topic of the Context and associated social issues. The introductory section ends with a "central question" that motivates study within the Context. The final section of each chapter is a "Context Connection," which discusses how the specific material in the chapter relates to the Context and to the central question. The final chapter in each Context is followed by a "Context Conclusion." Each Conclusion applies a combination of the principles learned in the various chapters of the Context to respond fully to the central question. Each chapter, as well as the Context Conclusions, includes problems related to the context material.

Text Features

Most instructors believe that the textbook selected for a course should be the student's primary guide for understanding and learning the subject matter. Furthermore, the textbook should be easily accessible and should be styled and written to facilitate instruction and learning. With these points in mind, we have included many pedagogical features, listed below, that are intended to enhance its usefulness to both students and instructors.

Problem Solving and Conceptual Understanding

General Problem-Solving Strategy. A general strategy outlined at the end of Chapter 1 (pages 25–26) provides students with a structured process for solving problems. In all remaining chapters, the strategy is employed explicitly in every example so that students learn how it is applied. Students are encouraged to follow this strategy when working end-of-chapter problems.

In most chapters, more specific strategies and suggestions are included for solving the types of problems featured in the end-of-chapter problems. This feature helps students identify the essential steps in solving problems and increases their skills as problem solvers.

Thinking Physics. We have included many Thinking Physics examples throughout each chapter. These questions relate the physics concepts to common experiences or extend the concepts beyond what is discussed in the textual material. Immediately following each of these questions is a "Reasoning" section that responds to the question. Ideally, the student will use these features to better understand physical concepts before being presented with quantitative examples and working homework problems.

MCAT Test Preparation Guide. Located at the front of the book, this guide outlines 12 concept-based study courses for the physics part of the MCAT exam. Students can use the guide to prepare for the MCAT exam, class tests, or homework assignments.

Active Figures. Many diagrams from the text have been animated to become Active Figures (identified in the figure legend), part of the Enhanced WebAssign online homework system. By viewing animations of phenomena and processes that cannot be fully represented on a static page, students greatly increase their conceptual understanding. In addition to viewing animations of the figures, students can see the outcome of changing variables, conduct suggested explorations of the principles involved in the figure, and take and receive feedback on quizzes related to the figure.

Quick Quizzes. Students are provided an opportunity to test their understanding of the physical concepts presented through Quick Quizzes. The questions require students to make decisions on the basis of sound reasoning, and some of the questions have been written to help students overcome common misconceptions. Quick Quizzes have been cast in an objective format, including multiple choice, true–false, and ranking. Answers to all Quick Quiz questions are found at the end of the text. Many instructors choose to use such questions in a "peer instruction" teaching style or with the use of personal response system "clickers," but they can be used in standard quiz format as well. An example of a Quick Quiz follows below.

QUICK QUIZ 6.5 A dart is inserted into a spring-loaded dart gun by pushing the spring in by a distance x. For the next loading, the spring is compressed a distance $2x$. How much faster does the second dart leave the gun compared with the first? (**a**) four times as fast (**b**) two times as fast (**c**) the same (**d**) half as fast (**e**) one-fourth as fast

Pitfall Preventions. Over 150 Pitfall Preventions (such as the one to the right) are provided to help students avoid common mistakes and misunderstandings. These features, which are placed in the margins of the text, address both common student misconceptions and situations in which students often follow unproductive paths.

> **Pitfall Prevention | 13.2**
> **Two Kinds of Speed/Velocity**
> Do not confuse v, the speed of the wave as it propagates along the string, with v_y, the transverse velocity of a point on the string. The speed v is constant for a uniform medium, whereas v_y varies sinusoidally.

Summaries. Each chapter contains a summary that reviews the important concepts and equations discussed in that chapter. New for the Fifth Edition is the Analysis Models for Problem Solving section of the Summary, which highlights the relevant analysis models presented in a given chapter.

Questions. As mentioned previously, the previous edition's Questions section is now divided into two sections: *Objective Questions* and *Conceptual Questions*. The instructor may select items to assign as homework or use in the classroom, possibly with "peer instruction" methods and possibly with personal response systems. More than seven hundred Objective and Conceptual Questions are included in this edition. Answers for selected questions are included in the *Student Solutions Manual/Study Guide*, and answers for all questions are found in the *Instructor's Solutions Manual*.

Problems. An extensive set of problems is included at the end of each chapter; in all, this edition contains over 2 200 problems. Answers for odd-numbered problems are provided at the end of the book. Full solutions for approximately 20% of the problems are included in the *Student Solutions Manual/Study Guide*, and solutions for all problems are found in the *Instructor's Solutions Manual*.

In addition to the new problem types mentioned previously, there are several other kinds of problems featured in this text:

- **Biomedical problems.** We added a number of problems related to biomedical situations in this edition (each indentified with a **BIO** icon), to highlight the relevance of physics principles to those students taking this course who are majoring in one of the life sciences.

- **Paired Problems.** As an aid for students learning to solve problems symbolically, paired numerical and symbolic problems are included in all chapters of the text. Paired problems are identified by a common background screen.
- **Review problems.** Many chapters include review problems requiring the student to combine concepts covered in the chapter with those discussed in previous chapters. These problems (marked **Review**) reflect the cohesive nature of the principles in the text and verify that physics is not a scattered set of ideas. When facing a real-world issue such as global warming or nuclear weapons, it may be necessary to call on ideas in physics from several parts of a textbook such as this one.
- **"Fermi problems."** One or more problems in most chapters ask the student to reason in order-of-magnitude terms.
- **Design problems.** Several chapters contain problems that ask the student to determine design parameters for a practical device so that it can function as required.
- **Calculus-based problems.** Most chapters contain at least one problem applying ideas and methods from differential calculus and one problem using integral calculus.

The instructor's Web site, **www.cengage.com/physics/serway,** provides lists of all the various problem types, including problems most often assigned in Enhanced WebAssign, symbolic problems, quantitative/conceptual problems, Master It tutorials, Watch It solution videos, impossibility problems, paired problems, problems using calculus, problems encouraging or requiring computer use, problems with **What If?** parts, problems referred to in the chapter text, problems based on experimental data, order-of-magnitude problems, problems about biological applications, design problems, review problems, problems reflecting historical reasoning, and ranking questions.

Alternative Representations. We emphasize alternative representations of information, including mental, pictorial, graphical, tabular, and mathematical representations. Many problems are easier to solve if the information is presented in alternative ways, to reach the many different methods students use to learn.

Math Appendix. The math appendix (Appendix B), a valuable tool for students, shows the math tools in a physics context. This resource is ideal for students who need a quick review on topics such as algebra, trigonometry, and calculus.

Helpful Features

Style. To facilitate rapid comprehension, we have written the book in a clear, logical, and engaging style. We have chosen a writing style that is somewhat informal and relaxed so that students will find the text appealing and enjoyable to read. New terms are carefully defined, and we have avoided the use of jargon.

Important Definitions and Equations. Most important definitions are set in **boldface** or are set off from the paragraph in centered text for added emphasis and ease of review. Similarly, important equations are highlighted with a background screen to facilitate location.

Marginal Notes. Comments and notes appearing in the margin with a ▶ icon can be used to locate important statements, equations, and concepts in the text.

Pedagogical Use of Color. Readers should consult the **pedagogical color chart** (inside the front cover) for a listing of the color-coded symbols used in the text diagrams. This system is followed consistently throughout the text.

Mathematical Level. We have introduced calculus gradually, keeping in mind that students often take introductory courses in calculus and physics concurrently. Most steps are shown when basic equations are developed, and reference is often made to mathematical appendices near the end of the textbook. Although vectors are discussed in detail in

Chapter 1, vector products are introduced later in the text, where they are needed in physical applications. The dot product is introduced in Chapter 6, which addresses energy of a system; the cross product is introduced in Chapter 10, which deals with angular momentum.

Significant Figures. In both worked examples and end-of-chapter problems, significant figures have been handled with care. Most numerical examples are worked to either two or three significant figures, depending on the precision of the data provided. End-of-chapter problems regularly state data and answers to three-digit precision. When carrying out estimation calculations, we shall typically work with a single significant figure. (More discussion of significant figures can be found in Chapter 1, pages 10–12.)

Units. The international system of units (SI) is used throughout the text. The U.S. customary system of units is used only to a limited extent in the chapters on mechanics and thermodynamics.

Appendices and Endpapers. Several appendices are provided near the end of the textbook. Most of the appendix material represents a review of mathematical concepts and techniques used in the text, including scientific notation, algebra, geometry, trigonometry, differential calculus, and integral calculus. Reference to these appendices is made throughout the text. Most mathematical review sections in the appendices include worked examples and exercises with answers. In addition to the mathematical reviews, the appendices contain tables of physical data, conversion factors, and the SI units of physical quantities as well as a periodic table of the elements. Other useful information—fundamental constants and physical data, planetary data, a list of standard prefixes, mathematical symbols, the Greek alphabet, and standard abbreviations of units of measure—appears on the endpapers.

TextChoice Custom Options for *Principles of Physics*

Cengage Learning's digital library, TextChoice, enables you to build your custom version of Serway and Jewett's *Principles of Physics* from scratch. You may pick and choose the content you want to include in your text and even add your own original materials creating a unique, all-in-one learning solution. This all happens from the convenience of your desktop. Visit **www.textchoice.com** to start building your book today.

Cengage Learning offers the fastest and easiest way to create unique customized learning materials delivered the way you want. For more information about custom publishing options, visit **www.cengage.com/custom** or contact your local Cengage Learning representative.

Course Solutions That Fit Your Teaching Goals and Your Students' Learning Needs

Recent advances in educational technology have made homework management systems and audience response systems powerful and affordable tools to enhance the way you teach your course. Whether you offer a more traditional text-based course, are interested in using or are currently using an online homework management system such as Enhanced WebAssign, or are ready to turn your lecture into an interactive learning environment with JoinIn, you can be confident that the text's proven content provides the foundation for each and every component of our technology and ancillary package.

Homework Management Systems

Enhanced WebAssign for *Principles of Physics*, Fifth Edition. Exclusively from Cengage Learning, Enhanced WebAssign offers an extensive online program for physics to encourage the practice that's so critical for concept mastery. The meticulously crafted pedagogy and exercises in our proven texts become even more effective in Enhanced WebAssign. Enhanced WebAssign includes the Cengage YouBook, a highly customizable, interactive eBook. WebAssign includes:

- All of the quantitative end-of-chapter problems
- Selected problems enhanced with targeted feedback. An example of targeted feedback appears below:

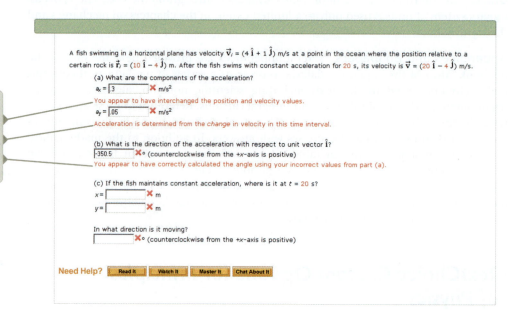

Selected problems include feedback to address common mistakes that students make. This feedback was developed by professors with years of classroom experience.

- Master It tutorials (indicated in the text by a **M** icon), to help students work through the problem one step at a time. An example of a Master It tutorial appears below:

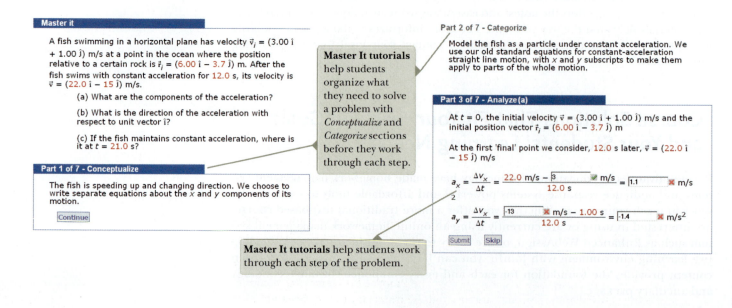

Master It tutorials help students organize what they need to solve a problem with *Conceptualize* and *Categorize* sections before they work through each step.

Master It tutorials help students work through each step of the problem.

- Watch It solution videos (indicated in the text by a **W** icon) that explain fundamental problem-solving strategies, to help students step through the problem. In addition, instructors can choose to include video hints of problem-solving strategies. A screen shot from a Watch It solution video appears below:

A projectile is launched at some angle to the horizontal with some initial speed v_b and air resistance is negligible.

(a) Is the projectile a freely falling body?
(b) What is its acceleration in the vertical direction?
(c) What is its acceleration in the horizontal direction?

Watch It solution videos help students visualize the steps needed to solve a problem.

- Concept Checks
- Active Figure simulation tutorials
- PhET simulations
- Most worked examples, enhanced with hints and feedback, to help strengthen students' problem-solving skills
- Every Quick Quiz, giving your students ample opportunity to test their conceptual understanding
- The Cengage YouBook

WebAssign has a customizable and interactive eBook, the **Cengage YouBook,** that lets you tailor the textbook to fit your course and connect with your students. You can remove and rearrange chapters in the table of contents and tailor assigned readings that match your syllabus exactly. Powerful editing tools let you change as much as you'd like—or leave it just like it is. You can highlight key passages or add sticky notes to pages to comment on a concept in the reading, and then share any of these individual notes and highlights with your students, or keep them personal. You can also edit narrative content in the textbook by adding a text box or striking out text. With a handy link tool, you can drop in an icon at any point in the eBook that lets you link to your own lecture notes, audio summaries, video lectures, or other files on a personal Web site or anywhere on the web. A simple YouTube widget lets you easily find and embed videos from YouTube directly into eBook pages. There is a light discussion board that lets students and instructors find others in their class and start a chat session. The Cengage YouBook helps students go beyond just reading the textbook. Students can also highlight the text, add their own notes, and bookmark the text. Animations play right on the page at the point of learning so that they're not speed bumps to reading but true enhancements. Please visit **www.webassign .net/brookscole** to view an interactive demonstration of Enhanced WebAssign.

- Offered exclusively in WebAssign, **Quick Prep** for physics is algebra and trigonometry math remediation within the context of physics applications and principles. Quick Prep helps students succeed by using narratives illustrated throughout with video examples. The Master It tutorial problems allow students to assess and retune their understanding of the material. The Practice Problems that go along with each tutorial allow both the student and the instructor to test the student's understanding of the material.

Quick Prep includes the following features:

- 67 interactive tutorials
- 67 additional practice problems
- Thorough overview of each topic that includes video examples
- Taken before the semester begins or during the first few weeks of the course
- Can also be assigned alongside each chapter for "just in time" remediation

Topics include: units, scientific notation, and significant figures; the motion of objects along a line; functions; approximation and graphing; probability and error; vectors, displacement, and velocity; spheres; force and vector projections.

CengageBrain.com

On **CengageBrain.com** students will be able to save up to 60% on their course materials through our full spectrum of options. Students will have the option to rent their textbooks, purchase print textbooks, e-textbooks, or individual e-chapters and audio books all for substantial savings over average retail prices. **CengageBrain.com** also includes access to Cengage Learning's broad range of homework and study tools and features a selection of free content.

Lecture Presentation Resources

PowerLecture with ExamView® and JoinIn for *Principles of Physics*, Fifth Edition. Bringing physics principles and concepts to life in your lectures has never been easier! The full-featured, two-volume **PowerLecture** Instructor's Resource DVD-ROM (Volume 1: Chapters 1–15; Volume 2: Chapters 16–31) provides everything you need for *Principles of Physics*, fifth edition. Key content includes the *Instructor's Solutions Manual* solutions, art and images from the text, pre-made chapter-specific PowerPoint lectures, ExamView test generator software with pre-loaded test questions, JoinIn response-system "clickers," Active Figures animations, and a physics movie library.

JoinIn. *Assessing to Learn in the Classroom* questions developed at the University of Massachusetts Amherst. This collection of 250 advanced conceptual questions has been tested in the classroom for more than ten years and takes peer learning to a new level. JoinIn helps you turn your lectures into an interactive learning environment that promotes conceptual understanding. Available exclusively for higher education from our partnership with Turning Technologies, JoinIn™ is the easiest way to turn your lecture hall into a personal, fully interactive experience for your students!

Assessment and Course Preparation Resources

A number of resources listed below will assist with your assessment and preparation processes.

Instructor's Solutions Manual by Vahe Peroomian (University of California at Los Angeles). Thoroughly revised for this edition, the *Instructor's Solutions Manual* contains complete worked solutions to all end-of-chapter problems in the textbook as well as answers to the even-numbered problems and all the questions. The solutions to problems new to the fifth edition are marked for easy identification. Volume 1 contains Chapters 1 through 15, and Volume 2 contains Chapters 16 through 31. Electronic files of the *Instructor's Solutions Manual* are available on the *PowerLecture*™ DVD-ROM.

Test Bank by Susan English (Durham Technical College). The test bank is available on the two-volume *PowerLecture*™ DVD-ROM via the ExamView® test software. This two-volume test bank contains approximately 2 000 multiple-choice questions. Instructors may print and duplicate pages for distribution to students. Volume 1 contains Chapters 1 through 15, and Volume 2 contains Chapters 16 through 31. WebCT and Blackboard versions of the test bank are available on the instructor's companion site at **www.cengage/physics/serway.**

Instructor's Companion Web Site. Consult the instructor's site by pointing your browser to **www.cengage.com/physics/serway** for a problem correlation guide, PowerPoint lectures, and JoinIn audience response content. Instructors adopting the fifth edition of *Principles of Physics* may download these materials after securing the appropriate password from their local sales representative.

Supporting Materials for the Instructor

Supporting instructor materials are available to qualified adopters. Please consult your local Cengage Learning, Brooks/Cole representative for details. Visit **www.cengage.com/physics/serway** to:

- request a desk copy
- locate your local representative
- download electronic files of select support materials

Student Resources

Visit the *Principles of Physics* Web site at **www.cengagebrain.com/shop/ISBN/9781133104261** to see samples of select student supplements. Go to **CengageBrain.com** to purchase and access this product at Cengage Learning's preferred online store.

CENGAGE **brain**.com

Student Solutions Manual/Study Guide by John R. Gordon, Vahe Peroomian, Raymond A. Serway, and John W. Jewett, Jr. This two-volume manual features detailed solutions to 20% of the end-of-chapter problems from the text. The manual also features a list of important equations, concepts, and notes from key sections of the text in addition to answers to selected end-of-chapter questions. Volume 1 contains Chapters 1 through 15, and Volume 2 contains Chapters 16 through 31.

Physics Laboratory Manual, Third Edition by David Loyd (Angelo State University) supplements the learning of basic physical principles while introducing laboratory procedures and equipment. Each chapter includes a prelaboratory assignment, objectives, an equipment list, the theory behind the experiment, experimental procedures, graphing exercises, and questions. A laboratory report form is included with each experiment so that the student can record data, calculations, and experimental results. Students are encouraged to apply statistical analysis to their data. A complete *Instructor's Manual* is also available to facilitate use of this lab manual.

Physics Laboratory Experiments, Seventh Edition by Jerry D. Wilson (Lander College) and Cecilia A. Hernández (American River College). This market-leading manual for the first-year physics laboratory course offers a wide range of class-tested experiments designed specifically for use in small to midsize lab programs. A series of integrated experiments emphasizes the use of computerized instrumentation and includes a set of "computer-assisted experiments" to allow students and instructors to gain experience with modern equipment. This option also enables instructors to determine the appropriate balance between traditional and computer-based experiments for their courses. By analyzing data through two different methods, students gain a greater understanding of the concepts behind the experiments. The seventh edition is updated with the latest information and techniques involving state-of-the-art equipment and a new Guided Learning feature addresses the growing interest in guided-inquiry pedagogy. Fourteen additional experiments are also available through custom printing.

Teaching Options

Although some topics found in traditional textbooks have been omitted from this textbook, instructors may find that the current text still contains more material than can be covered in a two-semester sequence. For this reason, we would like to offer the following suggestions. If you wish to place more emphasis on contemporary topics in physics, you should consider omitting parts or all of Chapters 15, 16, 17, 18, 24, 25, and 26. On the other hand, if you wish to follow a more traditional approach that places more emphasis on classical physics, you could omit Chapters 9, 11, 28, 29, 30, and 31. Either approach can be used without any loss in continuity. Other teaching options would fall somewhere between these two extremes by choosing to omit some or all of the following sections, which can be considered optional:

3.6	Relative Velocity and Relative Acceleration
6.10	Energy Diagrams and Equilibrium of a System
9.9	General Relativity
10.12	Rolling Motion of Rigid Objects
12.6	Damped Oscillations
12.7	Forced Oscillations
14.6	Nonsinusoidal Wave Patterns
14.7	The Ear and Theories of Pitch Perception
15.8	Other Applications of Fluid Dynamics
16.6	Distribution of Molecular Speeds
17.7	Molar Specific Heats of Ideal Gases
17.8	Adiabatic Processes for an Ideal Gas
17.9	Molar Specific Heats and the Equipartition of Energy
20.10	Capacitors with Dielectrics
22.11	Magnetism in Matter
26.5	The Eye
27.9	Diffraction of X-Rays by Crystals
28.13	Tunneling Through a Potential Energy Barrier

Acknowledgments

Prior to our work on this revision, we conducted two separate surveys of professors to gauge their textbook needs in the introductory calculus-based physics market. We were overwhelmed not only by the number of professors who wanted to take part in the survey but also by their insightful comments. Their feedback and suggestions helped shape the revision of this edition, and so we would like to thank the survey participants:

Anthony Abbott, *Willow-International Center;* Wagih Abdel Kader, *South Carolina State University;* Mikhail Agrest, *College of Charleston;* David Ailion, *University of Utah;* Rhett Allain, *Southeastern Louisiana University;* Bradley Antanaitis, *Lafayette College;* Vasudeva Rao Aravind, *Clarion University;* David Armstrong, *College of William & Mary;* Robert Arts, *University of Pikeville;* Robert Astalos, *Adams State College;* Charles Atchley, *Sauk Valley Community College;* Terry Austin, *LaGrange College;* Eric Ayars, *California State University, Chico;* Cristian Bahrim, *Lamar University;* Mirley Balasubramanya, *Texas A&M University–Corpus Christi;* Kenneth Balazovich, *University of Michigan;* David Balogh, *Fresno City College;* Alan Bardsley, *Salve Regina University;* Joseph Barranco, *San Francisco State University;* Perry Baskin, *Valdosta State University;* Celso Batalha, *Evergreen Valley College;* George Baxter, *Penn State Erie, The Behrend College;* Raymond Benge, *Tarrant County College;* Joel Berlinghieri, *The Citadel;* Matt Bigelow, *St. Cloud State University;* Raymond Bigliani, *Farmingdale State College, The State University of New York;* Mark S. Boley, *Western Illinois University;* Ken Bolland, *The Ohio State University;* Patrick Briggs, *The Citadel;* Douglas Brownell, *MiraCosta Community College;* William Bryant; Michael Butros, *Victor Valley College;* Gavin Buxton, *Robert Morris University;* Ralph Calhoun, *Northwest Florida State College;* Bruce Callen, *Drury University;* Bradley Carroll, *Weber State University;* Brian Carter, *Grossmont College;* Brian Carter, *San Diego City College;* Tom Carter, *College of DuPage;* Jennifer Cash, *South Carolina State University;* Cliff Castle, *Jefferson College;* Soumitra Chattopadhyay, *Georgia Highlands College;* Albert Chen, *Oklahoma Baptist University;* Li Chen, *Rhode Island College;* Norbert Chencinski, *College of Staten Island;* Kelvin Chu, *University of Vermont;* Darwin Church, *University of Cincinnati;* Michael Cohen, *Shippensburg University;* Robert Cohen, *East Stroudsburg University;* Stan Converse, *Wake Technical Community College;* S. Marie Cooper, *Immaculata University;* Susan Coppersmith, *University of Wisconsin–Madison;* Volker Crede,

Florida State University; Demetra Czegan, *Seton Hill University;* Deborah Damcott, *Harper College;* Chris Davis, *University of Louisville;* Lynn Delker, *Diablo Valley College;* Todd Devore, *University of Alabama at Birmingham;* Alfonso Diaz Jimenez, *ADJOIN Research Center;* Susan DiFranzo, *Hudson Valley Community College;* Edward Dingler, *Southwest Virginia Community College;* Gregory Dolise, *Harrisburg Area Community College;* Sandra Doty, *Ohio University Lancaster;* Diana Driscoll, *Case Western Reserve University;* Mike Durren, *Lake Michigan College;* Ephraim Eisenberger, *Stevenson University;* Eleazer Ekwue, *College of Southern Maryland;* Terry Ellis, *Jacksonville University;* Mark Engebretson, *Augsburg College;* Tim Farris, *Volunteer State Community College;* Michael Fauerbach, *Florida Gulf Coast University;* Nail Fazleev, *University of Texas at Arlington;* Jason Ferguson, *Wichita State University;* Michael Ferralli, *Gannon University;* Chrisopher Fischer, *University of Kansas;* Kent Fisher, *Columbus State Community College;* Matthew Fleenor, *Roanoke College;* Richard Fleming, *Midwestern State University;* Terrence Flower, *St. Catherine University;* Marco Fornari, *Central Michigan University;* Juhan Frank, *Louisiana State University;* Allan Franklin, *University of Colorado;* Carl Frederickson, *University of Central Arkansas;* Rica French, *MiraCosta College;* Jerry Fuller, *Kilgore College;* Sambandamurthy Ganapathy, *University at Buffalo, The State University of New York;* Mark Gealy, *Concordia College;* Benjamin Geinstein, *University of California, San Diego;* Brian Geislinger, *Gadsden State Community College;* Dan Gibson, *Denison University;* Svetlana Gladycheva, *University of Washington;* Paul Goains, *San Jacinto College;* John Goehl, *Barry University;* John Goff, *Lynchburg College;* Charles Goodman, *Pitt Community College;* Christopher Gould, *University of Southern California;* Michael Graf, *Boston College;* Morris Greenwood, *San Jacinto College;* Allan Greer, *Gonzaga University;* Elena Gregg, *Oral Roberts University;* Alec Habig, *University of Minnesota Duluth;* Robert Hallock, *University of Massachusetts Amherst;* Dean Hamden, *Montclair State University;*

Carlos Handy, *Texas Southern University;* Baher Hanna, *Owens Community College;* Wayne Hayes, *Greenville Technical College;* Charles Henderson, *Western Michigan University;* Paul Henriksen, *James Madison University;* Thomas Herring, *Western Nevada College;* Yoshinao Hirai, *Central Lakes College;* Dean Hirschi, *Washington State Community College;* Stanley Hirschi, *Central Michigan University;* Pei-Chun Ho, *California State University, Fresno;* Roy Hoffer, *Bloomsburg University;* Mikel Holcomb, *West Virginia University;* Dawn Hollenbeck, *Rochester Institute of Technology;* Zdeslav Hrepic, *Columbus State University;* James Hudgings, *Camden County College;* Yung Huh, *South Dakota State University;* David Ingram, *Ohio University;* M. Islam, *The State University of New York at Potsdam;* Howard Jackson, *University of Cincinnati;* Shawn Jackson, *University of Arizona;* Sa-Han Jang, *Franklin and Marshall College;* Tim Jenkins, *University of Oregon;* Erik Jensen, *Chemeketa Community College;* Corinna Jobe, *College of the Canyons;* Charles Johnson, *South Georgia College;* Adam Johnston, *Weber State University;* Edwin Jones, *University of South Carolina;* Rex Joyner, *Indiana Institute of Technology;* Michael Kaplan, *Simmons College;* David Kardelis, *Utah State University–College of Eastern Utah;* Debora Katz, *United States Naval Academy;* Edward Kelsey, *Monmouth University;* Leonard Khazan, *Camden County College;* Nikolaos Kidonakis, *Kennesaw State University;* Derrick Kiley, *University of California, Merced;* Wayne Kinnison, *Texas A&M University–Kingsville;* Terence Kite, *Pepperdine University;* Joseph Klarfeld, *Queens College, the City University of New York;* Mario Klaric, *Midlands Technical College;* Suja Kochat, *Palo Alto College;* Michael Korth, *University of Minnesota Morris;* Aneta Koynova, *University of Texas at San Antonio;* Tatiana Krentsel, *Binghamton University, State University of New York;* George Kuck, *California State University, Long Beach;* Fred Kuttner, *University of California, Santa Cruz;* John LaBrasca, *Clark College;* Dan Lawrence, *Northwest Nazarene University;* Lynne Lawson, *Providence College;* Geoffrey Lenters, *Grand Valley State University;* John Lestrade, *Mississippi State University;* Rudi Lindner, *University of Michigan;* Beth Lindsey, *Penn State Greater Allegheny;* Kehfei Liu, *University of Kentucky;* Zengqiang Liu, *St. Cloud State University;* James Lockhart, *San Francisco State University;* Jorge Lopez, *University of Texas at El Paso;* Donald Luttermoser, *East Tennessee State University;* Ntungwa Maasha, *College of Coastal Georgia;* Terrence Maher, *Fayetteville Technical Community College;* Rizwan Mahmood, *Slippery Rock University;* Kingshuk Majumdar, *Grand Valley State University;* Igor Makasyuk, *Stanford University;* Jay Mancini; David Mandelbaum, *DeVry University;* David Marasco, *Foothill College;* Jelena Maricic, *Drexel University;* Pete Maritato, *Suffolk Community College;* Collette Marsh, *Harper College;* Devon Mason, *Albright College;* Martin Mason, *Mt. San Antonio College;* Sylvio May, *North Dakota State University;* Richard McCorkle, *University of Rhode Island;* Jimmy McCoy, *Tarleton State University;* Laura McCullough, *University of Wisconsin–Stout;* Ralph McGrew, *Broome Community College;* Rahul Mehta, *University of Central Arkansas;* Albert Menard, *Saginaw Valley State University;* William Mendoza, *Florida State College at Jacksonville;* Michael Meyer, *Michigan Technological University;* Karie Meyers, *Pima Community College;* Stanley Micklavzina, *University of Oregon;* Sudipa Mitra-Kirtley, *Rose-Hulman Institute of Technology;* Bob Moltaji, *Northeastern Illinois University;* Tamar More, *University of Portland;* Muhammad Asim Mubeen, *University at Albany, State University of New York;* Carl Mungan, *United States Naval Academy;* Mim Nakarmi, *Brooklyn College;* Majid Noori, *Cumberland County College;* Irina Novikova, *College of William and Mary;* Richard Olenick, *University of Dallas;* Grant O'Rielly, *University of Massachusetts Dartmouth;* Nicola Orsini, *Marshall University;* Edward

Osagie, *Lane College;* Eric Page, *University of San Diego;* Bruce Palmquist, *Central Washington University;* Dr. James Pazun, *Pfeiffer University;* David Pengra, *University of Washington;* Dorn Peterson, *James Madison University;* Anna Petrova-Mayor, *California State University, Chico;* Chris Pettit, *Emporia State University;* Ronald Phaneuf, *University of Nevada, Reno;* Daniel Phillips, *Ohio University;* Mark Pickar, *Minnesota State University, Mankato;* Alberto Pinkas, *New Jersey City University;* Dale Pleticha, *Gordon College;* William Powell, *Texas Lutheran University;* Norris Preyer, *College of Charleston;* Steve Quon, *Ventura College;* Andrew Rader, *Indiana University–Purdue University Indianapolis;* Stanley Radford, *The College at Brockport, State University of New York;* Bayani Ramirez, *San Jacinto College South;* Michael Ramsdell, *Clarkson University;* Steven Rehse, *Wayne State University;* Michael Richmond, *Rochester Institute of Technology;* Shadow Robinson, *Millsaps College;* Stephen Roeder, *San Diego State University;* John Rollino, *Rutgers University, Newark;* Alvin Rosenthal, *Western Michigan University;* Richard Ross, *Rock Valley College;* Louis Rubbo, *Coastal Carolina University;* Nanjundiah Sadanand, *Central Connecticut State University;* Mik Sawicki, *John A. Logan College;* Joseph Scanio, *University of Cincinnati;* Beth Schaefer, *Georgian Court University;* Scott Schultz, *Delta College;* Leon Scott, *Borough of Manhattan Community College;* Naidu Seetala, *Grambling State University;* Dastgeer Shaikh, *The University of Alabama in Huntsville;* Nimmi Sharma, *Central Connecticut State University;* Peter Sheldon, *Randolph College;* Douglas Sherman, *San Jose State University;* Paresh Shettigar, *Hawkeye Community College;* Earl Skelton, *Georgetown University;* Steve Spicklemire, *University of Indianapolis;* Philip Spickler, *Bridgewater College;* Glenn Spiczak, *University of Wisconsin–River Falls;* Susan Spillane, *McHenry County College;* Phillip Sprunger, *Louisiana State University;* Sudha Srinivas, *Northeastern Illinois University;* Jason Stalnaker, *Oberlin College;* Lawrence Staunton, *Drake University;* Gay Stewart, *University of Arkansas;* Chris Stockdale, *Marquette University;* Glenn Stracher, *East Georgia College;* Igor Strakovsky, *The George Washington University;* Dave Stuva, *Creighton University;* K. V. Sudhakar, *Montana Tech of the University of Montana;* Edward Sunderhaus, *Cincinnati State Technical and Community College;* Richard Swanson, *Sandhills Community College;* Steven Sweeney, *King's College;* Douglas Szper, *Lakeland College;* Nacira Tache, *Santa Fe College;* Kerry Tanimoto, *Honolulu Community College;* Vahe Tatoian, *Mt. San Antonio College;* Greg Thompson, *Adrian College;* Kristen Thompson, *Loras College;* Ionel Tifrea, *California State University, Fullerton;* Ramon Tirado; Daniela Topasna, *Virginia Military Institute;* Jim Tressel, *Massasoit Community College;* Som Tyagi, *Drexel University;* Bob Tyndall, *Carteret Community College;* Toshiya Ueta, *University of Denver;* Trina Van Ausdal, *Salt Lake Community College;* Bira van Kolck, *University of Arizona;* Brian Vermillion, *University of Indianapolis;* Joan Vogtman, *Potomac State College;* Sergei Voloshin, *Wayne State University;* Judy Vondruska, *South Dakota State University;* William Waggoner, *San Antonio College;* Fa-chung (Fred) Wang, *Prairie View A&M University;* Michael Weber, *Brigham Young University-Hawaii;* Margaret Wessling, *Los Angeles Pierce College;* James White, *Juniata College;* Daniel Willey, *Allegheny College;* Suzanne Willis, *Northern Illinois University;* Keith Willson, *Geneva College;* Stephen Wimpenny, *University of California, Riverside;* Jeff Winger, *Mississippi State University;* Krista Wood, *University of Cincinnati;* Hai-Sheng Wu, *Minnesota State University, Mankato;* Qilin Wu, *College of Southern Nevada;* Scott Yost, *The Citadel;* Anne Young, *Rochester Institute of Technology;* Hashim Yousif, *University of Pittsburgh at Bradford;* Chen Zeng, *The George Washington University;* Michael Ziegler, *The Ohio State University*

We also thank the following people for their suggestions and assistance during the preparation of earlier editions of this textbook:

Edward Adelson, *Ohio State University;* Anthony Aguirre, *University of California at Santa Cruz;* Yildirim M. Aktas, *University of North Carolina–Charlotte;* Alfonso M. Albano, *Bryn Mawr College;* Royal Albridge, *Vanderbilt University;* Subash Antani, *Edgewood College;* Michael Bass, *University of Central Florida;* Harry Bingham, *University of California, Berkeley;* Billy E. Bonner, *Rice University;* Anthony Buffa, *California Polytechnic State University,* *San Luis Obispo;* Richard Cardenas, *St. Mary's University;* James Carolan, *University of British Columbia;* Kapila Clara Castoldi, *Oakland University;* Ralph V. Chamberlin, *Arizona State University;* Christopher R. Church, *Miami University (Ohio);* Gary G. DeLeo, *Lehigh University;* Michael Dennin, *University of California, Irvine;* Alan J. DeWeerd, *Creighton University;* Madi Dogariu, *University of Central Florida;* Gordon Emslie, *University of Alabama at Huntsville;*

Donald Erbsloe, *United States Air Force Academy;* William Fairbank, *Colorado State University;* Marco Fatuzzo, *University of Arizona;* Philip Fraundorf, *University of Missouri–St. Louis;* Patrick Gleeson, *Delaware State University;* Christopher M. Gould, *University of Southern California;* James D. Gruber, *Harrisburg Area Community College;* John B. Gruber, *San Jose State University;* Todd Hann, *United States Military Academy;* Gail Hanson, *Indiana University;* Gerald Hart, *Moorhead State University;* Dieter H. Hartmann, *Clemson University;* Richard W. Henry, *Bucknell University;* Athula Herat, *Northern Kentucky University;* Laurent Hodges, *Iowa State University;* Michael J. Hones, *Villanova University;* Huan Z. Huang, *University of California at Los Angeles;* Joey Huston, *Michigan State University;* George Igo, *University of California at Los Angeles;* Herb Jaeger, *Miami University;* David Judd, *Broward Community College;* Thomas H. Keil, *Worcester Polytechnic Institute;* V. Gordon Lind, *Utah State University;* Edwin Lo; Michael J. Longo, *University of Michigan;* Rafael Lopez-Mobilia, *University of Texas at San Antonio;* Roger M. Mabe, *United States Naval Academy;* David Markowitz, *University of Connecticut;* Thomas P. Marvin, *Southern Oregon University;* Bruce Mason, *University of Oklahoma at Norman;* Martin S. Mason, *College of the Desert;* Wesley N. Mathews, Jr., *Georgetown University;* Ian S. McLean, *University of California at Los Angeles;* John W. McClory, *United States Military Academy;* L. C. McIntyre, Jr., *University of Arizona;* Alan S. Meltzer, *Rensselaer Polytechnic Institute;* Ken Mendelson, *Marquette University;* Roy Middleton, *University of Pennsylvania;* Allen Miller, *Syracuse University;* Clement J. Moses, *Utica College of Syracuse University;* John W. Norbury, *University of Wisconsin–Milwaukee;* Anthony Novaco, *Lafayette College;* Romulo Ochoa, *The College of New Jersey;* Melvyn Oremland, *Pace University;* Desmond Penny, *Southern Utah University;* Steven J. Pollock, *University of Colorado–Boulder;* Prabha Ramakrishnan, *North Carolina State University;* Rex D. Ramsier, *The University of Akron;* Ralf Rapp, *Texas A&M University;* Rogers Redding, *University of North Texas;* Charles R. Rhyner, *University of Wisconsin–Green Bay;* Perry Rice, *Miami University;* Dennis Rioux, *University of Wisconsin–Oshkosh;* Richard Rolleigh, *Hendrix College;* Janet E. Seger, *Creighton University;* Gregory D. Severn, *University of San Diego;* Satinder S. Sidhu, *Washington College;* Antony Simpson, *Dalhousie University;* Harold Slusher, *University of Texas at El Paso;* J. Clinton Sprott, *University of Wisconsin at Madison;* Shirvel Stanislaus, *Valparaiso University;* Randall Tagg, *University of Colorado at Denver;* Cecil Thompson, *University of Texas at Arlington;* Harry W. K. Tom, *University of California at Riverside;* Chris Vuille, *Embry–Riddle Aeronautical University;* Fiona Waterhouse, *University of California at Berkeley;* Robert Watkins, *University of Virginia;* James Whitmore, *Pennsylvania State University*

Principles of Physics, fifth edition, was carefully checked for accuracy by Grant Hart (Brigham Young University), James E. Rutledge (University of California at Irvine), and Som Tyagi (Drexel University).

We are indebted to the developers of the IUPP models "A Particles Approach to Introductory Physics" and "Physics in Context," upon which much of the pedagogical approach in this textbook is based.

Vahe Peroomian wrote the initial draft of the new context on Heart Attacks, and we are very grateful for his efforts. He provided further assistance by reviewing early drafts of questions and problems sets.

We are grateful to John R. Gordon and Vahe Peroomian for writing the *Student Solutions Manual and Study Guide,* and to Vahe Peroomian for preparing an excellent *Instructor's Solutions Manual.* During the development of this text, the authors benefited from many useful discussions with colleagues and other physics instructors, including Robert Bauman, William Beston, Don Chodrow, Jerry Faughn, John R. Gordon, Kevin Giovanetti, Dick Jacobs, Harvey Leff, John Mallinckrodt, Clem Moses, Dorn Peterson, Joseph Rudmin, and Gerald Taylor.

Special thanks and recognition go to the professional staff at the Brooks/Cole Publishing Company—in particular, Charles Hartford, Ed Dodd, Brandi Kirksey, Rebecca Berardy-Schwartz, Jack Cooney, Cathy Brooks, Cate Barr, and Brendan Killion—for their fine work during the development and production of this textbook. We recognize the skilled production service provided by Jill Traut and the staff at Macmillan Solutions and the dedicated photo research efforts of Josh Garvin of the Bill Smith Group.

Finally, we are deeply indebted to our wives and children for their love, support, and long-term sacrifices.

RAYMOND A. SERWAY
St. Petersburg, Florida

JOHN W. JEWETT, JR.
Anaheim, California

It is appropriate to offer some words of advice that should be of benefit to you, the student. Before doing so, we assume you have read the Preface, which describes the various features of the text and support materials that will help you through the course.

How to Study

Instructors are often asked, "How should I study physics and prepare for examinations?" There is no simple answer to this question, but we can offer some suggestions based on our own experiences in learning and teaching over the years.

First and foremost, maintain a positive attitude toward the subject matter, keeping in mind that physics is the most fundamental of all natural sciences. Other science courses that follow will use the same physical principles, so it is important that you understand and are able to apply the various concepts and theories discussed in the text.

Concepts and Principles

It is essential that you understand the basic concepts and principles before attempting to solve assigned problems. You can best accomplish this goal by carefully reading the textbook before you attend your lecture on the covered material. When reading the text, you should jot down those points that are not clear to you. Also be sure to make a diligent attempt at answering the questions in the Quick Quizzes as you come to them in your reading. We have worked hard to prepare questions that help you judge for yourself how well you understand the material. Study the **What If?** features that appear in many of the worked examples carefully. They will help you extend your understanding beyond the simple act of arriving at a numerical result. The Pitfall Preventions will also help guide you away from common misunderstandings about physics. During class, take careful notes and ask questions about those ideas that are unclear to you. Keep in mind that few people are able to absorb the full meaning of scientific material after only one reading; several readings of the text and your notes may be necessary. Your lectures and laboratory work supplement the textbook and should clarify some of the more difficult material. You should minimize your memorization of material. Successful memorization of passages from the text, equations, and derivations does not necessarily indicate that you understand the material. Your understanding of the material will be enhanced through a combination of efficient study habits, discussions with other students and with instructors, and your ability to solve the problems presented in the textbook. Ask questions whenever you believe that clarification of a concept is necessary.

Study Schedule

It is important that you set up a regular study schedule, preferably a daily one. Make sure that you read the syllabus for the course and adhere to the schedule set by your instructor. The lectures will make much more sense if you read the corresponding text material *before* attending them. As a general rule, you should devote about two hours of study time for each hour you are in class. If you are having trouble with the course, seek the advice of the instructor or other students who have taken the course. You may find it necessary to seek further instruction from experienced students. Very often, instructors offer review sessions in addition to regular class periods. Avoid the practice of delaying study until a day or two before an exam. More often than not, this approach has disastrous results. Rather than undertake an all-night study session before a test, briefly review the basic concepts and equations, and then get a good night's rest. If you believe that you need additional help in understanding the concepts, in preparing for exams, or in problem solving, we suggest that you acquire a copy of the *Student Solutions Manual/Study Guide* that accompanies this textbook.

Visit the *Principles of Physics* Web site at **www.cengagebrain.com/shop/ISBN/9781133104261** to see samples of select student supplements. You can purchase any Cengage Learning product at your local college store or at our preferred online store **CengageBrain.com.**

Use the Features

You should make full use of the various features of the text discussed in the Preface. For example, marginal notes are useful for locating and describing important equations and concepts, and **boldface** indicates important definitions. Many useful tables are contained in the appendices, but most are incorporated in the text where they are most often referenced. Appendix B is a convenient review of mathematical tools used in the text.

Answers to Quick Quizzes and odd-numbered problems are given at the end of the textbook, and solutions to selected end-of-chapter questions and problems are provided in the *Student Solutions Manual/Study Guide.* The table of contents provides an overview of the entire text, and the index enables you to locate specific material quickly. Footnotes are sometimes used to supplement the text or to cite other references on the subject discussed.

After reading a chapter, you should be able to define any new quantities introduced in that chapter and discuss the principles and assumptions that were used to arrive at certain key relations. The chapter summaries and the review sections of the *Student Solutions Manual/Study Guide* should help you in this regard. In some cases, you may find it necessary to refer to the textbook's index to locate certain topics. You should be able to associate with each physical quantity the correct symbol used to represent that quantity and the unit in which the quantity is specified. Furthermore, you should be able to express each important equation in concise and accurate prose.

Problem Solving

R. P. Feynman, Nobel laureate in physics, once said, "You do not know anything until you have practiced." In keeping with this statement, we strongly advise you to develop the skills necessary to solve a wide range of problems. Your ability to solve problems will be one of the main tests of your knowledge of physics; therefore, you should try to solve as many problems as possible. It is essential that you understand basic concepts and principles before attempting to solve problems. It is good practice to try to find alternate solutions to the same problem. For example, you can solve problems in mechanics using Newton's laws, but very often an alternative method that draws on energy considerations is more direct. You should not deceive yourself into thinking that you understand a problem merely because you have seen it solved in class. You must be able to solve the problem and similar problems on your own.

The approach to solving problems should be carefully planned. A systematic plan is especially important when a problem involves several concepts. First, read the problem several times until you are confident you understand what is being asked. Look for any key words that will help you interpret the problem and perhaps allow you to make certain assumptions. Your ability to interpret a question properly is an integral part of problem solving. Second, you should acquire the habit of writing down the information given in a problem and those quantities that need to be found; for example, you might construct a table listing both the quantities given and the quantities to be found. This procedure is sometimes used in the worked examples of the textbook. Finally, after you have decided on the method you believe is appropriate for a given problem, proceed with your solution. The General Problem-Solving Strategy will guide you through complex problems. If you follow the steps of this procedure (*Conceptualize, Categorize, Analyze, Finalize*), you will find it easier to come up with a solution and gain more from your efforts. This strategy, located at the end of Chapter 1 (pages 25–26), is used in all worked examples in the remaining chapters so that you can learn how to apply it. Specific problem-solving strategies for certain types of situations are included in the text and appear with a special heading. These specific strategies follow the outline of the General Problem-Solving Strategy.

Often, students fail to recognize the limitations of certain equations or physical laws in a particular situation. It is very important that you understand and remember the assumptions that underlie a particular theory or formalism. For example, certain equations in kinematics apply only to a particle moving with constant acceleration. These equations are not valid for describing motion whose acceleration is not constant, such as the motion of an object connected to a spring or the motion of an object through a fluid. Study the Analysis Models for Problem Solving in the chapter summaries carefully so that you know how each model can be applied to a specific situation. The analysis models provide you with a logical structure for solving problems and help you develop your thinking skills to become more like those of a physicist. Use the analysis model approach to save you hours of looking for the correct equation and to make you a faster and more efficient problem solver.

Experiments

Physics is a science based on experimental observations. Therefore, we recommend that you try to supplement the text by performing various types of "hands-on" experiments either at home or in the laboratory. These experiments can be used to test ideas and models discussed in class or in the textbook. For example, the common Slinky toy is excellent for studying traveling waves, a ball swinging on the end of a long string can be used to investigate pendulum motion, various masses attached to the end of a vertical spring or rubber band can be used to determine its elastic nature, an old pair of polarized sunglasses and some discarded lenses and a magnifying glass are the components of various experiments in optics, and an approximate measure of the free-fall acceleration can be determined simply by measuring with a stopwatch the time interval required for a ball to drop from a known height. The list of such experiments is endless. When physical models are not available, be imaginative and try to develop models of your own.

New Media

If available, we strongly encourage you to use the **Enhanced WebAssign** product that is available with this textbook. It is far easier to understand physics if you see it in action, and the materials available in Enhanced WebAssign will enable you to become a part of that action.

It is our sincere hope that you will find physics an exciting and enjoyable experience and that you will benefit from this experience, regardless of your chosen profession. Welcome to the exciting world of physics!

The scientist does not study nature because it is useful; he studies it because he delights in it, and he delights in it because it is beautiful. If nature were not beautiful, it would not be worth knowing, and if nature were not worth knowing, life would not be worth living.

—**Henri Poincaré**

Life Science Applications and Problems

Global Warming

Numerous scientific studies have detailed the effects of an increase in temperature of the Earth, including melting of ice from the polar ice caps and changes in climate and the corresponding effects on vegetation. Data taken over the past few decades show a measurable global temperature increase. Life on this planet depends on a delicate balance that keeps the global temperature in a narrow range necessary for our survival. How is this temperature determined? What factors need to be in balance to keep the temperature constant? If we can devise an adequate structural model that predicts the correct surface temperature of the Earth, we can use the model to predict changes in the temperature as we vary the parameters.

You most likely have an intuitive sense for the temperature of an object, and as long as the object is small (and the object is not undergoing combustion or some other rapid process) no significant temperature variation occurs between different points on the object. What about a huge object like the Earth, though? It is clear that no single temperature describes the entire planet; we know that it is summer in Australia when it is winter in Canada. The polar ice caps clearly have different temperatures from the tropical regions. Variations also occur in temperature within a single large body of water such as an ocean. Temperature varies greatly with altitude in a relatively local region, such as in and near Palm Springs, California, as shown in Figure 1. Therefore, when we speak of the temperature of the Earth, we will refer to an *average* surface temperature, taking into account all the variations across the surface. It is this average temperature that we would like to calculate by building a structural model of the atmosphere and comparing its prediction with the measured surface temperature.

A primary factor in determining the surface temperature of the Earth is the existence of our atmosphere. The atmosphere is a relatively thin (compared with the radius of the Earth) layer of gas above the surface that provides us with life-supporting oxygen. In addition to providing this important element for life, the atmosphere plays a major role in the energy balance that determines the average temperature. As we proceed with

Figure 1 Temperature variations with altitude can exist in a local region on the Earth. Here in Palm Springs, California, palm trees grow in the city while snow is present at the top of the local mountains.

this Context, we shall focus on the physics of gases and apply the principles we learn to the atmosphere.

One important component of the global warming problem is the concentration of carbon dioxide in the atmosphere. Carbon dioxide plays an important role in absorbing energy and raising the temperature of the atmosphere. As seen in Figure 2, the

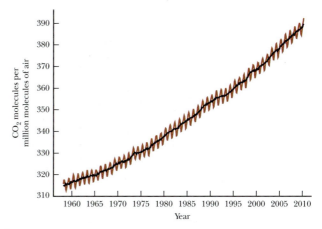

Figure 2 The concentration of atmospheric carbon dioxide in parts per million (ppm) of dry air as a function of time. These data were recorded at the Mauna Loa Observatory in Hawaii. The yearly variations (*red-brown curve*) coincide with growing seasons because vegetation absorbs carbon dioxide from the air. The steady increase in the average concentration (*black curve*) is of concern to scientists.

Courtesy of NASA

Figure 3 In this Context, we explore the energy balance of the Earth, which is a large, nonisolated system that interacts with its environment solely by means of electromagnetic radiation.

The Intergovernmental Panel on Climate Change (IPCC) is a scientific body that assesses the available information related to global warming and associated effects related to climate change. It was originally established in 1988 by two United Nations organizations, the World Meteorological Organization and the United Nations Environment Programme. The IPCC has published four assessment reports on climate change, the most recent in 2007, and a fifth report is scheduled to be released in 2014. The 2007 report concludes that there is a probability of greater than 90% that the increased global temperature measured by scientists is due to the placement of greenhouse gases such as carbon dioxide in the atmosphere by humans. The report also predicts a global temperature increase between 1 and 6°C in the 21st century, sea level rises from 18 to 59 cm, and very high probabilities of weather extremes, including heat waves, droughts, cyclones, and heavy rainfall.

In addition to its scientific aspects, global warming is a social issue with many facets. These facets encompass international politics and economics, because global warming is a worldwide problem. Changing our policies requires real costs to solve the problem. Global warming also has technological aspects, and new methods of manufacturing, transportation, and energy supply must be designed to slow down or reverse the increase in temperature. We shall restrict our attention to the physical aspects of global warming as we address this central question:

> **Can we build a structural model of the atmosphere that predicts the average temperature at the Earth's surface?**

amount of carbon dioxide in the atmosphere has been steadily increasing since the middle of the 20th century. This graph shows hard data that indicate that the atmosphere is undergoing a distinct change, although not all scientists agree on the interpretation of what that change means in terms of global temperatures.

Chapter 16

Temperature and the Kinetic Theory of Gases

Chapter Outline

A bubble in one of the many mud pots in Yellowstone National Park is caught just at the moment of popping. A mud pot is a pool of bubbling hot mud that demonstrates the existence of high temperatures below the Earth's surface.

Our study thus far has focused mainly on Newtonian mechanics, which explains a wide range of phenomena such as the motion of baseballs, rockets, and planets. We have applied these principles to isolated and nonisolated systems, oscillating systems, the propagation of mechanical waves through a medium, and the properties of fluids at rest and in motion. In Chapter 6, we introduced the notions of temperature and internal energy. We now extend the study of such notions as we focus on **thermodynamics,** which is concerned with the concepts of energy transfers between a system and its environment and the resulting variations in temperature or changes of state. As we shall see, thermodynamics explains the bulk properties of matter and the correlation between these properties and the mechanics of atoms and molecules.

Have you ever wondered how a refrigerator cools or what types of transformations occur in an automobile engine or why a bicycle pump becomes warm as you inflate a tire? The laws of thermodynamics enable us to answer such questions. In general, thermodynamics deals with the physical and chemical transformations of matter in various states: solid, liquid, and gas.

This chapter concludes with a study of ideal gases, which we shall approach on two levels. The first examines ideal gases on the macroscopic scale. Here we shall be concerned with the relationships among such quantities as pressure, volume, and temperature of the gas. On the second level, we shall examine gases on a microscopic (molecular) scale, using a structural model that treats the gas as a collection of particles. The latter approach will help us understand how behavior on the atomic level affects such macroscopic properties as pressure and temperature.

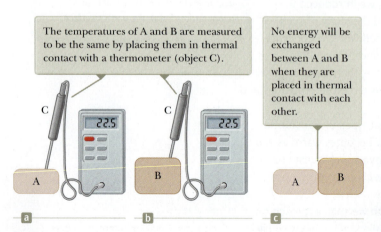 16.1 | Temperature and the Zeroth Law of Thermodynamics

We often associate the concept of temperature with how hot or cold an object feels to the touch. Our sense of touch provides us with a qualitative indication of temperature. Our senses are unreliable and often misleading, however. For example, if you stand with one bare foot on a tile floor and the other on an adjacent carpeted floor, the tile floor feels colder to your foot than the carpet even though the two are at the same temperature. That is because the properties of the tile are such that the transfer of energy (by heat) to the tile floor from your foot is more rapid than to the carpet. Your skin is sensitive to the *rate* of energy transfer—power—not the temperature of the object. Of course, the larger the difference in temperature between that of the object and that of your hand, the faster the energy transfer, so temperature and your sense of touch are related in *some* way. Recent research suggests that part of the temperature sensation in the skin is related to the TRPV3 protein present in sensory neurons that terminate in the skin. What we need is a reliable and reproducible method for establishing the relative "hotness" or "coldness" of objects that is related solely to the temperature of the object. Scientists have developed a variety of thermometers for making such quantitative measurements.

BIO Sense of warm and cold

We are all familiar with experiences in which two objects at different initial temperatures eventually reach some intermediate temperature when placed in contact with each other. For example, if you combine hot water and cold water in a bathtub from separate faucets, the combined water reaches an equilibrium temperature between the temperatures of the hot water and the cold water. Likewise, if an ice cube is placed in a cup of hot coffee, the ice eventually melts and the temperature of the coffee decreases.

We shall use these familiar examples to develop the scientific notion of temperature. Imagine two objects placed in an insulated container so that they form an isolated system. If the objects are at different temperatures, energy can be exchanged between them by, for example, heat or electromagnetic radiation. Objects that can exchange energy with each other in this way are said to be in **thermal contact.** Eventually, the temperatures of the two objects will become the same, one becoming warmer and the other cooler, as in our preceding examples. **Thermal equilibrium** is the situation in which two objects in thermal contact cease to have any net exchange of energy by heat or electromagnetic radiation.

Using these ideas, we can develop a formal definition of temperature. Consider two objects A and B that are not in thermal contact and a third object C that will be our **thermometer,** a device calibrated to measure the temperature of an object. We wish to determine whether A and B would be in thermal equilibrium if they were placed in thermal contact. The thermometer is first placed in thermal contact with A and its reading is recorded, as shown in Figure 16.1a. The thermometer is then placed in thermal contact with B and its reading is recorded (Fig. 16.1b). If the two readings are the same, A and B are in thermal equilibrium with each other. If they are placed in thermal contact with each other, as in Figure 16.1c, there is no net transfer of energy between them.

The temperatures of A and B are measured to be the same by placing them in thermal contact with a thermometer (object C).

No energy will be exchanged between A and B when they are placed in thermal contact with each other.

Figure 16.1 The zeroth law of thermodynamics.

We can summarize these results in a statement known as the **zeroth law of thermodynamics:**

> If objects A and B are separately in thermal equilibrium with a third object C, then A and B are in thermal equilibrium with each other.

▶ Zeroth law of thermodynamics

This statement, elementary as it may seem, is very important because it can be used to define the notion of temperature and is easily proved experimentally. We can think of temperature as the property that determines whether an object is in thermal equilibrium with other objects: Two objects in thermal equilibrium with each other are at the same temperature.

16.2 | Thermometers and Temperature Scales

In our discussion of the zeroth law, we mentioned a thermometer. Thermometers are devices used to measure the temperature of an object or a system with which the thermometer is in thermal equilibrium. All thermometers make use of some physical property that exhibits a change with temperature that can be calibrated to make the temperature measurable. Some of the physical properties used are (1) the volume of a liquid, (2) the length of a solid, (3) the pressure of a gas held at constant volume, (4) the volume of a gas held at constant pressure, (5) the electric resistance of a conductor, and (6) the color of a hot object.

A common thermometer in everyday use consists of a liquid—usually mercury or alcohol—that expands into a glass capillary tube when its temperature rises (Fig. 16.2). In this case, the physical property that changes is the volume of a liquid. Because the cross-sectional area of the capillary tube is uniform, the change in volume of the liquid varies linearly with its length along the tube. We can then define a temperature to be related to the length of the liquid column.

The thermometer can be calibrated by placing it in thermal contact with some environments that remain at constant temperature and marking the end of the liquid column on the thermometer. One such environment is a mixture of water and ice in thermal equilibrium with each other at atmospheric pressure. Once we have marked the ends of the liquid column for our chosen environments on our thermometer, we need to define a scale of numbers associated with various temperatures. One such scale is the **Celsius temperature scale.** On the Celsius scale, the temperature of the ice–water mixture is defined as zero degrees Celsius, written 0°C; this temperature is called the **ice point** or **freezing point** of water. Another commonly used environment is a mixture of water and steam in thermal equilibrium with each other at atmospheric pressure. On the Celsius scale, this temperature is defined as 100°C, the **steam point** or **boiling point** of water. Once the ends of the liquid column in the thermometer have been marked at these two points, the distance between the marks is divided into 100 equal segments, each denoting a change in temperature of one degree Celsius.

Thermometers calibrated in this way present problems when extremely accurate readings are needed. For instance, an alcohol thermometer calibrated at the ice and steam points of water might agree with a mercury thermometer only at the calibration points. Because mercury and alcohol have different thermal expansion properties (the expansion may not be perfectly linear with temperature), when one indicates a given temperature, the other may indicate a slightly different value. The discrepancies between different types of thermometers are especially large when the temperatures to be measured are far from the calibration points.

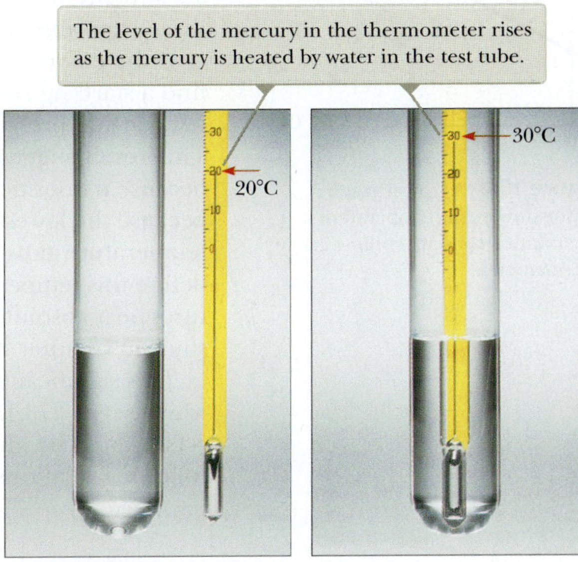

The level of the mercury in the thermometer rises as the mercury is heated by water in the test tube.

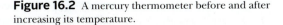

20°C

30°C

Figure 16.2 A mercury thermometer before and after increasing its temperature.

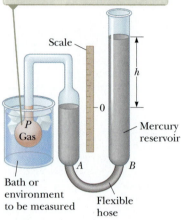

The volume of gas in the flask is kept constant by raising or lowering reservoir B to keep the mercury level in column A constant.

Figure 16.3 A constant-volume gas thermometer measures the pressure of the gas contained in the flask immersed in the bath.

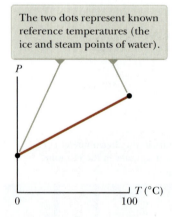

The two dots represent known reference temperatures (the ice and steam points of water).

Figure 16.4 A typical graph of pressure versus temperature taken with a constant-volume gas thermometer.

The Constant-Volume Gas Thermometer and the Kelvin Scale

Although practical devices such as the mercury thermometer can measure temperature, they do not define it in a fundamental way. Only one thermometer, the **gas thermometer,** offers a way to define temperature and relate it to internal energy directly. In a gas thermometer, the temperature readings are nearly independent of the substance used in the thermometer. One type of gas thermometer is the constant-volume example shown in Figure 16.3. The behavior observed in this device is the pressure variation with temperature of a fixed volume of gas.

When the constant-volume gas thermometer was developed, it was calibrated using the ice and steam points of water as follows. (A different calibration procedure, to be discussed shortly, is now used.) The gas flask is inserted into an ice bath, and mercury reservoir B is raised or lowered until the volume of the confined gas is at some value, indicated by the zero point on the scale. The height h, the difference between the levels in the reservoir and column A, indicates the pressure in the flask at 0°C, according to Equation 15.4. The flask is inserted into water at the steam point, and reservoir B is readjusted until the height in column A is again brought to zero on the scale, ensuring that the gas volume is the same as it had been in the ice bath (hence the designation "constant-volume"). A measure of the new value for h gives a value for the pressure at 100°C. These pressure and temperature values are then plotted on a graph, as in Figure 16.4. Based on experimental observations that the pressure of a gas varies linearly with its temperature, which is discussed in more detail in Section 16.4, we draw a straight line through our two points. The line connecting the two points serves as a calibration curve for measuring unknown temperatures. If we want to measure the temperature of a substance, we place the gas flask in thermal contact with the substance and adjust the column of mercury until the level in column A again returns to zero. The height of the mercury column tells us the pressure of the gas, and we can then find the temperature of the substance from the calibration curve.

Now suppose temperatures are measured with various gas thermometers containing different gases. Experiments show that the thermometer readings are nearly independent of the type of gas used as long as the gas pressure is low and the temperature is well above the point at which the gas liquefies.

We can also perform the temperature measurements with the gas in the flask at different starting pressures at 0°C. As long as the pressure is low, we will generate straight-line calibration curves for each different starting pressure, as shown for three experimental trials (solid lines) in Figure 16.5.

If the curves in Figure 16.5 are extended back toward negative temperatures, we find a startling result. In every case, regardless of the type of gas or the value of the low starting pressure, *the pressure extrapolates to zero when the temperature is* −273.15°C! This result suggests that this particular temperature is universal in its importance because it does not depend on the substance used in the thermometer. In addition, because the lowest possible pressure is $P = 0$, which would be a perfect vacuum, this temperature must represent a lower bound for physical processes. Therefore, we define this temperature as **absolute zero.** Some interesting effects occur at temperatures near absolute zero, such as the phenomenon of *superconductivity*, which we shall study in Chapter 21.

This significant temperature is used as the basis for the **Kelvin temperature scale,** which sets −273.15°C as its zero point (0 K). The size of a degree on the Kelvin scale is chosen to be identical to the size of a degree on the Celsius scale. Therefore, the following relationship enables conversion between these temperatures:

$$T_C = T - 273.15 \qquad \qquad \textbf{16.1} \blacktriangleleft$$

where T_C is the Celsius temperature and T is the Kelvin temperature (sometimes called the **absolute temperature**). The primary difference between these two temperature scales is a shift in the zero of the scale. The zero of the Celsius scale is arbitrary; it depends on a property associated with only one substance, water. The zero on the Kelvin scale is not arbitrary because it is characteristic of a behavior associated with all substances. Consequently, when an equation contains T as a variable, the absolute temperature must be used. Similarly, a ratio of temperatures is only meaningful if the temperatures are expressed on the Kelvin scale.

Equation 16.1 shows that the Celsius temperature T_C is shifted from the absolute temperature T by 273.15°. Because the size of a degree is the same on the two scales, a temperature difference of 5°C is equal to a temperature difference of 5 K. The two scales differ only in the choice of the zero point. Therefore, the ice point (273.15 K) corresponds to 0.00°C, and the steam point (373.15 K) is equivalent to 100.00°C.

Early gas thermometers made use of ice and steam points according to the procedure just described. These points are experimentally difficult to duplicate, however. For this reason, a new procedure based on two new points was adopted in 1954 by the International Committee on Weights and Measures. The first point is absolute zero. The second point is the **triple point of water,** which corresponds to the single temperature and pressure at which water, water vapor, and ice can coexist in equilibrium. This point is a convenient and reproducible reference temperature for the Kelvin scale. It occurs at a temperature of 0.01°C and a very low pressure of 4.58 mm of mercury. The temperature at the triple point of water on the Kelvin scale has a value of 273.16 K. Therefore, the SI unit of temperature, the **kelvin,** is defined as **1/273.16 of the temperature of the triple point of water.**

Figure 16.6 shows the Kelvin temperatures for various physical processes and conditions. As the figure reveals, absolute zero has never been achieved, although laboratory experiments have created conditions that are very close to absolute zero.

What would happen to a gas if its temperature could reach 0 K? As Figure 16.5 indicates (if we ignore the liquefaction and solidification of the substance), the pressure it would exert on the container's walls would be zero. In Section 16.5, we shall show that the pressure of a gas is proportional to the kinetic energy of the molecules of that gas. Therefore, according to classical physics, the kinetic energy of the gas would go to zero and there would be no motion at all of the individual components of the gas; hence, the molecules would settle out on the bottom of the container. Quantum theory, to be discussed in Chapter 28, modifies this statement to indicate that there would be some residual energy, called the *zero-point energy*, at this low temperature.

The Fahrenheit Scale

The most common temperature scale in everyday use in the United States is the **Fahrenheit scale.** This scale sets the temperature of the ice point at 32°F and the temperature of the steam point at 212°F. The relationship between the Celsius and Fahrenheit temperature scales is

$$T_F = \tfrac{9}{5} T_C + 32°F \qquad \textbf{16.2} \blacktriangleleft$$

Equation 16.2 can easily be used to find a relationship between changes in temperature on the Celsius and Fahrenheit scales. It is left as a problem for you to show that if the Celsius temperature changes by ΔT_C, the Fahrenheit temperature changes by an amount ΔT_F given by

$$\Delta T_F = \tfrac{9}{5} \Delta T_C \qquad \textbf{16.3} \blacktriangleleft$$

QUICK QUIZ 16.1 Consider the following pairs of materials. Which pair represents two materials, one of which is twice as hot as the other? (a) boiling water at 100°C, a glass of water at 50°C (b) boiling water at 100°C, frozen methane at −50°C (c) an ice cube at −20°C, flames from a circus fire-eater at 233°C (d) none of those pairs

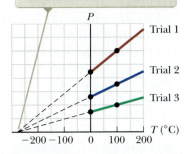

For all three trials, the pressure extrapolates to zero at the temperature −273.15°C.

Figure 16.5 Pressure versus temperature for experimental trials in which gases have different pressures in a constant volume gas thermometer.

Pitfall Prevention | 16.1
A Matter of Degree
Note that notations for temperatures in the Kelvin scale do not use the degree sign. The unit for a Kelvin temperature is simply "kelvins" and not "degrees Kelvin."

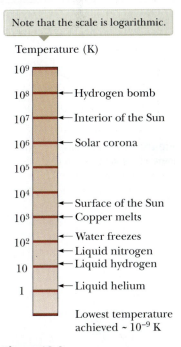

Note that the scale is logarithmic.

Temperature (K)

- 10^9
- 10^8 ← Hydrogen bomb
- 10^7 ← Interior of the Sun
- 10^6 ← Solar corona
- 10^5
- 10^4
- 10^3 ← Surface of the Sun
  ← Copper melts
- 10^2 ← Water freezes
  ← Liquid nitrogen
- 10 ← Liquid hydrogen
- 1 ← Liquid helium

Lowest temperature achieved ~ 10^{-9} K

Figure 16.6 Absolute temperatures at which various selected physical processes occur.

> ### THINKING PHYSICS 16.1
>
> A group of future astronauts lands on an inhabited planet. The astronauts strike up a conversation with the aliens about temperature scales. It turns out that the inhabitants of this planet have a temperature scale based on the freezing and boiling points of water, which are separated by 100 of the inhabitants' degrees. Would these two temperatures on this planet be the same as those on the Earth? Would the size of the aliens' degrees be the same as ours? Suppose the aliens have also devised a scale similar to the Kelvin scale. Would their absolute zero be the same as ours?
>
> **Reasoning** The values of 0°C and 100°C for the freezing and boiling points of water are defined at atmospheric pressure. On another planet, it is unlikely that atmospheric pressure would be exactly the same as that on the Earth. Therefore, water would freeze and boil at different temperatures on the alien planet. The aliens may call these temperatures 0° and 100°, but they would not be the same temperatures as our 0°C and 100°C. If the aliens did assign values of 0° and 100° for these temperatures, their degrees would not be the same size as our Celsius degrees (unless their atmospheric pressure were the same as ours). For an alien version of the Kelvin scale, the absolute zero would be the same as ours because it is based on a natural, universal definition rather than being associated with a particular substance or a given atmospheric pressure. ◀

Example 16.1 | Converting Temperatures

On a day when the temperature reaches 50°F, what is the temperature in degrees Celsius and in kelvins?

SOLUTION

Conceptualize In the United States, a temperature of 50°F is well understood. In many other parts of the world, however, this temperature might be meaningless because people are familiar with the Celsius temperature scale.

Categorize This example is a simple substitution problem.

Substitute the given temperature into Equation 16.2:

$$T_C = \tfrac{5}{9}(T_F - 32) = \tfrac{5}{9}(50 - 32) = \boxed{10°C}$$

Use Equation 16.1 to find the Kelvin temperature:

$$T = T_C + 273.15 = 10°C + 273.15 = \boxed{283 \text{ K}}$$

A convenient set of weather-related temperature equivalents to keep in mind is that 0°C is (literally) freezing at 32°F, 10°C is cool at 50°F, 20°C is room temperature, 30°C is warm at 86°F, and 40°C is a hot day at 104°F.

16.3 | Thermal Expansion of Solids and Liquids

Our discussion of the liquid thermometer makes use of one of the best known changes that occur in most substances, that as the temperature of a substance increases, its volume increases. This phenomenon, known as **thermal expansion,** plays an important role in numerous applications. For example, thermal expansion joints (Fig. 16.7) must

Figure 16.7 Thermal-expansion joints in (a) bridges and (b) walls.

Without these joints to separate sections of roadway on bridges, the surface would buckle due to thermal expansion on very hot days or crack due to contraction on very cold days.

The long, vertical joint is filled with a soft material that allows the wall to expand and contract as the temperature of the bricks changes.

a

b

© Cengage Learning/George Semple

© Cengage Learning/George Semple

be included in buildings, concrete highways, railroad tracks, and bridges to compensate for changes in dimensions with temperature variations.

The overall thermal expansion of an object is a consequence of the change in the average separation between its constituent atoms or molecules. To understand this concept, consider how the atoms in a solid substance behave. These atoms are located at fixed equilibrium positions; if an atom is pulled away from its position, a restoring force pulls it back. We can build a structural model in which we imagine that the atoms are particles at their equilibrium positions connected by springs to their neighboring atoms (Fig. 16.8). If an atom is pulled away from its equilibrium position, the distortion of the springs provides a restoring force. If the atom is released, it oscillates, and we can apply the particle in simple harmonic motion model to it. A number of macroscopic properties of the substance can be understood with this type of structural model on the atomic level.

In Chapter 6, we introduced the notion of internal energy and pointed out that it is related to the temperature of a system. For a solid, the internal energy is associated with the kinetic and potential energy of the vibrations of the atoms around their equilibrium positions. At ordinary temperatures, the atoms vibrate with an amplitude of about 10^{-11} m, and the average spacing between the atoms is about 10^{-10} m. As the temperature of the solid increases, the average separation between atoms increases. The increase in average separation with increasing temperature (and subsequent thermal expansion) is the result of a breakdown in the model of simple harmonic motion. Active Figure 6.23a in Chapter 6 shows the potential energy curve for an ideal simple harmonic oscillator. The potential energy curve for atoms in a solid is similar but not exactly the same as that one; it is slightly asymmetric around the equilibrium position. It is this asymmetry that leads to thermal expansion.

If the thermal expansion of an object is sufficiently small compared with the object's initial dimensions, the change in any dimension is, to a good approximation, dependent on the first power of the temperature change. For most situations, we can adopt a simplification model in which this dependence is true. Suppose an object has an initial length L_i along some direction at some temperature. The length increases by ΔL for a change in temperature ΔT. Experiments show that when ΔT is small enough, ΔL is proportional to ΔT and to L_i:

$$\Delta L = \alpha L_i \, \Delta T \qquad \qquad \text{16.4} \blacktriangleleft$$

or

$$L_f - L_i = \alpha L_i (T_f - T_i) \qquad \qquad \text{16.5} \blacktriangleleft$$

where L_f is the final length, T_f is the final temperature, and the proportionality constant α is called the **average coefficient of linear expansion** for a given material and has units of inverse degrees Celsius, or $(°C)^{-1}$.

It may be helpful to think of thermal expansion as a magnification or a photographic enlargement. For example, as a metal washer is heated (Active Fig. 16.9), all dimensions, including the radius of the hole, increase according to Equation 16.4. Because the linear dimensions of an object change with temperature, it follows that volume and surface area also change with temperature. Consider a cube having an initial edge length L_i and therefore an initial volume $V_i = L_i^3$. As the temperature is increased, the length of each side increases to

$$L_f = L_i + \alpha L_i \, \Delta T$$

The new volume, $V_f = L_f{}^3$, is

$$L_f{}^3 = (L_i + \alpha L_i \, \Delta T)^3 = L_i^3 + 3\alpha L_i^3 \Delta T + 3\alpha^2 L_i^3 (\Delta T)^2 + \alpha^3 L_i^3 (\Delta T)^3$$

The last two terms in this expression contain the quantity $\alpha \, \Delta T$ raised to the second and third powers. Because $\alpha \, \Delta T$ is a pure number much less than 1, raising it

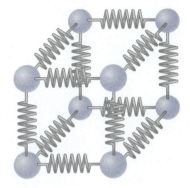

Figure 16.8 A structural model of the atomic configuration in a solid. The atoms (spheres) are imagined to be connected to one another by springs that reflect the elastic nature of the interatomic forces.

As the washer is heated, all dimensions increase, including the radius of the hole.

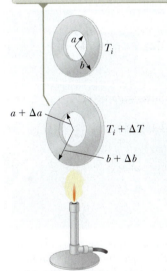

Active Figure 16.9 Thermal expansion of a homogeneous metal washer. (The expansion is exaggerated in this figure.)

Pitfall Prevention | 16.2
Do Holes Become Larger or Smaller?
When an object's temperature is raised, every linear dimension increases in size. Included are any holes in the material, which expand in the same way as if the hole were filled with the material, as shown in Active Figure 16.9. Keep in mind the notion of thermal expansion as being similar to a photographic enlargement.

TABLE 16.1 | Average Expansion Coefficients for Some Materials Near Room Temperature

Material	Average Coefficient of Linear Expansion $(\alpha)(°C^{-1})$	Material	Average Coefficient of Volume Expansion $(\beta)(°C^{-1})$
Aluminum	24×10^{-6}	Acetone	1.5×10^{-4}
Brass and bronze	19×10^{-6}	Alcohol, ethyl	1.12×10^{-4}
Concrete	12×10^{-6}	Benzene	1.24×10^{-4}
Copper	17×10^{-6}	Gasoline	9.6×10^{-4}
Glass (ordinary)	9×10^{-6}	Glycerin	4.85×10^{-4}
Glass (Pyrex)	3.2×10^{-6}	Mercury	1.82×10^{-4}
Invar (Ni–Fe alloy)	0.9×10^{-6}	Turpentine	9.0×10^{-4}
Lead	29×10^{-6}	Air[a] at 0°C	3.67×10^{-3}
Steel	11×10^{-6}	Helium[a]	3.665×10^{-3}

[a]Gases do not have a specific value for the volume expansion coefficient because the amount of expansion depends on the type of process through which the gas is taken. The values given here assume the gas undergoes an expansion at constant pressure.

to a power makes it even smaller. Therefore, we can ignore these terms to obtain a simpler expression:

$$V_f = L_f^{\,3} = L_i^{\,3} + 3\alpha L_i^{\,3}\,\Delta T = V_i + 3\alpha V_i\,\Delta T$$

or

$$\Delta V = V_f - V_i = \beta V_i\,\Delta T \qquad \text{16.6} \blacktriangleleft$$

where $\beta = 3\alpha$. The quantity β is called the **average coefficient of volume expansion.** We considered a cubic shape in deriving this equation, but Equation 16.6 describes a sample of any shape as long as the average coefficient of linear expansion is the same in all directions.

By a similar procedure, we can show that the *increase in area* of an object accompanying an increase in temperature is

$$\Delta A = \gamma A_i\,\Delta T \qquad \text{16.7} \blacktriangleleft$$

where γ, the **average coefficient of area expansion,** is given by $\gamma = 2\alpha$.

Table 16.1 lists the average coefficient of linear expansion for various materials. Note that for these materials α is positive, indicating an increase in length with increasing temperature, but that is not always the case. For example, some substances, such as calcite ($CaCO_3$), expand along one dimension (positive α) and contract along another (negative α) with increasing temperature.

Thermal expansion. The extremely high temperature of a July day in Asbury Park, New Jersey, caused these railroad tracks to buckle.

AP Photo

QUICK QUIZ 16.2 Two spheres are made of the same metal and have the same radius, but one is hollow and the other is solid. The spheres are taken through the same temperature increase. Which sphere expands more? (**a**) The solid sphere expands more. (**b**) The hollow sphere expands more. (**c**) They expand by the same amount. (**d**) There is not enough information to say.

THINKING PHYSICS 16.2

As a homeowner is painting a ceiling, a drop of paint falls from the brush onto an operating incandescent lightbulb. The bulb breaks. Why?

Reasoning The glass envelope of an incandescent lightbulb receives energy on the inside surface by electromagnetic radiation from the very hot filament. In addition, because the bulb contains gas, the glass envelope receives energy by matter transfer related to the movement of the hot gas near the filament to the colder glass. Therefore, the glass can become very hot. If a drop of relatively cold paint falls onto the glass, that portion of the glass envelope suddenly becomes colder than the other portions, and the contraction of this region can cause thermal stresses that might break the glass. ◄

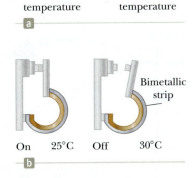

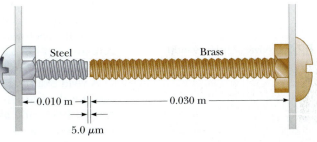

Figure 16.10 (a) A bimetallic strip bends as the temperature changes because the two metals have different expansion coefficients. (b) A bimetallic strip used in a thermostat to make or break electrical contact.

As Table 16.1 indicates, each substance has its own characteristic coefficients of expansion. For example, when the temperatures of a brass rod and a steel rod of equal length are raised by the same amount from some common initial value, the brass rod expands more than the steel rod because brass has a larger coefficient of expansion than steel. A simple device called a bimetallic strip that demonstrates this principle is found in practical devices such as thermostats in home furnace systems. The strip is made by securely bonding two different metals together along their surfaces. As the temperature of the strip increases, the two metals expand by different amounts and the strip bends as in Figure 16.10.

Example 16.2 | The Thermal Electrical Short

A poorly designed electronic device has two bolts attached to different parts of the device that almost touch each other in its interior as in Figure 16.11. The steel and brass bolts are at different electric potentials, and if they touch, a short circuit will develop, damaging the device. (We will study electric potential in Chapter 20.) The initial gap between the ends of the bolts is 5.0 μm at 27°C. At what temperature will the bolts touch? Assume the distance between the walls of the device is not affected by the temperature change.

Figure 16.11 (Example 16.2) Two bolts attached to different parts of an electrical device are almost touching when the temperature is 27°C. As the temperature increases, the ends of the bolts move toward each other.

SOLUTION

Conceptualize Imagine the ends of both bolts expanding into the gap between them as the temperature rises.

Categorize We categorize this example as a thermal expansion problem in which the *sum* of the changes in length of the two bolts must equal the length of the initial gap between the ends.

Analyze Set the sum of the length changes equal to the width of the gap:

$$\Delta L_{br} + \Delta L_{st} = \alpha_{br} L_{i,br} \Delta T + \alpha_{st} L_{i,st} \Delta T = 5.0 \times 10^{-6} \text{ m}$$

Solve for ΔT:

$$\Delta T = \frac{5.0 \times 10^{-6} \text{ m}}{\alpha_{br} L_{i,br} + \alpha_{st} L_{i,st}}$$

$$= \frac{5.0 \times 10^{-6} \text{ m}}{[19 \times 10^{-6} \text{ (°C)}^{-1}](0.030 \text{ m}) + [11 \times 10^{-6} \text{ (°C)}^{-1}](0.010 \text{ m})} = 7.4\text{°C}$$

Find the temperature at which the bolts touch:

$$T = 27\text{°C} + 7.4\text{°C} = \boxed{34\text{°C}}$$

Finalize This temperature is possible if the air conditioning in the building housing the device fails for a long period on a very hot summer day.

The Unusual Behavior of Water

Liquids generally increase in volume with increasing temperature and have volume expansion coefficients on the order ten times greater than those of solids. Water is an exception to this rule over a small temperature range, as we can see from its

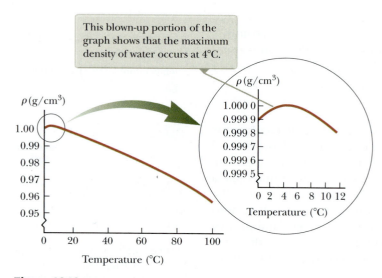

This blown-up portion of the graph shows that the maximum density of water occurs at 4°C.

Figure 16.12 The variation of density with temperature for water at atmospheric pressure.

density versus temperature curve in Figure 16.12. As the temperature increases from 0°C to 4°C, water contracts and therefore its density increases. Above 4°C, water exhibits the expected expansion with increasing temperature. Therefore, the density of water reaches a maximum value of 1 000 kg/m³ at 4°C.

We can use this unusual thermal expansion behavior of water to explain why a pond freezes at the surface. When the atmospheric temperature drops from 7°C to 6°C, for example, the water at the surface of the pond also cools and consequently decreases in volume. Hence, the surface water is denser than the water below it, which has not cooled and has not decreased in volume. As a result, the surface water sinks and warmer water from below moves to the surface to be cooled in a process called upwelling. When the atmospheric temperature is between 4°C and 0°C, however, the surface water expands as it cools, becoming less dense than the water below it. The sinking process stops, and eventually the surface water freezes. As the water freezes, the ice remains on the surface because ice is less dense than water. The ice continues to build up on the surface, while water near the bottom of the pool remains at 4°C. If that did not happen, fish and other forms of marine life would not survive through the winter.

A vivid example of the dangers of the absence of the upwelling and mixing processes is the sudden and deadly release of carbon dioxide gas from Lake Monoun in August 1984 and Lake Nyos in August 1986 (Fig. 16.13). Both lakes are located in the rain forest country of Cameroon in Africa. More than 1 700 natives of Cameroon died in these events.

In a lake located in a temperate zone such as the United States, significant temperature variations occur during the day and during the entire year. For example, imagine the Sun going down in the evening. As the temperature of the surface water drops because of the absence of sunlight, the sinking process tends to mix the upper and lower layers of water.

This mixing process does not normally occur in Lake Monoun and Lake Nyos because of two characteristics that contributed significantly to the disasters. First, the lakes are very deep, so mixing the various layers of water over such a large vertical distance is difficult. This factor also results in such very large pressure at the bottom

Figure 16.13 (a) Lake Nyos, in Cameroon, after an explosive outpouring of carbon dioxide. (b) The carbon dioxide caused many deaths, both of humans and animals, such as the cattle shown here.

of the lake that a large amount of carbon dioxide from local rocks and deep springs dissolves into the water. Second, both lakes are located in an equatorial rain forest region where the temperature variation is much smaller than in temperate zones, which results in little driving force to mix the layers of water in the lakes. Water near the bottom of the lake stays there for a long time and collects a large amount of dissolved carbon dioxide. In the absence of a mixing process, this carbon dioxide cannot be brought to the surface and released safely. It simply continues to increase in concentration.

The situation described is explosive. If the carbon dioxide–laden water is brought to the surface where the pressure is much lower, the gas expands and comes out of the solution rapidly. Once the carbon dioxide comes out of the solution, bubbles of carbon dioxide rise through the water and cause more mixing of layers.

Suppose the temperature of the surface water were to decrease; this water would become denser and sink, possibly triggering the release of carbon dioxide and the beginning of the explosive situation just described. The monsoon season in Cameroon occurs in August. Monsoon clouds block the sunlight, resulting in lower surface water temperatures, which may be the reason the disasters occurred in August. Climate data for Cameroon show lower than normal temperatures and higher than normal rainfall in the mid-1980s. The resulting decrease in surface temperature could explain why these events occurred in 1984 and 1986. The exact reasons for the sudden release of carbon dioxide are unknown and remain an area of active research.

Finally, once the carbon dioxide was released from the lakes, it stayed near the ground because carbon dioxide is denser than air. Therefore, a layer of carbon dioxide gas spread out over the land around the lake, representing a deadly suffocating gas for all humans and animals in its path.

BIO Suffocation by explosive release of carbon dioxide

◀ 16.4 | Macroscopic Description of an Ideal Gas

The properties of gases are very important in a number of thermal processes. Our everyday weather is a perfect example of the types of processes that depend on the behavior of gases.

If we introduce a gas into a container, it expands to fill the container uniformly. Therefore, the gas does not have a fixed volume or pressure. Its volume is that of the container, and its pressure depends on the size of the container. In this section, we shall be concerned with the properties of a gas with pressure P and temperature T, confined to a container of volume V. It is useful to know how these quantities are related. In general, the equation that interrelates these quantities, called the **equation of state,** can be complicated. If the gas is maintained at a very low pressure (or low density), however, the equation of state is found experimentally to be relatively simple. Such a low-density gas is commonly referred to as an **ideal gas.** Most gases at room temperature and atmospheric pressure behave approximately as ideal gases. We shall adopt a simplification model, called the **ideal gas model,** for these types of studies. In this model, an ideal gas is a collection of atoms or molecules that (1) move randomly, (2) exert no long-range forces on one another, and (3) are so small that they occupy a negligible fraction of the volume of their container.

It is convenient to express the amount of gas in a given volume in terms of the number of moles. One **mole** of any substance is that mass of the substance that contains **Avogadro's number,** $N_A = 6.022 \times 10^{23}$, of molecules. The number of moles n of a substance in a sample is related to its mass m through the expression

$$n = \frac{m}{M} \qquad \textbf{16.8} \blacktriangleleft$$

where M is the **molar mass** of the substance, usually expressed in grams per mole. For example, the molar mass of molecular oxygen O_2 is 32.0 g/mol. The mass of one mole of oxygen is therefore 32.0 g. We can calculate the mass m_0 of one molecule by

Active Figure 16.14 An ideal gas confined to a cylinder whose volume can be varied with a movable piston.

▶ Ideal gas law

▶ The universal gas constant

dividing the molar mass by the number of molecules, which is Avogadro's number. Therefore, for oxygen,

$$m_0 = \frac{M}{N_A} = \frac{32.0 \times 10^{-3} \text{ kg/mol}}{6.02 \times 10^{23} \text{ molecule/mol}} = 5.32 \times 10^{-26} \text{ kg/molecule}$$

Now suppose an ideal gas is confined to a cylindrical container whose volume can be varied by means of a movable piston, as in Active Figure 16.14. We shall assume that the cylinder does not leak, so the number of moles of gas remains constant. For such a system, experiments provide the following information:

- When the gas is kept at a constant temperature, its pressure is inversely proportional to the volume. (This behavior is described historically as Boyle's law.)
- When the pressure of the gas is kept constant, the volume is directly proportional to the temperature. (This behavior is described historically as Charles's law.)
- When the volume of the gas is kept constant, the pressure is directly proportional to the temperature. (This behavior is described historically as Gay-Lussac's law.)

These observations can be summarized by the following equation of state, known as the **ideal gas law:**

$$PV = nRT \qquad \textbf{16.9} \blacktriangleleft$$

In this expression, R is a constant for a specific gas that can be determined from experiments and T is the absolute temperature in kelvins. Experiments on several gases show that as the pressure approaches zero, the quantity PV/nT approaches the same value of R for *all* gases. For this reason, R is called the **universal gas constant.** In SI units, where pressure is expressed in pascals and volume in cubic meters, R has the value

$$R = 8.314 \text{ J/mol} \cdot \text{K} \qquad \textbf{16.10} \blacktriangleleft$$

If the pressure is expressed in atmospheres and the volume in liters ($1 \text{ L} = 10^3 \text{ cm}^3 = 10^{-3} \text{ m}^3$), R has the value

$$R = 0.082 \, 1 \text{ L} \cdot \text{atm/mol} \cdot \text{K}$$

Using this value of R and Equation 16.9, one finds that the volume occupied by 1 mol of any gas at atmospheric pressure and 0°C (273 K) is 22.4 L.

The ideal gas law is often expressed in terms of the total number of molecules N rather than the number of moles n. Because the total number of molecules equals the product of the number of moles and Avogadro's number N_A, we can write Equation 16.9 as

$$PV = nRT = \frac{N}{N_A} RT$$

$$PV = Nk_B T \qquad \textbf{16.11} \blacktriangleleft$$

where k_B is called **Boltzmann's constant** and has the value

$$k_B = \frac{R}{N_A} = 1.38 \times 10^{-23} \text{ J/K} \qquad \textbf{16.12} \blacktriangleleft$$

QUICK QUIZ 16.3 A common material for cushioning objects in packages is made by trapping bubbles of air between sheets of plastic. Is this material more effective at keeping the contents of the package from moving around inside the package on (**a**) a hot day, (**b**) a cold day, or (**c**) either hot or cold days?

QUICK QUIZ 16.4 On a winter day, you turn on your furnace and the temperature of the air inside your home increases. Assume that your home has the normal amount of leakage between inside air and outside air. Is the number of moles of air in your room at the higher temperature (**a**) larger than before, (**b**) smaller than before, or (**c**) the same as before?

Example 16.3 | **Heating a Spray Can**

A spray can containing a propellant gas at twice atmospheric pressure (202 kPa) and having a volume of 125.00 cm³ is at 22°C. It is then tossed into an open fire. (*Warning:* Do not do this experiment; it is very dangerous.) When the temperature of the gas in the can reaches 195°C, what is the pressure inside the can? Assume any change in the volume of the can is negligible.

SOLUTION

Conceptualize Intuitively, you should expect that the pressure of the gas in the container increases because of the increasing temperature.

Categorize We model the gas in the can as ideal and use the ideal gas law to calculate the new pressure.

Analyze Rearrange Equation 16.9:

$$(1)\quad \frac{PV}{T} = nR$$

No air escapes during the compression, so n, and therefore nR, remains constant. Hence, set the initial value of the left side of Equation (1) equal to the final value:

$$(2)\quad \frac{P_i V_i}{T_i} = \frac{P_f V_f}{T_f}$$

Because the initial and final volumes of the gas are assumed to be equal, cancel the volumes:

$$(3)\quad \frac{P_i}{T_i} = \frac{P_f}{T_f}$$

Solve for P_f:

$$P_f = \left(\frac{T_f}{T_i}\right) P_i = \left(\frac{468 \text{ K}}{295 \text{ K}}\right)(202 \text{ kPa}) = \boxed{320 \text{ kPa}}$$

Finalize The higher the temperature, the higher the pressure exerted by the trapped gas as expected. If the pressure increases sufficiently, the can may explode. Because of this possibility, you should never dispose of spray cans in a fire.

What If? Suppose we include a volume change due to thermal expansion of the steel can as the temperature increases. Does that alter our answer for the final pressure significantly?

Answer Because the thermal expansion coefficient of steel is very small, we do not expect much of an effect on our final answer.

Find the change in the volume of the can using Equation 16.6 and the value for α for steel from Table 16.1:

$$\Delta V = \beta V_i \Delta T = 3\alpha V_i \Delta T$$
$$= 3[11 \times 10^{-6} \, (°C)^{-1}](125.00 \text{ cm}^3)(173°C) = 0.71 \text{ cm}^3$$

Start from Equation (2) again and find an equation for the final pressure:

$$P_f = \left(\frac{T_f}{T_i}\right)\left(\frac{V_i}{V_f}\right) P_i$$

This result differs from Equation (3) only in the factor V_i/V_f. Evaluate this factor:

$$\frac{V_i}{V_f} = \frac{125.00 \text{ cm}^3}{(125.00 \text{ cm}^3 + 0.71 \text{ cm}^3)} = 0.994 = 99.4\%$$

Therefore, the final pressure will differ by only 0.6% from the value calculated without considering the thermal expansion of the can. Taking 99.4% of the previous final pressure, the final pressure including thermal expansion is 318 kPa.

16.5 | The Kinetic Theory of Gases

In the preceding section, we discussed the macroscopic properties of an ideal gas using such quantities as pressure, volume, number of moles, and temperature. From a *macroscopic* point of view, the mathematical representation of the ideal gas model is the ideal gas law. In this section, we consider the *microscopic* point of view of the ideal gas model. We shall show that the macroscopic properties can be understood on the basis of what is happening on the atomic scale.

Using the ideal gas model, we shall build a structural model of a gas enclosed in a container. The mathematical structure and the predictions made by this model

Ludwig Boltzmann
Austrian physicist (1844–1906)
Boltzmann made many important
contributions to the development of the
kinetic theory of gases, electromagnetism,
and thermodynamics. His pioneering work in
the field of kinetic theory led to the branch of
physics known as statistical mechanics.

constitute what is known as the **kinetic theory of gases.** With this theory, we shall interpret the pressure and temperature of an ideal gas in terms of microscopic variables. Our structural model will include the following components:

1. *A description of the physical components of the system*: The gas consists of a number of identical molecules within a cubic container of side length *d*. The number of molecules in the gas is large, and the average separation between them is large compared with their dimensions. Therefore, the molecules occupy a negligible volume in the container. This assumption is consistent with the ideal gas model, in which we imagine the molecules to be point-like.

2. *A description of where the components are located relative to one another and how they interact*: The molecules are distributed uniformly throughout the container and behave as follows:
   (a) The molecules obey Newton's laws of motion, but as a whole their motion is isotropic: any molecule can move in any direction with any speed.
   (b) The molecules interact only by short-range forces during elastic collisions. This assumption is consistent with the ideal gas model, in which the molecules exert no long-range forces on one another.
   (c) The molecules make elastic collisions with the walls.

3. *A description of the time evolution of the system*: The system has reached a steady-state situation so that macroscopic descriptions of the gas (volume, temperature, pressure, etc.) remain fixed. The velocities of individual molecules are constantly changing.

4. *A description of the agreement between predictions of the model and actual observations and, possibly, predictions of new effects that have not yet been observed*: Our structural model should make some specific predictions relating macroscopic measurements to microscopic behavior. In particular, we would like to predict how pressure and temperature are related to the microscopic parameters associated with the molecules.

Although we often picture an ideal gas as consisting of single atoms, molecular gases exhibit equally good approximations to ideal gas behavior at low pressures. Effects associated with molecular structure have no influence on the motions considered here. Therefore, we can apply the results of the following development to molecular gases as well as to monatomic gases.

Molecular Interpretation of the Pressure of an Ideal Gas

For our first application of kinetic theory, let us derive an expression for the pressure of *N* molecules of an ideal gas in a container of volume *V* in terms of microscopic quantities. As outlined in our structural model, the container is a cube with edges of length *d* (Fig. 16.15). We shall focus our attention on one of these molecules of mass m_0 and assumed to be moving so that its component of velocity in the *x* direction is v_{xi} as in Active Figure 16.16. (The subscript *i* here refers to the *i*th molecule, not to an initial value. We will combine the effects of all of the molecules shortly.) As the molecule collides elastically with any wall, as proposed in structural model component 2(c), its velocity component perpendicular to the wall is reversed because the mass of the wall is far greater than the mass of the molecule. Because the momentum component p_{xi} of the molecule is $m_0 v_{xi}$ before the collision and $-m_0 v_{xi}$ after the collision, the change in momentum of the molecule in the *x* direction is

$$\Delta p_{xi} = -m_0 v_{xi} - (m_0 v_{xi}) = -2m_0 v_{xi}$$

Applying the impulse–momentum theorem (Eq. 8.11) to the molecule gives

$$\overline{F}_{i,\text{on molecule}}\, \Delta t_{\text{collision}} = \Delta p_{xi} = -2m_0 v_{xi}$$

where $\overline{F}_{i,\text{on molecule}}$ is the average force component,[1] perpendicular to the wall, for the force that the wall exerts on the molecule during the collision and $\Delta t_{\text{collision}}$ is the

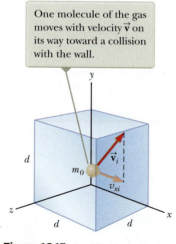

One molecule of the gas moves with velocity $\vec{v}$ on its way toward a collision with the wall.

Figure 16.15 A cubical box with sides of length *d* containing an ideal gas.

---

[1]For this discussion, we will use a bar over a variable to represent the average value of the variable, such as $\overline{F}$ for the average force, rather than the subscript "avg" that we have used before. This notation saves confusion because we will already have a number of subscripts on variables.

duration of the collision. For the molecule to make another collision with the same wall after this first collision, it must travel a distance of $2d$ in the x direction (across the container and back). The time interval between two collisions with the same wall is therefore

$$\Delta t = \frac{2d}{v_{xi}}$$

The force that causes the change in momentum of the molecule in the collision with the wall occurs only during the collision. We can, however, average the force over the time interval for the molecule to move across the cube and back. Sometime during this time interval the collision occurs, so the change in momentum for this time interval is the same as that for the short duration of the collision. Therefore, we can rewrite the impulse–momentum theorem as

$$\overline{F}_i \Delta t = -2m_0 v_{xi}$$

where $\overline{F}_i$ is interpreted as the average force component on the molecule over the time for the molecule to move across the cube and back. Because exactly one collision occurs for each such time interval, it is also the long-term average force component on the molecule, over long time intervals containing any number of multiples of Δt.

The substitution of Δt into the impulse–momentum equation enables us to express the long-term average force component of the wall on the molecule:

$$\overline{F}_i = \frac{-2m_0 v_{xi}}{\Delta t} = \frac{-2m_0 v_{xi}{}^2}{2d} = \frac{-m_0 v_{xi}{}^2}{d}$$

Now, by Newton's third law, the force component of the molecule on the wall is equal in magnitude and opposite in direction:

$$\overline{F}_{i,\text{on wall}} = -\overline{F}_i = -\left(\frac{-m_0 v_{xi}{}^2}{d}\right) = \frac{m_0 v_{xi}{}^2}{d}$$

The magnitude of the total average force $\overline{F}$ exerted on the wall by the gas is found by adding the average force components exerted by the individual molecules. We add terms such as those shown in the preceding equation for all molecules:

$$\overline{F} = \sum_{i=1}^{N} \frac{m_0 v_{xi}{}^2}{d} = \frac{m_0}{d} \sum_{i=1}^{N} v_{xi}{}^2$$

where we have factored out the length of the box and the mass m_0 because structural model component 1 tells us that all the molecules are the same. We now impose the condition that the number of molecules is large. For a small number of molecules, the actual force on the wall would vary with time. It would be nonzero during the short interval of a collision of a molecule with the wall and zero when no molecule happens to be hitting the wall. For a very large number of molecules, however, such as Avogadro's number, these variations in force are smoothed out, so the average force is the same over *any* time interval. Therefore, the *constant* force F on the wall due to the molecular collisions is the same as the average force $\overline{F}$ and is of magnitude

$$F = \frac{m_0}{d} \sum_{i=1}^{N} v_{xi}{}^2$$

To proceed further, let us consider how we express the average value of the square of the x component of the velocity for the N molecules. The traditional average of a value is the sum of the values over the number of values:

$$\overline{v_x{}^2} = \frac{\displaystyle\sum_{i=1}^{N} v_{xi}{}^2}{N}$$

The numerator of this expression is contained in the right-hand side of the previous equation. Therefore, by combining the two expressions the total force on the wall can be written

$$F = \frac{m_0}{d} N \overline{v_x{}^2}$$

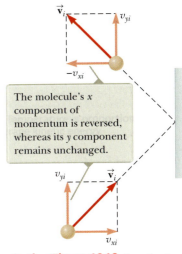

The molecule's x component of momentum is reversed, whereas its y component remains unchanged.

Active Figure 16.16 A molecule makes an elastic collision with the wall of the container. In this construction, we assume that the molecule moves in the xy plane.

9. A sample of lead has a mass of 20.0 kg and a density of 11.3×10^3 kg/m^3 at 0°C. (a) What is the density of lead at 90.0°C? (b) What is the mass of the sample of lead at 90.0°C?

10. **S** A sample of a solid substance has a mass m and a density ρ_0 at a temperature T_0. (a) Find the density of the substance if its temperature is increased by an amount ΔT in terms of the coefficient of volume expansion b. (b) What is the mass of the sample if the temperature is raised by an amount ΔT?

11. **BIO** Each year thousands of children are badly burned by hot tap water. Figure P16.11 shows a cross-sectional view of an antiscalding faucet attachment designed to prevent such accidents. Within the device, a spring made of material with a high coefficient of thermal expansion controls a movable plunger. When the water temperature rises above a preset safe value, the expansion of the spring causes the plunger to shut off the water flow. Assuming that the initial length L of the unstressed spring is 2.40 cm and its coefficient of linear expansion is 22.0×10^{-6} (°C)$^{-1}$, determine the increase in length of the spring when the water temperature rises by 30.0°C. (You will find the increase in length to be small. Therefore, to provide a greater variation in valve opening for the temperature change anticipated, actual devices have a more complicated mechanical design.)

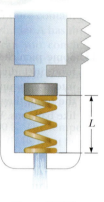

Figure P16.11

12. **BIO** A pair of eyeglass frames is made of epoxy plastic. At room temperature (20.0°C), the frames have circular lens holes 2.20 cm in radius. To what temperature must the frames be heated if lenses 2.21 cm in radius are to be inserted in them? The average coefficient of linear expansion for epoxy is 1.30×10^{-4} (°C)$^{-1}$.

13. **M** A copper telephone wire has essentially no sag between poles 35.0 m apart on a winter day when the temperature is −20.0°C. How much longer is the wire on a summer day when the temperature is 35.0°C?

14. **Q|C** **W** At 20.0°C, an aluminum ring has an inner diameter of 5.000 0 cm and a brass rod has a diameter of 5.050 0 cm. (a) If only the ring is warmed, what temperature must it reach so that it will just slip over the rod? (b) **What If?** If both the ring and the rod are warmed together, what temperature must they both reach so that the ring barely slips over the rod? (c) Would this latter process work? Explain. *Hint:* Consult Table 17.2 in the next chapter.

15. **M** The active element of a certain laser is made of a glass rod 30.0 cm long and 1.50 cm in diameter. Assume the average coefficient of linear expansion of the glass is equal to 9.00×10^{-6} (°C)$^{-1}$. If the temperature of the rod increases by 65.0°C, what is the increase in (a) its length, (b) its diameter, and (c) its volume?

16. **Review.** Inside the wall of a house, an L-shaped section of hot-water pipe consists of three parts: a straight, horizontal piece $h = $ 28.0 cm long; an elbow; and a straight, vertical piece $\ell = $ 134 cm long (Fig. P16.16).

Figure P16.16

A stud and a second-story floorboard hold the ends of this section of copper pipe stationary. Find the magnitude and direction of the displacement of the pipe elbow when the water flow is turned on, raising the temperature of the pipe from 18.0°C to 46.5°C.

17. A square hole 8.00 cm along each side is cut in a sheet of copper. (a) Calculate the change in the area of this hole resulting when the temperature of the sheet is increased by 50.0 K. (b) Does this change represent an increase or a decrease in the area enclosed by the hole?

18. The average coefficient of volume expansion for carbon tetrachloride is 5.81×10^{-4} (°C)$^{-1}$. If a 50.0-gal steel container is filled completely with carbon tetrachloride when the temperature is 10.0°C, how much will spill over when the temperature rises to 30.0°C?

19. A hollow aluminum cylinder 20.0 cm deep has an internal capacity of 2.000 L at 20.0°C. It is completely filled with turpentine at 20.0°C. The turpentine and the aluminum cylinder are then slowly warmed together to 80.0°C. (a) How much turpentine overflows? (b) What is the volume of turpentine remaining in the cylinder at 80.0°C? (c) If the combination with this amount of turpentine is then cooled back to 20.0°C, how far below the cylinder's rim does the turpentine's surface recede?

Section 16.4 Macroscopic Description of an Ideal Gas

20. Use the definition of Avogadro's number to find the mass of a helium atom.

21. **W** Gas is contained in an 8.00-L vessel at a temperature of 20.0°C and a pressure of 9.00 atm. (a) Determine the number of moles of gas in the vessel. (b) How many molecules are in the vessel?

22. A rigid tank having a volume of 0.100 m^3 contains helium gas at 150 atm. How many balloons can be inflated by opening the valve at the top of the tank? Each filled balloon is a sphere 0.300 m in diameter at an absolute pressure of 1.20 atm.

23. On your wedding day, your spouse gives you a gold ring of mass 3.80 g. Fifty years later its mass is 3.35 g. On the average, how many atoms were abraded from the ring during each second of your marriage? The molar mass of gold is 197 g/mol.

24. **W** A cook puts 9.00 g of water in a 2.00-L pressure cooker that is then warmed to 500°C. What is the pressure inside the container?

25. **M** An automobile tire is inflated with air originally at 10.0°C and normal atmospheric pressure. During the process, the air is compressed to 28.0% of its original volume and the temperature is increased to 40.0°C. (a) What is the tire pressure? (b) After the car is driven at high speed, the tire's air temperature rises to 85.0°C and the tire's interior volume increases by 2.00%. What is the new tire pressure (absolute)?

26. **Q|C** Your father and your younger brother are confronted with the same puzzle. Your father's garden sprayer and your brother's water cannon both have tanks with a capacity of 5.00 L (Fig. P16.26). Your father puts a negligible amount of concentrated fertilizer into his tank. They both pour in 4.00 L

of water and seal up their tanks, so the tanks also contain air at atmospheric pressure. Next, each uses a hand-operated pump to inject more air until the absolute pressure in the tank reaches 2.40 atm. Now each uses his device to spray out water—not air—until the stream becomes feeble, which it does when the pressure in the tank reaches 1.20 atm. To accomplish spraying out all the water, each finds he must pump up the tank three times. Here is the puzzle: most of the water sprays out after the second pumping. The first and the third pumping-up processes seem just as difficult as the second but result in a much smaller amount of water coming out. Account for this phenomenon.

Figure P16.26

27. **M** An auditorium has dimensions 10.0 m × 20.0 m × 30.0 m. How many molecules of air fill the auditorium at 20.0°C and a pressure of 101 kPa (1.00 atm)?

28. **S** A room of volume V contains air having equivalent molar mass M (in g/mol). If the temperature of the room is raised from T_1 to T_2, what mass of air will leave the room? Assume that the air pressure in the room is maintained at P_0.

29. **M** **Review.** The mass of a hot-air balloon and its cargo (not including the air inside) is 200 kg. The air outside is at 10.0°C and 101 kPa. The volume of the balloon is 400 m³. To what temperature must the air in the balloon be warmed before the balloon will lift off? (Air density at 10.0°C is 1.244 kg/m³.)

30. **Q|C** A container in the shape of a cube 10.0 cm on each edge contains air (with equivalent molar mass 28.9 g/mol) at atmospheric pressure and temperature 300 K. Find (a) the mass of the gas, (b) the gravitational force exerted on it, and (c) the force it exerts on each face of the cube. (d) Why does such a small sample exert such a great force?

31. A popular brand of cola contains 6.50 g of carbon dioxide dissolved in 1.00 L of soft drink. If the evaporating carbon dioxide is trapped in a cylinder at 1.00 atm and 20.0°C, what volume does the gas occupy?

32. Estimate the mass of the air in your bedroom. State the quantities you take as data and the value you measure or estimate for each.

33. **W** **Review.** At 25.0 m below the surface of the sea, where the temperature is 5.00°C, a diver exhales an air bubble having a volume of 1.00 cm³. If the surface temperature of the sea is 20.0°C, what is the volume of the bubble just before it breaks the surface?

34. **Review.** To measure how far below the ocean surface a bird dives to catch a fish, a scientist uses a method originated

by Lord Kelvin. He dusts the interiors of plastic tubes with powdered sugar and then seals one end of each tube. He captures the bird at nighttime in its nest and attaches a tube to its back. He then catches the same bird the next night and removes the tube. In one trial, using a tube 6.50 cm long, water washes away the sugar over a distance of 2.70 cm from the open end of the tube. Find the greatest depth to which the bird dived, assuming the air in the tube stayed at constant temperature.

35. **W** In state-of-the-art vacuum systems, pressures as low as 1.00×10^{-9} Pa are being attained. Calculate the number of molecules in a 1.00-m³ vessel at this pressure and a temperature of 27.0°C.

36. **M** The pressure gauge on a tank registers the gauge pressure, which is the difference between the interior pressure and exterior pressure. When the tank is full of oxygen (O_2), it contains 12.0 kg of the gas at a gauge pressure of 40.0 atm. Determine the mass of oxygen that has been withdrawn from the tank when the pressure reading is 25.0 atm. Assume the temperature of the tank remains constant.

Section 16.5 The Kinetic Theory of Gases

37. **W** A 2.00-mol sample of oxygen gas is confined to a 5.00-L vessel at a pressure of 8.00 atm. Find the average translational kinetic energy of the oxygen molecules under these conditions.

38. A 5.00-L vessel contains nitrogen gas at 27.0°C and 3.00 atm. Find (a) the total translational kinetic energy of the gas molecules and (b) the average kinetic energy per molecule.

39. **W** In a 30.0-s interval, 500 hailstones strike a glass window of area 0.600 m² at an angle of 45.0° to the window surface. Each hailstone has a mass of 5.00 g and a speed of 8.00 m/s. Assuming the collisions are elastic, find (a) the average force and (b) the average pressure on the window during this interval.

40. **M** A cylinder contains a mixture of helium and argon gas in equilibrium at 150°C. (a) What is the average kinetic energy for each type of gas molecule? (b) What is the rms speed of each type of molecule?

41. **M** A spherical balloon of volume 4.00×10^3 cm³ contains helium at a pressure of 1.20×10^5 Pa. How many moles of helium are in the balloon if the average kinetic energy of the helium atoms is 3.60×10^{-22} J?

42. **S** A spherical balloon of volume V contains helium at a pressure P. How many moles of helium are in the balloon if the average kinetic energy of the helium atoms is $\overline{K}$?

43. **W** In a period of 1.00 s, 5.00×10^{23} nitrogen molecules strike a wall with an area of 8.00 cm². Assume the molecules move with a speed of 300 m/s and strike the wall head-on in elastic collisions. What is the pressure exerted on the wall? *Note:* The mass of one N_2 molecule is 4.65×10^{-26} kg.

44. **M** (a) How many atoms of helium gas fill a spherical balloon of diameter 30.0 cm at 20.0°C and 1.00 atm? (b) What is the average kinetic energy of the helium atoms? (c) What is the rms speed of the helium atoms?

Section 16.6 Distribution of Molecular Speeds

45. **M** Fifteen identical particles have various speeds: one has a speed of 2.00 m/s, two have speeds of 3.00 m/s, three have speeds of 5.00 m/s, four have speeds of 7.00 m/s, three have speeds of 9.00 m/s, and two have speeds of 12.0 m/s. Find (a) the average speed, (b) the rms speed, and (c) the most probable speed of these particles.

46. **S** From the Maxwell–Boltzmann speed distribution, show that the most probable speed of a gas molecule is given by Equation 16.23. *Note:* The most probable speed corresponds to the point at which the slope of the speed distribution curve dN_v/dv is zero.

47. Review. At what temperature would the average speed of helium atoms equal (a) the escape speed from the Earth, 1.12×10^4 m/s, and (b) the escape speed from the Moon, 2.37×10^3 m/s? *Note:* The mass of a helium atom is 6.64×10^{-27} kg.

48. Helium gas is in thermal equilibrium with liquid helium at 4.20 K. Even though it is on the point of condensation, model the gas as ideal and determine the most probable speed of a helium atom (mass = 6.64×10^{-27} kg) in it.

Section 16.7 Context Connection: The Atmospheric Lapse Rate

49. The summit of Mount Whitney, in California, is 3 660 m higher than a point in the foothills. Assume that the atmospheric lapse rate in the Mount Whitney area is the same as the global average of −6.5°C/km. What is the temperature of the summit of Mount Whitney when eager hikers depart from the foothill location at a temperature of 30°C?

50. The theoretical lapse rate for dry air (no water vapor) in an atmosphere is given by

$$\frac{dT}{dy} = -\frac{\gamma - 1}{\gamma}\frac{gM}{R}$$

where g is the acceleration due to gravity, M is the molar mass of the uniform ideal gas in the atmosphere, R is the gas constant, and γ is the ratio of molar specific heats, which we will study in Chapter 17. (a) Calculate the theoretical lapse rate on the Earth given that $\gamma = 1.40$ and the effective molar mass of air is 28.9 g/mol. (b) Why is this value different from the value of −6.5°C/km given in the text? (c) The atmosphere of Mars is mostly dry carbon dioxide, with a molar mass of 44.0 g/mol and a ratio of molar specific heats of $\gamma = 1.30$. The mass of Mars is 6.42×10^{23} kg and the radius is 3.37×10^6 m. What is the lapse rate for the Martian troposphere? (d) A typical surface atmospheric temperature on Mars is −40.0°C. Using the lapse rate calculated in part (c), find the height in the Martian troposphere at which the temperature is −60.0°C. (e) Data from the *Mariner* flights in 1969 indicated a lapse rate in the Martian troposphere of about −1.5°C/km. The *Viking* missions in 1976 gave measured lapse rates of about −2°C/km. These values deviate from the ideal value calculated in part (c) because of *dust* in the Martian atmosphere. Why would dust affect the lapse rate? Which mission occurred in dustier conditions, *Mariner* or *Viking*?

Additional Problems

51. A student measures the length of a brass rod with a steel tape at 20.0°C. The reading is 95.00 cm. What will the tape indicate for the length of the rod when the rod and the tape are at (a) −15.0°C and (b) 55.0°C?

52. The density of gasoline is 730 kg/m³ at 0°C. Its average coefficient of volume expansion is 9.60×10^{-4} (°C)$^{-1}$. Assume 1.00 gal of gasoline occupies 0.003 80 m³. How many extra kilograms of gasoline would you receive if you bought 10.0 gal of gasoline at 0°C rather than at 20.0°C from a pump that is not temperature compensated?

53. **M** A mercury thermometer is constructed as shown in Figure P16.53. The Pyrex glass capillary tube has a diameter of 0.004 00 cm, and the bulb has a diameter of 0.250 cm. Find the change in height of the mercury column that occurs with a temperature change of 30.0°C.

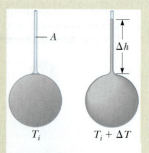

Figure P16.53 Problems 53 and 54.

54. **S** A liquid with a coefficient of volume expansion β just fills a spherical shell of volume V (Fig. P16.53). The shell and the open capillary of area A projecting from the top of the sphere are made of a material with an average coefficient of linear expansion α. The liquid is free to expand into the capillary. Assuming the temperature increases by ΔT, find the distance Δh the liquid rises in the capillary.

55. Review. A clock with a brass pendulum has a period of 1.000 s at 20.0°C. If the temperature increases to 30.0°C, (a) by how much does the period change and (b) how much time does the clock gain or lose in one week?

56. **GP** **S** A vertical cylinder of cross-sectional area A is fitted with a tight-fitting, frictionless piston of mass m (Fig. P16.56). The piston is not restricted in its motion in any way and is supported by the gas at pressure P below it. Atmospheric pressure is P_0. We wish to find the height h in Figure P16.56. (a) What analysis model is appropriate to describe the piston? (b) Write an appropriate force equation for the piston from this analysis model in terms of P, P_0, m, A, and g. (c) Suppose n moles of an ideal gas are in the cylinder at a temperature of T. Substitute for P in your answer to part (b) to find the height h of the piston above the bottom of the cylinder.

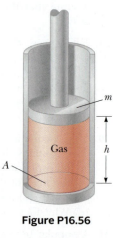

Figure P16.56

57. **BIO** Long-term space missions require reclamation of the oxygen in the carbon dioxide exhaled by the crew. In one method of reclamation, 1.00 mol of carbon dioxide produces 1.00 mol of oxygen and 1.00 mol of methane as a byproduct. The methane is stored in a tank under pressure and is available to control the attitude of the spacecraft by controlled venting. A single astronaut exhales 1.09 kg of carbon dioxide each day. If the methane generated in the respiration recycling of three astronauts during one week of flight is stored in an originally empty 150-L tank at −45.0°C, what is the final pressure in the tank?

58. **Q|C** **S** A bimetallic strip of length L is made of two ribbons of different metals bonded together. (a) First assume the strip is originally straight. As the strip is warmed, the metal with the greater average coefficient of expansion expands more than the other, forcing the strip into an arc with the outer radius having a greater circumference (Fig. P16.58). Derive an expression for the angle of bending θ as a function of the initial length of the strips, their average coefficients of linear expansion, the change in temperature, and the separation of the centers of the strips ($\Delta r = r_2 - r_1$). (b) Show that the angle of bending decreases to zero when ΔT decreases to zero and also when the two average coefficients of expansion become equal. (c) **What If?** What happens if the strip is cooled?

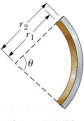

Figure P16.58

59. **BIO** **Review.** Oxygen at pressures much greater than 1 atm is toxic to lung cells. Assume a deep-sea diver breathes a mixture of oxygen (O_2) and helium (He). By weight, what ratio of helium to oxygen must be used if the diver is at an ocean depth of 50.0 m?

60. **Q|C** **S** The rectangular plate shown in Figure P16.60 has an area A_i equal to ℓw. If the temperature increases by ΔT, each dimension increases according to Equation 16.4, where α is the average coefficient of linear expansion. (a) Show that the increase in area is $\Delta A = 2\alpha A_i \Delta T$. (b) What approximation does this expression assume?

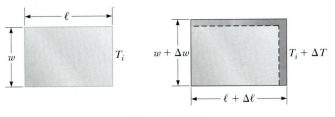

Figure P16.60

61. In a chemical processing plant, a reaction chamber of fixed volume V_0 is connected to a reservoir chamber of fixed volume $4V_0$ by a passage containing a thermally insulating porous plug. The plug permits the chambers to be at different temperatures. The plug allows gas to pass from either chamber to the other, ensuring that the pressure is the same in both. At one point in the processing, both chambers contain gas at a pressure of 1.00 atm and a temperature of 27.0°C. Intake and exhaust valves to the pair of chambers are closed. The reservoir is maintained at 27.0°C while the reaction chamber is heated to 400°C. What is the pressure in both chambers after that is done?

62. **Q|C** A liquid has a density ρ. (a) Show that the fractional change in density for a change in temperature ΔT is $\Delta\rho/\rho = -\beta\,\Delta T$. (b) What does the negative sign signify? (c) Fresh water has a maximum density of 1.000 0 g/cm³ at 4.0°C. At 10.0°C, its density is 0.999 7 g/cm³. What is β for water over this temperature interval? (d) At 0°C, the density of water is 0.999 9 g/cm³. What is the value for β over the temperature range 0°C to 4.00°C?

63. Two concrete spans of a 250-m-long bridge are placed end to end so that no room is allowed for expansion (Fig. P16.63a). If a temperature increase of 20.0°C occurs,

what is the height y to which the spans rise when they buckle (Fig. P16.63b)?

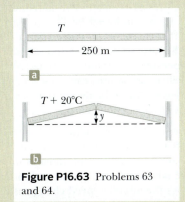

Figure P16.63 Problems 63 and 64.

64. **S** Two concrete spans that form a bridge of length L are placed end to end so that no room is allowed for expansion (Fig. P16.63a). If a temperature increase of ΔT occurs, what is the height y to which the spans rise when they buckle (Fig. P16.63b)?

65. A 1.00-km steel railroad rail is fastened securely at both ends when the temperature is 20.0°C. As the temperature increases, the rail buckles, taking the shape of an arc of a vertical circle. Find the height h of the center of the rail when the temperature is 25.0°C. (You will need to solve a transcendental equation.)

66. **Q|C** (a) Take the definition of the coefficient of volume expansion to be

$$\beta = \frac{1}{V}\frac{dV}{dT}\bigg]_{P=\text{constant}} = \frac{1}{V}\frac{\partial V}{\partial T}$$

Use the equation of state for an ideal gas to show that the coefficient of volume expansion for an ideal gas at constant pressure is given by $\beta = 1/T$, where T is the absolute temperature. (b) What value does this expression predict for β at 0°C? State how this result compares with the experimental values for (c) helium and (d) air in Table 16.1. *Note:* These values are much larger than the coefficients of volume expansion for most liquids and solids.

67. For a Maxwellian gas, use a computer or programmable calculator to find the numerical value of the ratio $N_v(v)/N_v(v_{mp})$ for the following values of v: (a) $v = (v_{mp}/50.0)$, (b) $(v_{mp}/10.0)$, (c) $(v_{mp}/2.00)$, (d) v_{mp}, (e) $2.00v_{mp}$, (f) $10.0v_{mp}$, and (g) $50.0v_{mp}$. Give your results to three significant figures.

68. (a) Show that the density of an ideal gas occupying a volume V is given by $\rho = PM/RT$, where M is the molar mass. (b) Determine the density of oxygen gas at atmospheric pressure and 20.0°C.

69. **Q|C** **Review.** Consider an object with any one of the shapes displayed in Table 10.2. What is the percentage increase in the moment of inertia of the object when it is warmed from 0°C to 100°C if it is composed of (a) copper or (b) aluminum? Assume the average linear expansion coefficients shown in Table 16.1 do not vary between 0°C and 100°C. (c) Why are the answers for parts (a) and (b) the same for all the shapes?

70. *Why is the following situation impossible?* An apparatus is designed so that steam initially at $T = 150$°C, $P = 1.00$ atm, and $V = 0.500$ m³ in a piston–cylinder apparatus undergoes a process in which (1) the volume remains constant and the pressure drops to 0.870 atm, followed by (2) an expansion in which the pressure remains constant and the volume increases to 1.00 m³, followed by (3) a return to the initial conditions. It is important that the pressure of the gas never fall below 0.850 atm so that the piston will support a delicate and very expensive part of the apparatus. Without such support, the delicate apparatus can be severely damaged and rendered useless. When the design is turned into a working prototype, it operates perfectly.

71. **Q|C** **Review.** Following a collision in outer space, a copper disk at 850°C is rotating about its axis with an angular speed of 25.0 rad/s. As the disk radiates infrared light, its temperature falls to 20.0°C. No external torque acts on the disk. (a) Does the angular speed change as the disk cools? Explain how it changes or why it does not. (b) What is its angular speed at the lower temperature?

72. A vessel contains 1.00×10^4 oxygen molecules at 500 K. (a) Make an accurate graph of the Maxwell speed distribution function versus speed with points at speed intervals of 100 m/s. (b) Determine the most probable speed from this graph. (c) Calculate the average and rms speeds for the molecules and label these points on your graph. (d) From the graph, estimate the fraction of molecules with speeds in the range 300 m/s to 600 m/s.

73. **W** A cylinder is closed by a piston connected to a spring of constant 2.00×10^3 N/m (see Fig. P16.73). With the spring relaxed, the cylinder is filled with 5.00 L of gas at a pressure of 1.00 atm and a temperature of 20.0°C. (a) If the piston has a cross-sectional area of 0.010 0 m² and negligible mass, how high will it rise when the temperature is raised to 250°C? (b) What is the pressure of the gas at 250°C?

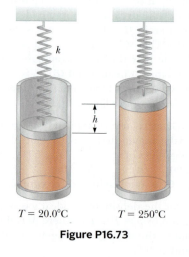

$T = 20.0°C$ $T = 250°C$

Figure P16.73

74. A cylinder that has a 40.0-cm radius and is 50.0 cm deep is filled with air at 20.0°C and 1.00 atm (Fig. P16.74a). A 20.0-kg piston is now lowered into the cylinder, compressing the air trapped inside as it takes equilibrium height h_i (Fig. P16.74b). Finally, a 25.0-kg dog stands on the piston, further compressing the air, which remains at 20°C (Fig. P16.74c). (a) How far down (Δh) does the piston move when the dog steps onto it? (b) To what temperature should the gas be warmed to raise the piston and dog back to h_i?

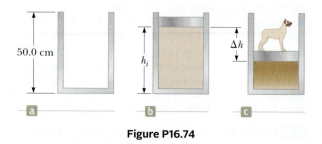

Figure P16.74

75. (a) Derive an expression for the buoyant force on a spherical balloon, submerged in a freshwater lake, as a function of the depth below the surface, the volume of the balloon at the surface, the pressure at the surface, and the density of the water. (Assume that the water temperature does not change with depth.) (b) Does the buoyant force increase or decrease as the balloon is submerged? (c) At what depth is the buoyant force one-half the surface value?

Chapter 17

Energy in Thermal Processes: The First Law of Thermodynamics

Chapter Outline

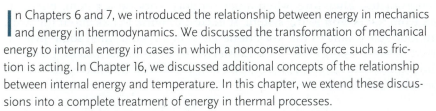

© iStockphoto.com/KingWu

In this photograph of the Mt. Baker area near Bellingham, Washington, we see evidence of water in all three phases. In the lake is liquid water, and solid water in the form of snow appears on the ground. The clouds in the sky consist of liquid water droplets that have condensed from the gaseous water vapor in the air. Changes of a substance from one phase to another are a result of energy transfer.

In Chapters 6 and 7, we introduced the relationship between energy in mechanics and energy in thermodynamics. We discussed the transformation of mechanical energy to internal energy in cases in which a nonconservative force such as friction is acting. In Chapter 16, we discussed additional concepts of the relationship between internal energy and temperature. In this chapter, we extend these discussions into a complete treatment of energy in thermal processes.

Until around 1850, the fields of thermodynamics and mechanics were considered to be two distinct branches of science, and the law of conservation of energy seemed to describe only certain kinds of mechanical systems. Mid-19th-century experiments performed by English physicist James Joule (1818–1889) and others showed that energy may enter or leave a system by heat and by work. Today, as we discussed in

James Prescott Joule
British Physicist (1818–1889)
Joule received some formal education in mathematics, philosophy, and chemistry from John Dalton but was in large part self-educated. Joule's research led to the establishment of the principle of conservation of energy. His study of the quantitative relationship among electrical, mechanical, and chemical effects of heat culminated in his announcement in 1843 of the amount of work required to produce a unit of energy, called the mechanical equivalent of heat.

Pitfall Prevention | 17.1
Heat, Temperature, and Internal Energy Are Different
As you read the newspaper or listen to the radio, be alert for incorrectly used phrases including the word *heat* and think about the proper word to be used in place of it. "As the truck braked to a stop, a large amount of heat was generated by friction" and "The heat of a hot summer day . . ." are two examples.

Chapter 6, internal energy is treated as a form of energy that can be transformed into mechanical energy and vice versa. Once the concept of energy was broadened to include internal energy, the law of conservation of energy emerged as a universal law of nature.

This chapter focuses on developing the concept of heat, extending our concept of work to thermal processes, introducing the first law of thermodynamics, and investigating some important applications.

❮17.1 | Heat and Internal Energy

A major distinction must be made between internal energy and heat because these terms tend to be used interchangeably in everyday communication. You should read the following descriptions carefully and try to use these terms correctly because they are not interchangeable. They have very different meanings.

We introduced internal energy in Chapter 6, and we formally define it here:

Internal energy E_{int} is the energy associated with the microscopic components of a system—atoms and molecules—when viewed from a reference frame at rest with respect to the system. It includes kinetic and potential energy associated with the random translational, rotational, and vibrational motion of the atoms or molecules that make up the system as well as intermolecular potential energy.

In Chapter 16, we showed that the internal energy of a monatomic ideal gas is associated with the translational motion of its atoms. In this special case, the internal energy is simply the total translational kinetic energy of the atoms; the higher the temperature of the gas, the greater the kinetic energy of the atoms and the greater the internal energy of the gas. For more complicated diatomic and polyatomic gases, internal energy includes other forms of molecular energy, such as rotational kinetic energy and the kinetic and potential energy associated with molecular vibrations.

Heat was introduced in Chapter 7 as one possible method of energy transfer, and we provide a formal definition here:

Heat is a mechanism by which energy is transferred between a system and its environment because of a temperature difference between them. It is also the amount of energy Q transferred by this mechanism.

Figure 17.1 shows a pan of water in contact with a gas flame. Energy enters the water by heat from the hot gases in the flame, and the internal energy of the water increases as a result. It is *incorrect* to say that the water has more heat as time goes by.

As further clarification of the use of the word *heat*, consider the distinction between work and energy. The work done on (or by) a system is a measure of the amount of energy transferred between the system and its surroundings, whereas the mechanical energy of the system (kinetic or potential) is a consequence of its motion and coordinates. Therefore, when a person does work on a system, energy is transferred from the person to the system. It makes no sense to talk about the work *in* a system; one refers only to the work done *on* or *by* a system when some process has occurred in which energy has been transferred to or from the system. Likewise, it makes no sense to use the term *heat* unless energy has been transferred as a result of a temperature difference.

Units of Heat

Early in the development of thermodynamics, before scientists recognized the connection between thermodynamics and mechanics, heat was defined in terms of the temperature changes it produced in an object, and a separate unit of energy, the calorie, was used for heat. The **calorie** (cal) was defined as the amount of energy

Figure 17.1 A pan of boiling water is warmed by a gas flame. Energy enters the water through the bottom of the pan by heat.

transfer necessary to raise the temperature of 1 g of water[1] from 14.5°C to 15.5°C. (The "Calorie," with a capital C, used in describing the energy content of foods, is actually a kilocalorie.) Likewise, the unit of heat in the U.S. customary system, the **British thermal unit** (Btu), was defined as the amount of energy transfer required to raise the temperature of 1 lb of water from 63°F to 64°F.

In 1948, scientists agreed that because heat (like work) is a measure of the transfer of energy, its SI unit should be the joule. The calorie is now defined to be exactly 4.186 J:

$$1 \text{ cal} \equiv 4.186 \text{ J}$$ **17.1◀**   ▶ Mechanical equivalent of heat

Note that this definition makes no reference to the heating of water. The calorie is a general energy unit. We could have used it in Chapter 6 for the kinetic energy of an object, for example. It is introduced here for historical reasons, but we shall make little use of it as an energy unit. The definition in Equation 17.1 is known historically as the **mechanical equivalent of heat.**

Example 17.1 | Losing Weight the Hard Way BIO

A student eats a dinner rated at 2 000 Calories. He wishes to do an equivalent amount of work in the gymnasium by lifting a 50.0-kg barbell. How many times must he raise the barbell to expend this much energy? Assume he raises the barbell 2.00 m each time he lifts it and he regains no energy when he lowers the barbell.

SOLUTION

Conceptualize Imagine the student raising the barbell. He is doing work on the system of the barbell and the Earth, so energy is leaving his body. The total amount of work that the student must do is 2 000 Calories.

Categorize We model the system of the barbell and the Earth as a nonisolated system.

Analyze Reduce the conservation of energy equation, Equation 7.2, to the appropriate expression for the system of the barbell and the Earth:

(1) $\Delta U_{\text{total}} = W_{\text{total}}$

Express the change in gravitational potential energy of the system after the barbell is raised once:

$\Delta U = mgh$

Express the total amount of energy that must be transferred into the system by work for lifting the barbell n times, assuming energy is not regained when the barbell is lowered:

(2) $\Delta U_{\text{total}} = nmgh$

Substitute Equation (2) into Equation (1):

$nmgh = W_{\text{total}}$

Solve for n:

$$n = \frac{W_{\text{total}}}{mgh}$$

$$= \frac{(2\ 000 \text{ Cal})}{(50.0 \text{ kg})(9.80 \text{ m/s}^2)(2.00 \text{ m})}\left(\frac{1.00 \times 10^3 \text{ cal}}{\text{Calorie}}\right)\left(\frac{4.186 \text{ J}}{1 \text{ cal}}\right)$$

$$= \boxed{8.54 \times 10^3 \text{ times}}$$

Finalize If the student is in good shape and lifts the barbell once every 5 s, it will take him about 12 h to perform this feat. Clearly, it is much easier for this student to lose weight by dieting.

In reality, the human body is not 100% efficient. Therefore, not all the energy transformed within the body from the dinner transfers out of the body by work done on the barbell. Some of this energy is used to pump blood and perform other functions within the body. Therefore, the 2 000 Calories can be worked off in less time than 12 h when these other energy processes are included.

---

[1]Originally, the calorie was defined as the heat necessary to raise the temperature of 1 g of water by 1°C at any initial temperature. Careful measurements, however, showed that the energy required depends somewhat on temperature; hence, a more precise definition evolved.

◀ 17.2 | Specific Heat

The definition of the calorie indicates the amount of energy necessary to raise the temperature of 1 g of a specific substance—water—by 1°C, which is 4.186 J. To raise the temperature of 1 kg of water by 1°C, we need to transfer 4 186 J of energy to it from the environment. The quantity of energy required to raise the temperature of 1 kg of an arbitrary substance by 1°C varies with the substance. For example, the energy required to raise the temperature of 1 kg of copper by 1°C is 387 J, which is significantly less than that required for water. Every substance requires a unique amount of energy per unit mass to change the temperature of that substance by 1°C.

Suppose a quantity of energy Q is transferred to a mass m of a substance, thereby changing its temperature by ΔT. The **specific heat** c of the substance is defined as

$$c \equiv \frac{Q}{m \, \Delta T}$$

17.2 ◀

The units of specific heat are joules per kilogram-degree Celsius, or $J/kg \cdot °C$. Table 17.1 lists specific heats for several substances. From the definition of the calorie, the specific heat of water is 4 186 $J/kg \cdot °C$.

From this definition, we can express the energy Q transferred between a system of mass m and its surroundings in terms of the resulting temperature change ΔT as

$$Q = mc \, \Delta T$$

17.3 ◀

For example, the energy required to raise the temperature of 0.500 kg of water by 3.00°C is $Q = (0.500 \text{ kg})(4 \text{ } 186 \text{ J/kg} \cdot °C)(3.00°C) = 6.28 \times 10^3$ J. Note that when the temperature increases, ΔT and Q are taken to be *positive*, corresponding to energy flowing *into* the system. When the temperature decreases, ΔT and Q are *negative* and energy flows *out* of the system. These sign conventions are consistent with those in our discussion of the conservation of energy equation, Equation 7.2.

Table 17.1 shows that water has a high specific heat relative to most other common substances (the specific heats of hydrogen and helium are higher). The high specific heat of water is responsible for the moderate temperatures found in regions near large bodies of water. As the temperature of a body of water decreases during the winter, the water transfers energy to the air, which carries the energy landward when prevailing winds are toward the land. For example, the prevailing winds off the western coast of the United States are toward the land, and the energy liberated by the Pacific Ocean as it cools keeps coastal areas much warmer than they would be otherwise. Therefore, the western coastal states generally have warmer winter weather than the eastern coastal states, where the winds do not transfer energy toward land.

That the specific heat of water is higher than that of sand accounts for the pattern of air flow at a beach. During the day, the Sun adds roughly equal amounts of energy to beach and water, but the lower specific heat of sand causes the beach to reach a higher temperature than the water. As a result, the air above the land reaches a higher temperature than the air above the water. The denser cold air pushes the less dense hot air upward (due to Archimedes's principle), which results in a breeze from ocean to land during the day. During the night, the sand cools more quickly than the water, and the direction of the breeze reverses because the hotter air is now over the water. These offshore and onshore breezes are well known to sailors.

◀ **TABLE 17.1** | Specific Heats of Some Substances at 25°C and Atmospheric Pressure

Substance	Specific Heat c	
	$J/kg \cdot °C$	$cal/g \cdot °C$
Elemental Solids		
Aluminum	900	0.215
Beryllium	1 830	0.436
Cadmium	230	0.055
Copper	387	0.092 4
Germanium	322	0.077
Gold	129	0.030 8
Iron	448	0.107
Lead	128	0.030 5
Silicon	703	0.168
Silver	234	0.056
Other Solids		
Brass	380	0.092
Glass	837	0.200
Ice (−5°C)	2 090	0.50
Marble	860	0.21
Wood	1 700	0.41
Liquids		
Alcohol (ethyl)	2 400	0.58
Mercury	140	0.033
Water (15°C)	4 186	1.00
Gas		
Steam (100°C)	2 010	0.48

QUICK QUIZ 17.1 Imagine you have 1 kg each of iron, glass, and water and that all three samples are at 10°C. **(a)** Rank the samples from highest to lowest temperature after 100 J of energy is added to each sample. **(b)** Rank the samples from greatest to least amount of energy transferred by heat if each sample increases in temperature by 20.0°C.

Calorimetry

One technique for measuring the specific heat of a solid or liquid is to raise the temperature of the substance to some value, place it into a vessel containing water of known mass and temperature, and measure the temperature of the combination after equilibrium is reached. Let us define the system as the substance and the water. If the vessel is assumed to be a good insulator so that energy does not leave the system by heat (nor by any other means), we can use the isolated system model. A vessel having this property is called a **calorimeter,** and the analysis performed by using such a vessel is called **calorimetry.** Figure 17.2 shows the hot sample in the cold water and the resulting energy transfer by heat from the high-temperature part of the system to the low-temperature part.

The principle of conservation of energy for this isolated system requires that the energy leaving by heat from the warmer substance (of unknown specific heat) equals the energy entering the water.[2] Therefore, we can write

$$Q_{cold} = -Q_{hot} \qquad \text{17.4} \blacktriangleleft$$

To see how to set up a calorimetry problem, suppose m_x is the mass of a substance whose specific heat we wish to determine, c_x its specific heat, and T_x its initial temperature. Let m_w, c_w, and T_w represent corresponding values for the water. If T is the final equilibrium temperature after the substance and the water are combined, from Equation 17.3 we find that the energy gained by the water is $m_w c_w (T - T_w)$ and that the energy transferred from the substance of unknown specific heat is $m_x c_x (T - T_x)$. Substituting these values into Equation 17.4, we have

$$m_w c_w (T - T_w) = -m_x c_x (T - T_x)$$

This equation can be solved for the unknown specific heat c_x.

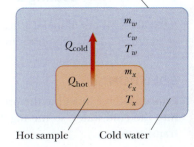

Figure 17.2 In a calorimetry experiment, a hot sample whose specific heat is unknown is placed in cold water in a container that isolates the system from the environment.

> **Pitfall Prevention | 17.3**
> **Remember the Negative Sign**
> It is *critical* to include the negative sign in Equation 17.4. The negative sign in the equation is necessary for consistency with our sign convention for energy transfer. The energy transfer Q_{hot} is negative because energy is leaving the hot substance. The negative sign in the equation ensures that the right-hand side is a positive number, consistent with the left-hand side, which is positive because energy is entering the cold substance.

> **Pitfall Prevention | 17.4**
> **Celsius versus Kelvin**
> In equations in which T appears (e.g., the ideal gas law), the Kelvin temperature *must* be used. In equations involving ΔT, such as calorimetry equations, it is *possible* to use Celsius temperatures because a change in temperature is the same on both scales. It is *safest,* however, to use Kelvin temperatures *consistently* in all equations involving T or ΔT.

> ▶ **THINKING PHYSICS 17.1**
>
> The equation $Q = mc\,\Delta T$ indicates the relationship between energy Q transferred to an object of mass m and specific heat c by means of heat and the resulting temperature change ΔT. In reality, the energy transfer on the left-hand side of the equation could be made by any method, not just heat. Give a few examples in which this equation could be used to calculate a temperature change of an object due to an energy transfer process other than heat.
>
> **Reasoning** The following are a few of several possible examples.
>
> During the first few seconds after turning on a toaster, the temperature of the electrical coils rises. The transfer mechanism here is *electrical transmission* of energy through the power cord.
>
> The temperature of a potato in a microwave oven increases due to the absorption of microwaves. In this case, the energy transfer mechanism is by *electromagnetic radiation,* the microwaves.
>
> A carpenter attempts to use a dull drill bit to bore a hole in a piece of wood. The bit fails to make much headway but becomes very warm. The increase in temperature in this case is due to *work* done on the bit by the wood.
>
> In each of these cases, as well as many other possibilities, the Q on the left of the equation of interest is not a measure of heat but, rather, is replaced with the energy transferred or transformed by other means. Even though heat is not involved, the equation can still be used to calculate the temperature change. ◀

---

[2]For precise measurements, the container holding the water should be included in the calculations because it also changes temperature. Doing so would require a knowledge of its mass and composition. If, however, the mass of the water is large compared with that of the container, we can adopt a simplification model in which we ignore the energy gained by the container.

Example 17.2 | Cooling a Hot Ingot

A 0.050 0-kg ingot of metal is heated to 200.0°C and then dropped into a calorimeter containing 0.400 kg of water initially at 20.0°C. The final equilibrium temperature of the mixed system is 22.4°C. Find the specific heat of the metal.

SOLUTION

Conceptualize Imagine the process occurring in the isolated system of Figure 17.2. Energy leaves the hot ingot and goes into the cold water, so the ingot cools off and the water warms up. Once both are at the same temperature, the energy transfer stops.

Categorize We use an equation developed in this section, so we categorize this example as a substitution problem.

Use Equation 17.3 to evaluate each side of Equation 17.4:

$$m_w c_w (T_f - T_w) = -m_x c_x (T_f - T_x)$$

Solve for c_x:

$$c_x = \frac{m_w c_w (T_f - T_w)}{m_x (T_x - T_f)}$$

Substitute numerical values:

$$c_x = \frac{(0.400 \text{ kg})(4\,186 \text{ J/kg} \cdot °\text{C})(22.4°\text{C} - 20.0°\text{C})}{(0.050\,0 \text{ kg})(200.0°\text{C} - 22.4°\text{C})}$$

$$= 453 \text{ J/kg} \cdot °\text{C}$$

The ingot is most likely iron as you can see by comparing this result with the data given in Table 17.1. The temperature of the ingot is initially above the steam point. Therefore, some of the water may vaporize when the ingot is dropped into the water. We assume the system is sealed and this steam cannot escape. Because the final equilibrium temperature is lower than the steam point, any steam that does result recondenses back into water.

What If? Suppose you are performing an experiment in the laboratory that uses this technique to determine the specific heat of a sample and you wish to decrease the overall uncertainty in your final result for c_x. Of the data given in this example, changing which value would be most effective in decreasing the uncertainty?

Answer The largest experimental uncertainty is associated with the small difference in temperature of 2.4°C for the water. For example, using the rules for propagation of uncertainty in Appendix Section B.8, an uncertainty of 0.1°C in each of T_f and T_w leads to an 8% uncertainty in their difference. For this temperature difference to be larger experimentally, the most effective change is to *decrease the amount of water.*

17.3 | Latent Heat

As we have seen in the preceding section, a substance can undergo a change in temperature when energy is transferred between it and its surroundings. In some situations, however, the transfer of energy does not result in a change in temperature. That is the case whenever the physical characteristics of the substance change from one form to another; such a change is commonly referred to as a **phase change.** Two common phase changes are from solid to liquid (melting) and from liquid to gas (boiling); another is a change in the crystalline structure of a solid. All such phase changes involve a change in the system's internal energy but no change in its temperature. The increase in internal energy in boiling, for example, is represented by the breaking of bonds between molecules in the liquid state; this bond breaking allows the molecules to move farther apart in the gaseous state, with a corresponding increase in intermolecular potential energy.

As you might expect, different substances respond differently to the addition or removal of energy as they change phase because their internal molecular arrangements vary. Also, the amount of energy transferred during a phase change depends on the amount of substance involved. (It takes less energy to melt an ice cube than it does to thaw a frozen lake.) When discussing two phases of a material, we will use the term *higher-phase material* to mean the material existing at the higher temperature. So, for example, if we discuss water and ice, water is the higher-phase material, whereas steam is the higher-phase material in a discussion of steam and water. Consider a system containing a substance in two phases in equilibrium such as water and ice. The initial amount of the higher-phase material, water, in the system is m_i. Now imagine that energy Q enters the system. As a result, the final amount of water is m_f due to the melting of some of the ice. Therefore, the amount of ice that melted, equal to the amount of *new* water, is $\Delta m = m_f - m_i$. We define the **latent heat** for this phase change as

$$L \equiv \frac{Q}{\Delta m}$$ 17.5 ◄

This parameter is called latent heat (literally, the "hidden" heat) because this added or removed energy does not result in a temperature change. The value of L for a substance depends on the nature of the phase change as well as on the properties of the substance. If the entire amount of the lower-phase material undergoes a phase change, the change in mass Δm of the higher-phase material is equal to the initial mass of the lower-phase material. For example, if an ice cube of mass m on a plate melts completely, the change in mass of the water is $m_f - 0 = m$, which is the mass of new water and is also equal to the initial mass of the ice cube.

From the definition of latent heat, and again choosing heat as our energy transfer mechanism, the energy required to change the phase of a pure substance is

$$Q = L \Delta m$$ 17.6 ◄

▶ Energy transferred to a substance during a phase change

where Δm is the change in mass of the higher-phase material.

Latent heat of fusion L_f is the term used when the phase change is from solid to liquid (*to fuse* means "to combine by melting"), and **latent heat of vaporization** L_v is the term used when the phase change is from liquid to gas (the liquid "vaporizes").[3] The latent heats of various substances vary considerably as data in Table 17.2 show. When energy enters a system, causing melting or vaporization, the amount of the higher-phase material increases, so Δm is positive and Q is positive, consistent with our sign convention. When energy is extracted from a system, causing freezing or condensation, the amount of the higher-phase material decreases, so Δm is negative and Q is negative, again consistent with our sign convention. Keep in mind that Δm in Equation 17.6 always refers to the higher-phase material.

Pitfall Prevention | 17.5
Signs Are Critical
Sign errors occur very often when students apply calorimetry equations. For phase changes, remember that Δm in Equation 17.6 is always the change in mass of the higher-phase material. In Equation 17.3, be sure your ΔT is *always* the final temperature minus the initial temperature. In addition, you must *always* include the negative sign on the right side of Equation 17.4.

◀ **TABLE 17.2** | **Latent Heats of Fusion and Vaporization**

Substance	Melting Point (°C)	Latent Heat of Fusion (J/kg)	Boiling Point (°C)	Latent Heat of Vaporization (J/kg)
Helium[a]	−272.2	5.23×10^3	−268.93	2.09×10^4
Oxygen	−218.79	1.38×10^4	−182.97	2.13×10^5
Nitrogen	−209.97	2.55×10^4	−195.81	2.01×10^5
Ethyl alcohol	−114	1.04×10^5	78	8.54×10^5
Water	0.00	3.33×10^5	100.00	2.26×10^6
Sulfur	119	3.81×10^4	444.60	3.26×10^5
Lead	327.3	2.45×10^4	1 750	8.70×10^5
Aluminum	660	3.97×10^5	2 450	1.14×10^7
Silver	960.80	8.82×10^4	2 193	2.33×10^6
Gold	1 063.00	6.44×10^4	2 660	1.58×10^6
Copper	1 083	1.34×10^5	1 187	5.06×10^6

[a]Helium does not solidify at atmospheric pressure. Therefore, its melting point is given under the conditions that the pressure is 2.5 MPa.

[3]When a gas cools, it eventually *condenses;* that is, it returns to the liquid phase. The energy given up per unit mass is called the *latent heat of condensation* and is numerically equal to the latent heat of vaporization. Likewise, when a liquid cools, it eventually solidifies, and the *latent heat of solidification* is numerically equal to the latent heat of fusion.

Figure 17.3 A plot of temperature versus energy added when 1.00 g of ice initially at −30.0°C is converted to steam at 120.0°C.

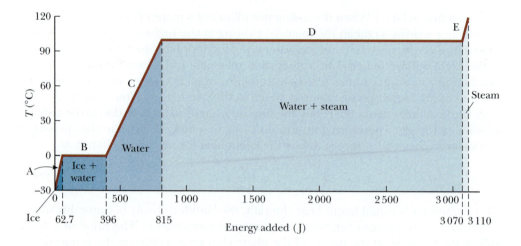

To understand the role of latent heat in phase changes, consider the energy required to convert a 1.00-g cube of ice at −30.0°C to steam at 120.0°C. Figure 17.3 indicates the experimental results obtained when energy is gradually added to the ice. The results are presented as a graph of temperature of the system of the ice cube versus energy added to the system. Let's examine each portion of the red-brown curve, which is divided into parts A through E.

Part A. On this portion of the curve, the temperature of the ice changes from −30.0°C to 0.0°C. Equation 17.3 indicates that the temperature varies linearly with the energy added, so the experimental result is a straight line on the graph. Because the specific heat of ice is 2 090 J/kg · °C, we can calculate the amount of energy added by using Equation 17.3:

$$Q = m_i c_i \, \Delta T = (1.00 \times 10^{-3} \text{ kg})(2\ 090 \text{ J/kg} \cdot °\text{C})(30.0°\text{C}) = 62.7 \text{ J}$$

Part B. When the temperature of the ice reaches 0.0°C, the ice–water mixture remains at this temperature—even though energy is being added—until all the ice melts. The energy required to melt 1.00 g of ice at 0.0°C is, from Equation 17.6,

$$Q = L_f \Delta m_w = L_f m_i = (3.33 \times 10^5 \text{ J/kg})(1.00 \times 10^{-3} \text{ kg}) = 333 \text{ J}$$

At this point, we have moved to the 396 J (= 62.7 J + 333 J) mark on the energy axis in Figure 17.3.

Part C. Between 0.0°C and 100.0°C, nothing surprising happens. No phase change occurs, and so all energy added to the water is used to increase its temperature. The amount of energy necessary to increase the temperature from 0.0°C to 100.0°C is

$$Q = m_w c_w \, \Delta T = (1.00 \times 10^{-3} \text{ kg})(4.19 \times 10^3 \text{ J/kg} \cdot °\text{C})(100.0°\text{C}) = 419 \text{ J}$$

Part D. At 100.0°C, another phase change occurs as the water changes from water at 100.0°C to steam at 100.0°C. Similar to the ice–water mixture in part B, the water–steam mixture remains at 100.0°C—even though energy is being added—until all the liquid has been converted to steam. The energy required to convert 1.00 g of water to steam at 100.0°C is

$$Q = L_v \Delta m_s = L_v m_w = (2.26 \times 10^6 \text{ J/kg})(1.00 \times 10^{-3} \text{ kg}) = 2.26 \times 10^3 \text{ J}$$

Part E. On this portion of the curve, as in parts A and C, no phase change occurs; therefore, all energy added is used to increase the temperature of the steam. The energy that must be added to raise the temperature of the steam from 100.0°C to 120.0°C is

$$Q = m_s c_s \, \Delta T = (1.00 \times 10^{-3} \text{ kg})(2.01 \times 10^3 \text{ J/kg} \cdot °\text{C})(20.0°\text{C}) = 40.2 \text{ J}$$

The total amount of energy that must be added to change 1 g of ice at −30.0°C to steam at 120.0°C is the sum of the results from all five parts of the curve, which is 3.11 × 10³ J. Conversely, to cool 1 g of steam at 120.0°C to ice at −30.0°C, we must remove 3.11 × 10³ J of energy.

Notice in Figure 17.3 the relatively large amount of energy that is transferred into the water to vaporize it to steam. Imagine reversing this process, with a large amount of energy transferred out of steam to condense it into water. That is why a burn to your skin from steam at 100°C is much more damaging than exposure of your skin to water at 100°C. A very large amount of energy enters your skin from the steam, and the steam remains at 100°C for a long time while it condenses. Conversely, when your skin makes contact with water at 100°C, the water immediately begins to drop in temperature as energy transfers from the water to your skin.

If liquid water is held perfectly still in a very clean container, it is possible for the water to drop below 0°C without freezing into ice. This phenomenon, called **supercooling,** arises because the water requires a disturbance of some sort for the molecules to move apart and start forming the large, open ice structure that makes the density of ice lower than that of water as discussed in Section 16.3. If supercooled water is disturbed, it suddenly freezes. The system drops into the lower-energy configuration of bound molecules of the ice structure, and the energy released raises the temperature back to 0°C.

Commercial hand warmers consist of liquid sodium acetate in a sealed plastic pouch. The solution in the pouch is in a stable supercooled state. When a disk in the pouch is clicked by your fingers, the liquid solidifies and the temperature increases, just like the supercooled water just mentioned. In this case, however, the freezing point of the liquid is higher than body temperature, so the pouch feels warm to the touch. To reuse the hand warmer, the pouch must be boiled until the solid liquefies. Then, as it cools, it passes below its freezing point into the supercooled state.

It is also possible to create **superheating.** For example, clean water in a very clean cup placed in a microwave oven can sometimes rise in temperature beyond 100°C without boiling because the formation of a bubble of steam in the water requires scratches in the cup or some type of impurity in the water to serve as a nucleation site. When the cup is removed from the microwave oven, the superheated water can become explosive as bubbles form immediately and the hot water is forced upward out of the cup.

A classic prank related to phase changes is to fashion a spoon made of pure gallium. The melting point of gallium is 29.8°C. Therefore, when the spoon is used to stir hot tea, the submerged portion of the spoon turns into a liquid and drops to the bottom of the cup. Picking up the spoon and beginning to stir must be done quickly, because the melting point of the gallium is lower than normal body temperature, so the spoon will melt in one's hand!

QUICK QUIZ 17.2 Suppose the same process of adding energy to the ice cube is performed as discussed above, but instead we graph the internal energy of the system as a function of energy input. What would this graph look like?

QUICK QUIZ 17.3 Calculate the slopes for the A, C, and E portions of Figure 17.3. Rank the slopes from least steep to steepest, and explain what this ordering means.

Example 17.3 | Cooling the Steam

What mass of steam initially at 130°C is needed to warm 200 g of water in a 100-g glass container from 20.0°C to 50.0°C?

SOLUTION

Conceptualize Imagine placing water and steam together in a closed insulated container. The system eventually reaches a uniform state of water with a final temperature of 50.0°C.

Categorize Based on our conceptualization of this situation, we categorize this example as one involving calorimetry in which a phase change occurs.

Analyze Write Equation 17.4 to describe the calorimetry process:

$$(1) \quad Q_{cold} = -Q_{hot}$$

continued

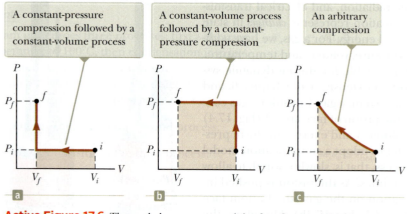

Active Figure 17.6 The work done on a gas as it is taken from an initial state to a final state depends on the path between these states.

from P_i to P_f by heating at constant volume V_f. The work done on the gas along this path is $-P_i(V_f - V_i)$. In Active Figure 17.6b, the pressure of the gas is increased from P_i to P_f at constant volume V_i and then the volume of the gas is reduced from V_i to V_f at constant pressure P_f. The work done on the gas along this path is $-P_f(V_f - V_i)$, which is greater in magnitude than that for the process described in Active Figure 17.6a because the piston is displaced through the same distance by a larger force than for the situation in Active Figure 17.6a. Finally, for the process described in Active Figure 17.6c, where both P and V change continuously, the work done on the gas has some value intermediate between the values obtained in the first two processes.

The energy transfer Q into or out of a system by heat also depends on the process. Consider the situations depicted in Figure 17.7. In both processes that are illustrated, the gas has the same initial volume, temperature, and pressure, and is assumed to be ideal. In Figure 17.7a, the gas is thermally insulated from its surroundings except at the bottom of the gas-filled region, where it is in thermal contact with an energy reservoir. An *energy reservoir* is a source of energy that is considered to be so great that a finite transfer of energy to or from the reservoir does not change its temperature. The piston is held at its initial position by an external agent such as a hand. When the force holding the piston is reduced slightly, the piston rises very slowly to its final position shown in Figure 17.7b. Because the piston is moving upward, the gas is doing work on the piston. During this expansion to the final volume V_f, just enough energy is transferred by heat from the reservoir to the gas to maintain a constant temperature T_i.

Now consider the completely thermally insulated system shown in Figure 17.7c. When the membrane is broken, the gas expands rapidly into the vacuum until it occupies a volume V_f and is at a pressure P_f. The final state of the gas is shown in Figure 17.7d. In this case, the gas does no work because it does not apply a force; no force is required to expand into a vacuum. Furthermore, no energy is transferred by heat through the insulating wall.

As we discuss in Section 17.6, experiments show that the temperature of the ideal gas does not change in the process indicated in Figures 17.7c and 17.7d. Therefore,

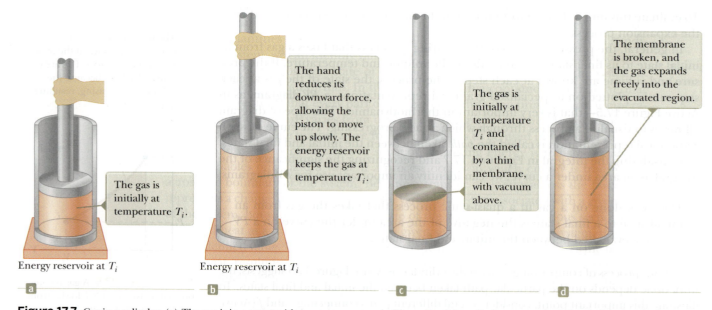

Figure 17.7 Gas in a cylinder. (a) The gas is in contact with an energy reservoir. The walls of the cylinder are perfectly insulating, but the base in contact with the reservoir is conducting. (b) The gas expands slowly to a larger volume. (c) The gas is contained by a membrane in half of a volume, with vacuum in the other half. The entire cylinder is perfectly insulating. (d) The gas expands freely into the larger volume.

the initial and final states of the ideal gas in Figures 17.7a and 17.7b are identical to the initial and final states in Figures 17.7c and 17.7d, but the paths are different. In the first case, the gas does work on the piston and energy is transferred slowly to the gas by heat. In the second case, no energy is transferred by heat and the value of the work done is zero. Therefore, energy transfer by heat, like work done, depends on the initial, final, and intermediate states of the system. In other words, because heat and work depend on the path, neither quantity is determined solely by the end-points of a thermodynamic process.

> ### Example 17.4 | Comparing Processes
>
> An ideal gas is taken through two processes in which $P_f = 1.00 \times 10^5$ Pa, $V_f = 2.00$ m^3, $P_i = 0.200 \times 10^5$ Pa, and $V_i = 10.0$ m^3. For process 1 shown in Active Figure 17.6c, the temperature remains constant. For process 2 shown in Active Figure 17.6a, the pressure remains constant and then the volume remains constant. What is the ratio of the work W_1 done on the gas in the first process to the work W_2 done in the second process?

SOLUTION

Conceptualize In Figure 17.6a (process 2), the displacement occurs at a fixed pressure equal to the initial pressure. In Figure 17.6c (process 1), the displacement occurs at an ever increasing pressure as the piston is moved inward. As a consequence, the force pushing the piston in during process 1 will become larger as the piston moves inward. We therefore expect the work to be larger for process 1 than for process 2.

Categorize We can categorize process 1 as taking place at constant temperature. We categorize process 2 as a combination of processes taking place at constant pressure and then at constant volume. In Section 17.6, we will discuss names for these types of processes.

Analyze For process 1, express the pressure as a function of volume using the ideal gas law:

$$P = \frac{nRT}{V}$$

For process 2, no work is done during the portion at constant volume because the piston does not move through a displacement. During the first part of the process, the pressure is constant at $P = P_i$. Use these results and set up the ratio of the work done in the two processes:

$$\frac{W_1}{W_2} = \frac{-\displaystyle\int_{\text{process 1}} P\,dV}{-\displaystyle\int_{\text{process 2}} P\,dV} = \frac{\displaystyle\int_{V_i}^{V_f} \frac{nRT}{V}\,dV}{\displaystyle\int_{V_i}^{V_f} P_i\,dV} = \frac{nRT\displaystyle\int_{V_i}^{V_f}\frac{dV}{V}}{P_i\displaystyle\int_{V_i}^{V_f}dV}$$

$$= \frac{nRT\ln\left(\dfrac{V_f}{V_i}\right)}{P_i(V_f - V_i)} = \frac{P_iV_i\ln\left(\dfrac{V_f}{V_i}\right)}{P_i(V_f - V_i)} = \frac{V_i\ln\left(\dfrac{V_f}{V_i}\right)}{V_f - V_i}$$

Substitute the numerical values for the initial and final volumes:

$$\frac{W_1}{W_2} = \frac{(10.0\text{ m}^3)\,\ln\left(\dfrac{2.00\text{ m}^3}{10.0\text{ m}^3}\right)}{(2.00\text{ m}^3 - 10.0\text{ m}^3)} = 2.01$$

Finalize As we expected, the work done in process 1 is larger, by about a factor of 2. How do you think the work done in process 1 would compare to that done in process 3, shown in Figure 17.6b?

17.5 | The First Law of Thermodynamics

In Chapter 7, we discussed the conservation of energy equation, Equation 7.2. Let us consider a special case of this general principle in which the only change in the energy of a system is in its internal energy E_{int} and the only transfer mechanisms are heat Q and work W, which we have discussed in this chapter. This case leads to an equation that can be used to analyze many problems in thermodynamics.

This special case of the conservation of energy equation, called the **first law of thermodynamics,** can be written as

$$\Delta E_{int} = Q + W \qquad \text{17.8}$$

◀ First law of thermodynamics

This equation indicates that the change in the internal energy of a system is equal to the sum of the energy transferred across the system boundary by heat and the energy transferred by work.

Active Figure 17.8 shows the energy transfers and change in internal energy for a gas in a cylinder consistent with the first law. Equation 17.8 can be used in a variety of problems in which the only energy considerations are internal energy, heat, and work. We shall consider several examples shortly. Some problems may not fit the conditions of the first law. For example, the internal energy of the coils in your toaster does not increase due to heat or work, but rather due to electrical transmission. Keep in mind that the first law is a special case of the conservation of energy equation, and the latter is the more general equation that covers the widest range of possible situations.

When a system undergoes an infinitesimal change in state, such that a small amount of energy dQ is transferred by heat and a small amount of work dW is done on the system, the internal energy also changes by a small amount dE_{int}. Therefore, for infinitesimal processes we can express the first law as[4]

$$dE_{int} = dQ + dW \qquad \text{17.9} \blacktriangleleft$$

No practical distinction exists between the results of heat and work on a microscopic scale. Each can produce a change in the internal energy of a system. Although the macroscopic quantities Q and W are *not* properties of a system, they are related to changes of the internal energy of a stationary system through the first law of thermodynamics. Once a process or path is defined, Q and W can be either calculated or measured, and the change in internal energy can be found from the first law.

Active Figure 17.8 The first law of thermodynamics equates the change ΔE_{int} in internal energy in a system to the net energy transfer to the system by heat Q and work W. In the situation shown here, the internal energy of the gas increases.

▶ **QUICK QUIZ 17.4** In the last three columns of the following table, fill in the boxes with the correct signs ($-$, $+$, or 0) for Q, W, and ΔE_{int}. For each situation, the system to be considered is identified.

Situation	System	Q	W	ΔE_{int}
(a) Rapidly pumping up a bicycle tire	Air in the pump			
(b) Pan of room-temperature water sitting on a hot stove	Water in the pan			
(c) Air quickly leaking out of a balloon	Air originally in the balloon			

▶ **THINKING PHYSICS 17.2**

In the late 1970s, casino gambling was approved in Atlantic City, New Jersey, which can become quite cold in the winter. Energy projections that were performed for the design of the casinos showed that the air-conditioning system would need to operate in the casino even in the middle of a very cold January. Why?

Reasoning If we consider the air in the casino to be the gas to which we apply the first law, imagine a simplification model in which there is no air conditioning and no ventilation so that this gas simply stays in the room. No work is being done on the gas, so we focus on the energy transferred by heat. A casino contains a large number of people, many of whom are active (throwing dice, cheering, etc.) and many of whom are in excited states (celebration, frustration, panic, etc.). As a result, these people have large rates of energy flow by heat from their bodies into the air. This energy results in an increase in internal energy of the air in the casino. With the large number of excited people in a casino (along with the large number of machines and incandescent lights), the temperature of the gas can rise quickly, and to a very high value. To keep the temperature at a comfortable level, energy must be transferred out of the air to compensate for the energy input. Calculations show that energy transfer by heat through the walls even on a 10°F January day is not sufficient to provide the required energy transfer, so the air-conditioning system must be in almost continuous use throughout the year. ◀

---

[4]It should be noted that dQ and dW are not true differential quantities because Q and W are not state variables, although dE_{int} is a true differential. For further details on this point, see R. P. Bauman, *Modern Thermodynamics and Statistical Mechanics* (New York: Macmillan, 1992).

17.6 | Some Applications of the First Law of Thermodynamics

To apply the first law of thermodynamics to specific systems, it is useful to first define some common thermodynamic processes. We shall identify several special processes used as simplification models to approximate real processes. For each of the following processes, we build a mental representation by imagining that the process occurs for the gas in Active Figure 17.8.

During an **adiabatic process,** no energy enters or leaves the system by heat; that is, $Q = 0$. For the piston in Active Figure 17.8, imagine that all surfaces of the piston are perfect insulators so that energy transfer by heat does not exist. (Another way to achieve an adiabatic process is to perform the process very rapidly because energy transfer by heat tends to be relatively slow.) Applying the first law in this case, we see that

$$\Delta E_{int} = W \qquad\qquad 17.10 \blacktriangleleft$$

From this result, we see that when a gas is compressed adiabatically, both W and ΔE_{int} are positive; work is done on the gas, representing a transfer of energy into the system, so the internal energy increases. Conversely, when the gas expands adiabatically, ΔE_{int} is negative.

Adiabatic processes are very important in engineering practice. Common applications include the expansion of hot gases in an internal combustion engine, the liquefaction of gases in a cooling system, and the compression stroke in a diesel engine. We study adiabatic processes in more detail in Section 17.8.

The **free expansion** depicted in Figures 17.7c and 17.7d is a unique adiabatic process in which no work is done on the gas. Because $Q = 0$ and $W = 0$, we see from the first law that $\Delta E_{int} = 0$ for this process. That is, the initial and final internal energies of a gas are equal in a free expansion. As we saw in Chapter 16, the internal energy of an ideal gas depends only on its temperature. Therefore, we expect no change in temperature during an adiabatic free expansion, which is in accord with experiments performed at low pressures. Experiments with real gases at high pressures show a slight increase or decrease in temperature after the expansion because of interactions between molecules.

A process that occurs at constant pressure is called an **isobaric process.** In Active Figure 17.8, as long as the piston is perfectly free to move, the pressure of the gas inside the cylinder is due to atmospheric pressure and the weight of the piston. Hence, the piston can be modeled as a particle in equilibrium. When such a process occurs, the work done on the gas is simply the negative of the constant pressure multiplied by the change in volume, or $-P(V_f - V_i)$. On a PV diagram, an isobaric process appears as a horizontal line, such as the first portion of the process in Active Figure 17.6a or the second portion of the process in Active Figure 17.6b.

A process that takes place at constant volume is called an **isovolumetric process.** In Active Figure 17.8, an isovolumetric process is created by locking the piston in place so that it cannot move. In such a process, the work done is zero because the volume does not change. Hence, the first law applied to an isovolumetric process gives

$$\Delta E_{int} = Q \qquad\qquad 17.11 \blacktriangleleft$$

This equation tells us that if energy is added by heat to a system kept at constant volume, all the energy goes into increasing the internal energy of the system and none enters or leaves the system by work. For example, when an aerosol can is thrown into a fire, energy enters the system (the gas in the can) by heat through the metal walls of the can. Consequently, the temperature and pressure of the gas rise until the can possibly explodes. On a PV diagram, an isovolumetric process appears as a vertical line, such as the second portion of the process in Active Figure 17.6a or the first portion of the process in Active Figure 17.6b.

A process that occurs at constant temperature is called an **isothermal process.** Because the internal energy of an ideal gas is a function of temperature only, in an

17.8 *cont.*

Categorize We categorize this example as a problem involving an adiabatic process.

Analyze Use Equation 17.25 to find the final pressure:

$$P_f = P_i \left(\frac{V_i}{V_f}\right)^\gamma = (1.00 \text{ atm}) \left(\frac{800.0 \text{ cm}^3}{60.0 \text{ cm}^3}\right)^{1.40}$$

$$= 37.6 \text{ atm}$$

Use the ideal gas law to find the final temperature:

$$\frac{P_i V_i}{T_i} = \frac{P_f V_f}{T_f}$$

$$T_f = \frac{P_f V_f}{P_i V_i} T_i = \frac{(37.6 \text{ atm})(60.0 \text{ cm}^3)}{(1.00 \text{ atm})(800.0 \text{ cm}^3)} (293 \text{ K})$$

$$= 826 \text{ K} = 553°C$$

Finalize The temperature of the gas increases by a factor of 826 K/293 K = 2.82. The high compression in a diesel engine raises the temperature of the gas enough to cause the combustion of fuel without the use of spark plugs.

❮17.9❯ Molar Specific Heats and the Equipartition of Energy

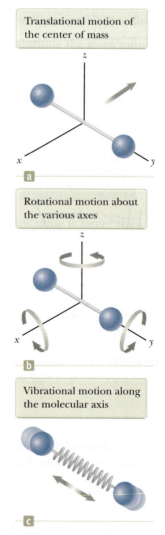

Translational motion of the center of mass

Rotational motion about the various axes

Vibrational motion along the molecular axis

Figure 17.15 Possible motions of a diatomic molecule.

We have found that predictions of molar specific heats based on kinetic theory agree quite well with the behavior of monatomic gases but not with the behavior of complex gases (Table 17.3). To explain the variations in C_V and C_P between monatomic gases and more complex gases, let us explore the origin of specific heat by extending our structural model of kinetic theory in Chapter 16. In Section 16.5, we discussed that the sole contribution to the internal energy of a monatomic gas is the translational kinetic energy of the molecules. We also discussed the theorem of equipartition of energy, which states that, at equilibrium, each degree of freedom contributes, on the average, $\frac{1}{2} k_B T$ of energy per molecule. The monatomic gas has three degrees of freedom, one associated with each of the independent directions of translational motion.

For more complex molecules, other types of motion exist in addition to translation. The internal energy of a diatomic or polyatomic gas includes contributions from the vibrational and rotational motion of the molecules in addition to translation. The rotational and vibrational motions of molecules with structure can be activated by collisions and therefore are "coupled" to the translational motion of the molecules. The branch of physics known as *statistical mechanics* suggests that the average energy for each of these additional degrees of freedom is the same as that for translation, which in turn suggests that the determination of a gas's internal energy is a simple matter of counting the degrees of freedom. We will find that this process works well, although the model must be modified with some notions from quantum physics for us to explain the experimental data completely.

Let us consider a diatomic gas, which we can model as consisting of dumbbell-shaped molecules (Fig. 17.15), and apply concepts that we studied in Chapter 10. In this model, the center of mass of the molecule can translate in the x, y, and z directions (Fig. 17.15a). For this motion, the molecule behaves as a particle, just like an atom in a monatomic gas. In addition, if we consider the molecule as a rigid object, it can rotate about three mutually perpendicular axes (Fig. 17.15b). We can ignore the rotation about the y axis because the moment of inertia and the rotational energy $\frac{1}{2} I \omega^2$ about this axis are negligible compared with those associated with the x and z axes. Therefore, there are five degrees of freedom: three associated with the translational motion and two associated with the rotational motion. Because each degree of freedom contributes, on average, $\frac{1}{2} k_B T$ of energy per molecule, the total internal energy for a diatomic gas consisting of N molecules and considering both translation and rotation is

$$E_{int} = 3N(\tfrac{1}{2} k_B T) + 2N(\tfrac{1}{2} k_B T) = \tfrac{5}{2} N k_B T = \tfrac{5}{2} nRT$$

We can use this result and Equation 17.19 to predict the molar specific heat at constant volume:

$$C_V = \frac{1}{n}\frac{dE_{int}}{dT} = \frac{1}{n}\frac{d}{dT}\left(\tfrac{5}{2}nRT\right) = \tfrac{5}{2}R = 20.8\,\text{J/mol}\cdot\text{K} \qquad \textbf{17.28} \blacktriangleleft$$

From Equations 17.21 and 17.22, we find that the model predicts that

$$C_P = C_V + R = \tfrac{7}{2}R \qquad \textbf{17.29} \blacktriangleleft$$

$$\gamma = \frac{C_P}{C_V} = \frac{\tfrac{7}{2}R}{\tfrac{5}{2}R} = \frac{7}{5} = 1.40 \qquad \textbf{17.30} \blacktriangleleft$$

Let us now incorporate the vibration of the molecule in the model. We use the structural model for the diatomic molecule in which the two atoms are joined by an imaginary spring (Fig. 17.15c) and apply concepts from the particle in simple harmonic motion model in Chapter 12. The vibrational motion has two types of energy associated with the vibrations along the length of the molecule—kinetic energy of the atoms and potential energy in the model spring—which adds two more degrees of freedom for a total of seven for translation, rotation, and vibration. Because each degree of freedom contributes $\tfrac{1}{2}k_B T$ of energy per molecule, the total internal energy for a diatomic gas consisting of N molecules and considering all types of motion is

$$E_{int} = 3N\left(\tfrac{1}{2}k_B T\right) + 2N\left(\tfrac{1}{2}k_B T\right) + 2N\left(\tfrac{1}{2}k_B T\right) = \tfrac{7}{2}Nk_B T = \tfrac{7}{2}nRT$$

Therefore, the molar specific heat at constant volume is predicted to be

$$C_V = \frac{1}{n}\frac{dE_{int}}{dT} = \frac{1}{n}\frac{d}{dT}\left(\tfrac{7}{2}nRT\right) = \tfrac{7}{2}R = 29.1\,\text{J/mol}\cdot\text{K} \qquad \textbf{17.31} \blacktriangleleft$$

From Equations 17.21 and 17.22,

$$C_P = C_V + R = \tfrac{9}{2}R \qquad \textbf{17.32} \blacktriangleleft$$

$$\gamma = \frac{C_P}{C_V} = \frac{\tfrac{9}{2}R}{\tfrac{7}{2}R} = \frac{9}{7} = 1.29 \qquad \textbf{17.33} \blacktriangleleft$$

When we compare our predictions with the section of Table 17.3 corresponding to diatomic gases, we find a curious result. For the first four gases—hydrogen, nitrogen, oxygen, and carbon monoxide—the value of C_V is close to that predicted by Equation 17.28, which includes rotation but not vibration. The value for the fifth gas, chlorine, lies between the prediction including rotation and the prediction that includes rotation and vibration. None of the values agrees with Equation 17.31, which is based on the most complete model for motion of the diatomic molecule!

It might seem that our model is a failure for predicting molar specific heats for diatomic gases. We can claim some success for our model, however, if measurements of molar specific heat are made over a wide temperature range rather than at the single temperature that gives us the values in Table 17.3. Figure 17.16 shows the molar specific heat of hydrogen as a function of temperature. The curve has three plateaus and they are at the values of the molar specific heat predicted by Equations 17.17, 17.28, and 17.31! For low temperatures, the diatomic hydrogen gas behaves like a monatomic gas. As the temperature rises to room temperature, its molar specific heat rises to a value for a diatomic gas, consistent with the inclusion of rotation but not vibration. For high temperatures, the molar specific heat is consistent with a model including all types of motion.

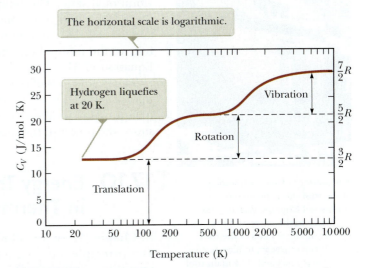

Figure 17.16 The molar specific heat of hydrogen as a function of temperature.

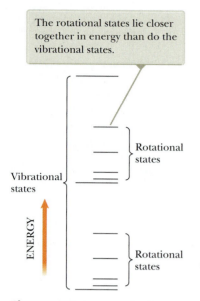

The rotational states lie closer together in energy than do the vibrational states.

Vibrational states

Rotational states

ENERGY

Rotational states

Figure 17.17 An energy level diagram for vibrational and rotational states of a diatomic molecule.

Before addressing the reason for this mysterious behavior, let us make a brief remark about polyatomic gases. For molecules with more than two atoms, the number of degrees of freedom is even larger and the vibrations are more complex than for diatomic molecules. These considerations result in an even higher predicted molar specific heat, which is in qualitative agreement with experiment. For the polyatomic gases shown in Table 17.3, we see that the molar specific heats are higher than those for diatomic gases. The more degrees of freedom available to a molecule, the more "ways" of storing energy are available, resulting in a higher molar specific heat.

A Hint of Energy Quantization

Our model for molar specific heats has been based so far on purely classical notions. It predicts a value of the specific heat for a diatomic gas that, according to Figure 17.16, only agrees with experimental measurements made at high temperatures. To explain why this value is only true at high temperatures and why the plateaus exist in Figure 17.16, we must go beyond classical physics and introduce some quantum physics into the model. In Section 11.5, we discussed energy quantization for the hydrogen atom. Only certain energies are allowed for the system, and an energy level diagram can be drawn to illustrate those allowed energies. For a molecule, quantum physics tells us that the rotational and vibrational energies are quantized. Figure 17.17 shows an energy level diagram for the rotational and vibrational quantum states of a diatomic molecule. Notice that vibrational states are separated by larger energy gaps than rotational states.

At low temperatures, the energy that a molecule gains in collisions with its neighbors is generally not large enough to raise it to the first excited state of either rotation or vibration. All molecules are in the ground state for rotation and vibration. Therefore, at low temperatures, the only contribution to the molecules' average energy is from translation, and the specific heat is that predicted by Equation 17.17.

As the temperature is raised, the average energy of the molecules increases. In some collisions, a molecule may have enough energy transferred to it from another molecule to excite the first rotational state. As the temperature is raised further, more molecules can be excited to this state. The result is that rotation begins to contribute to the internal energy and the molar specific heat rises. At about room temperature in Figure 17.16, the second plateau is reached and rotation contributes fully to the molar specific heat. The molar specific heat is now equal to the value predicted by Equation 17.28.

Vibration contributes nothing at room temperature because the vibrational states are farther apart in energy than the rotational states; the molecules are in the ground vibrational state. The temperature must be even higher to raise the molecules to the first excited vibrational state. That happens in Figure 17.16 between 1 000 K and 10 000 K. At 10 000 K on the right side of the figure, vibration is contributing fully to the internal energy and the molar specific heat has the value predicted by Equation 17.31.

The predictions of this structural model are supportive of the theorem of equipartition of energy. In addition, the inclusion in the model of energy quantization from quantum physics allows a full understanding of Figure 17.16. This excellent example shows the power of the modeling approach.

The absence of snow on some parts of the roof shows that energy is conducted from the inside of the residence to the exterior more rapidly on those parts of the roof. The dormer appears to have been added and insulated. The main roof does not appear to be well insulated.

17.10 | Energy Transfer Mechanisms in Thermal Processes

In Chapter 7, we introduced the conservation of energy equation $\Delta E_{system} = \Sigma T$ as a principle allowing a global approach to energy considerations in physical processes. Earlier in this chapter, we discussed two of the terms on the right-hand side of that equation: work and heat. In this section, we consider more details of heat and two other energy transfer methods that are often related to temperature changes: convection (a form of matter transfer) and electromagnetic radiation.

Conduction

The process of energy transfer by heat can also be called **conduction** or **thermal conduction.** In this process, the transfer mechanism can be viewed on an atomic scale as an exchange of kinetic energy between molecules in which the less energetic molecules gain energy by colliding with the more energetic molecules. For example, if you hold one end of a long metal bar and insert the other end into a flame, the temperature of the metal in your hand soon increases. The energy reaches your hand through conduction. How that happens can be understood by examining what is happening to the atoms in the metal. Initially, before the rod is inserted into the flame, the atoms are vibrating about their equilibrium positions. As the flame provides energy to the rod, those atoms near the flame begin to vibrate with larger and larger amplitudes. These atoms in turn collide with their neighbors and transfer some of their energy in the collisions. Slowly, metal atoms farther and farther from the flame increase their amplitude of vibration until eventually those in the metal near your hand are affected. This increased vibration represents an increase in temperature of the metal (and possibly a burned hand).

Although the transfer of energy through a material can be partially explained by atomic vibrations, the rate of conduction also depends on the properties of the substance. For example, it is possible to hold a piece of asbestos in a flame indefinitely, which implies that very little energy is being conducted through the asbestos. In general, metals are good thermal conductors because they contain large numbers of electrons that are relatively free to move through the metal and can transport energy from one region to another. Therefore, in a good thermal conductor, such as copper, conduction takes place via the vibration of atoms and via the motion of free electrons. Materials such as asbestos, cork, paper, and fiberglass are poor thermal conductors. Gases also are poor thermal conductors because of the large distance between the molecules.

Conduction occurs only if the temperatures differ between two parts of the conducting medium. This temperature difference drives the flow of energy. Consider a slab of material of thickness Δx and cross-sectional area A with its opposite faces at different temperatures T_c and T_h, where $T_h > T_c$ (Fig. 17.18). The slab allows energy to transfer from the region of high temperature to that of low temperature by thermal conduction. The rate of energy transfer by heat, $P = Q/\Delta t$, is proportional to the cross-sectional area of the slab and the temperature difference and inversely proportional to the thickness of the slab:

$$P = \frac{Q}{\Delta t} \propto A\frac{\Delta T}{\Delta x}$$

Note that P has units of watts when Q is in joules and Δt is in seconds. That is not surprising because P is *power*, the rate of transfer of energy by heat. For a slab of infinitesimal thickness dx and temperature difference dT, we can write the **law of conduction** as

$$P = kA\left|\frac{dT}{dx}\right|$$

17.34◀

where the proportionality constant k is called the **thermal conductivity** of the material and dT/dx is the **temperature gradient** (the variation of temperature with position). It is the higher thermal conductivity of tile relative to carpet that makes the tile floor feel colder than the carpeted floor in the discussion at the beginning of Chapter 16.

Suppose a substance is in the shape of a long uniform rod of length L as in Figure 17.19 and is insulated so that energy cannot escape by heat from its surface

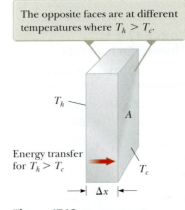

The opposite faces are at different temperatures where $T_h > T_c$.

Figure 17.18 Energy transfer through a conducting slab with cross-sectional area A and thickness Δx.

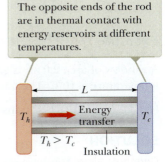

The opposite ends of the rod are in thermal contact with energy reservoirs at different temperatures.

Figure 17.19 Conduction of energy through a uniform, insulated rod of length L.

 ▶ Law of conduction

TABLE 17.4 |
Thermal Conductivities

Substance	Thermal Conductivity (W/m · °C)
Metals (at 25°C)	
Aluminum	238
Copper	397
Gold	314
Iron	79.5
Lead	34.7
Silver	427
Nonmetals (approximate values)	
Asbestos	0.08
Concrete	0.8
Diamond	2 300
Glass	0.8
Ice	2
Rubber	0.2
Water	0.6
Wood	0.08
Gases (at 20°C)	
Air	0.023 4
Helium	0.138
Hydrogen	0.172
Nitrogen	0.023 4
Oxygen	0.023 8

except at the ends, which are in thermal contact with reservoirs having temperatures T_c and T_h. When steady state is reached, the temperature at each point along the rod is constant in time. In this case, the temperature gradient is the same everywhere along the rod and is

$$\left| \frac{dT}{dx} \right| = \frac{T_h - T_c}{L}$$

Therefore, the rate of energy transfer by heat is

$$P = kA \frac{(T_h - T_c)}{L} \qquad \text{17.35} \blacktriangleleft$$

Substances that are good thermal conductors have large thermal conductivity values, whereas good thermal insulators have low thermal conductivity values. Table 17.4 lists thermal conductivities for various substances.

QUICK QUIZ 17.7 You have two rods of the same length and diameter, but they are formed from different materials. The rods will be used to connect two regions of different temperature such that energy will transfer through the rods by heat. They can be connected in series, as in Figure 17.20a, or in parallel, as in Figure 17.20b. In which case is the rate of energy transfer by heat larger? **(a)** The rate is larger when the rods are in series. **(b)** The rate is larger when the rods are in parallel. **(c)** The rate is the same in both cases.

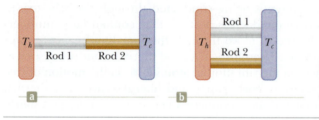

Figure 17.20 (Quick Quiz 17.7) In which case is the rate of energy transfer larger?

Example 17.9 | The Leaky Window

A window of area 2.0 m² is glazed with glass of thickness 4.0 mm. The window is in the wall of a house, and the outside temperature is 10°C. The temperature inside the house is 25°C.

(A) How much energy transfers through the window by heat in 1.0 h?

SOLUTION

Conceptualize You have several windows in your home. By placing your hand on the glass of a window on a cold winter day, you may have noticed that the glass is cold compared to the room temperature. The outside surface of the glass is even colder, resulting in a transfer of energy by heat through the glass.

Categorize We categorize the problem as one involving thermal conduction as well as our definition of power from Chapter 7.

Analyze Use Equation 17.35 to find the rate of energy transfer by heat:

$$P = kA \frac{(T_h - T_c)}{L}$$

Substitute numerical values, using the value for k for glass from Table 17.4:

$$P = (0.8 \text{ W/m} \cdot \text{°C})(2.0 \text{ m}^2) \frac{(25\text{°C} - 10\text{°C})}{4.0 \times 10^{-3} \text{ m}}$$

$$= 6 \times 10^3 \text{ W}$$

From the definition of power as the rate of energy transfer, find the energy transferred at this rate in 1.0 h:

$$Q = P \Delta t = (6 \times 10^3 \text{ W})(3.6 \times 10^3 \text{ s}) = 2 \times 10^7 \text{ J}$$

17.9 *cont.*

(B) If electrical energy costs 12¢/kWh, how much does the transfer of energy in part (A) cost to replace with electrical heating?

SOLUTION

Cast the answer to part (A) in units of kilowatt-hours:

$$Q = P \Delta t = (6 \times 10^3 \, \text{W})(1.0 \, \text{h}) = 6 \times 10^3 \, \text{Wh} = 6 \, \text{kWh}$$

Therefore, the cost to replace the energy transferred through the window is $(6 \, \text{kWh})(12¢/\text{kWh}) \approx$ **72¢**.

Finalize If you imagine paying this much for each hour for each window in your home, your electric bill will be extremely high! For example, for ten such windows, your electric bill would be over $5 000 for one month. It seems like something is wrong here because electric bills are not that high. In reality, a thin layer of air adheres to each of the two surfaces of the window. This air provides additional insulation to that of the glass. As seen in Table 17.4, air is a much poorer thermal conductor than glass, so most of the insulation is performed by the air, not the glass, in a window!

Convection

At one time or another you may have warmed your hands by holding them over an open flame. In this situation, the air directly above the flame is heated and expands. As a result, the density of the air decreases and the air rises. This warmed mass of air transfers energy by heat into your hands as it flows by. The transfer of energy from the flame to your hands is performed by matter transfer because the energy travels with the air. Energy transferred by the movement of a fluid is a process called **convection.** When the movement results from differences in density, as in the example of air around a fire, the process is called **natural convection.** When the fluid is forced to move by a fan or pump, as in some air and water heating systems, the process is called **forced convection.**

The circulating pattern of air flow at a beach (Section 17.2) is an example of convection in nature. The mixing that occurs as water is cooled and eventually freezes at its surface (Section 16.3) is another example.

If it were not for convection currents, it would be very difficult to boil water. As water is heated in a teakettle, the lower layers are warmed first. These regions expand and rise to the top because their density is lower than that of the cooler water. At the same time, the denser cool water falls to the bottom of the kettle so that it can be heated.

The same process occurs near the surface of the Sun. Figure 17.21 shows a close-up view of the solar surface. The granulation that appears is because of *convection cells.* The bright center of a cell is the location at which hot gases rise to the surface, just like the hot water rises to the surface in a pan of boiling water. As the gases cool, they sink back downward along the edges of the cell, forming the darker outline of each cell. The sinking gases appear dark because they are cooler than the gases in the center of the cell. Although the sinking gases emit a tremendous amount of radiation, the filter used to take the photograph in Figure 17.21 makes these areas appear dark relative to the warmer center of the cell.

Convection occurs when a room is heated by a radiator. The radiator warms the air in the lower regions of the room by heat at the interface between the radiator surface and the air. The warm air expands and floats to the ceiling because of its lower density, setting up the continuous air current pattern shown in Figure 17.22.

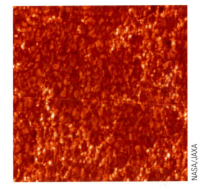

Figure 17.21 The surface of the Sun shows *granulation*, due to the existence of separate convection cells, each carrying energy to the surface by convection.

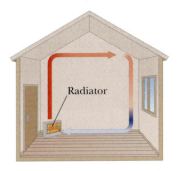

Radiator

Figure 17.22 Convection currents are set up in a room heated by a radiator.

NASA/JAXA

Radiation

Another method of transferring energy that can be related to a temperature change is **electromagnetic radiation.** All objects radiate energy continuously in the form of electromagnetic waves. As we shall find out in Chapter 24, electromagnetic radiation arises from accelerating electric charges. From our discussion of temperature, we know that temperature corresponds to random motion of molecules that are

is reradiated out into space according to Stefan's law, Equation 17.36. The only type of energy in the system that can change due to radiation is internal energy. Let us assume that any change in temperature of the Earth is so small over a time interval that we can approximate the change in internal energy as zero. This assumption leads to the following reduction of the conservation of energy equation, Equation 7.2:

$$0 = T_{ER}$$

Two energy transfer mechanisms occur by electromagnetic radiation, so we can write this equation as

$$0 = T_{ER} \text{ (in)} + T_{ER} \text{ (out)} \quad \rightarrow \quad T_{ER} \text{ (in)} = -T_{ER} \text{ (out)} \qquad \textbf{17.38} \blacktriangleleft$$

where "in" and "out" refer to energy transfers across the boundary of the system of the Earth. The energy coming into the system is from the Sun, and the energy going out of the system is by thermal radiation emitted from the Earth's surface. Figure 17.23 depicts these energy exchanges. The energy coming in from the Sun comes from only one direction, but the energy radiated out from the Earth's surface leaves in all directions. This distinction will be important in setting up our calculation of the equilibrium temperature.

As mentioned in Section 17.10, the rate of energy transfer per unit area from the Sun is approximately 1 370 W/m² at the top of the atmosphere. The rate of energy transfer per area is called **intensity,** and the intensity of radiation from the Sun at the top of the atmosphere is called the **solar constant** $I_S = 1\ 370$ W/m². A large amount of this energy is in the form of visible radiation, to which the atmosphere is transparent. The radiation emitted from the Earth's surface, however, is not visible. For a radiating object at the temperature of the Earth's surface, the radiation peaks in the infrared, with greatest intensity at a wavelength of about 10 μm. In general, objects with typical household temperatures have wavelength distributions in the infrared, so we do not see them glowing visibly. Only much hotter items emit enough radiation to be seen visibly. An example is a household electric stove burner. When turned off, it emits a small amount of radiation, mostly in the infrared. When turned to its highest setting, its much higher temperature results in significant radiation, with much of it in the visible. As a result, it appears to glow with a reddish color and is described as *red-hot*.

Let us divide Equation 17.38 by the time interval Δt during which the energy transfer occurs, which gives us

$$P_{ER} \text{ (in)} = -P_{ER} \text{ (out)} \qquad \textbf{17.39} \blacktriangleleft$$

We can express the rate of energy transfer into the top of the atmosphere of the Earth in terms of the solar constant I_S:

$$P_{ER} \text{ (in)} = I_S A_c$$

where A_c is the circular cross-sectional area of the Earth. Not all the radiation arriving at the top of the atmosphere reaches the ground. A fraction of it is reflected from clouds and the ground and escapes back into space. For the Earth, this fraction is about 30%, so only 70% of the incident radiation reaches the surface. Using this fact, we modify the input power, assuming that 70.0% reaches the surface:

$$P_{ER} \text{ (in)} = (0.700) I_S A_c$$

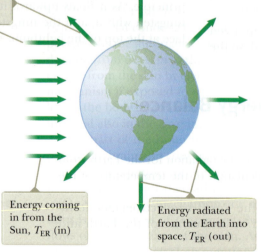

When the rates of incoming and outgoing energy transfer are equal, the temperature of the Earth's surface remains constant.

Energy coming in from the Sun, T_{ER} (in)

Energy radiated from the Earth into space, T_{ER} (out)

Figure 17.23 Energy exchanges by electromagnetic radiation for the Earth. The Sun is far to the left of the diagram and is not visible.

Stefan's law can be used to express the outgoing power, assuming that the Earth is a perfect emitter ($e = 1$):

$$P_{ER} \text{ (out)} = -\sigma A T^4$$

In this expression, A is the surface area of the Earth, T is its surface temperature, and the negative sign indicates that energy is leaving the Earth. Substituting the expressions for the input and output power into Equation 17.39, we have

$$(0.700) I_S A_c = -(-\sigma A T^4)$$

Solving for the temperature of the Earth's surface gives

$$T = \left(\frac{(0.700) I_S A_c}{\sigma A}\right)^{1/4}$$

Substituting the numbers, we find that

$$T = \left(\frac{(0.700)(1\ 370\ \text{W/m}^2)(\pi R_E{}^2)}{(5.67 \times 10^{-8}\ \text{W/m}^2 \cdot \text{K}^4)(4\pi R_E{}^2)}\right)^{1/4} = 255\ \text{K} \qquad \textbf{17.40} \blacktriangleleft$$

Measurements show that the average global temperature at the surface of the Earth is 288 K, about 33 K warmer than the temperature from our calculation. This difference indicates that a major factor was left out of our analysis. The major factor is the thermodynamic effects of the atmosphere, which result in additional energy from the Sun being "trapped" in the system of the Earth and raising the temperature. This effect is not included in the simple energy balance calculation we performed. To evaluate it, we must incorporate into our model the principles of thermodynamics of gases for the air in the atmosphere. The details of this incorporation are explored in the Context Conclusion.

❯ SUMMARY

The **internal energy** E_{int} of a system is the total of the kinetic and potential energies of the system associated with its microscopic components. **Heat** is a process by which energy is transferred as a consequence of a temperature difference. It is also the amount of energy Q transferred by this process.

The energy required to change the temperature of a substance by ΔT is

$$Q = mc\,\Delta T \qquad \textbf{17.3} \blacktriangleleft$$

where m is the mass of the substance and c is its **specific heat.**

The energy required to change the phase of a pure substance is

$$Q = L\,\Delta m \qquad \textbf{17.5} \blacktriangleleft$$

where L is **latent heat,** which depends on the substance and the nature of the phase change, and Δm is the change in mass of the higher-phase material.

A **state variable** of a system is a quantity that is defined for a given condition of the system. State variables for a gas include pressure, volume, temperature, and internal energy.

A **quasi-static process** is one that proceeds slowly enough to allow the system to always be in a state of thermal equilibrium.

The **work** done on a gas as its volume changes from some initial value V_i to some final value V_f is

$$W = -\int_{V_i}^{V_f} P\,dV \qquad \textbf{17.7} \blacktriangleleft$$

where P is the pressure, which may vary during the process.

The **first law of thermodynamics** is a special case of the conservation of energy equation, relating the internal energy of a system to energy transfer by heat and work:

$$\Delta E_{int} = Q + W \qquad \textbf{17.8} \blacktriangleleft$$

where Q is the energy transferred across the boundary of the system by heat and W is the work done on the system. Although Q and W both depend on the path taken from the initial state to the final state, internal energy is a state variable, so the quantity ΔE_{int} is independent of the path taken between given initial and final states.

An **adiabatic process** is one in which no energy is transferred by heat between the system and its surroundings ($Q = 0$). In this case, the first law gives $\Delta E_{int} = W$.

An **isobaric process** is one that occurs at constant pressure. The work done on the gas in such a process is $-P(V_f - V_i)$.

An **isovolumetric process** is one that occurs at constant volume. No work is done on the gas in such a process.

An **isothermal process** is one that occurs at constant temperature. The work done on an ideal gas during an isothermal process is

$$W = -nRT \ln\left(\frac{V_f}{V_i}\right) \qquad \textbf{17.12} \blacktriangleleft$$

In a **cyclic process** (one that originates and terminates at the same state), $\Delta E_{int} = 0$, and therefore $Q = -W$.

We define the **molar specific heats** of an ideal gas with the following equations:

$$Q = nC_V \Delta T \quad \text{(constant volume)} \qquad \textbf{17.13} \blacktriangleleft$$

$$Q = nC_P \Delta T \quad \text{(constant pressure)} \qquad \textbf{17.14} \blacktriangleleft$$

where C_V is the **molar specific heat at constant volume** and C_P is the **molar specific heat at constant pressure.**

The change in internal energy of an ideal gas for *any* process in which the temperature change is ΔT is

$$\Delta E_{int} = nC_V \Delta T \qquad \textbf{17.18} \blacktriangleleft$$

The molar specific heat at constant volume is related to internal energy as follows:

$$C_V = \frac{1}{n}\frac{dE_{int}}{dT} \qquad \textbf{17.19} \blacktriangleleft$$

The molar specific heat at constant volume and molar specific heat at constant pressure for all ideal gases are related as follows:

$$C_P - C_V = R \qquad \textbf{17.21} \blacktriangleleft$$

For an ideal gas undergoing an adiabatic process, where

$$\gamma = \frac{C_P}{C_V} \qquad \textbf{17.22} \blacktriangleleft$$

the pressure and volume are related as

$$PV^\gamma = \text{constant} \qquad \textbf{17.24} \blacktriangleleft$$

The theorem of equipartition of energy can be used to predict the molar specific heat at constant volume for various types of gases. Monatomic gases can only store energy by means of translational motion of the molecules of the gas. Diatomic and polyatomic gases can also store energy by means of rotation and vibration of the molecules. For a given molecule, the rotational and vibrational energies are quantized, so their contribution does not enter into the internal energy until the temperature is raised to a sufficiently high value.

Thermal conduction is the transfer of energy by molecular collisions. It is driven by a temperature difference, and the rate of energy transfer is

$$P = kA \left| \frac{dT}{dx} \right| \qquad \textbf{17.34} \blacktriangleleft$$

where the constant k is called the **thermal conductivity** of the material and dT/dx is the **temperature gradient** (the variation of temperature with position).

Convection is energy transfer by means of a moving fluid.

All objects emit **electromagnetic radiation** continuously in the form of **thermal radiation,** which depends on temperature according to **Stefan's law:**

$$P = \sigma A e T^4 \qquad \textbf{17.36} \blacktriangleleft$$

❯ OBJECTIVE QUESTIONS

1. An amount of energy is added to ice, raising its temperature from $-10°C$ to $-5°C$. A larger amount of energy is added to the same mass of water, raising its temperature from $15°C$ to $20°C$. From these results, what would you conclude? (a) Overcoming the latent heat of fusion of ice requires an input of energy. (b) The latent heat of fusion of ice delivers some energy to the system. (c) The specific heat of ice is less than that of water. (d) The specific heat of ice is greater than that of water. (e) More information is needed to draw any conclusion.

2. A 100-g piece of copper, initially at $95.0°C$, is dropped into 200 g of water contained in a 280-g aluminum can; the water and can are initially at $15.0°C$. What is the final temperature of the system? (Specific heats of copper and aluminum are 0.092 and 0.215 cal/g · °C, respectively.) (a) $16°C$ (b) $18°C$ (c) $24°C$ (d) $26°C$ (e) none of those answers

3. How long would it take a 1 000-W heater to melt 1.00 kg of ice at $-20.0°C$, assuming all the energy from the heater is absorbed by the ice? (a) 4.18 s (b) 41.8 s (c) 5.55 min (d) 6.25 min (e) 38.4 min

4. The specific heat of substance A is greater than that of substance B. Both A and B are at the same initial temperature when equal amounts of energy are added to them. Assuming no melting or vaporization occurs, which of the following can be concluded about the final temperature T_A of substance A and the final temperature T_B of substance B? (a) $T_A > T_B$ (b) $T_A < T_B$ (c) $T_A = T_B$ (d) More information is needed.

5. How much energy is required to raise the temperature of 5.00 kg of lead from $20.0°C$ to its melting point of $327°C$? The specific heat of lead is 128 J/kg · °C. (a) 4.04×10^5 J (b) 1.07×10^5 J (c) 8.15×10^4 J (d) 2.13×10^4 J (e) 1.96×10^5 J

6. If a gas undergoes an isobaric process, which of the following statements is true? (a) The temperature of the gas doesn't change. (b) Work is done on or by the gas. (c) No energy is transferred by heat to or from the gas. (d) The volume of the gas remains the same. (e) The pressure of the gas decreases uniformly.

7. Assume you are measuring the specific heat of a sample of originally hot metal by using a calorimeter containing water. Because your calorimeter is not perfectly insulating, energy can transfer by heat between the contents of the calorimeter and the room. To obtain the most accurate result for the specific heat of the metal, you should use water with which initial temperature? (a) slightly lower than room temperature (b) the same as room temperature (c) slightly higher than room temperature (d) whatever you like because the initial temperature makes no difference

8. Beryllium has roughly one-half the specific heat of water (H_2O). Rank the quantities of energy input required to produce the following changes from the largest to the smallest. In your ranking, note any cases of equality. (a) raising the temperature of 1 kg of H_2O from $20°C$ to $26°C$ (b) raising the temperature of 2 kg of H_2O from $20°C$ to $23°C$ (c) raising

the temperature of 2 kg of H_2O from 1°C to 4°C (d) raising the temperature of 2 kg of beryllium from −1°C to 2°C (e) raising the temperature of 2 kg of H_2O from −1°C to 2°C

9. A poker is a stiff, nonflammable rod used to push burning logs around in a fireplace. For safety and comfort of use, should the poker be made from a material with (a) high specific heat and high thermal conductivity, (b) low specific heat and low thermal conductivity, (c) low specific heat and high thermal conductivity, or (d) high specific heat and low thermal conductivity?

10. A person shakes a sealed insulated bottle containing hot coffee for a few minutes. (i) What is the change in the temperature of the coffee? (a) a large decrease (b) a slight decrease (c) no change (d) a slight increase (e) a large increase (ii) What is the change in the internal energy of the coffee? Choose from the same possibilities.

11. Star *A* has twice the radius and twice the absolute surface temperature of star *B*. The emissivity of both stars can be assumed to be 1. What is the ratio of the power output of star *A* to that of star *B*? (a) 4 (b) 8 (c) 16 (d) 32 (e) 64

12. If a gas is compressed isothermally, which of the following statements is true? (a) Energy is transferred into the gas by heat. (b) No work is done on the gas. (c) The temperature of the gas increases. (d) The internal energy of the gas remains constant. (e) None of those statements is true.

13. When a gas undergoes an adiabatic expansion, which of the following statements is true? (a) The temperature of the gas does not change. (b) No work is done by the gas. (c) No energy is transferred to the gas by heat. (d) The internal energy of the gas does not change. (e) The pressure increases.

14. Ethyl alcohol has about one-half the specific heat of water. Assume equal amounts of energy are transferred by heat into equal-mass liquid samples of alcohol and water in separate insulated containers. The water rises in temperature by 25°C. How much will the alcohol rise in temperature? (a) It will rise by 12°C. (b) It will rise by 25°C. (c) It will rise by 50°C. (d) It depends on the rate of energy transfer. (e) It will not rise in temperature.

15. An ideal gas is compressed to half its initial volume by means of several possible processes. Which of the following processes results in the most work done on the gas? (a) isothermal (b) adiabatic (c) isobaric (d) The work done is independent of the process.

CONCEPTUAL QUESTIONS

☐ denotes answer available in *Student Solutions Manual/Study Guide*

1. What is wrong with the following statement? "Given any two bodies, the one with the higher temperature contains more heat."

2. In 1801, Humphry Davy rubbed together pieces of ice inside an icehouse. He made sure that nothing in the environment was at a higher temperature than the rubbed pieces. He observed the production of drops of liquid water. Make a table listing this and other experiments or processes to illustrate each of the following situations. (a) A system can absorb energy by heat, increase in internal energy, and increase in temperature. (b) A system can absorb energy by heat and increase in internal energy without an increase in temperature. (c) A system can absorb energy by heat without increasing in temperature or in internal energy. (d) A system can increase in internal energy and in temperature without absorbing energy by heat. (e) A system can increase in internal energy without absorbing energy by heat or increasing in temperature.

3. Pioneers stored fruits and vegetables in underground cellars. In winter, why did the pioneers place an open barrel of water alongside their produce?

4. Why is a person able to remove a piece of dry aluminum foil from a hot oven with bare fingers, whereas a burn results if there is moisture on the foil?

5. Using the first law of thermodynamics, explain why the *total* energy of an isolated system is always constant.

6. Is it possible to convert internal energy to mechanical energy? Explain with examples.

7. It is the morning of a day that will become hot. You just purchased drinks for a picnic and are loading them, with ice, into a chest in the back of your car. (a) You wrap a wool blanket around the chest. Does doing so help to keep the beverages cool, or should you expect the wool blanket to warm them up? Explain your answer. (b) Your younger sister suggests you wrap her up in another wool blanket to keep her cool on the hot day like the ice chest. Explain your response to her.

8. You need to pick up a very hot cooking pot in your kitchen. You have a pair of cotton oven mitts. To pick up the pot most comfortably, should you soak them in cold water or keep them dry?

9. Rub the palm of your hand on a metal surface for about 30 seconds. Place the palm of your other hand on an unrubbed portion of the surface and then on the rubbed portion. The rubbed portion will feel warmer. Now repeat this process on a wood surface. Why does the temperature difference between the rubbed and unrubbed portions of the wood surface seem larger than for the metal surface?

10. When camping in a canyon on a still night, a camper notices that as soon as the sun strikes the surrounding peaks, a breeze begins to stir. What causes the breeze?

11. Suppose you pour hot coffee for your guests, and one of them wants it with cream. He wants the coffee to be as warm as possible several minutes later when he drinks it. To have the warmest coffee, should the person add the cream just after the coffee is poured or just before drinking? Explain.

12. In usually warm climates that experience a hard freeze, fruit growers will spray the fruit trees with water, hoping that a layer of ice will form on the fruit. Why would such a layer be advantageous?

PROBLEMS

Section 17.1 Heat and Internal Energy

1. On his honeymoon, James Joule traveled from England to Switzerland. He attempted to verify his idea of the interconvertibility of mechanical energy and internal energy by measuring the increase in temperature of water that fell in a waterfall. For the waterfall near Chamonix in the French Alps, which has a 120-m drop, what maximum temperature rise could Joule expect? He did not succeed in measuring it, partly because evaporation cooled the falling water and also because his thermometer was not sufficiently sensitive.

2. **W** Consider Joule's apparatus described in Figure P17.2. The mass of each of the two blocks is 1.50 kg, and the insulated tank is filled with 200 g of water. What is the increase in the water's temperature after the blocks fall through a distance of 3.00 m?

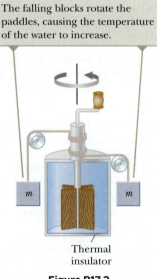

The falling blocks rotate the paddles, causing the temperature of the water to increase.

Thermal insulator

Figure P17.2

3. [**BIO**] A 55.0-kg woman cheats on her diet and eats a 540-Calorie (540 kcal) jelly doughnut for breakfast. (a) How many joules of energy are the equivalent of one jelly doughnut? (b) How many steps must the woman climb on a very tall stairway to change the gravitational potential energy of the woman–Earth system by a value equivalent to the food energy in one jelly doughnut? Assume the height of a single stair is 15.0 cm. (c) If the human body is only 25.0% efficient in converting chemical potential energy to mechanical energy, how many steps must the woman climb to work off her breakfast?

Section 17.2 Specific Heat

4. **M** The temperature of a silver bar rises by 10.0°C when it absorbs 1.23 kJ of energy by heat. The mass of the bar is 525 g. Determine the specific heat of silver from these data.

5. A 50.0-g sample of copper is at 25.0°C. If 1 200 J of energy is added to it by heat, what is the final temperature of the copper?

6. **Q|C** An electric drill with a steel drill bit of mass $m = 27.0$ g and diameter 0.635 cm is used to drill into a cubical steel block of mass $M = 240$ g. Assume steel has the same properties as iron. The cutting process can be modeled as happening at one point on the circumference of the bit. This point moves in a helix at constant tangential speed 40.0 m/s and exerts a force of constant magnitude 3.20 N on the block. As shown in Figure P17.6, a groove in the bit carries the chips up to the top of the block, where they form a pile around the hole. The drill is turned on and drills into the block for a time interval of 15.0 s. Let's assume this time interval is long enough for conduction within the steel to bring it all to a uniform temperature. Furthermore, assume the steel objects lose a negligible amount of energy by conduction, convection, and radiation into their environment. (a) Suppose the drill bit cuts three-quarters of the way through the block during 15.0 s. Find the temperature change of the whole quantity of steel. (b) **What If?** Now suppose the drill bit is dull and cuts only one-eighth of the way through the block in 15.0 s. Identify the temperature change of the whole quantity of steel in this case. (c) What pieces of data, if any, are unnecessary for the solution? Explain.

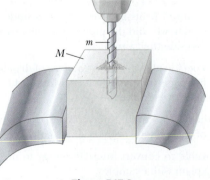

m
M

Figure P17.6

7. In cold climates, including the northern United States, a house can be built with very large windows facing south to

take advantage of solar heating. Sunlight shining in during the daytime is absorbed by the floor, interior walls, and objects in the room, raising their temperature to 38.0°C. If the house is well insulated, you may model it as losing energy by heat steadily at the rate 6 000 W on a day in April when the average exterior temperature is 4°C and when the conventional heating system is not used at all. During the period between 5:00 p.m. and 7:00 a.m., the temperature of the house drops and a sufficiently large "thermal mass" is required to keep it from dropping too far. The thermal mass can be a large quantity of stone (with specific heat 850 J/kg · °C) in the floor and the interior walls exposed to sunlight. What mass of stone is required if the temperature is not to drop below 18.0°C overnight?

8. **Q|C** **W** An aluminum calorimeter with a mass of 100 g contains 250 g of water. The calorimeter and water are in thermal equilibrium at 10.0°C. Two metallic blocks are placed into the water. One is a 50.0-g piece of copper at 80.0°C. The other has a mass of 70.0 g and is originally at a temperature of 100°C. The entire system stabilizes at a final temperature of 20.0°C. (a) Determine the specific heat of the unknown sample. (b) Using the data in Table 17.1, can you make a positive identification of the unknown material? Can you identify a possible material? (c) Explain your answers for part (b).

9. A combination of 0.250 kg of water at 20.0°C, 0.400 kg of aluminum at 26.0°C, and 0.100 kg of copper at 100°C is mixed in an insulated container and allowed to come to thermal equilibrium. Ignore any energy transfer to or from the container. What is the final temperature of the mixture?

10. **S** If water with a mass m_h at temperature T_h is poured into an aluminum cup of mass m_{Al} containing mass m_c of water at T_c, where $T_h > T_c$, what is the equilibrium temperature of the system?

11. **M** A 1.50-kg iron horseshoe initially at 600°C is dropped into a bucket containing 20.0 kg of water at 25.0°C. What is the final temperature of the water–horseshoe system? Ignore the heat capacity of the container and assume a negligible amount of water boils away.

12. An aluminum cup of mass 200 g contains 800 g of water in thermal equilibrium at 80.0°C. The combination of cup and water is cooled uniformly so that the temperature decreases by 1.50°C per minute. At what rate is energy being removed by heat? Express your answer in watts.

Section 17.3 Latent Heat

13. **W** How much energy is required to change a 40.0-g ice cube from ice at −10.0°C to steam at 110°C?

14. A 50.0-g copper calorimeter contains 250 g of water at 20.0°C. How much steam must be condensed into the water if the final temperature of the system is to reach 50.0°C?

15. In an insulated vessel, 250 g of ice at 0°C is added to 600 g of water at 18.0°C. (a) What is the final temperature of the system? (b) How much ice remains when the system reaches equilibrium?

16. **GP** **Review.** Two speeding lead bullets, one of mass 12.0 g moving to the right at 300 m/s and one of mass 8.00 g moving to the left at 400 m/s, collide head-on, and all the material sticks together. Both bullets are originally at temperature 30.0°C. Assume the change in kinetic energy of the system appears entirely as increased internal energy. We would like to determine the temperature and phase of the bullets after the collision. (a) What two analysis models are appropriate for the system of two bullets for the time interval from before to after the collision? (b) From one of these models, what is the speed of the combined bullets after the collision? (c) How much of the initial kinetic energy has transformed to internal energy in the system after the collision? (d) Does all the lead melt due to the collision? (e) What is the temperature of the combined bullets after the collision? (f) What is the phase of the combined bullets after the collision?

17. **M** A 3.00-g lead bullet at 30.0°C is fired at a speed of 240 m/s into a large block of ice at 0°C, in which it becomes embedded. What quantity of ice melts?

18. **Q|C** An automobile has a mass of 1 500 kg, and its aluminum brakes have an overall mass of 6.00 kg. (a) Assume all the mechanical energy that transforms into internal energy when the car stops is deposited in the brakes and no energy is transferred out of the brakes by heat. The brakes are originally at 20.0°C. How many times can the car be stopped from 25.0 m/s before the brakes start to melt? (b) Identify some effects ignored in part (a) that are important in a more realistic assessment of the warming of the brakes.

19. **W** A 1.00-kg block of copper at 20.0°C is dropped into a large vessel of liquid nitrogen at 77.3 K. How many kilograms of nitrogen boil away by the time the copper reaches 77.3 K? (The specific heat of copper is 0.092 4 cal/g · °C, and the latent heat of vaporization of nitrogen is 48.0 cal/g.)

20. **BIO** A resting adult of average size converts chemical energy in food into internal energy at the rate 120 W, called her *basal metabolic rate*. To stay at constant temperature, the body must put out energy at the same rate. Several processes exhaust energy from your body. Usually, the most important is thermal conduction into the air in contact with your exposed skin. If you are not wearing a hat, a convection current of warm air rises vertically from your head like a plume from a smokestack. Your body also loses energy by electromagnetic radiation, by your exhaling warm air, and by evaporation of perspiration. In this problem, consider still another pathway for energy loss: moisture in exhaled breath. Suppose you breathe out 22.0 breaths per minute, each with a volume of 0.600 L. Assume that you inhale dry air and exhale air at 37°C containing water vapor with a vapor pressure of 3.20 kPa. The vapor came from evaporation of liquid water in your body. Model the water vapor as an ideal gas. Assume that its latent heat of evaporation at 37°C is the same as its heat of vaporization at 100°C. Calculate the rate at which you lose energy by exhaling humid air.

Section 17.4 Work in Thermodynamic Processes

21. **M** An ideal gas is taken through a quasi-static process described by $P = \alpha V^2$, with $\alpha = 5.00$ atm/m^6, as shown in Figure P17.21. The gas is expanded to twice its original volume of 1.00 m^3. How much work is done on the expanding gas in this process?

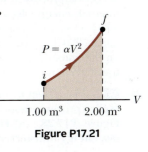

Figure P17.21

22. **QC** **S** One mole of an ideal gas is warmed slowly so that it goes from the PV state (P_i, V_i) to $(3P_i, 3V_i)$ in such a way that the pressure of the gas is directly proportional to the volume. (a) How much work is done on the gas in the process? (b) How is the temperature of the gas related to its volume during this process?

23. An ideal gas is enclosed in a cylinder with a movable piston on top of it. The piston has a mass of 8 000 g and an area of 5.00 cm² and is free to slide up and down, keeping the pressure of the gas constant. How much work is done on the gas as the temperature of 0.200 mol of the gas is raised from 20.0°C to 300°C?

24. **S** An ideal gas is enclosed in a cylinder that has a movable piston on top. The piston has a mass m and an area A and is free to slide up and down, keeping the pressure of the gas constant. How much work is done on the gas as the temperature of n mol of the gas is raised from T_1 to T_2?

25. **W** (a) Determine the work done on a gas that expands from i to f as indicated in Figure P17.25. (b) **What If?** How much work is done on the gas if it is compressed from f to i along the same path?

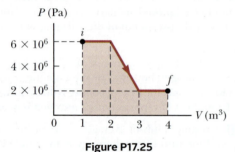

Figure P17.25

Section 17.5 The First Law of Thermodynamics

26. **W** A sample of an ideal gas goes through the process shown in Figure P17.26. From A to B, the process is adiabatic; from B to C, it is isobaric with 100 kJ of energy entering the system by heat; from C to D, it is isothermal; and from D to A, it is isobaric with 150 kJ of energy leaving the system by heat. Determine the difference in internal energy $E_{int,B} - E_{int,A}$.

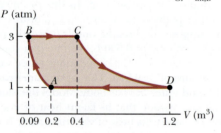

Figure P17.26

27. A thermodynamic system undergoes a process in which its internal energy decreases by 500 J. Over the same time interval, 220 J of work is done on the system. Find the energy transferred from it by heat.

28. **W** A gas is taken through the cyclic process described in Figure P17.28. (a) Find the net energy transferred to the

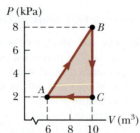

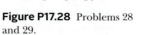

Figure P17.28 Problems 28 and 29.

system by heat during one complete cycle. (b) **What If?** If the cycle is reversed—that is, the process follows the path $ACBA$—what is the net energy input per cycle by heat?

29. Consider the cyclic process depicted in Figure P17.28. If Q is negative for the process BC and ΔE_{int} is negative for the process CA, what are the signs of Q, W, and ΔE_{int} that are associated with each of the three processes?

30. *Why is the following situation impossible?* An ideal gas undergoes a process with the following parameters: $Q = 10.0$ J, $W = 12.0$ J, and $\Delta T = -2.00°C$.

Section 17.6 Some Applications of the First Law of Thermodynamics

31. **M** An ideal gas initially at 300 K undergoes an isobaric expansion at 2.50 kPa. If the volume increases from 1.00 m³ to 3.00 m³ and 12.5 kJ is transferred to the gas by heat, what are (a) the change in its internal energy and (b) its final temperature?

32. In Figure P17.32, the change in internal energy of a gas that is taken from A to C along the blue path is +800 J. The work done on the gas along the red path ABC is −500 J. (a) How much energy must be added to the system by heat as it goes from A through B to C? (b) If the pressure at point A is five times that of point C, what is the work done on the system in going from C to D? (c) What is the energy exchanged with the surroundings by heat as the gas goes from C to A along the green path? (d) If the change in internal energy in going from point D to point A is +500 J, how much energy must be added to the system by heat as it goes from point C to point D?

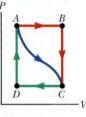

Figure P17.32

33. A 1.00-kg block of aluminum is warmed at atmospheric pressure so that its temperature increases from 22.0°C to 40.0°C. Find (a) the work done on the aluminum, (b) the energy added to it by heat, and (c) the change in its internal energy.

34. (a) How much work is done on the steam when 1.00 mol of water at 100°C boils and becomes 1.00 mol of steam at 100°C at 1.00 atm pressure? Assume the steam to behave as an ideal gas. (b) Determine the change in internal energy of the system of the water and steam as the water vaporizes.

35. **W** An ideal gas initially at P_i, V_i, and T_i is taken through a cycle as shown in Figure P17.35. (a) Find the net work done on the gas per cycle for 1.00 mol of gas initially at 0°C. (b) What is the net energy added by heat to the gas per cycle?

36. **S** An ideal gas initially at P_i, V_i, and T_i is taken through a cycle as shown in Figure P17.35. (a) Find the net work done on the gas per cycle. (b) What is the net energy added by heat to the system per cycle?

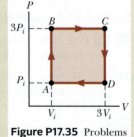

Figure P17.35 Problems 35 and 36.

37. **M** A 2.00-mol sample of helium gas initially at 300 K and 0.400 atm is compressed isothermally to 1.20 atm. Noting that the helium behaves as an ideal gas, find (a) the final volume of the gas, (b) the work done on the gas, and (c) the energy transferred by heat.

38. **W** One mole of an ideal gas does 3 000 J of work on its surroundings as it expands isothermally to a final pressure of 1.00 atm and volume of 25.0 L. Determine (a) the initial volume and (b) the temperature of the gas.

Section 17.7 Molar Specific Heats of Ideal Gases

Note: You may use data in Table 17.3 about particular gases. Here we define a "monatomic ideal gas" to have molar specific heats $C_V = \frac{3}{2}R$ and $C_P = \frac{5}{2}R$, and a "diatomic ideal gas" to have $C_V = \frac{5}{2}R$ and $C_P = \frac{7}{2}R$.

39. **M** A 1.00-mol sample of hydrogen gas is heated at constant pressure from 300 K to 420 K. Calculate (a) the energy transferred to the gas by heat, (b) the increase in its internal energy, and (c) the work done on the gas.

40. **S** A sample of a diatomic ideal gas has pressure *P* and volume *V*. When the gas is warmed, its pressure triples and its volume doubles. This warming process includes two steps, the first at constant pressure and the second at constant volume. Determine the amount of energy transferred to the gas by heat.

41. Calculate the change in internal energy of 3.00 mol of helium gas when its temperature is increased by 2.00 K.

42. A 1.00-L insulated bottle is full of tea at 90.0°C. You pour out one cup of tea and immediately screw the stopper back on the bottle. Make an order-of-magnitude estimate of the change in temperature of the tea remaining in the bottle that results from the admission of air at room temperature. State the quantities you take as data and the values you measure or estimate for them.

43. A vertical cylinder with a heavy piston contains air at 300 K. The initial pressure is 2.00×10^5 Pa, and the initial volume is 0.350 m³. Take the molar mass of air as 28.9 g/mol and assume $C_V = \frac{5}{2}R$. (a) Find the specific heat of air at constant volume in units of J/kg · °C. (b) Calculate the mass of the air in the cylinder. (c) Suppose the piston is held fixed. Find the energy input required to raise the temperature of the air to 700 K. (d) **What If?** Assume again the conditions of the initial state and assume the heavy piston is free to move. Find the energy input required to raise the temperature to 700 K.

44. **Review.** This problem is a continuation of Problem 16.29 in Chapter 16. A hot-air balloon consists of an envelope of constant volume 400 m³. Not including the air inside, the balloon and cargo have mass 200 kg. The air outside and originally inside is a diatomic ideal gas at 10.0°C and 101 kPa, with density 1.25 kg/m³. A propane burner at the center of the spherical envelope injects energy into the air inside. The air inside stays at constant pressure. Hot air, at just the temperature required to make the balloon lift off, starts to fill the envelope at its closed top, rapidly enough so that negligible energy flows by heat to the cool air below it or out through the wall of the balloon. Air at 10°C leaves through an opening at the bottom of the envelope until the whole balloon is filled with hot air at uniform temperature. Then the burner is shut off and the balloon rises from the ground. (a) Evaluate the quantity of energy the burner must transfer to the air in the balloon. (b) The "heat value" of propane—the internal energy released by burning each kilogram—is 50.3 MJ/kg. What mass of propane must be burned?

45. **W** In a constant-volume process, 209 J of energy is transferred by heat to 1.00 mol of an ideal monatomic gas initially at 300 K. Find (a) the work done on the gas, (b) the increase in internal energy of the gas, and (c) its final temperature.

Section 17.8 Adiabatic Processes for an Ideal Gas

46. **M** A 2.00-mol sample of a diatomic ideal gas expands slowly and adiabatically from a pressure of 5.00 atm and a volume of 12.0 L to a final volume of 30.0 L. (a) What is the final pressure of the gas? (b) What are the initial and final temperatures? Find (c) *Q*, (d) ΔE_{int}, and (e) *W* for the gas during this process.

47. A 4.00-L sample of a diatomic ideal gas with specific heat ratio 1.40, confined to a cylinder, is carried through a closed cycle. The gas is initially at 1.00 atm and 300 K. First, its pressure is tripled under constant volume. Then, it expands adiabatically to its original pressure. Finally, the gas is compressed isobarically to its original volume. (a) Draw a *PV* diagram of this cycle. (b) Determine the volume of the gas at the end of the adiabatic expansion. (c) Find the temperature of the gas at the start of the adiabatic expansion. (d) Find the temperature at the end of the cycle. (e) What was the net work done on the gas for this cycle?

48. **S** An ideal gas with specific heat ratio γ confined to a cylinder is put through a closed cycle. Initially, the gas is at P_i, V_i, and T_i. First, its pressure is tripled under constant volume. It then expands adiabatically to its original pressure and finally is compressed isobarically to its original volume. (a) Draw a *PV* diagram of this cycle. (b) Determine the volume at the end of the adiabatic expansion. Find (c) the temperature of the gas at the start of the adiabatic expansion and (d) the temperature at the end of the cycle. (e) What was the net work done on the gas for this cycle?

49. **W** During the compression stroke of a certain gasoline engine, the pressure increases from 1.00 atm to 20.0 atm. If the process is adiabatic and the air–fuel mixture behaves as a diatomic ideal gas, (a) by what factor does the volume change and (b) by what factor does the temperature change? Assuming the compression starts with 0.016 0 mol of gas at 27.0°C, find the values of (c) *Q*, (d) ΔE_{int}, and (e) *W* that characterize the process.

50. *Why is the following situation impossible?* A new diesel engine that increases fuel economy over previous models is designed. Automobiles fitted with this design become incredible best sellers. Two design features are responsible for the increased fuel economy: (1) the engine is made entirely of aluminum to reduce the weight of the automobile, and (2) the exhaust of the engine is used to prewarm the air to 50°C before it enters the cylinder to increase the final temperature of the compressed gas. The engine has a *compression ratio*—that is, the ratio of the initial volume of the air to its final volume after compression—of 14.5. The compression process is adiabatic, and the air behaves as a diatomic ideal gas with $\gamma = 1.40$.

51. **M** Air in a thundercloud expands as it rises. If its initial temperature is 300 K and no energy is lost by thermal conduction on expansion, what is its temperature when the initial volume has doubled?

52. **W** How much work is required to compress 5.00 mol of air at 20.0°C and 1.00 atm to one-tenth of the original volume (a) by an isothermal process? (b) **What If?** How much work is required to produce the same compression in an adiabatic process? (c) What is the final pressure in part (a)? (d) What is the final pressure in part (b)?

53. **GP** Air (a diatomic ideal gas) at 27.0°C and atmospheric pressure is drawn into a bicycle pump (Figure P17.53) that has a cylinder with an inner diameter of 2.50 cm and length 50.0 cm. The downstroke adiabatically compresses the air, which reaches a gauge pressure of 8.00×10^5 Pa before entering the tire. We wish to investigate the temperature increase of the pump. (a) What is the initial volume of the air in the pump? (b) What is the number of moles of air in the pump? (c) What is the absolute pressure of the compressed air? (d) What is the volume of the compressed air? (e) What is the temperature of the compressed air? (f) What is the increase in internal energy of the gas during the compression? **What If?** The pump is made of steel that is 2.00 mm thick. Assume 4.00 cm of the cylinder's length is allowed to come to thermal equilibrium with the air. (g) What is the volume of steel in this 4.00-cm length? (h) What is the mass of steel in this 4.00-cm length? (i) Assume the pump is compressed once. After the adiabatic expansion, conduction results in the energy increase in part (f) being shared between the gas and the 4.00-cm length of steel. What will be the increase in temperature of the steel after one compression?

Figure P17.53

54. During the power stroke in a four-stroke automobile engine, the piston is forced down as the mixture of combustion products and air undergoes an adiabatic expansion. Assume (1) the engine is running at 2 500 cycles/min; (2) the gauge pressure immediately before the expansion is 20.0 atm; (3) the volumes of the mixture immediately before and after the expansion are 50.0 cm³ and 400 cm³, respectively (Fig. P17.54); (4) the time interval for the expansion is one-fourth that of the total cycle; and (5) the mixture behaves like an ideal gas with specific heat ratio 1.40. Find the average power generated during the power stroke.

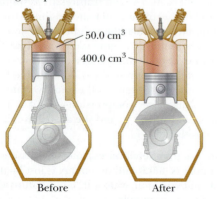

Figure P17.54

Section 17.9 Molar Specific Heats and the Equipartition of Energy

55. In a crude model (Fig. P17.55) of a rotating diatomic chlorine molecule (Cl_2), the two Cl atoms are 2.00×10^{-10} m apart and rotate about their center of mass with angular speed $\omega = 2.00 \times 10^{12}$ rad/s. What is the rotational kinetic energy of one molecule of Cl_2, which has a molar mass of 70.0 g/mol?

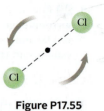

Figure P17.55

56. **S** A certain molecule has f degrees of freedom. Show that an ideal gas consisting of such molecules has the following properties: (a) its total internal energy is $fnRT/2$, (b) its molar specific heat at constant volume is $fR/2$, (c) its molar specific heat at constant pressure is $(f + 2)R/2$, and (d) its specific heat ratio is $\gamma = C_P/C_V = (f + 2)/f$.

57. **M** The relationship between the heat capacity of a sample and the specific heat of the sample material is discussed in Section 17.2. Consider a sample containing 2.00 mol of an ideal diatomic gas. Assuming the molecules rotate but do not vibrate, find (a) the total heat capacity of the sample at constant volume and (b) the total heat capacity at constant pressure. (c) **What If?** Repeat parts (a) and (b), assuming the molecules both rotate and vibrate.

58. *Why is the following situation impossible?* A team of researchers discovers a new gas, which has a value of $\gamma = C_P/C_V$ of 1.75.

Section 17.10 Energy Transfer Mechanisms in Thermal Processes

59. A bar of gold (Au) is in thermal contact with a bar of silver (Ag) of the same length and area (Fig. P17.59). One end of the compound bar is maintained at 80.0°C, and the opposite end is at 30.0°C. When the energy transfer reaches steady state, what is the temperature at the junction?

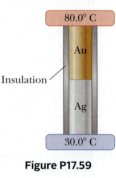

Figure P17.59

60. **BIO** The human body must maintain its core temperature inside a rather narrow range around 37°C. Metabolic processes, notably muscular exertion, convert chemical energy into internal energy deep in the interior. From the interior, energy must flow out to the skin or lungs to be expelled to the environment. During moderate exercise, an 80-kg man can metabolize food energy at the rate 300 kcal/h, do 60 kcal/h of mechanical work, and put out the remaining 240 kcal/h of energy by heat. Most of the energy is carried from the body interior out to the skin by forced convection (as a plumber would say), whereby blood is warmed in the interior and then cooled at the skin, which is a few degrees cooler than the body core. Without blood flow, living tissue is a good thermal insulator, with thermal conductivity about 0.210 W/m · °C. Show that blood flow is essential to cool the man's body by calculating the rate of energy conduction in kcal/h through the tissue layer under his skin. Assume that its area is 1.40 m², its thickness is

2.50 cm, and it is maintained at 37.0°C on one side and at 34.0°C on the other side.

61. **BIO** A student is trying to decide what to wear. His bedroom is at 20.0°C. His skin temperature is 35.0°C. The area of his exposed skin is 1.50 m². People all over the world have skin that is dark in the infrared, with emissivity about 0.900. Find the net energy transfer from his body by radiation in 10.0 min.

62. A box with a total surface area of 1.20 m² and a wall thickness of 4.00 cm is made of an insulating material. A 10.0-W electric heater inside the box maintains the inside temperature at 15.0°C above the outside temperature. Find the thermal conductivity k of the insulating material.

63. The surface of the Sun has a temperature of about 5 800 K. The radius of the Sun is 6.96×10^8 m. Calculate the total energy radiated by the Sun each second. Assume the emissivity of the Sun is 0.986.

Section 17.11 Context Connection: Energy Balance for the Earth

64. **Q|C** At our distance from the Sun, the intensity of solar radiation is 1 370 W/m². The temperature of the Earth is affected by the *greenhouse effect* of the atmosphere. This phenomenon describes the effect of absorption of infrared light emitted by the surface so as to make the surface temperature of the Earth higher than if it were airless. For comparison, consider a spherical object of radius r with no atmosphere at the same distance from the Sun as the Earth. Assume its emissivity is the same for all kinds of electromagnetic waves and its temperature is uniform over its surface. (a) Explain why the projected area over which it absorbs sunlight is πr^2 and the surface area over which it radiates is $4\pi r^2$. (b) Compute its steady-state temperature. Is it chilly?

65. At high noon, the Sun delivers 1 000 W to each square meter of a blacktop road. If the hot asphalt transfers energy only by radiation, what is its steady-state temperature?

66. **S** *A theoretical atmospheric lapse rate.* Section 16.7 described experimental data on the decrease in temperature with altitude in the Earth's atmosphere. Model the troposphere as an ideal gas, everywhere with equivalent molar mass M and ratio of specific heats γ. Absorption of sunlight at the Earth's surface warms the troposphere from below, so vertical convection currents are continually mixing the air. As a parcel of air rises, its pressure drops and it expands. The parcel does work on its surroundings, so its internal energy decreases and it drops in temperature. Assume that the vertical mixing is so rapid as to be adiabatic. (a) Show that the quantity $TP^{(1-\gamma)/\gamma}$ has a uniform value through the layers of the troposphere. (b) By differentiating with respect to altitude y, show that the lapse rate is given by

$$\frac{dT}{dy} = \frac{T}{P}\left(1 - \frac{1}{\gamma}\right)\frac{dP}{dy}$$

(c) A lower layer of air must support the weight of the layers above. From Equation 15.4, observe that mechanical equilibrium of the atmosphere requires that the pressure decrease with altitude according to $dP/dy = -\rho g$.

The depth of the troposphere is small compared with the radius of the Earth, so you may assume that the free-fall acceleration is uniform. Proceed to prove that the lapse rate is

$$\frac{dT}{dy} = -\left(1 - \frac{1}{\gamma}\right)\frac{Mg}{R}$$

Problem 16.50 in Chapter 16 calls for evaluation of this theoretical lapse rate on the Earth and on Mars and for comparison with experimental results.

Additional Problems

67. On a cold winter day, you buy roasted chestnuts from a street vendor. Into the pocket of your down parka you put the change he gives you: coins constituting 9.00 g of copper at $-12.0°C$. Your pocket already contains 14.0 g of silver coins at 30.0°C. A short time later the temperature of the copper coins is 4.00°C and is increasing at a rate of 0.500°C/s. At this time, (a) what is the temperature of the silver coins and (b) at what rate is it changing?

68. **Q|C** A sample of a monatomic ideal gas occupies 5.00 L at atmospheric pressure and 300 K (point A in Fig. P17.68). It is warmed at constant volume to 3.00 atm (point B). Then it is allowed to expand isothermally to 1.00 atm (point C) and at last compressed isobarically to its original state. (a) Find the number of moles in the sample. Find (b) the temperature at point B, (c) the temperature at point C, and (d) the volume at point C. (e) Now consider the processes $A \rightarrow B$, $B \rightarrow C$, and $C \rightarrow A$. Describe how to carry out each process experimentally. (f) Find Q, W, and ΔE_{int} for each of the processes. (g) For the whole cycle $A \rightarrow B \rightarrow C \rightarrow A$, find Q, W, and ΔE_{int}.

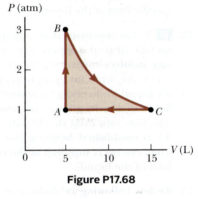

Figure P17.68

69. **M** An aluminum rod 0.500 m in length and with a cross-sectional area of 2.50 cm² is inserted into a thermally insulated vessel containing liquid helium at 4.20 K. The rod is initially at 300 K. (a) If one-half of the rod is inserted into the helium, how many liters of helium boil off by the time the inserted half cools to 4.20 K? Assume the upper half does not yet cool. (b) If the circular surface of the upper end of the rod is maintained at 300 K, what is the approximate boil-off rate of liquid helium in liters per second after the lower half has reached 4.20 K? (Aluminum has thermal conductivity of 3 100 W/m · K at 4.20 K; ignore its temperature variation. The density of liquid helium is 125 kg/m³.)

70. **BIO** **Q|C** For bacteriological testing of water supplies and in medical clinics, samples must routinely be incubated for 24 h at 37°C. Peace Corps volunteer and MIT engineer Amy Smith invented a low-cost, low-maintenance incubator. The incubator consists of a foam-insulated box containing a waxy material that melts at 37.0°C interspersed among tubes, dishes, or bottles containing the test samples and

18.9 | Context Connection: The Atmosphere as a Heat Engine

In Chapter 17, we predicted a global temperature based on the notion of energy balance between incoming visible radiation from the Sun and outgoing infrared radiation from the Earth. This model leads to a global temperature that is well below the measured temperature. This discrepancy results because atmospheric effects are not included in our model. In this section, we shall introduce some of these effects and show that the atmosphere can be modeled as a heat engine. In the Context Conclusion, we shall use concepts learned in the thermodynamics chapters to build a model that is more successful at predicting the correct temperature of the Earth.

What happens to the energy that enters the atmosphere by radiation from the Sun? Figure 18.13 helps answer this question by showing how the input energy undergoes various processes. If we identify the incoming energy as 100%, we find that 30% of it is reflected back into space, as we mentioned in Chapter 17. This 30% includes 6% back-scattered from air molecules, 20% reflected from clouds, and 4% reflected from the surface of the Earth. The remaining 70% is absorbed by either the air or the surface. Before reaching the surface, 20% of the original radiation is absorbed in the air; 4% by clouds; and 16% by water, dust particles, and ozone in the atmosphere. Of the original radiation striking the top of the atmosphere, the ground absorbs 50%.

The ground emits radiation upward and transfers energy to the atmosphere by several processes. Of the original 100% of the incoming energy, 6% simply passes back through the atmosphere into space (at the right in Fig. 18.13). In addition, 14% of the original incoming energy emitted as radiation from the ground is absorbed by water and carbon dioxide molecules. The air warmed by the surface rises

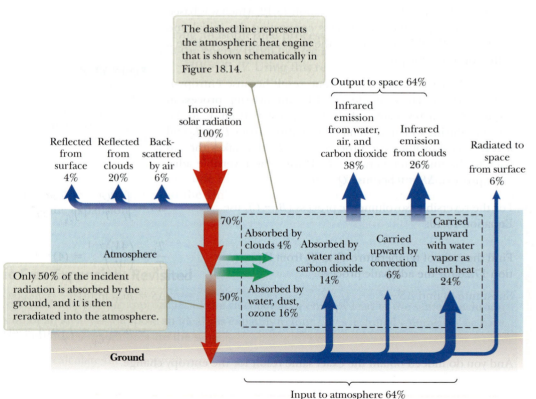

Figure 18.13 Energy input to the atmosphere from the Sun is divided into several components.

upward by convection, carrying 6% of the original energy into the atmosphere. The hydrological cycle results in 24% of the original energy being carried upward as water vapor and released into the atmosphere when the water vapor condenses into liquid water.

These processes result in a total of 64% of the original energy being absorbed in the atmosphere, with another 6% from the surface passing back through into space. Because the atmosphere is in steady state, this 64% is also emitted from the atmosphere into space. The emission is divided into two types. The first is infrared radiation from molecules in the atmosphere, including water vapor, carbon dioxide, and the nitrogen and oxygen molecules of the air, which accounts for emission of 38% of the original energy. The remaining 26% is emitted as infrared radiation from clouds.

Figure 18.13 accounts for all the energy; the amount of energy input equals the amount of energy output, which is the premise used in the Context Connection of Chapter 17. A major difference from our discussion in that chapter, however, is the notion of absorption of the energy by the atmosphere. It is this absorption that creates thermodynamic processes in the atmosphere to raise the surface temperature above the value we determined in Chapter 17. We shall explore more about these processes and the temperature profile of the atmosphere in the Context Conclusion.

To close this chapter, let us discuss one more process that is not included in Figure 18.13. The various processes depicted in that figure result in a small amount of work done on the air, which appears as the kinetic energy of the prevailing winds in the atmosphere.

The amount of the original solar energy that is converted to kinetic energy of prevailing winds is about 0.5%. The process of generating the winds does not change the energy balance shown in Figure 18.13. The kinetic energy of the wind is converted to internal energy as masses of air move past one another. This internal energy produces an increased infrared emission of the atmosphere into space, so the 0.5% is only temporarily in the form of kinetic energy before being emitted as radiation.

We can model the atmosphere as a heat engine, which is indicated in Figure 18.13 by the dotted rectangle. A schematic diagram of this heat engine is shown in Figure 18.14. The warm reservoir is the surface and the atmosphere, and the cold reservoir is empty space. We can calculate the efficiency of the atmospheric engine using Equation 18.2:

$$e = \frac{W_{\text{eng}}}{|Q_h|} = \frac{0.5\%}{64\%} = 0.008 = 0.8\%$$

which is a very low efficiency. Keep in mind, however, that a tremendous amount of energy enters the atmosphere from the Sun, so even a very small fraction of it can create a very complex and powerful wind system. Hurricanes represent a vivid example of the energy output of the atmospheric heat engine.

Notice that the output energy in Figure 18.14 is less than that in Figure 18.13 by 0.5%. As noted previously, the 0.5% transferred to the atmosphere by generating winds is eventually transformed to internal energy in the atmosphere by friction and then radiated into space as thermal radiation. We cannot separate the heat engine and the winds in the atmosphere in a diagram because the atmosphere *is* the heat engine and the winds are generated *in* the atmosphere!

We now have all the pieces we need to put together the puzzle of the temperature of the Earth. We shall discuss this subject in the Context Conclusion.

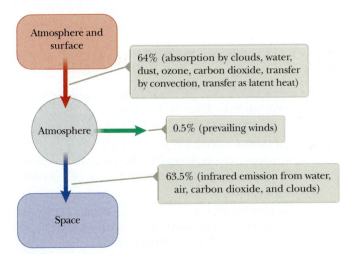

Figure 18.14 A schematic representation of the atmosphere as a heat engine.

SUMMARY

A **heat engine** is a device that takes in energy by heat and, operating in a cycle, expels a fraction of that energy by work. The net work done by a heat engine is

$$W_{\text{eng}} = |Q_h| - |Q_c| \qquad \textbf{18.1} \blacktriangleleft$$

where Q_h is the energy absorbed from a hot reservoir and Q_c is the energy expelled to a cold reservoir.

The **thermal efficiency** e of a heat engine is defined as the ratio of the net work done to the energy absorbed per cycle from the higher temperature reservoir:

$$e = \frac{W_{\text{eng}}}{|Q_h|} = 1 - \frac{|Q_c|}{|Q_h|} \qquad \textbf{18.2} \blacktriangleleft$$

The **Kelvin–Planck statement of the second law of thermodynamics** can be stated as follows:

- It is impossible to construct a heat engine that, operating in a cycle, produces no effect other than the absorption of energy by heat from a reservoir and the performance of an equal amount of work.

A **reversible** process is one for which the system can be returned to its initial conditions along the same path and for which every point along the path is an equilibrium state. A process that does not satisfy these requirements is **irreversible.**

The thermal efficiency of a heat engine operating in a **Carnot cycle** is given by

$$e_C = 1 - \frac{T_c}{T_h} \qquad \textbf{18.4} \blacktriangleleft$$

where T_c is the absolute temperature of the cold reservoir and T_h is the absolute temperature of the hot reservoir. No real heat engine operating between the temperatures T_c and T_h can be more efficient than an engine operating reversibly in a Carnot cycle between the same two temperatures.

The **Clausius statement of the second law** states that

- Energy will not transfer spontaneously by heat from a cold object to a hot object.

The **second law of thermodynamics** states that when real (irreversible) processes occur, the degree of disorder in the system plus the surroundings increases. The measure of disorder in a system is called **entropy** S.

The **change in entropy** dS of a system moving through an infinitesimal process between two equilibrium states is

$$dS = \frac{dQ_r}{T} \qquad \textbf{18.8} \blacktriangleleft$$

where dQ_r is the energy transferred by heat in a reversible process between the same states.

From a microscopic viewpoint, the entropy S associated with a macrostate of a system is defined as

$$S \equiv k_B \ln W \qquad \textbf{18.9} \blacktriangleleft$$

where k_B is Boltzmann's constant and W is the number of microstates corresponding to the macrostate whose entropy is S. Therefore, **entropy is a measure of microscopic disorder.** Because of the statistical tendency of systems to proceed toward states of greater probability and greater disorder, all natural processes are irreversible and result in an increase in entropy. Therefore, the **entropy statement of the second law of thermodynamics** is as follows:

- The entropy of the Universe increases in all real processes.

The change in entropy of a system moving between two general equilibrium states is

$$\Delta S = \int_i^f \frac{dQ_r}{T} \qquad \textbf{18.10} \blacktriangleleft$$

The value of ΔS is the same for all paths connecting the initial and final states.

The change in entropy for any reversible, cyclic process is zero, and when such a process occurs, the entropy of the Universe remains constant.

OBJECTIVE QUESTIONS

1. An engine does 15.0 kJ of work while exhausting 37.0 kJ to a cold reservoir. What is the efficiency of the engine? (a) 0.150 (b) 0.288 (c) 0.333 (d) 0.450 (e) 1.20

2. A steam turbine operates at a boiler temperature of 450 K and an exhaust temperature of 300 K. What is the maximum theoretical efficiency of this system? (a) 0.240 (b) 0.500 (c) 0.333 (d) 0.667 (e) 0.150

3. A refrigerator has 18.0 kJ of work done on it while 115 kJ of energy is transferred from inside its interior. What is its coefficient of performance? (a) 3.40 (b) 2.80 (c) 8.90 (d) 6.40 (e) 5.20

4. A compact air-conditioning unit is placed on a table inside a well-insulated apartment and is plugged in and turned on. What happens to the average temperature of the apartment?

(a) It increases. (b) It decreases. (c) It remains constant. (d) It increases until the unit warms up and then decreases. (e) The answer depends on the initial temperature of the apartment.

5. Consider cyclic processes completely characterized by each of the following net energy inputs and outputs. In each case, the energy transfers listed are the *only* ones occurring. Classify each process as (a) possible, (b) impossible according to the first law of thermodynamics, (c) impossible according to the second law of thermodynamics, or (d) impossible according to both the first and second laws. (i) Input is 5 J of work, and output is 4 J of work. (ii) Input is 5 J of work, and output is 5 J of energy transferred by heat. (iii) Input is 5 J of energy transferred by electrical transmission, and output is 6 J of work. (iv) Input is 5 J of energy transferred by heat,

and output is 5 J of energy transferred by heat. **(v)** Input is 5 J of energy transferred by heat, and output is 5 J of work. **(vi)** Input is 5 J of energy transferred by heat, and output is 3 J of work plus 2 J of energy transferred by heat.

6. Of the following, which is *not* a statement of the second law of thermodynamics? (a) No heat engine operating in a cycle can absorb energy from a reservoir and use it entirely to do work. (b) No real engine operating between two energy reservoirs can be more efficient than a Carnot engine operating between the same two reservoirs. (c) When a system undergoes a change in state, the change in the internal energy of the system is the sum of the energy transferred to the system by heat and the work done on the system. (d) The entropy of the Universe increases in all natural processes. (e) Energy will not spontaneously transfer by heat from a cold object to a hot object.

7. The arrow *OA* in the *PV* diagram shown in Figure OQ18.7 represents a reversible adiabatic expansion of an ideal gas. The same sample of gas, starting from the same state *O*, now undergoes an adiabatic free expansion to the same final volume. What point on the diagram could represent the final state of the gas? (a) the same point *A* as for the reversible expansion (b) point *B* (c) point *C* (d) any of those choices (e) none of those choices

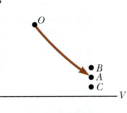

Figure OQ18.7

8. A thermodynamic process occurs in which the entropy of a system changes by −8 J/K. According to the second law of

thermodynamics, what can you conclude about the entropy change of the environment? (a) It must be +8 J/K or less. (b) It must be between +8 J/K and 0. (c) It must be equal to +8 J/K. (d) It must be +8 J/K or more. (e) It must be zero.

9. A sample of a monatomic ideal gas is contained in a cylinder with a piston. Its state is represented by the dot in the *PV* diagram shown in Figure OQ18.9. Arrows *A* through *E* represent isobaric, isothermal, adiabatic, and isovolumetric processes that the sample can undergo. In each process except *D*, the volume changes by a factor of 2. All five processes are reversible. Rank the processes according to the change in entropy of the gas from the largest positive value to the largest-magnitude negative value. In your rankings, display any cases of equality.

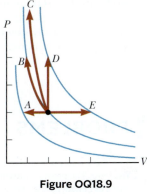

Figure OQ18.9

10. Assume a sample of an ideal gas is at room temperature. What action will *necessarily* make the entropy of the sample increase? (a) Transfer energy into it by heat. (b) Transfer energy into it irreversibly by heat. (c) Do work on it. (d) Increase either its temperature or its volume, without letting the other variable decrease. (e) None of those choices is correct.

11. The second law of thermodynamics implies that the coefficient of performance of a refrigerator must be what? (a) less than 1 (b) less than or equal to 1 (c) greater than or equal to 1 (d) finite (e) greater than 0

CONCEPTUAL QUESTIONS

☐ denotes answer available in *Student Solutions Manual/Study Guide*

1. The device shown in Figure CQ18.1, called a thermoelectric converter, uses a series of semiconductor cells to transform internal energy to electric potential energy, which we will study in Chapter 20. In the photograph on the left, both legs of the device are at the same temperature and no electric potential energy is produced. When one leg is at a higher temperature than the other as shown in the photograph on the right, however, electric potential energy is produced as the device extracts energy from the hot reservoir and drives a small electric motor. (a) Why is the difference in temperature necessary to produce electric potential energy in this demonstration? (b) In what sense does this intriguing experiment demonstrate the second law of thermodynamics?

Courtesy of PASCO Scientific Company.

Figure CQ18.1

2. (a) Give an example of an irreversible process that occurs in nature. (b) Give an example of a process in nature that is nearly reversible.

3. Does the second law of thermodynamics contradict or correct the first law? Argue for your answer.

4. Is it possible to construct a heat engine that creates no thermal pollution? Explain.

5. "The first law of thermodynamics says you can't really win, and the second law says you can't even break even." Explain how this statement applies to a particular device or process; alternatively, argue against the statement.

6. A steam-driven turbine is one major component of an electric power plant. Why is it advantageous to have the temperature of the steam as high as possible?

7. What are some factors that affect the efficiency of automobile engines?

8. (a) If you shake a jar full of jelly beans of different sizes, the larger beans tend to appear near the top and the smaller ones tend to fall to the bottom. Why? (b) Does this process violate the second law of thermodynamics?

9. Discuss the change in entropy of a gas that expands (a) at constant temperature and (b) adiabatically.

10. Suppose your roommate cleans and tidies up your messy room after a big party. Because she is creating more order, does this process represent a violation of the second law of thermodynamics?

11. "Energy is the mistress of the Universe, and entropy is her shadow." Writing for an audience of general readers, argue for this statement with at least two examples. Alternatively, argue for the view that entropy is like an executive who instantly determines what will happen, whereas energy is like a bookkeeper telling us how little we can afford. (Arnold Sommerfeld suggested the idea for this question.)

12. Discuss three different common examples of natural processes that involve an increase in entropy. Be sure to account for all parts of each system under consideration.

13. The energy exhaust from a certain coal-fired electric generating station is carried by "cooling water" into Lake Ontario. The water is warm from the viewpoint of living things in the lake. Some of them congregate around the outlet port and can impede the water flow. (a) Use the theory of heat engines to explain why this action can reduce the electric power output of the station. (b) An engineer says that the electric output is reduced because of "higher back pressure on the turbine blades." Comment on the accuracy of this statement.

▶ PROBLEMS

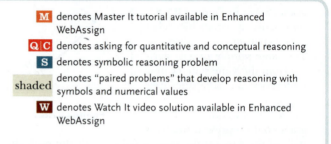

ENHANCED WebAssign The problems found in this chapter may be assigned online in Enhanced WebAssign.

  1. denotes straightforward problem; **2.** denotes intermediate problem; **3.** denotes challenging problem

  1. denotes full solution available in the *Student Solutions Manual/ Study Guide*

  1. denotes problems most often assigned in Enhanced WebAssign.

  BIO denotes biomedical problem

  GP denotes guided problem

  M denotes Master It tutorial available in Enhanced WebAssign

  Q|C denotes asking for quantitative and conceptual reasoning

  S denotes symbolic reasoning problem

  shaded denotes "paired problems" that develop reasoning with symbols and numerical values

  W denotes Watch It video solution available in Enhanced WebAssign

Section 18.1 Heat Engines and the Second Law of Thermodynamics

1. **W** A heat engine takes in 360 J of energy from a hot reservoir and performs 25.0 J of work in each cycle. Find (a) the efficiency of the engine and (b) the energy expelled to the cold reservoir in each cycle.

2. A multicylinder gasoline engine in an airplane, operating at 2.50×10^3 rev/min, takes in energy 7.89×10^3 J and exhausts 4.58×10^3 J for each revolution of the crankshaft. (a) How many liters of fuel does it consume in 1.00 h of operation if the heat of combustion of the fuel is equal to 4.03×10^7 J/L? (b) What is the mechanical power output of the engine? Ignore friction and express the answer in horsepower. (c) What is the torque exerted by the crankshaft on the load? (d) What power must the exhaust and cooling system transfer out of the engine?

3. **W** Suppose a heat engine is connected to two energy reservoirs, one a pool of molten aluminum (660°C) and the other a block of solid mercury (−38.9°C). The engine runs by freezing 1.00 g of aluminum and melting 15.0 g of mercury during each cycle. The heat of fusion of aluminum is 3.97×10^5 J/kg; the heat of fusion of mercury is 1.18×10^4 J/kg. What is the efficiency of this engine?

4. A gun is a heat engine. In particular, it is an internal combustion piston engine that does not operate in a cycle, but comes apart during its adiabatic expansion process. A certain gun consists of 1.80 kg of iron. It fires one 2.40-g bullet at 320 m/s with an energy efficiency of 1.10%. Assume the body of the gun absorbs all the energy exhaust—the other 98.9%—and increases uniformly in temperature for a short time interval before it loses any energy by heat into the environment. Find its temperature increase.

5. **M** A particular heat engine has a mechanical power output of 5.00 kW and an efficiency of 25.0%. The engine expels 8.00×10^3 J of exhaust energy in each cycle. Find (a) the energy taken in during each cycle and (b) the time interval for each cycle.

Section 18.2 Reversible and Irreversible Processes
Section 18.3 The Carnot Engine

6. **Q|C** An electric generating station is designed to have an electric output power of 1.40 MW using a turbine with two-thirds the efficiency of a Carnot engine. The exhaust energy is transferred by heat into a cooling tower at 110°C. (a) Find the rate at which the station exhausts energy by heat as a function of the fuel combustion temperature T_h. (b) If the firebox is modified to run hotter by using more advanced combustion technology, how does the amount of energy exhaust change? (c) Find the exhaust power for $T_h = 800°C$. (d) Find the value of T_h for which the exhaust power would be only half as large as in part (c). (e) Find the value of T_h for which the exhaust power would be one-fourth as large as in part (c).

7. **M** One of the most efficient heat engines ever built is a coal-fired steam turbine in the Ohio River valley, operating between 1 870°C and 430°C. (a) What is its maximum theoretical efficiency? (b) The actual efficiency of the engine is 42.0%. How much mechanical power does the engine deliver if it absorbs 1.40×10^5 J of energy each second from its hot reservoir?

8. Q|C S Suppose you build a two-engine device with the exhaust energy output from one heat engine supplying the input energy for a second heat engine. We say that the two engines are running *in series*. Let e_1 and e_2 represent the efficiencies of the two engines. (a) The overall efficiency of the two-engine device is defined as the total work output divided by the energy put into the first engine by heat. Show that the overall efficiency e is given by

$$e = e_1 + e_2 - e_1 e_2$$

What If? For parts (b) through (e) that follow, assume the two engines are Carnot engines. Engine 1 operates between temperatures T_h and T_i. The gas in engine 2 varies in temperature between T_i and T_c. In terms of the temperatures, (b) what is the efficiency of the combination engine? (c) Does an improvement in net efficiency result from the use of two engines instead of one? (d) What value of the intermediate temperature T_i results in equal work being done by each of the two engines in series? (e) What value of T_i results in each of the two engines in series having the same efficiency?

9. W A Carnot engine has a power output of 150 kW. The engine operates between two reservoirs at 20.0°C and 500°C. (a) How much energy enters the engine by heat per hour? (b) How much energy is exhausted by heat per hour?

10. S A Carnot engine has a power output P. The engine operates between two reservoirs at temperature T_c and T_h. (a) How much energy enters the engine by heat in a time interval Δt? (b) How much energy is exhausted by heat in the time interval Δt?

11. M An ideal gas is taken through a Carnot cycle. The isothermal expansion occurs at 250°C, and the isothermal compression takes place at 50.0°C. The gas takes in 1.20×10^3 J of energy from the hot reservoir during the isothermal expansion. Find (a) the energy expelled to the cold reservoir in each cycle and (b) the net work done by the gas in each cycle.

12. *Why is the following situation impossible?* An inventor comes to a patent office with the claim that her heat engine, which employs water as a working substance, has a thermodynamic efficiency of 0.110. Although this efficiency is low compared with typical automobile engines, she explains that her engine operates between an energy reservoir at room temperature and a water–ice mixture at atmospheric pressure and therefore requires no fuel other than that to make the ice. The patent is approved, and working prototypes of the engine prove the inventor's efficiency claim.

13. A power plant operates at a 32.0% efficiency during the summer when the seawater used for cooling is at 20.0°C. The plant uses 350°C steam to drive turbines. If the plant's efficiency changes in the same proportion as the ideal efficiency, what would be the plant's efficiency in the winter, when the seawater is at 10.0°C?

14. Q|C W An electric power plant that would make use of the temperature gradient in the ocean has been proposed. The system is to operate between 20.0°C (surface-water temperature) and 5.00°C (water temperature at a depth of about 1 km). (a) What is the maximum efficiency of such a system? (b) If the electric power output of the plant is 75.0 MW, how much energy is taken in from the warm reservoir per hour?

(c) In view of your answer to part (a), explain whether you think such a system is worthwhile. Note that the "fuel" is free.

15. W Argon enters a turbine at a rate of 80.0 kg/min, a temperature of 800°C, and a pressure of 1.50 MPa. It expands adiabatically as it pushes on the turbine blades and exits at pressure 300 kPa. (a) Calculate its temperature at exit. (b) Calculate the (maximum) power output of the turning turbine. (c) The turbine is one component of a model closed-cycle gas turbine engine. Calculate the maximum efficiency of the engine.

16. At point A in a Carnot cycle, 2.34 mol of a monatomic ideal gas has a pressure of 1 400 kPa, a volume of 10.0 L, and a temperature of 720 K. The gas expands isothermally to point B and then expands adiabatically to point C, where its volume is 24.0 L. An isothermal compression brings it to point D, where its volume is 15.0 L. An adiabatic process returns the gas to point A. (a) Determine all the unknown pressures, volumes, and temperatures as you fill in the following table:

	P	V	T
A	1 400 kPa	10.0 L	720 K
B			
C		24.0 L	
D		15.0 L	

(b) Find the energy added by heat, the work done by the engine, and the change in internal energy for each of the steps $A \rightarrow B$, $B \rightarrow C$, $C \rightarrow D$, and $D \rightarrow A$. (c) Calculate the efficiency $W_{net}/|Q_h|$. (d) Show that the efficiency is equal to $1 - T_C/T_A$, the Carnot efficiency.

Section 18.4 Heat Pumps and Refrigerators

17. W A refrigerator has a coefficient of performance equal to 5.00. The refrigerator takes in 120 J of energy from a cold reservoir in each cycle. Find (a) the work required in each cycle and (b) the energy expelled to the hot reservoir.

18. A refrigerator has a coefficient of performance of 3.00. The ice tray compartment is at −20.0°C, and the room temperature is at 22.0°C. The refrigerator can convert 30.0 g of water at 22.0°C to 30.0 g of ice at −20.0°C each minute. What input power is required? Give your answer in watts.

19. If a 35.0%-efficient Carnot heat engine (Active Fig. 18.1) is run in reverse so as to form a refrigerator (Active Fig. 18.7), what would be this refrigerator's coefficient of performance?

20. Q|C In 1993, the U.S. government instituted a requirement that all room air conditioners sold in the United States must have an energy efficiency ratio (EER) of 10 or higher. The EER is defined as the ratio of the cooling capacity of the air conditioner, measured in British thermal units per hour, or Btu/h, to its electrical power requirement in watts. (a) Convert the EER of 10.0 to dimensionless form, using the conversion 1 Btu = 1 055 J. (b) What is the appropriate name for this dimensionless quantity? (c) In the 1970s, it was common to find room air conditioners with EERs of 5 or lower. State how the operating costs compare for 10 000-Btu/h air conditioners with EERs of 5.00 and 10.0. Assume each air conditioner operates for 1 500 h during the summer in a city where electricity costs 17.0¢ per kWh.

21. What is the coefficient of performance of a refrigerator that operates with Carnot efficiency between temperatures $-3.00°C$ and $+27.0°C$?

22. **S** An ideal refrigerator or ideal heat pump is equivalent to a Carnot engine running in reverse. That is, energy $|Q_c|$ is taken in from a cold reservoir and energy $|Q_h|$ is rejected to a hot reservoir. (a) Show that the work that must be supplied to run the refrigerator or heat pump is

$$W = \frac{T_h - T_c}{T_c} |Q_c|$$

(b) Show that the coefficient of performance (COP) of the ideal refrigerator is

$$\text{COP} = \frac{T_c}{T_h - T_c}$$

23. What is the maximum possible coefficient of performance of a heat pump that brings energy from outdoors at $-3.00°C$ into a 22.0°C house? *Note:* The work done to run the heat pump is also available to warm the house.

24. A heat pump has a coefficient of performance of 3.80 and operates with a power consumption of 7.03×10^3 W. (a) How much energy does it deliver into a home during 8.00 h of continuous operation? (b) How much energy does it extract from the outside air?

25. A heat pump used for heating shown in Figure P18.25 is essentially an air conditioner installed backward. It extracts energy from colder air outside and deposits it in a warmer room. Suppose the ratio of the actual energy entering the room to the work done by the device's motor is 10.0% of the theoretical maximum ratio. Determine the energy entering the room per joule of work done by the motor given that the inside temperature is 20.0°C and the outside temperature is $-5.00°C$.

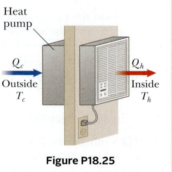

Heat pump

Q_c Outside T_c

Q_h Inside T_h

Figure P18.25

26. **M** How much work does an ideal Carnot refrigerator require to remove 1.00 J of energy from liquid helium at 4.00 K and expel this energy to a room-temperature (293-K) environment?

Section 18.6 Entropy

Section 18.7 Entropy and the Second Law of Thermodynamics

27. In making raspberry jelly, 900 g of raspberry juice is combined with 930 g of sugar. The mixture starts at room temperature, 23.0°C, and is slowly heated on a stove until it reaches 220°F. It is then poured into heated jars and allowed to cool. Assume that the juice has the same specific heat as water. The specific heat of sucrose is 0.299 cal/g · °C. Consider the heating process. (a) Which of the following terms describe(s) this process: adiabatic, isobaric, isothermal, isovolumetric, cyclic, reversible, isentropic?

(b) How much energy does the mixture absorb? (c) What is the minimum change in entropy of the jelly while it is heated?

28. **W** An ice tray contains 500 g of liquid water at 0°C. Calculate the change in entropy of the water as it freezes slowly and completely at 0°C.

29. Calculate the change in entropy of 250 g of water warmed slowly from 20.0°C to 80.0°C. (*Suggestion:* Note that $dQ = mc\,dT$.)

30. What change in entropy occurs when a 27.9-g ice cube at $-12°C$ is transformed into steam at 115°C?

31. Prepare a table like Table 18.1 by using the same procedure (a) for the case in which you draw three marbles from your bag rather than four and (b) for the case in which you draw five marbles rather than four.

32. (a) Prepare a table like Table 18.1 for the following occurrence. You toss four coins into the air simultaneously and then record the results of your tosses in terms of the numbers of heads (H) and tails (T) that result. For example, HHTH and HTHH are two possible ways in which three heads and one tail can be achieved. (b) On the basis of your table, what is the most probable result recorded for a toss? In terms of entropy, (c) what is the most ordered macrostate, and (d) what is the most disordered?

33. When an aluminum bar is connected between a hot reservoir at 725 K and a cold reservoir at 310 K, 2.50 kJ of energy is transferred by heat from the hot reservoir to the cold reservoir. In this irreversible process, calculate the change in entropy of (a) the hot reservoir, (b) the cold reservoir, and (c) the Universe, neglecting any change in entropy of the aluminum rod.

34. **S** When a metal bar is connected between a hot reservoir at T_h and a cold reservoir at T_c, the energy transferred by heat from the hot reservoir to the cold reservoir is Q. In this irreversible process, find expressions for the change in entropy of (a) the hot reservoir, (b) the cold reservoir, and (c) the Universe, neglecting any change in entropy of the metal rod.

35. If you roll two dice, what is the total number of ways in which you can obtain (a) a 12 and (b) a 7?

Section 18.8 Entropy Changes in Irreversible Processes

36. **W** The temperature at the surface of the Sun is approximately 5 800 K, and the temperature at the surface of the Earth is approximately 290 K. What entropy change of the Universe occurs when 1.00×10^3 J of energy is transferred by radiation from the Sun to the Earth?

37. A 2.00-L container has a center partition that divides it into two equal parts as shown in Figure P18.37. The left side contains 0.044 0 mol of H_2 gas, and the right side contains 0.044 0 mol of O_2 gas. Both gases are at room temperature and at atmospheric pressure. The partition is

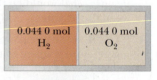

| 0.044 0 mol H_2 | 0.044 0 mol O_2 |

Figure P18.37

removed, and the gases are allowed to mix. What is the entropy increase of the system?

38. A 1.00-kg iron horseshoe is taken from a forge at 900°C and dropped into 4.00 kg of water at 10.0°C. Assuming that no energy is lost by heat to the surroundings, determine the total entropy change of the horseshoe-plus-water system.

39. A 1 500-kg car is moving at 20.0 m/s. The driver brakes to a stop. The brakes cool off to the temperature of the surrounding air, which is nearly constant at 20.0°C. What is the total entropy change?

40. How fast are you personally making the entropy of the Universe increase right now? Compute an order-of-magnitude estimate, stating what quantities you take as data and the values you measure or estimate for them.

41. A 1.00-mol sample of H_2 gas is contained in the left side of the container shown in Figure P18.41, which has equal volumes on the left and right. The right side is evacuated. When the valve is opened, the gas streams into the right side. (a) What is the entropy change of the gas? (b) Does the temperature of the gas change? Assume the container is so large that the hydrogen behaves as an ideal gas.

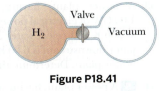

Figure P18.41

Section 18.9 Context Connection: The Atmosphere as a Heat Engine

42. We found the efficiency of the atmospheric heat engine to be about 0.8%. Taking the intensity of incoming solar radiation to be 1 370 W/m² and assuming that 64% of this energy is absorbed in the atmosphere, find the "wind power," that is, the rate at which energy becomes available for driving the winds.

43. (a) Find the kinetic energy of the moving air in a hurricane, modeled as a disk 600 km in diameter and 11 km thick, with wind blowing at a uniform speed of 60 km/h. (b) Consider sunlight with an intensity of 1 000 W/m² falling perpendicularly on a circular area 600 km in diameter. During what time interval would the sunlight deliver the amount of energy computed in part (a)?

Additional Problems

44. **Q|C** A firebox is at 750 K, and the ambient temperature is 300 K. The efficiency of a Carnot engine doing 150 J of work as it transports energy between these constant-temperature baths is 60.0%. The Carnot engine must take in energy 150 J/0.600 = 250 J from the hot reservoir and must put out 100 J of energy by heat into the environment. To follow Carnot's reasoning, suppose some other heat engine S could have an efficiency of 70.0%. (a) Find the energy input and exhaust energy output of engine S as it does 150 J of work. (b) Let engine S operate as in part (a) and run the Carnot engine in reverse between the same reservoirs. The output work of engine S is the input work for the Carnot refrigerator. Find the total energy transferred to or from the firebox and the total energy transferred to or from the environment as both

engines operate together. (c) Explain how the results of parts (a) and (b) show that the Clausius statement of the second law of thermodynamics is violated. (d) Find the energy input and work output of engine S as it puts out exhaust energy of 100 J. Let engine S operate as in part (c) and contribute 150 J of its work output to running the Carnot engine in reverse. Find (e) the total energy the firebox puts out as both engines operate together, (f) the total work output, and (g) the total energy transferred to the environment. (h) Explain how the results show that the Kelvin–Planck statement of the second law is violated. Therefore, our assumption about the efficiency of engine S must be false. (i) Let the engines operate together through one cycle as in part (d). Find the change in entropy of the Universe. (j) Explain how the result of part (i) shows that the entropy statement of the second law is violated.

45. **M** Energy transfers by heat through the exterior walls and roof of a house at a rate of 5.00×10^3 J/s = 5.00 kW when the interior temperature is 22.0°C and the outside temperature is −5.00°C. (a) Calculate the electric power required to maintain the interior temperature at 22.0°C if the power is used in electric resistance heaters that convert all the energy transferred in by electrical transmission into internal energy. (b) **What If?** Calculate the electric power required to maintain the interior temperature at 22.0°C if the power is used to drive an electric motor that operates the compressor of a heat pump that has a coefficient of performance equal to 60.0% of the Carnot-cycle value.

46. *Why is the following situation impossible?* Two samples of water are mixed at constant pressure inside an insulated container: 1.00 kg of water at 10.0°C and 1.00 kg of water at 30.0°C. Because the container is insulated, there is no exchange of energy by heat between the water and the environment. Furthermore, the amount of energy that leaves the warm water by heat is equal to the amount that enters the cool water by heat. Therefore, the entropy change of the Universe is zero for this process.

47. **GP S** In 1816, Robert Stirling, a Scottish clergyman, patented the *Stirling engine*, which has found a wide variety of applications ever since. Fuel is burned externally to warm one of the engine's two cylinders. A fixed quantity of inert gas moves cyclically between the cylinders, expanding in the hot one and contracting in the cold one. Figure P18.47 represents a model for its thermodynamic cycle. Consider *n* moles of an ideal monatomic gas being taken once through the cycle, consisting of two isothermal processes at temperatures $3T_i$ and T_i and two constant-volume processes. Let us find the efficiency of this engine. (a) Find the energy transferred by heat into the gas during the isovolumetric process *AB*. (b) Find the energy transferred by heat into the gas during the isothermal process *BC*. (c) Find the energy transferred by heat into the gas during the isovolumetric process *CD*.

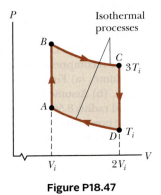

Figure P18.47

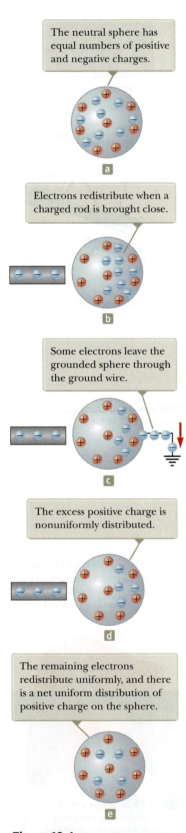

The neutral sphere has equal numbers of positive and negative charges.

a

Electrons redistribute when a charged rod is brought close.

b

Some electrons leave the grounded sphere through the ground wire.

c

The excess positive charge is nonuniformly distributed.

d

The remaining electrons redistribute uniformly, and there is a net uniform distribution of positive charge on the sphere.

e

Figure 19.4 Charging a metallic object by *induction*. (a) A neutral metallic sphere. (b) A charged rubber rod is placed near the sphere. (c) The sphere is grounded. (d) The ground connection is removed. (e) The rod is removed.

QUICK QUIZ 19.1 Three objects are brought close to one another, two at a time. When objects A and B are brought together, they repel. When objects B and C are brought together, they also repel. Which of the following statements are true? (**a**) Objects A and C possess charges of the same sign. (**b**) Objects A and C possess charges of opposite sign. (**c**) All three objects possess charges of the same sign. (**d**) One object is neutral. (**e**) Additional experiments must be performed to determine the signs of the charges.

19.3 | Insulators and Conductors

We have discussed the transfer of charge from one object to another. It is also possible for electric charges to move from one location to another within an object; such motion of charge is called **electrical conduction.** It is convenient to classify substances in terms of the ability of charges to move within the substance:

Electrical **conductors** are materials in which some of the electrons are free electrons[1] that are not bound to atoms and can move relatively freely through the material; electrical **insulators** are materials in which all electrons are bound to atoms and cannot move freely through the material.

Materials such as glass, rubber, and dry wood are insulators. When such materials are charged by rubbing, only the rubbed area becomes charged; the charge does not tend to move to other regions of the material. In contrast, materials such as copper, aluminum, and silver are good conductors. When such materials are charged in some small region, the charge readily distributes itself over the entire surface of the material. If you hold a copper rod in your hand and rub it with wool or fur, it will not attract a small piece of paper, which might suggest that a metal cannot be charged. If you hold the copper rod by an insulating handle and then rub, however, the rod remains charged and attracts the piece of paper. In the first case, the electric charges produced by rubbing readily move from the copper through your body, which is a conductor, and finally to the Earth. In the second case, the insulating handle prevents the flow of charge to your hand.

Semiconductors are a third class of materials, and their electrical properties are somewhere between those of insulators and those of conductors. Charges can move somewhat freely in a semiconductor, but far fewer charges are moving through a semiconductor than in a conductor. Silicon and germanium are well-known examples of semiconductors that are widely used in the fabrication of a variety of electronic devices. The electrical properties of semiconductors can be changed over many orders of magnitude by adding controlled amounts of certain foreign atoms to the materials.

Charging by Induction

When a conductor is connected to the Earth by means of a conducting wire or pipe, it is said to be **grounded.** For present purposes, the Earth can be modeled as an infinite reservoir for electrons, which means that it can accept or supply an unlimited number of electrons. In this context, the Earth serves a purpose similar to our energy reservoirs introduced in Chapter 17. With that in mind, we can understand how to charge a conductor by a process known as **charging by induction.**

To understand how to charge a conductor by induction, consider a neutral (uncharged) metallic sphere insulated from the ground as shown in Figure 19.4a. There are an equal number of electrons and protons in the sphere if the charge on the sphere is exactly zero. When a negatively charged rubber rod is brought near

[1]A metal atom contains one or more outer electrons, which are weakly bound to the nucleus. When many atoms combine to form a metal, the *free electrons* are these outer electrons, which are not bound to any one atom. These electrons move about the metal in a manner similar to that of gas molecules moving in a container.

the sphere, electrons in the region nearest the rod experience a repulsive force and migrate to the opposite side of the sphere. This migration leaves the side of the sphere near the rod with an effective positive charge because of the diminished number of electrons as in Figure 19.4b. (The left side of the sphere in Figure 19.4b is positively charged *as if* positive charges moved into this region, but in a metal it is only electrons that are free to move.) This migration occurs even if the rod never actually touches the sphere. If the same experiment is performed with a conducting wire connected from the sphere to the Earth (Fig. 19.4c), some of the electrons in the conductor are so strongly repelled by the presence of the negative charge in the rod that they move out of the sphere through the wire and into the Earth. The symbol ⏚ at the end of the wire in Figure 19.4c indicates that the wire is connected to **ground,** which means a reservoir such as the Earth. If the wire to ground is then removed (Fig. 19.4d), the conducting sphere contains an excess of *induced* positive charge because it has fewer electrons than it needs to cancel out the positive charge of the protons. When the rubber rod is removed from the vicinity of the sphere (Fig. 19.4e), this induced positive charge remains on the ungrounded sphere. Note that the rubber rod loses none of its negative charge during this process.

Charging an object by induction requires no contact with the object inducing the charge. This behavior is in contrast to charging an object by rubbing, which does require contact between the two objects.

A process similar to the first step in charging by induction in conductors takes place in insulators. In most neutral atoms and molecules, the average position of the positive charge coincides with the average position of the negative charge. In the presence of a charged object, however, these positions may shift slightly because of the attractive and repulsive forces from the charged object, resulting in more positive charge on one side of the molecule than on the other. This effect is known as **polarization.** The polarization of individual molecules produces a layer of charge on the surface of the insulator as shown in Figure 19.5a, in which a charged balloon on the left is placed against a wall on the right. In the figure, the negative charge layer in the wall is closer to the positively charged balloon than the positive charges at the other ends of the molecules. Therefore, the attractive force between the positive and negative charges is larger than the repulsive force between the positive charges. The result is a net attractive force between the charged balloon and the neutral insulator. Your knowledge of induction in insulators should help you explain why a charged rod attracts bits of electrically neutral paper (Fig. 19.5b) or why a balloon that has been rubbed against your hair can stick to a neutral wall.

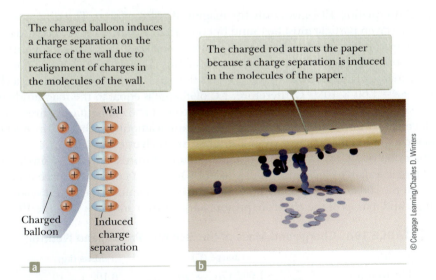

The charged balloon induces a charge separation on the surface of the wall due to realignment of charges in the molecules of the wall.

The charged rod attracts the paper because a charge separation is induced in the molecules of the paper.

Wall

Charged balloon   Induced charge separation

© Cengage Learning/Charles D. Winters

a   b

Figure 19.5 (a) A charged balloon is brought near an insulating wall. (b) A charged rod is brought close to bits of paper.

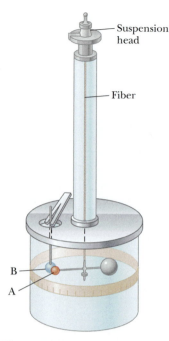

Figure 19.6 Coulomb's torsion balance, which was used to establish the inverse-square law for the electrostatic force between two charges.

Charles Coulomb
French Physicist (1736–1806)
Coulomb's major contributions to science were in the areas of electrostatics and magnetism. During his lifetime, he also investigated the strengths of materials and determined the forces that affect objects on beams, thereby contributing to the field of structural mechanics. In the field of ergonomics, his research provided a fundamental understanding of the ways in which people and animals can best do work.

© INTERFOTO/Alamy

▶ **QUICK QUIZ 19.2** Three objects are brought close to one another, two at a time. When objects A and B are brought together, they attract. When objects B and C are brought together, they repel. Which of the following are necessarily true? (a) Objects A and C possess charges of the same sign. (b) Objects A and C possess charges of opposite sign. (c) All three of the objects possess charges of the same sign. (d) One object is neutral. (e) Additional experiments must be performed to determine information about the charges on the objects.

19.4 | Coulomb's Law

Electric forces between charged objects were measured quantitatively by Charles Coulomb using the torsion balance, which he invented (Fig. 19.6). Coulomb confirmed that the electric force between two small charged spheres is proportional to the inverse square of their separation distance r, that is, $F_e \propto 1/r^2$. The operating principle of the torsion balance is the same as that of the apparatus used by Sir Henry Cavendish to measure the density of the Earth (Section 11.1), with the electrically neutral spheres replaced by charged ones. The electric force between charged spheres A and B in Figure 19.6 causes the spheres to either attract or repel each other, and the resulting motion causes the suspended fiber to twist. Because the restoring torque of the twisted fiber is proportional to the angle through which it rotates, a measurement of this angle provides a quantitative measure of the electric force of attraction or repulsion. Once the spheres are charged by rubbing, the electric force between them is very large compared with the gravitational attraction, and so the gravitational force can be ignored.

In Chapter 5, we introduced **Coulomb's law,** which describes the magnitude of the electrostatic force between two charged particles with charges q_1 and q_2 and separated by a distance r:

$$F_e = k_e \frac{|q_1||q_2|}{r^2}$$

▶ **19.1**

where k_e (= $8.987\ 6 \times 10^9$ N $\cdot$ m^2/C^2) is the **Coulomb constant** and the force is in newtons if the charges are in coulombs and if the separation distance is in meters. The constant k_e is also written as

$$k_e = \frac{1}{4\pi\epsilon_0}$$

where the constant ϵ_0 (Greek letter epsilon), known as the **permittivity of free space,** has the value

$$\epsilon_0 = 8.854\ 2 \times 10^{-12}\ \text{C}^2/\text{N} \cdot \text{m}^2$$

Note that Equation 19.1 gives only the magnitude of the force. The direction of the force on a given particle must be found by considering where the particles are located with respect to one another and the sign of each charge. Therefore, a pictorial representation of a problem in electrostatics is very important in analyzing the problem.

The charge of an electron is $q = -e = -1.60 \times 10^{-19}$ C, and the proton has a charge of $q = +e = 1.60 \times 10^{-19}$ C; therefore, 1 C of charge is equal to the magnitude of the charge of $(1.60 \times 10^{-19})^{-1} = 6.25 \times 10^{18}$ electrons. (The elementary charge e was introduced in Section 5.5.) Note that 1 C is a substantial amount of charge. In typical electrostatic experiments, where a rubber or glass rod is charged by friction, a net charge on the order of 10^{-6} C (= 1 μC) is obtained. In other words, only a very small fraction of the total available electrons (on the order of 10^{23} in a 1-cm^3 sample) are transferred between the rod and the rubbing material. The experimentally measured values of the charges and masses of the electron, proton, and neutron are given in Table 19.1.

▶ **TABLE 19.1** | Charge and Mass of the Electron, Proton, and Neutron

Particle	Charge (C)	Mass (kg)
Electron (e)	$-1.602\ 176\ 5 \times 10^{-19}$	$9.109\ 4 \times 10^{-31}$
Proton (p)	$+1.602\ 176\ 5 \times 10^{-19}$	$1.672\ 62 \times 10^{-27}$
Neutron (n)	0	$1.674\ 93 \times 10^{-27}$

When dealing with Coulomb's law, remember that force is a *vector* quantity and must be treated accordingly. Furthermore, Coulomb's law applies exactly only to particles.[2] The electrostatic force exerted by q_1 on q_2, written $\vec{\mathbf{F}}_{12}$, can be expressed in vector form as[3]

$$\vec{\mathbf{F}}_{12} = k_e \frac{q_1 q_2}{r^2} \hat{\mathbf{r}}_{12} \qquad \textbf{19.2} \blacktriangleleft$$

where $\hat{\mathbf{r}}_{12}$ is a unit vector directed from q_1 toward q_2 as in Active Figure 19.7a. Equation 19.2 can be used to find the direction of the force in space, although a carefully drawn pictorial representation is needed to clearly identify the direction of $\hat{\mathbf{r}}_{12}$. From Newton's third law, we see that the electric force exerted by q_2 on q_1 is equal in magnitude to the force exerted by q_1 on q_2 and in the opposite direction; that is, $\vec{\mathbf{F}}_{21} = -\vec{\mathbf{F}}_{12}$. From Equation 19.2, we see that if q_1 and q_2 have the same sign, the product $q_1 q_2$ is positive and the force is repulsive as in Active Figure 19.7a. The force on q_2 is in the same direction as $\hat{\mathbf{r}}_{12}$ and is directed away from q_1. If q_1 and q_2 are of opposite sign as in Active Figure 19.7b, the product $q_1 q_2$ is negative and the force is attractive. In this case, the force on q_2 is in the direction opposite to $\hat{\mathbf{r}}_{12}$, directed toward q_1.

When more than two charged particles are present, the force between any pair is given by Equation 19.2. Therefore, the resultant force on any one particle equals the *vector* sum of the individual forces due to all other particles. This **principle of superposition** as applied to electrostatic forces is an experimentally observed fact and simply represents the traditional vector sum of forces introduced in Chapter 4. As an example, if four charged particles are present, the resultant force on particle 1 due to particles 2, 3, and 4 is given by the vector sum

$$\vec{\mathbf{F}}_1 = \vec{\mathbf{F}}_{21} + \vec{\mathbf{F}}_{31} + \vec{\mathbf{F}}_{41}$$

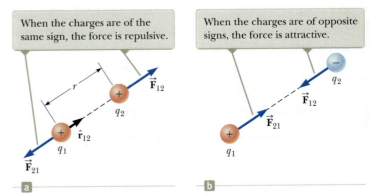

When the charges are of the same sign, the force is repulsive.

When the charges are of opposite signs, the force is attractive.

a   **b**

Active Figure 19.7 Two point charges separated by a distance r exert a force on each other given by Coulomb's law. Note that the force $\vec{\mathbf{F}}_{21}$ exerted by q_2 on q_1 is equal in magnitude and opposite in direction to the force $\vec{\mathbf{F}}_{12}$ exerted by q_1 on q_2.

> **QUICK QUIZ 19.3** Object A has a charge of $+2\ \mu C$, and object B has a charge of $+6\ \mu C$. Which statement is true about the electric forces on the objects?
> (a) $\vec{\mathbf{F}}_{AB} = -3\vec{\mathbf{F}}_{BA}$ (b) $\vec{\mathbf{F}}_{AB} = -\vec{\mathbf{F}}_{BA}$ (c) $3\vec{\mathbf{F}}_{AB} = -\vec{\mathbf{F}}_{BA}$ (d) $\vec{\mathbf{F}}_{AB} = 3\vec{\mathbf{F}}_{BA}$
> (e) $\vec{\mathbf{F}}_{AB} = \vec{\mathbf{F}}_{BA}$ (f) $3\vec{\mathbf{F}}_{AB} = \vec{\mathbf{F}}_{BA}$

Example 19.1 | **Where Is the Net Force Zero?**

Three point charges lie along the x axis as shown in Figure 19.8. The positive charge $q_1 = 15.0\ \mu C$ is at $x = 2.00$ m, the positive charge $q_2 = 6.00\ \mu C$ is at the origin, and the net force acting on q_3 is zero. What is the x coordinate of q_3?

Figure 19.8 (Example 19.1) Three point charges are placed along the x axis. If the resultant force acting on q_3 is zero, the force $\vec{\mathbf{F}}_{13}$ exerted by q_1 on q_3 must be equal in magnitude and opposite in direction to the force $\vec{\mathbf{F}}_{23}$ exerted by q_2 on q_3.

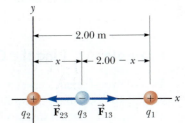

SOLUTION

Conceptualize Because q_3 is near two other charges, it experiences two electric forces. The forces lie along the same line as indicated in Figure 19.8. Because q_3 is negative and q_1 and q_2 are positive, the forces $\vec{\mathbf{F}}_{13}$ and $\vec{\mathbf{F}}_{23}$ are both attractive.

Categorize Because the net force on q_3 is zero, we model the point charge as a particle in equilibrium.

Analyze Write an expression for the net force on charge q_3 when it is in equilibrium:

$$\vec{\mathbf{F}}_3 = \vec{\mathbf{F}}_{23} + \vec{\mathbf{F}}_{13} = -k_e \frac{|q_2||q_3|}{x^2}\hat{\mathbf{i}} + k_e \frac{|q_1||q_3|}{(2.00-x)^2}\hat{\mathbf{i}} = 0$$

continued

---

[2]Coulomb's law can also be used for larger objects to which the particle model can be applied.

[3]Notice that we use "q_2" as shorthand notation for "the particle with charge q_2." This usage is common when discussing charged particles, similar to the use in mechanics of "m_2" for "the particle with mass m_2." The context of the sentence will tell you whether the symbol represents an amount of charge or a particle with that charge.

19.1 cont.

Move the second term to the right side of the equation and set the coefficients of the unit vector $\hat{\mathbf{i}}$ equal:

$$k_e \frac{|q_2||q_3|}{x^2} = k_e \frac{|q_1||q_3|}{(2.00 - x)^2}$$

Eliminate k_e and $|q_3|$ and rearrange the equation:

$$(2.00 - x)^2|q_2| = x^2|q_1|$$

$$(4.00 - 4.00x + x^2)(6.00 \times 10^{-6}\ C) = x^2(15.0 \times 10^{-6}\ C)$$

Reduce the quadratic equation to a simpler form:

$$3.00x^2 + 8.00x - 8.00 = 0$$

Solve the quadratic equation for the positive root:

$$x = 0.775\ m$$

Finalize The second root to the quadratic equation is $x = -3.44$ m. That is another location where the *magnitudes* of the forces on q_3 are equal, but both forces are in the same direction, so they do not cancel.

Example 19.2 | The Hydrogen Atom

The electron and proton of a hydrogen atom are separated (on the average) by a distance of approximately 5.3×10^{-11} m. Find the magnitudes of the electric force and the gravitational force between the two particles.

SOLUTION

Conceptualize Think about the two particles separated by the very small distance given in the problem statement. In Chapter 5, we mentioned that the gravitational force between an electron and a proton is very small compared to the electric force between them, so we expect this to be the case with the results of this example.

Categorize The electric and gravitational forces will be evaluated from universal force laws, so we categorize this example as a substitution problem.

Use Coulomb's law to find the magnitude of the electric force:

$$F_e = k_e \frac{|e||-e|}{r^2} = (8.99 \times 10^9\ N \cdot m^2/C^2) \frac{(1.60 \times 10^{-19}\ C)^2}{(5.3 \times 10^{-11}\ m)^2}$$

$$= 8.2 \times 10^{-8}\ N$$

Use Newton's law of universal gravitation and Table 19.1 (for the particle masses) to find the magnitude of the gravitational force:

$$F_g = G \frac{m_e m_p}{r^2}$$

$$= (6.67 \times 10^{-11}\ N \cdot m^2/kg^2) \frac{(9.11 \times 10^{-31}\ kg)(1.67 \times 10^{-27}\ kg)}{(5.3 \times 10^{-11}\ m)^2}$$

$$= 3.6 \times 10^{-47}\ N$$

The ratio $F_e/F_g \approx 2 \times 10^{39}$. Therefore, the gravitational force between charged atomic particles is negligible when compared with the electric force. Notice the similar forms of Newton's law of universal gravitation and Coulomb's law of electric forces. Other than the magnitude of the forces between elementary particles, what is a fundamental difference between the two forces?

Example 19.3 | Find the Charge on the Spheres

Two identical small charged spheres, each having a mass of 3.00×10^{-2} kg, hang in equilibrium as shown in Figure 19.9a. The length L of each string is 0.150 m, and the angle θ is 5.00°. Find the magnitude of the charge on each sphere.

SOLUTION

Conceptualize Figure 19.9a helps us conceptualize this example. The two spheres exert repulsive forces on each other. If they are held close to each other and released, they move outward from the center and settle into the configuration in Figure 19.9a after the oscillations have vanished due to air resistance.

Categorize The key phrase "in equilibrium" helps us model each sphere as a particle in equilibrium. This example is similar

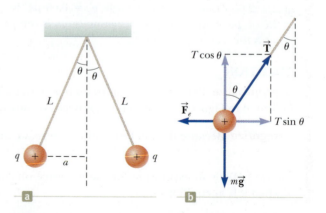

Figure 19.9 (Example 19.3) (a) Two identical spheres, each carrying the same charge q, suspended in equilibrium. (b) Diagram of the forces acting on the sphere on the left part of (a).

19.3 *cont.*

to the particle in equilibrium problems in Chapter 4 with the added feature that one of the forces on a sphere is an electric force.

....

Analyze The force diagram for the left-hand sphere is shown in Figure 19.9b. The sphere is in equilibrium under the application of the force $\vec{\mathbf{T}}$ from the string, the electric force $\vec{\mathbf{F}}_e$ from the other sphere, and the gravitational force $m\vec{\mathbf{g}}$.

Write Newton's second law for the left-hand sphere in component form:

(1) $\sum F_x = T \sin \theta - F_e = 0 \quad \rightarrow \quad T \sin \theta = F_e$

(2) $\sum F_y = T \cos \theta - mg = 0 \quad \rightarrow \quad T \cos \theta = mg$

Divide Equation (1) by Equation (2) to find F_e:

$\tan \theta = \dfrac{F_e}{mg} \quad \rightarrow \quad F_e = mg \tan \theta$

Use the geometry of the right triangle in Figure 19.9a to find a relationship between a, L, and θ:

$\sin \theta = \dfrac{a}{L} \quad \rightarrow \quad a = L \sin \theta$

Solve Coulomb's law (Eq. 19.1) for the charge $|q|$ on each sphere:

$|q| = \sqrt{\dfrac{F_e r^2}{k_e}} = \sqrt{\dfrac{F_e (2a)^2}{k_e}} = \sqrt{\dfrac{mg \tan \theta (2L \sin \theta)^2}{k_e}}$

Substitute numerical values:

$|q| = \sqrt{\dfrac{(3.00 \times 10^{-2} \text{ kg})(9.80 \text{ m/s}^2)\tan(5.00°)[2(0.150 \text{ m})\sin(5.00°)]^2}{8.99 \times 10^9 \text{ N} \cdot \text{m}^2/\text{C}^2}}$

$= 4.42 \times 10^{-8} \text{ C}$

....

Finalize If the sign of the charges were not given in Figure 19.9, we could not determine them. In fact, the sign of the charge is not important. The situation is the same whether both spheres are positively charged or negatively charged.

....

19.5 | Electric Fields

In Section 4.1, we discussed the differences between contact forces and field forces. Two field forces—the gravitational force in Chapter 11 and the electric force here—have been introduced into our discussions so far. As pointed out earlier, field forces can act through space, producing an effect even when no physical contact occurs between interacting objects. The gravitational field $\vec{\mathbf{g}}$ at a point in space due to a source particle was defined in Section 11.1 to be equal to the gravitational force $\vec{\mathbf{F}}_g$ acting on a test particle of mass m divided by that mass: $\vec{\mathbf{g}} \equiv \vec{\mathbf{F}}_g / m$. The concept of a field was developed by Michael Faraday (1791–1867) in the context of electric forces and is of such practical value that we shall devote much attention to it in the next several chapters. In this approach, an **electric field** is said to exist in the region of space around a charged object, the **source charge.** When another charged object—the **test charge**—enters this electric field, an electric force acts on it. As an example, consider Figure 19.10, which shows a small positive test charge q_0 placed near a second object carrying a much greater positive charge Q. We define the electric field due to the source charge at the location of the test charge to be the electric force on the test charge *per unit charge*, or, to be more specific, the **electric field vector** $\vec{\mathbf{E}}$ at a point in space is defined as the electric force $\vec{\mathbf{F}}_e$ acting on a positive test charge q_0 placed at that point divided by the test charge:[4]

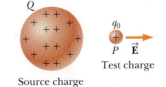

Q

q_0

P $\vec{\mathbf{E}}$

Test charge

Source charge

Figure 19.10 A small positive test charge q_0 placed at point P near an object carrying a much larger positive charge Q experiences an electric field $\vec{\mathbf{E}}$ at point P established by the source charge Q. We will *always* assume that the test charge is so small that the field of the source charge is unaffected by its presence.

$$\vec{\mathbf{E}} \equiv \frac{\vec{\mathbf{F}}_e}{q_0}$$

 19.3 ◀   ▶ Definition of electric field

The vector $\vec{\mathbf{E}}$ has the SI units of newtons per coulomb (N/C). The direction of $\vec{\mathbf{E}}$ as shown in Figure 19.10 is the direction of the force a positive test charge experiences when placed in the field. Note that $\vec{\mathbf{E}}$ is the field produced by some charge or charge distribution *separate from* the test charge; it is not the field produced by the

---

[4]When using Equation 19.3, we must assume the test charge q_0 is small enough that it does not disturb the charge distribution responsible for the electric field. If the test charge is great enough, the charge on the metallic sphere is redistributed and the electric field it sets up is different from the field it sets up in the presence of the much smaller test charge.

Active Figure 19.11 (a), (c) When a test charge q_0 is placed near a source charge q, the test charge experiences a force. (b), (d) At a point P near a source charge q, there exists an electric field.

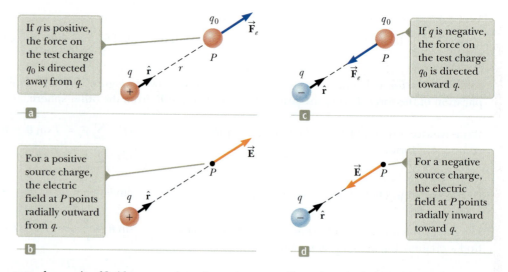

If q is positive, the force on the test charge q_0 is directed away from q.

If q is negative, the force on the test charge q_0 is directed toward q.

For a positive source charge, the electric field at P points radially outward from q.

For a negative source charge, the electric field at P points radially inward toward q.

test charge itself. Also note that the existence of an electric field is a property of its source; the presence of the test charge is not necessary for the field to exist. The test charge serves as a *detector* of the electric field: an electric field exists at a point if a test charge at that point experiences an electric force.

Once the electric field is known at some point, the force on *any* particle with charge q placed at that point can be calculated from a rearrangement of Equation 19.3:

$$\vec{\mathbf{F}}_e = q\vec{\mathbf{E}} \qquad\qquad \textbf{19.4} \blacktriangleleft$$

Once the electric force on a particle is evaluated, its motion can be determined from the particle under a net force model or the particle in equilibrium model (the electric force may have to be combined with other forces acting on the particle), and the techniques of earlier chapters can be used to find the motion of the particle.

Consider a point charge[5] q located a distance r from a test particle with charge q_0. According to Coulomb's law, the force exerted on the test particle by q is

$$\vec{\mathbf{F}}_e = k_e \frac{qq_0}{r^2}\hat{\mathbf{r}}$$

where $\hat{\mathbf{r}}$ is a unit vector directed from q toward q_0. This force in Active Figure 19.11a is directed away from the source charge q. Because the electric field at P, the position of the test charge, is defined by $\vec{\mathbf{E}} = \vec{\mathbf{F}}_e/q_0$, we find that at P, the electric field created by q is

$$\vec{\mathbf{E}} = k_e \frac{q}{r^2}\hat{\mathbf{r}} \qquad\qquad \textbf{19.5} \blacktriangleleft$$

▶ Electric field due to a point charge

If the source charge q is positive, Active Figure 19.11b shows the situation with the test charge removed; the source charge sets up an electric field at point P, directed away from q. If q is negative as in Active Figure 19.11c, the force on the test charge is toward the source charge, so the electric field at P is directed toward the source charge as in Active Figure 19.11d.

To calculate the electric field at a point P due to a group of point charges, we first calculate the electric field vectors at P individually using Equation 19.5 and then add them vectorially. In other words, the total electric field at a point in space due to a group of charged particles equals the vector sum of the electric fields at that point due to all the particles. This superposition principle applied to fields follows directly from the vector addition property of forces. Therefore, the electric field at point P of a group of source charges can be expressed as

▶ Electric field due to a finite number of point charges

$$\vec{\mathbf{E}} = k_e \sum_i \frac{q_i}{r_i^2}\hat{\mathbf{r}}_i \qquad\qquad \textbf{19.6} \blacktriangleleft$$

Pitfall Prevention | 19.1
Particles Only
Equation 19.4 is valid only for a *particle* of charge q, that is, an object of zero size. For a charged object of finite size in an electric field, the field may vary in magnitude and direction over the size of the object, so the corresponding force equation may be more complicated.

---

[5]We have used the phrase "charged particle" so far. The phrase "point charge" is somewhat misleading because charge is a property of a particle, not a physical entity. It is similar to misleading phrasing in mechanics such as "a mass m is placed . . ." (which we have avoided) rather than "a particle with mass m is placed. . . ." This phrase is so ingrained in physics usage, however, that we will use it and hope that this footnote suffices to clarify its use.

where r_i is the distance from the ith charge q_i to the point P (the location at which the field is to be evaluated) and $\hat{\mathbf{r}}_i$ is a unit vector directed from q_i toward P.

> **QUICK QUIZ 19.4** A test charge of $+3 \ \mu C$ is at a point P where an external electric field is directed to the right and has a magnitude of 4×10^6 N/C. If the test charge is replaced with another charge of $-3 \ \mu C$, what happens to the external electric field at P? (**a**) It is unaffected. (**b**) It reverses direction. (**c**) It changes in a way that cannot be determined.

Example 19.4 | Electric Field of a Dipole

An **electric dipole** consists of a point charge q and a point charge $-q$ separated by a distance of $2a$ as in Figure 19.12. As we shall see in later chapters, neutral atoms and molecules behave as dipoles when placed in an external electric field. Furthermore, many molecules, such as HCl, are permanent dipoles. (HCl can be effectively modeled as an H^+ ion combined with a Cl^- ion.) The effect of such dipoles on the behavior of materials subjected to electric fields is discussed in Chapter 20.

(A) Find the electric field $\vec{\mathbf{E}}$ due to the dipole along the y axis at the point P, which is a distance y from the origin.

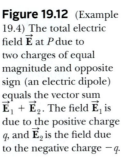

Figure 19.12 (Example 19.4) The total electric field $\vec{\mathbf{E}}$ at P due to two charges of equal magnitude and opposite sign (an electric dipole) equals the vector sum $\vec{\mathbf{E}}_1 + \vec{\mathbf{E}}_2$. The field $\vec{\mathbf{E}}_1$ is due to the positive charge q, and $\vec{\mathbf{E}}_2$ is the field due to the negative charge $-q$.

SOLUTION

Conceptualize In Example 19.1, we added vector forces to find the net force on a particle. Here, we add electric field vectors to find the net electric field at a point in space.

Categorize We have two source charges and wish to find the total electric field, so we categorize this example as one in which we can use the superposition principle represented by Equation 19.6.

. . . .

Analyze At P, the fields $\vec{\mathbf{E}}_1$ and $\vec{\mathbf{E}}_2$ due to the two particles are equal in magnitude because P is equidistant from the two charges. The total field at P is $\vec{\mathbf{E}} = \vec{\mathbf{E}}_1 + \vec{\mathbf{E}}_2$.

Find the magnitudes of the fields at P:

$$E_1 = E_2 = k_e \frac{q}{r^2} = k_e \frac{q}{y^2 + a^2}$$

The y components of $\vec{\mathbf{E}}_1$ and $\vec{\mathbf{E}}_2$ are equal in magnitude and opposite in sign, so they cancel. The x components are equal and add because they have the same sign. The total field $\vec{\mathbf{E}}$ is therefore parallel to the x axis.

Find an expression for the magnitude of the electric field at P:

$$(1) \quad E = 2k_e \frac{q}{y^2 + a^2} \cos \theta$$

From the geometry in Figure 19.12 we see that $\cos \theta = a/r = a/(y^2 + a^2)^{1/2}$. Substitute this result into Equation (1):

$$E = 2k_e \frac{q}{(y^2 + a^2)} \frac{a}{(y^2 + a^2)^{1/2}}$$

$$(2) \quad E = k_e \frac{2qa}{(y^2 + a^2)^{3/2}}$$

(B) Find the electric field for points $y \gg a$ far from the dipole.

SOLUTION

Equation (2) gives the value of the electric field on the y axis at all values of y. For points far from the dipole, for which $y \gg a$, neglect a^2 in the denominator and write the expression for E in this case:

$$(3) \quad E \approx k_e \frac{2qa}{y^3}$$

. . . .

Finalize Therefore, we see that along the y axis the field of a dipole at a distant point varies as $1/r^3$, whereas the more slowly varying field of a point charge varies as $1/r^2$. (*Note:* In the geometry of this example, $r = y$.) At distant points, the fields of the two charges in the dipole almost cancel each other. The $1/r^3$ variation in E for the dipole is also obtained for a distant point along the x axis (see Problem 20) and for a general distant point.

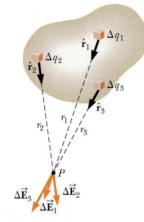

Figure 19.13 The electric field $\vec{\mathbf{E}}$ at P due to a continuous charge distribution is the vector sum of the fields $\Delta\vec{\mathbf{E}}$ due to all the elements Δq_i of the charge distribution. Three sample elements are shown.

Electric Field Due to Continuous Charge Distributions

In most practical situations (e.g., an object charged by rubbing), the average separation between source charges is small compared with their distances from the point at which the field is to be evaluated. In such cases, the system of source charges can be modeled as *continuous*. That is, we imagine that the system of closely spaced discrete charges is equivalent to a total charge that is continuously distributed through some volume or over some surface.

To evaluate the electric field of a continuous charge distribution, the following procedure is used. First, we divide the charge distribution into small elements, each of which contains a small amount of charge Δq as in Figure 19.13. Next, modeling the element as a point charge, we use Equation 19.5 to calculate the electric field $\Delta\vec{\mathbf{E}}$ at a point P due to one of these elements. Finally, we evaluate the total field at P due to the charge distribution by performing a vector sum of the contributions of all the charge elements (i.e., by applying the superposition principle).

The electric field at P in Figure 19.13 due to one element of charge Δq_i is given by

$$\Delta\vec{\mathbf{E}}_i = k_e \frac{\Delta q_i}{r_i^2}\hat{\mathbf{r}}_i$$

where the index i refers to the ith element in the distribution, r_i is the distance from the element to point P, and $\hat{\mathbf{r}}_i$ is a unit vector directed from the element toward P. The total electric field $\vec{\mathbf{E}}$ at P due to all elements in the charge distribution is approximately

$$\vec{\mathbf{E}} \approx k_e \sum_i \frac{\Delta q_i}{r_i^2}\hat{\mathbf{r}}_i$$

Now, we apply the model in which the charge distribution is continuous, and we let the elements of charge become infinitesimally small. With this model, the total field at P in the limit $\Delta q_i \to 0$ becomes

$$\vec{\mathbf{E}} = \lim_{\Delta q_i \to 0} k_e \sum_i \frac{\Delta q_i}{r_i^2}\hat{\mathbf{r}}_i = k_e \int \frac{dq}{r^2}\hat{\mathbf{r}} \qquad \textbf{19.7} \blacktriangleleft$$

where dq is an infinitesimal amount of charge and the integration is over all the charge creating the electric field. The integration is a *vector* operation and must be treated with caution. It can be evaluated in terms of individual components, or perhaps symmetry arguments can be used to reduce it to a scalar integral. We shall illustrate this type of calculation with several examples in which we assume that the charge is *uniformly* distributed on a line or a surface or throughout some volume. When performing such calculations, it is convenient to use the concept of a *charge density* along with the following notations:

- If a total charge Q is uniformly distributed throughout a volume V, the **volume charge density** ρ is defined by

▶ Volume charge density

$$\rho \equiv \frac{Q}{V} \qquad \textbf{19.8} \blacktriangleleft$$

where ρ has units of coulombs per cubic meter.

- If Q is uniformly distributed on a surface of area A, the **surface charge density** σ is defined by

▶ Surface charge density

$$\sigma \equiv \frac{Q}{A} \qquad \textbf{19.9} \blacktriangleleft$$

where σ has units of coulombs per square meter.

- If Q is uniformly distributed along a line of length ℓ, the **linear charge density** λ is defined by

▶ Linear charge density

$$\lambda \equiv \frac{Q}{\ell} \qquad \textbf{19.10} \blacktriangleleft$$

where λ has units of coulombs per meter.

> ### PROBLEM-SOLVING STRATEGY: Calculating the Electric Field

The following procedure is recommended for solving problems that involve the determination of an electric field due to individual charges or a charge distribution.

1. **Conceptualize.** Establish a mental representation of the problem: think carefully about the individual charges or the charge distribution and imagine what type of electric field it would create. Appeal to any symmetry in the arrangement of charges to help you visualize the electric field.

2. **Categorize.** Are you analyzing a group of individual charges or a continuous charge distribution? The answer to this question tells you how to proceed in the Analyze step.

3. **Analyze.**

   (a) If you are analyzing a group of individual charges, use the superposition principle: when several point charges are present, the resultant field at a point in space is the *vector sum* of the individual fields due to the individual charges (Eq. 19.6). Be very careful in the manipulation of vector quantities. It may be useful to review the material on vector addition in Chapter 1. Example 19.4 demonstrated this procedure.

   (b) If you are analyzing a continuous charge distribution, replace the vector sums for evaluating the total electric field from individual charges by vector integrals. The charge distribution is divided into infinitesimal pieces, and the vector sum is carried out by integrating over the entire charge distribution (Eq. 19.7). Examples 19.5 and 19.6 demonstrate such procedures.

   Consider symmetry when dealing with either a distribution of point charges or a continuous charge distribution. Take advantage of any symmetry in the system you observed in the Conceptualize step to simplify your calculations. The cancellation of field components perpendicular to the axis in Example 19.6 is an example of the application of symmetry.

4. **Finalize.** Check to see if your electric field expression is consistent with the mental representation and if it reflects any symmetry that you noted previously. Imagine varying parameters such as the distance of the observation point from the charges or the radius of any circular objects to see if the mathematical result changes in a reasonable way.

Example 19.5 | The Electric Field Due to a Charged Rod

A rod of length ℓ has a uniform positive charge per unit length λ and a total charge Q. Calculate the electric field at a point P that is located along the long axis of the rod and a distance a from one end (Fig. 19.14).

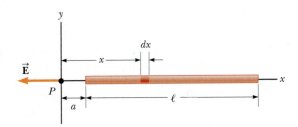

Figure 19.14 (Example 19.5) The electric field at P due to a uniformly charged rod lying along the x axis.

SOLUTION

Conceptualize The field $d\vec{E}$ at P due to each segment of charge on the rod is in the negative x direction because every segment carries a positive charge.

Categorize Because the rod is continuous, we are evaluating the field due to a continuous charge distribution rather than a group of individual charges. Because every segment of the rod produces an electric field in the negative x direction, the sum of their contributions can be handled without the need to add vectors.

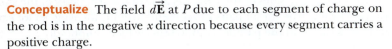

Analyze Let's assume the rod is lying along the x axis, dx is the length of one small segment, and dq is the charge on that segment. Because the rod has a charge per unit length λ, the charge dq on the small segment is $dq = \lambda\, dx$.

Find the magnitude of the electric field at P due to one segment of the rod having a charge dq:

$$dE = k_e \frac{dq}{x^2} = k_e \frac{\lambda\, dx}{x^2}$$

Find the total field at P using[6] Equation 19.7:

$$E = \int_{a}^{\ell+a} k_e \lambda \frac{dx}{x^2}$$

continued

---

[6]To carry out integrations such as this one, first express the charge element dq in terms of the other variables in the integral. (In this example, there is one variable, x, so we made the change $dq = \lambda\, dx$.) The integral must be over scalar quantities; therefore, express the electric field in terms of components, if necessary. (In this example, the field has only an x component, so this detail is of no concern.) Then, reduce your expression to an integral over a single variable (or to multiple integrals, each over a single variable). In examples that have spherical or cylindrical symmetry, the single variable is a radial coordinate.

19.5 *cont.*

Noting that k_e and $\lambda = Q/\ell$ are constants and can be removed from the integral, evaluate the integral:

$$E = k_e \lambda \int_a^{\ell+a} \frac{dx}{x^2} = k_e \lambda \left[-\frac{1}{x} \right]_a^{\ell+a}$$

$$(1) \quad E = k_e \frac{Q}{\ell} \left(\frac{1}{a} - \frac{1}{\ell + a} \right) = \frac{k_e Q}{a(\ell + a)}$$

Finalize If $a \rightarrow 0$, which corresponds to sliding the bar to the left until its left end is at the origin, then $E \rightarrow \infty$. That represents the condition in which the observation point P is at zero distance from the charge at the end of the rod, so the field becomes infinite. We explore large values of a below.

What If? Suppose point P is very far away from the rod. What is the nature of the electric field at such a point?

Answer If P is far from the rod ($a \gg \ell$), then ℓ in the denominator of Equation (1) can be neglected and $E \approx k_e Q / a^2$. That is exactly the form you would expect for a point charge. Therefore, at large values of a/ℓ, the charge distribution appears to be a point charge of magnitude Q; the point P is so far away from the rod we cannot distinguish that it has a size. The use of the limiting technique ($a/\ell \rightarrow \infty$) is often a good method for checking a mathematical expression.

Example 19.6 | The Electric Field of a Uniform Ring of Charge

A ring of radius a carries a uniformly distributed positive total charge Q. Calculate the electric field due to the ring at a point P lying a distance x from its center along the central axis perpendicular to the plane of the ring (Fig. 19.15a).

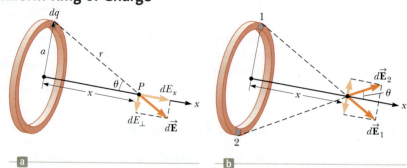

SOLUTION

Conceptualize Figure 19.15a shows the electric field contribution $d\vec{E}$ at P due to a single segment of charge at the top of the ring. This field vector can be resolved into components dE_x parallel to the axis of the ring and $dE_\perp$ perpendicular to the axis. Figure 19.15b

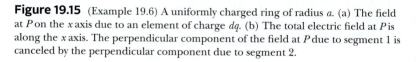

Figure 19.15 (Example 19.6) A uniformly charged ring of radius a. (a) The field at P on the x axis due to an element of charge dq. (b) The total electric field at P is along the x axis. The perpendicular component of the field at P due to segment 1 is canceled by the perpendicular component due to segment 2.

shows the electric field contributions from two segments on opposite sides of the ring. Because of the symmetry of the situation, the perpendicular components of the field cancel. That is true for all pairs of segments around the ring, so we can ignore the perpendicular component of the field and focus solely on the parallel components, which simply add.

Categorize Because the ring is continuous, we are evaluating the field due to a continuous charge distribution rather than a group of individual charges.

Analyze Evaluate the parallel component of an electric field contribution from a segment of charge dq on the ring:

$$(1) \quad dE_x = k_e \frac{dq}{r^2} \cos \theta = k_e \frac{dq}{a^2 + x^2} \cos \theta$$

From the geometry in Figure 19.15a, evaluate $\cos \theta$:

$$(2) \quad \cos \theta = \frac{x}{r} = \frac{x}{(a^2 + x^2)^{1/2}}$$

Substitute Equation (2) into Equation (1):

$$dE_x = k_e \frac{dq}{a^2 + x^2} \frac{x}{(a^2 + x^2)^{1/2}} = \frac{k_e x}{(a^2 + x^2)^{3/2}} dq$$

All segments of the ring make the same contribution to the field at P because they are all equidistant from this point. Integrate to obtain the total field at P:

$$E_x = \int \frac{k_e x}{(a^2 + x^2)^{3/2}} dq = \frac{k_e x}{(a^2 + x^2)^{3/2}} \int dq$$

$$(3) \quad E = \frac{k_e x}{(a^2 + x^2)^{3/2}} Q$$

19.6 *cont.*

Finalize This result shows that the field is zero at $x = 0$. Is that consistent with the symmetry in the problem? Furthermore, notice that Equation (3) reduces to $k_e Q / x^2$ if $x \gg a$, so the ring acts like a point charge for locations far away from the ring.

What If? Suppose a negative charge is placed at the center of the ring in Figure 19.15 and displaced slightly by a distance $x \ll a$ along the x axis. When the charge is released, what type of motion does it exhibit?

Answer In the expression for the field due to a ring of charge, let $x \ll a$, which results in

$$E_x = \frac{k_e Q}{a^3} x$$

Therefore, from Equation 19.4, the force on a charge $-q$ placed near the center of the ring is

$$F_x = -\frac{k_e q Q}{a^3} x$$

Because this force has the form of Hooke's law (Eq. 12.1), the motion of the negative charge is *simple harmonic!*

19.6 | Electric Field Lines

A convenient specialized pictorial representation for visualizing electric field patterns is created by drawing lines showing the direction of the electric field vector at any point. These lines, called **electric field lines,** are related to the electric field in any region of space in the following manner:

- The electric field vector $\vec{E}$ is *tangent* to the electric field line at each point.
- The number of electric field lines per unit area through a surface that is perpendicular to the lines is proportional to the magnitude of the electric field in that region. Therefore, E is large where the field lines are close together and small where they are far apart.

These properties are illustrated in Figure 19.16. The density of lines through surface A is greater than the density of lines through surface B. Therefore, the magnitude of the electric field on surface A is larger than on surface B. Furthermore, the field drawn in Figure 19.16 is nonuniform because the lines at different locations point in different directions.

Some representative electric field lines for a single positive point charge are shown in Figure 19.17a. Note that in this two-dimensional drawing we show only the field lines that lie in the plane of the page. The lines are actually directed radially outward in *all* directions from the charge, somewhat like the needles of a porcupine. Because a positively charged test particle placed in this field would be repelled by the charge q, the lines are directed radially away from q. Similarly, the electric field lines for a single negative point charge are directed toward the charge (Fig. 19.17b). In either case, the lines are radial and extend to infinity. Note that the lines are closer together as they come nearer to the charge, indicating that the magnitude of the field is increasing. The electric field lines end in Figure 19.17a and begin in Figure 19.17b on hypothetical charges we assume to be located infinitely far away.

Is this visualization of the electric field in terms of field lines consistent with Equation 19.5? To answer this question, consider an imaginary spherical surface of radius r, concentric with the charge. From

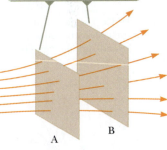

The magnitude of the field is greater on surface A than on surface B.

Figure 19.16 Electric field lines penetrating two surfaces.

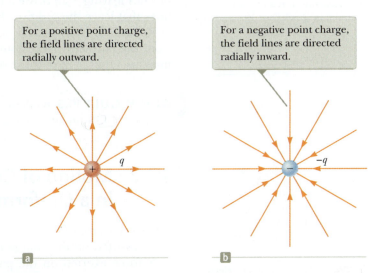

For a positive point charge, the field lines are directed radially outward.

For a negative point charge, the field lines are directed radially inward.

Figure 19.17 The electric field lines for a point charge. Notice that the figures show only those field lines that lie in the plane of the page.

The number of field lines leaving the positive charge equals the number terminating at the negative charge.

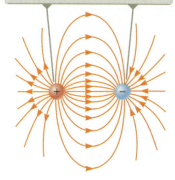

Figure 19.18 The electric field lines for two charges of equal magnitude and opposite sign (an electric dipole).

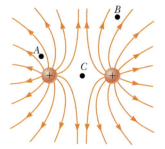

Figure 19.19 The electric field lines for two positive point charges. (The locations A, B, and C are discussed in Quick Quiz 19.5.)

Two field lines leave $+2q$ for every one that terminates on $-q$.

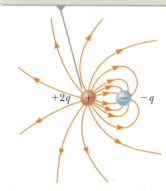

Active Figure 19.20 The electric field lines for a point charge $+2q$ and a second point charge $-q$.

symmetry, we see that the magnitude of the electric field is the same everywhere on the surface of the sphere. The number of lines N emerging from the charge is equal to the number penetrating the spherical surface. Hence, the number of lines per unit area on the sphere is $N/4\pi r^2$ (where the surface area of the sphere is $4\pi r^2$). Because E is proportional to the number of lines per unit area, we see that E varies as $1/r^2$. This result is consistent with that obtained from Equation 19.5; that is, $E = k_e q / r^2$.

The rules for drawing electric field lines for any charge distribution are as follows:

- The lines must begin on a positive charge and terminate on a negative charge. In the case of an excess of one type of charge, some lines will begin or end infinitely far away.
- The number of lines drawn leaving a positive charge or approaching a negative charge is proportional to the magnitude of the charge.
- No two field lines can cross.

Because charge is quantized, the number of lines leaving any positively charged object must be 0, ae, $2ae$, . . ., where a is an arbitrary (but fixed) proportionality constant chosen by the person drawing the lines. Once a is chosen, the number of lines is no longer arbitrary. For example, if object 1 has charge Q_1 and object 2 has charge Q_2, the ratio of the number of lines connected to object 2 to those connected to object 1 is $N_2/N_1 = Q_2/Q_1$.

The electric field lines for two point charges of equal magnitude but opposite signs (the electric dipole) are shown in Figure 19.18. In this case, the number of lines that begin at the positive charge must equal the number that terminate at the negative charge. At points very near the charges, the lines are nearly radial. The high density of lines between the charges indicates a region of strong electric field. The attractive nature of the force between the particles is also suggested by Figure 19.18, with the lines from one particle ending on the other particle.

Figure 19.19 shows the electric field lines in the vicinity of two equal positive point charges. Again, close to either charge the lines are nearly radial. The same number of lines emerges from each particle, because the charges are equal in magnitude, and end on hypothetical charges infinitely far away. At great distances from the particles, the field is approximately equal to that of a single point charge of magnitude $2q$. The repulsive nature of the electric force between particles of like charge is suggested in the figure in that no lines connect the particles and that the lines bend away from the region between the charges.

Finally, we sketch the electric field lines associated with a positive point charge $+2q$ and a negative point charge $-q$ in Active Figure 19.20. In this case, the number of lines leaving $+2q$ is twice the number terminating on $-q$. Hence, only half the lines that leave the positive charge end at the negative charge. The remaining half terminate on hypothetical negative charges infinitely far away. At large distances from the particles (large compared with the particle separation), the electric field lines are equivalent to those of a single point charge $+q$.

QUICK QUIZ 19.5 Rank the magnitudes of the electric field at points *A*, *B*, and *C* in Figure 19.19 (greatest magnitude first).

19.7 | Motion of Charged Particles in a Uniform Electric Field

When a particle of charge q and mass m is placed in an electric field $\vec{\mathbf{E}}$, the electric force exerted on the charge is given by Equation 19.4, $\vec{\mathbf{F}}_e = q\vec{\mathbf{E}}$. If this force is the only force exerted on the particle, it is the net force. If other forces also act on the particle, the electric force is simply added to the other forces vectorially to determine the net force. According to the particle under a net force model from

Chapter 4, the net force causes the particle to accelerate. If the electric force is the only force on the particle, Newton's second law applied to the particle gives

$$\vec{F}_e = q\vec{E} = m\vec{a}$$

The acceleration of the particle is therefore

$$\vec{a} = \frac{q\vec{E}}{m}$$ **19.11** ◀

If $\vec{E}$ is uniform (i.e., constant in magnitude and direction), the acceleration is constant and the particle under constant acceleration analysis model can be used to describe the motion of the particle. If the particle has a positive charge, its acceleration is in the direction of the electric field. If the particle has a negative charge, its acceleration is in the direction opposite the electric field.

Example 19.7 | An Accelerating Positive Charge: Two Models

A uniform electric field $\vec{E}$ is directed along the x axis between parallel plates of charge separated by a distance d as shown in Figure 19.21. A positive point charge q of mass m is released from rest at a point Ⓐ next to the positive plate and accelerates to a point Ⓑ next to the negative plate.

(A) Find the speed of the particle at Ⓑ by modeling it as a particle under constant acceleration.

SOLUTION

Conceptualize When the positive charge is placed at Ⓐ, it experiences an electric force toward the right in Figure 19.21 due to the electric field directed toward the right.

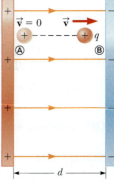

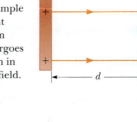

Figure 19.21 (Example 19.7) A positive point charge q in a uniform electric field $\vec{E}$ undergoes constant acceleration in the direction of the field.

Categorize Because the electric field is uniform, a constant electric force acts on the charge. Therefore, as suggested in the problem statement, the point charge can be modeled as a charged particle under constant acceleration.

Analyze Use Equation 2.14 to express the velocity of the particle as a function of position:

$$v_f^2 = v_i^2 + 2a(x_f - x_i) = 0 + 2a(d - 0) = 2ad$$

Solve for v_f and substitute for the magnitude of the acceleration from Equation 19.11:

$$v_f = \sqrt{2ad} = \sqrt{2\left(\frac{qE}{m}\right)d} = \sqrt{\frac{2qEd}{m}}$$

(B) Find the speed of the particle at Ⓑ by modeling it as a nonisolated system.

SOLUTION

Categorize The problem statement tells us that the charge is a nonisolated system. Energy is transferred to this charge by work done by the electric force exerted on the charge. The initial configuration of the system is when the particle is at Ⓐ, and the final configuration is when it is at Ⓑ.

Analyze Write the appropriate reduction of the conservation of energy equation, Equation 7.2, for the system of the charged particle:

$$W = \Delta K$$

Replace the work and kinetic energies with values appropriate for this situation:

$$F_e \Delta x = K_Ⓑ - K_Ⓐ = \tfrac{1}{2}mv_f^2 - 0 \quad \rightarrow \quad v_f = \sqrt{\frac{2F_e \Delta x}{m}}$$

Substitute for the electric force F_e and the displacement Δx:

$$v_f = \sqrt{\frac{2(qE)(d)}{m}} = \sqrt{\frac{2qEd}{m}}$$

Finalize The answer to part (B) is the same as that for part (A), as we expect.

Example **19.8** | **An Accelerated Electron**

An electron enters the region of a uniform electric field as shown in Active Figure 19.22, with $v_i = 3.00 \times 10^6$ m/s and $E = 200$ N/C. The horizontal length of the plates is $\ell = 0.100$ m.

(A) Find the acceleration of the electron while it is in the electric field.

SOLUTION

Conceptualize This example differs from the preceding one because the velocity of the charged particle is initially perpendicular to the electric field lines. (In Example 19.7, the velocity of the charged particle is always parallel to the electric field lines.) As a result, the electron in this example follows a curved path as shown in Active Figure 19.22.

Categorize Because the electric field is uniform, a constant electric force is exerted on the electron. To find the acceleration of the electron, we can model it as a particle under a net force.

· ·

Analyze The direction of the electron's acceleration is downward in Active Figure 19.22, opposite the direction of the electric field lines.

The particle under a net force model was used to develop Equation 19.11 in the case in which the electric force on a particle is the only force. Use this equation to evaluate the *y* component of the acceleration of the electron:

$$a_y = -\frac{eE}{m_e}$$

Substitute numerical values:

$$a_y = -\frac{(1.60 \times 10^{-19}\ \text{C})(200\ \text{N/C})}{9.11 \times 10^{-31}\ \text{kg}} = -3.51 \times 10^{13}\ \text{m/s}^2$$

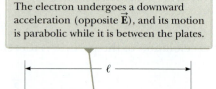

The electron undergoes a downward acceleration (opposite $\vec{E}$), and its motion is parabolic while it is between the plates.

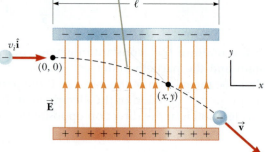

Active Figure 19.22 (Example 19.8) An electron is projected horizontally into a uniform electric field produced by two charged plates.

(B) Assuming the electron enters the field at time $t = 0$, find the time at which it leaves the field.

SOLUTION

Categorize Because the electric force acts only in the vertical direction in Active Figure 19.22, the motion of the particle in the horizontal direction can be analyzed by modeling it as a particle under constant velocity.

· ·

Analyze Solve Equation 2.5 for the time at which the electron arrives at the right edges of the plates:

$$x_f = x_i + v_x t \;\rightarrow\; t = \frac{x_f - x_i}{v_x}$$

Substitute numerical values:

$$t = \frac{\ell - 0}{v_x} = \frac{0.100\ \text{m}}{3.00 \times 10^6\ \text{m/s}} = 3.33 \times 10^{-8}\ \text{s}$$

· ·

Finalize We have neglected the gravitational force acting on the electron, which represents a good approximation when dealing with atomic particles. For an electric field of 200 N/C, the ratio of the magnitude of the electric force eE to the magnitude of the gravitational force mg is on the order of 10^{12} for an electron and on the order of 10^9 for a proton.

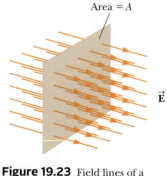

Area = A

Figure 19.23 Field lines of a uniform electric field penetrating a plane of area A perpendicular to the field. The electric flux Φ_E through this area is equal to EA.

19.8 | Electric Flux

Now that we have described the concept of electric field lines qualitatively, let us use a new concept, *electric flux,* to approach electric field lines on a quantitative basis. Electric flux is a quantity proportional to the number of electric field lines penetrating some surface. (We can define only a proportionality because the number of lines we choose to draw is arbitrary.)

First consider an electric field that is uniform in both magnitude and direction as in Figure 19.23. The field lines penetrate a plane rectangular surface of area A, which is perpendicular to the field. Recall that the number of lines per unit area is proportional to the magnitude of the electric field. The number of lines penetrating the surface of area A is therefore proportional to the product EA. The product of the electric field magnitude E and a surface area A perpendicular to the field is called the **electric flux** Φ_E:

$$\Phi_E \equiv EA \qquad \text{19.12} \blacktriangleleft$$

From the SI units of E and A, we see that electric flux has the units $\text{N} \cdot \text{m}^2/\text{C}$.

If the surface under consideration is not perpendicular to the field, the number of lines through it must be less than that given by Equation 19.12. This concept can be understood by considering Figure 19.24, where the normal to the surface of area A is at an angle of θ to the uniform electric field. Note that the number of lines that cross this area is equal to the number that cross the projected area $A_\perp$, which is perpendicular to the field. From Figure 19.24, we see that the two areas are related by $A_\perp = A \cos \theta$. Because the flux through area A equals the flux through $A_\perp$, we conclude that the desired flux is

$$\Phi_E = EA \cos \theta \qquad \text{19.13} \blacktriangleleft$$

From this result, we see that the flux through a surface of fixed area has the maximum value EA when the angle θ between the normal to the surface and the electric field is zero. This situation occurs when the normal is parallel to the field and the surface is perpendicular to the field. The flux is zero when the surface is parallel to the field because the angle θ in Equation 19.13 is then $90°$.

In more general situations, the electric field may vary in both magnitude and direction over the surface in question. Unless the field is uniform, our definition of flux given by Equation 19.13 therefore has meaning only over a small element of area. Consider a general surface divided up into a large number of small elements, each of area ΔA. The variation in the electric field over the element can be ignored if the element is small enough. It is convenient to define a vector $\Delta \vec{A}_i$ whose magnitude represents the area of the ith element and whose direction is defined to be perpendicular to the surface as in Figure 19.25. The electric flux $\Delta \Phi_E$ through this small element is

$$\Delta \Phi_E = E_i \, \Delta A_i \cos \theta_i = \vec{E}_i \cdot \Delta \vec{A}_i$$

where we have used the definition of the scalar product of two vectors $(\vec{A} \cdot \vec{B} = AB \cos \theta)$. By summing the contributions of all elements, we obtain the total flux through the surface. If we let the area of each element approach zero, the number of elements approaches infinity and the sum is replaced by an integral. The general definition of electric flux is therefore

$$\Phi_E \equiv \lim_{\Delta A_i \to 0} \sum \vec{E}_i \cdot \Delta \vec{A}_i = \int_{\text{surface}} \vec{E} \cdot d\vec{A} \qquad \text{19.14} \blacktriangleleft \qquad \blacktriangleright \text{Electric flux}$$

Equation 19.14 is a surface integral, which must be evaluated over the surface in question. In general, the value of Φ_E depends both on the field pattern and on the specified surface.

We shall often be interested in evaluating electric flux through a *closed surface*. A closed surface is defined as one that completely divides space into an inside region and an outside region so that movement cannot take place from one region to the other without penetrating the surface. This definition is similar to that of the system boundary in system models, in which the boundary divides space into a region inside the system and the outer region, the environment. The surface of a sphere is an example of a closed surface, whereas a drinking glass is an open surface.

Consider the closed surface in Active Figure 19.26 (page 638). Note that the vectors $\Delta \vec{A}_i$ point in different directions for the various surface elements. At each point, these vectors are *perpendicular* to the surface and, by convention, always point *outward* from the inside region. At the element labeled ①, $\vec{E}$ is outward and $\theta_i < 90°$; hence, the flux $\Delta \Phi_E = \vec{E} \cdot \Delta \vec{A}_i$ through this element is positive. For element ②, the field lines graze the surface (perpendicular to the vector $\Delta \vec{A}_i$); therefore, $\theta_i = 90°$ and the

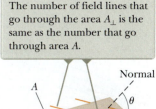

The number of field lines that go through the area $A_\perp$ is the same as the number that go through area A.

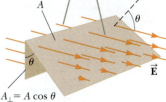

$A_\perp = A \cos \theta$

Figure 19.24 Field lines representing a uniform electric field penetrating an area A that is at an angle θ to the field.

The electric field makes an angle θ_i with the vector $\Delta \vec{A}_i$, defined as being normal to the surface element.

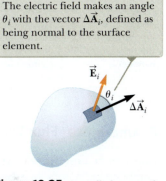

Figure 19.25 A small element of a surface of area ΔA_i.

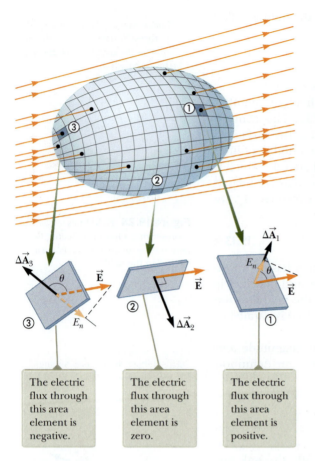

The electric flux through this area element is negative.

The electric flux through this area element is zero.

The electric flux through this area element is positive.

Active Figure 19.26 A closed surface in an electric field. The area vectors are, by convention, normal to the surface and point outward.

flux is zero. For elements such as ③, where the field lines are crossing the surface from the outside to the inside, $180° > \theta_i > 90°$ and the flux is negative because $\cos \theta_i$ is negative.

The net flux through the surface is proportional to the net number of lines penetrating the surface, where the net number means the number leaving the volume surrounded by the surface minus the number entering the volume. If more lines are leaving the surface than entering, the net flux is positive. If more lines enter than leave the surface, the net flux is negative. Using the symbol $\oint$ to represent an integral over a closed surface, we can write the net flux Φ_E through a closed surface as

$$\Phi_E = \oint \vec{E} \cdot d\vec{A} = \oint E_n \, dA \qquad \textbf{19.15} \blacktriangleleft$$

where E_n represents the component of the electric field normal to the surface.

Evaluating the net flux through a closed surface can be very cumbersome. If the field is perpendicular or parallel to the surface at each point and constant in magnitude, however, the calculation is straightforward. The following example illustrates this point.

Example 19.9 | Flux Through a Cube

Consider a uniform electric field $\vec{E}$ oriented in the x direction in empty space. A cube of edge length ℓ is placed in the field, oriented as shown in Figure 19.27. Find the net electric flux through the surface of the cube.

SOLUTION

Conceptualize Examine Figure 19.27 carefully. Notice that the electric field lines pass through two faces perpendicularly and are parallel to four other faces of the cube.

Categorize We evaluate the flux from its definition, so we categorize this example as a substitution problem.

The flux through four of the faces (③, ④, and the unnumbered faces) is zero because $\vec{E}$ is parallel to the four faces and therefore perpendicular to $d\vec{A}$ on these faces.

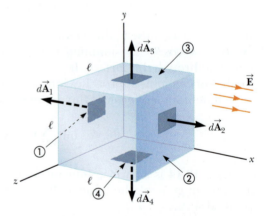

Figure 19.27 (Example 19.9) A closed surface in the shape of a cube in a uniform electric field oriented parallel to the x axis. Side ④ is the bottom of the cube, and side ① is opposite side ②.

Write the integrals for the net flux through faces ① and ②:

$$\Phi_E = \int_1 \vec{E} \cdot d\vec{A} + \int_2 \vec{E} \cdot d\vec{A}$$

For face ①, $\vec{E}$ is constant and directed inward but $d\vec{A}_1$ is directed outward ($\theta = 180°$). Find the flux through this face:

$$\int_1 \vec{E} \cdot d\vec{A} = \int_1 E(\cos 180°) \, dA = -E \int_1 dA = -EA = -E\ell^2$$

For face ②, $\vec{E}$ is constant and outward and in the same direction as $d\vec{A}_2$ ($\theta = 0°$). Find the flux through this face:

$$\int_2 \vec{E} \cdot d\vec{A} = \int_2 E(\cos 0°) \, dA = E \int_2 dA = +EA = E\ell^2$$

Find the net flux by adding the flux over all six faces:

$$\Phi_E = -E\ell^2 + E\ell^2 + 0 + 0 + 0 + 0 = 0$$

19.9 | Gauss's Law

In this section, we describe a general relation between the net electric flux through a closed surface and the charge *enclosed* by the surface. This relation, known as **Gauss's law,** is of fundamental importance in the study of electrostatic fields.

First, let us consider a positive point charge q located at the center of a spherical surface of radius r as in Figure 19.28. The field lines radiate outward and hence are perpendicular (or normal) to the surface at each point. That is, at each point on the surface, $\vec{E}$ is parallel to the vector $\Delta\vec{A}_i$ representing the local element of area ΔA_i. Therefore, at all points on the surface,

$$\vec{E} \cdot \Delta\vec{A}_i = E_n \Delta A_i = E \Delta A_i$$

and, from Equation 19.15, we find that the net flux through the surface is

$$\Phi_E = \oint E_n \, dA = \oint E \, dA = E \oint dA = EA$$

because E is constant over the surface. From Equation 19.5, we know that the magnitude of the electric field everywhere on the surface of the sphere is $E = k_e q/r^2$. Furthermore, for a spherical surface, $A = 4\pi r^2$ (the surface area of a sphere). Hence, the net flux through the surface is

$$\Phi_E = EA = \left(\frac{k_e q}{r^2}\right)(4\pi r^2) = 4\pi k_e q$$

Recalling that $k_e = 1/4\pi\epsilon_0$, we can write this expression in the form

$$\Phi_E = \frac{q}{\epsilon_0} \qquad\qquad \textbf{19.16} \blacktriangleleft$$

This result, which is independent of r, says that the net flux through a spherical surface is proportional to the charge q at the center *inside* the surface. This result mathematically represents that (1) the net flux is proportional to the number of field lines, (2) the number of field lines is proportional to the charge inside the surface, and (3) every field line from the charge must pass through the surface. That the net flux is independent of the radius is a consequence of the inverse-square dependence of the electric field given by Equation 19.5. That is, E varies as $1/r^2$, but the area of the sphere varies as r^2. Their combined effect produces a flux that is independent of r.

Now consider several closed surfaces surrounding a charge q as in Figure 19.29. Surface S_1 is spherical, whereas surfaces S_2 and S_3 are nonspherical. The flux that passes through surface S_1 has the value q/ϵ_0. As we discussed in Section 19.8, the flux is proportional to the number of electric field lines passing through that surface. The construction in Figure 19.29 shows that the number of electric field lines through the spherical surface S_1 is equal to the number of electric field lines through the nonspherical surfaces S_2 and S_3. It is therefore reasonable to conclude that the net flux through any closed surface is independent of the shape of that surface. (One can prove that conclusion using $E \propto 1/r^2$.) In fact,

the net flux through any closed surface surrounding the point charge q is given by q/ϵ_0 and is independent of the position of the charge within the surface.

Now consider a point charge located *outside* a closed surface of arbitrary shape as in Figure 19.30. As you can see from this construction, electric field lines enter the surface and then leave it. Therefore, the number of electric field lines entering the surface equals the number leaving the surface. Consequently, we conclude that the net electric flux through a closed surface that surrounds no net charge is zero. If we apply this result to Example 19.9, we see that the net flux through the cube is zero because there was no charge inside the cube. If there were charge in the cube, the electric field could not be uniform throughout the cube as specified in the example.

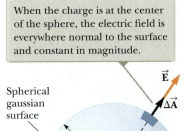

When the charge is at the center of the sphere, the electric field is everywhere normal to the surface and constant in magnitude.

Figure 19.28 A spherical surface of radius r surrounding a point charge q.

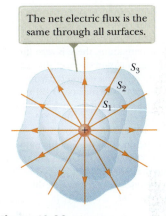

The net electric flux is the same through all surfaces.

Figure 19.29 Closed surfaces of various shapes surrounding a positive charge.

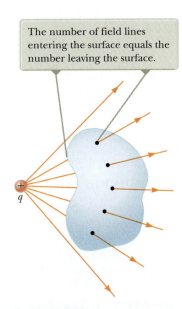

The number of field lines entering the surface equals the number leaving the surface.

Figure 19.30 A point charge located *outside* a closed surface.

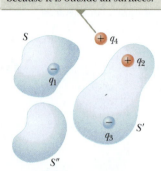

Charge q_4 does not contribute to the flux through any surface because it is outside all surfaces.

Active Figure 19.31 The net electric flux through any closed surface depends only on the charge *inside* that surface. The net flux through surface S is q_1/ϵ_0, the net flux through surface S' is $(q_2 + q_3)/\epsilon_0$, and the net flux through surface S'' is zero.

Pitfall Prevention | 19.3
Zero Flux Is Not Zero Field
In two situations, there is zero flux through a closed surface: either (1) there are no charged particles enclosed by the surface or (2) there are charged particles enclosed, but the net charge inside the surface is zero. For either situation, it is *incorrect* to conclude that the electric field on the surface is zero. Gauss's law states that the electric *flux* is proportional to the enclosed charge, not the electric *field*.

Let us extend these arguments to the generalized case of many point charges. We shall again make use of the superposition principle. That is, we can express the net flux through any closed surface as

$$\oint \vec{E} \cdot d\vec{A} = \oint (\vec{E}_1 + \vec{E}_2 + \cdots) \cdot d\vec{A}$$

where $\vec{E}$ is the total electric field at any point on the surface and $\vec{E}_1, \vec{E}_2, \ldots$ are the fields produced by the individual charges at that point. Consider the system of charges shown in Active Figure 19.31. The surface S surrounds only one charge, q_1; hence, the net flux through S is q_1/ϵ_0. The flux through S due to the charges outside it is zero because each electric field line from these charges that enters S at one point leaves it at another. The surface S' surrounds charges q_2 and q_3; hence, the net flux through S' is $(q_2 + q_3)/\epsilon_0$. Finally, the net flux through surface S'' is zero because no charge exists inside this surface. That is, *all* electric field lines that enter S'' at one point leave S'' at another. Notice that charge q_4 does not contribute to the net flux through any of the surfaces because it is outside all the surfaces.

Gauss's law, which is a generalization of the foregoing discussion, states that the net flux through *any* closed surface is

$$\Phi_E = \oint \vec{E} \cdot d\vec{A} = \frac{q_{in}}{\epsilon_0} \qquad \textbf{19.17} \blacktriangleleft$$

where q_{in} represents the *net charge inside the surface* and $\vec{E}$ represents the electric field at any point on the surface. In words, Gauss's law states that the net electric flux through any closed surface is equal to the net charge inside the surface divided by ϵ_0. The closed surface used in Gauss's law is called a **gaussian surface.**

Gauss's law is valid for the electric field of any system of charges or continuous distribution of charge. In practice, however, the technique is useful for calculating the electric field only in situations where the degree of symmetry is high. As we shall see in the next section, Gauss's law can be used to evaluate the electric field for charge distributions that have spherical, cylindrical, or plane symmetry. We do so by choosing an appropriate gaussian surface that allows $\vec{E}$ to be removed from the integral in Gauss's law and performing the integration. Note that a gaussian surface is a mathematical surface and need not coincide with any real physical surface.

QUICK QUIZ 19.6 If the net flux through a gaussian surface is zero, the following four statements could be true. Which of the statements must be true? **(a)** There are no charges inside the surface. **(b)** The net charge inside the surface is zero. **(c)** The electric field is zero everywhere on the surface. **(d)** The number of electric field lines entering the surface equals the number leaving the surface.

QUICK QUIZ 19.7 Consider the charge distribution shown in Active Figure 19.31. **(i)** What are the charges contributing to the total electric *flux* through surface S'? **(a)** q_1 only **(b)** q_4 only **(c)** q_2 and q_3 **(d)** all four charges **(e)** none of the charges **(ii)** What are the charges contributing to the total electric *field* at a chosen point on the surface S'? **(a)** q_1 only **(b)** q_4 only **(c)** q_2 and q_3 **(d)** all four charges **(e)** none of the charges

---

THINKING PHYSICS 19.1

A spherical gaussian surface surrounds a point charge q. Describe what happens to the net flux through the surface if (a) the charge is tripled, (b) the volume of the sphere is doubled, (c) the surface is changed to a cube, and (d) the charge is moved to another location *inside* the surface.

Reasoning (a) If the charge is tripled, the flux through the surface is also tripled because the net flux is proportional to the charge inside the surface. (b) The net flux remains constant when the volume changes because the surface surrounds the same amount of charge, regardless of its volume. (c) The net flux does not change when the shape of the closed surface changes. (d) The net flux through the closed surface remains unchanged as the charge inside the surface is moved to another location as long as the new location remains inside the surface. ◄

19.10 | Application of Gauss's Law to Various Charge Distributions

As mentioned earlier, Gauss's law is useful in determining electric fields when the charge distribution has a high degree of symmetry. The following examples show ways of choosing the gaussian surface over which the surface integral given by Equation 19.17 can be simplified and the electric field determined. The surface should always be chosen to take advantage of the symmetry of the charge distribution so that we can remove E from the integral and solve for it. The crucial step in applying Gauss's law is to determine a useful gaussian surface. Such a surface should be a closed surface for which each portion of the surface satisfies one or more of the following conditions:

1. The value of the electric field can be argued by symmetry to be constant over the portion of the surface.
2. The dot product in Equation 19.17 can be expressed as a simple algebraic product $E \, dA$ because $\vec{\mathbf{E}}$ and $d\vec{\mathbf{A}}$ are parallel.
3. The dot product in Equation 19.17 is zero because $\vec{\mathbf{E}}$ and $d\vec{\mathbf{A}}$ are perpendicular.
4. The electric field is zero over the portion of the surface.

Note that different portions of the gaussian surface can satisfy different conditions as long as every portion satisfies at least one condition. We will see all four of these conditions used in the examples and discussions in the remainder of this chapter. If the charge distribution does not have sufficient symmetry such that a gaussian surface that satisfies these conditions can be found, Gauss's law is not useful for determining the electric field for that charge distribution.

Example 19.10 | A Spherically Symmetric Charge Distribution

An insulating solid sphere of radius a has a uniform volume charge density ρ and carries a total positive charge Q (Fig. 19.32).

(A) Calculate the magnitude of the electric field at a point outside the sphere.

SOLUTION

Conceptualize Notice how this problem differs from our previous discussion of Gauss's law. The electric field due to point charges was discussed in Section 19.9. Now we are considering the electric field due to a distribution of charge. We found the field for various distributions of charge in Section 19.5 by integrating over the distribution. This example demonstrates a difference from our discussions in Section 19.5. In this example, we find the electric field using Gauss's law.

Categorize Because the charge is distributed uniformly throughout the sphere, the charge distribution has spherical symmetry and we can apply Gauss's law to find the electric field.

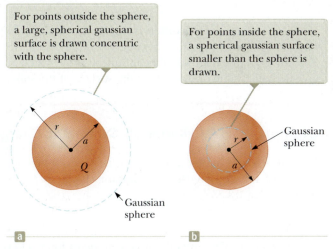

For points outside the sphere, a large, spherical gaussian surface is drawn concentric with the sphere.

For points inside the sphere, a spherical gaussian surface smaller than the sphere is drawn.

Figure 19.32 (Example 19.10) A uniformly charged insulating sphere of radius a and total charge Q. In diagrams such as this one, the dotted line represents the intersection of the gaussian surface with the plane of the page.

Analyze To reflect the spherical symmetry, let's choose a spherical gaussian surface of radius r, concentric with the sphere, as shown in Figure 19.32a. For this choice, condition (2) is satisfied everywhere on the surface and $\vec{\mathbf{E}} \cdot d\vec{\mathbf{A}} = E \, dA$.

Replace $\vec{\mathbf{E}} \cdot d\vec{\mathbf{A}}$ in Gauss's law with $E \, dA$:

$$\Phi_E = \oint \vec{\mathbf{E}} \cdot d\vec{\mathbf{A}} = \oint E \, dA = \frac{Q}{\epsilon_0}$$

continued

19.10 *cont.*

By symmetry, E is constant everywhere on the surface, which satisfies condition (1), so we can remove E from the integral:

$$\oint E \, dA = E \oint dA = E(4\pi r^2) = \frac{Q}{\epsilon_0}$$

Solve for E:

$$(1) \quad E = \frac{Q}{4\pi\epsilon_0 r^2} = k_e \frac{Q}{r^2} \quad (\text{for } r > a)$$

Finalize This field is identical to that for a point charge. Therefore, **the electric field due to a uniformly charged sphere in the region external to the sphere is *equivalent* to that of a point charge located at the center of the sphere.**

(B) Find the magnitude of the electric field at a point inside the sphere.

SOLUTION

Analyze In this case, let's choose a spherical gaussian surface having radius $r < a$, concentric with the insulating sphere (Fig. 19.32b). Let V' be the volume of this smaller sphere. To apply Gauss's law in this situation, recognize that the charge q_{in} within the gaussian surface of volume V' is less than Q.

Calculate q_{in} by using $q_{in} = \rho V'$:

$$q_{in} = \rho V' = \rho\left(\tfrac{4}{3}\pi r^3\right)$$

Notice that conditions (1) and (2) are satisfied everywhere on the gaussian surface in Figure 19.32b. Apply Gauss's law in the region $r < a$:

$$\oint E \, dA = E \oint dA = E(4\pi r^2) = \frac{q_{in}}{\epsilon_0}$$

Solve for E and substitute for q_{in}:

$$E = \frac{q_{in}}{4\pi\epsilon_0 r^2} = \frac{\rho\left(\tfrac{4}{3}\pi r^3\right)}{4\pi\epsilon_0 r^2} = \frac{\rho}{3\epsilon_0} r$$

Substitute $\rho = Q/\tfrac{4}{3}\pi a^3$ and $\epsilon_0 = 1/4\pi k_e$:

$$(2) \quad E = \frac{Q/\tfrac{4}{3}\pi a^3}{3(1/4\pi k_e)} r = k_e \frac{Q}{a^3} r \quad (\text{for } r < a)$$

Finalize This result for E differs from the one obtained in part (A). It shows that $E \rightarrow 0$ as $r \rightarrow 0$. Therefore, the result eliminates the problem that would exist at $r = 0$ if E varied as $1/r^2$ inside the sphere as it does outside the sphere. That is, if $E \propto 1/r^2$ for $r < a$, the field would be infinite at $r = 0$, which is physically impossible. Notice also that Equations (1) and (2) both give the same value of the field at the surface of the sphere ($r = a$), showing that the field is continuous.

Example **19.11** | **A Cylindrically Symmetric Charge Distribution**

Find the electric field a distance r from a line of positive charge of infinite length and constant charge per unit length λ (Fig. 19.33a).

SOLUTION

Conceptualize The line of charge is *infinitely* long. Therefore, the field is the same at all points equidistant from the line, regardless of the vertical position of the point in Figure 19.33a.

Categorize Because the charge is distributed uniformly along the line, the charge distribution has cylindrical symmetry and we can apply Gauss's law to find the electric field.

Analyze The symmetry of the charge distribution requires that $\vec{E}$ be perpendicular to the line charge and directed outward as shown in Figure 19.33b. To reflect the symmetry

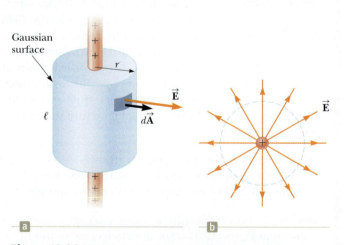

Figure 19.33 (Example 19.11) (a) An infinite line of charge surrounded by a cylindrical gaussian surface concentric with the line. (b) An end view shows that the electric field at the cylindrical surface is constant in magnitude and perpendicular to the surface.

19.11 *cont.*

of the charge distribution, let's choose a cylindrical gaussian surface of radius r and length ℓ that is coaxial with the line charge. For the curved part of this surface, $\vec{\mathbf{E}}$ is constant in magnitude and perpendicular to the surface at each point, satisfying conditions (1) and (2). Furthermore, the flux through the ends of the gaussian cylinder is zero because $\vec{\mathbf{E}}$ is parallel to these surfaces. That is the first application we have seen of condition (3).

We must take the surface integral in Gauss's law over the entire gaussian surface. Because $\vec{\mathbf{E}} \cdot d\vec{\mathbf{A}}$ is zero for the flat ends of the cylinder, however, we restrict our attention to only the curved surface of the cylinder.

Apply Gauss's law and conditions (1) and (2) for the curved surface, noting that the total charge inside our gaussian surface is $\lambda\ell$:

$$\Phi_E = \oint \vec{\mathbf{E}} \cdot d\vec{\mathbf{A}} = E \oint dA = EA = \frac{q_{\text{in}}}{\epsilon_0} = \frac{\lambda\ell}{\epsilon_0}$$

Substitute the area $A = 2\pi r\ell$ of the curved surface:

$$E(2\pi r\ell) = \frac{\lambda\ell}{\epsilon_0}$$

Solve for the magnitude of the electric field:

$$E = \frac{\lambda}{2\pi\epsilon_0 r} = \boxed{2k_e \frac{\lambda}{r}}$$

19.18 ◀

Finalize This result shows that the electric field due to a cylindrically symmetric charge distribution varies as $1/r$, whereas the field external to a spherically symmetric charge distribution varies as $1/r^2$. Equation 19.18 can also be derived by direct integration over the charge distribution. (See Problem 18.)

What If? What if the line segment in this example were not infinitely long?

Answer If the line charge in this example were of finite length, the electric field would not be given by Equation 19.18. A finite line charge does not possess sufficient symmetry to make use of Gauss's law because the magnitude of the electric field is no longer constant over the surface of the gaussian cylinder: the field near the ends of the line would be different from that far from the ends. Therefore, condition (1) would not be satisfied in this situation. Furthermore, $\vec{\mathbf{E}}$ is not perpendicular to the cylindrical surface at all points: the field vectors near the ends would have a component parallel to the line. Therefore, condition (2) would not be satisfied. For points close to a finite line charge and far from the ends, Equation 19.18 gives a good approximation of the value of the field.

It is left for you to show (see Problem 48) that the electric field inside a uniformly charged rod of finite radius and infinite length is proportional to r.

Example 19.12 | A Plane of Charge

Find the electric field due to an infinite plane of positive charge with uniform surface charge density σ.

SOLUTION

Conceptualize Notice that the plane of charge is *infinitely* large. Therefore, the electric field should be the same at all points equidistant from the plane.

Categorize Because the charge is distributed uniformly on the plane, the charge distribution is symmetric; hence, we can use Gauss's law to find the electric field.

Analyze By symmetry, $\vec{\mathbf{E}}$ must be perpendicular to the plane at all points. The direction of $\vec{\mathbf{E}}$ is away from positive charges, indicating that the direction of $\vec{\mathbf{E}}$ on one side of the plane must be opposite its direction on the other side as shown in Figure 19.34. A gaussian surface that reflects the symmetry is a small cylinder whose axis is perpendicular to the plane and whose ends each have an area A and are equidistant from the plane. Because $\vec{\mathbf{E}}$ is parallel to the curved surface of the cylinder—and therefore perpendicular to $d\vec{\mathbf{A}}$ at all points on this surface—condition (3) is satisfied and there is no contribution to the surface integral from this surface. For the flat ends of the cylinder, conditions (1) and (2) are satisfied. The flux through each end of the cylinder is EA; hence, the total flux through the entire gaussian surface is just that through the ends, $\Phi_E = 2EA$.

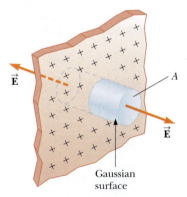

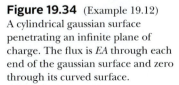

Figure 19.34 (Example 19.12) A cylindrical gaussian surface penetrating an infinite plane of charge. The flux is EA through each end of the gaussian surface and zero through its curved surface.

continued

19.12 *cont.*

Write Gauss's law for this surface, noting that the enclosed charge is $q_{in} = \sigma A$:

$$\Phi_E = 2EA = \frac{q_{in}}{\epsilon_0} = \frac{\sigma A}{\epsilon_0}$$

Solve for E:

$$E = \frac{\sigma}{2\epsilon_0}$$

19.19◄

Finalize Because the distance from each flat end of the cylinder to the plane does not appear in Equation 19.19, we conclude that $E = \sigma/2\epsilon_0$ at *any* distance from the plane. That is, the field is uniform everywhere.

What If? Suppose two infinite planes of charge are parallel to each other, one positively charged and the other negatively charged. Both planes have the same surface charge density. What does the electric field look like in this situation?

Answer The electric fields due to the two planes add in the region between the planes, resulting in a uniform field of magnitude σ/ϵ_0, and cancel elsewhere to give a field of zero. This method is a practical way to achieve uniform electric fields with finite-sized planes placed close to each other.

� 19.11 | Conductors in Electrostatic Equilibrium

A good electrical conductor, such as copper, contains charges (electrons) that are not bound to any atom and are free to move about within the material. When no motion of charge occurs within the conductor (other than thermal motion), the conductor is in **electrostatic equilibrium.** As we shall see, an isolated conductor (one that is insulated from ground) in electrostatic equilibrium has the following properties:

1. The electric field is zero everywhere inside the conductor, whether the conductor is solid or hollow.
2. If the conductor is isolated and carries a charge, the charge resides on its surface.
3. The electric field at a point just outside a charged conductor is perpendicular to the surface of the conductor and has a magnitude σ/ϵ_0, where σ is the surface charge density at that point.
4. On an irregularly shaped conductor, the surface charge density is greatest at locations where the radius of curvature of the surface is smallest.

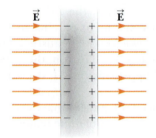

Figure 19.35 A conducting slab in an external electric field $\vec{E}$. The charges induced on the two surfaces of the slab produce an electric field that opposes the external field, giving a resultant field of zero *inside* the slab.

We will verify the first three properties in the following discussion. The fourth property is presented here so that we have a complete list of properties for conductors in electrostatic equilibrium. The verification of it, however, requires concepts from Chapter 20, so we will postpone its verification until then.

The first property can be understood by considering a conducting slab placed in an external field $\vec{E}$ (Fig. 19.35). The electric field inside the conductor *must* be zero under the assumption that we have electrostatic equilibrium. If the field were not zero, free charges in the conductor would accelerate under the action of the electric force. This motion of electrons, however, would mean that the conductor is not in electrostatic equilibrium. Therefore, the existence of electrostatic equilibrium is consistent only with a zero field in the conductor.

Let us investigate how this zero field is accomplished. Before the external field is applied, free electrons are uniformly distributed throughout the conductor. When the external field is applied, the free electrons accelerate to the left in Figure 19.35, causing a plane of negative charge to be present on the left surface. The movement of electrons to the left results in a plane of positive charge on the right surface. These planes of charge create an additional electric field inside the conductor that opposes the external field. As the electrons move, the surface charge density increases until the magnitude of the internal field equals that of the external field, giving a net field of zero inside the conductor.

We can use Gauss's law to verify the second property of a conductor in electrostatic equilibrium. Figure 19.36 shows an arbitrarily shaped conductor. A gaussian surface is drawn just inside the conductor and can be as close to the surface as we wish. As we have

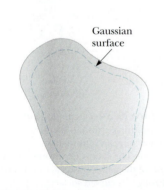

Gaussian surface

Figure 19.36 A conductor of arbitrary shape. The broken line represents a gaussian surface that can be just inside the conductor's surface.

just shown, the electric field everywhere inside a conductor in electrostatic equilibrium is zero. Therefore, the electric field must be zero at every point on the gaussian surface (condition 4 in Section 19.10). From this result and Gauss's law, we conclude that the net charge inside the gaussian surface is zero. Because there can be no net charge inside the gaussian surface (which is arbitrarily close to the conductor's surface), any net charge on the conductor must reside on its surface. Gauss's law does not tell us how this excess charge is distributed on the surface, only that it must reside on the surface.

Conceptually, we can understand the location of the charges on the surface by imagining placing many charges at the center of the conductor. The mutual repulsion of the charges causes them to move apart. They will move as far as they can, which is to various points on the surface.

To verify the third property, we can also use Gauss's law. We draw a gaussian surface in the shape of a small cylinder having its end faces parallel to the surface (Fig. 19.37). Part of the cylinder is just outside the conductor and part is inside. The field is normal to the surface because the conductor is in electrostatic equilibrium. If $\vec{E}$ had a component parallel to the surface, an electric force would be exerted on the charges parallel to the surface, free charges would move along the surface, and so the conductor would not be in equilibrium. Therefore, we satisfy condition 3 in Section 19.10 for the curved part of the cylinder in that no flux exists through this part of the gaussian surface because $\vec{E}$ is parallel to this part of the surface. No flux exists through the flat face of the cylinder inside the conductor because $\vec{E} = 0$ (condition 4). Hence, the net flux through the gaussian surface is the flux through the flat face outside the conductor where the field is perpendicular to the surface. Using conditions 1 and 2 for this face, the flux is EA, where E is the electric field just outside the conductor and A is the area of the cylinder's face. Applying Gauss's law to this surface gives

$$\Phi_E = \oint E \, dA = EA = \frac{q_{in}}{\epsilon_0} = \frac{\sigma A}{\epsilon_0}$$

where we have used that $q_{in} = \sigma A$. Solving for E gives

$$E = \frac{\sigma}{\epsilon_0} \qquad\qquad \textbf{19.20}\blacktriangleleft$$

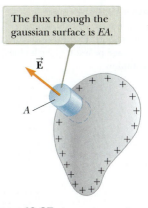

The flux through the gaussian surface is EA.

Figure 19.37 A gaussian surface in the shape of a small cylinder is used to calculate the electric field just outside a charged conductor.

> **THINKING PHYSICS 19.2**

Suppose a point charge $+Q$ is in empty space. We surround the charge with a spherical, uncharged conducting shell so that the charge is at the center of the shell. What effect does that have on the field lines from the charge?

Reasoning When the spherical shell is placed around the charge, the free charges in the shell adjust so as to satisfy the rules for a conductor in equilibrium and Gauss's law. A net charge of $-Q$ moves to the interior surface of the conductor, so the electric field within the conductor is zero (a spherical gaussian surface totally within the shell encloses no *net* charge). A net charge of $+Q$ resides on the outer surface, so a gaussian surface outside the sphere encloses a net charge of $+Q$, just as if the shell were not there. Therefore, the only change in the field lines from the initial situation is the absence of field lines over the thickness of the conducting shell. ◀

❰19.12 | Context Connection: The Atmospheric Electric Field

In this chapter, we discussed the electric field due to various charge distributions. On the surface of the Earth and in the atmosphere, a number of processes create charge distributions, resulting in an electric field in the atmosphere. These processes include cosmic rays entering the atmosphere, radioactive decay at the Earth's surface, and lightning, the focus of our study in this Context.

The result of these processes is an average negative charge distributed over the surface of the Earth of about 5×10^5 C, which is a tremendous amount of charge. (The Earth is neutral overall; the positive charges corresponding to this negative surface charge are spread through the atmosphere, as we shall discuss in

Figure 19.38 A typical tripolar charge distribution in a thundercloud. The dots indicate the average position of each charge distribution.

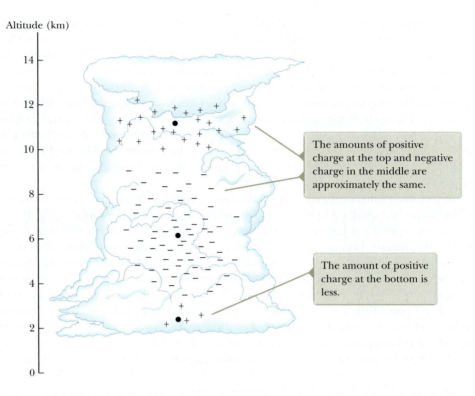

Altitude (km)

The amounts of positive charge at the top and negative charge in the middle are approximately the same.

The amount of positive charge at the bottom is less.

Chapter 20.) We can calculate the average surface charge density over the surface of the Earth:

$$\sigma_{avg} = \frac{Q}{A} = \frac{Q}{4\pi r^2} = \frac{5 \times 10^5 \, \text{C}}{4\pi (6.37 \times 10^6 \, \text{m})^2} \sim 10^{-9} \, \text{C/m}^2$$

Throughout this Context, we will be adopting a number of simplification models. Consequently, we will consider our calculations to be order-of-magnitude estimates of the actual values, as suggested by the $\sim$ sign above.

The Earth is a good conductor. Therefore, we can use the third property of conductors in Section 19.11 to find the average magnitude of the electric field at the surface of the Earth:

$$E_{avg} = \frac{\sigma_{avg}}{\epsilon_0} = \frac{10^{-9} \, \text{C/m}^2}{8.85 \times 10^{-12} \, \text{C}^2/\text{N} \cdot \text{m}^2} \sim 10^2 \, \text{N/C}$$

which is a typical value of the **fair-weather electric field** that exists in the absence of a thunderstorm. The direction of the field is downward because the charge on the Earth's surface is negative. During a thunderstorm, the electric field under the thundercloud is significantly higher than the fair-weather electric field, because of the charge distribution in the thundercloud.

Figure 19.38 shows a typical charge distribution in a thundercloud. The charge distribution can be modeled as a *tripole*, although the positive charge at the bottom of the cloud tends to be smaller than the other two charges. The mechanism of charging in thunderclouds is not well understood and continues to be an active area of research.

It is this high concentration of charge in the thundercloud that is responsible for the very strong electric fields that cause lightning discharge between the cloud and the ground. Typical electric fields during a thunderstorm are as high as 25 000 N/C. The distribution of negative charges in the center of the cloud in Figure 19.38 is the source of negative charge that moves downward in a lightning strike.

Transient Luminous Events

Normal lightning is related to atmospheric electric fields in the troposphere between a thundercloud and the ground. Let us consider the effects of electric fields *above* thunderclouds as shown in Figure 19.39. We find a number of visual effects

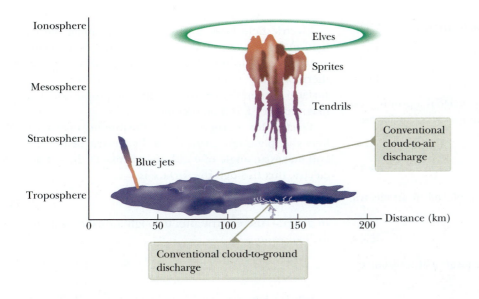

Figure 19.39 A representation of several types of transient luminous events in the atmosphere above thunderclouds.

associated with storms and lightning occurring in this region of the atmosphere. In general, these phenomena are called *transient luminous events*. One type of event is called a *sprite*, which occurs above thunderstorm clouds, with the light from the event originating between 90 and 100 km above the earth's surface. A sprite is triggered by normal tropospheric lightning from the thundercloud below it and appears as a luminous red flash, possibly with vertical tendrils hanging below. These displays last less than a second and are not easily seen with the naked eye. A sprite was first photographed by accident in 1989. Since then, additional photographic evidence has been made available and several astronauts at the International Space Station have reported seeing sprites while they were above a violent storm. Scientists believe that the electric fields at these altitudes are strong enough to ionize air molecules. Red light is released when the electrons recombine with molecular nitrogen ions, in a manner similar to the source of light in a fluorescent lamp.

Other lightning-induced transient luminous events are called *elves*. The light associated with these events lasts for less than 1 ms and has been observed using high speed photometers and CCD cameras. These events, which precede the onset of sprites, appear as expanding halos of light in the ionosphere at altitudes between 75 and 105 km. The expansion of the halos is faster than the speed of light, but there is no violation of the principles of relativity that we studied in Chapter 9 because no particles travel that fast. Current theories relate to an expanding spherical electromagnetic pulse from a lightning strike that interacts with the ionosphere to create the luminous display.

Also seen in Figure 19.39 is an optical event in the stratosphere called a *blue jet*. These displays occur as an upward-propagating ejection from the top of a thundercloud, disappearing at about 40–50 km from the ground. Blue jets are associated with storm clouds but do not appear to be directly triggered by lightning flashes as are the sprites.

Other types of luminous events include *blue starters, trolls, gnomes,* and *pixies*. Research on the origin of transient luminous events is ongoing.

▶ SUMMARY

Electric charges have the following important properties:

1. Two kinds of charges exist in nature, **positive** and **negative,** with the property that charges of opposite sign attract each other and charges of the same sign repel each other.
2. The force between charged particles varies as the inverse square of their separation distance.
3. Charge is conserved.
4. Charge is quantized.

Conductors are materials in which charges move relatively freely. **Insulators** are materials in which charges do not move freely.

Coulomb's law states that the electrostatic force between two stationary, charged particles separated by a distance r has the magnitude

$$F_e = k_e \frac{|q_1||q_2|}{r^2}$$ **19.1** ◀

where the Coulomb constant $k_e = 8.99 \times 10^9 \text{ N} \cdot \text{m}^2/\text{C}^2$. The vector form of Coulomb's law is

$$\vec{F}_{12} = k_e \frac{q_1 q_2}{r^2} \hat{r}_{12} \qquad \textbf{19.2} \blacktriangleleft$$

An **electric field** exists at a point in space if a positive test charge q_0 placed at that point experiences an electric force. The electric field is defined as

$$\vec{E} \equiv \frac{\vec{F}_e}{q_0} \qquad \textbf{19.3} \blacktriangleleft$$

The force on a particle with charge q placed in an electric field $\vec{E}$ is

$$\vec{F}_e = q\vec{E} \qquad \textbf{19.4} \blacktriangleleft$$

The electric field due to the point charge q at a distance r from the charge is

$$\vec{E} = k_e \frac{q}{r^2} \hat{r} \qquad \textbf{19.5} \blacktriangleleft$$

where $\hat{r}$ is a unit vector directed from the charge toward the point in question. The electric field is directed radially outward from a positive charge and is directed toward a negative charge.

The electric field due to a group of charges can be obtained using the superposition principle. That is, the total electric field equals the vector sum of the electric fields of all the charges at some point:

$$\vec{E} = k_e \sum_i \frac{q_i}{r_i^2} \hat{r}_i \qquad \textbf{19.6} \blacktriangleleft$$

Similarly, the electric field of a continuous charge distribution at some point is

$$\vec{E} = k_e \int \frac{dq}{r^2} \hat{r} \qquad \textbf{19.7} \blacktriangleleft$$

where dq is the charge on one element of the charge distribution and r is the distance from the element to the point in question.

Electric field lines are useful for describing the electric field in any region of space. The electric field vector $\vec{E}$ is always tangent to the electric field lines at every point. Furthermore, the number of lines per unit area through a surface perpendicular to the lines is proportional to the magnitude of $\vec{E}$ in that region.

Electric flux is proportional to the number of electric field lines that penetrate a surface. If the electric field is uniform and makes an angle of θ with the normal to the surface, the electric flux through the surface is

$$\Phi_E = EA \cos \theta \qquad \textbf{19.13} \blacktriangleleft$$

In general, the electric flux through a surface is defined by the expression

$$\Phi_E \equiv \int_{\text{surface}} \vec{E} \cdot d\vec{A} \qquad \textbf{19.14} \blacktriangleleft$$

Gauss's law says that the net electric flux Φ_E through any closed gaussian surface is equal to the *net* charge *inside* the surface divided by ϵ_0:

$$\Phi_E = \oint \vec{E} \cdot d\vec{A} = \frac{q_{\text{in}}}{\epsilon_0} \qquad \textbf{19.17} \blacktriangleleft$$

Using Gauss's law, one can calculate the electric field due to various symmetric charge distributions.

A conductor in **electrostatic equilibrium** has the following properties:

1. The electric field is zero everywhere inside the conductor, whether the conductor is solid or hollow.
2. If the conductor is isolated and carries a charge, the charge resides on its surface.
3. The electric field at a point just outside a charged conductor is perpendicular to the surface of the conductor and has a magnitude σ/ϵ_0, where σ is the surface charge density at that point.
4. On an irregularly shaped conductor, the surface charge density is greatest at locations where the radius of curvature of the surface is smallest.

▶ OBJECTIVE QUESTIONS

☐ denotes answer available in *Student Solutions Manual/Study Guide*

1. A point charge of -4.00 nC is located at $(0, 1.00)$ m. What is the x component of the electric field due to the point charge at $(4.00, -2.00)$ m? (a) 1.15 N/C (b) -0.864 N/C (c) 1.44 N/C (d) -1.15 N/C (e) 0.864 N/C

2. Charges of 3.00 nC, -2.00 nC, -7.00 nC, and 1.00 nC are contained inside a rectangular box with length 1.00 m, width 2.00 m, and height 2.50 m. Outside the box are charges of 1.00 nC and 4.00 nC. What is the electric flux through the surface of the box? (a) 0 (b) -5.64×10^2 N $\cdot$ m^2/C (c) -1.47×10^3 N $\cdot$ m^2/C (d) 1.47×10^3 N $\cdot$ m^2/C (e) 5.64×10^2 N $\cdot$ m^2/C

3. An object with negative charge is placed in a region of space where the electric field is directed vertically upward. What is the direction of the electric force exerted on this charge? (a) It is up. (b) It is down. (c) There is no force. (d) The force can be in any direction.

4. A particle with charge q is located inside a cubical gaussian surface. No other charges are nearby. **(i)** If the particle is at the center of the cube, what is the flux through each one of the faces of the cube? (a) 0 (b) $q/2\epsilon_0$ (c) $q/6\epsilon_0$ (d) $q/8\epsilon_0$ (e) depends on the size of the cube **(ii)** If the particle can be moved to any point within the cube, what maximum value can the flux through one face approach? Choose from the same possibilities as in part (i).

5. The magnitude of the electric force between two protons is 2.30×10^{-26} N. How far apart are they? (a) 0.100 m (b) 0.022 0 m (c) 3.10 m (d) 0.005 70 m (e) 0.480 m

6. Estimate the magnitude of the electric field due to the proton in a hydrogen atom at a distance of 5.29×10^{-11} m, the expected position of the electron in the atom. (a) 10^{-11} N/C (b) 10^8 N/C (c) 10^{14} N/C (d) 10^6 N/C (e) 10^{12} N/C

7. Rank the electric fluxes through each gaussian surface shown in Figure OQ19.7 from largest to smallest. Display any cases of equality in your ranking.

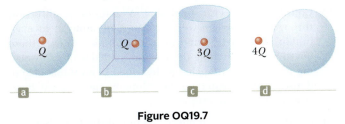

Figure OQ19.7

8. A circular ring of charge with radius b has total charge q uniformly distributed around it. What is the magnitude of the electric field at the center of the ring? (a) 0 (b) $k_e q/b^2$ (c) $k_e q^2/b^2$ (d) $k_e q^2/b$ (e) none of those answers

9. Two solid spheres, both of radius 5 cm, carry identical total charges of 2 μC. Sphere A is a good conductor. Sphere B is an insulator, and its charge is distributed uniformly throughout its volume. **(i)** How do the magnitudes of the electric fields they separately create at a radial distance of 6 cm compare? (a) $E_A > E_B = 0$ (b) $E_A > E_B > 0$ (c) $E_A = E_B > 0$ (d) $0 < E_A < E_B$ (e) $0 = E_A < E_B$ **(ii)** How do the magnitudes of the electric fields they separately create at radius 4 cm compare? Choose from the same possibilities as in part (i).

10. An electron with a speed of 3.00×10^6 m/s moves into a uniform electric field of magnitude 1.00×10^3 N/C. The field lines are parallel to the electron's velocity and pointing in the same direction as the velocity. How far does the electron travel before it is brought to rest? (a) 2.56 cm (b) 5.12 cm (c) 11.2 cm (d) 3.34 m (e) 4.24 m

11. A very small ball has a mass of 5.00×10^{-3} kg and a charge of 4.00 μC. What magnitude electric field directed upward will balance the weight of the ball so that the ball is suspended motionless above the ground? (a) 8.21×10^2 N/C (b) 1.22×10^4 N/C (c) 2.00×10^{-2} N/C (d) 5.11×10^6 N/C (e) 3.72×10^3 N/C

12. In which of the following contexts can Gauss's law *not* be readily applied to find the electric field? (a) near a long, uniformly charged wire (b) above a large, uniformly charged plane (c) inside a uniformly charged ball (d) outside a uniformly charged sphere (e) Gauss's law can be readily applied to find the electric field in all these contexts.

13. Two point charges attract each other with an electric force of magnitude F. If the charge on one of the particles is reduced to one-third its original value and the distance between the particles is doubled, what is the resulting magnitude of the electric force between them? (a) $\frac{1}{12}F$ (b) $\frac{1}{3}F$ (c) $\frac{1}{6}F$ (d) $\frac{3}{4}F$ (e) $\frac{3}{2}F$

14. Three charged particles are arranged on corners of a square as shown in Figure OQ19.14, with charge $-Q$ on both the particle at the upper left corner and the particle at the lower right corner and with charge $+2Q$ on the particle at the lower left corner. **(i)** What is the direction of the electric field at the upper right corner, which is a point in empty space? (a) It is upward and to the right. (b) It is straight to the right. (c) It is straight downward. (d) It is downward and to the left. (e) It is perpendicular to the plane of the picture and outward. **(ii)** Suppose the $+2Q$ charge at the lower left corner is removed. Then does the magnitude of the field at the upper right corner (a) become larger, (b) become smaller, (c) stay the same, or (d) change unpredictably?

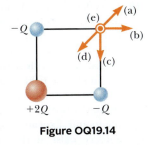

Figure OQ19.14

15. Assume the charged objects in Figure OQ19.15 are fixed. Notice that there is no sight line from the location of q_2 to the location of q_1. If you were at q_1, you would be unable to see q_2 because it is behind q_3. How would you calculate the electric force exerted on the object with charge q_1? (a) Find only the force exerted by q_2 on charge q_1. (b) Find only the force exerted by q_3 on charge q_1. (c) Add the force that q_2 would exert by itself on charge q_1 to the force that q_3 would exert by itself on charge q_1. (d) Add the force that q_3 would exert by itself to a certain fraction of the force that q_2 would exert by itself. (e) There is no definite way to find the force on charge q_1.

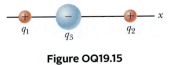

Figure OQ19.15

▶ CONCEPTUAL QUESTIONS

1. A uniform electric field exists in a region of space containing no charges. What can you conclude about the net electric flux through a gaussian surface placed in this region of space?

2. A glass object receives a positive charge by rubbing it with a silk cloth. In the rubbing process, have protons been added to the object or have electrons been removed from it?

3. If more electric field lines leave a gaussian surface than enter it, what can you conclude about the net charge enclosed by that surface?

4. Why must hospital personnel wear special conducting shoes while working around oxygen in an operating room? What might happen if the personnel wore shoes with rubber soles?

5. Would life be different if the electron were positively charged and the proton was negatively charged? (b) Does the choice of signs have any bearing on physical and chemical interactions? Explain your answers.

6. A student who grew up in a tropical country and is studying in the United States may have no experience with static electricity sparks and shocks until his or her first American winter. Explain.

7. If a suspended object A is attracted to a charged object B, can we conclude that A is charged? Explain.

8. A cubical surface surrounds a point charge q. Describe what happens to the total flux through the surface if (a) the charge is doubled, (b) the volume of the cube is doubled, (c) the surface is changed to a sphere, (d) the charge is moved to another location inside the surface, and (e) the charge is moved outside the surface.

9. A person is placed in a large, hollow, metallic sphere that is insulated from ground. (a) If a large charge is placed on the sphere, will the person be harmed upon touching the inside of the sphere? (b) Explain what will happen if the person also has an initial charge whose sign is opposite that of the charge on the sphere.

10. Consider point A in Figure CQ19.10 located an arbitrary distance from two positive point charges in otherwise empty space. (a) Is it possible for an electric field to exist at point A in empty space? Explain. (b) Does charge exist at this point? Explain. (c) Does a force exist at this point? Explain.

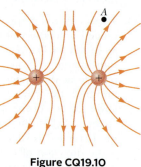

Figure CQ19.10

11. Consider two identical conducting spheres whose surfaces are separated by a small distance. One sphere is given a large net positive charge, and the other is given a small net positive charge. It is found that the force between the spheres is attractive even though they both have net charges of the same sign. Explain how this attraction is possible.

12. If the total charge inside a closed surface is known but the distribution of the charge is unspecified, can you use Gauss's law to find the electric field? Explain.

13. Consider an electric field that is uniform in direction throughout a certain volume. Can it be uniform in magnitude? Must it be uniform in magnitude? Answer these questions (a) assuming the volume is filled with an insulating material carrying charge described by a volume charge density and (b) assuming the volume is empty space. State reasoning to prove your answers.

14. On the basis of the repulsive nature of the force between like charges and the freedom of motion of charge within a conductor, explain why excess charge on an isolated conductor must reside on its surface.

15. A common demonstration involves charging a rubber balloon, which is an insulator, by rubbing it on your hair and then touching the balloon to a ceiling or wall, which is also an insulator. Because of the electrical attraction between the charged balloon and the neutral wall, the balloon sticks to the wall. Imagine now that we have two infinitely large, flat sheets of insulating material. One is charged, and the other is neutral. If these sheets are brought into contact, does an attractive force exist between them as there was for the balloon and the wall?

PROBLEMS

Section 19.2 Properties of Electric Charges

1. Find to three significant digits the charge and the mass of the following particles. *Suggestion:* Begin by looking up the mass of a neutral atom on the periodic table of the elements in Appendix C. (a) an ionized hydrogen atom, represented as H^+ (b) a singly ionized sodium atom, Na^+ (c) a chloride ion Cl^- (d) a doubly ionized calcium atom, $Ca^{++} = Ca^{2+}$ (e) the center of an ammonia molecule, modeled as an N^{3-} ion (f) quadruply ionized nitrogen atoms, N^{4+}, found in plasma in a hot star (g) the nucleus of a nitrogen atom (h) the molecular ion H_2O^-

2. **W** (a) Calculate the number of electrons in a small, electrically neutral silver pin that has a mass of 10.0 g. Silver has 47 electrons per atom, and its molar mass is 107.87 g/mol. (b) Imagine adding electrons to the pin until the negative charge has the very large value 1.00 mC. How many electrons are added for every 10^9 electrons already present?

Section 19.4 Coulomb's Law

3. Nobel laureate Richard Feynman (1918–1988) once said that if two persons stood at arm's length from each other and each person had 1% more electrons than protons, the force of repulsion between them would be enough to lift a "weight" equal to that of the entire Earth. Carry out an order-of-magnitude calculation to substantiate this assertion.

4. Two protons in an atomic nucleus are typically separated by a distance of 2×10^{-15} m. The electric repulsion force between the protons is huge, but the attractive nuclear force is even stronger and keeps the nucleus from bursting apart. What is the magnitude of the electric force between two protons separated by 2.00×10^{-15} m?

5. **W** Two identical conducting small spheres are placed with their centers 0.300 m apart. One is given a charge of 12.0 nC and the other a charge of −18.0 nC. (a) Find the electric force exerted by one sphere on the other. (b) **What If?** The spheres are connected by a conducting wire. Find the electric force each exerts on the other after they have come to equilibrium.

6. *Why is the following situation impossible?* Two identical dust particles of mass 1.00 μg are floating in empty space, far from any external sources of large gravitational or electric fields, and at rest with respect to each other. Both particles carry electric charges that are identical in magnitude and sign. The gravitational and electric forces between the particles happen to have the same magnitude, so each particle experiences zero net force and the distance between the particles remains constant.

7. **Q|C** **W** Two small beads having positive charges $q_1 = 3q$ and $q_2 = q$ are fixed at the opposite ends of a horizontal insulating rod of length $d = 1.50$ m.

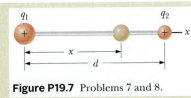

Figure P19.7 Problems 7 and 8.

The bead with charge q_1 is at the origin. As shown in Figure P19.7, a third small, charged bead is free to slide on the rod. (a) At what position x is the third bead in equilibrium? (b) Can the equilibrium be stable?

8. **Q|C** **S** Two small beads having charges q_1 and q_2 of the same sign are fixed at the opposite ends of a horizontal insulating rod of length d. The bead with charge q_1 is at the origin. As shown in Figure P19.7, a third small, charged bead is free to slide on the rod. (a) At what position x is the third bead in equilibrium? (b) Can the equilibrium be stable?

9. **M** Three charged particles are located at the corners of an equilateral triangle as shown in Figure P19.9. Calculate the total electric force on the 7.00-μC charge.

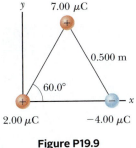

Figure P19.9

10. **GP** Particle A of charge 3.00×10^{-4} C is at the origin, particle B of charge -6.00×10^{-4} C is at (4.00 m, 0), and particle C of charge 1.00×10^{-4} C is at (0, 3.00 m). We wish to find the net electric force on C. (a) What is the x component of the electric force exerted by A on C? (b) What is the y component of the force exerted by A on C? (c) Find the magnitude of the force exerted by B on C. (d) Calculate the x component of the force exerted by B on C. (e) Calculate the y component of the force exerted by B on C. (f) Sum the two x components from parts (a) and (d) to obtain the resultant x component of the electric force acting on C. (g) Similarly, find the y component of the resultant force vector acting on C. (h) Find the magnitude and direction of the resultant electric force acting on C.

11. **Review.** In the Bohr theory of the hydrogen atom, an electron moves in a circular orbit about a proton, where the radius of the orbit is 5.29×10^{-11} m. (a) Find the magnitude of the electric force exerted on each particle. (b) If this force causes the centripetal acceleration of the electron, what is the speed of the electron?

12. A charged particle A exerts a force of 2.62 μN to the right on charged particle B when the particles are 13.7 mm apart. Particle B moves straight away from A to make the distance between them 17.7 mm. What vector force does it then exert on A?

13. **BIO** **Review.** A molecule of DNA (deoxyribonucleic acid) is 2.17 μm long. The ends of the molecule become singly ionized: negative on one end, positive on the other. The helical molecule acts like a spring and compresses 1.00% upon becoming charged. Determine the effective spring constant of the molecule.

Section 19.5 Electric Fields

14. **Q|C** **S** Two charged particles are located on the x axis. The first is a charge $+Q$ at $x = -a$. The second is an unknown charge located at $x = +3a$. The net electric field these charges produce at the origin has a magnitude of $2k_eQ/a^2$. Explain how many values are possible for the unknown charge and find the possible values.

15. **M** A uniformly charged ring of radius 10.0 cm has a total charge of 75.0 μC. Find the electric field on the axis of the ring at (a) 1.00 cm, (b) 5.00 cm, (c) 30.0 cm, and (d) 100 cm from the center of the ring.

16. What are the magnitude and direction of the electric field that will balance the weight of (a) an electron and (b) a proton? (You may use the data in Table 19.1.)

17. **M** In Figure P19.17, determine the point (other than infinity) at which the electric field is zero.

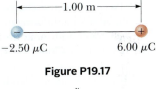

Figure P19.17

18. **S** A thin rod of length ℓ and uniform charge per unit length λ lies along the x axis as shown in Figure P19.18. (a) Show that the electric field at P, a distance y from the rod along its perpendicular bisector, has no x component and is given by $E = 2k_e\lambda \sin \theta_0 /y$. (b) Using your result to part (a), show that the field of a rod of infinite length is $E = 2k_e\lambda/y$. (*Suggestion:* First, calculate the field at P due to an element of length dx, which has a charge $\lambda\, dx$.

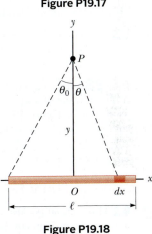

Figure P19.18

Then, change variables from x to θ, using the relationships $x = y \tan \theta$ and $dx = y \sec^2\theta\, d\theta$, and integrate over θ.)

19. Three point charges are arranged as shown in Figure P19.19. (a) Find the vector electric field that the 6.00-nC and -3.00-nC charges together create at the origin. (b) Find the vector force on the 5.00-nC charge.

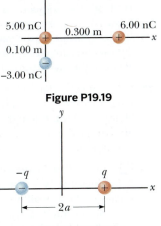

Figure P19.19

20. **S** Consider the electric dipole shown in Figure P19.20. Show that the electric field at a *distant* point on the +x axis is $E_x \approx 4k_eqa/x^3$.

Figure P19.20

21. A uniformly charged insulating rod of length 14.0 cm is bent into the shape of a semicircle as shown in Figure P19.21. The rod has a total charge of -7.50 μC. Find (a) the magnitude and (b) the direction of the electric field at O, the center of the semicircle.

• O

Figure P19.21

22. Two 2.00-μC point charges are located on the x axis. One is at $x = 1.00$ m, and the other is at $x = -1.00$ m. (a) Determine the electric field on the y axis at $y = 0.500$ m. (b) Calculate the electric force on a -3.00-μC charge placed on the y axis at $y = 0.500$ m.

23. [W] A rod 14.0 cm long is uniformly charged and has a total charge of -22.0 μC. Determine (a) the magnitude and (b) the direction of the electric field along the axis of the rod at a point 36.0 cm from its center.

24. [S] [W] A continuous line of charge lies along the x axis, extending from $x = +x_0$ to positive infinity. The line carries positive charge with a uniform linear charge density λ_0. What are (a) the magnitude and (b) the direction of the electric field at the origin?

25. Three solid plastic cylinders all have radius 2.50 cm and length 6.00 cm. Find the charge of each cylinder given the following additional information about each one. Cylinder (a) carries charge with uniform density 15.0 nC/m² everywhere on its surface. Cylinder (b) carries charge with uniform density 15.0 nC/m² on its curved lateral surface only. Cylinder (c) carries charge with uniform density 500 nC/m³ throughout the plastic.

26. [S] Show that the maximum magnitude E_{max} of the electric field along the axis of a uniformly charged ring occurs at $x = a/\sqrt{2}$ (see Fig. 19.15) and has the value $Q/(6\sqrt{3}\pi\epsilon_0 a^2)$.

27. [Q|C] [S] Four charged particles are at the corners of a square of side a as shown in Figure P19.27. Determine (a) the electric field at the location of charge q and (b) the total electric force exerted on q.

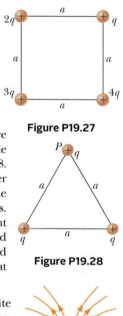

Figure P19.27

Section 19.6 Electric Field Lines

28. [S] Three equal positive charges q are at the corners of an equilateral triangle of side a as shown in Figure P19.28. Assume the three charges together create an electric field. (a) Sketch the field lines in the plane of the charges. (b) Find the location of one point (other than ∞) where the electric field is zero. What are (c) the magnitude and (d) the direction of the electric field at P due to the two charges at the base?

Figure P19.28

29. A negatively charged rod of finite length carries charge with a uniform charge per unit length. Sketch the electric field lines in a plane containing the rod.

30. [W] Figure P19.30 shows the electric field lines for two charged particles separated by a small distance. (a) Determine the ratio q_1/q_2. (b) What are the signs of q_1 and q_2?

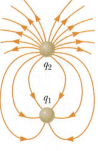

Figure P19.30

Section 19.7 Motion of Charged Particles in a Uniform Electric Field

31. [W] A proton is projected in the positive x direction into a region of a uniform electric field $\vec{E} = (-6.00 \times 10^5)\hat{i}$ N/C at $t = 0$. The proton travels 7.00 cm as it comes to rest. Determine (a) the acceleration of the proton, (b) its initial speed, and (c) the time interval over which the proton comes to rest.

32. [GP] Protons are projected with an initial speed $v_i = 9.55$ km/s from a field-free region through a plane and into a region where a uniform electric field $\vec{E} = -720\hat{j}$ N/C is present above the plane as shown in Figure P19.32. The initial velocity vector of the protons makes an angle θ with the plane. The protons are to hit a target that lies at a horizontal distance of $R = 1.27$ mm from the point where the protons cross the plane and enter the electric field. We wish to find the angle θ at which the protons must pass through the plane to strike the target. (a) What analysis model describes the horizontal motion of the protons above the plane? (b) What analysis model describes the vertical motion of the protons above the plane? (c) Argue that Equation 3.16 would be applicable to the protons in this situation. (d) Use Equation 3.16 to write an expression for R in terms of v_i, E, the charge and mass of the proton, and the angle θ. (e) Find the two possible values of the angle θ. (f) Find the time interval during which the proton is above the plane in Figure P19.32 for each of the two possible values of θ.

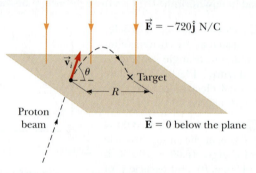

Figure P19.32

33. [M] A proton accelerates from rest in a uniform electric field of 640 N/C. At one later moment, its speed is 1.20 Mm/s (nonrelativistic because v is much less than the speed of light). (a) Find the acceleration of the proton. (b) Over what time interval does the proton reach this speed? (c) How far does it move in this time interval? (d) What is its kinetic energy at the end of this interval?

34. [S] The electrons in a particle beam each have a kinetic energy K. What are (a) the magnitude and (b) the direction of the electric field that will stop these electrons in a distance d?

35. [M] A proton moves at 4.50×10^5 m/s in the horizontal direction. It enters a uniform vertical electric field with a magnitude of 9.60×10^3 N/C. Ignoring any gravitational effects, find (a) the time interval required for the proton to travel 5.00 cm horizontally, (b) its vertical displacement during the time interval in which it travels 5.00 cm horizontally, and (c) the horizontal and vertical components of its velocity after it has traveled 5.00 cm horizontally.

Section 19.8 Electric Flux

36. [W] A vertical electric field of magnitude 2.00×10^4 N/C exists above the Earth's surface on a day when a thunderstorm is brewing. A car with a rectangular size of 6.00 m by 3.00 m is traveling along a dry gravel roadway sloping downward at 10.0°. Determine the electric flux through the bottom of the car.

37. **M** A 40.0-cm-diameter circular loop is rotated in a uniform electric field until the position of maximum electric flux is found. The flux in this position is measured to be 5.20×10^5 N · m²/C. What is the magnitude of the electric field?

Section 19.9 Gauss's Law

38. **S** A particle with charge Q is located a small distance δ immediately above the center of the flat face of a hemisphere of radius R as shown in Figure P19.38. What is the electric flux (a) through the curved surface and (b) through the flat face as $\delta \to 0$?

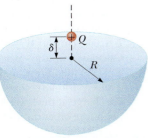

Figure P19.38

39. **M** The following charges are located inside a submarine: 5.00 μC, -9.00 μC, 27.0 μC, and -84.0 μC. (a) Calculate the net electric flux through the hull of the submarine. (b) Is the number of electric field lines leaving the submarine greater than, equal to, or less than the number entering it?

40. **Q|C** **W** A particle with charge of 12.0 μC is placed at the center of a spherical shell of radius 22.0 cm. What is the total electric flux through (a) the surface of the shell and (b) any hemispherical surface of the shell? (c) Do the results depend on the radius? Explain.

41. A particle with charge $Q = 5.00$ μC is located at the center of a cube of edge $L = 0.100$ m. In addition, six other identical charged particles having $q = -1.00$ μC are positioned symmetrically around Q as shown in Figure P19.41. Determine the electric flux through one face of the cube.

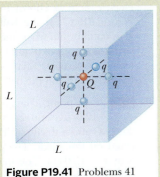

Figure P19.41 Problems 41 and 42.

42. **S** A particle with charge Q is located at the center of a cube of edge L. In addition, six other identical charged particles q are positioned symmetrically around Q as shown in Figure P19.41. For each of these particles, q is a negative number. Determine the electric flux through one face of the cube.

43. **W** The electric field everywhere on the surface of a thin, spherical shell of radius 0.750 m is of magnitude 890 N/C and points radially toward the center of the sphere. (a) What is the net charge within the sphere's surface? (b) What is the distribution of the charge inside the spherical shell?

44. **Q|C** A charge of 170 μC is at the center of a cube of edge 80.0 cm. No other charges are nearby. (a) Find the flux through each face of the cube. (b) Find the flux through the whole surface of the cube. (c) **What If?** Would your answers to either part (a) or part (b) change if the charge were not at the center? Explain.

Section 19.10 Application of Gauss's Law to Various Charge Distributions

45. **W** A solid sphere of radius 40.0 cm has a total positive charge of 26.0 μC uniformly distributed throughout its volume. Calculate the magnitude of the electric field (a) 0 cm, (b) 10.0 cm, (c) 40.0 cm, and (d) 60.0 cm from the center of the sphere.

46. **Q|C** **W** A nonconducting wall carries charge with a uniform density of 8.60 μC/cm². (a) What is the electric field 7.00 cm in front of the wall if 7.00 cm is small compared with the dimensions of the wall? (b) Does your result change as the distance from the wall varies? Explain.

47. In nuclear fission, a nucleus of uranium-238, which contains 92 protons, can divide into two smaller spheres, each having 46 protons and a radius of 5.90×10^{-15} m. What is the magnitude of the repulsive electric force pushing the two spheres apart?

48. **S** Consider a long, cylindrical charge distribution of radius R with a uniform charge density ρ. Find the electric field at distance r from the axis, where $r < R$.

49. A 10.0-g piece of Styrofoam carries a net charge of -0.700 μC and is suspended in equilibrium above the center of a large, horizontal sheet of plastic that has a uniform charge density on its surface. What is the charge per unit area on the plastic sheet?

50. **S** An insulating solid sphere of radius a has a uniform volume charge density and carries a total positive charge Q. A spherical gaussian surface of radius r, which shares a common center with the insulating sphere, is inflated starting from $r = 0$. (a) Find an expression for the electric flux passing through the surface of the gaussian sphere as a function of r for $r < a$. (b) Find an expression for the electric flux for $r > a$. (c) Plot the flux versus r.

51. **M** A large, flat, horizontal sheet of charge has a charge per unit area of 9.00 μC/m². Find the electric field just above the middle of the sheet.

52. **W** A cylindrical shell of radius 7.00 cm and length 2.40 m has its charge uniformly distributed on its curved surface. The magnitude of the electric field at a point 19.0 cm radially outward from its axis (measured from the midpoint of the shell) is 36.0 kN/C. Find (a) the net charge on the shell and (b) the electric field at a point 4.00 cm from the axis, measured radially outward from the midpoint of the shell.

53. **M** Consider a thin, spherical shell of radius 14.0 cm with a total charge of 32.0 μC distributed uniformly on its surface. Find the electric field (a) 10.0 cm and (b) 20.0 cm from the center of the charge distribution.

54. **M** A uniformly charged, straight filament 7.00 m in length has a total positive charge of 2.00 μC. An uncharged cardboard cylinder 2.00 cm in length and 10.0 cm in radius surrounds the filament at its center, with the filament as the axis of the cylinder. Using reasonable approximations, find (a) the electric field at the surface of the cylinder and (b) the total electric flux through the cylinder.

Section 19.11 Conductors in Electrostatic Equilibrium

55. **M** A long, straight metal rod has a radius of 5.00 cm and a charge per unit length of 30.0 nC/m. Find the electric field (a) 3.00 cm, (b) 10.0 cm, and (c) 100 cm from the axis of the rod, where distances are measured perpendicular to the rod's axis.

56. *Why is the following situation impossible?* A solid copper sphere of radius 15.0 cm is in electrostatic equilibrium and carries a charge of 40.0 nC. Figure P19.56 shows the magnitude of the electric field as a function of radial position r measured from the center of the sphere.

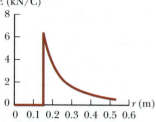

Figure P19.56

57. A solid conducting sphere of radius 2.00 cm has a charge 8.00 μC. A conducting spherical shell of inner radius 4.00 cm and outer radius 5.00 cm is concentric with the solid sphere and has a total charge −4.00 μC. Find the electric field at (a) $r = 1.00$ cm, (b) $r = 3.00$ cm, (c) $r = 4.50$ cm, and (d) $r = 7.00$ cm from the center of this charge configuration.

58. **S** A very large, thin, flat plate of aluminum of area A has a total charge Q uniformly distributed over its surfaces. Assuming the same charge is spread uniformly over the *upper* surface of an otherwise identical glass plate, compare the electric fields just above the center of the upper surface of each plate.

59. **M** A thin, square, conducting plate 50.0 cm on a side lies in the xy plane. A total charge of 4.00×10^{-8} C is placed on the plate. Find (a) the charge density on each face of the plate, (b) the electric field just above the plate, and (c) the electric field just below the plate. You may assume the charge density is uniform.

60. **S** A long, straight wire is surrounded by a hollow metal cylinder whose axis coincides with that of the wire. The wire has a charge per unit length of λ, and the cylinder has a net charge per unit length of 2λ. From this information, use Gauss's law to find (a) the charge per unit length on the inner surface of the cylinder, (b) the charge per unit length on the outer surface of the cylinder, and (c) the electric field outside the cylinder a distance r from the axis.

61. A square plate of copper with 50.0-cm sides has no net charge and is placed in a region of uniform electric field of 80.0 kN/C directed perpendicularly to the plate. Find (a) the charge density of each face of the plate and (b) the total charge on each face.

Section 19.12 Context Connection: The Atmospheric Electric Field

62. **Q|C** **Review.** In fair weather, the electric field in the air at a particular location immediately above the Earth's surface is 120 N/C directed downward. (a) What is the surface charge density on the ground? Is it positive or negative? (b) Imagine the surface charge density is uniform over the planet. What then is the charge of the whole surface of the Earth? (c) Imagine the Moon had a charge 27.3% as large as that on the surface of the Earth, with the same sign. Find the electric force the Earth would then exert on the Moon. (d) State how the answer to part (e) compares with the gravitational force the Earth exerts on the Moon.

63. In the air over a particular region at an altitude of 500 m above the ground, the electric field is 120 N/C directed downward. At 600 m above the ground, the electric field is 100 N/C downward. What is the average volume charge density in the layer of air between these two elevations? Is it positive or negative?

Additional Problems

64. **S** An infinitely long line charge having a uniform charge per unit length λ lies a distance d from point O as shown in Figure P19.64. Determine the total electric flux through the surface of a sphere of radius R centered at O resulting from this line charge. Consider both cases, where (a) $R < d$ and (b) $R > d$.

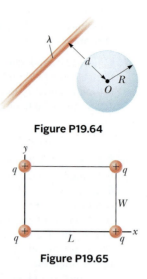

Figure P19.64

65. Four identical charged particles ($q = +10.0$ μC) are located on the corners of a rectangle as shown in Figure P19.65. The dimensions of the rectangle are $L = 60.0$ cm and $W = 15.0$ cm. Calculate (a) the magnitude and (b) the direction of the total electric force exerted on the charge at the lower left corner by the other three charges.

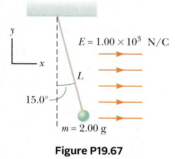

Figure P19.65

66. *Why is the following situation impossible?* An electron enters a region of uniform electric field between two parallel plates. The plates are used in a cathode-ray tube to adjust the position of an electron beam on a distant fluorescent screen. The magnitude of the electric field between the plates is 200 N/C. The plates are 0.200 m in length and are separated by 1.50 cm. The electron enters the region at a speed of 3.00×10^6 m/s, traveling parallel to the plane of the plates in the direction of their length. It leaves the plates heading toward its correct location on the fluorescent screen.

67. A small, 2.00-g plastic ball is suspended by a 20.0-cm-long string in a uniform electric field as shown in Figure P19.67. If the ball is in equilibrium when the string makes a 15.0° angle with the vertical, what is the net charge on the ball?

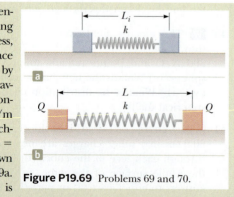

Figure P19.67

68. **Q|C** Two point charges $q_A = -12.0$ μC and $q_B = 45.0$ μC and a third particle with unknown charge q_C are located on the x axis. The particle q_A is at the origin, and q_B is at $x = 15.0$ cm. The third particle is to be placed so that each particle is in equilibrium under the action of the electric forces exerted by the other two particles. (a) Is this situation possible? If so, is it possible in more than one way? Explain. Find (b) the required location and (c) the magnitude and the sign of the charge of the third particle.

69. **Review.** Two identical blocks resting on a frictionless, horizontal surface are connected by a light spring having a spring constant $k = 100$ N/m and an unstretched length $L_i = 0.400$ m as shown in Figure P19.69a. A charge Q is slowly placed on each block, causing the spring to stretch to an

Figure P19.69 Problems 69 and 70.

equilibrium length $L = 0.500$ m as shown in Figure P19.69b. Determine the value of Q, modeling the blocks as charged particles.

70. **S** **Review.** Two identical blocks resting on a frictionless, horizontal surface are connected by a light spring having a spring constant k and an unstretched length L_i as shown in Figure P19.69a. A charge Q is slowly placed on each block, causing the spring to stretch to an equilibrium length L as shown in Figure P19.69b. Determine the value of Q, modeling the blocks as charged particles.

71. A line of charge with uniform density 35.0 nC/m lies along the line $y = -15.0$ cm, between the points with coordinates $x = 0$ and $x = 40.0$ cm. Find the electric field it creates at the origin.

72. **QC** **C** **S** Two small spheres of mass m are suspended from strings of length ℓ that are connected at a common point. One sphere has charge Q and the other charge $2Q$. The strings make angles θ_1 and θ_2 with the vertical. (a) Explain how θ_1 and θ_2 are related. (b) Assume θ_1 and θ_2 are small. Show that the distance r between the spheres is approximately

$$r \approx \left(\frac{4k_e Q^2 \ell}{mg}\right)^{1/3}$$

73. **S** Two infinite, nonconducting sheets of charge are parallel to each other as shown in Figure P19.73. The sheet on the left has a uniform surface charge density σ, and the one on the right has a uniform charge density $-\sigma$. Calculate the electric field at points (a) to the left of, (b) in between, and (c) to the right of the two sheets. (d) **What If?** Find the electric fields in all three regions if both sheets have *positive* uniform surface charge densities of value σ.

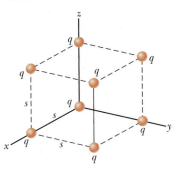

Figure P19.73

74. **S** Consider the charge distribution shown in Figure P19.74. (a) Show that the magnitude of the electric field at the center of any face of the cube has a value of $2.18k_e q/s^2$. (b) What is the direction of the electric field at the center of the top face of the cube?

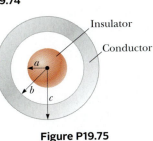

Figure P19.74

75. **GP** **S** A solid, insulating sphere of radius a has a uniform charge density throughout its volume and a total charge Q. Concentric with this sphere is an uncharged, conducting, hollow sphere whose inner and outer radii are b and c as shown in Figure P19.75. We

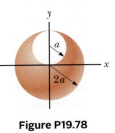

Figure P19.75

wish to understand completely the charges and electric fields at all locations. (a) Find the charge contained within a sphere of radius $r < a$. (b) From this value, find the magnitude of the electric field for $r < a$. (c) What charge is contained within a sphere of radius r when $a < r < b$? (d) From this value, find the magnitude of the electric field for r when $a < r < b$. (e) Now consider r when $b < r < c$. What is the magnitude of the electric field for this range of values of r? (f) From this value, what must be the charge on the inner surface of the hollow sphere? (g) From part (f), what must be the charge on the outer surface of the hollow sphere? (h) Consider the three spherical surfaces of radii a, b, and c. Which of these surfaces has the largest magnitude of surface charge density?

76. **S** **Review.** A negatively charged particle $-q$ is placed at the center of a uniformly charged ring, where the ring has a total positive charge Q as shown in Figure P19.76. The particle, confined to move along the x axis, is moved a small distance x along the axis (where $x \ll a$) and released. Show that the particle oscillates in simple harmonic motion with a frequency given by

$$f = \frac{1}{2\pi}\left(\frac{k_e qQ}{ma^3}\right)^{1/2}$$

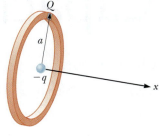

Figure P19.76

77. Inez is putting up decorations for her sister's quinceañera (fifteenth birthday party). She ties three light silk ribbons together to the top of a gateway and hangs a rubber balloon from each ribbon (Fig. P19.77). To include the effects of the gravitational and buoyant forces on it, each balloon can be modeled as a particle of mass 2.00 g, with its center 50.0 cm from the point of support. Inez rubs the whole surface of each balloon with her woolen scarf, making the balloons hang separately with gaps between them. Looking directly upward from below the balloons, Inez notices that the centers of the hanging balloons form a horizontal equilateral triangle with sides 30.0 cm long. What is the common charge each balloon carries?

Figure P19.77

78. **S** A sphere of radius $2a$ is made of a nonconducting material that has a uniform volume charge density ρ. (Assume the material does not affect the electric field.) A spherical cavity of radius a is now removed from the sphere as shown in Figure P19.78. Show that the electric field within the cavity is uniform and is given by $E_x = 0$ and $E_y = \rho a/3\epsilon_0$ (*Suggestion:* The field within the cavity is the superposition of the field due to the original uncut sphere plus the field due to a sphere the size of the cavity with a uniform negative charge density $-\rho$.)

Figure P19.78

Chapter 20

Electric Potential and Capacitance

Chapter Outline

© Cengage Learning/George Semple

This device is a *variable capacitor*, used to tune radios to a selected station. When one set of metal plates is rotated so as to lie between a fixed set of plates, the *capacitance* of the device changes. Capacitance is a parameter that depends on *electric potential*, the primary topic of this chapter.

The concept of potential energy was introduced in Chapter 6 in connection with such conservative forces as gravity and the elastic force of a spring. By using the principle of conservation of mechanical energy in an isolated system, we are often able to avoid working directly with forces when solving mechanical problems. In this chapter, we shall use the energy concept in our study of electricity. Because the electrostatic force (given by Coulomb's law) is conservative, electrostatic phenomena can conveniently be described in terms of an *electric* potential energy function. This concept enables us to define a quantity called *electric potential*, which is a scalar quantity and which therefore leads to a simpler means of describing some electrostatic phenomena than the electric field method. As we shall see in subsequent chapters, the concept of electric potential is of great practical value in many applications.

This chapter also addresses the properties of capacitors, devices that store charge. The ability of a capacitor to store charge is measured by its *capacitance*. Capacitors are used in common applications such as frequency tuners in radio receivers, filters in power supplies, and energy-storing devices in electronic flash units.

20.1 | Electric Potential and Potential Difference

When a test charge q_0 is placed in an electric field $\vec{\mathbf{E}}$ created by some source charge distribution, the electric force acting on the test charge is $q_0\vec{\mathbf{E}}$. The force $\vec{\mathbf{F}}_e = q_0\vec{\mathbf{E}}$ is conservative because the force between charges described by Coulomb's law is conservative. When the test charge is moved in the field at constant velocity by some external agent, the work done by the field on the charge is equal to the negative of the work done by the external agent causing the displacement. This situation is analogous to that of lifting an object with mass in a gravitational field: the work done by the external agent is mgh, and the work done by the gravitational force is $-mgh$.

When analyzing electric and magnetic fields, it is common practice to use the notation $d\vec{\mathbf{s}}$ to represent an infinitesimal displacement vector that is oriented tangent to a path through space. This path may be straight or curved, and an integral performed along this path is called either a *path integral* or a *line integral* (the two terms are synonymous).

For an infinitesimal displacement $d\vec{\mathbf{s}}$ of a point charge q_0 immersed in an electric field, the work done within the charge–field system by the electric field on the charge is $W_{int} = \vec{\mathbf{F}}_e \cdot d\vec{\mathbf{s}} = q_0\vec{\mathbf{E}} \cdot d\vec{\mathbf{s}}$. As this amount of work is done by the field, the potential energy of the charge–field system is changed by an amount $dU = -W_{int} = -q_0\vec{\mathbf{E}} \cdot d\vec{\mathbf{s}}$. For a finite displacement of the charge from point Ⓐ to point Ⓑ, the change in potential energy of the system $\Delta U = U_{Ⓑ} - U_{Ⓐ}$ is

$$\Delta U = -q_0 \int_{Ⓐ}^{Ⓑ} \vec{\mathbf{E}} \cdot d\vec{\mathbf{s}} \qquad \textbf{20.1} ◄$$

► Change in electric potential energy of a charge–field system

The integration is performed along the path that q_0 follows as it moves from Ⓐ to Ⓑ. Because the force $q_0\vec{\mathbf{E}}$ is conservative, this line integral does not depend on the path taken from Ⓐ to Ⓑ.

For a given position of the test charge in the field, the charge–field system has a potential energy U relative to the configuration of the system that is defined as $U = 0$. Dividing the potential energy by the test charge gives a physical quantity that depends only on the source charge distribution and has a value at every point in an electric field. This quantity is called the **electric potential** (or simply the **potential**) V:

$$V = \frac{U}{q_0} \qquad \textbf{20.2} ◄$$

Because potential energy is a scalar quantity, electric potential also is a scalar quantity.

As described by Equation 20.1, if the test charge is moved between two positions Ⓐ and Ⓑ in an electric field, the charge–field system experiences a change in potential energy. The **potential difference** $\Delta V = V_{Ⓑ} - V_{Ⓐ}$ between two points Ⓐ and Ⓑ in an electric field is defined as the change in potential energy of the system when a test charge q_0 is moved between the points divided by the test charge:

$$\Delta V \equiv \frac{\Delta U}{q_0} = -\int_{Ⓐ}^{Ⓑ} \vec{\mathbf{E}} \cdot d\vec{\mathbf{s}} \qquad \textbf{20.3} ◄$$

► Potential difference between two points

In this definition, the infinitesimal displacement $d\vec{\mathbf{s}}$ is interpreted as the displacement between two points in space rather than the displacement of a point charge as in Equation 20.1.

Just as with potential energy, only *differences* in electric potential are meaningful. We often take the value of the electric potential to be zero at some convenient point in an electric field.

Potential difference should not be confused with difference in potential energy. The potential difference between Ⓐ and Ⓑ exists solely because of a source charge and depends on the source charge distribution (consider points Ⓐ and Ⓑ *without* the presence of the test charge). For a potential energy to exist, we must have a *system* of two or more charges. The potential energy belongs to the system and changes only if a charge is moved relative to the rest of the system.

Pitfall Prevention | 20.1
Potential and Potential Energy
The *potential is characteristic of the field only*, independent of a charged test particle that may be placed in the field. *Potential energy is characteristic of the charge–field system* due to an interaction between the field and a charged particle placed in the field.

If an external agent moves a test charge from Ⓐ to Ⓑ without changing the kinetic energy of the test charge, the agent performs work that changes the potential energy of the system: $W = \Delta U$. Imagine an arbitrary charge q located in an electric field. From Equation 20.3, the work done by an external agent in moving a charge q through an electric field at constant velocity is

$$W = q\,\Delta V \qquad \textbf{20.4} \blacktriangleleft$$

Because electric potential is a measure of potential energy per unit charge, the SI unit of both electric potential and potential difference is joules per coulomb, which is defined as a **volt** (V):

$$1\ V \equiv 1\ J/C$$

That is, 1 J of work must be done to move a 1-C charge through a potential difference of 1 V.

Equation 20.3 shows that potential difference also has units of electric field times distance. It follows that the SI unit of electric field (N/C) can also be expressed in volts per meter:

$$1\ N/C = 1\ V/m$$

Therefore, we can interpret the electric field as a measure of the rate of change of the electric potential with respect to position.

As discussed in Section 9.7, a unit of energy commonly used in atomic and nuclear physics is the **electron volt** (eV), which is defined as the energy a charge–field system gains or loses when a charge of magnitude e (that is, an electron or a proton) is moved through a potential difference of 1 V. Because $1\ V = 1\ J/C$ and the fundamental charge is equal to 1.602×10^{-19} C, the electron volt is related to the joule as follows:

$$1\ eV = 1.602 \times 10^{-19}\ C \cdot V = 1.602 \times 10^{-19}\ J \qquad \textbf{20.5} \blacktriangleleft$$

For instance, an electron in the beam of a typical dental x-ray machine may have a speed of 1.4×10^{8} m/s. This speed corresponds to a kinetic energy 1.1×10^{-14} J (using relativistic calculations as discussed in Chapter 9), which is equivalent to 6.7×10^{4} eV. Such an electron has to be accelerated from rest through a potential difference of 67 kV to reach this speed.

QUICK QUIZ 20.1 In Figure 20.1, two points Ⓐ and Ⓑ are located within a region in which there is an electric field. **(i)** How would you describe the potential difference $\Delta V = V_{Ⓑ} - V_{Ⓐ}$? (a) It is positive. (b) It is negative. (c) It is zero. **(ii)** A negative charge is placed at Ⓐ and then moved to Ⓑ. How would you describe the change in potential energy of the charge–field system for this process? Choose from the same possibilities.

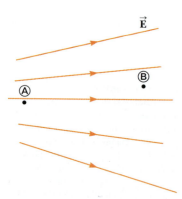

Figure 20.1 (Quick Quiz 20.1) Two points in an electric field.

20.2 | Potential Difference in a Uniform Electric Field

Equations 20.1 and 20.3 hold in all electric fields, whether uniform or varying, but they can be simplified for a uniform field. First, consider a uniform electric field directed along the negative y axis as shown in Active Figure 20.2a. Let's calculate the potential difference between two points Ⓐ and Ⓑ separated by a distance d, where the displacement $\vec{s}$ points from Ⓐ toward Ⓑ and is parallel to the field lines. Equation 20.3 gives

$$V_{Ⓑ} - V_{Ⓐ} = \Delta V = -\int_{Ⓐ}^{Ⓑ} \vec{E} \cdot d\vec{s} = -\int_{Ⓐ}^{Ⓑ} E\,ds\,(\cos 0°) = -\int_{Ⓐ}^{Ⓑ} E\,ds$$

Because E is constant, it can be removed from the integral sign, which gives

▶ Potential difference between two points in a uniform electric field

$$\Delta V = -E \int_{Ⓐ}^{Ⓑ} ds = -Ed \qquad \textbf{20.6} \blacktriangleleft$$

The negative sign indicates that the electric potential at point Ⓑ is lower than at point Ⓐ; that is, $V_Ⓑ < V_Ⓐ$. Electric field lines *always* point in the direction of decreasing electric potential as shown in Active Figure 20.2a.

Now suppose a test charge q_0 moves from Ⓐ to Ⓑ. We can calculate the change in the potential energy of the charge–field system from Equations 20.3 and 20.6:

$$\Delta U = q_0 \Delta V = -q_0 Ed \qquad \mathbf{20.7} \blacktriangleleft$$

This result shows that if q_0 is positive, then ΔU is negative. Therefore, in a system consisting of a positive charge and an electric field, the electric potential energy of the system decreases when the charge moves in the direction of the field. Equivalently, an electric field does work on a positive charge when the charge moves in the direction of the electric field. That is analogous to the work done by the gravitational field on a falling object as shown in Active Figure 20.2b. If a positive test charge is released from rest in this electric field, it experiences an electric force $q_0\vec{\mathbf{E}}$ in the direction of $\vec{\mathbf{E}}$ (downward in Active Fig. 20.2a). Therefore, it accelerates downward, gaining kinetic energy. As the charged particle gains kinetic energy, the potential energy of the charge–field system decreases by an equal amount. This equivalence should not be surprising; it is simply conservation of mechanical energy in an isolated system as introduced in Chapter 7.

The comparison between a system of an electric field with a positive test charge and a gravitational field with a test mass in Active Figure 20.2 is useful for conceptualizing electrical behavior. The electrical situation, however, has one feature that the gravitational situation does not: the test charge can be negative. If q_0 is negative, then ΔU in Equation 20.7 is positive and the situation is reversed. A system consisting of a negative charge and an electric field gains electric potential energy when the charge moves in the direction of the field. If a negative charge is released from rest in an electric field, it accelerates in a direction opposite the direction of the field. For the negative charge to move in the direction of the field, an external agent must apply a force and do positive work on the charge.

Now consider the more general case of a charged particle that moves between Ⓐ and Ⓑ in a uniform electric field such that the vector $\vec{\mathbf{s}}$ is *not* parallel to the field lines as shown in Figure 20.3. In this case, Equation 20.3 gives

$$\Delta V = -\int_Ⓐ^Ⓑ \vec{\mathbf{E}} \cdot d\vec{\mathbf{s}} = -\vec{\mathbf{E}} \cdot \int_Ⓐ^Ⓑ d\vec{\mathbf{s}} = -\vec{\mathbf{E}} \cdot \vec{\mathbf{s}} \qquad \mathbf{20.8} \blacktriangleleft$$

▶ Change in potential between two points in a uniform electric field

where again $\vec{\mathbf{E}}$ was removed from the integral because it is constant. The change in potential energy of the charge–field system is

$$\Delta U = q_0 \Delta V = -q_0 \vec{\mathbf{E}} \cdot \vec{\mathbf{s}} \qquad \mathbf{20.9} \blacktriangleleft$$

Finally, we conclude from Equation 20.8 that all points in a plane perpendicular to a uniform electric field are at the same electric potential. We can see that in Figure 20.3, where the potential difference $V_Ⓑ - V_Ⓐ$ is equal to the potential difference $V_Ⓒ - V_Ⓐ$. (Prove this fact to yourself by working out two dot products for $\vec{\mathbf{E}} \cdot \vec{\mathbf{s}}$: one for $\vec{\mathbf{s}}_{Ⓐ\to Ⓑ}$, where the angle θ between $\vec{\mathbf{E}}$ and $\vec{\mathbf{s}}$ is arbitrary as shown in Figure 20.3, and one for $\vec{\mathbf{s}}_{Ⓐ\to Ⓒ}$, where $\theta = 0$.) Therefore, $V_Ⓑ = V_Ⓒ$. The name **equipotential surface** is given to any surface consisting of a continuous distribution of points having the same electric potential.

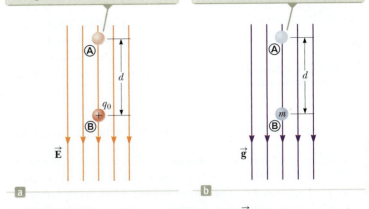

When a positive test charge moves from point Ⓐ to point Ⓑ, the electric potential energy of the charge–field system decreases.

When an object with mass moves from point Ⓐ to point Ⓑ, the gravitational potential energy of the object–field system decreases.

Active Figure 20.2 (a) When the electric field $\vec{\mathbf{E}}$ is directed downward, point Ⓑ is at a lower electric potential than point Ⓐ. (b) An object of mass m moving downward in a gravitational field $\vec{\mathbf{g}}$.

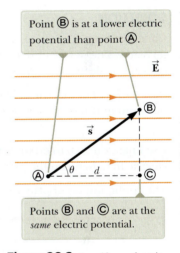

Point Ⓑ is at a lower electric potential than point Ⓐ.

Points Ⓑ and Ⓒ are at the *same* electric potential.

Figure 20.3 A uniform electric field directed along the positive x axis.

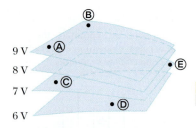

9 V
8 V
7 V
6 V

Figure 20.4 (Quick Quiz 20.2) Four equipotential surfaces.

The equipotential surfaces associated with a uniform electric field consist of a family of parallel planes that are all perpendicular to the field. Equipotential surfaces associated with fields having other symmetries are described in later sections.

QUICK QUIZ 20.2 The labeled points in Figure 20.4 are on a series of equipotential surfaces associated with an electric field. Rank (from greatest to least) the work done by the electric field on a positively charged particle that moves from Ⓐ to Ⓑ, from Ⓑ to Ⓒ, from Ⓒ to Ⓓ, and from Ⓓ to Ⓔ.

Example 20.1 | The Electric Field Between Two Parallel Plates of Opposite Charge

A battery has a specified potential difference ΔV between its terminals and establishes that potential difference between conductors attached to the terminals. A 12-V battery is connected between two parallel plates as shown in Figure 20.5. The separation between the plates is $d = 0.30$ cm, and we assume the electric field between the plates to be uniform. (This assumption is reasonable if the plate separation is small relative to the plate dimensions and we do not consider locations near the plate edges.) Find the magnitude of the electric field between the plates.

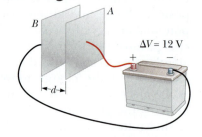

Figure 20.5 (Example 20.1) A 12-V battery connected to two parallel plates. The electric field between the plates has a magnitude given by the potential difference ΔV divided by the plate separation d.

SOLUTION

Conceptualize In Chapter 19 we investigated the uniform electric field between parallel plates. The new feature to this problem is that the electric field is related to the new concept of electric potential.

Categorize The electric field is evaluated from a relationship between field and potential given in this section, so we categorize this example as a substitution problem.

Use Equation 20.6 to evaluate the magnitude of the electric field between the plates:

$$E = \frac{|V_B - V_A|}{d} = \frac{12 \text{ V}}{0.30 \times 10^{-2} \text{ m}} = 4.0 \times 10^3 \text{ V/m}$$

The configuration of plates in Figure 20.5 is called a *parallel-plate capacitor* and is examined in greater detail in Section 20.7.

Example 20.2 | Motion of a Proton in a Uniform Electric Field

A proton is released from rest at point Ⓐ in a uniform electric field that has a magnitude of 8.0×10^4 V/m (Fig. 20.6). The proton undergoes a displacement of magnitude $d = 0.50$ m to point Ⓑ in the direction of $\vec{E}$. Find the speed of the proton after completing the displacement.

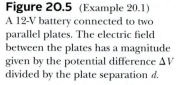

SOLUTION

Conceptualize Visualize the proton in Figure 20.6 moving downward through the potential difference. The situation is analogous to an object falling through a gravitational field.

Figure 20.6 (Example 20.2) A proton accelerates from Ⓐ to Ⓑ in the direction of the electric field.

Categorize The system of the proton and the two plates in Figure 20.6 does not interact with the environment, so we model it as an isolated system.

Analyze Use Equation 20.6 to find the potential difference between points Ⓐ and Ⓑ:

$$\Delta V = -Ed = -(8.0 \times 10^4 \text{ V/m})(0.50 \text{ m}) = -4.0 \times 10^4 \text{ V}$$

Write the appropriate reduction of Equation 7.2, the conservation of energy equation, for the isolated system of the charge and the electric field:

$$\Delta K + \Delta U = 0$$

20.2 *cont.*

Substitute the changes in energy for both terms:

$$(\tfrac{1}{2}mv^2 - 0) + e\,\Delta V = 0$$

Solve for the final speed of the proton:

$$v = \sqrt{\frac{-2e\,\Delta V}{m}}$$

Substitute numerical values:

$$v = \sqrt{\frac{-2(1.6 \times 10^{-19}\,\text{C})(-4.0 \times 10^4\,\text{V})}{1.67 \times 10^{-27}\,\text{kg}}}$$

$$= \boxed{2.8 \times 10^6\,\text{m/s}}$$

Finalize Because ΔV is negative for the field, ΔU is also negative for the proton–field system. The negative value of ΔU means the potential energy of the system decreases as the proton moves in the direction of the electric field. As the proton accelerates in the direction of the field, it gains kinetic energy while the electric potential energy of the system decreases at the same time.

Figure 20.6 is oriented so that the proton moves downward. The proton's motion is analogous to that of an object falling in a gravitational field. Although the gravitational field is always downward at the surface of the Earth, an electric field can be in any direction, depending on the orientation of the plates creating the field. Therefore, Figure 20.6 could be rotated 90° or 180° and the proton could move horizontally or upward in the electric field!

20.3 | Electric Potential and Potential Energy Due to Point Charges

As discussed in Section 19.6, an isolated positive point charge q produces an electric field directed radially outward from the charge. To find the electric potential at a point located a distance r from the charge, let's begin with the general expression for potential difference,

$$V_{\text{\textcircled{B}}} - V_{\text{\textcircled{A}}} = -\int_{\text{\textcircled{A}}}^{\text{\textcircled{B}}} \vec{\textbf{E}} \cdot d\vec{\textbf{s}}$$

where Ⓐ and Ⓑ are the two arbitrary points shown in Figure 20.7. At any point in space, the electric field due to the point charge is $\vec{\textbf{E}} = (k_e q/r^2)\hat{\textbf{r}}$ (Eq. 19.5), where $\hat{\textbf{r}}$ is a unit vector directed radially outward from the charge. The quantity $\vec{\textbf{E}} \cdot d\vec{\textbf{s}}$ can be expressed as

$$\vec{\textbf{E}} \cdot d\vec{\textbf{s}} = k_e \frac{q}{r^2}\hat{\textbf{r}} \cdot d\vec{\textbf{s}}$$

Because the magnitude of $\hat{\textbf{r}}$ is 1, the dot product $\hat{\textbf{r}} \cdot d\vec{\textbf{s}} = ds\cos\theta$, where θ is the angle between $\hat{\textbf{r}}$ and $d\vec{\textbf{s}}$. Furthermore, $ds\cos\theta$ is the projection of $d\vec{\textbf{s}}$ onto $\hat{\textbf{r}}$; therefore, $ds\cos\theta = dr$. That is, any displacement $d\vec{\textbf{s}}$ along the path from point Ⓐ to point Ⓑ produces a change dr in the magnitude of $\vec{\textbf{r}}$, the position vector of the point relative to the charge creating the field. Making these substitutions, we find that $\vec{\textbf{E}} \cdot d\vec{\textbf{s}} = (k_e q/r^2)\,dr$; hence, the expression for the potential difference becomes

$$V_{\text{\textcircled{B}}} - V_{\text{\textcircled{A}}} = -k_e q \int_{r_{\text{\textcircled{A}}}}^{r_{\text{\textcircled{B}}}} \frac{dr}{r^2} = k_e \frac{q}{r}\Big|_{r_{\text{\textcircled{A}}}}^{r_{\text{\textcircled{B}}}}$$

$$V_{\text{\textcircled{B}}} - V_{\text{\textcircled{A}}} = k_e q\left[\frac{1}{r_{\text{\textcircled{B}}}} - \frac{1}{r_{\text{\textcircled{A}}}}\right] \qquad \text{20.10} \blacktriangleleft$$

Equation 20.10 shows us that the integral of $\vec{\textbf{E}} \cdot d\vec{\textbf{s}}$ is *independent* of the path between points Ⓐ and Ⓑ. Multiplying by a charge q_0 that moves between points Ⓐ and Ⓑ, we see that the integral of $q_0\vec{\textbf{E}} \cdot d\vec{\textbf{s}}$ is also independent of path. This latter integral, which is the work done by the electric force on the charge q_0, shows that the electric force is conservative (see Section 6.7). We define a field that is related to a

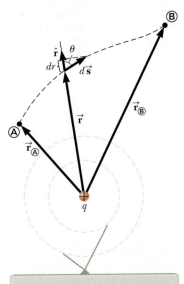

The two dashed circles represent intersections of spherical equipotential surfaces with the page.

Figure 20.7 The potential difference between points Ⓐ and Ⓑ due to a point charge q depends *only* on the initial and final radial coordinates $r_{\text{\textcircled{A}}}$ and $r_{\text{\textcircled{B}}}$.

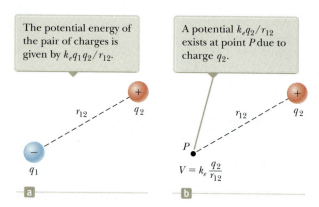

The potential energy of the pair of charges is given by $k_e q_1 q_2/r_{12}$.

A potential $k_e q_2/r_{12}$ exists at point P due to charge q_2.

$V = k_e \dfrac{q_2}{r_{12}}$

Active Figure 20.8 (a) Two point charges separated by a distance r_{12}. (b) Charge q_1 is removed.

▶ Electric potential due to several point charges

conservative force as a **conservative field.** Therefore, Equation 20.10 tells us that the electric field of a fixed point charge q is conservative. Furthermore, Equation 20.10 expresses the important result that the potential difference between any two points Ⓐ and Ⓑ in a field created by a point charge depends only on the radial coordinates $r_Ⓐ$ and $r_Ⓑ$. It is customary to choose the reference of electric potential for a point charge to be $V = 0$ at $r_Ⓐ = \infty$. With this reference choice, the electric potential due to a point charge at any distance r from the charge is

$$V = k_e \frac{q}{r} \qquad \text{20.11} ◀$$

We obtain the electric potential resulting from two or more point charges by applying the superposition principle. That is, the total electric potential at some point P due to several point charges is the sum of the potentials due to the individual charges. For a group of point charges, we can write the total electric potential at P as

$$V = k_e \sum_i \frac{q_i}{r_i} \qquad \text{20.12} ◀$$

where the potential is again taken to be zero at infinity and r_i is the distance from the point P to the charge q_i. Notice that the sum in Equation 20.12 is an algebraic sum of scalars rather than a vector sum (which we used to calculate the electric field of a group of charges in Eq. 19.6). Therefore, it is often much easier to evaluate V than $\vec{E}$.

Now consider the potential energy of a system of two charged particles. If V_2 is the electric potential at a point P due to charge q_2, the work an external agent must do to bring a second charge q_1 from infinity to P without acceleration is $q_1 V_2$. This work represents a transfer of energy into the system, and the energy appears in the system as potential energy U when the particles are separated by a distance r_{12} (Active Fig. 20.8a). Therefore, the potential energy of the system can be expressed as[1]

$$U = k_e \frac{q_1 q_2}{r_{12}} \qquad \text{20.13} ◀$$

If the charges are of the same sign, then U is positive. Positive work must be done by an external agent on the system to bring the two charges near each other (because charges of the same sign repel). If the charges are of opposite sign, then U is negative. Negative work is done by an external agent against the attractive force between the charges of opposite sign as they are brought near each other; a force must be applied opposite the displacement to prevent q_1 from accelerating toward q_2.

In Active Figure 20.8b, we have removed the charge q_1. At the position this charge previously occupied, point P, Equations 20.2 and 20.13 can be used to define a potential due to charge q_2 as $V = U/q_1 = k_e q_2/r_{12}$. This expression is consistent with Equation 20.11.

If the system consists of more than two charged particles, we can obtain the total potential energy of the system by calculating U for every *pair* of charges and summing the terms algebraically. The total electric potential energy of a system of point charges is equal to the work required to bring the charges, one at a time, from an infinite separation to their final positions.

QUICK QUIZ 20.3 A spherical balloon contains a positively charged object at its center. (**i**) As the balloon is inflated to a greater volume while the charged object remains at the center, does the electric potential at the surface of the balloon (a) increase, (b) decrease, or (c) remain the same? (**ii**) Does the electric flux through the surface of the balloon (a) increase, (b) decrease, or (c) remain the same?

---

[1]The expression for the electric potential energy of a system made up of two point charges, Equation 20.13, is of the *same* form as the equation for the gravitational potential energy of a system made up of two point masses, $-Gm_1 m_2/r$ (see Chapter 11). The similarity is not surprising considering that both expressions are derived from an inverse-square force law.

QUICK QUIZ 20.4 In Active Figure 20.8a, take q_1 to be a negative source charge and q_2 to be the test charge. (i) If q_2 is initially positive and is changed to a charge of the same magnitude but negative, what happens to the potential at the position of q_2 due to q_1? (a) It increases. (b) It decreases. (c) It remains the same. (ii) When q_2 is changed from positive to negative, what happens to the potential energy of the two-charge system? Choose from the same possibilities.

Example 20.3 | The Electric Potential Due to Two Point Charges

As shown in Figure 20.9a, a charge $q_1 = 2.00~\mu C$ is located at the origin and a charge $q_2 = -6.00~\mu C$ is located at $(0, 3.00)$ m.

(A) Find the total electric potential due to these charges at the point P, whose coordinates are $(4.00, 0)$ m.

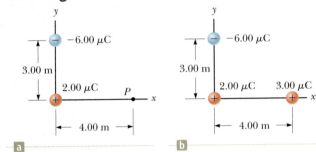

SOLUTION

Conceptualize Recognize first that the 2.00-μC and -6.00-μC charges are source charges and set up an electric field as well as a potential at all points in space, including point P.

Figure 20.9 (Example 20.3) (a) The electric potential at P due to the two charges q_1 and q_2 is the algebraic sum of the potentials due to the individual charges. (b) A third charge $q_3 = 3.00~\mu C$ is brought from infinity to point P.

Categorize The potential is evaluated using an equation developed in this chapter, so we categorize this example as a substitution problem.

Use Equation 20.12 for the system of two source charges:

$$V_P = k_e\left(\frac{q_1}{r_1} + \frac{q_2}{r_2}\right)$$

Substitute numerical values:

$$V_P = (8.99 \times 10^9~\text{N} \cdot \text{m}^2/\text{C}^2)\left(\frac{2.00 \times 10^{-6}~\text{C}}{4.00~\text{m}} + \frac{-6.00 \times 10^{-6}~\text{C}}{5.00~\text{m}}\right)$$

$$= -6.29 \times 10^3~\text{V}$$

(B) Find the change in potential energy of the system of two charges plus a third charge $q_3 = 3.00~\mu C$ as the latter charge moves from infinity to point P (Fig. 20.9b).

SOLUTION

Assign $U_i = 0$ for the system to the configuration in which the charge q_3 is at infinity. Use Equation 20.2 to evaluate the potential energy for the configuration in which the charge is at P:

$$U_f = q_3 V_P$$

Substitute numerical values to evaluate ΔU:

$$\Delta U = U_f - U_i = q_3 V_P - 0 = (3.00 \times 10^{-6}~\text{C})(-6.29 \times 10^3~\text{V})$$

$$= -1.89 \times 10^{-2}~\text{J}$$

Therefore, because the potential energy of the system has decreased, an external agent has to do positive work to remove the charge q_3 from point P back to infinity.

What If? You are working through this example with a classmate and she says, "Wait a minute! In part (B), we ignored the potential energy associated with the pair of charges q_1 and q_2!" How would you respond?

Answer Given the statement of the problem, it is not necessary to include this potential energy because part (B) asks for the *change* in potential energy of the system as q_3 is brought in from infinity. Because the configuration of charges q_1 and q_2 does not change in the process, there is no ΔU associated with these charges.

spherical conductor of radius R and charge Q. (Based on the shape of the field lines from a single spherical conductor, we can model the second conductor as a concentric spherical shell of infinite radius.) Because the potential of the sphere is simply $k_e Q/R$ (and $V = 0$ for the shell of infinite radius), the capacitance of the sphere is

$$C = \frac{Q}{\Delta V} = \frac{Q}{k_e Q/R} = \frac{R}{k_e} = 4\pi\epsilon_0 R \qquad \textbf{20.21} \blacktriangleleft$$

(Remember from Section 19.4 that the Coulomb constant $k_e = 1/4\pi\epsilon_0$.) Equation 20.21 shows that the capacitance of an isolated charged sphere is proportional to the sphere's radius and is independent of both the charge and the potential difference.

The capacitance of a pair of oppositely charged conductors can be calculated in the following manner. A convenient charge of magnitude Q is assumed, and the potential difference is calculated using the techniques described in Section 20.5. One then uses $C = Q/\Delta V$ to evaluate the capacitance. As you might expect, the calculation is relatively straightforward if the geometry of the capacitor is simple.

Let us illustrate with two familiar geometries: parallel plates and concentric cylinders. In these examples, we shall assume that the charged conductors are separated by a vacuum. (The effect of a material between the conductors will be treated in Section 20.10.)

The Parallel-Plate Capacitor

A parallel-plate capacitor consists of two parallel plates of equal area A separated by a distance d as in Figure 20.20. If the capacitor is charged, one plate has charge Q and the other, charge $-Q$. The magnitude of the charge per unit area on either plate is $\sigma = Q/A$. If the plates are very close together (compared with their length and width), we adopt a simplification model in which the electric field is uniform between the plates and zero elsewhere, as we discussed in Example 19.12. According to Example 19.12, the magnitude of the electric field between the plates is

$$E = \frac{\sigma}{\epsilon_0} = \frac{Q}{\epsilon_0 A}$$

Because the field is uniform, the potential difference across the capacitor can be found from Equation 20.6. Therefore,

$$\Delta V = Ed = \frac{Qd}{\epsilon_0 A}$$

Substituting this result into Equation 20.20, we find that the capacitance is

$$C = \frac{Q}{\Delta V} = \frac{Q}{Qd/\epsilon_0 A}$$

$$\boxed{C = \frac{\epsilon_0 A}{d}} \qquad \textbf{20.22} \blacktriangleleft$$

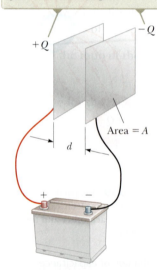

When the capacitor is connected to the terminals of a battery, electrons transfer between the plates and the wires so that the plates become charged.

$+Q$ $-Q$

Area $= A$

d

Figure 20.20 A parallel-plate capacitor consists of two parallel conducting plates, each of area A, separated by a distance d.

That is, the capacitance of a parallel-plate capacitor is proportional to the area of its plates and inversely proportional to the plate separation.

As you can see from the definition of capacitance, $C = Q/\Delta V$, the amount of charge a given capacitor can store for a given potential difference across its plates increases as the capacitance increases. It therefore seems reasonable that a capacitor constructed from plates having large areas should be able to store a large charge.

A careful inspection of the electric field lines for a parallel-plate capacitor reveals that the field is uniform in the central region between the plates, but is nonuniform at the edges of the plates. Figure 20.21 shows a drawing of the electric field pattern of a parallel-plate capacitor, showing the nonuniform field lines at the plates' edges. As long as the separation between the plates is small compared with the dimensions of the plates, the edge effects can be ignored and we can use the simplification model in which the electric field is uniform everywhere between the plates.

Active Figure 20.22 shows a battery connected to a single parallel-plate capacitor with a switch in the circuit. Let us identify the circuit as a system. When the switch is closed, the battery establishes an electric field in the wires and charges flow between the wires and the capacitor. As that occurs, energy is transformed within the system. Before the switch is closed, energy is stored as chemical potential energy in the battery. This type of energy is associated with chemical bonds and is transformed during the chemical reaction that occurs within the battery when it is operating in an electric circuit. When the switch is closed, some of the chemical potential energy in the battery is converted to electric potential energy related to the separation of positive and negative charges on the plates. As a result, we can describe a capacitor as a device that stores *energy* as well as *charge*. We will explore this energy storage in more detail in Section 20.9.

As a biological example of a parallel-plate capacitor, consider the plasma membrane for a neuron. The *plasma membrane* is a lipid bilayer containing a variety of types of molecules. This membrane contains a number of structures, including *ion channels* and *ion pumps*, which control concentrations of various ions on either side of the membrane. These ions include potassium, chlorine, calcium, and sodium. As a result of differences in these concentrations, there is an effective sheet of negative charge on the intracellular side of the membrane and a sheet of positive charge on the extracellular side. This results in a voltage of about 70 to 80 mV across the membrane. The sheets of charge act as parallel plates, so that the membrane can be modeled as a parallel-plate capacitor. The capacitance of the plasma membrane is about 2 μF for each cm^2 of membrane area.

When a neuron is carrying a signal, an event called an *action potential* occurs. Special structures in the cell membrane, called *voltage-gated ion channels*, are normally closed. If the voltage across the membrane capacitor falls in magnitude to a threshold value of about 50 mV, the ion channels open, allowing a flow of sodium ions into the cell. This flow reduces the voltage even further, allowing more sodium ions to enter the cell, thereby reversing the polarity of the voltage across the capacitor in a time interval measured in milliseconds. The voltage-gated ion channels then close and other channels open, allowing movement of ions until the neuron returns to its resting state.

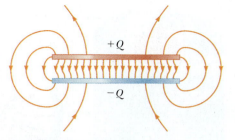

Figure 20.21 The electric field between the plates of a parallel-plate capacitor is uniform near the center but nonuniform near the edges.

BIO Capacitance of cell membranes

BIO Action potential

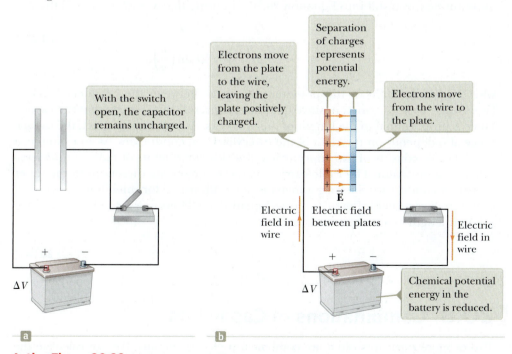

With the switch open, the capacitor remains uncharged.

Electrons move from the plate to the wire, leaving the plate positively charged.

Separation of charges represents potential energy.

Electrons move from the wire to the plate.

Electric field in wire

Electric field between plates

$\vec{E}$

Electric field in wire

ΔV

ΔV

Chemical potential energy in the battery is reduced.

a   **b**

Active Figure 20.22 (a) A circuit consisting of a capacitor, a battery, and a switch. (b) When the switch is closed, the battery establishes an electric field in the wire and the capacitor becomes charged.

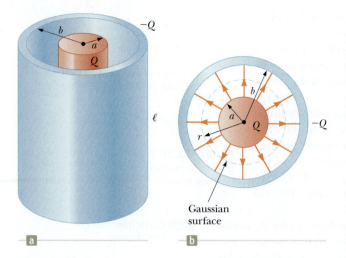

Figure 20.23 (a) A cylindrical capacitor consists of a solid cylindrical conductor of radius a and length ℓ surrounded by a coaxial cylindrical shell of radius b. (b) End view. The dashed line represents the end of the cylindrical gaussian surface of radius r and length ℓ.

This process can disturb neighboring regions of the plasma membrane so that the action potential is propagated along the neuron. In the next chapter, we will see how the capacitance of the plasma membrane combines with another electrical characteristic of the membrane to provide an electrical model for the conduction of a signal along the neuron.

The Cylindrical Capacitor

A cylindrical capacitor consists of a cylindrical conductor of radius a and charge Q coaxial with a larger cylindrical shell of radius b and charge $-Q$ (Fig. 20.23a). Let us find the capacitance of this device if its length is ℓ. If we assume that ℓ is large compared with a and b, we can adopt a simplification model in which we ignore end effects. In this case, the field is perpendicular to the axis of the cylinders and is confined to the region between them (Fig. 20.23b). We first calculate the potential difference between the two cylinders, which is given in general by

$$V_b - V_a = -\int_a^b \vec{\mathbf{E}} \cdot d\vec{\mathbf{s}}$$

where $\vec{\mathbf{E}}$ is the electric field in the region $a < r < b$. In Chapter 19, using Gauss's law, we showed that the electric field of a cylinder with charge per unit length λ has the magnitude $E = 2k_e\lambda/r$. The same result applies here because the outer cylinder does not contribute to the electric field inside it. Using this result and noting that the direction of $\vec{\mathbf{E}}$ is radially away from the inner cylinder in Figure 20.23b, we find that

$$V_b - V_a = -\int_a^b E_r\,dr = -2k_e\lambda\int_a^b \frac{dr}{r} = -2k_e\lambda \ln\left(\frac{b}{a}\right)$$

Substituting this result into Equation 20.20 and using that $\lambda = Q/\ell$, we find that

$$C = \frac{Q}{\Delta V} = \frac{Q}{\dfrac{2k_e Q}{\ell}\ln\left(\dfrac{b}{a}\right)} = \frac{\ell}{2k_e \ln\left(\dfrac{b}{a}\right)} \qquad \text{20.23} \blacktriangleleft$$

where the magnitude of the potential difference between the cylinders is $\Delta V = |V_a - V_b| = 2k_e\lambda \ln(b/a)$, a positive quantity. Our result for C shows that the capacitance is proportional to the length of the cylinders. As you might expect, the capacitance also depends on the radii of the two cylindrical conductors. As an example, a coaxial cable consists of two concentric cylindrical conductors of radii a and b separated by an insulator. The cable carries currents in opposite directions in the inner and outer conductors. Such a geometry is especially useful for shielding an electrical signal from external influences. From Equation 20.23, we see that the capacitance per unit length of a coaxial cable is

$$\frac{C}{\ell} = \frac{1}{2k_e \ln\left(\dfrac{b}{a}\right)} \qquad \text{20.24} \blacktriangleleft$$

20.8 | Combinations of Capacitors

Two or more capacitors often are combined in electric circuits. We can calculate the equivalent capacitance of certain combinations using methods described in this section. Throughout this section, we assume the capacitors to be combined are initially uncharged.

In studying electric circuits, we use a simplified pictorial representation called a **circuit diagram.** Such a diagram uses **circuit symbols** to represent various circuit elements. The circuit symbols are connected by straight lines that represent the wires between the circuit elements. The circuit symbols for capacitors, batteries, and switches as well as the color codes used for them in this text are given in Figure 20.24. The symbol for the capacitor reflects the geometry of the most common model for a capacitor, a pair of parallel plates. The positive terminal of the battery is at the higher potential and is represented in the circuit symbol by the longer line.

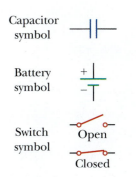

Capacitor symbol

Battery symbol

Switch symbol — Open — Closed

Figure 20.24 Circuit symbols for capacitors, batteries, and switches. Notice that capacitors are in blue, batteries are in green, and switches are in red. The closed switch can carry current, whereas the open one cannot.

Parallel Combination

Two capacitors connected as shown in Active Figure 20.25a are known as a **parallel combination** of capacitors. Active Figure 20.25b shows a circuit diagram for this combination of capacitors. The left plates of the capacitors are connected to the positive terminal of the battery by a conducting wire and are therefore both at the same electric potential as the positive terminal. Likewise, the right plates are connected to the negative terminal and so are both at the same potential as the negative terminal. Therefore, the individual potential differences across capacitors connected in parallel are the same and are equal to the potential difference applied across the combination. That is,

$$\Delta V_1 = \Delta V_2 = \Delta V$$

where ΔV is the battery terminal voltage.

After the battery is attached to the circuit, the capacitors quickly reach their maximum charge. Let's call the maximum charges on the two capacitors Q_1 and Q_2. The *total charge* Q_{tot} stored by the two capacitors is the sum of the charges on the individual capacitors:

$$Q_{tot} = Q_1 + Q_2 \qquad \textbf{20.25} \blacktriangleleft$$

Suppose you wish to replace these two capacitors by one *equivalent capacitor* having a capacitance C_{eq} as in Active Figure 20.25c. The effect this equivalent capacitor has on the circuit must be exactly the same as the effect of the combination of the two individual capacitors. That is, the equivalent capacitor must store charge Q_{tot} when connected to the battery. Active Figure 20.25c shows that the voltage across

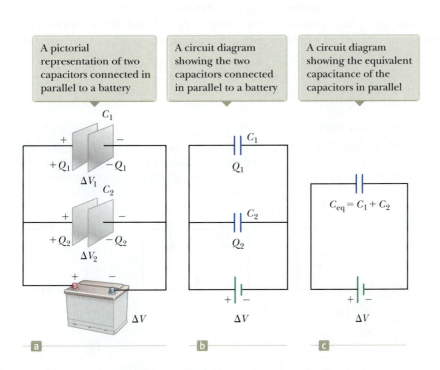

A pictorial representation of two capacitors connected in parallel to a battery

A circuit diagram showing the two capacitors connected in parallel to a battery

A circuit diagram showing the equivalent capacitance of the capacitors in parallel

C_1

$+Q_1$ $-Q_1$

ΔV_1 C_2

$+Q_2$ $-Q_2$

ΔV_2

ΔV

a

C_1

Q_1

C_2

Q_2

ΔV

b

$C_{eq} = C_1 + C_2$

ΔV

c

Active Figure 20.25 Two capacitors connected in parallel. All three diagrams are equivalent.

the equivalent capacitor is ΔV because the equivalent capacitor is connected directly across the battery terminals. Therefore, for the equivalent capacitor,

$$Q_{tot} = C_{eq} \Delta V$$

Substituting for the charges in Equation 20.25 gives

$$C_{eq} \Delta V = Q_1 + Q_2 = C_1 \Delta V_1 + C_2 \Delta V_2$$

$$C_{eq} = C_1 + C_2 \quad \text{(parallel combination)}$$

where we have canceled the voltages because they are all the same. If this treatment is extended to three or more capacitors connected in parallel, the **equivalent capacitance** is found to be

▶ Equivalent capacitance for capacitors in parallel

$$C_{eq} = C_1 + C_2 + C_3 + \cdots \quad \text{(parallel combination)} \qquad \textbf{20.26} \blacktriangleleft$$

Therefore, the equivalent capacitance of a parallel combination of capacitors is (1) the algebraic sum of the individual capacitances and (2) greater than any of the individual capacitances. Statement (2) makes sense because we are essentially combining the areas of all the capacitor plates when they are connected with conducting wire, and capacitance of parallel plates is proportional to area (Eq. 20.22).

Series Combination

Two capacitors connected as shown in Active Figure 20.26a and the equivalent circuit diagram in Active Figure 20.26b are known as a **series combination** of capacitors. The left plate of capacitor 1 and the right plate of capacitor 2 are connected to the terminals of a battery. The other two plates are connected to each other and to nothing else; hence, they form an isolated system that is initially uncharged and must continue to have zero net charge. To analyze this combination, let's first consider the uncharged capacitors and then follow what happens immediately after a battery is connected to the circuit. When the battery is connected, electrons are transferred to the leftmost wire out of the left plate of C_1 and into the right plate of C_2 from the rightmost wire. As this negative charge accumulates on the right plate of C_2, an equivalent amount of negative charge is forced off the left plate of C_2, and this left plate therefore has an excess positive charge. The negative charge leaving the left plate of C_2 causes negative charges to accumulate on the right plate of C_1. As a result, both right plates end up with a charge $-Q$ and both left plates end up with a charge $+Q$. Therefore, the charges on capacitors connected in series are the same:

$$Q_1 = Q_2 = Q$$

Active Figure 20.26 Two capacitors connected in series. All three diagrams are equivalent.

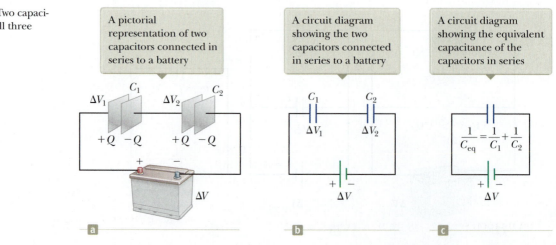

A pictorial representation of two capacitors connected in series to a battery

A circuit diagram showing the two capacitors connected in series to a battery

A circuit diagram showing the equivalent capacitance of the capacitors in series

where Q is the charge that moved between a wire and the connected outside plate of one of the capacitors.

Active Figure 20.26a shows that the total voltage ΔV_{tot} across the combination is split between the two capacitors:

$$\Delta V_{tot} = \Delta V_1 + \Delta V_2 \qquad \text{20.27} \blacktriangleleft$$

where ΔV_1 and ΔV_2 are the potential differences across capacitors C_1 and C_2, respectively. In general, the total potential difference across any number of capacitors connected in series is the sum of the potential differences across the individual capacitors.

Suppose the equivalent single capacitor in Active Figure 20.26c has the same effect on the circuit as the series combination when it is connected to the battery. After it is fully charged, the equivalent capacitor must have a charge of $-Q$ on its right plate and a charge of $+Q$ on its left plate. Applying the definition of capacitance to the circuit in Active Figure 20.26c gives

$$\Delta V_{tot} = \frac{Q}{C_{eq}}$$

Substituting for the voltages in Equation 20.27, we have

$$\frac{Q}{C_{eq}} = \Delta V_1 + \Delta V_2 = \frac{Q_1}{C_1} + \frac{Q_2}{C_2}$$

Canceling the charges because they are all the same gives

$$\frac{1}{C_{eq}} = \frac{1}{C_1} + \frac{1}{C_2} \quad \text{(series combination)}$$

When this analysis is applied to three or more capacitors connected in series, the relationship for the **equivalent capacitance** is

$$\frac{1}{C_{eq}} = \frac{1}{C_1} + \frac{1}{C_2} + \frac{1}{C_3} + \cdots \quad \text{(series combination)} \qquad \text{20.28} \blacktriangleleft$$

▶ Equivalent capacitance for capacitors in series

This expression shows that (1) the inverse of the equivalent capacitance is the algebraic sum of the inverses of the individual capacitances and (2) the equivalent capacitance of a series combination is always less than any individual capacitance in the combination.

QUICK QUIZ 20.7 Two capacitors are identical. They can be connected in series or in parallel. If you want the *smallest* equivalent capacitance for the combination, how should you connect them? **(a)** in series **(b)** in parallel **(c)** either way because both combinations have the same capacitance

Example 20.8 | Equivalent Capacitance

Find the equivalent capacitance between a and b for the combination of capacitors shown in Figure 20.27a. All capacitances are in microfarads.

SOLUTION

Conceptualize Study Figure 20.27a carefully and make sure you understand how the capacitors are connected.

Categorize Figure 20.27a shows that the circuit contains both series and parallel connections, so we use the rules for series and parallel combinations discussed in this section.

Analyze Using Equations 20.26 and 20.28, we reduce the combination step by step as indicated in the figure.

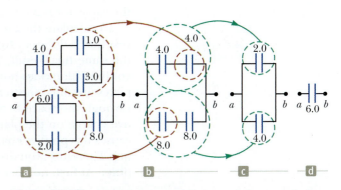

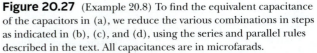

Figure 20.27 (Example 20.8) To find the equivalent capacitance of the capacitors in (a), we reduce the various combinations in steps as indicated in (b), (c), and (d), using the series and parallel rules described in the text. All capacitances are in microfarads.

continued

20.8 *cont.*

The 1.0-μF and 3.0-μF capacitors (upper red-brown circle in Fig. 20.27a) are in parallel. Find the equivalent capacitance from Equation 20.26:

$$C_{eq} = C_1 + C_2 = 4.0 \ \mu\text{F}$$

The 2.0-μF and 6.0-μF capacitors (lower red-brown circle in Fig. 20.27a) are also in parallel:

$$C_{eq} = C_1 + C_2 = 8.0 \ \mu\text{F}$$

The circuit now looks like Figure 20.27b. The two 4.0-μF capacitors (upper green circle in Fig. 20.27b) are in series. Find the equivalent capacitance from Equation 20.28:

$$\frac{1}{C_{eq}} = \frac{1}{C_1} + \frac{1}{C_2} = \frac{1}{4.0 \ \mu\text{F}} + \frac{1}{4.0 \ \mu\text{F}} = \frac{1}{2.0 \ \mu\text{F}}$$

$$C_{eq} = 2.0 \ \mu\text{F}$$

The two 8.0-μF capacitors (lower green circle in Fig. 20.27b) are also in series. Find the equivalent capacitance from Equation 20.28:

$$\frac{1}{C_{eq}} = \frac{1}{C_1} + \frac{1}{C_2} = \frac{1}{8.0 \ \mu\text{F}} + \frac{1}{8.0 \ \mu\text{F}} = \frac{1}{4.0 \ \mu\text{F}}$$

$$C_{eq} = 4.0 \ \mu\text{F}$$

The circuit now looks like Figure 20.27c. The 2.0-μF and 4.0-μF capacitors are in parallel:

$$C_{eq} = C_1 + C_2 = \boxed{6.0 \ \mu\text{F}}$$

Finalize This final value is that of the single equivalent capacitor shown in Figure 20.27d. For further practice in treating circuits with combinations of capacitors, imagine a battery is connected between points *a* and *b* in Figure 20.27a so that a potential difference ΔV is established across the combination. Can you find the voltage across and the charge on each capacitor?

20.9 | Energy Stored in a Charged Capacitor

Almost everyone who works with electronic equipment has at some time verified that a capacitor can store energy. If the plates of a charged capacitor are connected by a conductor, such as a wire, charge transfers between the plates and the wire until the two plates are uncharged. The discharge can often be observed as a visible spark. If you accidentally touch the opposite plates of a charged capacitor, your fingers act as pathways by which the capacitor discharges, resulting in an electric shock. The degree of shock depends on the capacitance and the voltage applied to the capacitor. When high voltages are present, such as in the power supply of an electronic instrument, the shock can be fatal.

Consider a parallel-plate capacitor that is initially uncharged so that the initial potential difference across the plates is zero. Now imagine that the capacitor is connected to a battery and develops a charge of Q. The final potential difference across the capacitor is $\Delta V = Q/C$.

To calculate the energy stored in the capacitor, we shall assume a charging process that is different from the actual process described in Section 20.7 but that gives the same final result. This assumption is justified because the energy in the final configuration does not depend on the actual charge-transfer process.[2] Imagine the plates are disconnected from the battery and you transfer the charge mechanically through the space between the plates as follows. You grab a small amount of positive charge on the plate connected to the negative terminal and apply a force that causes this positive charge to move over to the plate connected to the positive terminal. Therefore, you do work on the charge as it is transferred from one plate to the other. At first, no work is required to transfer a small amount of charge dq from one

---

[2]This discussion is similar to that of state variables in thermodynamics. The change in a state variable such as temperature is independent of the path followed between the initial and final states. The potential energy of a capacitor (or any system) is also a state variable, so its change does not depend on the process followed to charge the capacitor.

plate to the other,[3] but once this charge has been transferred, a small potential difference exists between the plates. Therefore, work must be done to move additional charge through this potential difference. As more and more charge is transferred from one plate to the other, the potential difference increases in proportion and more work is required.

The work required to transfer an increment of charge dq from one plate to the other is

$$dW = \Delta V \, dq = \frac{q}{C} \, dq$$

Therefore, the total work required to charge the capacitor from $q = 0$ to the final charge $q = Q$ is

$$W = \int_0^Q \frac{q}{C} \, dq = \frac{Q^2}{2C}$$

The capacitor can be modeled as a nonisolated system for this discussion. The work done by the external agent on the system in charging the capacitor appears as potential energy U stored in the capacitor. In reality, of course, this energy is not the result of mechanical work done by an external agent moving charge from one plate to the other, but is due to transformation of chemical energy in the battery. We have used a model of work done by an external agent that gives us a result that is also valid for the actual situation. Using $Q = C \Delta V$, the energy stored in a charged capacitor can be expressed in the following alternative forms:

$$U = \frac{Q^2}{2C} = \tfrac{1}{2} Q \Delta V = \tfrac{1}{2} C (\Delta V)^2 \qquad \textbf{20.29} \blacktriangleleft$$

► Energy stored in a charged capacitor

This result applies to *any* capacitor, regardless of its geometry. In practice, the maximum energy (or charge) that can be stored is limited because electric discharge ultimately occurs between the plates of the capacitor at a sufficiently large value of ΔV. For this reason, capacitors are usually labeled with a maximum operating voltage.

For an object on an extended spring, the elastic potential energy can be modeled as being stored *in the spring*. Internal energy of a substance associated with its temperature is located *throughout the substance*. Where is the energy in a capacitor located? The energy stored in a capacitor can be modeled as being stored *in the electric field between the plates of the capacitor*. For a parallel-plate capacitor, the potential difference is related to the electric field through the relationship $\Delta V = Ed$. Furthermore, the capacitance is $C = \epsilon_0 A / d$. Substituting these expressions into Equation 20.29 gives

$$U = \tfrac{1}{2} \left(\frac{\epsilon_0 A}{d} \right) (Ed)^2 = \tfrac{1}{2} (\epsilon_0 A d) E^2 \qquad \textbf{20.30} \blacktriangleleft$$

Because the volume of a parallel-plate capacitor that is occupied by the electric field is Ad, the energy per unit volume $u_E = U/Ad$, called the **energy density**, is

$$u_E = \tfrac{1}{2} \epsilon_0 E^2 \qquad \textbf{20.31} \blacktriangleleft$$

► Energy density in an electric field

Although Equation 20.31 was derived for a parallel-plate capacitor, the expression is generally valid. That is, the energy density in any electric field is proportional to the square of the magnitude of the electric field at a given point.

◄ **QUICK QUIZ 20.8** You have three capacitors and a battery. In which of the following combinations of the three capacitors is the maximum possible energy stored when the combination is attached to the battery? **(a)** series **(b)** parallel **(c)** no difference because both combinations store the same amount of energy

---

[3]We shall use lowercase q for the time-varying charge on the capacitor while it is charging to distinguish it from uppercase Q, which is the total charge on the capacitor after it is completely charged.

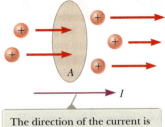

The direction of the current is the direction in which positive charges flow when free to do so.

Figure 21.1 Charges in motion through an area A. The time rate at which charge flows through the area is defined as the current I.

▶ Electric current

Pitfall Prevention | 21.1
"Current Flow" Is Redundant
The phrase *current flow* is commonly used, although it is technically incorrect because current is a flow (of charge). This wording is similar to the phrase *heat transfer*, which is also redundant because heat is a transfer (of energy). We will avoid this phrase and speak of *flow of charge* or *charge flow*.

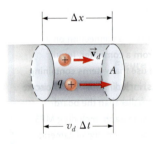

Figure 21.2 A segment of a uniform conductor of cross-sectional area A.

⟨21.1 | Electric Current

Whenever charge is flowing, an **electric current** is said to exist. To define current mathematically, suppose charged particles are moving perpendicular to a surface of area A as in Figure 21.1. (This area could be the cross-sectional area of a wire, for example.) The current is defined as **the rate at which electric charge flows through this surface.** If ΔQ is the amount of charge that passes through this area in a time interval Δt, the average current I_{avg} over the time interval is the ratio of the charge to the time interval:

$$I_{avg} = \frac{\Delta Q}{\Delta t} \qquad \textbf{21.1} ◀$$

It is possible for the rate at which charge flows to vary in time. We define the **instantaneous current** I as the limit of the preceding expression as Δt goes to zero:

$$I \equiv \lim_{\Delta t \to 0} \frac{\Delta Q}{\Delta t} = \frac{dQ}{dt} \qquad \textbf{21.2} ◀$$

The SI unit of current is the **ampere** (A):

$$1\,\text{A} = 1\,\text{C/s} \qquad \textbf{21.3} ◀$$

That is, 1 A of current is equivalent to 1 C of charge passing through a surface in 1 s.

The particles flowing through a surface as in Figure 21.1 can be charged positively or negatively, or we can have two or more types of particles moving, with charges of both signs in the flow. Conventionally, we define the direction of the current as the direction of flow of positive charge, regardless of the sign of the actual charged particles in motion.[1] In a common conductor such as copper, the current is physically due to the motion of the negatively charged electrons. Therefore, when we speak of current in such a conductor, the direction of the current is opposite the direction of flow of electrons. On the other hand, if one considers a beam of positively charged protons in a particle accelerator, the current is in the direction of motion of the protons. In some cases—gases and electrolytes, for example—the current is the result of the flow of both positive and negative charged particles. It is common to refer to a moving charged particle (whether it is positive or negative) as a mobile **charge carrier**. For example, the charge carriers in a metal are electrons.

We now build a structural model that will allow us to relate the macroscopic current to the motion of the charged particles. Consider identical charged particles moving in a conductor of cross sectional area A (Fig. 21.2). The volume of a segment of the conductor of length Δx (between the two circular cross sections shown in Fig. 21.2) is $A\,\Delta x$. If n represents the number of mobile charge carriers per unit volume (in other words, the charge carrier density), the number of carriers in the segment is $nA\,\Delta x$. Therefore, the total charge ΔQ in this segment is

$$\Delta Q = \text{number of carriers in section} \times \text{charge per carrier} = (nA\,\Delta x)q$$

where q is the charge on each carrier. If the carriers move with an average velocity component v_d in the x direction (along the wire), the displacement they experience in this direction in a time interval Δt is $\Delta x = v_d\,\Delta t$. The speed v_d of the charge carrier along the wire is an average speed called the **drift speed**. Let us choose Δt to be the time interval required for the charges in the segment to move through a displacement whose magnitude is equal to the length of the segment. This time interval is also that required for all the charges in the segment to pass through the circular area at one end. With this choice, we can write ΔQ in the form

$$\Delta Q = (nAv_d\,\Delta t)q$$

---

[1]Even though we discuss a direction for current, current is not a vector. As we shall see later in the chapter, currents add algebraically and not vectorially.

If we divide both sides of this equation by Δt, we see that the average current in the conductor is

$$I_{avg} = \frac{\Delta Q}{\Delta t} = nqv_d A \qquad \textbf{21.4} \blacktriangleleft$$

▶ Current in terms of microscopic parameters

Equation 21.4 relates a macroscopically measured average current to the microscopic origin of the current: the density of charge carriers n, the charge per carrier q, and the drift speed v_d.

QUICK QUIZ 21.1 Consider positive and negative charges moving horizontally through the four regions shown in Figure 21.3. Rank the currents in these four regions, from highest to lowest.

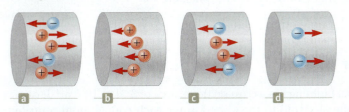

Figure 21.3 (Quick Quiz 21.1) Four groups of charges move through a region.

Let us investigate further the notion of drift speed. We have identified drift speed as an average speed along the wire, but the charge carriers are by no means moving in a straight line with speed v_d. Consider a conductor in which the charge carriers are free electrons. In the absence of a potential difference across the conductor, these electrons undergo random motion similar to that of gas molecules in the structural model of kinetic theory that we studied in Chapter 16. This random motion is related to the temperature of the conductor. The electrons undergo repeated collisions with the metal atoms, and the result is a complicated zigzag motion. When a potential difference is applied across the conductor, an electric field is established in the conductor. The electric field exerts an electric force on the electrons (Eq. 19.4). This force accelerates the electrons and hence produces a current. The motion of the electrons due to the electric force is superimposed on their random motion to provide an average velocity whose magnitude is the drift speed as shown in Active Figure 21.4.

When electrons make collisions with metal atoms during their motion, they transfer energy to the atoms. This energy transfer causes an increase in the vibrational energy of the atoms and a corresponding increase in the temperature of the conductor.[2] This process involves all three types of energy storage in the conservation of energy equation, Equation 7.2. If we consider the system to be the electrons, the metal atoms, and the electric field (which is established by an external source such as a battery), the energy at the instant when the potential difference is applied across the conductor is electric potential energy associated with the field and the electrons. This energy is transformed by work done within the system by the field on the electrons to kinetic energy of electrons. When the electrons strike the metal atoms, some of the kinetic energy is transferred to the atoms, which adds to the internal energy of the system.

The **current density** J in the conductor is defined as the current per unit area. From Equation 21.4, the current density is

$$J \equiv \frac{I}{A} = nqv_d \qquad \textbf{21.5} \blacktriangleleft$$

where J has the SI units amperes per square meter.

Pitfall Prevention | 21.2
Batteries Do Not Supply Electrons A battery does not supply electrons to the circuit. It establishes the electric field that exerts a force on electrons already in the wires and elements of the circuit.

The random motion of the charge carriers is modified by the field, and they have a drift velocity opposite the direction of the electric field.

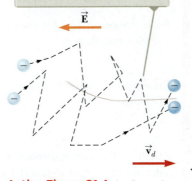

Active Figure 21.4 A schematic representation of the zigzag motion of negative charge carriers in a conductor. Because of the acceleration of the charge carriers due to the electric force, the paths are actually parabolic. The drift speed, however, is much smaller than the average speed, so the parabolic shape is not visible on this scale.

---

[2]This increase in temperature is sometimes called *Joule heating*, but that term is a misnomer because there is no heat involved. We will not use this wording.

> **THINKING PHYSICS 21.1**
>
> In Chapter 19, we claimed that the electric field inside a conductor is zero. In the preceding discussion, however, we have used the notion of an electric field in a conducting wire that exerts electric forces on electrons, causing them to move with a drift velocity. Is this notion inconsistent with Chapter 19?
>
> **Reasoning** The electric field is zero only in a conductor in *electrostatic equilibrium*, that is, a conductor in which the charges are at rest after having moved to equilibrium positions. In a current-carrying conductor, the charges are not at rest, so the requirement for a zero field is not imposed. The electric field in a conductor in a circuit is due to a distribution of charge over the surface of the conductor that can be quite complicated.[3] ◄

Example 21.1 | Drift Speed in a Copper Wire

The 12-gauge copper wire in a typical residential building has a cross-sectional area of 3.31×10^{-6} m^2. It carries a constant current of 10.0 A. What is the drift speed of the electrons in the wire? Assume each copper atom contributes one free electron to the current. The density of copper is 8.92 g/cm^3.

SOLUTION

Conceptualize Imagine electrons following a zigzag motion with a drift velocity parallel to the wire superimposed on the motion as in Active Figure 21.4. As mentioned earlier, the drift speed is small, and this example helps us quantify the speed.

Categorize We evaluate the drift speed using Equation 21.4. Because the current is constant, the average current during any time interval is the same as the constant current: $I_{avg} = I$.

Analyze The periodic table of the elements in Appendix C shows that the molar mass of copper is $M = 63.5$ g/mol. Recall that 1 mol of any substance contains Avogadro's number of atoms ($N_A = 6.02 \times 10^{23}$ mol^{-1}).

Use the molar mass and the density of copper to find the volume of 1 mole of copper:

$$V = \frac{M}{\rho}$$

From the assumption that each copper atom contributes one free electron to the current, find the electron density in copper:

$$n = \frac{N_A}{V} = \frac{N_A \rho}{M}$$

Solve Equation 21.4 for the drift speed and substitute for the electron density:

$$v_d = \frac{I_{avg}}{nqA} = \frac{I}{nqA} = \frac{IM}{qAN_A\rho}$$

Substitute numerical values:

$$v_d = \frac{(10.0 \text{ A})(0.063\ 5 \text{ kg/mol})}{(1.60 \times 10^{-19} \text{ C})(3.31 \times 10^{-6} \text{ m}^2)(6.02 \times 10^{23} \text{ mol}^{-1})(8\ 920 \text{ kg/m}^3)}$$

$$= 2.23 \times 10^{-4} \text{ m/s}$$

Finalize This result shows that typical drift speeds are very small. For instance, electrons traveling at 2.23×10^{-4} m/s would take about 75 min to travel 1 m! You might therefore wonder why a light turns on almost instantaneously when its switch is thrown. In a conductor, changes in the electric field that drives the free electrons travel through the conductor with a speed close to that of light. So, when you flip on a light switch, electrons already in the filament of an incandescent lightbulb experience electric forces and begin moving after a time interval on the order of nanoseconds.

---

[3]See Chapter 18 in R. Chabay and B. Sherwood, *Matter & Interactions II: Electric and Magnetic Interactions* (Hoboken: Wiley, 2007) for details on this charge distribution.

21.2 | Resistance and Ohm's Law

The drift speed of electrons in a current-carrying wire is related to the electric field in the wire. If the field is increased, the electric force on the electrons is stronger and the drift speed increases. We shall show in Section 21.4 that this relationship is linear and that the drift speed is directly proportional to the electric field. For a uniform field in a conductor of uniform cross-section, the potential difference across the conductor is proportional to the electric field as in Equation 20.6. Therefore, when a potential difference ΔV is applied across the ends of a metallic conductor as in Figure 21.5, the current in the conductor is found to be proportional to the applied voltage; that is, $I \propto \Delta V$. We can write this proportionality as $\Delta V = IR$, where R is called the **resistance** of the conductor. We define this resistance according to the equation we have just written, as the ratio of the voltage across the conductor to the current it carries:

$$R \equiv \frac{\Delta V}{I} \qquad \text{21.6} \blacktriangleleft$$

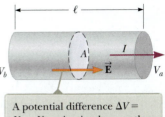

A potential difference $\Delta V = V_b - V_a$ maintained across the conductor sets up an electric field $\vec{\mathbf{E}}$, and this field produces a current I that is proportional to the potential difference.

Figure 21.5 A uniform conductor of length ℓ and cross-sectional area A.

▶ Definition of resistance

Resistance has the SI unit volt per ampere, called an **ohm** (Ω). Therefore, if a potential difference of 1 V across a conductor produces a current of 1 A, the resistance of the conductor is 1 Ω. As another example, if an electrical appliance connected to a 120-V source carries a current of 6.0 A, its resistance is 20 Ω.

Resistance is the quantity that determines the current that results due to a voltage in a simple circuit. For a fixed voltage, if the resistance increases, the current decreases. If the resistance decreases, the current increases.

It might be useful for you to build a mental representation for current, voltage, and resistance by comparing these concepts to analogous concepts for the flow of water in a river. As water flows downhill in a river of constant width and depth, the rate of flow of water (analogous to current) depends on the total vertical distance through which the water drops between two points (analogous to voltage) and on the width and depth as well as on the effects of rocks, the riverbank, and other obstructions (analogous to resistance). Likewise, electric current in a uniform conductor depends on the applied voltage, and the resistance of the conductor is caused by collisions of the electrons with atoms in the conductor.

For many materials, including most metals, experiments show that the resistance is constant over a wide range of applied voltages. This behavior is known as **Ohm's law** after Georg Simon Ohm (1787–1854), who was the first to conduct a systematic study of electrical resistance.

Many individuals call Equation 21.6 Ohm's law, but this terminology is incorrect. This equation is simply the definition of resistance, and it provides an important relationship between voltage, current, and resistance. Ohm's law is *not* a fundamental law of nature, but a behavior that is valid only for certain materials and devices, and only over a limited range of conditions. Materials or devices that obey Ohm's law, and hence that have a constant resistance over a wide range of voltages, are said to be **ohmic** (Fig. 21.6a). Materials or devices that do not obey Ohm's law are **nonohmic**. One common semiconducting device that is nonohmic is the *diode*, a circuit element that acts like a one-way valve for current. Its resistance is small for currents in one direction (positive ΔV) and large for currents in the reverse direction (negative ΔV)

Pitfall Prevention | 21.3
We've Seen Something Like Equation 21.6 Before
In Chapter 4, we introduced Newton's second law, $\Sigma F = ma$, for a net force on an object of mass m. It can be written as

$$m = \frac{\Sigma F}{a}$$

In Chapter 4, we defined mass as *resistance to a change in motion in response to an external force*. Mass as resistance to changes in motion is analogous to electrical resistance to charge flow, and Equation 21.6 is analogous to the form of Newton's second law above. Each equation states that the resistance (electrical or mechanical) is equal to (1) ΔV, the cause of current, or (2) ΣF, the cause of changes in motion, divided by the result, (1) a charge flow, quantified by current I, or (2) a change in motion, quantified by acceleration a.

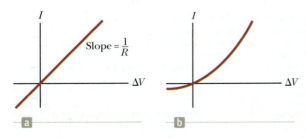

Figure 21.6 (a) The current–potential difference curve for an ohmic material. The curve is linear, and the slope is equal to the inverse of the resistance of the conductor. (b) A nonlinear current–potential difference curve for a semiconducting diode. This device does not obey Ohm's law.

as shown in Figure 21.6b. Most modern electronic devices have nonlinear current–voltage relationships; their operation depends on the particular ways they violate Ohm's law.

> **QUICK QUIZ 21.2** In Figure 21.6b, as the applied voltage increases, does the resistance of the diode **(a)** increase, **(b)** decrease, or **(c)** remain the same?

A **resistor** is a simple circuit element that provides a specified resistance in an electrical circuit. The symbol for a resistor in circuit diagrams is a zigzag red line (—W—). We can express Equation 21.6 in the form

$$\Delta V = IR \qquad \text{21.7} \blacktriangleleft$$

This equation tells us that the voltage across a resistor is the product of the resistance and the current in the resistor.

The resistance of an ohmic conducting wire such as that shown in Figure 21.5 is found to be proportional to its length ℓ and inversely proportional to its cross-sectional area A. That is,

▶ Resistance of a uniform material of resistivity ρ along a length ℓ

$$R = \rho \frac{\ell}{A} \qquad \text{21.8} \blacktriangleleft$$

where the constant of proportionality ρ is called the **resistivity** of the material,[4] which has the unit ohm · meter ($\Omega \cdot$ m). To understand this relationship between resistance and resistivity, note that every ohmic material has a characteristic resistivity, a parameter that depends on the properties of the material and on temperature. On the other hand, as you can see from Equation 21.8, the resistance of a particular conductor depends on its size and shape as well as on the resistivity of the material. Table 21.1 provides a list of resistivities for various materials measured at 20°C.

Pitfall Prevention | 21.4
Resistance and Resistivity
Resistivity is a property of a *substance*, whereas resistance is a property of an *object*. We have seen similar pairs of variables before. For example, density is a property of a substance, whereas mass is a property of an object. Equation 21.8 relates resistance to resistivity, and we have seen a previous equation (Eq. 1.1) that relates mass to density.

> **TABLE 21.1 | Resistivities and Temperature Coefficients of Resistivity for Various Materials**

Material	Resistivity[a] ($\Omega \cdot$ m)	Temperature Coefficient[b] α [(°C)$^{-1}$]
Silver	1.59×10^{-8}	3.8×10^{-3}
Copper	1.7×10^{-8}	3.9×10^{-3}
Gold	2.44×10^{-8}	3.4×10^{-3}
Aluminum	2.82×10^{-8}	3.9×10^{-3}
Tungsten	5.6×10^{-8}	4.5×10^{-3}
Iron	10×10^{-8}	5.0×10^{-3}
Platinum	11×10^{-8}	3.92×10^{-3}
Lead	22×10^{-8}	3.9×10^{-3}
Nichrome[c]	1.00×10^{-6}	0.4×10^{-3}
Carbon	3.5×10^{-5}	-0.5×10^{-3}
Germanium	0.46	-48×10^{-3}
Silicon[d]	2.3×10^{3}	-75×10^{-3}
Glass	10^{10} to 10^{14}	
Hard rubber	$\sim 10^{13}$	
Sulfur	10^{15}	
Quartz (fused)	75×10^{16}	

[a]All values at 20°C. All elements in this table are assumed to be free of impurities.
[b]The temperature coefficient of resistivity will be discussed later in this section.
[c]A nickel–chromium alloy commonly used in heating elements. The resistivity of Nichrome varies with composition and ranges between 1.00×10^{-6} and $1.50 \times 10^{-6} \Omega \cdot$ m.
[d]The resistivity of silicon is very sensitive to purity. The value can be changed by several orders of magnitude when it is doped with other atoms.

---

[4]The symbol ρ used for resistivity should not be confused with the same symbol used earlier in the text for mass density and volume charge density.

An assortment of resistors used in electric circuits.

The inverse of the resistivity is defined[5] as the **conductivity** σ. Hence, the resistance of an ohmic conductor can be expressed in terms of its conductivity as

$$R = \frac{\ell}{\sigma A} \qquad \blacktriangleleft \text{ 21.9}$$

where $\sigma = 1/\rho$.

Equation 21.9 shows that the resistance of a conductor is proportional to its length and inversely proportional to its cross-sectional area, similar to the flow of liquid through a pipe. As the length of the pipe is increased and the pressure difference between the ends of the pipe is held constant, the pressure difference between any two points separated by a fixed distance decreases and there is less force pushing the element of fluid between these points through the pipe. Therefore, less fluid flows for a given pressure difference between the ends of the pipe, representing an increased resistance. As its cross-sectional area is increased, the pipe can transport more fluid in a given time interval for a given pressure difference between the ends of the pipe, so its resistance drops.

As another analogy between electrical circuits and our previous studies, let us combine Equations 21.6 and 21.9:

$$R = \frac{\ell}{\sigma A} = \frac{\Delta V}{I} \;\rightarrow\; I = \sigma A \frac{\Delta V}{\ell} \;\rightarrow\; \frac{q}{\Delta t} = \sigma A \frac{\Delta V}{\ell}$$

where q is the amount of charge transferred in a time interval Δt. Let us compare this equation to Equation 17.35 for conduction of energy through a slab of material of area A, length ℓ, and thermal conductivity k, which we reproduce below:

$$P = kA \frac{(T_h - T_c)}{L} \;\rightarrow\; \frac{Q}{\Delta t} = kA \frac{\Delta T}{L}$$

In this equation, Q is the amount of energy transferred by heat in a time interval Δt. Notice the striking similarity between these last two equations.

Another analogy arises in an example that is important in biochemical applications. *Fick's law* describes the rate of transfer of a chemical solute through a solvent by the process of *diffusion*. This transfer occurs because of a difference in concentration of the solute (mass of solute per volume) between the two locations. Fick's law is as follows:

BIO Diffusion in biological systems

$$\frac{n}{\Delta t} = DA \frac{\Delta C}{L}$$

where $n/\Delta t$ is the rate of flow of the solute in moles per second, A is the area through which the solute moves, and L is the length over which the concentration difference is ΔC. The concentration is measured in moles per cubic meter. The parameter D is a diffusion constant (with units of meters squared per second) that describes the

---

[5]Do not confuse the symbol σ for conductivity with the same symbol used earlier for the Stefan–Boltzmann constant and surface charge density.

The colored bands on this resistor are yellow, violet, black, and gold.

dexns/Shutterstock.com

Figure 21.7 A close-up view of a circuit board shows the color coding on a resistor.

BIO Electrical activity in the heart

BIO Catheter ablation for atrial fibrillation

rate of diffusion of a solute through the solvent and is similar in nature to electrical or thermal conductivity. Fick's law has important applications in describing the transport of molecules across biological membranes.

All three of the preceding equations have exactly the same mathematical form. Each has a time rate of change on the left, and each has the product of a conductivity, an area, and a ratio of a difference in a variable to a length on the right. This type of equation is a *transport equation* used when we transport energy, charge, or moles of matter. The difference in the variable on the right side of each equation is what drives the transport. A temperature difference drives energy transfer by heat, an electric potential difference drives a transfer of charge, and a concentration difference drives a transfer of matter.

Most electric circuits use resistors to control the current level in the various parts of the circuit. Two common types of resistors are the *composition* resistor containing carbon and the *wire-wound* resistor, which consists of a coil of wire. Resistors are normally color-coded to give their values in ohms, as shown in Figure 21.7 and Table 21.2. As an example, the four colors on the resistor at the bottom of Figure 21.7 are yellow (= 4), violet (= 7), black (= 10^0), and gold (= 5%), and so the resistance value is $47 \times 10^0 \ \Omega = 47 \ \Omega$ with a tolerance value of $5\% = 2 \ \Omega$.

Let's consider the role of electrical resistance in maintaining proper beating of the human heart. The right atrium of the heart contains a specialized set of muscle fibers called the SA (sinoatrial) node that initiates the heartbeat. Electrical impulses that originate in these fibers gradually spread from cell to cell throughout the right and left atrial muscles, causing them to contract. When the impulses reach the atrioventricular (AV) node, the muscles of the atria begin to relax, and the impulses are directed to the ventricular muscles by an assembly of heart muscle cells called the *bundle of His* and the *Purkinje fibers*. After the resulting contraction of the ventricles, the heartbeat is complete and the cycle begins again.

The heart can experience a number of *arrhythmias*, in which the normal heartbeat rhythm is interrupted. Arrhythmias are generally caused by abnormal electrical activity in the heart. The most common cardiac arrhythmia is *atrial fibrillation* (AF). In this condition, the two upper chambers of the heart, the atria, undergo random quivering at a rate that can be greater than 300 per minute, rather than the usual coordinated contractions. In *paroxysmal* AF, the patient goes into episodes of atrial fibrillation that may last from a few minutes to a few days. It some cases, the condition may even become chronic. With episodes longer than a few days, blood can pool in the atria, due to the inefficiency of the quivering action in pumping blood out of the heart. This pooled blood can result in clots, which can then travel to the brain and cause a stroke. Patients with long-lasting episodes of AF are treated with anticoagulants to prevent clots, rate control medications to slow the rate of impulses conducted to the ventricles, and antiarrhythmics to return the heart to its normal rhythm. Defibrillator device paddles are sometimes used to deliver an electric shock to a patient's chest in order to restore normal heart rhythm.

In a large percentage of patients, the source of the chaotic activity is found in the four pulmonary veins leading into the left atrium. Atrial tissue has grown into these veins and can act as electrical triggers competing with the SA node. As a result, the atrial muscles receive electrical signals from a variety of sources rather than the SA node alone, leading to chaotic contractions. Patients whose arrhythmias can not be controlled with medications, as well as those who do not wish to take medications, have an additional option. A procedure known as *cardiac catheter ablation* may be performed by an *electrophysiologist* in an effort to restore normal sinus rhythm. In this procedure, the patient is anesthetized

◀ **TABLE 21.2 | Color Code for Resistors**

Color	Number	Multiplier	Tolerance
Black	0	1	
Brown	1	10^1	
Red	2	10^2	
Orange	3	10^3	
Yellow	4	10^4	
Green	5	10^5	
Blue	6	10^6	
Violet	7	10^7	
Gray	8	10^8	
White	9	10^9	
Gold		10^{-1}	5%
Silver		10^{-2}	10%
Colorless			20%

and catheters are inserted into a vein in the groin and guided into the right atrium of the heart. The catheter then punctures the septum and enters the left atrium. Figure 21.8 shows an ablation catheter passing into the heart from a vein. The electrophysiologist maps the atrium and then stimulates the heart to determine areas of abnormal electrical activity. Finally, the electrophysiologist *ablates* the tissue around the four pulmonary veins, usually with radiofrequency energy from the tip of one of the catheters. The resulting scar tissue represents a high-resistance path, through which the electrical signals from the AF triggers in the pulmonary veins cannot travel. As a result, the SA node alone again controls the electrical activity of the heart. Because the triggers in the pulmonary veins have been cut off electrically from the rest of the heart, this specific procedure is called a *pulmonary vein isolation*.

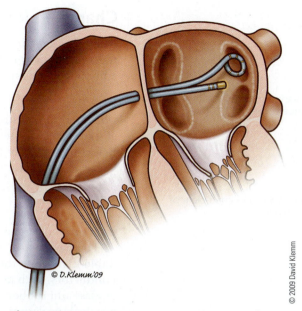

© D.Klemm'09

© 2009 David Klemm

Figure 21.8 During a cardiac catheter ablation procedure, catheters are guided into the left atrium through a vein from the groin. Radiofrequency energy is used to ablate tissue surrounding the pulmonary veins, where abnormal electrical activity is happening.

Example 21.2 | **The Resistance of Nichrome Wire**

The radius of 22-gauge Nichrome wire is 0.32 mm.

(A) Calculate the resistance per unit length of this wire.

SOLUTION

Conceptualize Table 21.1 shows that Nichrome has a resistivity two orders of magnitude larger than the best conductors in the table. Therefore, we expect it to have some special practical applications that the best conductors may not have.

Categorize We model the wire as a cylinder so that a simple geometric analysis can be applied to find the resistance.

Analyze Use Equation 21.8 and the resistivity of Nichrome from Table 21.1 to find the resistance per unit length:

$$\frac{R}{\ell} = \frac{\rho}{A} = \frac{\rho}{\pi r^2} = \frac{1.0 \times 10^{-6}\,\Omega \cdot \text{m}}{\pi(0.32 \times 10^{-3}\,\text{m})^2} = \boxed{3.1\ \Omega/\text{m}}$$

(B) If a potential difference of 10 V is maintained across a 1.0-m length of the Nichrome wire, what is the current in the wire?

SOLUTION

Analyze Use Equation 21.6 to find the current:

$$I = \frac{\Delta V}{R} = \frac{\Delta V}{(R/\ell)\ell} = \frac{10\ \text{V}}{(3.1\ \Omega/\text{m})(1.0\ \text{m})} = \boxed{3.2\ \text{A}}$$

Finalize Because of its high resistivity and resistance to oxidation, Nichrome is often used for heating elements in toasters, irons, and electric heaters.

What If? What if the wire were composed of copper instead of Nichrome? How would the values of the resistance per unit length and the current change?

Answer Table 21.1 shows us that copper has a resistivity two orders of magnitude smaller than that for Nichrome. Therefore, we expect the answer to part (A) to be smaller and the answer to part (B) to be larger. Calculations show that a copper wire of the same radius would have a resistance per unit length of only 0.053 Ω/m. A 1.0-m length of copper wire of the same radius would carry a current of 190 A with an applied potential difference of 10 V.

Change in Resistivity with Temperature

Resistivity depends on a number of factors, one of which is temperature. For most metals, resistivity increases approximately linearly with increasing temperature over a limited temperature range according to the expression

▶ Variation of resistivity with temperature

$$\rho = \rho_0[1 + \alpha(T - T_0)] \qquad \textbf{21.10} \blacktriangleleft$$

where ρ is the resistivity at some temperature T (in degrees Celsius), ρ_0 is the resistivity at some reference temperature T_0 (usually 20°C), and α is called the **temperature coefficient of resistivity** (not to be confused with the average coefficient of linear expansion α in Chapter 16). From Equation 21.10, we see that α can be expressed as

▶ Temperature coefficient of resistivity

$$\alpha = \frac{1}{\rho_0}\frac{\Delta\rho}{\Delta T} \qquad \textbf{21.11} \blacktriangleleft$$

where $\Delta\rho = \rho - \rho_0$ is the change in resistivity in the temperature interval $\Delta T = T - T_0$.

The resistivities and temperature coefficients of certain materials are listed in Table 21.1. Note the enormous range in resistivities, from very low values for good conductors, such as copper and silver, to very high values for good insulators, such as glass and rubber. An ideal, or "perfect," conductor would have zero resistivity, and an ideal insulator would have infinite resistivity.

Because resistance is proportional to resistivity according to Equation 21.8, the temperature variation of the resistance can be written as

▶ Variation of resistance with temperature

$$R = R_0[1 + \alpha(T - T_0)] \qquad \textbf{21.12} \blacktriangleleft$$

Precise temperature measurements are often made using this property.

> **QUICK QUIZ 21.3** When does an incandescent lightbulb carry more current? **(a)** immediately after it is turned on and the glow of the metal filament is increasing or **(b)** after it has been on for a few milliseconds and the glow is steady?

21.3 | Superconductors

For several metals, resistivity is nearly proportional to temperature as shown in Figure 21.9. In reality, however, there is always a nonlinear region at very low temperatures, and the resistivity usually approaches some finite value near absolute zero (see the magnified inset in Fig. 21.9). This residual resistivity near absolute zero is due primarily to collisions of electrons with impurities and to imperfections in the metal. In contrast, the high temperature resistivity (the linear region) is dominated by collisions of electrons with the vibrating metal atoms. We shall describe this process in more detail in Section 21.4.

There is a class of metals and compounds whose resistance decreases to zero when they are below a certain temperature T_c, known as the **critical temperature.** These materials are known as **superconductors.** The resistance–temperature graph for a superconductor follows that of a normal metal at temperatures above T_c (Fig. 21.10). When the temperature is at or below T_c, the resistivity drops suddenly

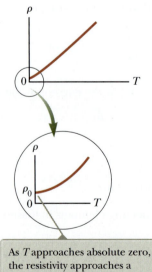

As T approaches absolute zero, the resistivity approaches a finite value ρ_0.

Figure 21.9 Resistivity versus temperature for a normal metal, such as copper. The curve is linear over a wide range of temperatures, and ρ increases with increasing temperature.

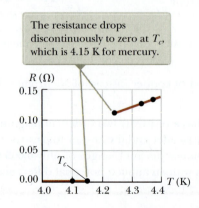

The resistance drops discontinuously to zero at T_c, which is 4.15 K for mercury.

Figure 21.10 Resistance versus temperature for a sample of mercury (Hg). The graph follows that of a normal metal above the critical temperature T_c.

to zero. This phenomenon was discovered in 1911 by Dutch physicist Heike Kamerlingh-Onnes (1853–1926) as he worked with mercury, which is a superconductor below 4.15 K. Measurements have shown that the resistivities of superconductors below their T_c values are less than 4×10^{-25} $\Omega \cdot$ m, or approximately 10^{17} times smaller than the resistivity of copper. In practice, these resistivities are considered to be zero.

Today, thousands of superconductors are known, and as Table 21.3 illustrates, the critical temperatures of recently discovered superconductors are substantially higher than initially thought possible. Two kinds of superconductors are recognized. The more recently identified ones are essentially ceramics with high critical temperatures, whereas superconducting materials such as those observed by Kamerlingh-Onnes are metals. If a room-temperature superconductor is ever identified, its effect on technology could be tremendous.

The value of T_c is sensitive to chemical composition, pressure, and molecular structure. Copper, silver, and gold, which are excellent conductors at room temperature, do not exhibit superconductivity.

One truly remarkable feature of superconductors is that once a current is set up in them, it persists *without any applied potential difference* (because $R = 0$). Steady currents have been observed to persist in superconducting loops for several years with no apparent decay!

An important and useful application of superconductivity is in the development of superconducting magnets, in which the magnitudes of the magnetic field are approximately ten times greater than those produced by the best normal electromagnets. Such superconducting magnets are being considered as a means of storing energy. Superconducting magnets are currently used in medical magnetic resonance imaging, or MRI, units, which produce high-quality images of internal organs without the need for excessive exposure of patients to x-rays or other harmful radiation.

21.4 | A Model for Electrical Conduction

In this section, we describe a classical model of electrical conduction in metals that was first proposed by Paul Drude (1863–1906) in 1900. This structural model leads to Ohm's law and shows that resistivity can be related to the motion of electrons in metals. Although the Drude model described here has limitations, it introduces concepts that are applied in more elaborate treatments.

Using the components of structural models introduced in Section 11.2, we can describe the Drude model as follows.

1. *A description of the physical components of the system:* Consider a conductor as a regular array of ionized atoms plus a collection of free electrons, which are sometimes called *conduction* electrons. The conduction electrons, although bound to their respective atoms when the atoms are not part of a solid, become free when the atoms condense into a solid.

2. *A description of where the components are located relative to one another and how they interact:* The conduction electrons fill the interior of the conductor. In the absence of an electric field, they move in random directions through the conductor. The situation is similar to the motion of gas molecules confined in a vessel. In fact, some scientists refer to conduction electrons in a metal as an *electron gas.* The conduction electrons experience no interaction with the array of ionized atoms except during a collision with one of those atoms.

3. *A description of the time evolution of the system:* When an electric field is applied to the conductor, the conduction electrons drift slowly in a direction opposite that of the electric field (Active Fig. 21.4), with an average drift speed v_d that is much smaller (typically 10^{-4} m/s) than their average speed between collisions

TABLE 21.3 | Critical Temperatures for Various Superconductors

Material	T_c (K)
$HgBa_2Ca_2Cu_3O_8$	134
Tl—Ba—Ca—Cu—O	125
Bi—Sr—Ca—Cu—O	105
$YBa_2Cu_3O_7$	92
Nb_3Ge	23.2
Nb_3Sn	18.05
Nb	9.46
Pb	7.18
Hg	4.15
Sn	3.72
Al	1.19
Zn	0.88

A small permanent magnet levitated above a disk of the superconductor $YBa_2Cu_3O_7$, which is in liquid nitrogen at 77 K.

(typically 10^6 m/s). An electron's motion after a collision is independent of its motion before the collision. The kinetic energy acquired by the electrons in the electric field is transferred to the ionized atoms of the conductor when the electrons and atoms collide. The energy transferred to the atoms increases their vibrational energy, which causes the temperature of the conductor to increase.

4. *A description of the agreement between predictions of the model and actual observations and, possibly, predictions of new effects that have not yet been observed:* The test of Drude's model will be this: Can we generate an expression for the resistivity of the conductor that agrees with experimental observations?

We begin to answer the question in (4) above by deriving an expression for the drift velocity. When a free electron of mass m_e and charge q $(= -e)$ is subjected to an electric field $\vec{E}$, it experiences a force $\vec{F} = q\vec{E}$. The electron is a particle under a net force, and its acceleration can be found from Newton's second law, $\Sigma \vec{F} = m\vec{a}$:

$$\vec{a} = \frac{\Sigma \vec{F}}{m} = \frac{q\vec{E}}{m_e} \qquad \textbf{21.13} \blacktriangleleft$$

Because the electric field is uniform, the electron's acceleration is constant, so the electron can be modeled as a particle under constant acceleration. If $\vec{v}_i$ is the electron's initial velocity the instant after a collision (which occurs at a time defined as $t = 0$), the velocity of the electron at a very short time t later (immediately before the next collision occurs) is, from Equation 3.8,

$$\vec{v}_f = \vec{v}_i + \vec{a}t = \vec{v}_i + \frac{q\vec{E}}{m_e} t \qquad \textbf{21.14} \blacktriangleleft$$

Let's now take the average value of $\vec{v}_f$ for all the electrons in the wire over all possible collision times t and all possible values of $\vec{v}_i$. Assuming the initial velocities are randomly distributed over all possible directions, the average value of $\vec{v}_i$ is zero. The average value of the second term of Equation 21.14 is $(q\vec{E}/m_e)\tau$, where τ is the *average time interval between successive collisions*. Because the average value of $\vec{v}_f$ is equal to the drift velocity,

$\blacktriangleright$ Drift velocity in terms of microscopic quantities

$$\vec{v}_{f,\text{avg}} = \vec{v}_d = \frac{q\vec{E}}{m_e} \tau \qquad \textbf{21.15} \blacktriangleleft$$

Substituting the magnitude of this drift velocity (the drift speed) into Equation 21.4, we have

$$I = nev_d A = ne\left(\frac{eE}{m_e}\tau\right) A = \frac{ne^2 E}{m_e} \tau A \qquad \textbf{21.16} \blacktriangleleft$$

According to Equation 21.6, the current is related to the macroscopic variables of potential difference and resistance:

$$I = \frac{\Delta V}{R}$$

Incorporating Equation 21.8, we can write this expression as

$$I = \frac{\Delta V}{\left(\rho \dfrac{\ell}{A}\right)} = \frac{\Delta V}{\rho \ell} A$$

In the conductor, the electric field is uniform, so we use Equation 20.6, $\Delta V = E\ell$, to substitute for the magnitude of the potential difference across the conductor:

$$I = \frac{E\ell}{\rho \ell} A = \frac{E}{\rho} A \qquad \textbf{21.17} \blacktriangleleft$$

Setting the two expressions for the current, Equations 21.16 and 21.17, equal, we solve for the resistivity:

$$I = \frac{ne^2 E}{m_e}\tau A = \frac{E}{\rho}A \;\rightarrow\; \rho = \frac{m_e}{ne^2\tau}$$

21.18 ◄   ► Resistivity in terms of microscopic parameters

According to this structural model, our prediction is that resistivity does not depend on the electric field or, equivalently, on the potential difference, but depends only on fixed parameters associated with the material and the electron. This feature is characteristic of a conductor obeying Ohm's law. The model shows that the resistivity can be calculated from a knowledge of the density of the electrons, their charge and mass, and the average time interval τ between collisions. This time interval is related to the average distance between collisions ℓ_{avg} (the *mean free path*) and the average speed v_{avg} through the expression[6]

$$\tau = \frac{\ell_{avg}}{v_{avg}}$$

21.19 ◄

EXAMPLE 21.3 | **Electron Collisions in Copper**

(A) Using the data from Example 21.1 and the structural model of electron conduction, estimate the average time interval between collisions for electrons in copper at 20°C.

SOLUTION

Conceptualize Imagine the conduction electrons moving in the conductor and making collisions with the array of ionized atoms. Because the speed of the electrons is high, we expect many collisions to occur per unit time interval, so the time interval between collisions should be short.

Categorize We will be using the results of our structural model, so we categorize this problem as a substitution problem.

Solve Equation 21.18 for the average time interval between collisions:

$$(1) \quad \tau = \frac{m_e}{ne^2\rho}$$

In Equation (1), ρ is the *resistivity* of the conductor. From Example 21.1, write the expression for the electron density in a conductor:

$$(2) \quad n = \frac{N_A\rho}{M}$$

In Equation (2), ρ is the *density* of the conductor and M is the molecular mass of the conductor. Substitute numerical values in Equation (2):

$$n = \frac{(6.022 \times 10^{23}\ \text{mol}^{-1})(8\,920\ \text{kg/m}^3)}{0.063\,5\ \text{kg/mol}} = 8.46 \times 10^{28}\ \text{m}^{-3}$$

Substitute this result and other numerical values into Equation (1):

$$\tau = \frac{9.109 \times 10^{-31}\ \text{kg}}{(8.46 \times 10^{28}\ \text{m}^{-3})(1.602 \times 10^{-19}\ \text{C})^2(1.7 \times 10^{-8}\ \Omega \cdot \text{m})}$$

$$= 2.5 \times 10^{-14}\ \text{s}$$

Note that this result is a very short time interval so that the electrons make a very large number of collisions per second.

(B) Assuming that the average speed for free electrons in copper is 1.6×10^6 m/s and using the result from part (A), calculate the mean free path for electrons in copper.

SOLUTION

Solve Equation 21.19 for the mean free path and substitute numerical values:

$$\ell_{avg} = v_{avg}\,\tau = (1.6 \times 10^6\ \text{m/s})(2.5 \times 10^{-14}\ \text{s}) = 4.0 \times 10^{-8}\ \text{m}$$

This result is equivalent to 40 nm (compared with atomic spacings of about 0.2 nm). Therefore, although the time interval between collisions is very short, the electrons travel about 200 atomic distances before colliding with an atom.

---

[6]Recall that the average speed of a group of particles depends on the temperature of the group (Chapter 16) and is not the same as the drift speed v_d.

Although this structural model of conduction is consistent with Ohm's law, it does not correctly predict the values of resistivity or the behavior of the resistivity with temperature. For example, the results of classical calculations for v_{avg} using the ideal gas model for the electrons are about a factor of ten smaller than the actual values, which results in incorrect predictions of values of resistivity from Equation 21.18. Furthermore, according to Equations 21.18 and 21.19, the resistivity is predicted to vary with temperature as does v_{avg}, which according to an ideal-gas model (Chapter 16, Eq. 16.22) is proportional to $\sqrt{T}$. This behavior is in disagreement with the linear dependence of resistivity with temperature for pure metals (Fig. 21.9). Because of these incorrect predictions, we must modify our structural model. We shall call the model that we have developed so far the *classical* model for electrical conduction. To account for the incorrect predictions of the classical model, we will develop it further into a *quantum mechanical* model, which we shall describe briefly.

We discussed two important simplification models in earlier chapters, the particle model and the wave model. Although we discussed these two simplification models separately, quantum physics tells us that this separation is not so clear-cut. As we shall discuss in detail in Chapter 28, particles have wave-like properties. The predictions of some models can only be matched to experimental results if the model includes the wave-like behavior of particles. The structural model for electrical conduction in metals is one of these cases.

Let us imagine that the electrons moving through the metal have wave-like properties. If the array of atoms in a conductor is regularly spaced (that is, periodic), the wave-like character of the electrons makes it possible for them to move freely through the conductor and a collision with an atom is unlikely. For an idealized conductor, no collisions would occur, the mean free path would be infinite, and the resistivity would be zero. Electrons are scattered only if the atomic arrangement is irregular (not periodic), as a result of structural defects or impurities, for example. At low temperatures, the resistivity of metals is dominated by scattering caused by collisions between the electrons and impurities. At high temperatures, the resistivity is dominated by scattering caused by collisions between the electrons and the atoms of the conductor, which are continuously displaced as a result of thermal agitation, destroying the perfect periodicity. The thermal motion of the atoms makes the structure irregular (compared with an atomic array at rest), thereby reducing the electron's mean free path.

Although it is beyond the scope of this text to show this modification in detail, the classical model modified with the wave-like character of the electrons results in predictions of resistivity values that are in agreement with measured values and predicts a linear temperature dependence. When discussing the hydrogen atom in Chapter 11, we had to introduce some quantum notions to understand experimental observations such as atomic spectra. Likewise, we had to introduce quantum notions in Chapter 17 to understand the temperature behavior of molar specific heats of gases. Here we have another case in which quantum physics is necessary for the model to agree with experiment. Although classical physics can explain a tremendous range of phenomena, we continue to see hints that quantum physics must be incorporated into our models. We shall study quantum physics in detail in Chapters 28 through 31.

21.5 | Energy and Power in Electric Circuits

In Section 21.1, we discussed the energy transformations occurring when current exists in a conductor. If a battery is used to establish an electric current in a conductor, there is a continuous transformation of chemical energy in the battery to kinetic energy of the electrons to internal energy in the conductor, resulting in an increase in the temperature of the conductor.

In typical electric circuits, energy is transferred from a source, such as a battery, to some device, such as a lightbulb or a radio receiver by electrical transmission (T_{ET} in Eq. 7.2). Let us determine an expression that will allow us to calculate the rate of this energy transfer. First, consider the simple circuit in Active Figure 21.11,

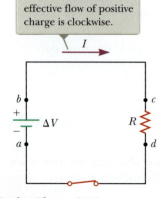

The direction of the effective flow of positive charge is clockwise.

Active Figure 21.11 A circuit consisting of a resistor of resistance R and a battery having a potential difference ΔV across its terminals.

where we imagine that energy is being delivered to a resistor. Because the connecting wires also have resistance, some energy is delivered to the wires and some energy to the resistor. Unless noted otherwise, we will adopt a simplification model in which the resistance of the wires is so small compared with the resistance of the circuit element that we ignore the energy delivered to the wires.

Let us now analyze the energetics of the circuit in which a battery is connected to a resistor of resistance R as in Active Figure 21.11. Imagine following a positive quantity of charge Q around the circuit from point a through the battery and resistor and back to a. Point a is a reference point at which the potential is defined as zero. We identify the entire circuit as our system. As the charge moves from a to b through the battery whose potential difference is ΔV, the electrical potential energy of the system increases by the amount $Q\,\Delta V$, whereas the chemical energy in the battery decreases by the same amount. (Recall from Chapter 20 that $\Delta U = q\,\Delta V$.) As the charge moves from c to d through the resistor, however, the system loses this electrical potential energy during collisions with atoms in the resistor. In this process, the energy is transformed to internal energy corresponding to increased vibrational motion of the atoms in the resistor. Because we have neglected the resistance of the interconnecting wires, no energy transformation occurs for paths bc and da. When the charge returns to point a, the net result is that some of the chemical energy in the battery has been delivered to the resistor and resides in the resistor as internal energy associated with molecular vibration.

The resistor is normally in contact with air, so its increased temperature results in a transfer of energy by heat into the air. In addition, there will be thermal radiation from the resistor, representing another means of escape for the energy. After some time interval has passed, the resistor remains at a constant temperature because the input of energy from the battery is balanced by the output of energy by heat and radiation. Some electrical devices include *heat sinks*[7] connected to parts of the circuit to prevent these parts from reaching dangerously high temperatures. Heat sinks are pieces of metal with many fins. The high thermal conductivity of the metal provides a rapid transfer of energy by heat away from the hot component and the large number of fins provides a large surface area in contact with the air, so energy can transfer by radiation and into the air by heat at a high rate.

Let us consider now the rate at which the system loses electric potential energy as the charge Q passes through the resistor:

$$\frac{dU}{dt} = \frac{d}{dt}\,(Q\,\Delta V) = \frac{dQ}{dt}\,\Delta V = I\,\Delta V$$

where I is the current in the circuit. Of course, the system regains this potential energy when the charge passes through the battery, at the expense of chemical energy in the battery. The rate at which the system loses potential energy as the charge passes through the resistor is equal to the rate at which the system gains internal energy in the resistor. Therefore, the **power** P, representing the rate at which energy is delivered to the resistor, is

$$P = I\,\Delta V \qquad\qquad \textbf{21.20} \blacktriangleleft \qquad \blacktriangleright \text{ Power delivered to a device}$$

We have developed this result by considering a battery delivering energy to a resistor. Equation 21.20, however, can be used to determine the power transferred from a voltage source to *any* device carrying a current I and having a potential difference ΔV between its terminals.

Using Equation 21.20 and that $\Delta V = IR$ for a resistor, we can express the power delivered to the resistor in the alternative forms

$$P = I^2 R = \frac{(\Delta V)^2}{R} \qquad\qquad \textbf{21.21} \blacktriangleleft$$

---

[7]This terminology is another misuse of the word *heat* that is ingrained in our common language.

Pitfall Prevention | 21.5
Misconceptions about Current
Several common misconceptions are associated with current in a circuit like that in Active Figure 21.11. One is that current comes out of one terminal of the battery and is then "used up" as it passes through the resistor. According to this approach, there is current in only one part of the circuit. The current is actually the same *everywhere* in the circuit. A related misconception has the current coming out of the resistor being smaller than that going in because some of the current is "used up." Yet another misconception has current coming out of both terminals of the battery, in opposite directions, and then "clashing" in the resistor, delivering the energy in this manner. We know that is not the case; charges flow in the same rotational sense at *all* points in the circuit.

Pitfall Prevention | 21.6
Charges Do Not Move All the Way Around a Circuit
Because of the very small magnitude of the drift velocity, it might take *hours* for a single electron to make one complete trip around the circuit. In terms of understanding the energy transfer in a circuit, however, it is useful to *imagine* a charge moving all the way around the circuit.

Pitfall Prevention | 21.7
Energy Is Not "Dissipated"
In some books, you may see Equation 21.21 described as the power "dissipated in" a resistor, suggesting that energy disappears. Instead, we say energy is "delivered to" a resistor. The notion of *dissipation* arises because a warm resistor expels energy by radiation and heat, and energy delivered by the battery leaves the circuit. (It does not disappear!)

The SI unit of power is the watt, introduced in Chapter 7. If you analyze the units in Equations 21.20 and 21.21, you will see that the result of the calculation provides a watt as the unit. The power delivered to a conductor of resistance R is often referred to as an I^2R *loss*.

As we learned in Section 7.6, the unit of energy your electric company uses to calculate energy transfer, the kilowatt-hour, is the amount of energy transferred in 1 h at the constant rate of 1 kW. We learned there that 1 kWh = 3.6×10^6 J.

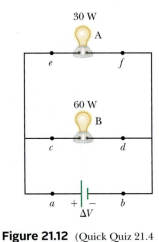

Figure 21.12 (Quick Quiz 21.4 and Thinking Physics 21.2) Two incandescent lightbulbs connected across the same potential difference.

QUICK QUIZ 21.4 For the two incandescent lightbulbs shown in Figure 21.12, rank the currents at points *a* through *f*, from greatest to least.

THINKING PHYSICS 21.2

Two incandescent lightbulbs A and B are connected across the same potential difference as in Figure 21.12. The electric input powers to the lightbulbs are shown. Which lightbulb has the higher resistance? Which carries the greater current?

Reasoning Because the voltage across each lightbulb is the same and the rate of energy delivered to a resistor is $P = (\Delta V)^2/R$, the lightbulb with the higher resistance exhibits the lower rate of energy transfer. In this case, the resistance of A is larger than that for B. Furthermore, because $P = I\,\Delta V$, we see that the current carried by B is larger than that of A. ◄

THINKING PHYSICS 21.3

When is an incandescent lightbulb more likely to fail, just after it is turned on or after it has been on for a while?

Reasoning When the switch is closed, the source voltage is immediately applied across the lightbulb. As the voltage is applied across the cold filament when the lightbulb is first turned on, the resistance of the filament is low. Therefore, the current is high and a relatively large amount of energy is delivered to the bulb per unit time

interval. This causes the temperature of the filament to rise rapidly, resulting in thermal stress on the filament that makes it likely to fail at that moment. As the filament warms up in the absence of failure, its resistance rises and the current falls. As a result, the rate of energy delivered to the lightbulb falls. The thermal stress on the filament is reduced so that the failure is less likely to occur after the bulb has been on for a while. ◄

Example 21.4 | Linking Electricity and Thermodynamics

An immersion heater must increase the temperature of 1.50 kg of water from 10.0°C to 50.0°C in 10.0 min while operating at 110 V.

(A) What is the required resistance of the heater?

SOLUTION

Conceptualize An immersion heater is a resistor that is inserted into a container of water. As energy is delivered to the immersion heater, raising its temperature, energy leaves the surface of the resistor by heat, going into the water. When the immersion heater reaches a constant temperature, the rate of energy delivered to the resistance by electrical transmission (T_{ET}) is equal to the rate of energy delivered by heat (Q) to the water.

Categorize This example allows us to link our new understanding of power in electricity with our experience with specific heat in thermodynamics (Chapter 17). The water is a nonisolated system. Its internal energy is rising because of energy transferred into the water by heat from the resistor: $\Delta E_{int} = Q$. In our model, we assume the energy that enters the water from the heater remains in the water.

21.4 *cont.*

Analyze To simplify the analysis, let's ignore the initial period during which the temperature of the resistor increases and also ignore any variation of resistance with temperature. Therefore, we imagine a constant rate of energy transfer for the entire 10.0 min.

Set the rate of energy delivered to the resistor equal to the rate of energy Q entering the water by heat:

$$P = \frac{(\Delta V)^2}{R} = \frac{Q}{\Delta t}$$

Use Equation 17.3, $Q = mc \, \Delta T$, to relate the energy input by heat to the resulting temperature change of the water and solve for the resistance:

$$\frac{(\Delta V)^2}{R} = \frac{mc \, \Delta T}{\Delta t} \;\rightarrow\; R = \frac{(\Delta V)^2 \, \Delta t}{mc \, \Delta T}$$

Substitute the values given in the statement of the problem:

$$R = \frac{(110 \text{ V})^2 (600 \text{ s})}{(1.50 \text{ kg})(4\,186 \text{ J/kg} \cdot {}^\circ\text{C})(50.0{}^\circ\text{C} - 10.0{}^\circ\text{C})} = \boxed{28.9 \ \Omega}$$

(B) Estimate the cost of heating the water.

SOLUTION

Multiply the power by the time interval to find the amount of energy transferred:

$$T_{ET} = P \, \Delta t = \frac{(\Delta V)^2}{R} \, \Delta t = \frac{(110 \text{ V})^2}{28.9 \ \Omega} (10.0 \text{ min}) \left(\frac{1 \text{ h}}{60.0 \text{ min}}\right)$$

$$= 69.8 \text{ Wh} = 0.069\,8 \text{ kWh}$$

Find the cost knowing that energy is purchased at an estimated price of 11¢ per kilowatt-hour:

Cost $= (0.069\,8 \text{ kWh})(\$0.11/\text{kWh}) = \$0.008 = \boxed{0.8¢}$

Finalize The cost to heat the water is very low, less than one cent. In reality, the cost is higher because some energy is transferred from the water into the surroundings by heat and electromagnetic radiation while its temperature is increasing. If you have electrical devices in your home with power ratings on them, use this power rating and an approximate time interval of use to estimate the cost for one use of the device.

21.6 | Sources of emf

The entity that maintains the constant voltage in Figure 21.13 is called a **source of emf**.[8] Sources of emf are any devices (such as batteries and generators) that increase the potential energy of a circuit system by maintaining a potential difference between points in the circuit while charges move through the circuit. One can think of a source of emf as a "charge pump." The emf $\mathcal{E}$ of a source describes the work done per unit charge, and hence the SI unit of emf is the volt.

At this point, you may wonder why we need to define a second quantity, emf, with the volt as a unit when we have already defined the potential difference. To see the need for this new quantity, consider the circuit shown in Figure 21.13, consisting of a battery connected to a resistor. We shall assume that the connecting wires have no resistance. We might be tempted to claim that the potential difference across the battery terminals (the terminal voltage) equals the emf of the battery. A real battery, however, always has some **internal resistance** r. As a result, the terminal voltage is not equal to the emf, as we shall show.

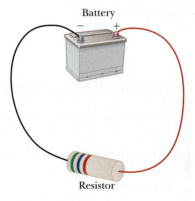

Battery

Resistor

Figure 21.13 A circuit consisting of a resistor connected to the terminals of a battery.

---

[8]The term *emf* was originally an abbreviation for *electromotive force*, but it is not a force, so the long form is discouraged. The name electromotive force was used early in the study of electricity before the understanding of batteries was as sophisticated as it is today.

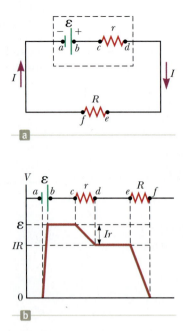

Active Figure 21.14 (a) Circuit diagram of a source of emf $\mathcal{E}$ (in this case, a battery) with internal resistance r, connected to an external resistor of resistance R. (b) Graphical representation showing how the potential changes as the circuit in (a) is traversed clockwise.

Pitfall Prevention | 21.8
What Is Constant in a Battery?
It is a common misconception that a battery is a source of constant current. Equation 21.24 shows that this is not true. The current in the circuit depends on the resistance R connected to the battery. It is also not true that a battery is a source of constant terminal voltage, as shown by Equation 21.22. **A battery is a source of constant emf.**

The circuit shown in Figure 21.13 can be described by the circuit diagram in Active Figure 21.14a. The battery within the dashed rectangle is modeled as an ideal, zero-resistance source of emf $\mathcal{E}$ in series with the internal resistance r. Now imagine moving from a to d in Active Figure 21.14a. As you pass from the negative to the positive terminal within the source of emf the potential increases by $\mathcal{E}$. As you move through the resistance r, however, the potential decreases by an amount Ir, where I is the current in the circuit. Therefore, the terminal voltage $\Delta V = V_d - V_a$ of the battery is[9]

$$\Delta V = \mathcal{E} - Ir \qquad \textbf{21.22}\blacktriangleleft$$

Note from this expression that $\mathcal{E}$ is equivalent to the **open-circuit voltage,** that is, the terminal voltage when the current is zero. Active Figure 21.14b is a graphical representation of the changes in potential as the circuit is traversed clockwise. By inspecting Active Figure 21.14a, we see that the terminal voltage ΔV must also equal the potential difference across the external resistance R, often called the **load resistance;** that is, $\Delta V = IR$. Combining this expression with Equation 21.22, we see that

$$\mathcal{E} = IR + Ir \qquad \textbf{21.23}\blacktriangleleft$$

Solving for the current gives

$$I = \frac{\mathcal{E}}{R + r} \qquad \textbf{21.24}\blacktriangleleft$$

which shows that the current in this simple circuit depends on both the resistance R external to the battery and the internal resistance r. If R is much greater than r, we can adopt a simplification model in which we neglect r in our analysis. In many circuits, we shall adopt this simplification model.

If we multiply Equation 21.23 by the current I, we have

$$I\mathcal{E} = I^2R + I^2r$$

This equation tells us that the total power output $I\mathcal{E}$ of the source of emf is equal to the rate I^2R at which energy is delivered to the load resistance plus the rate I^2r at which energy is delivered to the internal resistance. If $r \ll R$, much more of the energy from the battery is delivered to the load resistance than stays in the battery, although the amount of energy is relatively small because the load resistance is large, resulting in a small current. If $r \gg R$, a significant fraction of the energy from the source of emf stays in the battery package because it is delivered to the internal resistance. For example, if a wire is simply connected between the terminals of a flashlight battery, the battery becomes warm. This warming represents the transfer of energy from the source of emf to the internal resistance, where it appears as internal energy associated with temperature. Problem 73 explores the conditions under which the largest amount of energy is transferred from the battery to the load resistor.

Example 21.5 | Terminal Voltage of a Battery

A battery has an emf of 12.0 V and an internal resistance of 0.050 0 Ω. Its terminals are connected to a load resistance of 3.00 Ω.

(A) Find the current in the circuit and the terminal voltage of the battery.

SOLUTION

Conceptualize Study Active Figure 21.14a, which shows a circuit consistent with the problem statement. The battery delivers energy to the load resistor.

Categorize This example involves simple calculations from this section, so we categorize it as a substitution problem.

---

[9]The terminal voltage in this case is less than the emf by the amount Ir. In some situations, the terminal voltage may *exceed* the emf by the amount Ir. Such a situation occurs when the direction of the current is *opposite* that of the emf, as when a battery is being charged by another source of emf.

21.5 *cont.*

Use Equation 21.24 to find the current in the circuit:

$$I = \frac{\mathcal{E}}{R + r} = \frac{12.0 \text{ V}}{3.00 \ \Omega + 0.050 \ 0 \ \Omega} = 3.93 \text{ A}$$

Use Equation 21.22 to find the terminal voltage:

$$\Delta V = \mathcal{E} - Ir = 12.0 \text{ V} - (3.93 \text{ A})(0.050 \ 0 \ \Omega) = 11.8 \text{ V}$$

To check this result, calculate the voltage across the load resistance R:

$$\Delta V = IR = (3.93 \text{ A})(3.00 \ \Omega) = 11.8 \text{ V}$$

(B) Calculate the power delivered to the load resistor, the power delivered to the internal resistance of the battery, and the power delivered by the battery.

SOLUTION

Use Equation 21.21 to find the power delivered to the load resistor:

$$P_R = I^2R = (3.93 \text{ A})^2(3.00 \ \Omega) = 46.3 \text{ W}$$

Find the power delivered to the internal resistance:

$$P_r = I^2r = (3.93 \text{ A})^2(0.050 \ 0 \ \Omega) = 0.772 \text{ W}$$

Find the power delivered by the battery by adding these quantities:

$$P = P_R + P_r = 46.3 \text{ W} + 0.772 \text{ W} = 47.1 \text{ W}$$

What If? As a battery ages, its internal resistance increases. Suppose the internal resistance of this battery rises to $2.00 \ \Omega$ toward the end of its useful life. How does that alter the battery's ability to deliver energy?

Answer Let's connect the same 3.00-Ω load resistor to the battery.

Find the new current in the battery:

$$I = \frac{\mathcal{E}}{R + r} = \frac{12.0 \text{ V}}{3.00 \ \Omega + 2.00 \ \Omega} = 2.40 \text{ A}$$

Find the new terminal voltage:

$$\Delta V = \mathcal{E} - Ir = 12.0 \text{ V} - (2.40 \text{ A})(2.00 \ \Omega) = 7.2 \text{ V}$$

Find the new powers delivered to the load resistor and internal resistance:

$$P_R = I^2R = (2.40 \text{ A})^2(3.00 \ \Omega) = 17.3 \text{ W}$$
$$P_r = I^2r = (2.40 \text{ A})^2(2.00 \ \Omega) = 11.5 \text{ W}$$

The terminal voltage is only 60% of the emf. Notice that 40% of the power from the battery is delivered to the internal resistance when r is $2.00 \ \Omega$. When r is $0.050 \ 0 \ \Omega$ as in part (B), this percentage is only 1.6%. Consequently, even though the emf remains fixed, the increasing internal resistance of the battery significantly reduces the battery's ability to deliver energy to an external load.

21.7 | Resistors in Series and Parallel

When two or more resistors are connected together as are the incandescent light-bulbs in Active Figure 21.15a (page 716), they are said to be in a **series combination.** Active Figure 21.15b is the circuit diagram for the lightbulbs, shown as resistors, and the battery. In a series connection, if an amount of charge Q exits resistor R_1, charge Q must also enter the second resistor R_2. Otherwise, charge would accumulate on the wire between the resistors. Therefore, the same amount of charge passes through both resistors in a given time interval and the currents are the same in both resistors:

$$I = I_1 = I_2$$

where I is the current leaving the battery, I_1 is the current in resistor R_1, and I_2 is the current in resistor R_2.

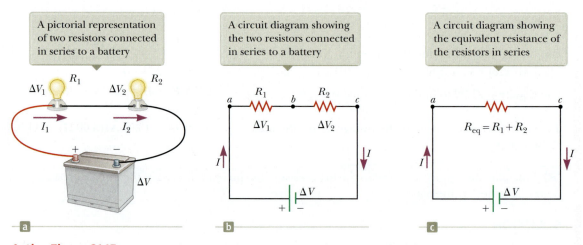

Active Figure 21.15 Two incandescent lightbulbs with resistances R_1 and R_2 connected in series. All three diagrams are equivalent.

The potential difference applied across the series combination of resistors divides between the resistors. In Active Figure 21.15b, because the voltage drop[10] from a to b equals I_1R_1 and the voltage drop from b to c equals I_2R_2, the voltage drop from a to c is

$$\Delta V = \Delta V_1 + \Delta V_2 = I_1R_1 + I_2R_2$$

The potential difference across the battery is also applied to the **equivalent resistance** R_{eq} in Active Figure 21.15c:

$$\Delta V = IR_{eq}$$

where the equivalent resistance has the same effect on the circuit as the series combination because it results in the same current I in the battery. Combining these equations for ΔV gives

$$\Delta V = IR_{eq} = I_1R_1 + I_2R_2 \quad \rightarrow \quad R_{eq} = R_1 + R_2 \qquad \textbf{21.25} \blacktriangleleft$$

where we have canceled the currents I, I_1, and I_2 because they are all the same. We see that we can replace the two resistors in series with a single equivalent resistance whose value is the *sum* of the individual resistances.

The equivalent resistance of three or more resistors connected in series is

> ▶ The equivalent resistance of a series combination of resistors

$$R_{eq} = R_1 + R_2 + R_3 + \cdots \qquad \textbf{21.26} \blacktriangleleft$$

This relationship indicates that the equivalent resistance of a series combination of resistors is the numerical sum of the individual resistances and is always greater than any individual resistance.

Looking back at Equation 21.24, we see that the denominator of the right-hand side is the simple algebraic sum of the internal and external resistances. That is consistent with the internal and external resistances being in series in Active Figure 21.14a.

If the filament of one lightbulb in Active Figure 21.15a were to fail, the circuit would no longer be complete (resulting in an open-circuit condition) and the second lightbulb would also go out. This fact is a general feature of a series circuit: if one device in the series creates an open circuit, all devices are inoperative.

Pitfall Prevention | 21.9
Lightbulbs Don't Burn
We will describe the end of the life of an incandescent lightbulb by saying *the filament fails* rather than by saying the lightbulb "burns out." The word *burn* suggests a combustion process, which is not what occurs in a lightbulb. The failure of a lightbulb results from the slow sublimation of tungsten from the very hot filament over the life of the lightbulb. The filament eventually becomes very thin because of this process. The mechanical stress from a sudden temperature increase when the lightbulb is turned on causes the thin filament to break.

---

[10]The term *voltage drop* is synonymous with a decrease in electric potential across a resistor. It is often used by individuals working with electric circuits.

QUICK QUIZ 21.5 With the switch in the circuit of Figure 21.16a closed, there is no current in R_2 because the current has an alternate zero-resistance path through the switch. There is current in R_1, and this current is measured with the ammeter (a device for measuring current) at the bottom of the circuit. If the switch is opened (Figure 21.16b), there is current in R_2. What happens to the reading on the ammeter when the switch is opened? **(a)** The reading goes up. **(b)** The reading goes down. **(c)** The reading does not change.

Now consider two resistors in a **parallel combination** as shown in Active Figure 21.17. Notice that both resistors are connected directly across the terminals of the battery. Therefore, the potential differences across the resistors are the same:

$$\Delta V = \Delta V_1 = \Delta V_2$$

where ΔV is the terminal voltage of the battery.

When charges reach point a in Active Figure 21.17b, they split into two parts, with some going toward R_1 and the rest going toward R_2. A **junction** is any such point in a circuit where a current can split. This split results in less current in each individual resistor than the current leaving the battery. Because electric charge is conserved, the current I that enters point a must equal the total current leaving that point:

$$I = I_1 + I_2 = \frac{\Delta V_1}{R_1} + \frac{\Delta V_2}{R_2}$$

where I_1 is the current in R_1 and I_2 is the current in R_2.

The current in the **equivalent resistance** R_{eq} in Active Figure 21.17c is

$$I = \frac{\Delta V}{R_{eq}}$$

where the equivalent resistance has the same effect on the circuit as the two resistors in parallel; that is, the equivalent resistance draws the same current I from the battery. Combining these equations for I, we see that the equivalent resistance of two resistors in parallel is given by

$$I = \frac{\Delta V}{R_{eq}} = \frac{\Delta V_1}{R_1} + \frac{\Delta V_2}{R_2} \quad \rightarrow \quad \frac{1}{R_{eq}} = \frac{1}{R_1} + \frac{1}{R_2} \qquad \textbf{21.27} \blacktriangleleft$$

where we have canceled ΔV, ΔV_1, and ΔV_2 because they are all the same.

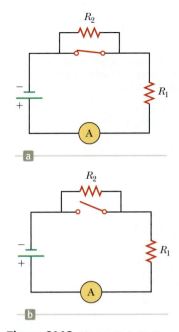

Figure 21.16 (Quick Quiz 21.5) What happens when the switch is opened?

Pitfall Prevention | 21.10
Local and Global Changes
A local change in one part of a circuit may result in a global change throughout the circuit. For example, if a single resistor is changed in a circuit containing several resistors and batteries, the currents in all resistors and batteries, the terminal voltages of all batteries, and the voltages across all resistors may change as a result.

Pitfall Prevention | 21.11
Current Does Not Take the Path of Least Resistance
You may have heard the phrase "current takes the path of least resistance" (or similar wording) in reference to a parallel combination of current paths such that there are two or more paths for the current to take. Such wording is incorrect. The current takes *all* paths. Those paths with lower resistance have larger currents, but even very high resistance paths carry *some* of the current. In theory, if current has a choice between a zero-resistance path and a finite resistance path, all the current takes the path of zero resistance; a path with zero resistance, however, is an idealization.

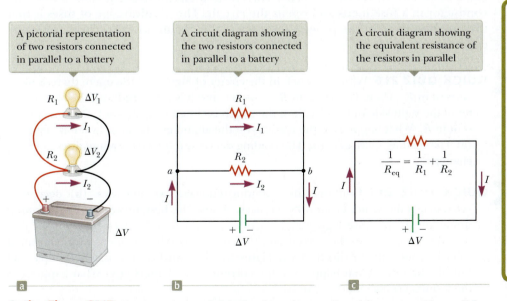

A pictorial representation of two resistors connected in parallel to a battery

A circuit diagram showing the two resistors connected in parallel to a battery

A circuit diagram showing the equivalent resistance of the resistors in parallel

$$\frac{1}{R_{eq}} = \frac{1}{R_1} + \frac{1}{R_2}$$

Active Figure 21.17 Two incandescent lightbulbs with resistances R_1 and R_2 connected in parallel. All three diagrams are equivalent.

An extension of this analysis to three or more resistors in parallel gives

▶ The equivalent resistance of a parallel combination of resistors

$$\frac{1}{R_{eq}} = \frac{1}{R_1} + \frac{1}{R_2} + \frac{1}{R_3} + \cdots$$

21.28 ◀

This expression shows that the inverse of the equivalent resistance of two or more resistors in a parallel combination is equal to the sum of the inverses of the individual resistances. Furthermore, the equivalent resistance is always less than the smallest resistance in the group.

A circuit consisting of resistors can often be reduced to a simple circuit containing only one resistor. To do so, examine the initial circuit and replace any resistors in series or any in parallel with equivalent resistances using Equations 21.26 and 21.28. Draw a sketch of the new circuit after these changes have been made. Examine the new circuit and replace any new series or parallel combinations that now exist. Continue this process until a single equivalent resistance is found for the entire circuit. (That may not be possible; if not, see the techniques of Section 21.8.)

If the current in or the potential difference across a resistor in the initial circuit is to be found, start with the final circuit and gradually work your way back through the equivalent circuits. Find currents and voltages across resistors using $\Delta V = IR$ and your understanding of series and parallel combinations.

Household circuits are always wired so that the electrical devices are connected in parallel as in Active Figure 21.17a. In this manner, each device operates independently of the others so that if one is switched off, the others remain on. For example, if one of the lightbulbs in Active Figure 21.17a were removed from its socket, the other would continue to operate. Equally important, each device operates on the same voltage. If devices were connected in series, the voltage applied to the combination would divide among the devices, so the voltage applied to any one device would depend on how many devices were in the combination.

In many household circuits, circuit breakers are used in series with other circuit elements for safety purposes. A circuit breaker is designed to switch off and open the circuit at some maximum current (typically 15 A or 20 A) whose value depends on the nature of the circuit. If a circuit breaker were not used, large currents caused by turning on many devices could result in excessive temperatures in wires and, perhaps, cause a fire. In older home construction, fuses were used in place of circuit breakers. When the current in a circuit exceeds some value, the conductor in a fuse melts and opens the circuit. The disadvantage of fuses is that they are destroyed in the process of opening the circuit, whereas circuit breakers can be reset.

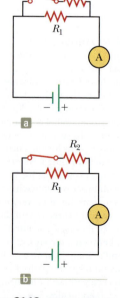

Figure 21.18 (Quick Quiz 21.6) What happens when the switch is closed?

QUICK QUIZ 21.6 With the switch in the circuit of Figure 21.18a open, there is no current in R_2. There is current in R_1, however, and it is measured with the ammeter at the right side of the circuit. If the switch is closed (Fig. 21.18b), there is current in R_2. What happens to the reading on the ammeter when the switch is closed? (a) The reading increases. (b) The reading decreases. (c) The reading does not change.

QUICK QUIZ 21.7 Consider the following choices: (a) increases, (b) decreases, (c) remains the same. From these choices, choose the best answer for the following situations. (i) In Active Figure 21.15, a third resistor is added in series with the first two. What happens to the current in the battery? (ii) What happens to the terminal voltage of the battery? (iii) In Active Figure 21.17, a third resistor is added in parallel with the first two. What happens to the current in the battery? (iv) What happens to the terminal voltage of the battery?

THINKING PHYSICS 21.4

Compare the brightnesses of the four identical lightbulbs in Figure 21.19. What happens if bulb A fails so that it cannot conduct? What if bulb C fails? What if bulb D fails?

Reasoning Bulbs A and B are connected in series across the battery, whereas bulb C is connected by itself. Therefore, the terminal voltage of the battery is split between bulbs A and B. As a result, bulb C will be brighter than bulbs A and B, which should be equally as bright as each other. Bulb D has a wire connected across it. Therefore, there is no potential difference across bulb D and it does not glow at all. If bulb A fails, bulb B goes out but bulb C stays lit. If bulb C fails, there is no effect on the other bulbs. If bulb D fails, the event is undetectable because bulb D was not glowing initially. ◀

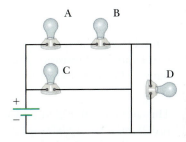

Figure 21.19 (Thinking Physics 21.4) What happens to the lightbulbs if one fails?

THINKING PHYSICS 21.5

Figure 21.20 illustrates how a three-way incandescent lightbulb is constructed to provide three levels of light intensity. The socket of the lamp is equipped with a three-way switch for selecting different light intensities. The bulb contains two filaments. Why are the filaments connected in parallel? Explain how the two filaments are used to provide three different light intensities.

Reasoning If the filaments were connected in series and one of them were to fail, there would be no current in the bulb and the bulb would give no illumination, regardless of the switch position. When the filaments are connected in parallel, however, and one of them (say the 75-W filament) fails, the bulb still operates in some switch positions because there is current in the other (100-W) filament. The three light intensities are made possible by selecting one of three values of filament resistance, using a single value of 120 V for the applied voltage. The 75-W filament offers one value of resistance, the 100-W filament offers a second value, and the third resistance is obtained by combining the two filaments in parallel. When switch S_1 is closed and switch S_2 is opened, only the 75-W filament carries current. When switch S_1 is open and switch S_2 is closed, only the 100-W filament carries current. When both switches are closed, both filaments carry current, and a total illumination corresponding to 175 W is obtained. ◀

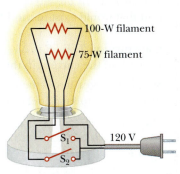

Figure 21.20 (Thinking Physics 21.5) A three-way incandescent lightbulb.

Example 21.6 | Find the Equivalent Resistance

Four resistors are connected as shown in Figure 21.21a.

(A) Find the equivalent resistance between points *a* and *c*.

SOLUTION

Conceptualize Imagine charges flowing into this combination from the left. All charges must pass through the first two resistors, but the charges split into two different paths when encountering the combination of the 6.0-Ω and the 3.0-Ω resistors.

Categorize Because of the simple nature of the combination of resistors in Figure 21.21, we categorize this example as one for which we can use the rules for series and parallel combinations of resistors.

Analyze The combination of resistors can be reduced in steps as shown in Figure 21.21.

Find the equivalent resistance between *a* and *b* of the 8.0-Ω and 4.0-Ω resistors, which are in series (left-hand red-brown circles):

$$R_{eq} = 8.0 \ \Omega + 4.0 \ \Omega = 12.0 \ \Omega$$

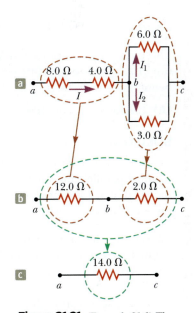

Figure 21.21 (Example 21.6) The original network of resistors is reduced to a single equivalent resistance.

continued

21.6 *cont.*

Find the equivalent resistance between b and c of the 6.0-Ω and 3.0-Ω resistors, which are in parallel (right-hand red-brown circles):

$$\frac{1}{R_{eq}} = \frac{1}{6.0 \ \Omega} + \frac{1}{3.0 \ \Omega} = \frac{3}{6.0 \ \Omega}$$

$$R_{eq} = \frac{6.0 \ \Omega}{3} = 2.0 \ \Omega$$

The circuit of equivalent resistances now looks like Figure 21.21b. The 12.0-Ω and 2.0-Ω resistors are in series (green circles). Find the equivalent resistance from a to c:

$$R_{eq} = 12.0 \ \Omega + 2.0 \ \Omega = \boxed{14.0 \ \Omega}$$

This resistance is that of the single equivalent resistor in Figure 21.21c.

(B) What is the current in each resistor if a potential difference of 42 V is maintained between a and c?

SOLUTION

The currents in the 8.0-Ω and 4.0-Ω resistors are the same because they are in series. In addition, they carry the same current that would exist in the 14.0-Ω equivalent resistor subject to the 42-V potential difference.

Use Equation 21.6 ($R = \Delta V/I$) and the result from part (A) to find the current in the 8.0-Ω and 4.0-Ω resistors:

$$I = \frac{\Delta V_{ac}}{R_{eq}} = \frac{42 \ V}{14.0 \ \Omega} = \boxed{3.0 \ A}$$

Set the voltages across the resistors in parallel in Figure 21.21a equal to find a relationship between the currents:

$$\Delta V_1 = \Delta V_2 \quad \rightarrow \quad (6.0 \ \Omega)I_1 = (3.0 \ \Omega)I_2 \quad \rightarrow \quad I_2 = 2I_1$$

Use $I_1 + I_2 = 3.0$ A to find I_1:

$$I_1 + I_2 = 3.0 \ A \quad \rightarrow \quad I_1 + 2I_1 = 3.0 \ A \quad \rightarrow \quad I_1 = \boxed{1.0 \ A}$$

Find I_2:

$$I_2 = 2I_1 = 2(1.0 \ A) = \boxed{2.0 \ A}$$

Finalize As a final check of our results, note that $\Delta V_{bc} = (6.0 \ \Omega)I_1 = (3.0 \ \Omega)I_2 = 6.0$ V and $\Delta V_{ab} = (12.0 \ \Omega)I = 36$ V; therefore, $\Delta V_{ac} = \Delta V_{ab} + \Delta V_{bc} = 42$ V, as it must.

Example 21.7 | Three Resistors in Parallel

Three resistors are connected in parallel as shown in Figure 21.22a. A potential difference of 18.0 V is maintained between points a and b.

(A) Calculate the equivalent resistance of the circuit.

SOLUTION

Conceptualize Figure 21.22a shows that we are dealing with a simple parallel combination of three resistors. Notice that the current I splits into three currents I_1, I_2, and I_3 in the three resistors.

Categorize Because the three resistors are connected in parallel, we can use Equation 21.28 to evaluate the equivalent resistance.

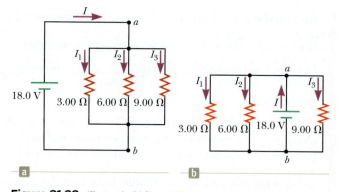

Figure 21.22 (Example 21.7) (a) Three resistors connected in parallel. The voltage across each resistor is 18.0 V. (b) Another circuit with three resistors and a battery. Is it equivalent to the circuit in (a)?

Analyze Use Equation 21.28 to find R_{eq}:

$$\frac{1}{R_{eq}} = \frac{1}{3.00 \ \Omega} + \frac{1}{6.00 \ \Omega} + \frac{1}{9.00 \ \Omega} = \frac{11.0}{18.0 \ \Omega}$$

$$R_{eq} = \frac{18.0 \ \Omega}{11.0} = \boxed{1.64 \ \Omega}$$

21.7 *cont.*

(B) Find the current in each resistor.

SOLUTION

The potential difference across each resistor is 18.0 V. Apply the relationship $\Delta V = IR$ to find the currents:

$$I_1 = \frac{\Delta V}{R_1} = \frac{18.0 \text{ V}}{3.00 \ \Omega} = 6.00 \text{ A}$$

$$I_2 = \frac{\Delta V}{R_2} = \frac{18.0 \text{ V}}{6.00 \ \Omega} = 3.00 \text{ A}$$

$$I_3 = \frac{\Delta V}{R_3} = \frac{18.0 \text{ V}}{9.00 \ \Omega} = 2.00 \text{ A}$$

(C) Calculate the power delivered to each resistor and the total power delivered to the combination of resistors.

SOLUTION

Apply the relationship $P = I^2 R$ to each resistor using the currents calculated in part (B):

3.00-Ω: $P_1 = I_1^2 R_1 = (6.00 \text{ A})^2 (3.00 \ \Omega) = 108 \text{ W}$

6.00-Ω: $P_2 = I_2^2 R_2 = (3.00 \text{ A})^2 (6.00 \ \Omega) = 54 \text{ W}$

9.00-Ω: $P_3 = I_3^2 R_3 = (2.00 \text{ A})^2 (9.00 \ \Omega) = 36 \text{ W}$

Finalize Part (C) shows that the smallest resistor receives the most power. Summing the three quantities gives a total power of 198 W. We could have calculated this final result from part (A) by considering the equivalent resistance as follows: $P = (\Delta V)^2 / R_{eq} = (18.0 \text{ V})^2 / 1.64 \ \Omega = 198 \text{ W}$.

What If? What if the circuit were as shown in Figure 21.22b instead of as in Figure 21.22a? How would that affect the calculation?

Answer There would be no effect on the calculation. The physical placement of the battery is not important. Only the electrical arrangement is important. In Figure 21.22b, the battery still maintains a potential difference of 18.0 V between points *a* and *b*, so the two circuits in the figure are electrically identical.

21.8 | Kirchhoff's Rules

As we saw in the preceding section, combinations of resistors can be simplified and analyzed using the expression $\Delta V = IR$ and the rules for series and parallel combinations of resistors. Very often, however, it is not possible to reduce a circuit to a single loop using these rules. The procedure for analyzing more complex circuits is made possible by using the two following principles, called **Kirchhoff's rules.**

1. **Junction rule.** At any junction, the sum of the currents must equal zero:

$$\sum_{\text{junction}} I = 0 \qquad \qquad \textbf{21.29} \blacktriangleleft$$

2. **Loop rule.** The sum of the potential differences across all elements around any closed circuit loop must be zero:

$$\sum_{\text{loop}} \Delta V = 0 \qquad \qquad \textbf{21.30} \blacktriangleleft$$

Kirchhoff's first rule is a statement of **conservation of electric charge.** All charges that enter a given point in a circuit must leave that point because charge cannot build up at a point. Currents directed into the junction are entered into the junction rule as $+I$, whereas currents directed out of a junction are entered as $-I$. Applying this rule to the junction in Figure 21.23a gives

$$I_1 - I_2 - I_3 = 0$$

Figure 21.23b represents a mechanical analog of this situation, in which water flows through a branched pipe having no leaks. Because water does not build up anywhere

The amount of charge flowing out of the branches on the right must equal the amount flowing into the single branch on the left.

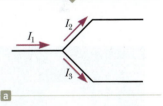

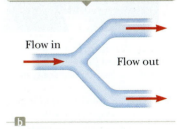

The amount of water flowing out of the branches on the right must equal the amount flowing into the single branch on the left.

Flow in

Flow out

Figure 21.23 (a) Kirchhoff's junction rule. (b) A mechanical analog of the junction rule.

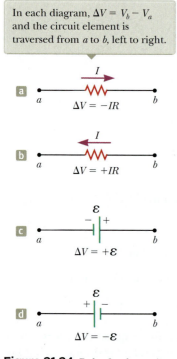

In each diagram, $\Delta V = V_b - V_a$ and the circuit element is traversed from a to b, left to right.

Figure 21.24 Rules for determining the potential differences across a resistor and a battery. (The battery is assumed to have no internal resistance.)

in the pipe, the flow rate into the pipe on the left equals the total flow rate out of the two branches on the right.

Kirchhoff's second rule follows from the law of **conservation of energy.** Let's imagine moving a charge around a closed loop of a circuit. When the charge returns to the starting point, the charge–circuit system must have the same total energy as it had before the charge was moved. The sum of the increases in energy as the charge passes through some circuit elements must equal the sum of the decreases in energy as it passes through other elements. The potential energy decreases whenever the charge moves through a potential drop $-IR$ across a resistor or whenever it moves in the reverse direction through a source of emf. The potential energy increases whenever the charge passes through a battery from the negative terminal to the positive terminal.

When applying Kirchhoff's second rule, imagine *traveling* around the loop and consider changes in *electric potential* rather than the changes in *potential energy* described in the preceding paragraph. Imagine traveling through the circuit elements in Figure 21.24 toward the right. The following sign conventions apply when using the second rule:

- Charges move from the high-potential end of a resistor toward the low-potential end, so if a resistor is traversed in the direction of the current, the potential difference ΔV across the resistor is $-IR$ (Fig. 21.24a).
- If a resistor is traversed in the direction *opposite* the current, the potential difference ΔV across the resistor is $+IR$ (Fig. 21.24b).
- If a source of emf (assumed to have zero internal resistance) is traversed in the direction of the emf (from negative to positive), the potential difference ΔV is $+\mathcal{E}$ (Fig. 21.24c).
- If a source of emf (assumed to have zero internal resistance) is traversed in the direction opposite the emf (from positive to negative), the potential difference ΔV is $-\mathcal{E}$ (Fig. 21.24d).

There are limits on the number of times you can usefully apply Kirchhoff's rules in analyzing a circuit. You can use the junction rule as often as you need as long as you include in it a current that has not been used in a preceding junction-rule equation. In general, the number of times you can use the junction rule is one fewer than the number of junction points in the circuit. You can apply the loop rule as often as needed as long as a new circuit element (resistor or battery) or a new current appears in each new equation. In general, to solve a particular circuit problem, the number of independent equations you need to obtain from the two rules equals the number of unknown currents.

Gustav Kirchhoff
German Physicist (1824–1887)
Kirchhoff, a professor at Heidelberg, and Robert Bunsen invented the spectroscope and founded the science of spectroscopy, which we discussed in Chapter 11. They discovered the elements cesium and rubidium and invented astronomical spectroscopy.

> **PROBLEM-SOLVING STRATEGY: Kirchhoff's Rules**
>
> The following procedure is recommended for solving problems that involve circuits that cannot be reduced by the rules for combining resistors in series or parallel.
>
> 1. **Conceptualize** Study the circuit diagram and make sure you recognize all elements in the circuit. Identify the polarity of each battery and try to imagine the directions in which the current would exist in the batteries.
>
> 2. **Categorize** Determine whether the circuit can be reduced by means of combining series and parallel resistors. If so, use the techniques of Section 21.7. If not, apply Kirchhoff's rules according to the *Analyze* step below.
>
> 3. **Analyze** Assign labels to all known quantities and symbols to all unknown quantities. You must assign *directions* to the currents in each part of the circuit. Although the assignment of current directions is arbitrary, you must adhere *rigorously* to the directions you assign when you apply Kirchhoff's rules.
>
>   Apply the junction rule (Kirchhoff's first rule) to all junctions in the circuit except one. Now apply the loop rule (Kirchhoff's second rule) to as many loops in the circuit as are needed to obtain, in combination with the

equations from the junction rule, as many equations as there are unknowns. To apply this rule, you must choose a direction in which to travel around the loop (either clockwise or counterclockwise) and correctly identify the change in potential as you cross each element. Be careful with signs!

Solve the equations simultaneously for the unknown quantities.

4. **Finalize** Check your numerical answers for consistency. Do not be alarmed if any of the resulting currents have a negative value. That only means you have guessed the direction of that current incorrectly, but *its magnitude will be correct.*

Example 21.8 | A Multiloop Circuit

Find the currents I_1, I_2, and I_3 in the circuit shown in Figure 21.25.

SOLUTION

Conceptualize Imagine physically rearranging the circuit while keeping it electrically the same. Can you rearrange it so that it consists of simple series or parallel combinations of resistors? You should find that you cannot.

Categorize We cannot simplify the circuit by the rules associated with combining resistances in series and in parallel. (If the 10.0-V battery were removed and replaced by a wire from b to the 6.0-Ω resistor, we could reduce the remaining circuit.) Because the circuit is not a simple series and parallel combination of resistances, this problem is one in which we must use Kirchhoff's rules.

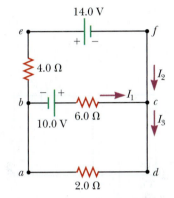

Figure 21.25 (Example 21.8) A circuit containing different branches.

Analyze We arbitrarily choose the directions of the currents as labeled in Figure 21.25.

Apply Kirchhoff's junction rule to junction c:

(1)　$I_1 + I_2 - I_3 = 0$

We now have one equation with three unknowns: I_1, I_2, and I_3. There are three loops in the circuit: *abcda*, *befcb*, and *aefda*. We need only two loop equations to determine the unknown currents. (The third equation for loop *aefda* would give no new information.) Let's choose to traverse these loops in the clockwise direction. Apply Kirchhoff's loop rule to loops *abcda* and *befcb*:

abcda: (2)　$10.0\text{ V} - (6.0\ \Omega)I_1 - (2.0\ \Omega)I_3 = 0$

befcb: $-(4.0\ \Omega)I_2 - 14.0\text{ V} + (6.0\ \Omega)I_1 - 10.0\text{ V} = 0$

(3)　$-24.0\text{ V} + (6.0\ \Omega)I_1 - (4.0\ \Omega)I_2 = 0$

Solve Equation (1) for I_3 and substitute into Equation (2):

$10.0\text{ V} - (6.0\ \Omega)I_1 - (2.0\ \Omega)(I_1 + I_2) = 0$

(4)　$10.0\text{ V} - (8.0\ \Omega)I_1 - (2.0\ \Omega)I_2 = 0$

Multiply each term in Equation (3) by 4 and each term in Equation (4) by 3:

(5)　$-96.0\text{ V} + (24.0\ \Omega)I_1 - (16.0\ \Omega)I_2 = 0$

(6)　$30.0\text{ V} - (24.0\ \Omega)I_1 - (6.0\ \Omega)I_2 = 0$

Add Equation (6) to Equation (5) to eliminate I_1 and find I_2:

$-66.0\text{ V} - (22.0\ \Omega)I_2 = 0$

$I_2 = \boxed{-3.0\text{ A}}$

Use this value of I_2 in Equation (3) to find I_1:

$-24.0\text{ V} + (6.0\ \Omega)I_1 - (4.0\ \Omega)(-3.0\text{ A}) = 0$

$-24.0\text{ V} + (6.0\ \Omega)I_1 + 12.0\text{ V} = 0$

$I_1 = \boxed{2.0\text{ A}}$

Use Equation (1) to find I_3:

$I_3 = I_1 + I_2 = 2.0\text{ A} - 3.0\text{ A} = \boxed{-1.0\text{ A}}$

Finalize Because our values for I_2 and I_3 are negative, the directions of these currents are opposite those indicated in Figure 21.25. The numerical values for the currents are correct. Despite the incorrect direction, we *must* continue to use these negative values in subsequent calculations because our equations were established with our original choice of direction. What would have happened had we left the current directions as labeled in Figure 21.25 but traversed the loops in the opposite direction?

⟨21.9 | *RC* Circuits

So far, we have analyzed direct-current circuits in which the current is constant. In DC circuits containing capacitors, the current is always in the same direction but may vary in time. A circuit containing a series combination of a resistor and a capacitor is called an **RC circuit.**

Charging a Capacitor

Active Figure 21.26 shows a simple series *RC* circuit. Let's assume the capacitor in this circuit is initially uncharged. There is no current while the switch is open (Active Fig. 21.26a). If the switch is thrown to position *a* at *t* = 0 (Active Fig. 21.26b), however, charge begins to flow, setting up a current in the circuit, and the capacitor begins to charge.[11] Notice that during charging, charges do not jump across the capacitor plates because the gap between the plates represents an open circuit. Instead, charge is transferred between each plate and its connecting wires due to the electric field established in the wires by the battery until the capacitor is fully charged. As the plates are being charged, the potential difference across the capacitor increases. The value of the maximum charge on the plates depends on the voltage of the battery. Once the maximum charge is reached, the current in the circuit is zero because the potential difference across the capacitor matches that supplied by the battery.

To analyze this circuit quantitatively, let's apply Kirchhoff's loop rule to the circuit after the switch is thrown to position *a*. Traversing the loop in Active Figure 21.26b clockwise gives

$$\mathcal{E} - \frac{q}{C} - IR = 0 \qquad\qquad \text{21.31} \blacktriangleleft$$

where q/C is the potential difference across the capacitor and IR is the potential difference across the resistor. We have used the sign conventions discussed earlier for the signs on $\mathcal{E}$ and IR. The capacitor is traversed in the direction from the positive plate to the negative plate, which represents a decrease in potential. Therefore, we use a negative sign for this potential difference in Equation 21.31. Note that q and I are *instantaneous* values that depend on time (as opposed to steady-state values) as the capacitor is being charged.

We can use Equation 21.31 to find the initial current in the circuit and the maximum charge on the capacitor. At the instant the switch is thrown to position *a* (*t* = 0), the charge on the capacitor is zero. Equation 21.31 shows that the initial current I_i in the circuit is a maximum and is given by

$$I_i = \frac{\mathcal{E}}{R} \quad \text{(current at } t = 0) \qquad \text{21.32} \blacktriangleleft$$

At this time, the potential difference from the battery terminals appears entirely across the resistor. Later, when the capacitor is charged to its maximum value Q, charges cease to flow, the current in the circuit is zero, and the potential difference from the battery terminals appears entirely across the capacitor. Substituting $I = 0$ into Equation 21.31 gives the maximum charge on the capacitor:

$$Q = C\mathcal{E} \quad \text{(maximum charge)} \qquad \text{21.33} \blacktriangleleft$$

To determine analytical expressions for the time dependence of the charge and current, we must solve Equation 21.31, a single equation containing two variables q and I. The current in all parts of the series circuit must be the same. Therefore,

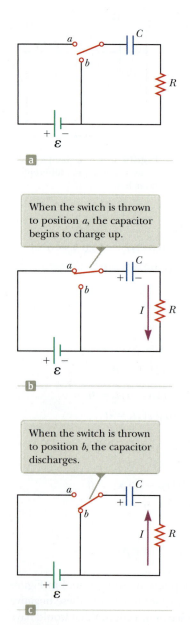

When the switch is thrown to position *a*, the capacitor begins to charge up.

When the switch is thrown to position *b*, the capacitor discharges.

Active Figure 21.26 A capacitor in series with a resistor, switch, and battery.

---

[11]In previous discussions of capacitors, we assumed a steady-state situation, in which no current was present in any branch of the circuit containing a capacitor. Now we are considering the case *before* the steady-state condition is realized; in this situation, charges are moving and a current exists in the wires connected to the capacitor.

the current in the resistance R must be the same as the current between each capacitor plate and the wire connected to it. This current is equal to the time rate of change of the charge on the capacitor plates. Therefore, we substitute $I = dq/dt$ into Equation 21.31 and rearrange the equation:

$$\frac{dq}{dt} = \frac{\mathcal{E}}{R} - \frac{q}{RC}$$

To find an expression for q, we solve this separable differential equation as follows. First combine the terms on the right-hand side:

$$\frac{dq}{dt} = \frac{C\mathcal{E}}{RC} - \frac{q}{RC} = -\frac{q - C\mathcal{E}}{RC}$$

Multiply this equation by dt and divide by $q - C\mathcal{E}$:

$$\frac{dq}{q - C\mathcal{E}} = -\frac{1}{RC} dt$$

Integrate this expression, using $q = 0$ at $t = 0$:

$$\int_0^q \frac{dq}{q - C\mathcal{E}} = -\frac{1}{RC} \int_0^t dt$$

$$\ln\left(\frac{q - C\mathcal{E}}{-C\mathcal{E}}\right) = -\frac{t}{RC}$$

From the definition of the natural logarithm, we can write this expression as

$$q(t) = C\mathcal{E}(1 - e^{-t/RC}) = Q(1 - e^{-t/RC}) \qquad \textbf{21.34} \blacktriangleleft$$

▶ Charge as a function of time for a capacitor being charged

where e is the base of the natural logarithm and we have made the substitution from Equation 21.33.

We can find an expression for the charging current by differentiating Equation 21.34 with respect to time. Using $I = dq/dt$, we find that

$$I(t) = \frac{\mathcal{E}}{R} e^{-t/RC} \qquad \textbf{21.35} \blacktriangleleft$$

▶ Current as a function of time for a capacitor being charged

Plots of capacitor charge and circuit current versus time are shown in Figure 21.27. Notice that the charge is zero at $t = 0$ and approaches the maximum value $C\mathcal{E}$ as $t \to \infty$. The current has its maximum value $I_i = \mathcal{E}/R$ at $t = 0$ and decays exponentially to zero as $t \to \infty$. The quantity RC, which appears in the exponents of Equations 21.34 and 21.35, is called the **time constant** τ of the circuit:

$$\tau = RC \qquad \textbf{21.36} \blacktriangleleft$$

The time constant represents the time interval during which the current decreases to $1/e$ of its initial value; that is, after a time interval τ, the current decreases to $I = e^{-1}I_i = 0.368I_i$. After a time interval 2τ, the current decreases to $I = e^{-2}I_i = 0.135I_i$, and so forth. Likewise, in a time interval τ, the charge increases from zero to $C\mathcal{E}[1 - e^{-1}] = 0.632C\mathcal{E}$.

The energy supplied by the battery during the time interval required to fully charge the capacitor is $Q\mathcal{E} = C\mathcal{E}^2$. After the capacitor is fully charged, the energy stored in the capacitor is $\frac{1}{2}Q\mathcal{E} = \frac{1}{2}C\mathcal{E}^2$, which is only half the energy output of the battery. It is left as a problem (Problem 68) to show that the remaining half of the energy supplied by the battery appears as internal energy in the resistor.

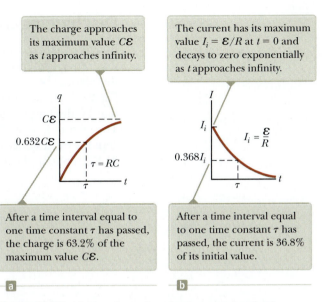

The charge approaches its maximum value $C\mathcal{E}$ as t approaches infinity.

The current has its maximum value $I_i = \mathcal{E}/R$ at $t = 0$ and decays to zero exponentially as t approaches infinity.

After a time interval equal to one time constant τ has passed, the charge is 63.2% of the maximum value $C\mathcal{E}$.

After a time interval equal to one time constant τ has passed, the current is 36.8% of its initial value.

Figure 21.27 (a) Plot of capacitor charge versus time for the circuit shown in Active Figure 21.26b. (b) Plot of current versus time for the circuit shown in Active Figure 21.26b.

Discharging a Capacitor

Imagine that the capacitor in Active Figure 21.26b is completely charged. A potential difference Q/C exists across the capacitor, and there is zero potential difference across the resistor because $I = 0$. If the switch is now thrown to position b at $t = 0$ (Active Fig. 21.26c), the capacitor begins to discharge through the resistor. At some time t during the discharge, the current in the circuit is I and the charge on the capacitor is q. The circuit in Active Figure 21.26c is the same as the circuit in Active Figure 21.26b except for the absence of the battery. Therefore, we eliminate the emf $\mathcal{E}$ from Equation 21.31 to obtain the appropriate loop equation for the circuit in Active Figure 21.26c:

$$-\frac{q}{C} - IR = 0 \qquad \text{21.37} \blacktriangleleft$$

When we substitute $I = dq/dt$ into this expression, it becomes

$$-R\frac{dq}{dt} = \frac{q}{C}$$

$$\frac{dq}{q} = -\frac{1}{RC}\,dt$$

Integrating this expression using $q = Q$ at $t = 0$ gives

$$\int_{Q}^{q} \frac{dq}{q} = -\frac{1}{RC}\int_{0}^{t} dt$$

$$\ln\left(\frac{q}{Q}\right) = -\frac{t}{RC}$$

▶ **Charge as a function of time for a discharging capacitor**

$$q(t) = Qe^{-t/RC} \qquad \text{21.38} \blacktriangleleft$$

Differentiating Equation 21.38 with respect to time gives the instantaneous current as a function of time:

▶ **Current as a function of time for a discharging capacitor**

$$I(t) = -\frac{Q}{RC}e^{-t/RC} \qquad \text{21.39} \blacktriangleleft$$

where $Q/RC = I_i$ is the initial current. The negative sign indicates that as the capacitor discharges, the current direction is opposite its direction when the capacitor was being charged. (Compare the current directions in Active Figs. 21.26b and 21.26c.) Both the charge on the capacitor and the current decay exponentially at a rate characterized by the time constant $\tau = RC$.

BIO Cable theory for propagation of an action potential along a nerve

In Section 20.7, we discussed the modeling of a patch of cell membrane as a capacitor. Let us call the capacitance of a given patch of membrane C_m. We also discussed the flow of ions through various ion channels and ion pumps in the membrane. This flow represents a current. The ions cannot move across the membrane unimpeded, so there is a resistance to the current, called the *membrane resistance* R_m. As a result, each small patch of the cell membrane can be modeled as an RC circuit as shown in Figure 21.28.

Figure 21.28 Modeling the cell membrane of a neuron using cable theory. Four small patches of the cell membrane are shown, with each patch being modeled electrically as an RC circuit consisting of resistance R_m and capacitance C_m. Adjacent patches are connected electrically by a resistance R_i in the cytoplasm of the cell interior.

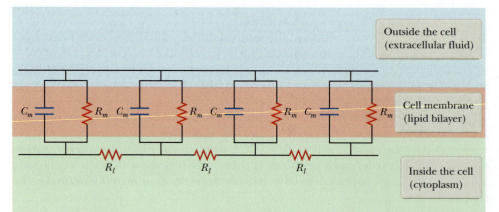

A given long structure in a neuron (such as a dendrite or an axon) can be modeled as a series of *RC* circuit modules connected by a longitudinal resistance as shown in Figure 21.28. The *longitudinal resistance* R_l represents resistance to the current along the axis of the neuron through the cytoplasm. This model of a neuron can be analyzed using *cable theory*, first used by Kelvin in the 1850s to analyze the decay of signals in underwater telegraphic cables. In a neuron, we consider the decay of the propagation of an action potential along the neuron.

Using cable theory, we can model the propagation of an action potential along a nerve cell and relate this model to the transfer of information within the human nervous system. The propagation of the action potential is governed by two primary parameters: the time constant and the length constant. The *time constant* $\tau = R_m C_m$ for the *RC* circuit associated with each patch of membrane is similar to the time constant discussed above and determines how rapidly the membrane capacitor can charge and discharge. For a given input at a point on the neuron, the membrane voltage along the neuron decays exponentially. The *length constant* $\lambda = (R_m / R_l)^{1/2}$ determines a characteristic length along the neuron through which the voltage decays to e^{-1} of its original value. Together, these two parameters describe how efficient the neuron is at transmitting a signal along its length.

The axons of some nerves are wrapped with sections of *myelin*, with each section separated from the next by intervals called the *nodes of Ranvier*. The myelin has the effect of shutting off the transfer of ions across the cell membrane. As a result, the relatively slow patch-to-patch propagation of an action potential as described above does not occur. Instead, the signal is carried primarily within the cell interior, such that an action potential at one node rapidly causes another action potential at the next node. As a result, the signal travels much faster along the neuron, in a process called *saltatory conduction*.

Some diseases cause damage to the myelin sheath around nerve cells, degrading the process of saltatory conduction. As a result, patients suffering from these diseases experience impaired movement, due to the slowness of signals traveling to the muscles. For example, *transverse myelitis* is an autoimmune disease, in which the body attacks the spinal cord, with the resulting inflammation damaging the myelin. In severe cases, patients are left wheelchair-bound and require assistance with daily activities. If the damage to the myelin occurs within the white matter of the brain, the disease is called *multiple sclerosis*, a highly debilitating disease.

BIO The role of myelin in nerve conduction

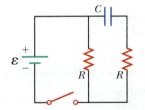

QUICK QUIZ 21.8 Consider the circuit in Figure 21.29 and assume the battery has no internal resistance. (**i**) Just after the switch is closed, what is the current in the battery? (a) 0 (b) $\mathcal{E}/2R$ (c) $2\mathcal{E}/R$ (d) $\mathcal{E}/R$ (e) impossible to determine (**ii**) After a very long time, what is the current in the battery? Choose from the same choices.

Figure 21.29 (Quick Quiz 21.8) How does the current vary after the switch is closed?

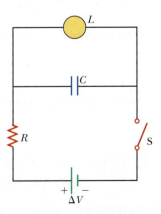

Figure 21.30 (Thinking Physics 21.6) The *RC* circuit in a roadway construction flasher. When the switch is closed, the charge on the capacitor increases until the voltage across the capacitor (and across the flash lamp) is high enough for the lamp to flash, discharging the capacitor.

▶ **THINKING PHYSICS 21.6**

Many roadway construction sites have flashing yellow lights to warn motorists of possible dangers. What causes the lightbulbs to flash?

Reasoning A typical circuit for such a flasher is shown in Figure 21.30. The lamp *L* is a gas-filled lamp that acts as an open circuit until a large potential difference causes an electrical discharge in the gas, which gives off a bright light. During this discharge, charges flow through the gas between the electrodes of the lamp. After switch S is closed, the battery charges up the capacitor of capacitance *C*. At the beginning, the current is high and the charge on the capacitor is low, so most of the potential difference appears across the resistance *R*. As the capacitor charges, more potential difference appears across it, reflecting the lower current and therefore lower potential difference across the resistor. Eventually, the potential difference across the capacitor reaches a value at which the lamp will conduct, causing a flash. This discharges the capacitor through the lamp and the process of charging begins again. The period between flashes can be adjusted by changing the time constant of the *RC* circuit. ◀

Example 21.9 | Charging a Capacitor in an *RC* Circuit

An uncharged capacitor and a resistor are connected in series to a battery as shown in Active Figure 21.26, where $\mathcal{E} = 12.0$ V, $C = 5.00$ μF, and $R = 8.00 \times 10^5$ Ω. The switch is thrown to position a. Find the time constant of the circuit, the maximum charge on the capacitor, the maximum current in the circuit, and the charge and current as functions of time.

SOLUTION

Conceptualize Study Active Figure 21.26 and imagine throwing the switch to position a as shown in Active Figure 21.26b. Upon doing so, the capacitor begins to charge.

Categorize We evaluate our results using equations developed in this section, so we categorize this example as a substitution problem.

Evaluate the time constant of the circuit from Equation 21.36:

$$\tau = RC = (8.00 \times 10^5 \ \Omega)(5.00 \times 10^{-6} \ \text{F}) = \boxed{4.00 \ \text{s}}$$

Evaluate the maximum charge on the capacitor from Equation 21.33:

$$Q = C\mathcal{E} = (5.00 \ \mu\text{F})(12.0 \ \text{V}) = \boxed{60.0 \ \mu\text{C}}$$

Evaluate the maximum current in the circuit from Equation 21.32:

$$I_i = \frac{\mathcal{E}}{R} = \frac{12.0 \ \text{V}}{8.00 \times 10^5 \ \Omega} = \boxed{15.0 \ \mu\text{A}}$$

Use these values in Equations 21.34 and 21.35 to find the charge and current as functions of time:

$$(1) \quad q(t) = \boxed{60.0(1 - e^{-t/4.00})}$$

$$(2) \quad I(t) = \boxed{15.0e^{-t/4.00}}$$

In Equations (1) and (2), q is in microcoulombs, I is in microamperes, and t is in seconds.

Example 21.10 | Discharging a Capacitor in an *RC* Circuit

Consider a capacitor of capacitance C that is being discharged through a resistor of resistance R as shown in Active Figure 21.26c.

(A) After how many time constants is the charge on the capacitor one-fourth its initial value?

SOLUTION

Conceptualize Study Active Figure 21.26 and imagine throwing the switch to position b as shown in Active Figure 21.26c. Upon doing so, the capacitor begins to discharge.

Categorize We categorize the example as one involving a discharging capacitor and use the appropriate equations.

Analyze Substitute $q(t) = Q/4$ into Equation 21.38:

$$\frac{Q}{4} = Qe^{-t/RC}$$

$$\frac{1}{4} = e^{-t/RC}$$

Take the logarithm of both sides of the equation and solve for t:

$$-\ln 4 = -\frac{t}{RC}$$

$$t = RC \ln 4 = 1.39RC = \boxed{1.39\tau}$$

(B) The energy stored in the capacitor decreases with time as the capacitor discharges. After how many time constants is this stored energy one-fourth its initial value?

SOLUTION

Use Equations 20.29 and 21.38 to express the energy stored in the capacitor at any time t:

$$(1) \quad U(t) = \frac{q^2}{2C} = \frac{Q^2}{2C}e^{-2t/RC}$$

Substitute $U(t) = \frac{1}{4}(Q^2/2C)$ into Equation (1):

$$\frac{1}{4}\frac{Q^2}{2C} = \frac{Q^2}{2C}e^{-2t/RC}$$

$$\frac{1}{4} = e^{-2t/RC}$$

21.10 *cont.*

Take the logarithm of both sides of the equation and solve for t:

$$-\ln 4 = -\frac{2t}{RC}$$

$$t = \tfrac{1}{2}RC \ln 4 = 0.693RC = \boxed{0.693\tau}$$

Finalize Notice that because the energy depends on the square of the charge, the energy in the capacitor drops more rapidly than the charge on the capacitor.

21.10 | Context Connection: The Atmosphere as a Conductor

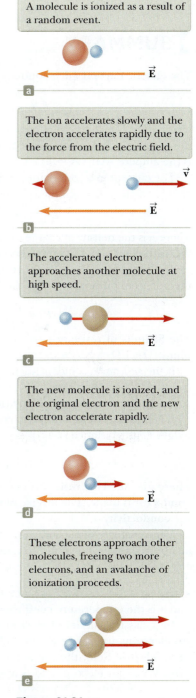

A molecule is ionized as a result of a random event.

The ion accelerates slowly and the electron accelerates rapidly due to the force from the electric field.

The accelerated electron approaches another molecule at high speed.

The new molecule is ionized, and the original electron and the new electron accelerate rapidly.

These electrons approach other molecules, freeing two more electrons, and an avalanche of ionization proceeds.

Figure 21.31 The anatomy of a spark.

When discussing capacitors with air between the plates in Chapter 20, we adopted the simplification model that air was a perfect insulator. Although that was a good model for typical potential differences encountered in capacitors, we know that it is possible for a current to exist in air. Lightning is a dramatic example of this possibility, but a more mundane example is the common spark that you might receive upon bringing your finger near a doorknob after rubbing your feet across a carpet.

Let us analyze the process that occurs in electrical discharge, which is the same for lightning and the doorknob spark except for the size of the current. Whenever a strong electric field exists in air, it is possible for the air to undergo electrical breakdown in which the effective resistivity of the air drops dramatically and the air becomes a conductor. At any given time, due to cosmic ray collisions and other events, air contains a number of ionized molecules (Fig. 21.31a). For a relatively weak electric field, such as the fair-weather electric field, these ions and freed electrons accelerate slowly due to the electric force. They collide with other molecules with no effect and eventually neutralize as a freed electron ultimately finds an ion and combines with it. In a strong electric field such as that associated with a thunderstorm, however, the freed electrons can accelerate to very high speeds (Fig. 21.31b) before making a collision with a molecule (Fig. 21.31c). If the field is strong enough, the electron may have enough energy to ionize the molecule in this collision (Fig. 21.31d). Now there are two electrons to be accelerated by the field, and each can strike another molecule at high speed (Fig. 21.31e). The result is a very rapid increase in the number of charge carriers available in the air and a corresponding decrease in resistance of the air. Therefore, there can be a large current in the air that tends to neutralize the charges that established the initial potential difference, such as the charges in the cloud and on the ground. When that happens, we have lightning.

Typical currents during lightning strikes can be very high. While the stepped leader is making its way toward the ground, the current is relatively modest, in the range of 200 to 300 A. This current is large compared with typical household currents but small compared with peak currents in lightning discharges. Once the connection is made between the stepped leader and the return stroke, the current rises rapidly to a typical value of 5×10^4 A. Considering that typical potential differences between cloud and ground in a thunderstorm can be measured in hundreds of thousands of volts, the power during a lightning stroke is measured in billions of watts. Much of the energy in the stroke is delivered to the air, resulting in a rapid temperature increase and the resultant flash of light and sound of thunder.

Even in the absence of a thundercloud, there is a flow of charge through the air. The ions in the air make the air a conductor, although not a very good one. Atmospheric measurements indicate a typical potential difference across our atmospheric capacitor (Section 20.11) of about 3×10^5 V. As we shall show in the Context Conclusion, the total resistance of the air between the plates in the atmospheric capacitor is about 300 Ω. Therefore, the average fair-weather current in the air is

$$I = \frac{\Delta V}{R} = \frac{3 \times 10^5 \,\text{V}}{300 \,\Omega} \approx 1 \times 10^3 \,\text{A}$$

A number of simplifying assumptions were made in these calculations, but this result is on the right order of magnitude for the global current. Although the result might seem surprisingly large, remember that this current is spread out over the entire surface area of the Earth. Therefore, the average fair-weather current density is

$$J = \frac{I}{A} = \frac{I}{4\pi R_E{}^2} = \frac{1 \times 10^3 \, \text{A}}{4\pi (6.4 \times 10^6 \, \text{m})^2} \approx 2 \times 10^{-12} \, \text{A/m}^2$$

In comparison, the current density in a lightning strike is on the order of $10^5 \, \text{A/m}^2$.

The fair-weather current and the lightning current are in opposite directions. The fair-weather current delivers positive charge to the ground, whereas lightning delivers negative charge. These two effects are in balance,[12] which is the principle that we shall use to estimate the average number of lightning strikes on the Earth in the Context Conclusion.

SUMMARY

The **electric current** I in a conductor is defined as

$$I \equiv \frac{dQ}{dt} \qquad \text{21.2} \blacktriangleleft$$

where dQ is the charge that passes through a cross-section of the conductor in the time interval dt. The SI unit of current is the ampere (A); $1 \, \text{A} = 1 \, \text{C/s}$.

The current in a conductor is related to the motion of the charge carriers through the relationship

$$I_{\text{avg}} = nqv_d A \qquad \text{21.4} \blacktriangleleft$$

where n is the density of charge carriers, q is their charge, v_d is the **drift speed**, and A is the cross-sectional area of the conductor.

The **resistance** R of a conductor is defined as the ratio of the potential difference across the conductor to the current:

$$R \equiv \frac{\Delta V}{I} \qquad \text{21.6} \blacktriangleleft$$

The SI units of resistance are volts per ampere, defined as ohms (Ω); $1 \, \Omega = 1 \, \text{V/A}$.

If the resistance is independent of the applied voltage, the conductor obeys **Ohm's law,** and conductors that have a constant resistance over a wide range of voltages are said to be **ohmic.**

If a conductor has a uniform cross-sectional area A and a length ℓ, its resistance is

$$R = \rho \frac{\ell}{A} \qquad \text{21.8} \blacktriangleleft$$

where ρ is called the **resistivity** of the material from which the conductor is made. The inverse of the resistivity is defined as the **conductivity** $\sigma = 1/\rho$.

The resistivity of a conductor varies with temperature in an approximately linear fashion; that is,

$$\rho = \rho_0[1 + \alpha(T - T_0)] \qquad \text{21.10} \blacktriangleleft$$

where ρ_0 is the resistivity at some reference temperature T_0 and α is the **temperature coefficient of resistivity.**

In a classical model of electronic conduction in a metal, the electrons are treated as molecules of a gas. In the absence of an electric field, the average velocity of the electrons is zero. When an electric field is applied, the electrons move (on the average) with a drift velocity $\vec{v}_d$, given by

$$\vec{v}_d = \frac{q\vec{E}}{m_e}\tau \qquad \text{21.15} \blacktriangleleft$$

where τ is the average time interval between collisions with the atoms of the metal. The resistivity of the material according to this model is

$$\rho = \frac{m_e}{ne^2\tau} \qquad \text{21.18} \blacktriangleleft$$

where n is the number of free electrons per unit volume.

If a potential difference ΔV is maintained across a circuit element, the **power,** or the rate at which energy is delivered to the circuit element, is

$$P = I\,\Delta V \qquad \text{21.20} \blacktriangleleft$$

Because the potential difference across a resistor is $\Delta V = IR$, we can express the power delivered to a resistor in the form

$$P = I^2 R = \frac{(\Delta V)^2}{R} \qquad \text{21.21} \blacktriangleleft$$

The **emf** of a battery is the voltage across its terminals when the current is zero. Because of the voltage drop across the **internal resistance** r of a battery, the **terminal voltage** of the battery is less than the emf when a current exists in the battery.

The **equivalent resistance** of a set of resistors connected in **series** is

$$R_{\text{eq}} = R_1 + R_2 + R_3 + \cdots \qquad \text{21.26} \blacktriangleleft$$

The **equivalent resistance** of a set of resistors connected in **parallel** is given by

$$\frac{1}{R_{\text{eq}}} = \frac{1}{R_1} + \frac{1}{R_2} + \frac{1}{R_3} + \cdots \qquad \text{21.28} \blacktriangleleft$$

Circuits involving more than one loop are analyzed using two simple rules called **Kirchhoff's rules:**

- At any junction, the sum of the currents must equal zero:

$$\sum_{\text{junction}} I = 0 \qquad \text{21.29} \blacktriangleleft$$

---

[12]There are a number of other effects, too, but we will adopt a simplification model in which these are the only two effects. For more information, see E. A. Bering, A. A. Few, and J. R. Benbrook, "The Global Electric Circuit," *Physics Today,* October 1998, pp. 24–30.

• The sum of the potential differences across each element around any closed circuit loop must be zero:

$$\sum_{\text{loop}} \Delta V = 0 \qquad \textbf{21.30} \blacktriangleleft$$

For the junction rule, current in a direction into a junction is $+I$, whereas current with a direction away from a junction is $-I$.

For the loop rule, when a resistor is traversed in the direction of the current, the change in potential ΔV across the resistor is $-IR$. If a resistor is traversed in the direction opposite the current, $\Delta V = +IR$.

If a source of emf is traversed in the direction of the emf (negative to positive), the change in potential is $+\mathcal{E}$. If it is traversed opposite the emf (positive to negative), the change in potential is $-\mathcal{E}$.

If a capacitor is charged with a battery of emf $\mathcal{E}$ through a resistance R, the charge on the capacitor and the current in the circuit vary in time according to the expressions

$$q(t) = Q(1 - e^{-t/RC}) \qquad \textbf{21.34} \blacktriangleleft$$

$$I(t) = \frac{\mathcal{E}}{R} e^{-t/RC} \qquad \textbf{21.35} \blacktriangleleft$$

where $Q = C\mathcal{E}$ is the maximum charge on the capacitor. The product RC is called the **time constant** of the circuit.

If a charged capacitor is discharged through a resistance R, the charge and current decrease exponentially in time according to the expressions

$$q(t) = Qe^{-t/RC} \qquad \textbf{21.38} \blacktriangleleft$$

$$I(t) = -\frac{Q}{RC} e^{-t/RC} \qquad \textbf{21.39} \blacktriangleleft$$

where Q is the initial charge on the capacitor.

> OBJECTIVE QUESTIONS |

1. If the terminals of a battery with zero internal resistance are connected across two identical resistors in series, the total power delivered by the battery is 8.00 W. If the same battery is connected across the same resistors in parallel, what is the total power delivered by the battery? (a) 16.0 W (b) 32.0 W (c) 2.00 W (d) 4.00 W (e) none of those answers

2. Wire B has twice the length and twice the radius of wire A. Both wires are made from the same material. If wire A has a resistance R, what is the resistance of wire B? (a) $4R$ (b) $2R$ (c) R (d) $\frac{1}{2}R$ (e) $\frac{1}{4}R$

3. The current-versus-voltage behavior of a certain electrical device is shown in Figure OQ21.3. When the potential difference across the device is 2 V, what is its resistance? (a) 1 Ω (b) $\frac{3}{4}$ Ω (c) $\frac{4}{3}$ Ω (d) undefined (e) none of those answers

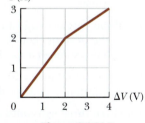

Figure OQ21.3

4. Several resistors are connected in parallel. Which of the following statements are correct? Choose all that are correct. (a) The equivalent resistance is greater than any of the resistances in the group. (b) The equivalent resistance is less than any of the resistances in the group. (c) The equivalent resistance depends on the voltage applied across the group. (d) The equivalent resistance is equal to the sum of the resistances in the group. (e) None of those statements is correct.

5. A potential difference of 1.00 V is maintained across a 10.0-Ω resistor for a period of 20.0 s. What total charge passes by a point in one of the wires connected to the resistor in this time interval? (a) 200 C (b) 20.0 C (c) 2.00 C (d) 0.005 00 C (e) 0.050 0 C

6. Several resistors are connected in series. Which of the following statements is correct? Choose all that are correct. (a) The equivalent resistance is greater than any of the resistances in the group. (b) The equivalent resistance is less than any of the resistances in the group. (c) The equivalent resistance depends on the voltage applied across the group. (d) The equivalent resistance is equal to the sum of the resistances in the group. (e) None of those statements is correct.

7. A metal wire of resistance R is cut into three equal pieces that are then placed together side by side to form a new cable with a length equal to one-third the original length. What is the resistance of this new cable? (a) $\frac{1}{9}R$ (b) $\frac{1}{3}R$ (c) R (d) $3R$ (e) $9R$

8. The terminals of a battery are connected across two resistors in parallel. The resistances of the resistors are not the same. Which of the following statements is correct? Choose all that are correct. (a) The resistor with the larger resistance carries more current than the other resistor. (b) The resistor with the larger resistance carries less current than the other resistor. (c) The potential difference across each resistor is the same. (d) The potential difference across the larger resistor is greater than the potential difference across the smaller resistor. (e) The potential difference is greater across the resistor closer to the battery.

9. A cylindrical metal wire at room temperature is carrying electric current between its ends. One end is at potential $V_A = 50$ V, and the other end is at potential $V_B = 0$ V. Rank the following actions in terms of the change that each one separately would produce in the current from the greatest increase to the greatest decrease. In your ranking, note any cases of equality. (a) Make $V_A = 150$ V with $V_B = 0$ V. (b) Adjust V_A to triple the power with which the wire converts electrically transmitted energy into internal energy. (c) Double the radius of the wire. (d) Double the length of the wire. (e) Double the Celsius temperature of the wire.

10. Two conducting wires A and B of the same length and radius are connected across the same potential difference. Conductor A has twice the resistivity of conductor B. What is the ratio of the power delivered to A to the power delivered to B? (a) 2 (b) $\sqrt{2}$ (c) 1 (d) $1/\sqrt{2}$ (e) $\frac{1}{2}$

11. When resistors with different resistances are connected in series, which of the following must be the same for each resistor? Choose all correct answers. (a) potential difference (b) current (c) power delivered (d) charge entering each resistor in a given time interval (e) none of those answers

12. When operating on a 120-V circuit, an electric heater receives 1.30×10^3 W of power, a toaster receives 1.00×10^3 W, and an

electric oven receives 1.54×10^3 W. If all three appliances are connected in parallel on a 120-V circuit and turned on, what is the total current drawn from an external source? (a) 24.0 A (b) 32.0 A (c) 40.0 A (d) 48.0 A (e) none of those answers

13. Car batteries are often rated in ampere-hours. Does this information designate the amount of (a) current, (b) power, (c) energy, (d) charge, or (e) potential the battery can supply?

14. The terminals of a battery are connected across two resistors in series. The resistances of the resistors are not the same. Which of the following statements are correct? Choose all that are correct. (a) The resistor with the smaller resistance carries more current than the other resistor. (b) The resistor with the larger resistance carries less current than the other resistor. (c) The current in each resistor is the same. (d) The potential difference across each resistor is the same. (e) The potential

difference is greatest across the resistor closest to the positive terminal.

15. In the circuit shown in Figure OQ21.15, each battery is delivering energy to the circuit by electrical transmission. All the resistors have equal resistance.

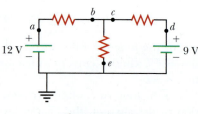

Figure OQ21.15

(i) Rank the electric potentials at points a, b, c, d, and e from highest to lowest, noting any cases of equality in the ranking. (ii) Rank the magnitudes of the currents at the same points from greatest to least, noting any cases of equality.

☐ denotes answer available in *Student Solutions Manual/Study Guide*

CONCEPTUAL QUESTIONS

1. Suppose a parachutist lands on a high-voltage wire and grabs the wire as she prepares to be rescued. (a) Will she be electrocuted? (b) If the wire then breaks, should she continue to hold onto the wire as she falls to the ground? Explain.

2. What factors affect the resistance of a conductor?

3. Newspaper articles often contain statements such as "10 000 volts of electricity surged through the victim's body." What is wrong with this statement?

4. Referring to Figure CQ21.4, describe what happens to the lightbulb after the switch is closed. Assume the capacitor has a large capacitance and is initially uncharged. Also assume the light illuminates when connected directly across the battery terminals.

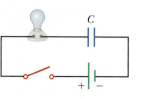

Figure CQ21.4

5. When the potential difference across a certain conductor is doubled, the current is observed to increase by a factor of 3. What can you conclude about the conductor?

6. Use the atomic theory of matter to explain why the resistance of a material should increase as its temperature increases.

7. So that your grandmother can listen to *A Prairie Home Companion*, you take her bedside radio to the hospital where she is staying. You are required to have a maintenance worker test the radio for electrical safety. Finding that it develops 120 V on one of its knobs, he does not let you take it to your grandmother's

room. Your grandmother complains that she has had the radio for many years and nobody has ever gotten a shock from it. You end up having to buy a new plastic radio. (a) Why is your grandmother's old radio dangerous in a hospital room? (b) Will the old radio be safe back in her bedroom?

8. (a) What advantage does 120-V operation offer over 240 V? (b) What disadvantages does it have?

9. How does the resistance for copper and for silicon change with temperature? Why are the behaviors of these two materials different?

10. If charges flow very slowly through a metal, why does it not require several hours for a light to come on when you throw a switch?

11. If you were to design an electric heater using Nichrome wire as the heating element, what parameters of the wire could you vary to meet a specific power output such as 1 000 W?

12. Is the direction of current in a battery always from the negative terminal to the positive terminal? Explain.

13. Given three lightbulbs and a battery, sketch as many different electric circuits as you can.

14. A student claims that the second of two lightbulbs in series is less bright than the first because the first lightbulb uses up some of the current. How would you respond to this statement?

15. Why is it possible for a bird to sit on a high-voltage wire without being electrocuted?

PROBLEMS

Section 21.1 Electric Current

1. In a particular cathode-ray tube, the measured beam current is 30.0 μA. How many electrons strike the tube screen every 40.0 s?

2. **S** Suppose the current in a conductor decreases exponentially with time according to the equation $I(t) = I_0 e^{-t/\tau}$, where I_0 is the initial current (at $t = 0$) and τ is a constant having dimensions of time. Consider a fixed observation point within the conductor. (a) How much charge passes this point between $t = 0$ and $t = \tau$? (b) How much charge passes this point between $t = 0$ and $t = 10\tau$? (c) **What If?** How much charge passes this point between $t = 0$ and $t = \infty$?

3. **W** The quantity of charge q (in coulombs) that has passed through a surface of area 2.00 cm^2 varies with time according to the equation $q = 4t^3 + 5t + 6$, where t is in seconds. What is the instantaneous current through the surface at $t = 1.00$ s? (b) What is the value of the current density?

4. **S** A small sphere that carries a charge q is whirled in a circle at the end of an insulating string. The angular frequency of revolution is ω. What average current does this revolving charge represent?

5. **M** The electron beam emerging from a certain high-energy electron accelerator has a circular cross section of radius 1.00 mm. (a) The beam current is 8.00 μA. Find the current density in the beam assuming it is uniform throughout. (b) The speed of the electrons is so close to the speed of light that their speed can be taken as 300 Mm/s with negligible error. Find the electron density in the beam. (c) Over what time interval does Avogadro's number of electrons emerge from the accelerator?

6. **QC** **W** Figure P21.6 represents a section of a conductor of nonuniform diameter carrying a current of $I = 5.00$ A. The radius of cross-section A_1 is $r_1 = 0.400$ cm. (a) What is the magnitude of the current density across A_1? The radius r_2 at A_2 is larger than the radius r_1 at A_1. (b) Is the current at A_2 larger, smaller, or the same? (c) Is the current density at A_2 larger, smaller, or the same? Assume $A_2 = 4A_1$. Specify the (d) radius, (e) current, and (f) current density at A_2.

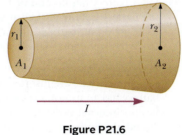

Figure P21.6

7. **W** An aluminum wire having a cross-sectional area equal to 4.00×10^{-6} m^2 carries a current of 5.00 A. The density of aluminum is 2.70 g/cm^3. Assume each aluminum atom supplies one conduction electron per atom. Find the drift speed of the electrons in the wire.

Section 21.2 Resistance and Ohm's Law

8. A 0.900-V potential difference is maintained across a 1.50-m length of tungsten wire that has a cross-sectional area of 0.600 mm^2. What is the current in the wire?

9. **M** An aluminum wire with a diameter of 0.100 mm has a uniform electric field of 0.200 V/m imposed along its entire length. The temperature of the wire is 50.0°C. Assume

one free electron per atom. (a) Use the information in Table 21.1 to determine the resistivity of aluminum at this temperature. (b) What is the current density in the wire? (c) What is the total current in the wire? (d) What is the drift speed of the conduction electrons? (e) What potential difference must exist between the ends of a 2.00-m length of the wire to produce the stated electric field?

10. **W** A lightbulb has a resistance of 240 Ω when operating with a potential difference of 120 V across it. What is the current in the lightbulb?

11. **M** Suppose you wish to fabricate a uniform wire out of 1.00 g of copper. If the wire is to have a resistance of $R = 0.500$ Ω and all the copper is to be used, what must be (a) the length and (b) the diameter of this wire?

12. **S** Suppose you wish to fabricate a uniform wire from a mass m of a metal with density ρ_m and resistivity ρ. If the wire is to have a resistance of R and all the metal is to be used, what must be (a) the length and (b) the diameter of this wire?

13. While taking photographs in Death Valley on a day when the temperature is 58.0°C, Bill Hiker finds that a certain voltage applied to a copper wire produces a current of 1.000 A. Bill then travels to Antarctica and applies the same voltage to the same wire. What current does he register there if the temperature is −88.0°C? Assume that no change occurs in the wire's shape and size.

14. **Review.** An aluminum rod has a resistance of 1.234 Ω at 20.0°C. Calculate the resistance of the rod at 120°C by accounting for the changes in both the resistivity and the dimensions of the rod. The coefficient of linear expansion for aluminum is 24.0×10^{-6} (°C)$^{-1}$.

Section 21.4 A Model for Electrical Conduction

15. If the current carried by a conductor is doubled, what happens to (a) the charge carrier density, (b) the current density, (c) the electron drift velocity, and (d) the average time interval between collisions?

16. **GP** **QC** An iron wire has a cross-sectional area equal to 5.00×10^{-6} m^2. Carry out the following steps to determine the drift speed of the conduction electrons in the wire if it carries a current of 30.0 A. (a) How many kilograms are there in 1.00 mole of iron? (b) Starting with the density of iron and the result of part (a), compute the molar density of iron (the number of moles of iron per cubic meter). (c) Calculate the number density of iron atoms using Avogadro's number. (d) Obtain the number density of conduction electrons given that there are two conduction electrons per iron atom. (e) Calculate the drift speed of conduction electrons in this wire.

17. **M** If the magnitude of the drift velocity of free electrons in a copper wire is 7.84×10^{-4} m/s, what is the electric field in the conductor?

Section 21.5 Energy and Power in Electric Circuits

18. **BIO** The potential difference across a resting neuron in the human body is about 75.0 mV and carries a current of about 0.200 mA. How much power does the neuron release?

19. In a hydroelectric installation, a turbine delivers 1 500 hp to a generator, which in turn transfers 80.0% of the mechanical

energy out by electrical transmission. Under these conditions, what current does the generator deliver at a terminal potential difference of 2 000 V?

20. **QC W** Residential building codes typically require the use of 12-gauge copper wire (diameter 0.205 cm) for wiring receptacles. Such circuits carry currents as large as 20.0 A. If a wire of smaller diameter (with a higher gauge number) carried that much current, the wire could rise to a high temperature and cause a fire. (a) Calculate the rate at which internal energy is produced in 1.00 m of 12-gauge copper wire carrying 20.0 A. (b) **What If?** Repeat the calculation for a 12-gauge aluminum wire. (c) Explain whether a 12-gauge aluminum wire would be as safe as a copper wire.

21. **M** A certain toaster has a heating element made of Nichrome wire. When the toaster is first connected to a 120-V source (and the wire is at a temperature of 20.0°C), the initial current is 1.80 A. The current decreases as the heating element warms up. When the toaster reaches its final operating temperature, the current is 1.53 A. (a) Find the power delivered to the toaster when it is at its operating temperature. (b) What is the final temperature of the heating element?

22. *Why is the following situation impossible?* A politician is decrying wasteful uses of energy and decides to focus on energy used to operate plug-in electric clocks in the United States. He estimates there are 270 million of these clocks, approximately one clock for each person in the population. The clocks transform energy taken in by electrical transmission at the average rate 2.50 W. The politician gives a speech in which he complains that, at today's electrical rates, the nation is losing $100 million every year to operate these clocks.

23. An 11.0-W energy-efficient fluorescent lightbulb is designed to produce the same illumination as a conventional 40.0-W incandescent lightbulb. Assuming a cost of $0.110/kWh for energy from the electric company, how much money does the user of the energy-efficient bulb save during 100 h of use?

24. Make an order-of-magnitude estimate of the cost of one person's routine use of a handheld hair dryer for 1 year. If you do not use a hair dryer yourself, observe or interview someone who does. State the quantities you estimate and their values.

25. A 100-W lightbulb connected to a 120-V source experiences a voltage surge that produces 140 V for a moment. By what percentage does its power output increase? Assume its resistance does not change.

26. The cost of energy delivered to residences by electrical transmission varies from $0.070/kWh to $0.258/kWh throughout the United States; $0.110/kWh is the average value. At this average price, calculate the cost of (a) leaving a 40.0-W porch light on for two weeks while you are on vacation, (b) making a piece of dark toast in 3.00 min with a 970-W toaster, and (c) drying a load of clothes in 40.0 min in a 5.20×10^3-W dryer.

27. **M** Assuming the cost of energy from the electric company is $0.110/kWh, compute the cost per day of operating a lamp that draws a current of 1.70 A from a 110-V line.

28. **Review.** An office worker uses an immersion heater to warm 250 g of water in a light, covered, insulated cup from 20.0°C to 100°C in 4.00 min. The heater is a Nichrome resistance wire connected to a 120-V power supply. Assume the wire is at 100°C throughout the 4.00-min time interval. Specify a relationship between a diameter and a length that the wire can have. (b) Can it be made from less than 0.500 cm³ of Nichrome?

29. A toaster is rated at 600 W when connected to a 120-V source. What current does the toaster carry, and what is its resistance?

30. **Review.** A rechargeable battery of mass 15.0 g delivers an average current of 18.0 mA to a portable DVD player at 1.60 V for 2.40 h before the battery must be recharged. The recharger maintains a potential difference of 2.30 V across the battery and delivers a charging current of 13.5 mA for 4.20 h. (a) What is the efficiency of the battery as an energy storage device? (b) How much internal energy is produced in the battery during one charge–discharge cycle? (c) If the battery is surrounded by ideal thermal insulation and has an effective specific heat of 975 J/kg · °C, by how much will its temperature increase during the cycle?

31. **M** An all-electric car (not a hybrid) is designed to run from a bank of 12.0-V batteries with total energy storage of 2.00×10^7 J. If the electric motor draws 8.00 kW as the car moves at a steady speed of 20.0 m/s, (a) what is the current delivered to the motor? (b) How far can the car travel before it is "out of juice"?

32. **M Review.** A well-insulated electric water heater warms 109 kg of water from 20.0°C to 49.0°C in 25.0 min. Find the resistance of its heating element, which is connected across a 240-V potential difference.

Section 21.6 Sources of emf

33. **M** A battery has an emf of 15.0 V. The terminal voltage of the battery is 11.6 V when it is delivering 20.0 W of power to an external load resistor R. (a) What is the value of R? (b) What is the internal resistance of the battery?

34. Two 1.50-V batteries—with their positive terminals in the same direction—are inserted in series into a flashlight. One battery has an internal resistance of 0.255 Ω, and the other has an internal resistance of 0.153 Ω. When the switch is closed, the bulb carries a current of 600 mA. (a) What is the bulb's resistance? (b) What fraction of the chemical energy transformed appears as internal energy in the batteries?

35. **W** An automobile battery has an emf of 12.6 V and an internal resistance of 0.080 0 Ω. The headlights together have an equivalent resistance of 5.00 Ω (assumed constant). What is the potential difference across the headlight bulbs (a) when they are the only load on the battery and (b) when the starter motor is operated, requiring an additional 35.0 A from the battery?

Section 21.7 Resistors in Series and Parallel

36. **BIO** For the purpose of measuring the electric resistance of shoes through the body of the wearer standing on a metal ground plate, the American National Standards Institute (ANSI) specifies the circuit shown in Figure P21.36. The potential

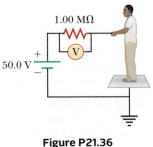

1.00 MΩ

50.0 V

Figure P21.36

difference ΔV across the 1.00-MΩ resistor is measured with an ideal voltmeter. (a) Show that the resistance of the footwear is

$$R_{shoes} = \frac{50.0\ V - \Delta V}{\Delta V}$$

(b) In a medical test, a current through the human body should not exceed 150 μA. Can the current delivered by the ANSI-specified circuit exceed 150 μA? To decide, consider a person standing barefoot on the ground plate.

37. (a) Find the equivalent resistance between points a and b in Figure P21.37. (b) A potential difference of 34.0 V is applied between points a and b. Calculate the current in each resistor.

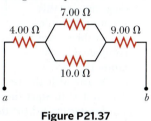

Figure P21.37

38. *Why is the following situation impossible?* A technician is testing a circuit that contains a resistance R. He realizes that a better design for the circuit would include a resistance $\frac{7}{3}R$ rather than R. He has three additional resistors, each with resistance R. By combining these additional resistors in a certain combination that is then placed in series with the original resistor, he achieves the desired resistance.

39. **M** Consider the circuit shown in Figure P21.39. Find (a) the current in the 20.0-Ω resistor and (b) the potential difference between points a and b.

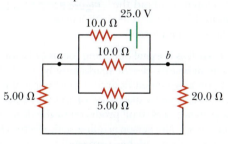

Figure P21.39

40. Four resistors are connected to a battery as shown in Figure P21.40. The current in the battery is I, the battery emf is $\mathcal{E}$, and the resistor values are $R_1 = R$, $R_2 = 2R$, $R_3 = 4R$, and $R_4 = 3R$. (a) Rank the resistors according to the potential difference across them, from largest to smallest. Note any cases of equal potential differences. (b) Determine the potential difference across each resistor in terms of $\mathcal{E}$. (c) Rank the resistors according to the current in them, from largest to smallest. Note any cases of equal currents. (d) Determine the current in each resistor in terms of I. (e) If R_3 is increased, what happens to the current in each of the resistors? (f) In the limit that $R_3 \to \infty$, what are the new values of the current in each resistor in terms of I, the original current in the battery?

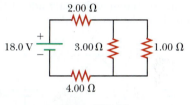

Figure P21.40

41. **W** Three 100-Ω resistors are connected as shown in Figure P21.41. The

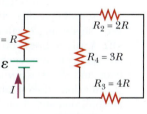

Figure P21.41

maximum power that can safely be delivered to any one resistor is 25.0 W. (a) What is the maximum potential difference that can be applied to the terminals a and b? (b) For the voltage determined in part (a), what is the power delivered to each resistor? (c) What is the total power delivered to the combination of resistors?

42. A young man owns a canister vacuum cleaner marked "535 W [at] 120 V" and a Volkswagen Beetle, which he wishes to clean. He parks the car in his apartment parking lot and uses an inexpensive extension cord 15.0 m long to plug in the vacuum cleaner. You may assume the cleaner has constant resistance. (a) If the resistance of each of the two conductors in the extension cord is 0.900 Ω, what is the actual power delivered to the cleaner? (b) If instead the power is to be at least 525 W, what must be the diameter of each of two identical copper conductors in the cord he buys? (c) Repeat part (b) assuming the power is to be at least 532 W.

43. **W** Calculate the power delivered to each resistor in the circuit shown in Figure P21.43.

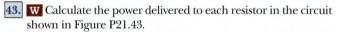

Figure P21.43

44. **Q C** A lightbulb marked "75 W @ 120 V" is screwed into a socket at one end of a long extension cord, in which each of the two conductors has resistance 0.800 Ω. The other end of the extension cord is plugged into a 120-V outlet. (a) Explain why the actual power delivered to the lightbulb cannot be 75 W in this situation. (b) Draw a circuit diagram. (c) Find the actual power delivered to the lightbulb in this circuit.

Section 21.8 Kirchhoff's Rules

> *Note:* The currents are not necessarily in the direction shown for some circuits.

45. **W** The ammeter shown in Figure P21.45 reads 2.00 A. Find I_1, I_2, and $\mathcal{E}$.

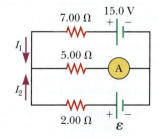

Figure P21.45

46. The following equations describe an electric circuit:

$$-I_1\ (220\ \Omega) + 5.80\ V - I_2\ (370\ \Omega) = 0$$

$$+I_2\ (370\ \Omega) + I_3\ (150\ \Omega) - 3.10\ V = 0$$

$$I_1 + I_3 - I_2 = 0$$

(a) Draw a diagram of the circuit. (b) Calculate the unknowns and identify the physical meaning of each unknown.

47. M Q|C The circuit shown in Figure P21.47 is connected for 2.00 min. (a) Determine the current in each branch of the circuit. (b) Find the energy delivered by each battery. (c) Find the energy delivered to each resistor. (d) Identify the type of energy storage transformation that occurs in the operation of the circuit. (e) Find the total amount of energy transformed into internal energy in the resistors.

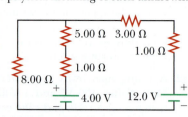

Figure P21.47 Problems 47 and 48.

48. In Figure P21.47, show how to add just enough ammeters to measure every different current. Show how to add just enough voltmeters to measure the potential difference across each resistor and across each battery.

49. Taking $R = 1.00$ kΩ and $\mathcal{E} = 250$ V in Figure P21.49, determine the direction and magnitude of the current in the horizontal wire between a and e.

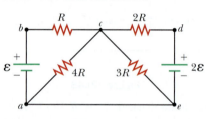

Figure P21.49

50. GP Q|C For the circuit shown in Figure P21.50, we wish to find the currents I_1, I_2, and I_3. Use Kirchhoff's rules to obtain equations for (a) the upper loop, (b) the lower loop, and (c) the junction on the left side. In each case, suppress units for clarity and simplify, combining the terms. (d) Solve the junction equation for I_3. (e) Using the equation found in part (d), eliminate I_3 from the equation found in part (b). (f) Solve the equations found in parts (a) and (e) simultaneously for the two unknowns I_1 and I_2. (g) Substitute the answers found in part (f) into the junction equation found in part (d), solving for I_3. (h) What is the significance of the negative answer for I_2?

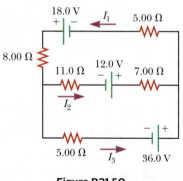

Figure P21.50

51. W In the circuit of Figure P21.51, determine (a) the current in each resistor and (b) the potential difference across the 200-Ω resistor.

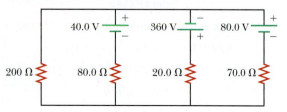

Figure P21.51

52. Q|C Jumper cables are connected from a fresh battery in one car to charge a dead battery in another car. Figure P21.52 shows the circuit diagram for this situation. While the cables are connected, the ignition switch of the car with the dead battery is closed and the starter is activated to start the engine. Determine the current in (a) the starter and (b) the dead battery. (c) Is the dead battery being charged while the starter is operating?

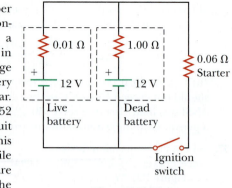

Figure P21.52

Section 21.9 RC Circuits

53. W Consider a series RC circuit as in Figure P21.53 for which $R = 1.00$ MΩ, $C = 5.00$ μF, and $\mathcal{E} = 30.0$ V. Find (a) the time constant of the circuit and (b) the maximum charge on the capacitor after the switch is thrown closed. (c) Find the current in the resistor 10.0 s after the switch is closed.

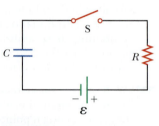

Figure P21.53 Problems 53, 67, and 68.

54. In places such as hospital operating rooms or factories for electronic circuit boards, electric sparks must be avoided. A person standing on a grounded floor and touching nothing else can typically have a body capacitance of 150 pF, in parallel with a foot capacitance of 80.0 pF produced by the dielectric soles of his or her shoes. The person acquires static electric charge from interactions with his or her surroundings. The static charge flows to ground through the equivalent resistance of the two shoe soles in parallel with each other. A pair of rubber-soled street shoes can present an equivalent resistance of 5.00×10^3 MΩ. A pair of shoes with special static-dissipative soles can have an equivalent resistance of 1.00 MΩ. Consider the person's body and shoes as forming an RC circuit with the ground. (a) How long does it take the rubber-soled shoes to reduce a person's potential from 3.00×10^3 V to 100 V? (b) How long does it take the static-dissipative shoes to do the same thing?

55. W A 2.00-nF capacitor with an initial charge of 5.10 μC is discharged through a 1.30-kΩ resistor. (a) Calculate the current in the resistor 9.00 μs after the resistor is connected across the terminals of the capacitor. (b) What charge remains on the capacitor after 8.00 μs? (c) What is the maximum current in the resistor?

56. A 10.0-μF capacitor is charged by a 10.0-V battery through a resistance R. The capacitor reaches a potential difference of 4.00 V in a time interval of 3.00 s after charging begins. Find R.

57. W In the circuit of Figure P21.57, the switch S has been open for a long time. It is then suddenly closed. Take $\mathcal{E} = 10.0$ V, $R_1 = 50.0$ kΩ, $R_2 = 100$ kΩ, and $C = 10.0$ μF. Determine the time constant (a) before the switch is closed and (b) after the switch is closed. (c) Let the switch be closed at $t = 0$. Determine the current in the switch as a function of time.

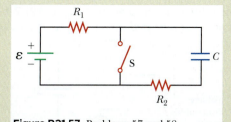

Figure P21.57 Problems 57 and 58.

58. **S** In the circuit of Figure P21.57, the switch S has been open for a long time. It is then suddenly closed. Determine the time constant (a) before the switch is closed and (b) after the switch is closed. (c) Let the switch be closed at $t = 0$. Determine the current in the switch as a function of time.

59. **M** The circuit in Figure P21.59 has been connected for a long time. (a) What is the potential difference across the capacitor? (b) If the battery is disconnected from the circuit, over what time interval does the capacitor discharge to one-tenth its initial voltage?

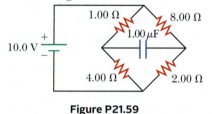

Figure P21.59

Section 21.10 Context Connection: The Atmosphere as a Conductor

60. Assume that global lightning on the Earth constitutes a constant current of 1.00 kA between the ground and an atmospheric layer at potential 300 kV. (a) Find the power of terrestrial lightning. (b) For comparison, find the power of sunlight falling on the Earth. Sunlight has an intensity of 1 370 W/m² above the atmosphere. Sunlight falls perpendicularly on the circular projected area that the Earth presents to the Sun.

61. A current density of 6.00×10^{-13} A/m² exists in the atmosphere at a location where the electric field is 100 V/m. Calculate the electrical conductivity of the Earth's atmosphere in this region.

Additional Problems

62. **Q|C** Lightbulb A is marked "25 W 120 V," and lightbulb B is marked "100 W 120 V." These labels mean that each lightbulb has its respective power delivered to it when it is connected to a constant 120-V source. (a) Find the resistance of each lightbulb. (b) During what time interval does 1.00 C pass into lightbulb A? (c) Is this charge different upon its exit versus its entry into the lightbulb? Explain. (d) In what time interval does 1.00 J pass into lightbulb A? (e) By what mechanisms does this energy enter and exit the lightbulb? Explain. (f) Find the cost of running lightbulb A continuously for 30.0 days, assuming the electric company sells its product at $0.110 per kWh.

63. A straight, cylindrical wire lying along the x axis has a length of 0.500 m and a diameter of 0.200 mm. It is made of a material described by Ohm's law with a resistivity of $\rho = 4.00 \times 10^{-8} \; \Omega \cdot$ m. Assume a potential of 4.00 V is maintained at the left end of the wire at $x = 0$. Also assume $V = 0$

at $x = 0.500$ m. Find (a) the magnitude and direction of the electric field in the wire, (b) the resistance of the wire, (c) the magnitude and direction of the electric current in the wire, and (d) the current density in the wire. (e) Show that $E = \rho J$.

64. **S** A straight, cylindrical wire lying along the x axis has a length L and a diameter d. It is made of a material described by Ohm's law with a resistivity ρ. Assume potential V is maintained at the left end of the wire at $x = 0$. Also assume the potential is zero at $x = L$. In terms of L, d, V, ρ, and physical constants, derive expressions for (a) the magnitude and direction of the electric field in the wire, (b) the resistance of the wire, (c) the magnitude and direction of the electric current in the wire, and (d) the current density in the wire. (e) Show that $E = \rho J$.

65. Four 1.50-V AA batteries in series are used to power a small radio. If the batteries can move a charge of 240 C, how long will they last if the radio has a resistance of 200 Ω?

66. **S** An oceanographer is studying how the ion concentration in seawater depends on depth. She makes a measurement by lowering into the water a pair of concentric metallic cylinders (Fig. P21.66) at the end of a cable and taking data to determine the resistance between these electrodes as a function of depth. The water between the two cylinders forms a cylindrical shell of inner radius r_a, outer radius r_b, and length L much larger than r_b. The scientist applies a potential difference ΔV between the inner and outer surfaces, producing an outward radial current I. Let ρ represent the resistivity of the water. (a) Find the resistance of the water between the cylinders in terms of L, ρ, r_a, an r_b. (b) Express the resistivity of the water in terms of the measured quantities L, r_a, r_b, ΔV, and I.

Figure P21.66

67. **M** The values of the components in a simple series RC circuit containing a switch (Fig. P21.53) are $C = 1.00 \; \mu$F, $R = 2.00 \times 10^6 \; \Omega$, and $\mathcal{E} = 10.0$ V. At the instant 10.0 s after the switch is closed, calculate (a) the charge on the capacitor, (b) the current in the resistor, (c) the rate at which energy is being stored in the capacitor, and (d) the rate at which energy is being delivered by the battery.

68. **S** A battery is used to charge a capacitor through a resistor as shown in Figure P21.53. Show that half the energy supplied by the battery appears as internal energy in the resistor and half is stored in the capacitor.

69. Switch S shown in Figure P21.69 has been closed for a long time, and the electric circuit carries a constant current. Take $C_1 = 3.00 \; \mu$F, $C_2 = 6.00 \; \mu$F, $R_1 = 4.00$ kΩ, and $R_2 = 7.00$ kΩ. The power delivered to R_2 is 2.40 W.

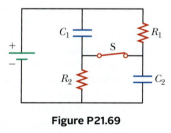

Figure P21.69

(a) Find the charge on C_1. (b) Now the switch is opened. After many milliseconds, by how much has the charge on C_2 changed?

70. *Why is the following situation impossible?* A battery has an emf of $\mathcal{E} = 9.20$ V and an internal resistance of $r = 1.20$ Ω. A resistance R is connected across the battery and extracts from it a power of $P = 21.2$ W.

71. The student engineer of a campus radio station wishes to verify the effectiveness of the lightning rod on the antenna mast (Fig. P21.71). The unknown resistance R_x is between points C and E. Point E is a true ground, but it is inaccessible for direct measurement because this stratum is several meters below the

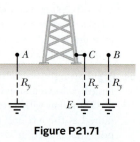

Figure P21.71

Earth's surface. Two identical rods are driven into the ground at A and B, introducing an unknown resistance R_y. The procedure is as follows. Measure resistance R_1 between points A and B, then connect A and B with a heavy conducting wire and measure resistance R_2 between points A and C. (a) Derive an equation for R_x in terms of the observable resistances, R_1 and R_2. (b) A satisfactory ground resistance would be $R_x < 2.00\ \Omega$. Is the grounding of the station adequate if measurements give $R_1 = 13.0\ \Omega$ and $R_2 = 6.00\ \Omega$? Explain.

72. The circuit shown in Figure P21.72 is set up in the laboratory to measure an unknown capacitance C in series with a resistance $R = 10.0\ M\Omega$ powered by a battery whose emf is 6.19 V. The data given in the table are the measured voltages across the capacitor as a function of time, where

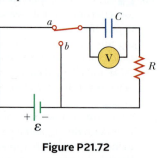

Figure P21.72

$t = 0$ represents the instant at which the switch is thrown to position b. (a) Construct a graph of $\ln(\mathcal{E}/\Delta V)$ versus t and perform a linear least-squares fit to the data. (b) From the slope of your graph, obtain a value for the time constant of the circuit and a value for the capacitance.

ΔV (V)	t (s)	$\ln(\mathcal{E}/\Delta V)$
6.19	0	
5.55	4.87	
4.93	11.1	
4.34	19.4	
3.72	30.8	
3.09	46.6	
2.47	67.3	
1.83	102.2	

73. **S** A battery has an emf $\mathcal{E}$ and internal resistance r. A variable load resistor R is connected across the terminals of the battery. (a) Determine the value of R such that the potential difference across the terminals is a maximum. (b) Determine the value of R so that the current in the circuit is a maximum. (c) Determine the value of R so that the power delivered to the load resistor is a maximum. Choosing the load resistance for maximum power transfer is a case of what is called *impedance matching* in general. Impedance matching is important in shifting gears on a bicycle, in connecting a loudspeaker to an audio amplifier, in connecting a battery charger to a bank of solar photoelectric cells, and in many other applications.

74. **S** The switch in Figure P21.74a closes when $\Delta V_c > \frac{2}{3}\Delta V$ and opens when $\Delta V_c < \frac{1}{3}\Delta V$. The ideal voltmeter reads a

potential difference as plotted in Figure P21.74b. What is the period T of the waveform in terms of R_1, R_2, and C?

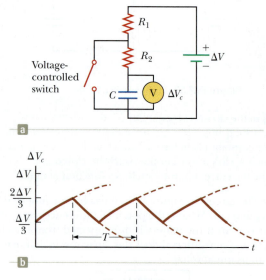

Figure P21.74

75. **M** **Q|C** An electric heater is rated at 1.50×10^3 W, a toaster at 750 W, and an electric grill at 1.00×10^3 W. The three appliances are connected to a common 120-V household circuit. (a) How much current does each draw? (b) If the circuit is protected with a 25.0-A circuit breaker, will the circuit breaker be tripped in this situation? Explain your answer.

76. **Q|C** An experiment is conducted to measure the electrical resistivity of Nichrome in the form of wires with different lengths and cross-sectional areas. For one set of measurements, a student uses 30-gauge wire, which has a cross-sectional area of $7.30 \times 10^{-8}\ m^2$. The student measures the potential difference across the wire and the current in the wire with a voltmeter and an ammeter, respectively. For each set of measurements given in the table taken on wires of three different lengths, calculate the resistance of the wires and the corresponding values of the resistivity. (b) What is the average value of the resistivity? (c) Explain how this value compares with the value given in Table 21.1.

L (m)	ΔV (V)	I (A)	R (Ω)	ρ ($\Omega \cdot m$)
0.540	5.22	0.72		
1.028	5.82	0.414		
1.543	5.94	0.281		

77. **Q|C** Four resistors are connected in parallel across a 9.20-V battery. They carry currents of 150 mA, 45.0 mA, 14.0 mA, and 4.00 mA. If the resistor with the largest resistance is replaced with one having twice the resistance, (a) what is the ratio of the new current in the battery to the original current? (b) **What If?** If instead the resistor with the smallest resistance is replaced with one having twice the resistance, what is the ratio of the new total current to the original current? (c) On a February night, energy leaves a house by several energy leaks, including 1.50×10^3 W by conduction through the ceiling, 450 W by infiltration (air-flow) around the windows, 140 W by conduction through the basement wall above the foundation sill, and 40.0 W by conduction through the plywood door to the attic. To produce the biggest saving in heating bills, which one of these energy transfers should be reduced first? Explain how you decide. Clifford Swartz suggested the idea for this problem.

Determining the Number of Lightning Strikes

N ow that we have investigated the principles of electricity, let us respond to our central question for the *Lightning* Context:

> **How can we determine the number of lightning strikes on the Earth in a typical day?**

We must combine several ideas from our knowledge of electricity to perform this calculation. In Chapter 20, the atmosphere was modeled as a capacitor. Such modeling was first done by Lord Kelvin, who modeled the ionosphere as the positive plate several tens of kilometers above the Earth's surface. More sophisticated models have shown the effective height of the positive plate to be the 5 km that we used in our earlier calculation.

The Atmospheric Capacitor Model

The plates of the atmospheric capacitor are separated by a layer of air containing a large number of free ions that can carry current. Air is a good insulator; measurements show that the resistivity of air is about $3 \times 10^{13} \, \Omega \cdot$ m. Let us calculate the resistance of the air between our capacitor plates. The shape of the resistor is that of a spherical shell between the plates of the atmospheric capacitor (Fig. 1a). The length of 5 km, however, is very short compared with the radius of 6 400 km. Therefore, we can ignore the spherical shape and approximate the resistor as a 5-km slab of flat material whose area is the surface area of the Earth. Using Equation 21.8,

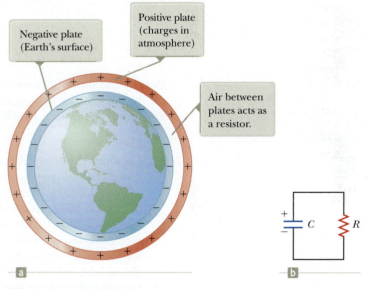

Figure 1 (a) The atmosphere can be modeled as a capacitor, with conductive air between the plates. (b) We can imagine an equivalent *RC* circuit for the atmosphere, with the natural discharge of the capacitor in balance with the charging of the capacitor by lightning.

$$R = \rho \frac{\ell}{A} = (3 \times 10^{13} \, \Omega \cdot \text{m}) \frac{5 \times 10^3 \, \text{m}}{4\pi(6.4 \times 10^6 \, \text{m})^2} \approx 3 \times 10^2 \, \Omega$$

The charge on the atmospheric capacitor can pass from the upper plate to the ground by electric current in the air between the plates. Therefore, we can model the atmosphere as an *RC* circuit (Fig. 1b), using the capacitance found in Chapter 20, and the resistance connecting the plates calculated above. The time constant for this *RC* circuit is

$$\tau = RC = (0.9 \, \text{F})(3 \times 10^2 \, \Omega) \approx 3 \times 10^2 \, \text{s} = 5 \, \text{min}$$

Therefore, the charge on the atmospheric capacitor should fall to $e^{-1} = 37\%$ of its original value after only 5 min! After 30 min, less than 0.3% of the charge would

739

remain! Why doesn't that happen? What keeps the atmospheric capacitor charged? The answer is *lightning*. The processes occurring in cloud charging result in lightning strikes that deliver negative charge to the ground to replace that neutralized by the flow of charge through the air. On the average, a net charge on the atmospheric capacitor results from a balance between these two processes.

Now, let's use this balance to numerically answer our central question. We first address the charge on the atmospheric capacitor. In Chapter 19, we mentioned a charge of 5×10^5 C that is spread over the surface of the Earth, which is the charge on the atmospheric capacitor.

A typical lightning strike delivers about 25 C of negative charge to the ground in the process of charging the capacitor. Dividing the charge on the capacitor by the charge per lightning strike tells us the number of lightning strikes required to charge the capacitor:

$$\text{Number of lightning strikes} = \frac{\text{total charge}}{\text{charge per lightning strike}}$$

$$= \frac{5 \times 10^5 \text{ C}}{25 \text{ C per strike}} \approx 2 \times 10^4 \text{ lightning strikes}$$

According to our calculation for the *RC* circuit, the atmospheric capacitor almost completely discharges through the air in about 30 min. Therefore, 2×10^4 lightning strikes must occur every 30 min, or 4×10^4/h, to keep the charging and discharging processes in balance. Multiplying by the number of hours in a day gives us

$$\text{Number of lightning strikes per day} = (4 \times 10^4 \text{ strikes/h})\left(\frac{24 \text{ h}}{1 \text{ d}}\right)$$

$$\approx 1 \times 10^6 \text{ strokes/day}$$

Despite the simplifications that we have adopted in our calculations, this number is on the right order of magnitude for the actual number of lightning strikes on the Earth in a typical day: 1 million!

Problems

1. Consider the atmospheric capacitor described in the text, with the ground as one plate and positive charges in the atmosphere as the other. On one particular day, the capacitance of the atmospheric capacitor is 0.800 F. The effective plate separation distance is 4.00 km, and the resistivity of the air between the plates is 2.00×10^{13} $\Omega \cdot$ m. If no lightning events occur, the capacitor will discharge through the air. If a charge of 4.00×10^4 C is on the atmospheric capacitor at time $t = 0$, at what later time is the charge reduced (a) to 2.00×10^4 C, (b) to 5.00×10^3 C, and (c) to zero?

2. Consider this alternative line of reasoning to estimate the number of lightning strikes on the Earth in one day. Using the charge on the Earth of 5.00×10^5 C and the atmospheric capacitance of 0.9 F, we find that the potential difference across the capacitor is $\Delta V = Q/C = 5.00 \times 10^5$ C/0.9 F $\approx 6 \times 10^5$ V. The leakage current in the air is $I = \Delta V/R = 6 \times 10^5$ V/300 $\Omega \approx 2$ kA. To keep the capacitor charged, lightning should deliver the same net current in the opposite direction. (a) If each lightning strike delivers 25 C of charge to the ground, what is the average time interval between lightning strikes so that the average current due to lightning is 2 kA? (b) Using this average time interval between lightning strikes, calculate the number of lightning strikes in one day.

3. Consider again the atmospheric capacitor discussed in the text. (a) Assume that atmospheric conditions are such that, for one complete day, the lower 2.50 km of the air between the capacitor plates has resistivity 2.00×10^{13} $\Omega \cdot$ m and the upper 2.50 km has resistivity 0.500×10^{13} $\Omega \cdot$ m. How many lightning strikes occur on this day? (b) Assume that atmospheric conditions are such that, for one complete day, resistivity of the air between the plates in the southern hemisphere is 2.00×10^{13} $\Omega \cdot$ m and the resistivity between the plates in the northern hemisphere is 0.200×10^{13} $\Omega \cdot$ m. How many lightning strikes occur on this day?

Magnetism in Medicine

Now that we have studied electricity, we turn our attention to the closely related topic of magnetism. Magnetism is prevalent in our everyday life. Magnets are essential for the operation of motors. Magnets in generators provide electricity to homes and businesses. Loudspeaker systems use magnets to convert electrical signals to sound waves. Magnets are also critical in keeping important data securely fixed to refrigerator doors.

Magnetism has also entered the field of medicine with a number of applications that can improve health and save lives. Various medical tests or procedures involve magnets. We will explore some of these important applications in this Context. We begin, however, by exploring some questionable applications of magnetism in medicine between the 18th century and the present day.

You may have heard advertisements or even own a magnetic bracelet, such as those shown in Figure 1. Such a bracelet is just one example of devices that provide purported *magnetic therapy*. Additional such devices include other magnetic jewelry, magnetic straps for various body parts, magnetic shoe inserts, magnetic blankets and mattresses, and magnetic creams. Despite having sales of a billion dollars a year, magnetic therapy has not been shown in any scientific studies to be effective.

The United States Food and Drug Administration prohibits marketing any magnetic therapy device as having proven medical advantages.

Let us now move back in time and investigate some earlier applications of magnetism in medicine. Some doctors who employed these applications actually believed that their magnetic instruments would help their patients. Other so-called *quack doctors* knew that the instruments would not work, but used them anyway.

Franz Anton Mesmer, from Vienna, was one of the earliest individuals to develop a theory of medicine involving magnetism. In his doctoral thesis (*The Influence of the Planets on the Human Body,* 1767), Mesmer suggested that a universal fluid that he called "animal gravitation" was responsible for all health and illnesses. In 1773, Mesmer began to use magnets to heal diseases. He claimed that he could "cure" some diseases with a combination of stroking the patient with magnets, various forms of wailing, and listening to music from the glass armonica, recently invented by Benjamin Franklin.

By 1776, Mesmer announced that the magnets were not necessary for his treatment—they were only serving as conductors of the universal fluid, which by now had become a magnetic fluid, which he called *animal magnetism.* Mesmer was very careful about the selection of diseases he attempted to cure. For organic diseases, he would refer the patient to a traditional doctor. He used magnets to treat only nervous or hysterical diseases. The startling aspect of Mesmer's practice is that he supposedly restored sight in a blind pianist and relieved many patients from chronic convulsions. Today we realize that his magnetic treatments were not really curing patients. Mesmer was actually *hypnotizing* patients, using stares, stroking, glass armonica music, power of suggestion, etc. In fact, his name is the root of the word *mesmerize.*

Figure 2 (page 742) shows an example of the *Davis and Kidder Magneto-Electric Machine for Nervous Disorders,* which was used from the 1850s into the latter part of the 19th century. It is simply an electromagnetic generator, developed shortly after the magnetic induction discoveries of Michael Faraday. A pair of wire coils was rotated in the vicinity of a permanent magnet. The patient held on to the two metal cylinders, which

Figure 1 Magnetic bracelets are sold to consumers to promote good health and pain relief. Do you think devices such as these bracelets work?

Figure 2 The Davis & Kidder Magneto-Electric Machine for Nervous Disorders. The patient would hold a brass tube in each hand, while the caretaker turned the crank. The patient received a shock from the voltage generated by the rotating coils in the presence of the magnetic field of the large permanent magnet in the back of the case.

were connected to the generator. The caretaker then turned the crank, providing a jolt of electricity to the patient. Providing jolts of electricity to the patient continued as a supposed treatment until well into the 20th century, and even has its supporters today. The hand-cranked Davis and Kidder device, however, was replaced by plug-in devices such as *violet ray machines* and the various devices of Albert Abrams. (Check out Abrams and his struggle with the American Medical Association on the Internet.)

Another magnetic quack device that arose in the 20th century was originally manufactured under the name of the IONACO, developed by Gaylord Wilshire. A large loop of wire, covered in leather, was plugged into an electrical socket. The goal was to magnetize the blood by wearing the loop around the body. Figure 3 shows a

Figure 3 The Theronoid magnetic device. The leather-coated loop of wire is worn around the body to "magnetize the blood."

later version of this device, the Theronoid, developed by Philip Ilsey. In its 1933 annual report, the United States Federal Trade Commission (FTC) states, "It was claimed by respondents that the use of said device or appliance . . . was a beneficial therapeutic agent in the aid, relief, prevention, or cure of . . . asthma, arthritis, bladder trouble, bronchitis, catarrh, constipation, diabetes, eczema, heart trouble, hemorrhoids, indigestion, insomnia, lumbago, nervous disorders, neuralgia, neuritis, rheumatism, sciatica, stomach trouble, varicose veins, and high blood pressure." The FTC closes this section of its report by banning advertising of the Theronoid: "the Commission issued an order to . . . cease and desist from representing in any manner whatsoever that the said belt or device or any similar device or appliance . . . has any physiotherapeutic effect upon such subject, or that it is calculated or likely to aid in the prevention, treatment, or cure of any human ailment, sickness, or disease."

In this Context, we will look at scientifically supported uses of magnetism in medicine today as opposed to the unsubstantiated and, in some cases, fraudulent uses discussed here. We will address the central question:

> **How has magnetism entered the field of medicine to diagnose and cure illnesses and save lives?**

Magnetic Forces and Magnetic Fields

Chapter Outline

CERN

An engineer performs a test on the electronics associated with one of the superconducting magnets in the Large Hadron Collider at the European Laboratory for Particle Physics, operated by the European Organization for Nuclear Research (CERN). The magnets are used to control the motion of charged particles in the accelerator. We will study the effects of magnetic fields on moving charged particles in this chapter.

The list of technological applications of magnetism is very long. For instance, large electromagnets are used to pick up heavy loads in scrap yards. Magnets are used in such devices as meters, motors, and loudspeakers. Magnetic tapes are routinely used in sound and video recording equipment. Magnetic stripes on the backs of credit cards allow our purchase to be completed quickly in a store. Intense magnetic fields generated by superconducting magnets are currently being used as a means of containing plasmas at temperatures on the order of 10^8 K used in controlled nuclear fusion research.

As we investigate magnetism in this chapter, we shall find that the subject cannot be divorced from electricity. For example, magnetic fields affect moving

Example 22.2 | A Proton Moving Perpendicular to a Uniform Magnetic Field

A proton is moving in a circular orbit of radius 14 cm in a uniform 0.35-T magnetic field perpendicular to the velocity of the proton. Find the speed of the proton.

SOLUTION

Conceptualize From our discussion in this section, we know the proton follows a circular path when moving perpendicular to a uniform magnetic field.

Categorize We evaluate the speed of the proton using an equation developed in this section, so we categorize this example as a substitution problem.

Solve Equation 22.3 for the speed of the particle:

$$v = \frac{qBr}{m_p}$$

Substitute numerical values:

$$v = \frac{(1.60 \times 10^{-19}\,\text{C})(0.35\,\text{T})(0.14\,\text{m})}{1.67 \times 10^{-27}\,\text{kg}}$$

$$= 4.7 \times 10^{6}\,\text{m/s}$$

What If? What if an electron, rather than a proton, moves in a direction perpendicular to the same magnetic field with this same speed? Will the radius of its orbit be different?

Answer An electron has a much smaller mass than a proton, so the magnetic force should be able to change its velocity much more easily than that for the proton. Therefore, we expect the radius to be smaller. Equation 22.3 shows that r is proportional to m with q, B, and v the same for the electron as for the proton. Consequently, the radius will be smaller by the same factor as the ratio of masses m_e/m_p.

Example 22.3 | Bending an Electron Beam

In an experiment designed to measure the magnitude of a uniform magnetic field, electrons are accelerated from rest through a potential difference of 350 V and then enter a uniform magnetic field that is perpendicular to the velocity vector of the electrons. The electrons travel along a curved path because of the magnetic force exerted on them, and the radius of the path is measured to be 7.5 cm. (Such a curved beam of electrons is shown in Fig. 22.10.)

(A) What is the magnitude of the magnetic field?

Figure 22.10
(Example 22.3)
The bending
of an electron
beam in a
magnetic field.

Henry Leap and Jim Lehman

SOLUTION

Conceptualize This example involves electrons accelerating from rest due to an electric force and then moving in a circular path due to a magnetic force. With the help of Active Figure 22.7 and Figure 22.10, visualize the circular motion of the electrons.

Categorize Equation 22.3 shows that we need the speed v of the electron to find the magnetic field magnitude, and v is not given. Consequently, we must find the speed of the electron based on the potential difference through which it is accelerated. To do so, we categorize the first part of the problem by modeling an electron and the electric field as an isolated system. Once the electron enters the magnetic field, we categorize the second part of the problem as one similar to those we have studied in this section.

Analyze Write the appropriate reduction of the conservation of energy equation, Equation 7.2, for the electron–electric field system:

$$\Delta K + \Delta U = 0$$

Substitute the appropriate initial and final energies:

$$(\tfrac{1}{2}m_e v^2 - 0) + (q\,\Delta V) = 0$$

Solve for the speed of the electron:

$$v = \sqrt{\frac{-2q\,\Delta V}{m_e}}$$

22.3 cont.

Substitute numerical values:

$$v = \sqrt{\frac{-2(-1.60 \times 10^{-19}\,\text{C})(350\,\text{V})}{9.11 \times 10^{-31}\,\text{kg}}} = 1.11 \times 10^{7}\,\text{m/s}$$

Now imagine the electron entering the magnetic field with this speed. Solve Equation 22.3 for the magnitude of the magnetic field:

$$B = \frac{m_e v}{er}$$

Substitute numerical values:

$$B = \frac{(9.11 \times 10^{-31}\,\text{kg})(1.11 \times 10^{7}\,\text{m/s})}{(1.60 \times 10^{-19}\,\text{C})(0.075\,\text{m})} = 8.4 \times 10^{-4}\,\text{T}$$

(B) What is the angular speed of the electrons?

SOLUTION

Use Equation 10.10:

$$\omega = \frac{v}{r} = \frac{1.11 \times 10^{7}\,\text{m/s}}{0.075\,\text{m}} = 1.5 \times 10^{8}\,\text{rad/s}$$

Finalize The angular speed can be represented as $\omega = (1.5 \times 10^{8}\,\text{rad/s})(1\,\text{rev}/2\pi\,\text{rad}) = 2.4 \times 10^{7}\,\text{rev/s}$. The electrons travel around the circle 24 million times per second! This answer is consistent with the very high speed found in part (A).

What If? What if a sudden voltage surge causes the accelerating voltage to increase to 400 V? How does that affect the angular speed of the electrons, assuming the magnetic field remains constant?

Answer The increase in accelerating voltage ΔV causes the electrons to enter the magnetic field with a higher speed v. This higher speed causes them to travel in a circle with a larger radius r. The angular speed is the ratio of v to r. Both v and r increase by the same factor, so the effects cancel and the angular speed remains the same. Equation 22.4 is an expression for the cyclotron frequency, which is the same as the angular speed of the electrons. The cyclotron frequency depends only on the charge q, the magnetic field B, and the mass m_e, none of which have changed. Therefore, the voltage surge has no effect on the angular speed. (In reality, however, the voltage surge may also increase the magnetic field if the magnetic field is powered by the same source as the accelerating voltage. In that case, the angular speed increases according to Eq. 22.4.)

22.4 | Applications Involving Charged Particles Moving in a Magnetic Field

A charge moving with velocity $\vec{v}$ in the presence of an electric field $\vec{E}$ and a magnetic field $\vec{B}$ experiences both an electric force $q\vec{E}$ and a magnetic force $q\vec{v} \times \vec{B}$. The total force, called the **Lorentz force,** acting on the charge is therefore the vector sum,

$$\vec{F} = q\vec{E} + q\vec{v} \times \vec{B} \qquad \text{22.6} \blacktriangleleft$$

In this section, we look at three applications involving particles experiencing the Lorentz force.

Velocity Selector

In many experiments involving moving charged particles, it is important that all particles move with essentially the same velocity, which can be achieved by applying a combination of an electric field and a magnetic field oriented as shown in Active Figure 22.11. A uniform electric field is directed to the right (in the plane of the page in Active Fig. 22.11), and a uniform magnetic field is applied in the direction perpendicular to the electric field (into the page in Active Fig. 22.11). If q is positive and the velocity $\vec{v}$ is upward, the magnetic force $q\vec{v} \times \vec{B}$ is to the left and the electric force $q\vec{E}$ is to the right. When the magnitudes of the two fields are chosen so that $qE = qvB$, the charged particle is modeled as a particle in equilibrium and moves in a straight vertical line through the region of the fields. From the expression $qE = qvB$, we find that

$$v = \frac{E}{B} \qquad \text{22.7} \blacktriangleleft$$

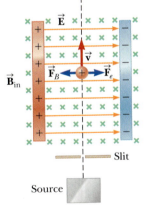

Active Figure 22.11 A velocity selector. When a positively charged particle is moving with velocity $\vec{v}$ in the presence of a magnetic field directed into the page and an electric field directed to the right, it experiences an electric force $q\vec{E}$ to the right and a magnetic force $q\vec{v} \times \vec{B}$ to the left.

Pitfall Prevention | 22.3
The Electron Does Not Spin
The electron is *not* physically spinning. It has an intrinsic angular momentum *as if it were spinning*, but the notion of rotation for a point particle is meaningless. Rotation applies only to a *rigid object*, with an extent in space, as in Chapter 10. Spin angular momentum is actually a relativistic effect.

TABLE 22.1 |
Magnetic Moments of Some Atoms and Ions

Atom or Ion	Magnet Moment per Atom or Ion (10^{-24} J/T)
H	9.27
He	0
Ne	0
Ce^{3+}	19.8
Yb^{3+}	37.1

an electron is an angular momentum separate from its orbital angular momentum, just as the spin of the Earth is separate from its orbital motion about the Sun. Even if the electron is at rest, it still has an angular momentum associated with spin. We shall investigate spin more deeply in Chapter 29.

In atoms or ions containing multiple electrons, many electrons are paired up with their spins in opposite directions, an arrangement that results in a cancellation of the spin magnetic moments. An atom with an odd number of electrons, however, must have at least one "unpaired" electron and a corresponding spin magnetic moment. The net magnetic moment of the atom leads to various types of magnetic behavior. The magnetic moments of several atoms and ions are listed in Table 22.1.

Ferromagnetic Materials

Iron, cobalt, nickel, gadolinium, and dysprosium are strongly magnetic materials and are said to be **ferromagnetic.** Ferromagnetic substances, used to fabricate permanent magnets, contain atoms with spin magnetic moments that tend to align parallel to each other even in a weak external magnetic field. Once the moments are aligned, the substance remains magnetized after the external field is removed. This permanent alignment is due to strong coupling between neighboring atoms, which can only be understood using quantum physics.

All ferromagnetic materials contain microscopic regions called **domains,** within which all magnetic moments are aligned. The domains range from about 10^{-12} to 10^{-8} m^3 in volume and contain 10^{17} to 10^{21} atoms. The boundaries between domains having different orientations are called **domain walls.** In an unmagnetized sample, the domains are randomly oriented so that the net magnetic moment is zero as in Figure 22.38a. When the sample is placed in an external magnetic field, domains with magnetic moment vectors initially oriented along the external field grow in size at the expense of other domains, which results in a magnetized sample, as in Figures 22.38b and 22.38c. When the external field is removed, the sample may retain most of its magnetism.

The extent to which a ferromagnetic substance retains its magnetism is described by its classification as being magnetically **hard** or **soft.** Soft magnetic materials, such as iron, are easily magnetized but also tend to lose their magnetism easily. When a soft magnetic material is magnetized and the external magnetic field is removed, thermal agitation produces domain motion and the material quickly returns to an unmagnetized state. In contrast, hard magnetic materials, such as cobalt and nickel, are difficult to magnetize but tend to retain their magnetism, and domain alignment persists in them after the external magnetic field is removed. Such hard magnetic materials are referred to as **permanent magnets.** Rare-earth permanent magnets, such as samarium–cobalt, are now regularly used in industry.

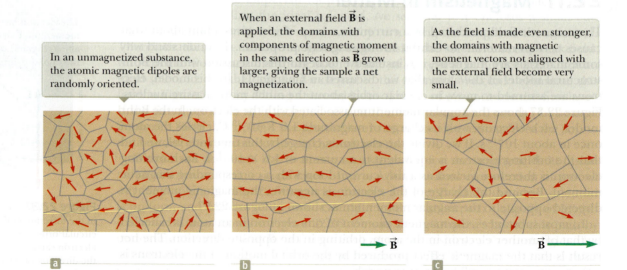

In an unmagnetized substance, the atomic magnetic dipoles are randomly oriented.

When an external field $\vec{B}$ is applied, the domains with components of magnetic moment in the same direction as $\vec{B}$ grow larger, giving the sample a net magnetization.

As the field is made even stronger, the domains with magnetic moment vectors not aligned with the external field become very small.

Figure 22.38 Orientation of magnetic dipoles before and after a magnetic field is applied to a ferromagnetic substance.

22.12 | Context Connection: Remote Magnetic Navigation for Cardiac Catheter Ablation Procedures

In the *Heart Attacks* Context, we studied the role of fluid flow in blood vessels and the dangerous effect of plaque buildup on delivery of blood to the heart. In Section 21.2, we looked at the heart again as we investigated the details of cardiac catheter ablation for a patient who suffers from atrial fibrillation. In this Context Connection, we return to atrial fibrillation in the heart, but consider a newer development in the ablation procedure.

There are a number of risks with a traditional cardiac catheter ablation procedure. One possible outcome is a perforation of the heart wall with one of the catheters. Because the esophagus passes right behind the heart, it is possible to burn through too much tissue during a particular ablation and create an esophageal fistula. Other risks come from exposure to x-rays. In order to observe the positions of the catheters, the electrophysiologist must use x-rays and a fluoroscope to make the heart and the catheters visible. As a result, the patient receives a relatively high dose of radiation during the procedure. In addition, despite the use of lead aprons, the electrophysiologist receives radiation from each ablation procedure during his or her entire career. In addition to the effects of this prolonged radiation exposure, studies have shown that a large percentage of electrophysiologists have been treated for back and neck pain due to the long hours of wearing lead aprons.

One possibility for reducing risks to both patient and doctor is the use of *remote magnetic navigation* in catheter ablation procedures. This procedure uses softer and more flexible catheters than the traditional approach, reducing the risk of perforation and allowing catheters to reach areas of the heart unavailable to the stiffer traditional catheters. The tips of the catheters are guided magnetically with the aid of a computer. The electrophysiologist can sit comfortably at a computer in another room and guide the catheters with a joystick, avoiding exposure to radiation. Figure 22.39 shows a typical computer display that helps the electrophysiologist guide the catheters.

During a catheter ablation procedure using remote magnetic navigation, the patient is located between two strong magnets as shown in Figure 22.40 (page 770). The magnets can be moved over a wide range of positions and orientations relative to the patient. The magnetic field from these magnets is strong, but only about 10% as strong as that used in magnetic resonance imaging (to be discussed in the Context Conclusion). The tip of the catheter includes ferromagnetic material so that its orientation can be precisely controlled by the positions of the external magnets. Once the tip is correctly oriented, it can be advanced mechanically as with the traditional approach.

In addition to the increased safety of the softer catheter and the precise magnetic orientation of its tip, the computer control of the procedure provides additional advantages. For example, locations of ablations can be "memorized" by the computer. The catheter tip can be quickly returned to this exact location for a repeated ablation by calling up the memorized location.

While there are many advantages to remote magnetic navigation, clinical evidence shows one disadvantage. The total procedure time with remote navigation has been measured to be significantly longer than that with the traditional approach.[2] Reasons for this longer time interval include the learning curve for the procedure, "interruption time" because the electrophysiologist is available to other staff in a room separate from the patient, and increased time for the more complicated mapping procedures.

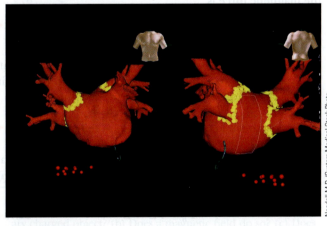

Needell M.D./Custom Medical Stock Photo

Figure 22.39 In remote magnetic navigation procedures for cardiac catheter ablations, the electrophysiologist views a computer model of the heart such as the front and back images shown here. The yellow dots are lesions around the pulmonary veins made by the ablation process.

[2] A. Arya, R. Zaker-Shahrak, P. Sommer, A. Bollmann, U. Wetzel, T. Gaspar, S. Richter, D. Husser, C. Piorkowski, and G. Hindricks, "Catheter Ablation of Atrial Fibrillation Using Remote Magnetic Catheter Navigation: A Case–Control Study," *Europace,* **13**, pp. 45–50 (2011).

Additional Problems

63. **M** A particle with positive charge $q = 3.20 \times 10^{-19}$ C moves with a velocity $\vec{v} = (2\hat{i} + 3\hat{j} - \hat{k})$ m/s through a region where both a uniform magnetic field and a uniform electric field exist. (a) Calculate the total force on the moving particle (in unit-vector notation), taking $\vec{B} = (2\hat{i} + 4\hat{j} + \hat{k})$ T and $\vec{E} = (4\hat{i} - \hat{j} - 2\hat{k})$ V/m. (b) What angle does the force vector make with the positive x axis?

64. **S** An infinite sheet of current lying in the yz plane carries a surface current of linear density J_s. The current is in the positive z direction, and J_s represents the current per unit length measured along the y axis. Figure P22.64 is an edge view of the sheet. Prove that the magnetic field near the sheet is parallel to the sheet and perpendicular to the current direction, with magnitude $\mu_0 J_s/2$.

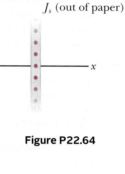

J_s (out of paper)

Figure P22.64

65. Carbon-14 and carbon-12 ions (each with charge of magnitude e) are accelerated in a cyclotron. If the cyclotron has a magnetic field of magnitude 2.40 T, what is the difference in cyclotron frequencies for the two ions?

66. The *Hall effect* finds important application in the electronics industry. It is used to find the sign and density of the carriers of electric current in semiconductor chips. The arrangement is shown in Figure P22.66. A semiconducting block of thickness t and width d carries a current I in the x direction. A uniform magnetic field B is applied in the y direction. If the charge carriers are positive, the magnetic force deflects them in the z direction. Positive charge accumulates on the top surface of the sample and negative charge on the bottom surface, creating a downward electric field. In equilibrium, the downward electric force on the charge carriers balances the upward magnetic force and the carriers move through the sample without deflection. The *Hall voltage* $\Delta V_H = V_c - V_a$ between the top and bottom surfaces is measured, and the density of the charge carriers can be calculated from it. (a) Demonstrate that if the charge carriers are negative the Hall voltage will be negative. Hence, the Hall effect reveals the sign of the charge carriers, so the sample can be classified as *p*-type (with positive majority charge carriers) or *n*-type (with negative). (b) Determine the number of charge carriers per unit volume n in terms of I, t, B, ΔV_H, and the magnitude q of the carrier charge.

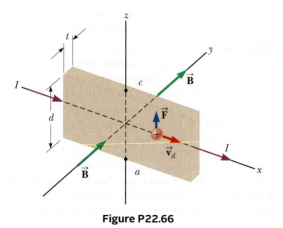

Figure P22.66

67. **Review.** A 0.200-kg metal rod carrying a current of 10.0 A glides on two horizontal rails 0.500 m apart. If the coefficient of kinetic friction between the rod and rails is 0.100, what vertical magnetic field is required to keep the rod moving at a constant speed?

68. **S** **Review.** A metal rod of mass m carrying a current I glides on two horizontal rails a distance d apart. If the coefficient of kinetic friction between the rod and rails is μ, what vertical magnetic field is required to keep the rod moving at a constant speed?

69. **Q|C** Two circular loops are parallel, coaxial, and almost in contact, with their centers 1.00 mm apart (Fig. P22.69). Each loop is 10.0 cm in radius. The top loop carries a clockwise current of $I = 140$ A. The bottom loop carries a counterclockwise current of $I = 140$ A. (a) Calculate the magnetic force exerted by the bottom loop on the top loop. (b) Suppose a student thinks the first step in solving part (a) is to use Equation 22.23 to find the magnetic field created by one of the loops. How would you argue for or against this idea? (c) The upper loop has a mass of 0.021 0 kg. Calculate its acceleration, assuming the only forces acting on it are the force in part (a) and the gravitational force.

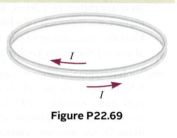

Figure P22.69

70. **S** Consider a thin, straight wire segment carrying a constant current I and placed along the x axis as shown in Figure P22.70. (a) Use the Biot–Savart law to show that the total magnetic field at the point P, located a distance a from the wire, is

$$B = \frac{\mu_0 I}{4\pi a}(\cos\theta_1 - \cos\theta_2)$$

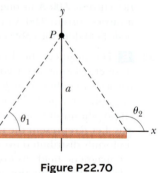

Figure P22.70

(b) Assuming that the wire is infinitely long, show that the result in part (a) gives a magnetic field that agrees with that obtained by using Ampère's law in Example 22.8.

71. Assume the region to the right of a certain plane contains a uniform magnetic field of magnitude 1.00 mT and the field is zero in the region to the left of the plane as shown in Figure P22.71. An electron, originally traveling perpendicular to the boundary plane, passes into the region of the field. (a) Determine the time interval required for the electron to leave the "field-filled" region, noting that the electron's path is a semicircle. (b) Assuming the maximum depth of penetration into the field is 2.00 cm, find the kinetic energy of the electron.

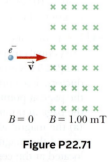

$B = 0$ $B = 1.00$ mT

Figure P22.71

72. **BIO** **Q|C** **S** Heart–lung machines and artificial kidney machines employ electromagnetic blood pumps. The blood is confined to an electrically insulating tube, cylindrical in practice but represented here for simplicity as a rectangle of interior width w and height h. Figure P22.72 shows a rectangular section of blood within the tube. Two electrodes fit into the top and the bottom of the tube. The potential difference between them establishes an electric current through the blood, with current density J over the section of length L

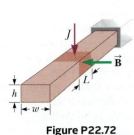

Figure P22.72

shown in Figure P22.72. A perpendicular magnetic field exists in the same region. (a) Explain why this arrangement produces on the liquid a force that is directed along the length of the pipe. (b) Show that the section of liquid in the magnetic field experiences a pressure increase JLB. (c) After the blood leaves the pump, is it charged? (d) Is it carrying current? (e) Is it magnetized? (The same electromagnetic pump can be used for any fluid that conducts electricity, such as liquid sodium in a nuclear reactor.)

73. **BIO** **Q|C** A heart surgeon monitors the flow rate of blood through an artery using an electromagnetic flowmeter (Fig. P22.73). Electrodes A and B make contact with the outer surface of the blood

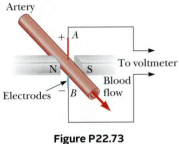

Figure P22.73

vessel, which has a diameter of 3.00 mm. (a) For a magnetic field magnitude of 0.040 0 T, an emf of 160 μV appears between the electrodes. Calculate the speed of the blood. (b) Explain why electrode A has to be positive as shown. (c) Does the sign of the emf depend on whether the mobile ions in the blood are predominantly positively or negatively charged? Explain.

74. *Why is the following situation impossible?* The magnitude of the Earth's magnetic field at either pole is approximately 7.00×10^{-5} T. Suppose the field fades away to zero before its next reversal. Several scientists propose plans for artificially generating a replacement magnetic field to assist with devices that depend on the presence of the field. The plan that is selected is to lay a copper wire around the equator and supply it with a current that would generate a magnetic field of magnitude 7.00×10^{-5} T at the poles. (Ignore magnetization of any materials inside the Earth.) The plan is implemented and is highly successful.

75. Protons having a kinetic energy of 5.00 MeV $(1 \text{ eV} = 1.60 \times 10^{-19} \text{ J})$ are moving in the positive x direction and enter a magnetic field $\vec{B} = 0.050\ 0\hat{k}$ T directed out of the plane of the page and extending from $x = 0$

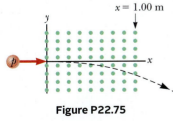

Figure P22.75

to $x = 1.00$ m as shown in Figure P22.75. (a) Ignoring relativistic effects, find the angle α between the initial velocity vector of the proton beam and the velocity vector after the beam emerges from the field. (b) Calculate the y component of the protons' momenta as they leave the magnetic field.

76. **GP** **Review.** Rail guns have been suggested for launching projectiles into space without chemical rockets. A tabletop model rail gun (Fig. P22.76) consists of two long, parallel, horizontal rails $\ell = 3.50$ cm apart, bridged by a bar of mass $m = 3.00$ g that is free to slide without friction. The rails and bar have low electric resistance, and the current is limited to a constant $I = 24.0$ A by a power supply that is far to the left of the figure, so it has no magnetic effect on the bar. Figure P22.76 shows the bar at rest at the midpoint of the rails at the moment the current is established. We wish to find the speed with which the bar leaves the rails after being released from the midpoint of the rails. (a) Find the magnitude of the magnetic field at a distance of 1.75 cm from a single long wire carrying a current of 2.40 A. (b) For purposes of evaluating the magnetic field, model the rails as infinitely long. Using the result of part (a), find the magnitude and direction of the magnetic field at the midpoint of the bar. (c) Argue that this value of the field will be the same at all positions of the bar to the right of the midpoint of the rails. At other points along the bar, the field is in the same direction as at the midpoint, but is larger in magnitude. Assume the average effective magnetic field along the bar is five times larger than the field at the midpoint. With this assumption, find (d) the magnitude and (e) the direction of the force on the bar. (f) Is the bar properly modeled as a particle under constant acceleration? (g) Find the velocity of the bar after it has traveled a distance $d = 130$ cm to the end of the rails.

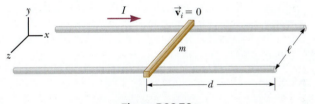

Figure P22.76

77. **M** A nonconducting ring of radius 10.0 cm is uniformly charged with a total positive charge 10.0 μC. The ring rotates at a constant angular speed 20.0 rad/s about an axis through its center, perpendicular to the plane of the ring. What is the magnitude of the magnetic field on the axis of the ring 5.00 cm from its center?

78. **S** A nonconducting ring of radius R is uniformly charged with a total positive charge q. The ring rotates at a constant angular speed ω about an axis through its center, perpendicular to the plane of the ring. What is the magnitude of the magnetic field on the axis of the ring a distance $\frac{1}{2}R$ from its center?

79. Model the electric motor in a handheld electric mixer as a single flat, compact, circular coil carrying electric current in a region where a magnetic field is produced by an external permanent magnet. You need consider only one instant in the operation of the motor. (We will consider motors again in Chapter 23.) Make order-of-magnitude estimates of (a) the magnetic field, (b) the torque on the coil, (c) the current in the coil, (d) the coil's area, and (e) the number of turns in the coil. The input power to the motor is electric, given by $P = I\Delta V$, and the useful output power is mechanical, $P = \tau\omega$.

80. *Why is the following situation impossible?* Figure P22.80 (page 780) shows an experimental technique for altering the

direction of travel for a charged particle. A particle of charge $q = 1.00\ \mu C$ and mass $m = 2.00 \times 10^{-13}$ kg enters the bottom of the region of uniform magnetic field at speed $v = 2.00 \times 10^5$ m/s, with a velocity vector perpendicular to the field lines. The magnetic force on the particle causes its direction of travel to change so that it leaves the region of the magnetic field at the top traveling at an angle from its original direction. The magnetic field has magnitude $B = 0.400$ T and is directed out of the page. The length h of the magnetic field region is 0.110 m.

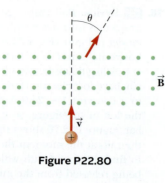

Figure P22.80

An experimenter performs the technique and measures the angle θ at which the particles exit the top of the field. She finds that the angles of deviation are exactly as predicted.

81. **S** A very long, thin strip of metal of width w carries a current I along its length as shown in Figure P22.81. The current is distributed uniformly across the width of the strip. Find the magnetic field at point P in the diagram. Point P is in the plane of the strip at distance b away from its edge.

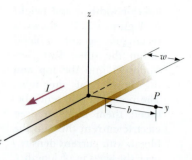

Figure P22.81

Faraday's Law and Inductance

Chapter Outline

Marine Current Turbines TM Ltd.

O ur studies in electromagnetism so far have been concerned with the electric fields due to stationary charges and the magnetic fields produced by moving charges. This chapter introduces a new type of electric field, one that is due to a changing magnetic field.

As we learned in Section 19.1, experiments conducted by Michael Faraday in England in the early 1800s and independently by Joseph Henry in the United States showed that an electric current can be induced in a circuit by a changing magnetic field. The results of those experiments led to a very basic and important law of electromagnetism known as *Faraday's law of induction*. Faraday's law explains how generators, as well as other practical devices, work.

Faraday's law is also the basis for a new circuit element, the *inductor*. This new circuit element combines with resistors and capacitors to allow for a variety of useful electric circuits.

An artist's impression of the Skerries SeaGen Array, a tidal energy generator under development near the island of Anglesey, North Wales. When it is brought on line, possibly in 2015, it will offer 10.5 MW of power from generators turned by tidal streams. The image shows the underwater blades that are driven by the tidal currents. The second blade system has been raised from the water for servicing. We will study generators in this chapter.

❮23.1❯ Faraday's Law of Induction

We begin discussing the concepts in this chapter by considering a simple experiment that builds on material presented in Chapter 22. Imagine that a straight metal conductor resides in a uniform magnetic field directed into the page as in Figure 23.1 (page 782). Within the conductor, there are free electrons. Suppose the conductor is now moved with a velocity $\vec{v}$ toward the right. Equation 22.1

A current is induced in the conductor due to the magnetic force on charged particles in the conductor.

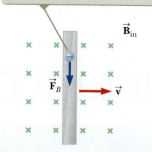

Figure 23.1 A straight electrical conductor moving with a velocity $\vec{v}$ through a uniform magnetic field $\vec{B}$ directed perpendicular to $\vec{v}$.

When a magnet is moved toward a loop of wire connected to a sensitive ammeter, the ammeter shows that a current is induced in the loop.

When the magnet is held stationary, there is no induced current in the loop, even when the magnet is inside the loop.

When the magnet is moved away from the loop, the ammeter shows that the induced current is opposite that shown in part **a**.

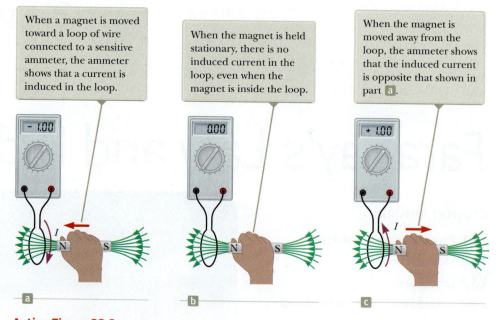

Active Figure 23.2 A simple experiment showing that a current is induced in a loop when a magnet is moved toward or away from the loop.

Michael Faraday
British Physicist and Chemist (1791–1867)
Faraday is often regarded as the greatest experimental scientist of the 1800s. His many contributions to the study of electricity include the invention of the electric motor, the electric generator, and the transformer as well as the discovery of electromagnetic induction and the laws of electrolysis. Greatly influenced by his religious beliefs, he refused to work on the development of poison gas for the British military.

tells us that a magnetic force acts on the electrons in the conductor. Using the right-hand rule, the force on the electrons is downward in Figure 23.1 (remember that the electrons carry a negative charge). Because this direction is along the conductor, the electrons move along the conductor in response to this force. Therefore, a *current* is produced in the conductor as it moves through a magnetic field!

Let us consider another simple experiment that demonstrates that an electric current can be produced by a magnetic field. Consider a loop of wire connected to a sensitive ammeter, a device that measures current, as illustrated in Active Figure 23.2. If a magnet is moved toward the loop, the ammeter display shows the existence of a current as in Active Figure 23.2a. When the magnet is held stationary as in Active Figure 23.2b, the ammeter shows no current. If the magnet is moved away from the loop as in Active Figure 23.2c, the ammeter display shows a current in the opposite direction from that caused by the motion of the magnet toward the ammeter. Finally, if the magnet is held stationary and the loop is moved either toward or away from it, the ammeter shows a current again. From these observations comes the conclusion that an electric current is set up in the loop as long as relative motion occurs between the magnet and the loop.

These results are quite remarkable when we consider that a current exists in a loop of wire even though no batteries are connected to the wire. We call such a current an **induced current**, and it is produced by an **induced emf.**

Another experiment, first conducted by Faraday, is illustrated in Active Figure 23.3. Part of the apparatus consists of a coil of insulated wire connected to a switch and a battery. We shall refer to this coil as the *primary coil* of wire and to the corresponding circuit as the primary circuit. The coil is wrapped around an iron ring to intensify the magnetic field produced by the current through the coil. A second coil of insulated wire at the right is also wrapped around the iron ring and is connected to a sensitive ammeter. We shall refer to this coil as the *secondary coil* and to the corresponding circuit as the secondary circuit. The secondary circuit has no battery, and the secondary coil is not electrically connected to the primary coil. The purpose of this apparatus is to detect any current that might be generated in the secondary circuit by a change in the magnetic field produced by the primary circuit.

Initially, you might guess that no current would ever be detected in the secondary circuit. Something quite surprising happens, however, when the switch in the primary circuit is opened or thrown closed. At the instant the switch is thrown closed, the ammeter display briefly shows a current and then returns to zero. When the switch is opened, the ammeter display shows a current in the opposite direction and then again returns to zero. Finally, the ammeter reads zero when the primary circuit carries a steady current.

As a result of these observations, Faraday concluded that an electric current can be produced by a time-varying magnetic field. A current cannot be produced by a steady magnetic field. In the experiment shown in Active Figure 23.2, the changing magnetic field is a result of the relative motion between the magnet and the loop of wire. As long as the motion persists, the current is maintained. In the experiment shown in Active Figure 23.3, the current produced in the secondary circuit occurs for only an instant after the switch is closed while the magnetic field acting on the secondary coil builds from its zero value to its final value. In effect, the secondary circuit behaves as though a source of emf were connected to it for an instant. It is customary to say that an emf is induced in the secondary circuit by the changing magnetic field produced by the current in the primary circuit.

To quantify such observations, we define a quantity called **magnetic flux.** The flux associated with a magnetic field is defined in a similar manner to the electric flux (Section 19.8) and is proportional to the number of magnetic field lines passing through an area. Consider an element of area dA on an arbitrarily shaped open surface as in Figure 23.4. If the magnetic field at the location of this element is $\vec{B}$, the magnetic flux through the element is $\vec{B} \cdot d\vec{A}$, where $d\vec{A}$ is a vector perpendicular to the surface whose magnitude equals the area dA. Hence, the total magnetic flux Φ_B through the surface is

$$\Phi_B = \int \vec{B} \cdot d\vec{A} \qquad \textbf{23.1} \blacktriangleleft$$

 ▶ Magnetic flux

The SI unit of magnetic flux is a tesla·meter squared, which is named the *weber* (Wb); 1 Wb = 1 T·m².

The two experiments illustrated in Figures 23.2 and 23.3 have one thing in common. In both cases, an emf is induced in a circuit when the magnetic flux through the surface bounded by the circuit changes with time. A general statement known as **Faraday's law of induction** summarizes such experiments involving induced emfs:

The emf induced in a circuit is equal to the time rate of change of magnetic flux through the circuit:

$$\mathcal{E} = -\frac{d\Phi_B}{dt} \qquad \textbf{23.2} \blacktriangleleft$$

In Equation 23.2, Φ_B is the magnetic flux through the surface bounded by the circuit and is given by Equation 23.1. The negative sign in Equation 23.2 will be discussed in Section 23.3. If the circuit is a coil consisting of N identical and concentric loops and if the field lines pass through all loops, the induced emf is

$$\mathcal{E} = -N\frac{d\Phi_B}{dt} \qquad \textbf{23.3} \blacktriangleleft$$

The emf is increased by the factor N because all the loops are in series, so the emfs in the individual loops add to give the total emf.

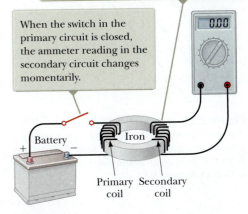

The emf induced in the secondary circuit is caused by the changing magnetic field through the secondary coil.

When the switch in the primary circuit is closed, the ammeter reading in the secondary circuit changes momentarily.

Active Figure 23.3 Faraday's experiment.

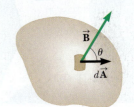

Figure 23.4 The magnetic flux through an area element $d\vec{A}$ is given by $\vec{B} \cdot d\vec{A} = B \, dA \cos\theta$. Note that vector $d\vec{A}$ is perpendicular to the surface.

▶ Faraday's law

Pitfall Prevention | 23.1
Induced emf Requires a Change
The *existence* of a magnetic flux through an area is not sufficient to create an induced emf. The magnetic flux must *change* to induce an emf.

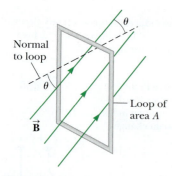

Figure 23.5 A conducting loop that encloses an area A in the presence of a uniform magnetic field $\vec{B}$. The angle between $\vec{B}$ and the normal to the loop is θ.

Figure 23.6 (a) In an electric guitar, a vibrating magnetized string induces an emf in a pickup coil. (b) The pickups (the circles beneath the metallic strings) of this electric guitar detect the vibrations of the strings and send this information through an amplifier and into speakers. (A switch on the guitar allows the musician to select which set of six pickups is used.)

Suppose a loop enclosing an area A lies in a uniform magnetic field $\vec{B}$ as in Figure 23.5. In this case, the magnetic flux through the loop is

$$\Phi_B = \int \vec{B} \cdot d\vec{A} = \int B\, dA \cos\theta = B\cos\theta \int dA = BA\cos\theta$$

Hence, the induced emf is

$$\mathcal{E} = -\frac{d}{dt}(BA\cos\theta) \qquad \textbf{23.4} \blacktriangleleft$$

This expression shows that an emf can be induced in a circuit by changing the magnetic flux in several ways: (1) the magnitude of $\vec{B}$ can vary with time, (2) the area A of the circuit can change with time, (3) the angle θ between $\vec{B}$ and the normal to the plane can change with time, and (4) any combination of these changes can occur.

An interesting application of Faraday's law is the production of sound in an electric guitar (Fig. 23.6). The coil in this case, called the *pickup coil*, is placed near the vibrating guitar string, which is made of a metal that can be magnetized. A permanent magnet inside the coil magnetizes the portion of the string nearest the coil. When the string vibrates at some frequency, its magnetized segment produces a changing magnetic flux through the coil. The changing flux induces an emf in the coil that is fed to an amplifier. The output of the amplifier is sent to the loudspeakers, which produce the sound waves we hear.

QUICK QUIZ 23.1 A circular loop of wire is held in a uniform magnetic field, with the plane of the loop perpendicular to the field lines. Which of the following will *not* cause a current to be induced in the loop? (a) crushing the loop (b) rotating the loop about an axis perpendicular to the field lines (c) keeping the orientation of the loop fixed and moving it along the field lines (d) pulling the loop out of the field

QUICK QUIZ 23.2 Figure 23.7 shows a graphical representation of the field magnitude versus time for a magnetic field that passes through a fixed loop and that is oriented perpendicular to the plane of the loop. The magnitude of the magnetic field at any time is uniform over the area of the loop. Rank the magnitudes of the emf generated in the loop at the five instants indicated, from largest to smallest.

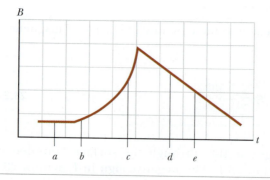

Figure 23.7 (Quick Quiz 23.2) The time behavior of a magnetic field through a loop.

THINKING PHYSICS 23.1

The ground fault circuit interrupter (GFCI) is a safety device that protects users of electric power against electric shock when they touch appliances. Its essential parts are shown in Figure 23.8. How does the operation of a GFCI make use of Faraday's law?

Reasoning Wire 1 leads from the wall outlet to the appliance being protected, and wire 2 leads from the appliance back to the wall outlet. An iron ring surrounds the two wires. A sensing coil wrapped around part of the iron ring activates a circuit breaker when changes in magnetic flux occur. Because the currents in the two wires are in opposite directions during normal operation of the appliance, the net magnetic field through the sensing coil due to the currents is zero. A change in magnetic flux through the sensing coil can happen, however, if one of the wires on the appliance loses its insulation and accidentally touches the metal case of the appliance, providing a direct path to ground. When such a short to ground occurs, a net magnetic flux occurs through the sensing coil that alternates in time because household current is alternating. This changing flux produces an induced voltage in the coil, which in turn triggers a circuit breaker, stopping the current before it reaches a level that might be harmful to the person using the appliance. ◄

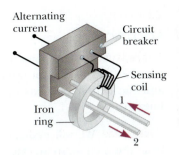

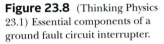

Figure 23.8 (Thinking Physics 23.1) Essential components of a ground fault circuit interrupter.

Example 23.1 | Inducing an emf in a Coil

A coil consists of 200 turns of wire. Each turn is a square of side $d = 18$ cm, and a uniform magnetic field directed perpendicular to the plane of the coil is turned on. If the field changes linearly from 0 to 0.50 T in 0.80 s, what is the magnitude of the induced emf in the coil while the field is changing?

SOLUTION

Conceptualize From the description in the problem, imagine magnetic field lines passing through the coil. Because the magnetic field is changing in magnitude, an emf is induced in the coil.

Categorize We will evaluate the emf using Faraday's law from this section, so we categorize this example as a substitution problem.

Evaluate Equation 23.3 for the situation described here, noting that the magnetic field changes linearly with time:

$$|\varepsilon| = N\frac{\Delta\Phi_B}{\Delta t} = N\frac{\Delta(BA)}{\Delta t} = NA\frac{\Delta B}{\Delta t} = Nd^2\frac{B_f - B_i}{\Delta t}$$

Substitute numerical values:

$$|\varepsilon| = (200)(0.18\text{ m})^2\frac{(0.50\text{ T} - 0)}{0.80\text{ s}} = 4.0\text{ V}$$

What If? What if you were asked to find the magnitude of the induced current in the coil while the field is changing? Can you answer that question?

Answer If the ends of the coil are not connected to a circuit, the answer to this question is easy: the current is zero! (Charges move within the wire of the coil, but they cannot move into or out of the ends of the coil.) For a steady current to exist, the ends of the coil must be connected to an external circuit. Let's assume the coil is connected to a circuit and the total resistance of the coil and the circuit is 2.0 Ω. Then, the magnitude of the induced current in the coil is

$$I = \frac{|\varepsilon|}{R} = \frac{4.0\text{ V}}{2.0\text{ }\Omega} = 2.0\text{ A}$$

Example 23.2 | An Exponentially Decaying Magnetic Field

A loop of wire enclosing an area A is placed in a region where the magnetic field is perpendicular to the plane of the loop. The magnitude of $\vec{B}$ varies in time according to the expression $B = B_{max}e^{-at}$, where a is some constant. That is, at $t = 0$, the field is B_{max}, and for $t > 0$, the field decreases exponentially as in Figure 23.9 (page 786). Find the induced emf in the loop as a function of time.

continued

23.2 *cont.*

SOLUTION

Conceptualize The physical situation is similar to that in Example 23.1 except for two things: there is only one loop, and the field varies exponentially with time rather than linearly.

Categorize We will evaluate the emf using Faraday's law from this section, so we categorize this example as a substitution problem.

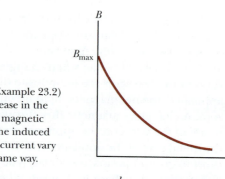

Figure 23.9 (Example 23.2) Exponential decrease in the magnitude of the magnetic field with time. The induced emf and induced current vary with time in the same way.

Evaluate Equation 23.2 for the situation described here:

$$\varepsilon = -\frac{d\Phi_B}{dt} = -\frac{d}{dt}(AB_{max}\,e^{-at}) = -AB_{max}\frac{d}{dt}e^{-at} = aAB_{max}\,e^{-at}$$

This expression indicates that the induced emf decays exponentially in time. The maximum emf occurs at $t = 0$, where $\varepsilon_{max} = aAB_{max}$. The plot of ε versus t is similar to the B-versus-t curve shown in Figure 23.9.

23.2 | Motional emf

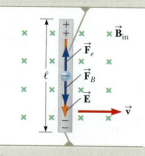

In steady state, the electric and magnetic forces on an electron in the conductor are balanced.

Due to the magnetic force on electrons, the ends of the conductor become oppositely charged, which establishes an electric field in the conductor.

Figure 23.10 A straight electrical conductor of length ℓ moving with a velocity $\vec{v}$ through a uniform magnetic field $\vec{B}$ directed perpendicular to $\vec{v}$.

Examples 23.1 and 23.2 are cases in which an emf is produced in a circuit when the magnetic field changes with time. In this section, we describe **motional emf,** in which an emf is induced in a conductor moving through a magnetic field. This is the situation described in Figure 23.1 at the beginning of Section 23.1.

Consider a straight conductor of length ℓ moving with constant velocity through a uniform magnetic field directed into the page as in Figure 23.10. For simplicity, we shall assume that the conductor is moving with a velocity that is perpendicular to the field. The electrons in the conductor experience a force along the conductor with magnitude $|\vec{F}_B| = |q\vec{v} \times \vec{B}| = qvB$. According to Newton's second law, the electrons accelerate in response to this force and move along the conductor. Once the electrons move to the lower end of the conductor, they accumulate there, leaving a net positive charge at the upper end. As a result of this charge separation, an electric field $\vec{E}$ is produced within the conductor. The charge at the ends of the conductor builds up until the magnetic force qvB on an electron in the conductor is balanced by the electric force qE on the electron as shown in Figure 23.10. At this point, charge stops flowing. In this situation, the zero net force on an electron allows us to relate the electric field to the magnetic field:

$$\sum\vec{F} = \vec{F}_e - \vec{F}_B = 0 \quad \rightarrow \quad qE = qvB \quad \rightarrow \quad E = vB$$

Because the electric field produced in the conductor is uniform, it is related to the potential difference across the ends of the conductor according to the relation $\Delta V = E\ell$ (Section 20.2). Therefore,

$$\Delta V = E\ell = B\ell v$$

where the upper end is at a higher potential than the lower end. Therefore, a potential difference is maintained as long as the conductor is moving through the magnetic field. If the motion is reversed, the polarity of ΔV is also reversed.

An interesting situation occurs if we now consider what happens when the moving conductor is part of a closed circuit. Consider a circuit consisting of a conducting bar of length ℓ sliding along two fixed parallel conducting rails as in Active Figure 23.11a. For simplicity, we assume that the moving bar has zero electrical resistance and that the stationary part of the circuit has a resistance R. A uniform and constant magnetic field $\vec{B}$ is applied perpendicular to the plane of the circuit.

As the bar is pulled to the right with a velocity $\vec{v}$ under the influence of an applied force $\vec{F}_{app}$, free charges in the bar experience a magnetic force along the length of the bar. Because the moving bar is part of a complete circuit, a continuous current is established in the circuit. In this case, the rate of change of magnetic flux through the loop and the accompanying induced emf across the moving bar are proportional to the change in loop area as the bar moves through the magnetic field.

Because the area of the circuit at any instant is ℓx, the magnetic flux through the circuit is

$$\Phi_B = B\ell x$$

where x is the width of the circuit, a parameter that changes with time. Using Faraday's law, we find that the induced emf is

$$\mathcal{E} = -\frac{d\Phi_B}{dt} = -\frac{d}{dt}(B\ell x) = -B\ell\frac{dx}{dt}$$

$$\mathcal{E} = -B\ell v \qquad \qquad \textbf{23.5} \blacktriangleleft$$

Because the resistance of the circuit is R, the magnitude of the induced current is

$$I = \frac{|\mathcal{E}|}{R} = \frac{B\ell v}{R} \qquad\qquad \textbf{23.6}\blacktriangleleft$$

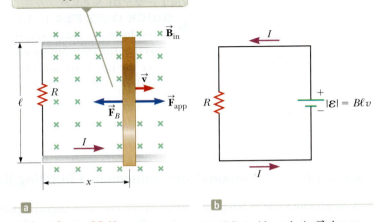

A counterclockwise current I is induced in the loop. The magnetic force $\vec{\mathbf{F}}_B$ on the bar carrying this current opposes the motion.

Active Figure 23.11 (a) A conducting bar sliding with a velocity $\vec{\mathbf{v}}$ along two conducting rails under the action of an applied force $\vec{\mathbf{F}}_{app}$. (b) The equivalent circuit diagram for the pictorial representation in (a).

The equivalent circuit diagram for this example is shown in Active Figure 23.11b. The moving bar is behaving like a battery in that it is a source of emf as long as the bar continues to move.

Let us examine this situation using energy considerations in the nonisolated system model, with the system being the entire circuit. Because the circuit has no battery, you might wonder about the origin of the induced current and the energy delivered to the resistor. Note that the external force $\vec{\mathbf{F}}_{app}$ does work on the conductor, thereby moving charges through a magnetic field, which causes the charges to move along the conductor with some average drift velocity. Hence, a current is established. From the viewpoint of the conservation of energy equation (Eq. 7.2), the total work done on the system by the applied force while the bar moves with constant speed must equal the increase in internal energy in the resistor during this time interval. (This statement assumes that the energy stays in the resistor; in reality, energy leaves the resistor by heat and electromagnetic radiation.)

As the conductor of length ℓ moves through the uniform magnetic field $\vec{\mathbf{B}}$, it experiences a magnetic force $\vec{\mathbf{F}}_B$ of magnitude $I\ell B$ (Eq. 22.10), where I is the current induced due to its motion. The direction of this force is opposite the motion of the bar, or to the left in Active Figure 23.11a.

If the bar is to move with a *constant* velocity, it is modeled as a particle in equilibrium, so the applied force $\vec{\mathbf{F}}_{app}$ must be equal in magnitude and opposite in direction to the magnetic force, or to the right in Active Figure 23.11a. (If the magnetic force acted in the direction of motion, it would cause the bar to accelerate once it was in motion, thereby increasing its speed. This state of affairs would represent a violation of the principle of energy conservation.) Using Equation 23.6 and that $F_{app} = F_B = I\ell B$, we find that the power delivered by the applied force is

$$P = F_{app}v = (I\ell B)v = \frac{B^2\ell^2 v^2}{R} = \left(\frac{B\ell v}{R}\right)^2 R = I^2R \qquad \textbf{23.7}\blacktriangleleft$$

This power is equal to the rate at which energy is delivered to the resistor, as we expect.

QUICK QUIZ 23.3 You wish to move a rectangular loop of wire into a region of uniform magnetic field at a given speed so as to induce an emf in the loop. The plane of the loop must remain perpendicular to the magnetic field lines. In which orientation

Reasoning As the magnet moves to the right toward the loop, the external magnetic flux through the loop increases with time. To counteract this increase in flux due to a field toward the right, the induced current produces its own magnetic field to the left as illustrated in Figure 23.18b; hence, the induced current is in the direction shown. Knowing that like magnetic poles repel each other, we conclude that the left face of the current loop acts like a north pole and the right face acts like a south pole.

(B) Find the direction of the induced current in the loop when the magnet is pulled away from the loop.

Reasoning If the magnet moves to the left as in Figure 23.18c, its flux through the area enclosed by the loop decreases in time. Now the induced current in the loop is in the direction shown in Figure 23.18d because this current direction produces a magnetic field in the same direction as the external field. In this case, the left face of the loop is a south pole and the right face is a north pole. ◄

❮ 23.4 | Induced emfs and Electric Fields

We have seen that a changing magnetic flux induces an emf and a current in a conducting loop. We can also interpret this phenomenon from another point of view. Because the normal flow of charges in a circuit is due to an electric field in the wires set up by a source such as a battery, we can interpret the changing magnetic field as creating an induced electric field. This electric field applies a force on the charges to cause them to move. With this approach, then, we see that an electric field is created in the conductor as a result of changing magnetic flux. In fact, the law of electromagnetic induction can be interpreted as follows: An electric field is always generated by a changing magnetic flux, even in free space where no charges are present. This induced electric field, however, has quite different properties from those of the electrostatic field produced by stationary charges.

Let us illustrate this point by considering a conducting loop of radius r, situated in a uniform magnetic field that is perpendicular to the plane of the loop as in Figure 23.19. If the magnetic field changes with time, Faraday's law tells us that an emf $\mathcal{E} = -d\Phi_B/dt$ is induced in the loop. The induced current thus produced implies the presence of an induced electric field $\vec{E}$ that must be tangent to the loop so as to provide an electric force on the charges around the loop. The work done by the electric field on the loop in moving a test charge q once around the loop is equal to $W = q\mathcal{E}$. Because the magnitude of the electric force on the charge is qE, the work done on the charge by the electric field can also be expressed from Equation 6.8 as $W = \int \vec{F} \cdot d\vec{r} = qE(2\pi r)$, where $2\pi r$ is the circumference of the loop. These two expressions for the work must be equal; therefore, we see that

$$q\mathcal{E} = qE(2\pi r)$$

$$E = \frac{\mathcal{E}}{2\pi r}$$

Using this result along with Faraday's law and that $\Phi_B = BA = B\pi r^2$ for a circular loop, we find that the induced electric field can be expressed as

$$E = -\frac{1}{2\pi r}\frac{d\Phi_B}{dt} = -\frac{1}{2\pi r}\frac{d}{dt}(B\pi r^2) = -\frac{r}{2}\frac{dB}{dt}$$

This expression can be used to calculate the induced electric field if the time variation of the magnetic field is specified. The negative sign indicates that the induced electric field $\vec{E}$ results in a current that opposes the change in the magnetic field. It is important to understand that this result is also valid in the absence of a conductor or charges. That is, the same electric field is induced by the changing magnetic field in empty space.

In general, the magnitude of the emf for any closed path can be expressed as the line integral of $\vec{E} \cdot d\vec{s}$ over that path: $\mathcal{E} = \oint \vec{E} \cdot d\vec{s}$ (Eq. 20.3). Hence, the general form of Faraday's law of induction is

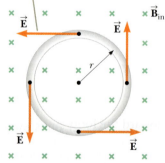

If $\vec{B}$ changes in time, an electric field is induced in a direction tangent to the circumference of the loop.

Figure 23.19 A conducting loop of radius r in a uniform magnetic field perpendicular to the plane of the loop.

▶ Faraday's law in general form

$$\oint \vec{E} \cdot d\vec{s} = -\frac{d\Phi_B}{dt}$$

23.9 ◄

It is important to recognize that the induced electric field $\vec{E}$ that appears in Equation 23.9 is a nonconservative field that is generated by a changing magnetic field. We call it a nonconservative field because the work done in moving a charge around a closed path (the loop in Fig. 23.19) is not zero. This type of electric field is very different from an electrostatic field.

> **QUICK QUIZ 23.6** In a region of space, a magnetic field is uniform over space but increases at a constant rate. This changing magnetic field induces an electric field that **(a)** increases in time, **(b)** is conservative, **(c)** is in the direction of the magnetic field, or **(d)** has a constant magnitude.

> ### THINKING PHYSICS 23.4
>
> In studying electric fields, we noted that electric field lines begin on positive charges and end on negative charges. Do *all* electric field lines begin and end on charges?
>
> **Reasoning** The statement that electric field lines begin and end on charges is true only for *electrostatic* fields, that is, electric fields due to stationary charges. Electric field lines due to changing magnetic fields form closed loops, with no beginning and no end, and are independent of the presence of charges. ◄

Example 23.5 | Electric Field Induced by a Changing Magnetic Field in a Solenoid

A long solenoid of radius R has n turns of wire per unit length and carries a time-varying current that varies sinusoidally as $I = I_{max} \cos \omega t$, where I_{max} is the maximum current and ω is the angular frequency of the alternating current source (Fig. 23.20).

(A) Determine the magnitude of the induced electric field outside the solenoid at a distance $r > R$ from its long central axis.

SOLUTION

Conceptualize Figure 23.20 shows the physical situation. As the current in the coil changes, imagine a changing magnetic field at all points in space as well as an induced electric field.

Categorize Because the current varies in time, the magnetic field is changing, leading to an induced electric field as opposed to the electrostatic electric fields due to stationary electric charges.

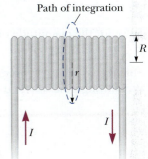

Path of integration

Figure 23.20 (Example 23.5) A long solenoid carrying a time-varying current given by $I = I_{max} \cos \omega t$. An electric field is induced both inside and outside the solenoid.

Analyze First consider an external point and take the path for the line integral to be a circle of radius r centered on the solenoid as illustrated in Figure 23.20.

Evaluate the right side of Equation 23.9, noting that the magnetic field $\vec{B}$ inside the solenoid is perpendicular to the circle bounded by the path of integration:

$$(1) \quad -\frac{d\Phi_B}{dt} = -\frac{d}{dt}(B\pi R^2) = -\pi R^2 \frac{dB}{dt}$$

Evaluate the magnetic field in the solenoid from Equation 22.32:

$$(2) \quad B = \mu_0 nI = \mu_0 nI_{max} \cos \omega t$$

Substitute Equation (2) into Equation (1):

$$(3) \quad -\frac{d\Phi_B}{dt} = -\pi R^2 \mu_0 nI_{max} \frac{d}{dt}(\cos \omega t) = \pi R^2 \mu_0 nI_{max}\omega \sin \omega t$$

Evaluate the left side of Equation 23.9, noting that the magnitude of $\vec{E}$ is constant on the path of integration and $\vec{E}$ is tangent to it:

$$(4) \quad \oint \vec{E} \cdot d\vec{s} = E(2\pi r)$$

23.5 *cont.*

Substitute Equations (3) and (4) into Equation 23.9: $E(2\pi r) = \pi R^2 \mu_0 n I_{max} \omega \sin \omega t$

Solve for the magnitude of the electric field: $E = \dfrac{\mu_0 n I_{max} \omega R^2}{2r} \sin \omega t \quad \text{(for } r > R\text{)}$

Finalize This result shows that the amplitude of the electric field outside the solenoid falls off as $1/r$ and varies sinusoidally with time. As we will learn in Chapter 24, the time-varying electric field creates an additional contribution to the magnetic field. The magnetic field can be somewhat stronger than we first stated, both inside and outside the solenoid. The correction to the magnetic field is small if the angular frequency ω is small. At high frequencies, however, a new phenomenon can dominate: The electric and magnetic fields, each re-creating the other, constitute an electromagnetic wave radiated by the solenoid as we will study in Chapter 24.

(B) What is the magnitude of the induced electric field inside the solenoid, a distance r from its axis?

SOLUTION

Analyze For an interior point $(r < R)$, the magnetic flux through an integration loop is given by $\Phi_B = B\pi r^2$.

Evaluate the right side of Equation 23.9:   (5) $-\dfrac{d\Phi_B}{dt} = -\dfrac{d}{dt}(B\pi r^2) = -\pi r^2 \dfrac{dB}{dt}$

Substitute Equation (2) into Equation (5):   (6) $-\dfrac{d\Phi_B}{dt} = -\pi r^2 \mu_0 \, n I_{max} \dfrac{d}{dt}(\cos \omega t) = \pi r^2 \mu_0 n I_{max} \omega \sin \omega t$

Substitute Equations (4) and (6) into Equation 23.9: $E(2\pi r) = \pi r^2 \mu_0 n I_{max} \omega \sin \omega t$

Solve for the magnitude of the electric field: $E = \dfrac{\mu_0 n I_{max} \omega}{2} r \sin \omega t \quad \text{(for } r < R\text{)}$

Finalize This result shows that the amplitude of the electric field induced inside the solenoid by the changing magnetic flux through the solenoid increases linearly with r and varies sinusoidally with time.

23.5 | Inductance

Consider an isolated circuit consisting of a switch, a resistor, and a source of emf as in Figure 23.21. The circuit diagram is represented in perspective so that we can see the orientations of some of the magnetic field lines due to the current in the circuit. When the switch is thrown to its closed position, the current doesn't immediately jump from zero to its maximum value $\mathcal{E}/R$; the law of electromagnetic induction (Faraday's law) describes the actual behavior. As the current increases with time, the magnetic flux through the loop of the circuit itself due to the current also increases with time. This increasing magnetic flux *from* the circuit induces an emf *in* the circuit (sometimes referred to as a *back emf*) that opposes the change in the net magnetic flux through the loop of the circuit. By Lenz's law, the induced electric field in the wires must therefore be opposite the direction of the current, and the opposing emf results in a *gradual* increase in the current. This effect is called *self-induction* because the changing magnetic flux through the circuit arises from the circuit itself. The emf set up in this case is called a **self-induced emf**.

To obtain a quantitative description of self-induction, we recall from Faraday's law that the induced emf is the negative time rate of change of the magnetic flux. The magnetic flux is proportional to the magnetic field, which in turn is proportional to the current in the circuit. Therefore, the self-induced emf is always proportional to the time rate of change of the current. For a closely spaced coil of N turns of fixed

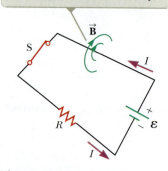

After the switch is closed, the current produces a magnetic flux through the area enclosed by the loop. As the current increases toward its equilibrium value, this magnetic flux changes in time and induces an emf in the loop.

Figure 23.21 Self-induction in a simple circuit

geometry (a toroidal coil or the ideal solenoid), we can express this proportionality as follows:

$$\mathcal{E}_L = -N\frac{d\Phi_B}{dt} = -L\frac{dI}{dt} \qquad \textbf{23.10} \blacktriangleleft \qquad \blacktriangleright \text{ Self-induced emf}$$

where L is a proportionality constant, called the **inductance** of the coil, that depends on the geometric features of the coil and other physical characteristics. From this expression, we see that the inductance of a coil containing N turns is

$$L = \frac{N\Phi_B}{I} \qquad \textbf{23.11} \blacktriangleleft$$

where it is assumed that the same magnetic flux passes through each turn. Later we shall use this equation to calculate the inductance of some special coil geometries.

From Equation 23.10, we can also write the inductance as the ratio

$$L = -\frac{\mathcal{E}_L}{dI/dt} \qquad \textbf{23.12} \blacktriangleleft$$

which is usually taken to be the defining equation for the inductance of any coil, regardless of its shape, size, or material characteristics. If we compare Equation 23.10 with Equation 21.6, $R = \Delta V/I$, we see that resistance is a measure of opposition to current, whereas inductance is a measure of opposition to the *change* in current.

The SI unit of inductance is the **henry (H),** which, from Equation 23.12, is seen to be equal to 1 volt·second per ampere:

$$1\text{ H} = 1\text{ V}\cdot\text{s/A}$$

As we shall see, the **inductance of a coil depends on its geometry.** Because inductance calculations can be quite difficult for complicated geometries, the examples we shall explore involve simple situations for which inductances are easily evaluated.

Joseph Henry
American Physicist (1797–1878)
Henry became the first director of the Smithsonian Institution and first president of the Academy of Natural Science. He improved the design of the electromagnet and constructed one of the first motors. He also discovered the phenomenon of self-induction, but he failed to publish his findings. The unit of inductance, the henry, is named in his honor.

Brady-Handy Collection, Library of Congress Prints and Photographs Division [LC-BH83-997]

Example 23.6 | Inductance of a Solenoid

Consider a uniformly wound solenoid having N turns and length ℓ. Assume ℓ is much longer than the radius of the windings and the core of the solenoid is air.

(A) Find the inductance of the solenoid.

SOLUTION

Conceptualize The magnetic field lines from each turn of the solenoid pass through all the turns, so an induced emf in each coil opposes changes in the current.

Categorize Because the solenoid is long, we can use the results for an ideal solenoid obtained in Chapter 22.

Analyze Find the magnetic flux through each turn of area A in the solenoid, using the expression for the magnetic field from Equation 22.32:

$$\Phi_B = BA = \mu_0 nIA = \mu_0\frac{N}{\ell}IA$$

Substitute this expression into Equation 23.11:

$$(1)\quad L = \frac{N\Phi_B}{I} = \mu_0\frac{N^2}{\ell}A$$

(B) Calculate the inductance of the solenoid if it contains 300 turns, its length is 25.0 cm, and its cross-sectional area is 4.00 cm².

SOLUTION

Substitute numerical values into Equation (1):

$$L = (4\pi \times 10^{-7}\text{ T}\cdot\text{m/A})\frac{300^2}{25.0 \times 10^{-2}\text{ m}}(4.00 \times 10^{-4}\text{ m}^2)$$

$$= 1.81 \times 10^{-4}\text{ T}\cdot\text{m}^2/\text{A} = 0.181\text{ mH}$$

continued

23.6 *cont.*

(C) Calculate the self-induced emf in the solenoid if the current it carries decreases at the rate of 50.0 A/s.

SOLUTION

Substitute $dI/dt = -50.0$ A/s and the answer to part (B) into Equation 23.10:

$$\varepsilon_L = -L\frac{dI}{dt} = -(1.81 \times 10^{-4}\ \text{H})(-50.0\ \text{A/s})$$

$$= 9.05\ \text{mV}$$

Finalize The result for part (A) shows that L depends on geometry and is proportional to the square of the number of turns. Because $N = n\ell$, we can also express the result in the form

$$L = \mu_0 \frac{(n\ell)^2}{\ell}A = \mu_0 n^2 A\ell = \mu_0 n^2 V$$

where $V = A\ell$ is the interior volume of the solenoid.

⟨ 23.6 | *RL* Circuits

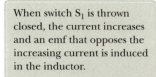

When switch S_1 is thrown closed, the current increases and an emf that opposes the increasing current is induced in the inductor.

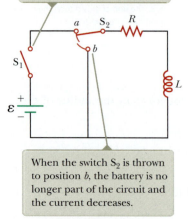

When the switch S_2 is thrown to position b, the battery is no longer part of the circuit and the current decreases.

Active Figure 23.22 An *RL* circuit. When switch S_2 is in position a, the battery is in the circuit.

If a circuit contains a coil such as a solenoid, the inductance of the coil prevents the current in the circuit from increasing or decreasing instantaneously. A circuit element that has a large inductance is called an **inductor** and has the circuit symbol —. We always assume the inductance of the remainder of a circuit is negligible compared with that of the inductor. Keep in mind, however, that even a circuit without a coil has some inductance that can affect the circuit's behavior.

Because the inductance of an inductor results in a back emf, an inductor in a circuit opposes changes in the current in that circuit. The inductor attempts to keep the current the same as it was before the change occurred. If the battery voltage in the circuit is increased so that the current rises, the inductor opposes this change and the rise is not instantaneous. If the battery voltage is decreased, the inductor causes a slow drop in the current rather than an immediate drop. Therefore, the inductor causes the circuit to be "sluggish" as it reacts to changes in the voltage.

Consider the circuit shown in Active Figure 23.22, which contains a battery of negligible internal resistance. This circuit is an **RL circuit** because the elements connected to the battery are a resistor and an inductor. The curved lines on switch S_2 suggest this switch can never be open; it is always set to either a or b. (If the switch is connected to neither a nor b, any current in the circuit suddenly stops.) Suppose S_2 is set to a and switch S_1 is open for $t < 0$ and then thrown closed at $t = 0$. The current in the circuit begins to increase, and a back emf (Eq. 23.10) that opposes the increasing current is induced in the inductor.

With this point in mind, let's apply Kirchhoff's loop rule to this circuit, traversing the circuit in the clockwise direction:

$$\varepsilon - IR - L\frac{dI}{dt} = 0 \qquad\qquad \text{23.13} \blacktriangleleft$$

where IR is the voltage drop across the resistor. (Kirchhoff's rules were developed for circuits with steady currents, but they can also be applied to a circuit in which the current is changing if we imagine them to represent the circuit at one *instant* of time.) Now let's find a solution to this differential equation, which is similar to that for the *RC* circuit (see Section 21.9).

A mathematical solution of Equation 23.13 represents the current in the circuit as a function of time. To find this solution, we change variables for convenience, letting $x = (\varepsilon/R) - I$, so $dx = -dI$. With these substitutions, Equation 23.13 becomes

$$x + \frac{L}{R}\frac{dx}{dt} = 0$$

Rearranging and integrating this last expression gives

$$\int_{x_0}^{x} \frac{dx}{x} = -\frac{R}{L} \int_{0}^{t} dt$$

$$\ln \frac{x}{x_0} = -\frac{R}{L} t$$

where x_0 is the value of x at time $t = 0$. Taking the antilogarithm of this result gives

$$x = x_0 e^{-Rt/L}$$

Because $I = 0$ at $t = 0$, note from the definition of x that $x_0 = \mathcal{E}/R$. Hence, this last expression is equivalent to

$$\frac{\mathcal{E}}{R} - I = \frac{\mathcal{E}}{R} e^{-Rt/L}$$

$$I = \frac{\mathcal{E}}{R} (1 - e^{-Rt/L})$$

This expression shows how the inductor affects the current. The current does not increase instantly to its final equilibrium value when the switch is closed, but instead increases according to an exponential function. If the inductance is removed from the circuit, which corresponds to letting L approach zero, the exponential term becomes zero and there is no time dependence of the current in this case; the current increases instantaneously to its final equilibrium value in the absence of the inductance.

We can also write this expression as

$$I = \frac{\mathcal{E}}{R} (1 - e^{-t/\tau}) \qquad \text{23.14} \blacktriangleleft$$

where the constant τ is the **time constant** of the *RL* circuit:

$$\tau = \frac{L}{R} \qquad \text{23.15} \blacktriangleleft$$

Physically, τ is the time interval required for the current in the circuit to reach $(1 - e^{-1}) = 0.632 = 63.2\%$ of its final value $\mathcal{E}/R$. The time constant is a useful parameter for comparing the time responses of various circuits.

Active Figure 23.23 shows a graph of the current versus time in the *RL* circuit. Notice that the equilibrium value of the current, which occurs as t approaches infinity, is $\mathcal{E}/R$. That can be seen by setting dI/dt equal to zero in Equation 23.13 and solving for the current I. (At equilibrium, the change in the current is zero.) Therefore, the current initially increases very rapidly and then gradually approaches the equilibrium value $\mathcal{E}/R$ as t approaches infinity.

Let's also investigate the time rate of change of the current. Taking the first time derivative of Equation 23.14 gives

$$\frac{dI}{dt} = \frac{\mathcal{E}}{L} e^{-t/\tau} \qquad \text{23.16} \blacktriangleleft$$

This result shows that the time rate of change of the current is a maximum (equal to $\mathcal{E}/L$) at $t = 0$ and falls off exponentially to zero as t approaches infinity (Fig. 23.24).

Now consider the *RL* circuit in Active Figure 23.22 again. Suppose switch S_2 has been set at position a long enough (and switch S_1 remains closed) to allow the current to reach its equilibrium value $\mathcal{E}/R$. In this situation, the circuit is described by the outer loop in Active Figure 23.22. If S_2 is thrown from a to b, the circuit is now described by only the right-hand loop in Active Figure 23.22. Therefore, the battery has been eliminated from the circuit. Setting $\mathcal{E} = 0$ in Equation 23.13 gives

$$IR + L \frac{dI}{dt} = 0 \qquad \text{23.17} \blacktriangleleft$$

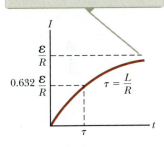

After switch S_1 is thrown closed at $t = 0$, the current increases toward its maximum value $\mathcal{E}/R$.

Active Figure 23.23 Plot of the current versus time for the *RL* circuit shown in Active Figure 23.22. The time constant τ is the time interval required for I to reach 63.2% of its maximum value.

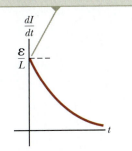

The time rate of change of current is a maximum at $t = 0$, which is the instant at which switch S_1 is thrown closed.

Figure 23.24 Plot of dI/dt versus time for the *RL* circuit shown in Active Figure 23.22. The rate decreases exponentially with time as I increases toward its maximum value.

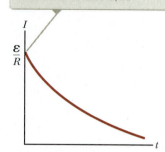

At $t = 0$, the switch is thrown to position b and the current has its maximum value $\mathcal{E}/R$.

Active Figure 23.25 Current versus time for the right-hand loop of the circuit shown in Active Figure 23.22. For $t < 0$, switch S_2 is at position a.

It is left as a problem (Problem 38) to show that the solution of this differential equation is

$$I = \frac{\mathcal{E}}{R} e^{-t/\tau} = I_i e^{-t/\tau}$$ 23.18 ◀

where $\mathcal{E}$ is the emf of the battery and $I_i = \mathcal{E}/R$ is the initial current at the instant the switch is thrown to b.

If the circuit did not contain an inductor, the current would immediately decrease to zero when the battery is removed. When the inductor is present, it opposes the decrease in the current and causes the current to decrease exponentially. A graph of the current in the circuit versus time (Active Fig. 23.25) shows that the current is continuously decreasing with time.

QUICK QUIZ 23.7 The circuit in Figure 23.26 includes a power source that provides a sinusoidal voltage. Therefore, the magnetic field in the inductor is constantly changing. The inductor is a simple air-core solenoid. The switch in the circuit is closed and the lightbulb glows steadily. An iron rod is inserted into the interior of the solenoid, which increases the magnitude of the magnetic field in the solenoid. As that happens, the brightness of the lightbulb (**a**) increases, (**b**) decreases, or (**c**) is unaffected.

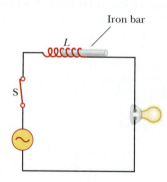

Iron bar

Figure 23.26 (Quick Quiz 23.7) A lightbulb is powered by an AC source with an inductor in the circuit. When the iron bar is inserted into the coil, what happens to the brightness of the lightbulb?

QUICK QUIZ 23.8 Two circuits like the one shown in Active Figure 23.22 are identical except for the value of L. In circuit A, the inductance of the inductor is L_A, and in circuit B, it is L_B. Switch S_2 has been in position b for both circuits for a long time. At $t = 0$, the switch is thrown to a in both circuits. At $t = 10$ s, the switch is thrown to b in both circuits. The resulting graphical representation of the current as a function of time is shown in Figure 23.27. Assuming that the time constant of each circuit is much less than 10 s, which of the following is true? (a) $L_A > L_B$. (b) $L_A < L_B$. (c) There is not enough information to determine the relative values.

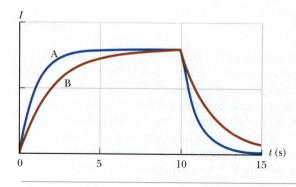

Figure 23.27 (Quick Quiz 23.8) Current–time graphs for two circuits with different inductances.

Example 23.7 | Time Constant of an *RL* Circuit

Consider the circuit in Active Figure 23.22 again. Suppose the circuit elements have the following values: $\mathcal{E} = 12.0$ V, $R = 6.00\ \Omega$, and $L = 30.0$ mH.

(A) Find the time constant of the circuit.

SOLUTION

Conceptualize You should understand the behavior of this circuit from the discussion in this section.

Categorize We evaluate the results using equations developed in this section, so this example is a substitution problem.

Evaluate the time constant from Equation 23.15:

$$\tau = \frac{L}{R} = \frac{30.0 \times 10^{-3}\ \text{H}}{6.00\ \Omega} = \boxed{5.00\ \text{ms}}$$

(B) Switch S_2 is at position a, and switch S_1 is thrown closed at $t = 0$. Calculate the current in the circuit at $t = 2.00$ ms.

SOLUTION

Evaluate the current at $t = 2.00$ ms from Equation 23.14:

$$I = \frac{\mathcal{E}}{R}(1 - e^{-t/\tau}) = \frac{12.0\ \text{V}}{6.00\ \Omega}(1 - e^{-2.00\ \text{ms}/5.00\ \text{ms}}) = 2.00\ \text{A}\ (1 - e^{-0.400})$$

$$= \boxed{0.659\ \text{A}}$$

(C) Compare the potential difference across the resistor with that across the inductor.

SOLUTION

At the instant the switch is closed, there is no current and therefore no potential difference across the resistor. At this instant, the battery voltage appears entirely across the inductor in the form of a back emf of 12.0 V as the inductor tries to maintain the zero-current condition. (The top end of the inductor in Active Fig. 23.22 is at a higher electric potential than the bottom end.) As time passes, the emf across the inductor decreases and the current in the resistor (and hence the voltage across it) increases as shown in Figure 23.28. The sum of the two voltages at all times is 12.0 V.

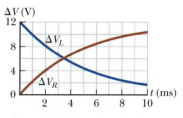

Figure 23.28 (Example 23.7) The time behavior of the voltages across the resistor and inductor in Active Figure 23.22 given the values provided in this example.

23.7 | Energy Stored in a Magnetic Field

In the preceding section, we found that the induced emf set up by an inductor prevents a battery from establishing an instantaneous current. Part of the energy supplied by the battery goes into internal energy in the resistor, and the remaining energy is stored in the inductor. If we multiply each term in Equation 23.13 by the current I and rearrange the expression, we have

$$I\mathcal{E} = I^2R + LI\frac{dI}{dt} \qquad \textbf{23.19} \blacktriangleleft$$

This expression tells us that the rate $I\mathcal{E}$ at which energy is supplied by the battery equals the sum of the rate I^2R at which energy is delivered to the resistor and the rate $LI\,(dI/dt)$ at which energy is delivered to the inductor. Therefore, Equation 23.19 is simply an expression of energy conservation for the isolated system of the circuit. (Actually, energy can leave the circuit by thermal conduction into the air and by electromagnetic radiation, so the system need not be completely isolated.) If we let U denote the energy stored in the inductor at any time, the rate dU/dt at which energy is delivered to the inductor can be written as

$$\frac{dU}{dt} = LI\frac{dI}{dt}$$

Pitfall Prevention | 23.3
Capacitors, Resistors, and Inductors Store Energy Differently
Different energy-storage mechanisms are at work in capacitors, inductors, and resistors. A charged capacitor stores energy as electrical potential energy. An inductor stores energy as what we could call magnetic potential energy when it carries current. Energy delivered to a resistor is transformed to internal energy.

To find the total energy stored in the inductor at any instant, we can rewrite this expression as $dU = LI\,dI$ and integrate:

$$U = \int_0^U dU = \int_0^I LI\,dI$$

▶ Energy stored in an inductor

$$U = \tfrac{1}{2}LI^2 \qquad\qquad\qquad \textbf{23.20}◀$$

where L is constant and so has been removed from the integral. Equation 23.20 represents the energy stored in the magnetic field of the inductor when the current is I.

Equation 23.20 is similar to the equation for the energy stored in the electric field of a capacitor, $U = \tfrac{1}{2}C(\Delta V)^2$ (Eq. 20.29). In either case, we see that energy from a battery is required to establish a field and that energy is stored in the field. In the case of the capacitor, we can conceptually relate the energy stored in the capacitor to the electric potential energy associated with the separated charge on the plates. We have not discussed a magnetic analogy to electric potential energy, so the storage of energy in an inductor is not as easy to conceptualize.

To argue that energy is stored in an inductor, consider the circuit in Figure 23.29a, which is the same circuit as in Active Figure 23.22, with the addition of a switch S_3 across the resistor R. With switch S_2 set to position a and S_3 closed as shown, a current is established in the inductor. Now, as in Figure 23.29b, switch S_2 is thrown to position b. The current persists in this (ideally) resistance-free and battery-free circuit (the right-hand loop in Fig. 23.29b), consisting of only the inductor and a conducting path between its ends. There is no current in the resistor (because the path around it through S_3 is resistance free), so no energy is being delivered to it. The next step is to open switch S_3 as shown in Figure 23.29c, which puts the resistor into the circuit. There is now current in the resistor, and energy is delivered to the resistor. Where is the energy coming from? The only other element in the circuit previous to opening switch S_3 was the inductor. Energy must therefore have been stored in the inductor and is now being delivered to the resistor.

Now let us determine the energy per unit volume, or energy density, stored in a magnetic field. For simplicity, consider a solenoid whose inductance is $L = \mu_0 n^2 A\ell$ (see Example 23.6). The magnetic field of the solenoid is $B = \mu_0 nI$. Substituting the expression for L and $I = B/\mu_0 n$ into Equation 23.20 gives

$$U = \tfrac{1}{2}LI^2 = \tfrac{1}{2}\mu_0 n^2 A\ell \left(\frac{B}{\mu_0 n}\right)^2 = \frac{B^2}{2\mu_0}(A\ell) \qquad \textbf{23.21}◀$$

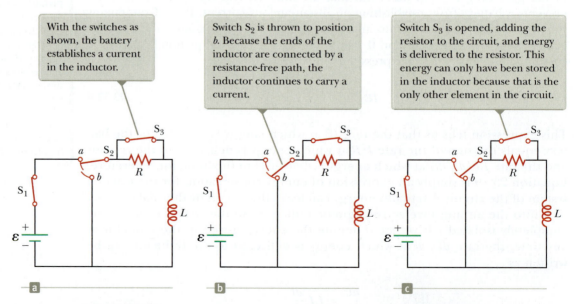

With the switches as shown, the battery establishes a current in the inductor.

Switch S_2 is thrown to position b. Because the ends of the inductor are connected by a resistance-free path, the inductor continues to carry a current.

Switch S_3 is opened, adding the resistor to the circuit, and energy is delivered to the resistor. This energy can only have been stored in the inductor because that is the only other element in the circuit.

Figure 23.29 An RL circuit used for conceptualizing energy storage in an inductor.

Because $A\ell$ is the volume of the solenoid, the energy stored per unit volume in a magnetic field—in other words, the *magnetic energy density*—is

$$u_B = \frac{U}{A\ell} = \frac{B^2}{2\mu_0}$$

23.22 ◀   ▶ Magnetic energy density

Although Equation 23.22 was derived for the special case of a solenoid, it is valid for any region of space in which a magnetic field exists. Note that it is similar to the equation for the energy per unit volume stored in an electric field, given by $\frac{1}{2}\epsilon_0 E^2$ (Eq. 20.31). In both cases, the energy density is proportional to the square of the magnitude of the field.

QUICK QUIZ 23.9 You are performing an experiment that requires the highest-possible magnetic energy density in the interior of a very long current-carrying solenoid. Which of the following adjustments increases the energy density? (More than one choice may be correct.) (**a**) increasing the number of turns per unit length on the solenoid (**b**) increasing the cross-sectional area of the solenoid (**c**) increasing only the length of the solenoid while keeping the number of turns per unit length fixed (**d**) increasing the current in the solenoid

---

Example 23.8 | What Happens to the Energy in the Inductor?

Consider once again the *RL* circuit shown in Active Figure 23.22, with switch S_2 at position *a* and the current having reached its steady-state value. When S_2 is thrown to position *b*, the current in the right-hand loop decays exponentially with time according to the expression $I = I_i e^{-t/\tau}$, where $I_i = \mathcal{E}/R$ is the initial current in the circuit and $\tau = L/R$ is the time constant. Show that all the energy initially stored in the magnetic field of the inductor appears as internal energy in the resistor as the current decays to zero.

SOLUTION

Conceptualize Before S_2 is thrown to *b*, energy is being delivered at a constant rate to the resistor from the battery and energy is stored in the magnetic field of the inductor. After $t = 0$, when S_2 is thrown to *b*, the battery can no longer provide energy and energy is delivered to the resistor only from the inductor.

Categorize We model the right-hand loop of the circuit as an isolated system so that energy is transferred between components of the system but does not leave the system.

Analyze The energy in the magnetic field of the inductor at any time is U. The rate dU/dt at which energy leaves the inductor and is delivered to the resistor is equal to $I^2 R$, where I is the instantaneous current.

Substitute the current given by Equation 23.18 into $dU/dt = I^2 R$:

$$\frac{dU}{dt} = I^2 R = (I_i e^{-Rt/L})^2 R = I_i^2 R e^{-2Rt/L}$$

Solve for dU and integrate this expression over the limits $t = 0$ to $t \rightarrow \infty$:

$$U = \int_0^\infty I_i^2 R e^{-2Rt/L}\, dt = I_i^2 R \int_0^\infty e^{-2Rt/L}\, dt$$

The value of the definite integral can be shown to be $L/2R$ (see Problem 74). Use this result to evaluate U:

$$U = I_i^2 R\left(\frac{L}{2R}\right) = \tfrac{1}{2}LI_i^2$$

Finalize This result is equal to the initial energy stored in the magnetic field of the inductor, given by Equation 23.20, as we set out to prove.

Example 23.9 | The Coaxial Cable

Coaxial cables are often used to connect electrical devices, such as your video system, and in receiving signals in television cable systems. Model a long coaxial cable as a thin, cylindrical conducting shell of radius *b* concentric with a solid cylinder of radius *a* as in Figure 23.30 (page 804). The conductors carry the same current *I* in opposite directions. Calculate the inductance *L* of a length ℓ of this cable.

continued

23.9 *cont.*

SOLUTION

Conceptualize Consider Figure 23.30. Although we do not have a visible coil in this geometry, imagine a thin, radial slice of the coaxial cable such as the light gold rectangle in Figure 23.30. If the inner and outer conductors are connected at the ends of the cable (above and below the figure), this slice represents one large conducting loop. The current in the loop sets up a magnetic field between the inner and outer conductors that passes through this loop. If the current changes, the magnetic field changes and the induced emf opposes the original change in the current in the conductors.

Categorize We categorize this situation as one in which we must return to the fundamental definition of inductance, Equation 23.11.

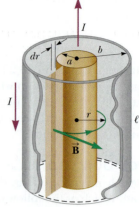

Figure 23.30 (Example 23.9) Section of a long coaxial cable. The inner and outer conductors carry equal currents in opposite directions.

Analyze We must find the magnetic flux through the light gold rectangle in Figure 23.30. Ampère's law (see Section 22.9) tells us that the magnetic field in the region between the conductors is due to the inner conductor alone and that its magnitude is $B = \mu_0 I / 2\pi r$, where r is measured from the common center of the cylinders. A sample circular field line is shown in Figure 23.30, along with a field vector tangent to the field line. The magnetic field is zero outside the outer shell because the net current passing through the area enclosed by a circular path surrounding the cable is zero; hence, from Ampère's law, $\oint \vec{\mathbf{B}} \cdot d\vec{\mathbf{s}} = 0$.

The magnetic field is perpendicular to the light gold rectangle of length ℓ and width $b - a$, the cross section of interest. Because the magnetic field varies with radial position across this rectangle, we must use calculus to find the total magnetic flux.

Divide the light gold rectangle into strips of width dr such as the darker strip in Figure 23.30. Evaluate the magnetic flux through such a strip:

$$d\Phi_B = B\,dA = B\ell\,dr$$

Substitute for the magnetic field and integrate over the entire light gold rectangle:

$$\Phi_B = \int_a^b \frac{\mu_0 I}{2\pi r}\,\ell\,dr = \frac{\mu_0 I\ell}{2\pi}\int_a^b \frac{dr}{r} = \frac{\mu_0 I\ell}{2\pi}\,\ln\left(\frac{b}{a}\right)$$

Use Equation 23.11 to find the inductance of the cable:

$$L = \frac{\Phi_B}{I} = \frac{\mu_0 \ell}{2\pi}\,\ln\left(\frac{b}{a}\right)$$

Finalize The inductance increases if ℓ increases, if b increases, or if a decreases. This result is consistent with our conceptualization: any of these changes increases the size of the loop represented by our radial slice and through which the magnetic field passes, increasing the inductance.

23.8 | Context Connection: The Use of Transcranial Magnetic Stimulation in Depression BIO

In Sections 20.7 and 21.9, we discussed the electrical characteristics of neurons and the propagation of an action potential along a neuron. In this Context Connection, we discuss a new treatment for depression that is early in its development and is closely related to the propagation of impulses along and between nerves.

Depression is a mental disorder in which patients exhibit decreased self-esteem, low moods, sadness, loss of interest in previously enjoyable activities, and increased possibility of suicidal thoughts and behaviors. It is a complicated disorder whose cause seems to include biological conditions, psychological effects, social interactions, drug and alcohol use, and even genetics. As a result of the large number of possible influences causing depression, the particular treatment plan for a given individual is not clear without extensive counseling of the patient and attempts to use a variety of treatments.

In our discussion, we will focus on the possible biological origins of depression. One hypothesis suggests that depression is related to low levels of *neurotransmitters* (particularly serotonin, norepinephrine, and dopamine) in the synapses between neurons. Antidepressant medications, such as *sertraline*, act to increase the levels of these neurotransmitters.

One of the more controversial treatments for severe depression that has not responded to other treatments is *electroconvulsive therapy* (ECT), in which seizures are induced in a patient under anesthesia. The seizures are induced by placing electrodes on the patient's head and passing a pulsed current between the electrodes. While the effects of this procedure on human brains cannot ethically be studied in detail, results from animal experiments suggest possible new synapse formation from the treatment. Because of the role of levels of neurotransmitters in depression, this growth of synapses may be the reason for the improvement in some depressed patients after undergoing electroconvulsive therapy. ECT was used in the 1940s and 1950s on severely disturbed patients in large mental institutions. Today, its main use is in psychiatric hospitals. There is still continuing controversy about the usage of ECT for patients suffering from mental disorders.

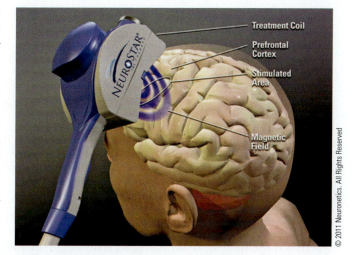

Figure 23.31 The magnetic coil of a Neurostar TMS apparatus is held near the head of a patient.

A newer method of introducing electric current into the brain is **transcranial magnetic stimulation** (TMS). This procedure induces currents in the brain by means of magnetic induction rather than by the application of a large voltage from contact electrodes. A large coil of wire is placed against the scalp of the patient. The coil carries an alternating current, creating an oscillating magnetic field, which induces currents in the nerve cells of the brain. Unlike electroconvulsive therapy, the patient is awake and does not experience seizures. Figure 23.31 shows the coil of a Neurostar TMS machine being applied to a patient's head. While the United States Food and Drug Administration has not approved TMS as a generic procedure at the time of this printing, the agency has cleared the specific Neurostar device that performs TMS.

BIO Transcranial magnetic stimulation

TMS has been used to perform motor cortex mapping, in which connections between the primary motor cortex and various muscles are measured to determine damage from spinal cord injuries, strokes, and motor neuron disease. With this technique, muscular responses in the index finger, forearm, biceps, jaw, and leg can be observed as the magnetic coil is moved to different locations of the cortex.

When the coil is moved over the occipital cortex, some patients report *magnetophosphenes*, which are flashes of light seen even though the eyes are closed. The most common method of inducing phosphenes is mechanical: by rubbing the closed eyes. Phosphenes resulting from a blow to the head are the origin of the phrase "seeing stars" associated with such trauma.

The use of TMS in depression is more recent. Figure 23.32 shows a patient being treated with a TMS apparatus. The patient sits in a chair and the coil is placed against the head. The coil is switched on and off at frequencies up to 10 Hz, inducing currents in the brain. The magnetic field is increased until the patient's fingers begin to twitch, indicating a therapeutic level. Once that level is attained, the treatment portion of the experience continues for about 40 minutes. These treatments are repeated on a daily basis for a period of several weeks.

Studies have reported some effectiveness of the procedure in treating depression, but more studies need to be performed to validate the evidence. One of the factors working against tests of effectiveness is the difficulty in establishing a "fake" TMS experience to be used as a placebo for comparison to the actual treatment. The treatment causes low-level neck pain, headache, and twitching in the scalp, which are difficult to reproduce in a placebo intervention.

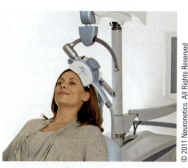

Figure 23.32 A patient is treated with the Neurostar TMS apparatus.

SUMMARY

The **magnetic flux** through a surface associated with a magnetic field $\vec{B}$ is

$$\Phi_B = \int \vec{B} \cdot d\vec{A} \qquad \text{23.1} \blacktriangleleft$$

where the integral is over the surface.

Faraday's law of induction states that the emf induced in a circuit is directly proportional to the time rate of change of magnetic flux through the circuit:

$$\varepsilon = -N \frac{d\Phi_B}{dt} \qquad \text{23.3} \blacktriangleleft$$

where N is the number of turns and Φ_B is the magnetic flux through each turn.

When a conducting bar of length ℓ moves through a magnetic field $\vec{B}$ with a velocity $\vec{v}$ so that $\vec{v}$ is perpendicular to $\vec{B}$, the emf induced in the bar (called the **motional emf**) is

$$\varepsilon = -B\ell v \qquad \text{23.5} \blacktriangleleft$$

Lenz's law states that the induced current and induced emf in a conductor are in such a direction as to oppose the change that produced them.

A general form of Faraday's law of induction is

$$\oint \vec{E} \cdot d\vec{s} = -\frac{d\Phi_B}{dt} \qquad \text{23.9} \blacktriangleleft$$

where $\vec{E}$ is a nonconservative electric field produced by the changing magnetic flux.

When the current in a coil changes with time, an emf is induced in the coil according to Faraday's law. The **self-induced emf** is described by the expression

$$\varepsilon_L = -L \frac{dI}{dt} \qquad \text{23.10} \blacktriangleleft$$

where L is the *inductance* of the coil. Inductance is a measure of the opposition of a device to a change in current.

The **inductance** of a coil is

$$L = \frac{N\Phi_B}{I} \qquad \text{23.11} \blacktriangleleft$$

where Φ_B is the magnetic flux through the coil and N is the total number of turns. Inductance has the SI unit the **henry** (H), where $1\ \text{H} = 1\ \text{V} \cdot \text{s/A}$.

If a resistor and inductor are connected in series to a battery of emf ε as shown in Active Figure 23.22, switch S_2 is set at position a, and switch S_1 is thrown closed at $t = 0$, the current in the circuit varies with time according to the expression

$$I(t) = \frac{\varepsilon}{R}(1 - e^{-t/\tau}) \qquad \text{23.14} \blacktriangleleft$$

where $\tau = L/R$ is the **time constant** of the RL circuit.

If switch S_2 in Active Figure 23.22 is thrown to position b, the current decays exponentially with time according to the expression

$$I(t) = \frac{\varepsilon}{R} e^{-t/\tau} \qquad \text{23.18} \blacktriangleleft$$

where ε/R is the initial current in the circuit.

The energy stored in the magnetic field of an inductor carrying a current I is

$$U = \tfrac{1}{2} L I^2 \qquad \text{23.20} \blacktriangleleft$$

The energy per unit volume (or energy density) at a point where the magnetic field is B is

$$u_B = \frac{B^2}{2\mu_0} \qquad \text{23.22} \blacktriangleleft$$

OBJECTIVE QUESTIONS

☐ denotes answer available in *Student Solutions Manual/Study Guide*

1. Figure OQ23.1 is a graph of the magnetic flux through a certain coil of wire as a function of time during an interval while the radius of the coil is increased, the coil is rotated through 1.5 revolutions, and the external source of the magnetic field is turned off, in that order. Rank the emf induced in the coil at the instants marked A through E from the largest positive value to the largest-magnitude negative value. In your ranking, note any cases of equality and also any instants when the emf is zero.

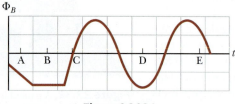

Figure OQ23.1

2. A long, fine wire is wound into a coil with inductance 5 mH. The coil is connected across the terminals of a battery, and

the current is measured a few seconds after the connection is made. The wire is unwound and wound again into a different coil with $L = 10$ mH. This second coil is connected across the same battery, and the current is measured in the same way. Compared with the current in the first coil, is the current in the second coil (a) four times as large, (b) twice as large, (c) unchanged, (d) half as large, or (e) one-fourth as large?

3. Two solenoids, A and B, are wound using equal lengths of the same kind of wire. The length of the axis of each solenoid is large compared with its diameter. The axial length of A is twice as large as that of B, and A has twice as many turns as B. What is the ratio of the inductance of solenoid A to that of solenoid B? (a) 4 (b) 2 (c) 1 (d) $\frac{1}{2}$ (e) $\frac{1}{4}$

4. A circular loop of wire with a radius of 4.0 cm is in a uniform magnetic field of magnitude 0.060 T. The plane of the loop is perpendicular to the direction of the magnetic field. In a time interval of 0.50 s, the magnetic field changes to

the opposite direction with a magnitude of 0.040 T. What is the magnitude of the average emf induced in the loop? (a) 0.20 V (b) 0.025 V (c) 5.0 mV (d) 1.0 mV (e) 0.20 mV

5. A rectangular conducting loop is placed near a long wire carrying a current I as shown in Figure OQ23.5. If I decreases in time, what can be said of the current induced in the loop? (a) The direction of the current depends on the size of the loop. (b) The current is clockwise. (c) The current is counterclockwise. (d) The current is zero. (e) Nothing can be said about the current in the loop without more information.

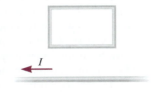

Figure OQ23.5

6. A flat coil of wire is placed in a uniform magnetic field that is in the y direction. **(i)** The magnetic flux through the coil is a maximum if the plane of the coil is where? More than one answer may be correct. (a) in the xy plane (b) in the yz plane (c) in the xz plane (d) in any orientation, because it is a constant **(ii)** For what orientation is the flux zero? Choose from the same possibilities as in part (i).

7. A solenoidal inductor for a printed circuit board is being redesigned. To save weight, the number of turns is reduced by one-half, with the geometric dimensions kept the same. By how much must the current change if the energy stored in the inductor is to remain the same? (a) It must be four times larger. (b) It must be two times larger. (c) It should be left the same. (d) It should be one-half as large. (e) No change in the current can compensate for the reduction in the number of turns.

8. If the current in an inductor is doubled, by what factor is the stored energy multiplied? (a) 4 (b) 2 (c) 1 (d) $\frac{1}{2}$ (e) $\frac{1}{4}$

9. A square, flat loop of wire is pulled at constant velocity through a region of uniform magnetic field directed perpendicular to the plane of the loop as shown in Figure OQ23.9. Which of the following statements are correct? More than one statement may be correct. (a) Current is induced in the loop in the clockwise direction. (b) Current is induced in the loop in the counterclockwise direction. (c) No current is induced in the loop. (d) Charge separation occurs in the loop, with the top edge positive. (e) Charge separation occurs in the loop, with the top edge negative.

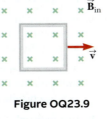

Figure OQ23.9

10. The bar in Figure OQ23.10 moves on rails to the right with a velocity $\vec{v}$, and a uniform, constant magnetic field is directed out of the page. Which of the following statements are correct? More than one statement may be correct. (a) The induced current in the loop is zero. (b) The induced current in the loop is clockwise. (c) The induced current in the loop is counterclockwise. (d) An external force is required to keep the bar moving at constant speed. (e) No force is required to keep the bar moving at constant speed.

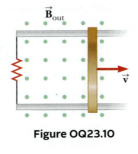

Figure OQ23.10

11. Initially, an inductor with no resistance carries a constant current. Then the current is brought to a new constant value twice as large. *After* this change, when the current is constant at its higher value, what has happened to the emf in the inductor? (a) It is larger than before the change by a factor of 4. (b) It is larger by a factor of 2. (c) It has the same nonzero value. (d) It continues to be zero. (e) It has decreased.

12. In Figure OQ23.12, the switch is left in position a for a long time interval and is then quickly thrown to position b. Rank the magnitudes of the voltages across the four circuit elements a short time thereafter from the largest to the smallest.

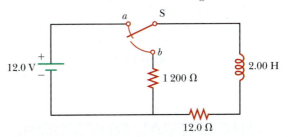

Figure OQ23.12

13. A bar magnet is held in a vertical orientation above a loop of wire that lies in the horizontal plane as shown in Figure OQ23.13. The south end of the magnet is toward the loop. After the magnet is dropped, what is true of the induced current in the loop as viewed from above? (a) It is clockwise as the magnet falls toward the loop. (b) It is counterclockwise as the magnet falls toward the loop. (c) It is clockwise after the magnet has moved through the loop and moves away from it. (d) It is always clockwise. (e) It is first counterclockwise as the magnet approaches the loop and then clockwise after it has passed through the loop.

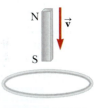

Figure OQ23.13

14. What happens to the amplitude of the induced emf when the rate of rotation of a generator coil is doubled? (a) It becomes four times larger. (b) It becomes two times larger. (c) It is unchanged. (d) It becomes one-half as large. (e) It becomes one-fourth as large.

15. Two coils are placed near each other as shown in Figure OQ23.15. The coil on the left is connected to a battery and a switch, and the coil on the right is connected to a resistor. What is the direction of the current in the resistor **(i)** at an instant immediately after the switch is thrown closed, **(ii)** after the switch has been closed for several seconds, and **(iii)** at an instant after the switch has then been thrown open? Choose each answer from the possibilities (a) left, (b) right, or (c) the current is zero.

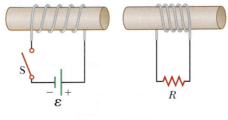

Figure OQ23.15

16. A circuit consists of a conducting movable bar and a light bulb connected to two conducting rails as shown in

Figure OQ23.16. An external magnetic field is directed perpendicular to the plane of the circuit. Which of the following actions will make the bulb light up? More than one statement may be correct. (a) The bar is moved to the left. (b) The bar is moved to the right. (c) The magnitude of the magnetic field is increased. (d) The magnitude of the magnetic field is decreased. (e) The bar is lifted off the rails.

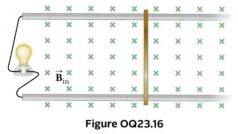

Figure OQ23.16

17. Two rectangular loops of wire lie in the same plane as shown in Figure OQ23.17. If the current I in the outer loop is counterclockwise and increases with time, what is true of the current induced in the inner loop? More than one statement may be correct. (a) It is zero. (b) It is clockwise. (c) It is counterclockwise. (d) Its magnitude depends on the dimensions of the loops. (e) Its direction depends on the dimensions of the loops.

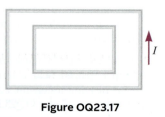

Figure OQ23.17

CONCEPTUAL QUESTIONS

☐ denotes answer available in *Student Solutions Manual/Study Guide*

1. A switch controls the current in a circuit that has a large inductance. The electric arc at the switch (Fig. CQ23.1) can melt and oxidize the contact surfaces, resulting in high resistivity of the contacts and eventual destruction of the switch. Is a spark more likely to be produced at the switch when the switch is being closed, when it is being opened, or does it not matter?

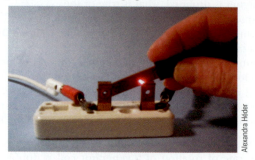

Figure CQ23.1

2. Consider the four circuits shown in Figure CQ23.2, each consisting of a battery, a switch, a lightbulb, a resistor, and either a capacitor or an inductor. Assume the capacitor has a large capacitance and the inductor has a large inductance but no resistance. The lightbulb has high efficiency, glowing whenever it carries electric current. **(i)** Describe what the lightbulb does in each of circuits (a) through (d) after the switch is thrown

closed. **(ii)** Describe what the lightbulb does in each of circuits (a) through (d) when, having been closed for a long time interval, the switch is opened.

3. What is the difference between magnetic flux and magnetic field?

4. Discuss the similarities between the energy stored in the electric field of a charged capacitor and the energy stored in the magnetic field of a current-carrying coil.

5. A spacecraft orbiting the Earth has a coil of wire in it. An astronaut measures a small current in the coil, although there is no battery connected to it and there are no magnets in the spacecraft. What is causing the current?

6. A circular loop of wire is located in a uniform and constant magnetic field. Describe how an emf can be induced in the loop in this situation.

7. A bar magnet is dropped toward a conducting ring lying on the floor. As the magnet falls toward the ring, does it move as a freely falling object? Explain.

8. In a hydroelectric dam, how is energy produced that is then transferred out by electrical transmission? That is, how is the energy of motion of the water converted to energy that is transmitted by AC electricity?

9. A piece of aluminum is dropped vertically downward between the poles of an electromagnet. Does the magnetic field affect the velocity of the aluminum?

10. The current in a circuit containing a coil, a resistor, and a battery has reached a constant value. (a) Does the coil have an inductance? (b) Does the coil affect the value of the current?

11. (a) What parameters affect the inductance of a coil? (b) Does the inductance of a coil depend on the current in the coil?

12. In Section 6.7, we defined conservative and nonconservative forces. In Chapter 19, we stated that an electric charge creates an electric field that produces a conservative force. Argue now that induction creates an electric field that produces a nonconservative force.

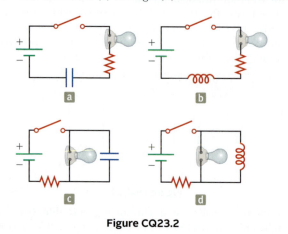

Figure CQ23.2

13. When the switch in Figure CQ23.13a is closed, a current is set up in the coil and the metal ring springs upward (Fig. CQ23.13b). Explain this behavior.

14. Assume the battery in Figure CQ23.13a is replaced by an AC source and the switch is held closed. If held down, the metal ring on top of the solenoid becomes hot. Why?

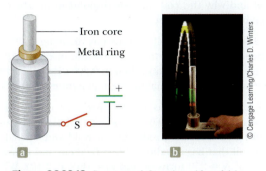

Figure CQ23.13 Conceptual Questions 13 and 14.

15. A loop of wire is moving near a long, straight wire carrying a constant current I as shown in Figure CQ23.15. (a) Determine the direction of the induced current in the loop as it moves away from the wire. (b) What would be the direction of the induced current in the loop if it were moving toward the wire?

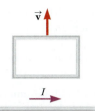

Figure CQ23.15

16. After the switch is closed in the LC circuit shown in Figure CQ23.16, the charge on the capacitor is sometimes zero, but at such instants the current in the circuit is not zero. How is this behavior possible?

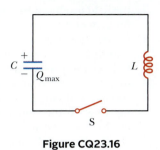

Figure CQ23.16

❯ PROBLEMS

Section 23.1 Faraday's Law of Induction

1. **W** A 30-turn circular coil of radius 4.00 cm and resistance 1.00 Ω is placed in a magnetic field directed perpendicular to the plane of the coil. The magnitude of the magnetic field varies in time according to the expression $B = 0.010\ 0t + 0.040\ 0t^2$, where B is in teslas and t is in seconds. Calculate the induced emf in the coil at $t = 5.00$ s.

2. An instrument based on induced emf has been used to measure projectile speeds up to 6 km/s. A small magnet is imbedded in the projectile as shown in Figure P23.2. The projectile passes through two coils separated by a distance d. As the projectile passes through each coil, a pulse of emf is induced in the coil. The time interval between pulses can be measured accurately with an oscilloscope, and thus the speed can be determined. (a) Sketch a graph of ΔV versus t for the arrangement shown. Consider a current that flows counterclockwise as viewed from the starting point of the projectile as positive. On your graph, indicate which pulse is from coil 1 and which is from coil 2. (b) If the pulse separation is 2.40 ms and $d = 1.50$ m, what is the projectile speed?

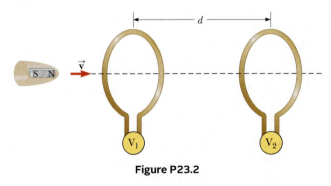

Figure P23.2

3. A flat loop of wire consisting of a single turn of cross-sectional area 8.00 cm² is perpendicular to a magnetic field that increases uniformly in magnitude from 0.500 T to 2.50 T in 1.00 s. What is the resulting induced current if the loop has a resistance of 2.00 Ω?

4. **BIO** Scientific work is currently underway to determine whether weak oscillating magnetic fields can affect human health. For example, one study found that drivers of trains had a higher incidence of blood cancer than other railway workers, possibly due to long exposure to mechanical devices in the train engine cab. Consider a magnetic field of magnitude 1.00×10^{-3} T, oscillating sinusoidally at 60.0 Hz. If the diameter of a red blood cell is 8.00 μm, determine the maximum emf that can be generated around the perimeter of a cell in this field.

5. **M** An aluminum ring of radius $r_1 = 5.00$ cm and resistance 3.00×10^{-4} Ω is placed around one end of a long air-core solenoid with 1 000 turns per meter and radius $r_2 = 3.00$ cm as shown in Figure P23.5. Assume the axial component of the field produced by the solenoid is one-half as strong over the area of the end of the solenoid as at the center of the solenoid. Also assume the solenoid produces negligible field outside its cross-sectional area. The current in the solenoid is increasing at a rate of 270 A/s. (a) What is the induced current in the ring? At the center of the ring, what are (b) the magnitude and (c) the direction of the magnetic field produced by the induced current in the ring?

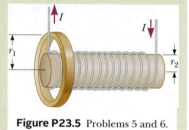

Figure P23.5 Problems 5 and 6.

6. **S** An aluminum ring of radius r_1 and resistance R is placed around one end of a long aircore solenoid with n turns per meter and smaller radius r_2 as shown in Figure P23.5. Assume the axial component of the field produced by the solenoid over the area of the end of the solenoid is one-half as strong as at the center of the solenoid. Also assume the solenoid produces negligible field outside its cross-sectional area. The current in the solenoid is increasing at a rate of $\Delta I/\Delta t$. (a) What is the induced current in the ring? (b) At the center of the ring, what is the magnetic field produced by the induced current in the ring? (c) What is the direction of this field?

7. **W** A loop of wire in the shape of a rectangle of width w and length L and a long, straight wire carrying a current I lie on a tabletop as shown in Figure P23.7. (a) Determine the magnetic flux through the loop due to the current I. (b) Suppose the current is changing with time according to $I = a + bt$, where a and b are constants. Determine the emf that is induced in the loop if $b = 10.0$ A/s, $h = 1.00$ cm, $w = 10.0$ cm, and $L = 1.00$ m. (c) What is the direction of the induced current in the rectangle?

Figure P23.7

8. **QC** **S** When a wire carries an AC current with a known frequency, you can use a *Rogowski coil* to determine the amplitude I_{max} of the current without disconnecting the wire to shunt the current through a meter. The Rogowski coil, shown in Figure P23.8, simply clips around the wire. It consists of a toroidal conductor wrapped

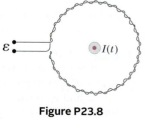

Figure P23.8

around a circular return cord. Let n represent the number of turns in the toroid per unit distance along it. Let A represent the cross-sectional area of the toroid. Let $I(t) = I_{max} \sin \omega t$ represent the current to be measured. (a) Show that the amplitude of the emf induced in the Rogowski coil is $\mathcal{E}_{max} = \mu_0 n A \omega I_{max}$. (b) Explain why the wire carrying the unknown current need not be at the center of the Rogowski coil and why the coil will not respond to nearby currents that it does not enclose.

9. A 25-turn circular coil of wire has diameter 1.00 m. It is placed with its axis along the direction of the Earth's magnetic field of 50.0 μT and then in 0.200 s is flipped 180°. An average emf of what magnitude is generated in the coil?

10. **M** A long solenoid has $n = 400$ turns per meter and carries a current given by $I = 30.0 (1 - e^{-1.60t})$, where I is in amperes and t is in seconds. Inside the solenoid and coaxial with it is a coil that has a radius of $R = 6.00$ cm and consists of a total of $N = 250$ turns of fine wire (Fig. P23.10). What emf is induced in the coil by the changing current?

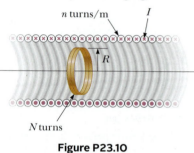

Figure P23.10

11. **W** A strong electromagnet produces a uniform magnetic field of 1.60 T over a cross-sectional area of 0.200 m^2. A coil having 200 turns and a total resistance of 20.0 Ω is placed around the electromagnet. The current in the electromagnet is then smoothly reduced until it reaches zero in 20.0 ms. What is the current induced in the coil?

12. A piece of insulated wire is shaped into a figure eight as shown in Figure P23.12. For simplicity, model the two halves of the figure eight as circles. The radius of the upper circle is 5.00 cm and that of the lower circle is 9.00 cm. The wire has a uniform resistance per unit length of 3.00 Ω/m. A uniform magnetic field is applied perpendicular to the plane of the two circles, in the direction shown. The magnetic field is increasing at a constant rate of 2.00 T/s. Find (a) the magnitude and (b) the direction of the induced current in the wire.

Figure P23.12

13. **W** A coil of 15 turns and radius 10.0 cm surrounds a long solenoid of radius 2.00 cm and 1.00×10^3 turns/meter (Fig. P23.13). The current in the solenoid changes as $I = 5.00 \sin 120t$, where I is in amperes and t is in seconds. Find the induced emf in the 15-turn coil as a function of time.

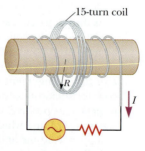

Figure P23.13

Section 23.2 Motional emf

Section 23.3 Lenz's Law

Note: Problem 73 in Chapter 22 can be assigned with this section.

14. A helicopter (Fig. P23.14) has blades of length 3.00 m, extending out from a central hub and rotating at 2.00 rev/s. If the vertical component of the Earth's magnetic field is 50.0 μT, what is the emf induced between the blade tip and the center hub?

Figure P23.14

Sascha Hahn/Shutterstock.com

15. **M** Figure P23.15 shows a top view of a bar that can slide on two frictionless rails. The resistor is $R = 6.00\ \Omega$, and a 2.50-T magnetic field is directed perpendicularly downward, into the paper. Let $\ell = 1.20$ m. (a) Calculate the applied force required to move the bar to the right at a constant speed of 2.00 m/s. (b) At what rate is energy delivered to the resistor?

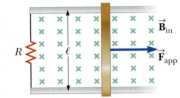

Figure P23.15 Problems 15 through 18.

16. Consider the arrangement shown in Figure P23.15. Assume that $R = 6.00\ \Omega$, $\ell = 1.20$ m, and a uniform 2.50-T magnetic field is directed into the page. At what speed should the bar be moved to produce a current of 0.500 A in the resistor?

17. A conducting rod of length ℓ moves on two horizontal, frictionless rails as shown in Figure P23.15. If a constant force of 1.00 N moves the bar at 2.00 m/s through a magnetic field $\vec{\mathbf{B}}$ that is directed into the page, (a) what is the current through the 8.00-Ω resistor R? (b) What is the rate at which energy is delivered to the resistor? (c) What is the mechanical power delivered by the force $\vec{\mathbf{F}}_{app}$?

18. **S** A metal rod of mass m slides without friction along two parallel horizontal rails, separated by a distance ℓ and connected by a resistor R, as shown in Figure P23.15. A uniform vertical magnetic field of magnitude B is applied perpendicular to the plane of the paper. The applied force shown in the figure acts only for a moment, to give the rod a speed v. In terms of m, ℓ, R, B, and v, find the distance the rod will then slide as it coasts to a stop.

19. **Review.** After removing one string while restringing his acoustic guitar, a student is distracted by a video game. His experimentalist roommate notices his inattention and attaches one end of the string, of linear density $\mu = 3.00 \times 10^{-3}$ kg/m, to a rigid support. The other end passes over a pulley, a distance $\ell = 64.0$ cm from the fixed end, and an object of mass $m = 27.2$ kg is attached to the hanging end of the string. The roommate places a magnet across the string as shown in Figure P23.19. The magnet does not touch the string, but produces a uniform field of 4.50 mT over a 2.00-cm length of the string and negligible field elsewhere. Strumming the string sets it vibrating vertically at its fundamental (lowest) frequency. The section of the string in the magnetic field

moves perpendicular to the field with a uniform amplitude of 1.50 cm. Find (a) the frequency and (b) the amplitude of the emf induced between the ends of the string.

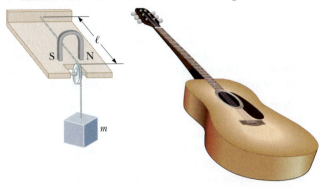

Figure P23.19

20. *Why is the following situation impossible?* An automobile has a vertical radio antenna of length $\ell = 1.20$ m. The automobile travels on a curvy, horizontal road where the Earth's magnetic field has a magnitude of $B = 50.0\ \mu$T and is directed toward the north and downward at an angle of $\theta = 65.0°$ below the horizontal. The motional emf developed between the top and bottom of the antenna varies with the speed and direction of the automobile's travel and has a maximum value of 4.50 mV.

21. The *homopolar generator,* also called the *Faraday disk,* is a low-voltage, high-current electric generator. It consists of a rotating conducting disk with one stationary brush (a sliding electrical contact) at its axle and another at a point on its circumference as shown in Figure P23.21. A uniform magnetic field is applied perpendicular to the plane of the disk. Assume the field is 0.900 T, the angular speed is 3.20×10^3 rev/min, and the radius of the disk is 0.400 m. Find the emf generated between the brushes. When superconducting coils are used to produce a large magnetic field, a homopolar generator can have a power output of several megawatts. Such a generator is useful, for example, in purifying metals by electrolysis. If a voltage is applied to the output terminals of the generator, it runs in reverse as a *homopolar motor* capable of providing great torque, useful in ship propulsion.

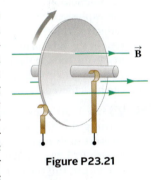

Figure P23.21

22. **S** A rectangular coil with resistance R has N turns, each of length ℓ and width w as shown in Figure P23.22. The coil moves into a uniform magnetic field $\vec{\mathbf{B}}$ with constant velocity $\vec{\mathbf{v}}$. What are the magnitude and direction of the total magnetic force on the coil (a) as it enters the magnetic field, (b) as it moves within the field, and (c) as it leaves the field?

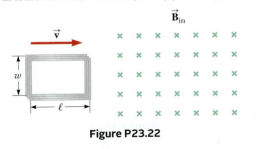

Figure P23.22

23. A long solenoid, with its axis along the x axis, consists of 200 turns per meter of wire that carries a steady current of 15.0 A. A coil is formed by wrapping 30 turns of thin wire around a circular frame that has a radius of 8.00 cm. The coil is placed inside the solenoid and mounted on an axis that is a diameter of the coil and coincides with the y axis. The coil is then rotated with an angular speed of 4.00π rad/s. The plane of the coil is in the yz plane at $t = 0$. Determine the emf generated in the coil as a function of time.

24. *Why is the following situation impossible?* A conducting rectangular loop of mass $M = 0.100$ kg, resistance $R = 1.00$ Ω, and dimensions $w = 50.0$ cm by $\ell = 90.0$ cm is held with its lower edge just above a region with a uniform magnetic field of magnitude $B = 1.00$ T as shown in Figure P23.24. The loop is released from rest. Just as the top edge of the loop reaches the region containing the field, the loop moves with a speed 4.00 m/s.

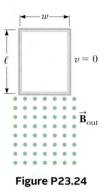

Figure P23.24

25. Very large magnetic fields can be produced using a procedure called *flux compression*. A metallic cylindrical tube of radius R is placed coaxially in a long solenoid of somewhat larger radius. The space between the tube and the solenoid is filled with a highly explosive material. When the explosive is set off, it collapses the tube to a cylinder of radius $r < R$. If the collapse happens very rapidly, induced current in the tube maintains the magnetic flux nearly constant inside the tube. If the initial magnetic field in the solenoid is 2.50 T and $R/r = 12.0$, what maximum value of magnetic field can be achieved?

26. W Use Lenz's law to answer the following questions concerning the direction of induced currents. Express your answers in terms of the letter labels a and b in each part of Figure P23.26. (a) What is the direction of the induced current in the resistor R in Figure P23.26a when the bar magnet is moved to the left? (b) What is the direction of the current induced in the resistor R immediately after the switch S in Figure P23.26b is closed? (c) What is the direction of the induced current in the resistor R when the current I in Figure P23.26c decreases rapidly to zero?

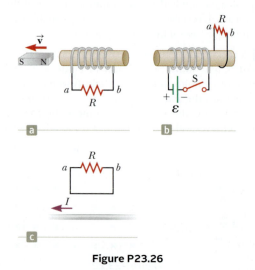

Figure P23.26

27. A coil of area 0.100 m² is rotating at 60.0 rev/s with the axis of rotation perpendicular to a 0.200-T magnetic field. (a) If the coil has 1 000 turns, what is the maximum emf generated in it? (b) What is the orientation of the coil with respect to the magnetic field when the maximum induced voltage occurs?

Section 23.4 Induced emfs and Electric Fields

28. M A magnetic field directed into the page changes with time according to $B = 0.030\ 0t^2 + 1.40$, where B is in teslas and t is in seconds. The field has a circular cross section of radius $R = 2.50$ cm (see Fig. P23.28). When $t = 3.00$ s and $r_2 = 0.020\ 0$ m, what are (a) the magnitude and (b) the direction of the electric field at point P_2?

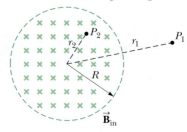

Figure P23.28 Problems 28 and 29.

29. W Within the green dashed circle shown in Figure P23.28, the magnetic field changes with time according to the expression $B = 2.00t^3 - 4.00t^2 + 0.800$, where B is in teslas, t is in seconds, and $R = 2.50$ cm. When $t = 2.00$ s, calculate (a) the magnitude and (b) the direction of the force exerted on an electron located at point P_1, which is at a distance $r_1 = 5.00$ cm from the center of the circular field region. (c) At what instant is this force equal to zero?

Section 23.5 Inductance

30. A coiled telephone cord forms a spiral with 70 turns, a diameter of 1.30 cm, and an unstretched length of 60.0 cm. Determine the self-inductance of one conductor in the unstretched cord.

31. A coil has an inductance of 3.00 mH, and the current in it changes from 0.200 A to 1.50 A in a time interval of 0.200 s. Find the magnitude of the average induced emf in the coil during this time interval.

32. S A toroid has a major radius R and a minor radius r and is tightly wound with N turns of wire on a hollow cardboard torus. Figure P23.32 shows half of this toroid, allowing us to see its cross section. If $R \gg r$, the magnetic field in the region enclosed by the wire is essentially the same as the magnetic field of a solenoid that has been bent into a large circle of radius R. Modeling the field as the uniform field of a long solenoid, show that the inductance of such a toroid is approximately

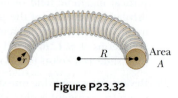

Figure P23.32

$$L \approx \tfrac{1}{2}\mu_0 N^2 \frac{r^2}{R}$$

33. M A 10.0-mH inductor carries a current $I = I_{max} \sin \omega t$, with $I_{max} = 5.00$ A and $f = \omega/2\pi = 60.0$ Hz. What is the self-induced emf as a function of time?

34. An emf of 24.0 mV is induced in a 500-turn coil when the current is changing at the rate of 10.0 A/s. What is the magnetic flux through each turn of the coil at an instant when the current is 4.00 A?

35. W The current in a 90.0-mH inductor changes with time as $I = 1.00t^2 - 6.00t$, where I is in amperes and t is in seconds. Find the magnitude of the induced emf at (a) $t = 1.00$ s and (b) $t = 4.00$ s. (c) At what time is the emf zero?

36. M An inductor in the form of a solenoid contains 420 turns and is 16.0 cm in length. A uniform rate of decrease of current through the inductor of 0.421 A/s induces an emf of 175 μV. What is the radius of the solenoid?

Section 23.6 *RL* Circuits

37. M A 12.0-V battery is connected into a series circuit containing a 10.0-Ω resistor and a 2.00-H inductor. In what time interval will the current reach (a) 50.0% and (b) 90.0% of its final value?

38. S Show that $I = I_i e^{-t/\tau}$ is a solution of the differential equation

$$IR + L\frac{dI}{dt} = 0$$

where I_i is the current at $t = 0$ and $\tau = L/R$.

39. Consider the circuit in Figure P23.39, taking $\mathcal{E} = 6.00$ V, $L = 8.00$ mH, and $R = 4.00$ Ω. (a) What is the inductive time constant of the circuit? (b) Calculate the current in the circuit 250 μs after the switch is closed. (c) What is the value of the final steady-state current? (d) After what time interval does the current reach 80.0% of its maximum value?

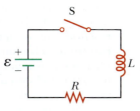

Figure P23.39 Problems 39, 40, 42 and 45.

40. For the *RL* circuit shown in Figure P23.39, let the inductance be 3.00 H, the resistance 8.00 Ω, and the battery emf 36.0 V. (a) Calculate $\Delta V_R/\mathcal{E}_L$, that is, the ratio of the potential difference across the resistor to the emf across the inductor when the current is 2.00 A. (b) Calculate the emf across the inductor when the current is 4.50 A.

41. A circuit consists of a coil, a switch, and a battery, all in series. The internal resistance of the battery is negligible compared with that of the coil. The switch is originally open. It is thrown closed, and after a time interval Δt, the current in the circuit reaches 80.0% of its final value. The switch remains closed for a time interval much longer than Δt. Then the battery is disconnected and the terminals of the coil are connected together to form a short circuit. (a) After an equal additional time interval Δt elapses, the current is what percentage of its maximum value? (b) At the moment $2\Delta t$ after the coil is short-circuited, the current in the coil is what percentage of its maximum value?

42. When the switch in Figure P23.39 is closed, the current takes 3.00 ms to reach 98.0% of its final value. If $R = 10.0$ Ω, what is the inductance?

43. The switch in Figure P23.43 is open for $t < 0$ and is then thrown closed at time $t = 0$. Assume $R = 4.00$ Ω, $L = 1.00$ H, and $\mathcal{E} = 10.0$ V. Find (a) the current in the inductor and (b) the current in the switch as functions of time thereafter.

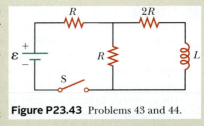

Figure P23.43 Problems 43 and 44.

44. S The switch in Figure P23.43 is open for $t < 0$ and is then thrown closed at time $t = 0$. Find (a) the current in the inductor and (b) the current in the switch as functions of time thereafter.

45. W In the circuit shown in Figure P23.39, let $L = 7.00$ H, $R = 9.00$ Ω, and $\mathcal{E} = 120$ V. What is the self-induced emf 0.200 s after the switch is closed?

46. One application of an *RL* circuit is the generation of time-varying high voltage from a low-voltage source as shown in Figure P23.46. (a) What is the current in the circuit a long time after the switch has been in position a? (b) Now the switch is thrown quickly from a to b. Compute the initial voltage across each resistor and across the inductor. (c) How much time elapses before the voltage across the inductor drops to 12.0 V?

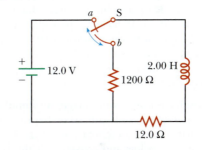

Figure P23.46

47. M A 140-mH inductor and a 4.90-Ω resistor are connected with a switch to a 6.00-V battery as shown in Figure P23.47. (a) After the switch is first thrown to a (connecting the battery), what time interval elapses before the current reaches 220 mA? (b) What is the current in the inductor 10.0 s after the switch is closed? (c) Now the switch is quickly thrown from a to b. What time interval elapses before the current in the inductor falls to 160 mA?

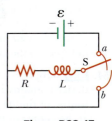

Figure P23.47

Section 23.7 Energy Stored in a Magnetic Field

48. The magnetic field inside a superconducting solenoid is 4.50 T. The solenoid has an inner diameter of 6.20 cm and a length of 26.0 cm. Determine (a) the magnetic energy density in the field and (b) the energy stored in the magnetic field within the solenoid.

49. M An air-core solenoid with 68 turns is 8.00 cm long and has a diameter of 1.20 cm. When the solenoid carries a current of 0.770 A, how much energy is stored in its magnetic field?

50. Q|C W A flat coil of wire has an inductance of 40.0 mH and a resistance of 5.00 Ω. It is connected to a 22.0-V battery at the instant $t = 0$. Consider the moment when the current is 3.00 A. (a) At what rate is energy being delivered by the battery? (b) What is the power being delivered to the resistance of the coil? (c) At what rate is energy being stored in the magnetic field of the coil? (d) What is the relationship among these three power values? (e) Is the relationship described in part (d) true at other instants as well? (f) Explain the relationship at the moment immediately after $t = 0$ and at a moment several seconds later.

51. On a clear day at a certain location, a 100-V/m vertical electric field exists near the Earth's surface. At the same place, the Earth's magnetic field has a magnitude of 0.500×10^{-4} T. Compute the energy densities of (a) the electric field and (b) the magnetic field.

Section 23.8 Context Connection: The Use of Transcranial Magnetic Stimulation in Depression

52. BIO Transcranial magnetic stimulation (TMS) is a noninvasive technique used to stimulate regions of the human brain. In TMS, a small coil is placed on the scalp and a brief burst of current in the coil produces a rapidly changing magnetic field inside the brain. The induced emf can stimulate neuronal activity. (a) One such device generates an upward magnetic field within the brain that rises from zero to 1.50 T in 120 ms. Determine the induced emf around a horizontal circle of tissue of radius 1.60 mm. (b) **What If?** The field next changes to 0.500 T downward in 80.0 ms. How does the emf induced in this process compare with that in part (a)?

53. BIO Consider a transcranial magnetic stimulation (TMS) device containing a coil with several turns of wire, each of radius 6.00 cm. In a circular area of the brain of radius 6.00 cm directly below and coaxial with the coil, the magnetic field changes at the rate of 1.00×10^4 T/s. Assume that this rate of change is the same everywhere inside the circular area. (a) What is the emf induced around the circumference of this circular area in the brain? (b) What electric field is induced on the circumference of this circular area?

54. BIO Suppose a transcranial magnetic stimulation (TMS) device contains a single coil of wire of radius 6.00 cm as the source of the magnetic field. A typical peak magnetic field at the center of the coil in such a device is 1.50 T. (a) If the field at the center of the coil has this peak value, what current exists in the coil? (b) The treatment area in the brain is about 2.50 cm below the center of the coil. If the current in the coil is changing at the rate of 1.00×10^7 A/s, what is the rate of change of the magnetic field at the treatment area and along the axis of the coil? (c) Some patients are advised against TMS treatment if they have magnetic-sensitive metals in their heads, such as aneurysm clips, stents, or bullet fragments. In addition, patients with electronic devices may also be advised against TMS: cochlear implants, pacemakers, cardioverter defibrillators, insulin pumps, and vagus nerve stimulators. The recommended magnetic field exclusion level for magnetic-sensitive metals and electronic devices is 5.00×10^{-4} T. For the current in part (a), at what distance from the coil, along the center axis, must these metals or devices be in order to experience a magnetic field

smaller than the exclusion level? (We will assume the magnetic field exists in empty space. Because of the effect of biological tissue, the exclusion level occurs in reality at about 30 cm from the coil.)

Additional Problems

55. A guitar's steel string vibrates (see Fig. 23.6). The component of magnetic field perpendicular to the area of a pickup coil nearby is given by

$$B = 50.0 + 3.20 \sin (1\,046\pi t)$$

where B is in milliteslas and t is in seconds. The circular pickup coil has 30 turns and radius 2.70 mm. Find the emf induced in the coil as a function of time.

56. GP Consider the apparatus shown in Figure P31.56 in which a conducting bar can be moved along two rails connected to a lightbulb. The whole system is immersed in a magnetic field of magnitude $B = 0.400$ T perpendicular and into the page. The distance between the horizontal rails is $\ell = 0.800$ m. The resistance of the lightbulb is $R = 48.0$ Ω, assumed to be constant. The bar and rails have negligible resistance. The bar is moved toward the right by a constant force of magnitude $F = 0.600$ N. We wish to find the maximum power delivered to the lightbulb. (a) Find an expression for the current in the lightbulb as a function of B, ℓ, R, and v, the speed of the bar. (b) When the maximum power is delivered to the lightbulb, what analysis model properly describes the moving bar? (c) Use the analysis model in part (b) to find a numerical value for the speed v of the bar when the maximum power is being delivered to the lightbulb. (d) Find the current in the lightbulb when maximum power is being delivered to it. (e) Using $P = I^2R$, what is the maximum power delivered to the lightbulb? (f) What is the maximum mechanical input power delivered to the bar by the force F? (g) We have assumed the resistance of the lightbulb is constant. In reality, as the power delivered to the lightbulb increases, the filament temperature increases and the resistance increases. Does the speed found in part (c) change if the resistance increases and all other quantities are held constant? (h) If so, does the speed found in part (c) increase or decrease? If not, explain. (i) With the assumption that the resistance of the lightbulb increases as the current increases, does the power found in part (f) change? (j) If so, is the power found in part (f) larger or smaller? If not, explain.

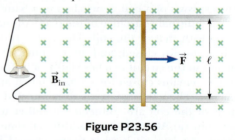

Figure P23.56

57. BIO Strong magnetic fields are used in such medical procedures as magnetic resonance imaging, or MRI. A technician wearing a brass bracelet enclosing area 0.005 00 m² places her hand in a solenoid whose magnetic field is 5.00 T directed perpendicular to the plane of the bracelet. The electrical resistance around the bracelet's circumference is

+Test

0.020 0 Ω. An unexpected power failure causes the field to drop to 1.50 T in a time interval of 20.0 ms. Find (a) the current induced in the bracelet and (b) the power delivered to the bracelet. *Note:* As this problem implies, you should not wear any metal objects when working in regions of strong magnetic fields.

58. Figure P23.58 is a graph of the induced emf versus time for a coil of N turns rotating with angular speed ω in a uniform magnetic field directed perpendicular to the coil's axis of rotation. **What If?** Copy this sketch (on a larger scale) and on the same set of axes show the graph of emf versus t (a) if the number of turns in the coil is doubled, (b) if instead the angular speed is doubled, and (c) if the angular speed is doubled while the number of turns in the coil is halved.

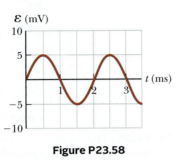

Figure P23.58

59. Suppose you wrap wire onto the core from a roll of cellophane tape to make a coil. Describe how you can use a bar magnet to produce an induced voltage in the coil. What is the order of magnitude of the emf you generate? State the quantities you take as data and their values.

60. **GP** **S** At $t = 0$, the open switch in Figure P23.60 is thrown closed. We wish to find a symbolic expression for the current in the inductor for time $t > 0$. Let this current be called I and choose it to be downward in the inductor in Figure P23.60. Identify I_1 as the current to the right through R_1 and I_2 as the current downward through R_2. (a) Use Kirchhoff's junction rule to find a relation among the three currents. (b) Use Kirchhoff's loop rule around the left loop to find another relationship. (c) Use Kirchhoff's loop rule around the outer loop to find a third relationship. (d) Eliminate I_1 and I_2 among the three equations to find an equation involving only the current I. (e) Compare the equation in part (d) with Equation 23.13 in the text. Use this comparison to rewrite Equation 23.14 in the text for the situation in this problem and show that

$$I(t) = \frac{\mathcal{E}}{R_1}[1 - e^{-(R'/L)t}]$$

where $R' = R_1 R_2/(R_1 + R_2)$.

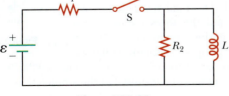

Figure P23.60

61. **M** The magnetic flux through a metal ring varies with time t according to $\Phi_B = at^3 - bt^2$, where Φ_B is in webers, $a = 6.00 \text{ s}^{-3}$, $b = 18.0 \text{ s}^{-2}$, and t is in seconds. The resistance of the ring is 3.00 Ω. For the interval from $t = 0$ to $t = 2.00$ s, determine the maximum current induced in the ring.

62. **Review.** A particle with a mass of 2.00×10^{-16} kg and a charge of 30.0 nC starts from rest, is accelerated through a

potential difference ΔV, and is fired from a small source in a region containing a uniform, constant magnetic field of magnitude 0.600 T. The particle's velocity is perpendicular to the magnetic field lines. The circular orbit of the particle as it returns to the location of the source encloses a magnetic flux of 15.0 μWb. (a) Calculate the particle's speed. (b) Calculate the potential difference through which the particle was accelerated inside the source.

63. **BIO** To monitor the breathing of a hospital patient, a thin belt is girded around the patient's chest. The belt is a 200-turn coil. When the patient inhales, the area encircled by the coil increases by 39.0 cm². The magnitude of the Earth's magnetic field is 50.0 μT and makes an angle of 28.0° with the plane of the coil. Assuming a patient takes 1.80 s to inhale, find the average induced emf in the coil during this time interval.

64. A *betatron* is a device that accelerates electrons to energies in the MeV range by means of electromagnetic induction. Electrons in a vacuum chamber are held in a circular orbit by a magnetic field perpendicular to the orbital plane. The magnetic field is gradually increased to induce an electric field around the orbit. (a) Show that the electric field is in the correct direction to make the electrons speed up. (b) Assume the radius of the orbit remains constant. Show that the average magnetic field over the area enclosed by the orbit must be twice as large as the magnetic field at the circle's circumference.

65. **M** A long, straight wire carries a current given by $I = I_{max} \sin (\omega t + \phi)$. The wire lies in the plane of a rectangular coil of N turns of wire as shown in Figure P23.65. The quantities I_{max}, ω, and ϕ are all constants. Assume $I_{max} = 50.0$ A, $\omega = 200\pi \text{ s}^{-1}$, $N = 100$, $h = w = 5.00$ cm, and $L = 20.0$ cm. Determine the emf induced in the coil by the magnetic field created by the current in the straight wire.

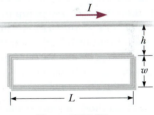

Figure P23.65

66. Figure P23.66 shows a stationary conductor whose shape is similar to the letter e. The radius of its circular portion is $a = 50.0$ cm. It is placed in a constant magnetic field of 0.500 T directed out of the page. A straight conducting rod, 50.0 cm long, is pivoted about point O and rotates with a constant angular speed of 2.00 rad/s. (a) Determine the induced emf in the loop POQ. Note that the area of the loop is $\theta a^2/2$. (b) If all the conducting material has a resistance per length of 5.00 Ω/m, what is the induced current in the loop POQ at the instant 0.250 s after point P passes point Q?

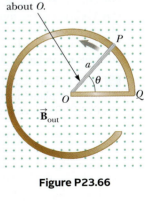

Figure P23.66

67. The toroid in Figure P23.67 consists of N turns and has a rectangular cross section. Its inner and outer radii are a and b, respectively. The figure shows half of the toroid to allow us to see its cross section. Compute the inductance of a 500-turn toroid for which $a = 10.0$ cm, $b = 12.0$ cm, and $h = 1.00$ cm.

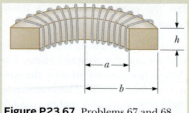

Figure P23.67 Problems 67 and 68.

68. **S** The toroid in Figure P23.67 consists of N turns and has a rectangular cross section. Its inner and outer radii are a and b, respectively. Find the inductance of the toroid.

69. (a) A flat, circular coil does not actually produce a uniform magnetic field in the area it encloses. Nevertheless, estimate the inductance of a flat, compact, circular coil with radius R and N turns by assuming the field at its center is uniform over its area. (b) A circuit on a laboratory table consists of a 1.50-volt battery, a 270-Ω resistor, a switch, and three 30.0-cm-long patch cords connecting them. Suppose the circuit is arranged to be circular. Think of it as a flat coil with one turn. Compute the order of magnitude of its inductance and (c) of the time constant describing how fast the current increases when you close the switch.

Review problems. Problems 70, 71, and 73 apply ideas from this and earlier chapters to some properties of superconductors, which were introduced in Section 21.3.

70. **Review.** In an experiment carried out by S. C. Collins between 1955 and 1958, a current was maintained in a superconducting lead ring for 2.50 yr with no observed loss, even though there was no energy input. If the inductance of the ring were 3.14×10^{-8} H and the sensitivity of the experiment were 1 part in 10^9, what was the maximum resistance of the ring? *Suggestion:* Treat the ring as an RL circuit carrying decaying current and recall that the approximation $e^{-x} \approx 1 - x$ is valid for small x.

71. **Review.** A novel method of storing energy has been proposed. A huge underground superconducting coil, 1.00 km in diameter, would be fabricated. It would carry a maximum current of 50.0 kA in each winding of a 150-turn Nb_3Sn solenoid. (a) If the inductance of this huge coil were 50.0 H, what would be the total energy stored? (b) What would be the compressive force per unit length acting between two adjacent windings 0.250 m apart?

72. **S** A bar of mass m and resistance R slides without friction in a horizontal plane, moving on parallel rails as shown in Figure P23.72. The rails are separated by a distance d. A battery that maintains a constant emf ε is connected between the rails, and a constant magnetic field $\vec{B}$ is directed perpendicularly out of the page. Assuming the bar starts from rest at time $t = 0$, show that at time t it moves with a speed

$$v = \frac{\varepsilon}{Bd}(1 - e^{-B^2 d^2 t / mR})$$

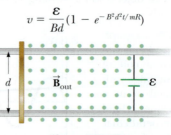

Figure P23.72

73. **Review.** The use of superconductors has been proposed for power transmission lines. A single coaxial cable (Fig. P23.73) could carry a power of 1.00×10^3 MW (the output of a large power plant) at 200 kV, DC, over a distance of 1.00×10^3 km without loss. An inner wire of

Figure P23.73

radius $a = 2.00$ cm, made from the superconductor Nb_3Sn, carries the current I in one direction. A surrounding superconducting cylinder of radius $b = 5.00$ cm would carry the return current I. In such a system, what is the magnetic field (a) at the surface of the inner conductor and (b) at the inner surface of the outer conductor? (c) How much energy would be stored in the magnetic field in the space between the conductors in a 1.00×10^3 km superconducting line? (d) What is the pressure exerted on the outer conductor due to the current in the inner conductor?

74. **S** Complete the calculation in Example 23.8 by proving that

$$\int_0^\infty e^{-2Rt/L}\, dt = \frac{L}{2R}$$

75. **M** The plane of a square loop of wire with edge length $a = 0.200$ m is oriented vertically and along an east–west axis. The Earth's magnetic field at this point is of magnitude $B = 35.0\ \mu T$ and is directed northward at 35.0° below the horizontal. The total resistance of the loop and the wires connecting it to a sensitive ammeter is 0.500 Ω. If the loop is suddenly collapsed by horizontal forces as shown in Figure P23.75, what total charge enters one terminal of the ammeter?

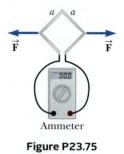

Figure P23.75

Nuclear Magnetic Resonance and Magnetic Resonance Imaging

In this Context Conclusion, we discuss an application that is now widely used as a noninvasive diagnostic tool in medical practice. This application is known as *MRI*, for *magnetic resonance imaging.*

In Section 22.11, we discussed the spin angular momentum of an electron and the associated magnetic moment of the electron. Spin is a general property of all particles. For example, protons and neutrons in the nucleus of an atom have spin and an associated magnetic dipole moment $\vec{\mu}$. As we saw in Section 22.6, the potential energy of a system consisting of a magnetic dipole moment in an external magnetic field is $U = -\vec{\mu} \cdot \vec{B}$.

When the magnetic moment $\vec{\mu}$ is aligned with the field as closely as quantum physics allows, the potential energy of the dipole–field system has its minimum value E_{min}. When $\vec{\mu}$ is as antiparallel to the field as possible, the potential energy has its maximum value E_{max}. Because the directions of the spin and the magnetic moment for a particle are quantized (see Chapter 29), the energies of the dipole–field system are also quantized. We introduced the concept of quantized states of energy in Chapter 11. In general, there are other allowed energy states between E_{min} and E_{max} corresponding to the quantized directions of the magnetic moment with respect to the field. These states are often called **spin states** because they differ in energy as a result of the direction of the spin.

The number of spin states depends on the spin of the nucleus. The simplest situation is shown in Figure 1, for a nucleus with only two possible spin states having energies E_{min} and E_{max}.

It is possible to observe transitions between these two spin states in a sample using a technique known as **nuclear magnetic resonance** (NMR). A constant magnetic field changes the energy associated with the spin states, splitting them apart in energy as shown in Figure 1. The sample is also exposed to electromagnetic waves in the radio range of the electromagnetic spectrum. When the frequency of the radio waves is adjusted such that the photon energy matches the separation energy between spin states, a resonance condition exists and the photon is absorbed by a nucleus in the ground state, raising the nucleus–magnetic field system to the higher energy spin state. This results in a net absorption of energy by the system, which is detected by the experimental control and measurement system. A diagram of the apparatus used to detect an NMR signal is illustrated in Figure 2 (page 818). The absorbed energy is supplied by the oscillator producing the radio waves. Nuclear magnetic resonance and a related technique called *electron spin resonance* are extremely important methods for studying nuclear and atomic systems and how these systems interact with their surroundings.

A widely used medical diagnostic technique called **MRI,** for **magnetic resonance imaging,** is based on nuclear magnetic resonance. In MRI, the patient is placed

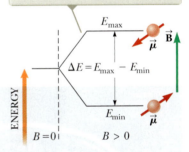

The magnetic field splits a single state of the nucleus into two states.

Figure 1 A nucleus with spin $\frac{1}{2}$ is placed in a magnetic field.

▶ Nuclear magnetic resonance

BIO Magnetic resonance imaging (MRI)

Electromagnetic Waves

Chapter Outline

NASA/CXC/SAO/F. Seward

Electromagnetic waves cover a broad spectrum of wavelengths, with waves in various wavelength ranges having distinct properties. This photo of the Crab Nebula is made with x-rays, which are like visible light, except for their very short wavelength. In this chapter, we shall study the common features of x-rays, visible light, and other forms of electromagnetic radiation.

Although we are not always aware of their presence, electromagnetic waves permeate our environment. In the form of visible light, they enable us to view the world around us with our eyes. Infrared waves from the surface of the Earth warm our environment, radio-frequency waves carry our favorite radio entertainment, microwaves cook our food and are used in radar communication systems, and the list goes on and on. The waves described in Chapter 13 are mechanical waves, which require a medium through which to propagate. Electromagnetic waves, in contrast, can propagate through a vacuum. Despite this difference between mechanical and electromagnetic waves, much of the behavior in the wave models of Chapters 13 and 14 is similar for electromagnetic waves.

The purpose of this chapter is to explore the properties of electromagnetic waves. The fundamental laws of electricity and magnetism—Maxwell's equations—form the basis of all electromagnetic phenomena. One of these equations predicts that a time-varying electric field produces a magnetic field just as a time-varying magnetic field produces an electric field. From this generalization, Maxwell provided the final important link between electric and magnetic fields. The most dramatic prediction of his equations is the existence of electromagnetic waves that propagate through empty space with the speed of light. This discovery led to many practical applications, such as communication by radio, television, and cellular telephone, and to the realization that light is one form of electromagnetic radiation.

24.1 | Displacement Current and the Generalized Form of Ampère's Law

We have seen that charges in motion, or currents, produce magnetic fields. When a current-carrying conductor has high symmetry, we can calculate the magnetic field using Ampère's law, given by Equation 22.29:

$$\oint \vec{B} \cdot d\vec{s} = \mu_0 I$$

where the line integral is over any closed path through which the conduction current passes and the conduction current is defined by $I = dq/dt$.

In this section, we shall use the term *conduction current* to refer to the type of current that we have already discussed, that is, current carried by charged particles in a wire. We use this term to differentiate this current from a different type of current we will introduce shortly. Ampère's law in this form is valid only if the conduction current is continuous in space. Maxwell recognized this limitation and modified Ampère's law to include all possible situations.

This limitation can be understood by considering a capacitor being charged as in Figure 24.1. When conduction current exists in the wires, the charge on the plates changes, but no conduction current exists between the plates. Consider the two surfaces S_1 (a circle, shown in blue) and S_2 (a paraboloid, in tan, passing between the plates) in Figure 24.1 bounded by the same path P. Ampère's law says that the line integral of $\vec{B} \cdot d\vec{s}$ around this path must equal $\mu_0 I$, where I is the conduction current through *any* surface bounded by the path P.

When the path P is considered as bounding S_1, the right-hand side of Equation 22.29 is $\mu_0 I$ because the conduction current passes through S_1 while the capacitor is charging. When the path bounds S_2, however, the right-hand side of Equation 22.29 is zero because no conduction current passes through S_2. Therefore, a contradictory situation arises because of the discontinuity of the current! Maxwell solved this problem by postulating an additional term on the right side of Equation 22.29, called the **displacement current** I_d, defined as

$$I_d \equiv \epsilon_0 \frac{d\Phi_E}{dt}$$

24.1 ◀   ▶ Displacement current

Recall that Φ_E is the flux of the electric field, defined as $\Phi_E \equiv \oint \vec{E} \cdot d\vec{A}$ (Eq. 19.14). (The word *displacement* here does not have the same meaning as in Chapter 2; it is historically entrenched in the language of physics, however, so we continue to use it.)

Equation 24.1 is interpreted as follows. As the capacitor is being charged (or discharged), the changing electric field between the plates may be considered as equivalent to a current between the plates that acts as a continuation of the conduction current in the wire. When the expression for the displacement current given by Equation 24.1 is added to the conduction current on the right side of Ampère's law, the difficulty represented in Figure 24.1 is resolved. No matter what surface bounded by the path P is chosen, either conduction current or displacement current passes through it. With this new notion of displacement current, we can express the general form of Ampère's law (sometimes called the **Ampère–Maxwell law**) as[1]

$$\oint \vec{B} \cdot d\vec{s} = \mu_0(I + I_d) = \mu_0 I + \mu_0 \epsilon_0 \frac{d\Phi_E}{dt}$$

24.2 ◀   ▶ Ampère–Maxwell law

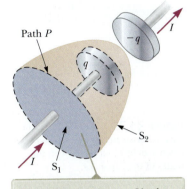

The conduction current I in the wire passes only through S_1, which leads to a contradiction in Ampère's law that is resolved only if one postulates a displacement current through S_2.

Figure 24.1 Two surfaces S_1 and S_2 near the plate of a capacitor are bounded by the same path P.

---

[1]Strictly speaking, this expression is valid only in a vacuum. If a magnetic material is present, a magnetizing current must also be included on the right side of Equation 24.2 to make Ampère's law fully general.

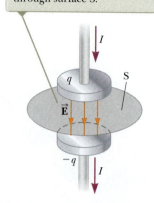

The electric field lines between the plates create an electric flux through surface S.

Figure 24.2 When a conduction current exists in the wires, a changing electric field $\vec{E}$ exists between the plates of the capacitor.

James Clerk Maxwell
Scottish Theoretical Physicist
(1831–1879)
Maxwell developed the electromagnetic theory of light and the kinetic theory of gases and explained the nature of Saturn's rings and color vision. Maxwell's successful interpretation of the electromagnetic field resulted in the field equations that bear his name. Formidable mathematical ability combined with great insight enabled him to lead the way in the study of electromagnetism and kinetic theory. He died of cancer before he was 50.

▶ Gauss's law

▶ Gauss's law in magnetism

▶ Faraday's law

▶ Ampère–Maxwell law

We can understand the meaning of this expression by referring to Figure 24.2. The electric flux through surface S is $\Phi_E = \int \vec{E} \cdot d\vec{A} = EA$, where A is the area of the capacitor plates and E is the magnitude of the uniform electric field between the plates. If q is the charge on the plates at any instant, then $E = \sigma/\epsilon_0 = q/(\epsilon_0 A)$ (see Example 19.12). Therefore, the electric flux through S is

$$\Phi_E = EA = \frac{q}{\epsilon_0}$$

Hence, the displacement current through S is

$$I_d = \epsilon_0 \frac{d\Phi_E}{dt} = \frac{dq}{dt} \qquad \textbf{24.3} \blacktriangleleft$$

That is, the displacement current I_d through S is precisely equal to the conduction current I in the wires connected to the capacitor!

By considering surface S, we can identify the displacement current as the source of the magnetic field on the surface boundary. The displacement current has its physical origin in the time-varying electric field. The central point of this formalism is that magnetic fields are produced *both* by conduction currents *and* by time-varying electric fields. This result was a remarkable example of theoretical work by Maxwell, and it contributed to major advances in the understanding of electromagnetism.

▶ **QUICK QUIZ 24.1** In an *RC* circuit, the capacitor begins to discharge. **(i)** During the discharge, in the region of space between the plates of the capacitor, is there (a) conduction current but no displacement current, (b) displacement current but no conduction current, (c) both conduction and displacement current, or (d) no current of any type? **(ii)** In the same region of space, is there (a) an electric field but no magnetic field, (b) a magnetic field but no electric field, (c) both electric and magnetic fields, or (d) no fields of any type?

◀24.2 | Maxwell's Equations and Hertz's Discoveries

We now present four equations that are regarded as the basis of all electrical and magnetic phenomena. These equations, developed by Maxwell, are as fundamental to electromagnetic phenomena as Newton's laws are to mechanical phenomena. In fact, the theory that Maxwell developed was more far-reaching than even he imagined because it turned out to be in agreement with the special theory of relativity, as Einstein showed in 1905.

Maxwell's equations represent the laws of electricity and magnetism that we have already discussed, but they have additional important consequences. For simplicity, we present **Maxwell's equations** as applied to free space, that is, in the absence of any dielectric or magnetic material. The four equations are

$$\oint \vec{E} \cdot d\vec{A} = \frac{q}{\epsilon_0} \qquad \textbf{24.4} \blacktriangleleft$$

$$\oint \vec{B} \cdot d\vec{A} = 0 \qquad \textbf{24.5} \blacktriangleleft$$

$$\oint \vec{E} \cdot d\vec{s} = -\frac{d\Phi_B}{dt} \qquad \textbf{24.6} \blacktriangleleft$$

$$\oint \vec{B} \cdot d\vec{s} = \mu_0 I + \epsilon_0 \mu_0 \frac{d\Phi_E}{dt} \qquad \textbf{24.7} \blacktriangleleft$$

Equation 24.4 is Gauss's law: the total electric flux through any closed surface equals the net charge inside that surface divided by ϵ_0. This law relates an electric field to the charge distribution that creates it.

Equation 24.5 is Gauss's law in magnetism, and it states that the net magnetic flux through a closed surface is zero. That is, the number of magnetic field lines that enter a closed volume must equal the number that leave that volume, which implies that magnetic field lines cannot begin or end at any point. If they did, it would mean that isolated magnetic monopoles existed at those points. That isolated magnetic monopoles have not been observed in nature can be taken as a confirmation of Equation 24.5.

Equation 24.6 is Faraday's law of induction, which describes the creation of an electric field by a changing magnetic flux. This law states that the emf, which is the line integral of the electric field around any closed path, equals the rate of change of magnetic flux through any surface bounded by that path. One consequence of Faraday's law is the current induced in a conducting loop placed in a time-varying magnetic field.

Equation 24.7 is the Ampère–Maxwell law, and it describes the creation of a magnetic field by a changing electric field and by electric current: the line integral of the magnetic field around any closed path is the sum of μ_0 multiplied by the net current through that path and $\epsilon_0\mu_0$ multiplied by the rate of change of electric flux through any surface bounded by that path.

Once the electric and magnetic fields are known at some point in space, the force acting on a particle of charge q can be calculated from the expression

$$\vec{F} = q\vec{E} + q\vec{v} \times \vec{B} \qquad \textbf{24.8}◀$$

▶ Lorentz force law

This relationship is called the **Lorentz force law.** (We saw this relationship earlier as Eq. 22.6.) Maxwell's equations, together with this force law, completely describe all classical electromagnetic interactions in a vacuum.

Notice the symmetry of Maxwell's equations. Equations 24.4 and 24.5 are symmetric, apart from the absence of the term for magnetic monopoles in Equation 24.5. Furthermore, Equations 24.6 and 24.7 are symmetric in that the line integrals of $\vec{E}$ and $\vec{B}$ around a closed path are related to the rate of change of magnetic flux and electric flux, respectively. Maxwell's equations are of fundamental importance not only to electromagnetism, but to all science. Hertz once wrote, "One cannot escape the feeling that these mathematical formulas have an independent existence and an intelligence of their own, that they are wiser than we are, wiser even than their discoverers, that we get more out of them than we put into them."

In the next section, we show that Equations 24.6 and 24.7 can be combined to obtain a wave equation for both the electric field and the magnetic field. In empty space, where $q = 0$ and $I = 0$, the solution to these two equations shows that the speed at which electromagnetic waves travel equals the measured speed of light. This result led Maxwell to predict that light waves are a form of electromagnetic radiation.

Hertz performed experiments that verified Maxwell's prediction. The experimental apparatus Hertz used to generate and detect electromagnetic waves is shown schematically in Figure 24.3. An induction coil is connected to a transmitter made up of two spherical electrodes separated by a narrow gap. The coil provides short voltage surges to the electrodes, making one positive and the other negative. A spark is generated between the spheres when the electric field near either electrode surpasses the dielectric strength for air (3×10^6 V/m; see Table 20.1). Free electrons in a strong electric field are accelerated and gain enough energy to ionize any molecules they strike. This ionization provides more electrons, which can accelerate and cause further ionizations. As the air in the gap is ionized, it becomes a much better conductor and the discharge between the electrodes exhibits an oscillatory behavior at a very high frequency. From an electric-circuit viewpoint, this experimental apparatus is equivalent to an LC circuit in which the inductance is that of the coil and the capacitance is due to the spherical electrodes. By applying Kirchhoff's loop rule to an LC circuit, similar to the way we applied it to an RC circuit in Section 21.9, we can show that the current in an LC circuit oscillates in simple harmonic motion at the frequency

$$\omega = \frac{1}{\sqrt{LC}} \qquad \textbf{24.9}◀$$

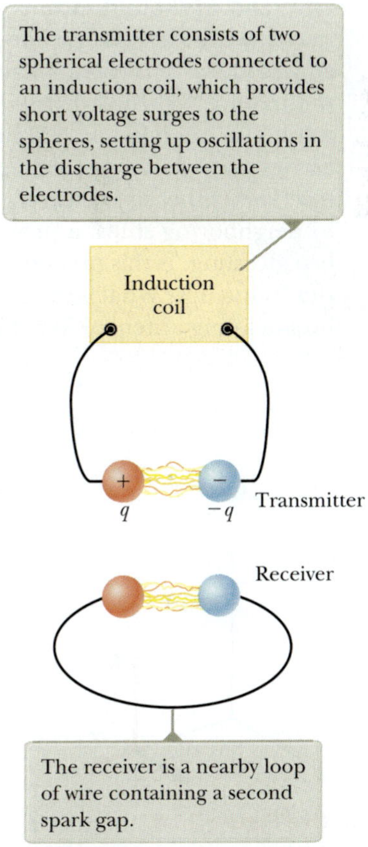

The transmitter consists of two spherical electrodes connected to an induction coil, which provides short voltage surges to the spheres, setting up oscillations in the discharge between the electrodes.

Induction coil

Transmitter

Receiver

The receiver is a nearby loop of wire containing a second spark gap.

Figure 24.3 Schematic diagram of Hertz's apparatus for generating and detecting electromagnetic waves.

© Bettmann/Corbis

Heinrich Rudolf Hertz
German Physicist (1857–1894)
Hertz made his most important discovery of electromagnetic waves in 1887. After finding that the speed of an electromagnetic wave was the same as that of light, Hertz showed that electromagnetic waves, like light waves, could be reflected, refracted, and diffracted. The hertz, equal to one complete vibration or cycle per second, is named after him.

Because L and C are small in Hertz's apparatus, the frequency of oscillation is high, on the order of 100 MHz. Electromagnetic waves are radiated at this frequency as a result of the oscillation (and hence acceleration) of free charges in the transmitter circuit. Hertz was able to detect these waves using a single loop of wire with its own spark gap (the receiver). Such a receiver loop, placed several meters from the transmitter, has its own effective inductance, capacitance, and natural frequency of oscillation. In Hertz's experiment, sparks were induced across the gap of the receiving electrodes when the receiver's frequency was adjusted to match that of the transmitter. In this way, Hertz demonstrated that the oscillating current induced in the receiver was produced by electromagnetic waves radiated by the transmitter. His experiment is analogous to the mechanical phenomenon in which a tuning fork responds to acoustic vibrations from an identical tuning fork that is oscillating.

In addition, Hertz showed in a series of experiments that the radiation generated by his spark-gap device exhibited the wave properties of interference, diffraction, reflection, refraction, and polarization, which are all properties exhibited by light as we shall see in this chapter and Chapters 25–27. Therefore, it became evident that the radio-frequency waves Hertz was generating had properties similar to those of light waves and that they differed only in frequency and wavelength. Perhaps his most convincing experiment was the measurement of the speed of this radiation. Waves of known frequency were reflected from a metal sheet and created a standing-wave interference pattern whose nodal points could be detected. The measured distance between the nodal points enabled determination of the wavelength λ. Using the relationship $v = \lambda f$ (Eq. 13.12), Hertz found that v was close to 3×10^8 m/s, the known speed c of visible light.

▷ THINKING PHYSICS 24.1

In radio transmission, a radio wave serves as a carrier wave and the sound wave is superimposed on the carrier wave. In amplitude modulation (AM radio), the amplitude of the carrier wave varies according to the sound wave. (The word *modulate* means "to change.") In frequency modulation (FM radio), the frequency of the carrier wave varies according to the sound wave. The navy sometimes uses flashing lights to send Morse code to neighboring ships, a process that is similar to radio broadcasting. Is this process AM or FM? What is the carrier frequency? What is the signal frequency? What is the broadcasting antenna? What is the receiving antenna?

Reasoning The flashing of the light according to Morse code is a drastic amplitude modulation because the amplitude is changing between a maximum value and zero. In this sense, it is similar to the on-and-off binary code used in computers and compact discs. The carrier frequency is that of the visible light, on the order of 10^{14} Hz. The signal frequency depends on the skill of the signal operator, but is on the order of a few hertz, as the light is flashed on and off. The broadcasting antenna for this modulated signal is the filament of the lightbulb in the signal source. The receiving antenna is the eye. ◀

▌24.3 | Electromagnetic Waves

In his unified theory of electromagnetism, Maxwell showed that time-dependent electric and magnetic fields satisfy a linear wave equation. (The linear wave equation for mechanical waves is Equation 13.20.) The most significant outcome of this theory is the prediction of the existence of **electromagnetic waves.**

Maxwell's equations predict that an electromagnetic wave consists of oscillating electric and magnetic fields. The changing fields induce each other, which maintains the propagation of the wave; a changing electric field induces a magnetic field, and a changing magnetic field induces an electric field. The $\vec{E}$ and $\vec{B}$ vectors are perpendicular to each other, and to the direction of propagation, as shown in Active Figure 24.4 at one instant of time and one point in space. The direction of propagation is the direction of the vector product $\vec{E} \times \vec{B}$, which we shall explore more fully in Section 24.4.

In Active Figure 24.4, we have chosen the direction of propagation of the wave to be the positive x axis. We have also chosen the y axis to be parallel to the electric field

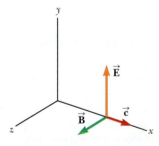

Active Figure 24.4 The fields in an electromagnetic wave traveling at velocity $\vec{c}$ in the positive x direction at one point on the x axis. These fields depend only on x and t.

vector. Given these choices, it is necessarily true that the magnetic field $\vec{\mathbf{B}}$ is in the z direction as in Active Figure 24.4. Waves in which the electric and magnetic fields are restricted to being parallel to certain directions are said to be **linearly polarized waves.** Furthermore, let us assume that at any point in space in Active Figure 24.4, the magnitudes E and B of the fields depend on x and t only, not on the y or z coordinates.

Let us also imagine that the source of the electromagnetic waves is such that a wave radiated from *any* position in the yz plane (not just from the origin as might be suggested by Active Fig. 24.4) propagates in the x direction and that all such waves are emitted in phase. If we define a **ray** as the line along which a wave travels, all rays for these waves are parallel. This whole collection of waves is often called a **plane wave.** A surface connecting points of equal phase on all waves, which we call a **wave front,** is a geometric plane. In comparison, a point source of radiation sends waves out in all directions. A surface connecting points of equal phase for this situation is a sphere, so we call the radiation from a point source a **spherical wave.**

To generate the prediction of electromagnetic waves, we start with Faraday's law, Equation 24.6:

$$\oint \vec{\mathbf{E}} \cdot d\vec{\mathbf{s}} = -\frac{d\Phi_B}{dt}$$

Let's again assume the electromagnetic wave is traveling in the x direction, with the electric field $\vec{\mathbf{E}}$ in the positive y direction and the magnetic field $\vec{\mathbf{B}}$ in the positive z direction.

Consider a rectangle of width dx and height ℓ lying in the xy plane as shown in Figure 24.5. To apply Equation 24.6, let's first evaluate the line integral of $\vec{\mathbf{E}} \cdot d\vec{\mathbf{s}}$ around this rectangle in the counterclockwise direction at an instant of time when the wave is passing through the rectangle. The contributions from the top and bottom of the rectangle are zero because $\vec{\mathbf{E}}$ is perpendicular to $d\vec{\mathbf{s}}$ for these paths. We can express the electric field on the right side of the rectangle as

$$E(x + dx) \approx E(x) + \frac{dE}{dx}\bigg]_{t\text{ constant}} dx = E(x) + \frac{\partial E}{\partial x} dx$$

where $E(x)$ is the field on the left side of the rectangle at this instant.[2] Therefore, the line integral over this rectangle is approximately

$$\oint \vec{\mathbf{E}} \cdot d\vec{\mathbf{s}} = [E(x + dx)]\ell - [E(x)]\ell \approx \ell\left(\frac{\partial E}{\partial x}\right) dx \qquad \textbf{24.10} \blacktriangleleft$$

Because the magnetic field is in the z direction, the magnetic flux through the rectangle of area $\ell\,dx$ is approximately $\Phi_B = B\ell\,dx$ (assuming dx is very small compared with the wavelength of the wave). Taking the time derivative of the magnetic flux gives

$$\frac{d\Phi_B}{dt} = \ell\,dx\,\frac{dB}{dt}\bigg]_{x\text{ constant}} = \ell\,dx\,\frac{\partial B}{\partial t} \qquad \textbf{24.11} \blacktriangleleft$$

Substituting Equations 24.10 and 24.11 into Equation 24.6 gives

$$\ell\left(\frac{\partial E}{\partial x}\right) dx = -\ell\,dx\,\frac{\partial B}{\partial t}$$

$$\frac{\partial E}{\partial x} = -\frac{\partial B}{\partial t} \qquad \textbf{24.12} \blacktriangleleft$$

In a similar manner, we can derive a second equation by starting with Maxwell's fourth equation in empty space (Eq. 24.7). In this case, the line integral of $\vec{\mathbf{B}} \cdot d\vec{\mathbf{s}}$ is evaluated around a rectangle lying in the xz plane and having width dx and length ℓ as in Figure 24.6 (page 828). Noting that the magnitude of the magnetic field

According to Equation 24.12, this spatial variation in $\vec{\mathbf{E}}$ gives rise to a time-varying magnetic field along the z direction.

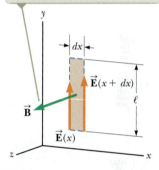

Figure 24.5 At an instant when a plane wave moving in the positive x direction passes through a rectangular path of width dx lying in the xy plane, the electric field in the y direction varies from $\vec{\mathbf{E}}(x)$ to $\vec{\mathbf{E}}(x + dx)$.

---

[2]Because dE/dx in this equation is expressed as the change in E with x at a given instant t, dE/dx is equivalent to the partial derivative $\partial E/\partial x$. Likewise, dB/dt means the change in B with time at a particular position x; therefore, in Equation 24.11, we can replace dB/dt with $\partial B/\partial t$.

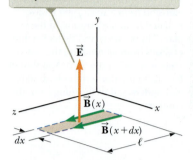

According to Equation 24.15, this spatial variation in $\vec{B}$ gives rise to a time-varying electric field along the y direction.

Figure 24.6 At an instant when a plane wave passes through a rectangular path of width dx lying in the xz plane, the magnetic field in the z direction varies from $\vec{B}(x)$ to $\vec{B}(x + dx)$.

changes from $B(x)$ to $B(x + dx)$ over the width dx and that the direction for taking the line integral is counterclockwise when viewed from above in Figure 24.6, the line integral over this rectangle is found to be approximately

$$\oint \vec{B} \cdot d\vec{s} = [B(x)]\ell - [B(x + dx)]\ell \approx -\ell \left(\frac{\partial B}{\partial x}\right) dx \qquad \text{24.13} \blacktriangleleft$$

The electric flux through the rectangle is $\Phi_E = E\ell\, dx$, which, when differentiated with respect to time, gives

$$\frac{\partial \Phi_E}{\partial t} = \ell\, dx \frac{\partial E}{\partial t} \qquad \text{24.14} \blacktriangleleft$$

Substituting Equations 24.13 and 24.14 into Equation 24.7 gives

$$-\ell \left(\frac{\partial B}{\partial x}\right) dx = \mu_0 \epsilon_0 \,\ell\, dx \left(\frac{\partial E}{\partial t}\right)$$

$$\frac{\partial B}{\partial x} = -\mu_0 \epsilon_0 \frac{\partial E}{\partial t} \qquad \text{24.15} \blacktriangleleft$$

Taking the derivative of Equation 24.12 with respect to x and combining the result with Equation 24.15 gives

$$\frac{\partial^2 E}{\partial x^2} = -\frac{\partial}{\partial x}\left(\frac{\partial B}{\partial t}\right) = -\frac{\partial}{\partial t}\left(\frac{\partial B}{\partial x}\right) = -\frac{\partial}{\partial t}\left(-\mu_0 \epsilon_0 \frac{\partial E}{\partial t}\right)$$

$$\frac{\partial^2 E}{\partial x^2} = \mu_0 \epsilon_0 \frac{\partial^2 E}{\partial t^2} \qquad \text{24.16} \blacktriangleleft$$

In the same manner, taking the derivative of Equation 24.15 with respect to x and combining it with Equation 24.12 gives

$$\frac{\partial^2 B}{\partial x^2} = \mu_0 \epsilon_0 \frac{\partial^2 B}{\partial t^2} \qquad \text{24.17} \blacktriangleleft$$

Equations 24.16 and 24.17 both have the form of the linear wave equation[3] with the wave speed v replaced by c, where

▶ Speed of electromagnetic waves

$$c = \frac{1}{\sqrt{\mu_0 \epsilon_0}} \qquad \text{24.18} \blacktriangleleft$$

Let's evaluate this speed numerically:

$$c = \frac{1}{\sqrt{(4\pi \times 10^{-7}\ \text{T} \cdot \text{m/A})(8.854\,19 \times 10^{-12}\ \text{C}^2/\text{N} \cdot \text{m}^2)}}$$

$$= 2.997\,92 \times 10^8\ \text{m/s}$$

Because this speed is precisely the same as the speed of light in empty space, we are led to believe (correctly) that light is an electromagnetic wave.

The simplest solution to Equations 24.16 and 24.17 is a sinusoidal wave for which the field magnitudes E and B vary with x and t according to the expressions

▶ Sinusoidal electric and magnetic fields

$$E = E_{max} \cos (kx - \omega t) \qquad \text{24.19} \blacktriangleleft$$

$$B = B_{max} \cos (kx - \omega t) \qquad \text{24.20} \blacktriangleleft$$

where E_{max} and B_{max} are the maximum values of the fields. The angular wave number is $k = 2\pi/\lambda$, where λ is the wavelength. The angular frequency is $\omega = 2\pi f$, where f is the wave frequency. The ratio ω/k equals the speed of an electromagnetic wave, c:

$$\frac{\omega}{k} = \frac{2\pi f}{2\pi/\lambda} = \lambda f = c$$

---

[3] The linear wave equation is of the form $(\partial^2 y/\partial x^2) = (1/v^2)(\partial^2 y/\partial t^2)$, where v is the speed of the wave and y is the wave function. The linear wave equation was introduced as Equation 13.20, and we suggest you review Section 13.2.

where we have used Equation 13.12, $v = c = \lambda f$, which relates the speed, frequency, and wavelength of any continuous wave. Therefore, for electromagnetic waves, the wavelength and frequency of these waves are related by

$$\lambda = \frac{c}{f} = \frac{3.00 \times 10^8 \text{ m/s}}{f}$$ 24.21 ◀

Active Figure 24.7 is a pictorial representation, at one instant, of a sinusoidal, linearly polarized electromagnetic wave moving in the positive x direction.

Taking partial derivatives of Equations 24.19 (with respect to x) and 24.20 (with respect to t) gives

$$\frac{\partial E}{\partial x} = -kE_{max} \sin (kx - \omega t)$$

$$\frac{\partial B}{\partial t} = \omega B_{max} \sin (kx - \omega t)$$

Substituting these results into Equation 24.12 shows that, at any instant,

$$kE_{max} = \omega B_{max}$$

$$\frac{E_{max}}{B_{max}} = \frac{\omega}{k} = c$$

Using these results together with Equations 24.19 and 24.20 gives

$$\frac{E_{max}}{B_{max}} = \frac{E}{B} = c$$ 24.22 ◀

That is, at every instant, the ratio of the magnitude of the electric field to the magnitude of the magnetic field in an electromagnetic wave equals the speed of light.

Finally, note that electromagnetic waves obey the superposition principle (which we discussed in Section 14.1 with respect to mechanical waves) because the differential equations involving E and B are linear equations. For example, we can add two waves with the same frequency and polarization simply by adding the magnitudes of the two electric fields algebraically.

Doppler Effect for Light

Another feature of electromagnetic waves is that there is a shift in the observed frequency of the waves when there is relative motion between the source of the waves and the observer. This phenomenon, known as the Doppler effect, was introduced in Chapter 13 as it pertains to sound waves. In the case of sound, the motion of the source with respect to the medium of propagation can be distinguished from the motion of the observer with respect to the medium. Light waves must be analyzed differently, however, because they require no medium of propagation, and no method exists for distinguishing the motion of a light source from the motion of the observer.

If a light source and an observer approach each other with a relative speed v, the frequency f' measured by the observer is

$$f' = \sqrt{\frac{c + v}{c - v}} f$$ 24.23 ◀

where f is the frequency of the source measured in its rest frame. This Doppler shift equation, unlike the Doppler shift equation for sound, depends only on the relative speed v of the source and observer and holds for relative speeds as great as c. As you might expect, the equation predicts that $f' > f$ when the source and observer approach each other. We obtain the expression for the case in which the source and observer recede from each other by substituting negative values for v in Equation 24.23.

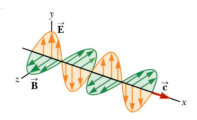

Active Figure 24.7 A sinusoidal electromagnetic wave moves in the positive x direction with a speed c.

Pitfall Prevention | 24.2
$\vec{E}$ Stronger Than $\vec{B}$?
Because the value of c is so large, some students incorrectly interpret Equation 24.22 as meaning that the electric field is much stronger than the magnetic field. Electric and magnetic fields are measured in different units, however, so they cannot be directly compared. In Section 24.4, we find that the electric and magnetic fields contribute equally to the wave's energy.

▶ Doppler effect for electromagnetic waves

When this expression is averaged over one or more cycles of an electromagnetic wave, we again obtain a factor of $\frac{1}{2}$. Therefore, the total average energy per unit volume of an electromagnetic wave is

▶ Average energy density of an
electromagnetic wave

$$u_{avg} = \epsilon_0 (E^2)_{avg} = \tfrac{1}{2}\epsilon_0 E^2_{max} = \frac{B^2_{max}}{2\mu_0}$$ 24.29 ◀

Comparing this result with Equation 24.26 for the average value of S, we see that

$$I = S_{avg} = cu_{avg}$$ 24.30 ◀

In other words, the intensity of an electromagnetic wave equals the average energy density multiplied by the speed of light.

QUICK QUIZ 24.2 An electromagnetic wave propagates in the negative y direction. The electric field at a point in space is momentarily oriented in the positive x direction. In which direction is the magnetic field at that point momentarily oriented? (a) the negative x direction (b) the positive y direction (c) the positive z direction (d) the negative z direction

QUICK QUIZ 24.3 Which of the following quantities does not vary in time for plane electromagnetic waves? (a) magnitude of the Poynting vector (b) energy density u_E (c) energy density u_B (d) intensity I

Example 24.2 | Fields on the Page

Estimate the maximum magnitudes of the electric and magnetic fields of the light that is incident on this page because of the visible light coming from your desk lamp. Treat the lightbulb as a point source of electromagnetic radiation that is 5% efficient at transforming energy coming in by electrical transmission to energy leaving by visible light.

SOLUTION

Conceptualize The filament in your lightbulb emits electromagnetic radiation. The brighter the light, the larger the magnitudes of the electric and magnetic fields.

Categorize Because the lightbulb is to be treated as a point source, it emits equally in all directions, so the outgoing electromagnetic radiation can be modeled as a spherical wave.

Analyze We mentioned that intensity is equivalent to the average power of radiation per unit area. For a point source emitting uniformly in all directions, the power is distributed evenly over the surface area $4\pi r^2$ of an expanding sphere of increasing radius r, centered on the source. Therefore, $I = P_{avg}/4\pi r^2$, where P represents power.

Set this expression for I equal to the intensity of an electromagnetic wave given by Equation 24.26:

$$I = \frac{P_{avg}}{4\pi r^2} = \frac{E^2_{max}}{2\mu_0 c}$$

Solve for the electric field magnitude:

$$E_{max} = \sqrt{\frac{\mu_0 c P_{avg}}{2\pi r^2}}$$

Let's make some assumptions about numbers to enter in this equation. The visible light output of a 60-W lightbulb operating at 5% efficiency is approximately 3.0 W by visible light. (The remaining energy transfers out of the lightbulb by conduction and invisible radiation.) A reasonable distance from the lightbulb to the page might be 0.30 m.

Substitute these values:

$$E_{max} = \sqrt{\frac{(4\pi \times 10^{-7}\ \text{T} \cdot \text{m/A})(3.00 \times 10^8\ \text{m/s})(3.0\ \text{W})}{2\pi(0.30\ \text{m})^2}}$$

$$= 45\ \text{V/m}$$

Use Equation 24.22 to find the magnetic field magnitude:

$$B_{max} = \frac{E_{max}}{c} = \frac{45\ \text{V/m}}{3.00 \times 10^8\ \text{m/s}} = 1.5 \times 10^{-7}\ \text{T}$$

Finalize This value of the magnetic field magnitude is two orders of magnitude smaller than the Earth's magnetic field.

❮24.5 | Momentum and Radiation Pressure

Electromagnetic waves transport linear momentum as well as energy. Hence, it follows that pressure is exerted on a surface when an electromagnetic wave impinges on it. In what follows, let us assume that the electromagnetic wave strikes a surface at normal incidence and transports a total energy T_{ER} to a surface in a time interval Δt. If the surface absorbs all the incident energy T_{ER} in this time, Maxwell showed that the total momentum $\vec{p}$ delivered to this surface has a magnitude

$$p = \frac{T_{ER}}{c} \qquad \text{(complete absorption)} \qquad \textbf{24.31} \blacktriangleleft$$

▶ Momentum delivered to a perfectly absorbing surface

The radiation pressure P exerted on the surface is defined as force per unit area F/A. Let us combine this definition with Newton's second law:

$$P = \frac{F}{A} = \frac{1}{A}\frac{dp}{dt}$$

If we now replace p, the momentum transported to the surface by radiation, from Equation 24.31, we have

$$P = \frac{1}{A}\frac{dp}{dt} = \frac{1}{A}\frac{d}{dt}\left(\frac{T_{ER}}{c}\right) = \frac{1}{c}\frac{(dT_{ER}/dt)}{A}$$

Pitfall Prevention | 24.4
So Many *p*'s
We have *p* for momentum and *P* for pressure, and they are both related to *P* for power! Be sure to keep all these symbols straight.

We recognize $(dT_{ER}/dt)/A$ as the rate at which energy is arriving at the surface per unit area, which is the magnitude of the Poynting vector. Therefore, the radiation pressure P exerted on the perfectly absorbing surface is

$$P = \frac{S}{c} \qquad \text{(complete absorption)} \qquad \textbf{24.32} \blacktriangleleft$$

▶ Radiation pressure exerted on a perfectly absorbing surface

An absorbing surface for which all the incident energy is absorbed (none is reflected) is called a **black body.** A more detailed discussion of a black body will be presented in Chapter 28.

As we found in the last section, the intensity of an electromagnetic wave I is equal to the average value of S (Eq. 24.26), so we can express the average radiation pressure as

$$P_{avg} = \frac{S_{avg}}{c} = \frac{I}{c} \qquad \text{(complete absorption)} \qquad \textbf{24.33} \blacktriangleleft$$

Furthermore, because S_{avg} represents power per unit area, we find that the average power delivered to a surface of area A is (using "*Power*" to represent power because we also have P for pressure here)

$$(Power)_{avg} = IA \qquad \text{(complete absorption)} \qquad \textbf{24.34} \blacktriangleleft$$

If the surface is a perfect reflector, the momentum delivered in a time interval Δt for normal incidence is twice that given by Equation 24.31, or $p = 2T_{ER}/c$. That is, a momentum T_{ER}/c is delivered first by the incident wave and then again by the reflected wave, a situation analogous to a ball colliding elastically with a wall.[4] Finally, the radiation pressure exerted on a perfect reflecting surface for normal incidence of the wave is twice that given by Equation 24.32, or

$$P = \frac{2S}{c} \qquad \text{(complete reflection)} \qquad \textbf{24.35} \blacktriangleleft$$

Although radiation pressures are very small (about 5×10^{-6} N/m² for direct sunlight), they have been measured using torsion balances such as the one shown in Figure 24.10. Light is allowed to strike either a mirror or a black disk, both of which are suspended from a fine fiber. Light striking the black disk is completely absorbed, so all its momentum is transferred to the disk. Light striking the mirror (normal

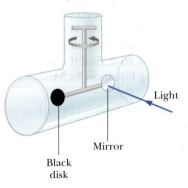

Figure 24.10 An apparatus for measuring the pressure exerted by light. In practice, the system is contained in a high vacuum.

---

[4]For *oblique* incidence, the momentum transferred is $2T_{ER}\cos\theta/c$ and the pressure is given by $P = 2S\cos^2\theta/c$, where θ is the angle between the normal to the surface and the direction of propagation.

The orange dot is the sample of trapped sodium atoms.

Courtesy of National Institute of Standards and Technology, U.S. Dept. of Commerce

Figure 24.19 A staff member of the National Institute of Standards and Technology views a sample of trapped sodium atoms cooled to a temperature measured in microkelvins.

Chu shared the 1997 Nobel Prize in Physics with two of his colleagues for the development of the techniques of optical trapping.

An extension of laser trapping, *laser cooling*, is possible because the normal high speeds of the atoms are reduced when they are restricted to the region of the trap. As a result, the temperature of the collection of atoms can be reduced to a few microkelvins. The technique of laser cooling allows scientists to study the behavior of atoms at extremely low temperatures (Fig. 24.19).

In the 1920s, Satyendra Nath Bose (1894–1974) was studying photons and investigating collections of identical photons, which can all be in the same quantum state. Einstein followed up on the work of Bose and predicted that a collection of atoms could all be in the same quantum state if the temperature were low enough. The proposed collection of atoms is called a *Bose–Einstein condensate*. In 1995, using laser cooling supplemented with evaporative cooling, the first Bose–Einstein condensate was created in the laboratory by Eric Cornell and Carl Wieman, who won the 2001 Nobel Prize in Physics for their work. Many laboratories are now creating Bose–Einstein condensates and studying their properties and possible applications. One interesting result was reported by a Harvard University group led by Lene Vestergaard Hau in 2001. She and her colleagues announced that they were able to bring a light pulse to a complete stop by using a Bose–Einstein condensate.[6]

More recently, scientists have discovered a new type of Bose–Einstein condensate based on a quasiparticle called the *polariton*.[7] The polariton, which is a superposition of a photon and an electronic excitation in a solid, exists typically for only a few picoseconds in an optical cavity. These condensates are unique because they are extremely light compared to atomic condensates and therefore exhibit quantum effects at higher temperatures.

We have explored general properties of laser light in this chapter. In the Context Connection of Chapter 25, we shall explore the technology of optical fibers, in which lasers are used in a variety of applications.

SUMMARY

Displacement current I_d is defined as

$$I_d \equiv \epsilon_0 \frac{d\Phi_E}{dt} \qquad \text{24.1} \blacktriangleleft$$

and represents an effective current through a region of space in which an electric field is changing in time.

When used with the Lorentz force law ($\vec{F} = q\vec{E} + q\vec{v} \times \vec{B}$), **Maxwell's equations** describe *all* electromagnetic phenomena:

$$\oint \vec{E} \cdot d\vec{A} = \frac{q}{\epsilon_0} \qquad \text{24.4} \blacktriangleleft$$

$$\oint \vec{B} \cdot d\vec{A} = 0 \qquad \text{24.5} \blacktriangleleft$$

$$\oint \vec{E} \cdot d\vec{s} = -\frac{d\Phi_B}{dt} \qquad \text{24.6} \blacktriangleleft$$

$$\oint \vec{B} \cdot d\vec{s} = \mu_0 I + \epsilon_0 \mu_0 \frac{d\Phi_E}{dt} \qquad \text{24.7} \blacktriangleleft$$

Electromagnetic waves, which are predicted by Maxwell's equations, have the following properties:

- The electric and magnetic fields satisfy the following wave equations, which can be obtained from Maxwell's third and fourth equations:

$$\frac{\partial^2 E}{\partial x^2} = \mu_0 \epsilon_0 \frac{\partial^2 E}{\partial t^2} \qquad \text{24.16} \blacktriangleleft$$

$$\frac{\partial^2 B}{\partial x^2} = \mu_0 \epsilon_0 \frac{\partial^2 B}{\partial t^2} \qquad \text{24.17} \blacktriangleleft$$

- Electromagnetic waves travel through a vacuum with the speed of light $c = 3.00 \times 10^8$ m/s, where

$$c = \frac{1}{\sqrt{\mu_0 \epsilon_0}} \qquad \text{24.18} \blacktriangleleft$$

- The electric and magnetic fields of an electromagnetic wave are perpendicular to each other and perpendicular to the direction of wave propagation; hence, electromagnetic waves are transverse waves. The electric and magnetic

[6]C. Liu, Z. Dutton, C. H. Behroozi, and L. V. Hau, "Observation of Coherent Optical Information Storage in an Atomic Medium Using Halted Light Pulses," *Nature*, 409, 490–493, January 25, 2001.

[7]D. Snoke and P. Littlewood, "Polariton Condensates," *Physics Today*, 42–47, August 2010.

fields of a sinusoidal plane electromagnetic wave propagating in the positive x direction can be written

$$E = E_{max} \cos(kx - \omega t) \qquad \textbf{24.19} \blacktriangleleft$$

$$B = B_{max} \cos(kx - \omega t) \qquad \textbf{24.20} \blacktriangleleft$$

where ω is the angular frequency of the wave and k is the angular wave number. These equations represent special solutions to the wave equations for $\vec{E}$ and $\vec{B}$.

- The instantaneous magnitudes of $\vec{E}$ and $\vec{B}$ in an electromagnetic wave are related by the expression

$$\frac{E}{B} = c \qquad \textbf{24.22} \blacktriangleleft$$

- Electromagnetic waves carry energy. The rate of flow of energy crossing a unit area is described by the **Poynting vector $\vec{S}$**, where

$$\vec{S} \equiv \frac{1}{\mu_0} \vec{E} \times \vec{B} \qquad \textbf{24.24} \blacktriangleleft$$

The average value of the Poynting vector for a plane electromagnetic wave has the magnitude

$$I = S_{avg} = \frac{E_{max} B_{max}}{2\mu_0} = \frac{E^2_{max}}{2\mu_0 c} = \frac{cB^2_{max}}{2\mu_0} \qquad \textbf{24.26} \blacktriangleleft$$

The average power per unit area (intensity) of a sinusoidal plane electromagnetic wave equals the average value of the Poynting vector taken over one or more cycles.

- Electromagnetic waves carry momentum and hence can exert pressure on surfaces. If an electromagnetic wave whose intensity is I is completely absorbed by a surface on which it is normally incident, the radiation pressure on that surface is

$$P = \frac{S}{c} \quad \text{(complete absorption)} \qquad \textbf{24.32} \blacktriangleleft$$

If the surface totally reflects a normally incident wave, the pressure is doubled.

The **electromagnetic spectrum** includes waves covering a broad range of frequencies and wavelengths.

When polarized light of intensity I_{max} is incident on a polarizing film, the light transmitted through the film has an intensity equal to $I_{max} \cos^2 \theta$, where θ is the angle between the transmission axis of the polarizing film and the electric field vector of the incident light.

▶ OBJECTIVE QUESTIONS |

1. Which of the following statements are true regarding electromagnetic waves traveling through a vacuum? More than one statement may be correct. (a) All waves have the same wavelength. (b) All waves have the same frequency. (c) All waves travel at 3.00×10^8 m/s. (d) The electric and magnetic fields associated with the waves are perpendicular to each other and to the direction of wave propagation. (e) The speed of the waves depends on their frequency.

2. An electromagnetic wave with a peak magnetic field magnitude of 1.50×10^{-7} T has an associated peak electric field of what magnitude? (a) 0.500×10^{-15} N/C (b) 2.00×10^{-5} N/C (c) 2.20×10^4 N/C (d) 45.0 N/C (e) 22.0 N/C

3. Assume you charge a comb by running it through your hair and then hold the comb next to a bar magnet. Do the electric and magnetic fields produced constitute an electromagnetic wave? (a) Yes they do, necessarily. (b) Yes they do because charged particles are moving inside the bar magnet. (c) They can, but only if the electric field of the comb and the magnetic field of the magnet are perpendicular. (d) They can, but only if both the comb and the magnet are moving. (e) They can, if either the comb or the magnet or both are accelerating.

4. If plane polarized light is sent through two polarizers, the first at 45° to the original plane of polarization and the second at 90° to the original plane of polarization, what fraction of the original polarized intensity passes through the last polarizer? (a) 0 (b) $\frac{1}{4}$ (c) $\frac{1}{2}$ (d) $\frac{1}{8}$ (e) $\frac{1}{10}$

5. A typical microwave oven operates at a frequency of 2.45 GHz. What is the wavelength associated with the electromagnetic waves in the oven? (a) 8.20 m (b) 12.2 cm (c) 1.20×10^8 m (d) 8.20×10^{-9} m (e) none of those answers

6. A student working with a transmitting apparatus like Heinrich Hertz's wishes to adjust the electrodes to generate electromagnetic waves with a frequency half as large as before. **(i)** How large should she make the effective capacitance of the pair of electrodes? (a) four times larger than before (b) two times larger than before (c) one-half as large as before (d) one-fourth as large as before (e) none of those answers **(ii)** After she makes the required adjustment, what will the wavelength of the transmitted wave be? Choose from the same possibilities as in part (i).

7. A small source radiates an electromagnetic wave with a single frequency into vacuum, equally in all directions. **(i)** As the wave moves, does its frequency (a) increase, (b) decrease, or (c) stay constant? Using the same choices, answer the same question about **(ii)** its wavelength, **(iii)** its speed, **(iv)** its intensity, and **(v)** the amplitude of its electric field.

8. A plane electromagnetic wave with a single frequency moves in vacuum in the positive x direction. Its amplitude is uniform over the yz plane. **(i)** As the wave moves, does its frequency (a) increase, (b) decrease, or (c) stay constant? Using the same choices, answer the same question about **(ii)** its wavelength, **(iii)** its speed, **(iv)** its intensity, and **(v)** the amplitude of its magnetic field.

44. A radar pulse returns to the transmitter-receiver after a total travel time of 4.00×10^{-4} s. How far away is the object that reflected the wave?

Section 24.7 Polarization of Light Waves

45. M Plane-polarized light is incident on a single polarizing disk with the direction of $\vec{E}_0$ parallel to the direction of the transmission axis. Through what angle should the disk be rotated so that the intensity in the transmitted beam is reduced by a factor of (a) 3.00, (b) 5.00, and (c) 10.0?

46. S In Figure P24.46, suppose the transmission axes of the left and right polarizing disks are perpendicular to each other. Also, let the center disk be rotated on the common axis with an angular speed ω. Show that if unpolarized light is incident on the left disk with an intensity I_{max}, the intensity of the beam emerging from the right disk is

$$I = \tfrac{1}{16} I_{max}(1 - \cos 4\omega t)$$

This result means that the intensity of the emerging beam is modulated at a rate four times the rate of rotation of the center disk. *Suggestion:* Use the trigonometric identities

$$\cos^2 \theta = \tfrac{1}{2}(1 + \cos 2\theta) \quad \text{and} \quad \sin^2 \theta = \tfrac{1}{2}(1 - \cos 2\theta)$$

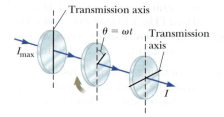

Figure P24.46

47. W Unpolarized light passes through two ideal Polaroid sheets. The axis of the first is vertical, and the axis of the second is at 30.0° to the vertical. What fraction of the incident light is transmitted?

48. Two handheld radio transceivers with dipole antennas are separated by a large fixed distance. If the transmitting antenna is vertical, what fraction of the maximum received power will appear in the receiving antenna when it is inclined from the vertical by (a) 15.0°, (b) 45.0°, and (c) 90.0°?

49. You use a sequence of ideal polarizing filters, each with its axis making the same angle with the axis of the previous filter, to rotate the plane of polarization of a polarized light beam by a total of 45.0°. You wish to have an intensity reduction no larger than 10.0%. (a) How many polarizers do you need to achieve your goal? (b) What is the angle between adjacent polarizers?

50. Two polarizing sheets are placed together with their transmission axes crossed so that no light is transmitted. A third sheet is inserted between them with its transmission axis at an angle of 45.0° with respect to each of the other axes. Find the fraction of incident unpolarized light intensity transmitted by the three-sheet combination. (Assume each polarizing sheet is ideal.)

Section 24.8 Context Connection: The Special Properties of Laser Light

51. W High-power lasers in factories are used to cut through cloth and metal (Fig. P24.51). One such laser has a beam diameter of 1.00 mm and generates an electric field having an amplitude of 0.700 MV/m at the target. Find (a) the amplitude of the magnetic field produced, (b) the intensity of the laser, and (c) the power delivered by the laser.

Figure P24.51 A laser cutting device mounted on a robot arm is being used to cut through a metallic plate.

52. Figure P24.52 shows portions of the energy-level diagrams of the helium and neon atoms. An electrical discharge excites the He atom from its ground state (arbitrarily assigned the energy $E_1 = 0$) to its excited state of 20.61 eV. The excited He atom collides with a Ne atom in its ground state and excites this atom to the state at 20.66 eV. Lasing action takes place for electron transitions from E_3^* to E_2 in the Ne atoms. From the data in the figure, show that the wavelength of the red He–Ne laser light is approximately 633 nm.

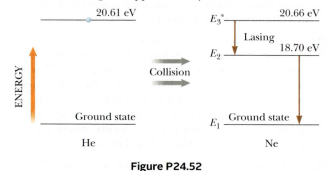

Figure P24.52

53. W A neodymium–yttrium–aluminum garnet laser used in eye surgery emits a 3.00-mJ pulse in 1.00 ns, focused to a spot 30.0 μm in diameter on the retina. (a) Find (in SI units) the power per unit area at the retina. (In the optics industry, this quantity is called the *irradiance.*) (b) What energy is delivered by the pulse to an area of molecular size, taken as a circular area 0.600 nm in diameter?

54. A pulsed ruby laser emits light at 694.3 nm. For a 14.0-ps pulse containing 3.00 J of energy, find (a) the physical length of the pulse as it travels through space and (b) the number of photons in it. (c) Assuming that the beam has a circular cross section of 0.600 cm diameter, find the number of photons per cubic millimeter.

55. The carbon dioxide laser is one of the most powerful developed. The energy difference between the two laser levels is 0.117 eV. Determine (a) the frequency and (b) the wavelength of the radiation emitted by this laser. (c) In what portion of the electromagnetic spectrum is this radiation?

56. **Review.** Figure 24.16 represents the light bouncing between two mirrors in a laser cavity as two traveling waves. These traveling waves moving in opposite directions constitute a standing wave. If the reflecting surfaces are metallic films, the electric field has nodes at both ends. The electromagnetic standing wave is analogous to the standing string wave represented in Active Figure 14.9. (a) Assume that a helium–neon laser has precisely flat and parallel mirrors 35.124 103 cm apart. Assume that the active medium can efficiently amplify only light with wavelengths between 632.808 40 nm and 632.809 80 nm. Find the number of components that constitute the laser light, and the wavelength of each component, precise to eight digits. (b) Find the root-mean-square speed for a neon atom at 120°C. (c) Show that at this temperature the Doppler effect for light emission by moving neon atoms should realistically make the bandwidth of the light amplifier larger than the 0.001 40 nm assumed in part (a).

57. **M** A ruby laser delivers a 10.0-ns pulse of 1.00-MW average power. If the photons have a wavelength of 694.3 nm, how many are contained in the pulse?

58. **Q|C** The number N of atoms in a particular state is called the population of that state. This number depends on the energy of that state and the temperature. In thermal equilibrium, the population of atoms in a state of energy E_n is given by a Boltzmann distribution expression

$$N = N_g e^{-(E_n - E_g)/k_B T}$$

where N_g is the population of the ground state of energy E_g, k_B is Boltzmann's constant, and T is the absolute temperature. For simplicity, assume each energy level has only one quantum state associated with it. (a) Before the power is switched on, the neon atoms in a laser are in thermal equilibrium at 27.0°C. Find the equilibrium ratio of the populations of the states E_4^* and E_3 shown for the red transition in Figure P24.58. Lasers operate by a clever artificial production of a "population inversion" between the upper and lower

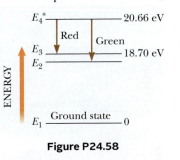

Figure P24.58

atomic energy states involved in the lasing transition. This term means that more atoms are in the upper excited state than in the lower one. Consider the $E_4^* - E_3$ transition in Figure P24.58. Assume 2% more atoms occur in the upper state than in the lower. (b) To demonstrate how unnatural such a situation is, find the temperature for which the Boltzmann distribution describes a 2.00% population inversion. (c) Why does such a situation not occur naturally?

Additional Problems

59. Assume the intensity of solar radiation incident on the cloud tops of the Earth is 1 370 W/m². (a) Taking the average Earth–Sun separation to be 1.496×10^{11} m, calculate the total power radiated by the Sun. Determine the maximum values of (b) the electric field and (c) the magnetic field in the sunlight at the Earth's location.

60. One goal of the Russian space program is to illuminate dark northern cities with sunlight reflected to the Earth from a 200-m diameter mirrored surface in orbit. Several smaller prototypes have already been constructed and put into orbit. (a) Assume that sunlight with intensity 1 370 W/m² falls on the mirror nearly perpendicularly and that the

atmosphere of the Earth allows 74.6% of the energy of sunlight to pass though it in clear weather. What is the power received by a city when the space mirror is reflecting light to it? (b) The plan is for the reflected sunlight to cover a circle of diameter 8.00 km. What is the intensity of light (the average magnitude of the Poynting vector) received by the city? (c) This intensity is what percentage of the vertical component of sunlight at St. Petersburg in January, when the sun reaches an angle of 7.00° above the horizon at noon?

61. The intensity of solar radiation at the top of the Earth's atmosphere is 1 370 W/m². Assuming 60% of the incoming solar energy reaches the Earth's surface and you absorb 50% of the incident energy, make an order-of-magnitude estimate of the amount of solar energy you absorb if you sunbathe for 60 minutes.

62. **S** **Review.** In the absence of cable input or a satellite dish, a television set can use a dipole-receiving antenna for VHF channels and a loop antenna for UHF channels. In Figure CQ24.1, the "rabbit ears" form the VHF antenna and the smaller loop of wire is the UHF antenna. The UHF antenna produces an emf from the changing magnetic flux through the loop. The television station broadcasts a signal with a frequency f, and the signal has an electric field amplitude E_{max} and a magnetic field amplitude B_{max} at the location of the receiving antenna. (a) Using Faraday's law, derive an expression for the amplitude of the emf that appears in a single-turn, circular loop antenna with a radius r that is small compared with the wavelength of the wave. (b) If the electric field in the signal points vertically, what orientation of the loop gives the best reception?

63. **M** A dish antenna having a diameter of 20.0 m receives (at normal incidence) a radio signal from a distant source as shown in Figure P24.63. The radio signal is a continuous sinusoidal wave with amplitude $E_{max} = 0.200\ \mu\text{V/m}$. Assume the antenna absorbs all the radiation that falls on the dish. (a) What is the amplitude of the magnetic field in this wave? (b) What is the intensity of the radiation received by this antenna? (c) What is the power received by the antenna? (d) What force is exerted by the radio waves on the antenna?

Figure P24.63

64. **GP** **Q|C** You may wish to review Section 13.5 on the transport of energy by sinusoidal waves on strings. Figure P24.13 is a graphical representation of an electromagnetic wave moving in the x direction. We wish to find an expression for the intensity of this wave by means of a different process from that by which Equation 24.26 was generated. (a) Sketch a graph of the electric field in this wave at the instant $t = 0$, letting your flat paper represent the xy plane. (b) Compute the energy density u_E in the electric field as a function of x at the instant $t = 0$. (c) Compute the energy density in the magnetic field u_B as a function of x at that instant. (d) Find the total energy density u as a function of x, expressed in terms of only the electric field amplitude. (e) The energy in a "shoebox" of length λ and frontal area A is $E_\lambda = \int_0^\lambda uA\, dx$. (The symbol E_λ for energy in a wavelength imitates the notation of Section 13.5.) Perform the integration to compute the amount of this energy in terms of A, λ, E_{max}, and universal constants. (f) We may think of the energy transport by the whole wave as a series of these shoeboxes going past as if carried on a conveyor belt. Each shoebox passes by a point in a time interval defined as the period $T = 1/f$ of the wave. Find the power the wave carries through area A. (g) The intensity of the wave is the power per unit area through which the wave passes. Compute this intensity in terms of E_{max} and universal constants. (h) Explain how your result compares with that given in Equation 24.26.

65. Consider a small, spherical particle of radius r located in space a distance $R = 3.75 \times 10^{11}$ m from the Sun. Assume the particle has a perfectly absorbing surface and a mass density of $\rho = 1.50$ g/cm^3. Use $S = 214$ W/m^2 as the value of the solar intensity at the location of the particle. Calculate the value of r for which the particle is in equilibrium between the gravitational force and the force exerted by solar radiation.

66. **S** Consider a small, spherical particle of radius r located in space a distance R from the Sun, of mass M_S. Assume the particle has a perfectly absorbing surface and a mass density ρ. The value of the solar intensity at the particle's location is S. Calculate the value of r for which the particle is in equilibrium between the gravitational force and the force exerted by solar radiation. Your answer should be in terms of S, R, ρ, and other constants.

67. **Review.** A 1.00-m-diameter circular mirror focuses the Sun's rays onto a circular absorbing plate 2.00 cm in radius, which holds a can containing 1.00 L of water at 20.0°C. (a) If the solar intensity is 1.00 kW/m^2, what is the intensity on the absorbing plate? At the plate, what are the maximum magnitudes of the fields (b) $\vec{E}$ and (c) $\vec{B}$? (d) If 40.0% of the energy is absorbed, what time interval is required to bring the water to its boiling point?

68. In 1965, Arno Penzias and Robert Wilson discovered the cosmic microwave radiation left over from the big bang expansion of the Universe. Suppose the energy density of this background radiation is 4.00×10^{-14} J/m^3. Determine the corresponding electric field amplitude.

Review problems. Section 17.10 discussed electromagnetic radiation as a mode of energy transfer. Problems 69 through 71 use ideas introduced both there and in this chapter.

69. **Review.** A 5.50-kg black cat and her four black kittens, each with mass 0.800 kg, sleep snuggled together on a mat on a cool night, with their bodies forming a hemisphere. Assume the hemisphere has a surface temperature of 31.0°C, an emissivity of 0.970, and a uniform density of 990 kg/m^3. Find (a) the radius of the hemisphere, (b) the area of its curved surface, (c) the radiated power emitted by the cats at their curved surface and, (d) the intensity of radiation at this surface. You may think of the emitted electromagnetic wave as having a single predominant frequency. Find (e) the amplitude of the electric field in the electromagnetic wave just outside the surface of the cozy pile and (f) the amplitude of the magnetic field. (g) **What If?** The next night, the kittens all sleep alone, curling up into separate hemispheres like their mother. Find the total radiated power of the family.

(For simplicity, ignore the cats' absorption of radiation from the environment.)

70. **Q|C Review.** Gliese 581c is the first Earth-like terrestrial planet discovered. Its parent star, Gliese 581, is a red dwarf that radiates electromagnetic waves with power 5.00×10^{24} W, which is only 1.30% of the power of the Sun. Assume the emissivity of the planet is equal for infrared and for visible light and the planet has a uniform surface temperature. Identify (a) the projected area over which the planet absorbs light from Gliese 581 and (b) the radiating area of the planet. (c) If an average temperature of 287 K is necessary for life to exist on Gliese 581c, what should the radius of the planet's orbit be?

71. **Review.** (a) A homeowner has a solar water heater installed on the roof of his house (Fig. P24.71). The heater is a flat, closed box with excellent thermal insulation. Its interior is painted black, and its front face is made of insulating glass. Its emissivity for visible light is 0.900, and its emissivity for infrared light is 0.700. Light from the noontime Sun is incident perpendicular to the glass with an intensity of 1 000 W/m^2, and no water enters or leaves the box. Find the steady-state temperature of the box's interior. (b) **What If?** The homeowner builds an identical box with no water tubes. It lies flat on the ground in front of the house. He uses it as a cold frame, where he plants seeds in early spring. Assuming the same noontime Sun is at an elevation angle of 50.0°, find the steady-state temperature of the interior of the box when its ventilation slots are tightly closed.

Figure P24.71

72. A microwave source produces pulses of 20.0-GHz radiation, with each pulse lasting 1.00 ns. A parabolic reflector with a face area of radius 6.00 cm is used to focus the microwaves into a parallel beam of radiation as shown in Figure P24.72. The average power during each pulse is 25.0 kW. (a) What is the wavelength of these microwaves? (b) What is the total energy contained in each pulse? (c) Compute the average energy density inside each pulse. (d) Determine the amplitude of the electric and magnetic fields in these microwaves. (e) Assuming that this pulsed beam strikes an absorbing surface, compute the force exerted on the surface during the 1.00-ns duration of each pulse.

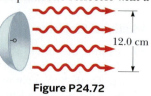

Figure P24.72

12.0 cm

73. A linearly polarized microwave of wavelength 1.50 cm is directed along the positive x axis. The electric field vector has a maximum value of 175 V/m and vibrates in the xy plane. Assuming the magnetic field component of the wave can be written in the form $B = B_{max} \sin (kx - \omega t)$, give values for (a) B_{max}, (b) k, and (c) ω. (d) Determine in which plane the magnetic field vector vibrates. (e) Calculate the average value of the Poynting vector for this wave. (f) If this wave were directed at normal incidence onto a perfectly reflecting sheet, what radiation pressure would it exert? (g) What acceleration would be imparted to a 500-g sheet (perfectly reflecting and at normal incidence) with dimensions of 1.00 m × 0.750 m?

74. The electromagnetic power radiated by a non-relativistic particle with charge q moving with acceleration a is

$$P = \frac{q^2 a^2}{6\pi\epsilon_0 c^3}$$

where ϵ_0 is the permittivity of free space (also called the permittivity of vacuum) and c is the speed of light in vacuum.

(a) Show that the right side of this equation has units of watts. An electron is placed in a constant electric field of magnitude 100 N/C. Determine (b) the acceleration of the electron and (c) the electromagnetic power radiated by this electron. (d) **What If?** If a proton is placed in a cyclotron with a radius of 0.500 m and a magnetic field of magnitude 0.350 T, what electromagnetic power does this proton radiate just before leaving the cyclotron?

75. **Review.** An astronaut, stranded in space 10.0 m from her spacecraft and at rest relative to it, has a mass (including equipment) of 110 kg. Because she has a 100-W flashlight that forms a directed beam, she considers using the beam as a photon rocket to propel herself continuously toward the spacecraft. (a) Calculate the time interval required for her to reach the spacecraft by this method. (b) **What If?** Suppose she throws the 3.00-kg flashlight in the direction away from the spacecraft instead. After being thrown, the flashlight moves at 12.0 m/s relative to the recoiling astronaut. After what time interval will the astronaut reach the spacecraft?

Chapter 25

Reflection and Refraction of Light

Chapter Outline

The preceding chapter serves as a bridge between electromagnetism and the area of physics called *optics*. Now that we have established the wave nature of electromagnetic radiation, we shall study the behavior of visible light and apply what we learn to all electromagnetic radiation. Our emphasis in this chapter will be on the behavior of light as it encounters an interface between two media.

So far, we have focused on the wave nature of light and discussed it in terms of our wave simplification model. As we learn more about the behavior of light, however, we shall return to our particle simplification model, especially as we incorporate the notions of quantum physics, beginning in Chapter 28. As we discuss in Section 25.1, a long historical debate took place between proponents of wave and particle models for light.

This photograph of a rainbow shows a distinct secondary rainbow with the colors reversed. The appearance of the rainbow depends on three optical phenomena discussed in this chapter: reflection, refraction, and dispersion.

Patrick J. Endres/Visuals Unlimited

25.1 | The Nature of Light

We encounter light every day, as soon as we open our eyes in the morning. This everyday experience involves a phenomenon that is actually quite complicated. Since the beginning of this book, we have discussed both the particle model and the wave model as simplification models to help us gain understanding of physical phenomena. Both of these models have been applied to the behavior of light. Until the beginning of the 19th century, most scientists thought that light was a stream of particles emitted by a light source. According to this model, the light particles stimulated the sense of sight on entering the eye. The chief architect of this particle model of light was Isaac Newton. The model provided a simple explanation of some known experimental facts concerning the nature of light—namely, the laws of reflection and refraction—to be discussed in this chapter.

Most scientists at the time accepted the particle model of light. During Newton's lifetime, however, another model was proposed—a model that views light as having wave-like properties. In 1678, a Dutch physicist and astronomer, Christian Huygens, showed that a wave model of light can also explain the laws of reflection and refraction. The wave model did not receive immediate acceptance for several reasons. All the waves known at the time (sound, water, and so on) traveled through a medium, but light from the Sun could travel to the Earth through empty space. Even though experimental evidence for the wave nature of light was discovered by Francesco Grimaldi (1618–1663) around 1660, most scientists rejected the wave model for more than a century and adhered to Newton's particle model due, for the most part, to Newton's great reputation as a scientist.

The first clear and convincing demonstration of the wave nature of light was provided in 1801 by Englishman Thomas Young (1773–1829), who showed that under appropriate conditions, light exhibits interference behavior. That is, light waves emitted by a single source and traveling along two different paths can arrive at some point, combine, and cancel each other by destructive interference. Such behavior could not be explained at that time by a particle model, because scientists could not imagine how two or more particles could come together and cancel one another. Additional developments during the 19th century led to the general acceptance of the wave model of light.

A critical development concerning the understanding of light was the work of James Clerk Maxwell, who in 1865 mathematically predicted that light is a form of high-frequency electromagnetic wave. As discussed in Chapter 24, Hertz in 1887 provided experimental confirmation of Maxwell's theory by producing and detecting other electromagnetic waves. Furthermore, Hertz and other investigators showed that these waves exhibited reflection, refraction, and all the other characteristic properties of waves.

Although the electromagnetic wave model seemed to be well established and could explain most known properties of light, some experiments could not be explained by the assumption that light was a wave. The most striking of these was the *photoelectric effect*, discovered by Hertz, in which electrons are ejected from a metal when its surface is exposed to light. We shall explore this experiment in detail in Chapter 28.

In view of these developments, light must be regarded as having a dual nature. In some cases, light acts like a wave, and in others, it acts like a particle. The classical electromagnetic wave model provides an adequate explanation of light propagation and interference, whereas the photoelectric effect and other experiments involving the interaction of light with matter are best explained by assuming that light is a particle. Light is light, to be sure. The question "Is light a wave or a particle?" is inappropriate; in some experiments, we measure its wave properties; in other experiments, we measure its particle properties. This curious dual nature of light may be unsettling at this point, but it will be clarified when we introduce the notion of a *quantum particle*. The photon, a particle of light, is our first example of a quantum particle, which we shall explore more fully in Chapter 28. Until then, we focus our attention on the properties of light that can be satisfactorily explained with the wave model.

25.2 | The Ray Model in Geometric Optics

In the beginning of our study of optics, we shall use a simplification model called the **ray model** or the **ray approximation.** A **ray** is a straight line drawn along the direction of propagation of a single wave, showing the path of the wave as it travels through space. The ray approximation involves geometric models based on these straight lines. Phenomena explained with the ray approximation do not depend explicitly on the wave nature of light, other than its propagation along a straight line.

A set of light waves can be represented by wave fronts (defined in Section 24.3) as illustrated in the pictorial representation in Figure 25.1 for a plane wave, which was

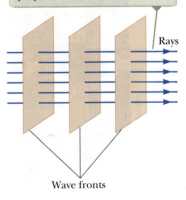

The rays, which always point in the direction of the wave propagation, are straight lines perpendicular to the wave fronts.

Rays

Wave fronts

Figure 25.1 A plane wave propagating to the right.

speed? (d) What is the maximum speed of the spot on the wall? (e) In what time interval does the spot change from its minimum to its maximum speed?

60. *Why is the following situation impossible?* While at the bottom of a calm freshwater lake, a scuba diver sees the Sun at an apparent angle of 38.0° above the horizontal.

61. **S** A material having an index of refraction n is surrounded by vacuum and is in the shape of a quarter circle of radius R (Fig. P25.61). A light ray parallel to the base of the material is incident from the left at a distance L above the base and emerges from the material at the angle θ. Determine an expression for θ in terms of n, R, and L.

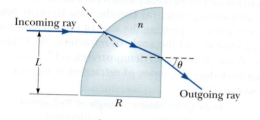

Incoming ray

n

L

θ

Outgoing ray

R

Figure P25.61

62. **Q|C** As sunlight enters the Earth's atmosphere, it changes direction due to the small difference between the speeds of light in vacuum and in air. The duration of an *optical* day is defined as the time interval between the instant when the top of the rising Sun is just visible above the horizon and the instant when the top of the Sun just disappears below the horizontal plane. The duration of the *geometric* day is defined as the time interval between the instant a mathematically straight line between an observer and the top of the Sun just clears the horizon and the instant this line just dips below the horizon. (a) Explain which is longer, an optical day or a geometric day. (b) Find the difference between these two time intervals. Model the Earth's atmosphere as uniform, with index of refraction 1.000 293, a sharply defined upper surface, and depth 8 614 m. Assume the observer is at the Earth's equator so that the apparent path of the rising and setting Sun is perpendicular to the horizon.

Image Formation by Mirrors and Lenses

Don Hammond Photography Ltd. RF

This chapter is concerned with the images formed when light interacts with flat and curved surfaces. We find that images of an object can be formed by reflection or by refraction and that mirrors and lenses work because of these phenomena.

Images formed by reflection and refraction are used in a variety of everyday devices, such as the rearview mirror in your car, a shaving or makeup mirror, a camera, your eyeglasses, and a magnifying glass. In addition, more scientific devices, such as telescopes and microscopes, take advantage of the image formation principles discussed in this chapter.

We shall make extensive use of geometric models developed from the principles of reflection and refraction. Such constructions allow us to develop mathematical representations for the image locations of various types of mirrors and lenses.

The light rays coming from the leaves in the background of this scene did not form a focused image on the film of the camera that took this photograph. Consequently, the background appears very blurry. Light rays passing though the raindrop, however, have been altered so as to form a focused image of the background leaves on the film. In this chapter, we investigate the formation of images as light rays reflect from mirrors and refract through lenses.

26.1 | Images Formed by Flat Mirrors

We begin by considering the simplest possible mirror, the flat mirror. Consider a point source of light[1] placed at O in Figure 26.1 (page 880), a distance p in front of a flat mirror. The distance p is called the **object distance**. Light rays leave the source and are reflected from the mirror. Upon reflection, the rays continue to diverge (spread apart). The dashed lines in Figure 26.1 are extensions of the diverging rays back to a point of intersection at I.

---

[1]We imagine the object to be a point source of light. It could actually *be* a point source, such as a very small lightbulb, but more often is a single point on some extended object that is illuminated from the exterior by a light source. Therefore, the reflected light leaves the point on the object as if the point were a source of light.

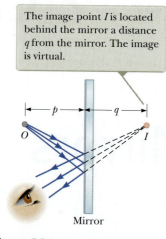

The image point *I* is located behind the mirror a distance *q* from the mirror. The image is virtual.

Figure 26.1 An image formed by reflection from a flat mirror.

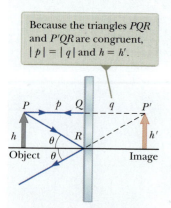

Because the triangles *PQR* and *P'QR* are congruent, |*p*| = |*q*| and *h* = *h'*.

Active Figure 26.2 A geometric construction used to locate the image of an object placed in front of a flat mirror.

▶ Magnification of an image

Pitfall Prevention | 26.1
Magnification Does Not Necessarily Imply Enlargement
For optical elements other than flat mirrors, the magnification defined in Equation 26.1 can result in a number with a magnitude larger *or* smaller than 1. Therefore, despite the cultural usage of the word *magnification* to mean *enlargement*, the image could be smaller than the object. We shall see examples of such a situation in this chapter.

The diverging rays appear to the viewer to come from the point *I* behind the mirror. Point *I* is called the **image** of the object at *O*. Regardless of the system under study, we always locate images by extending diverging rays back to a point at which they intersect.[2] Images are located either at a point from which rays of light *actually* diverge or at a point from which they *appear* to diverge. Because the rays in Figure 26.1 appear to originate at *I*, which is a distance *q* behind the mirror, that is the location of the image. The distance *q* is called the **image distance.**

Images are classified as **real** or **virtual.** A **real image** is formed when light rays pass through and diverge from the image point; a **virtual image** is formed when the light rays do not pass through the image point but only appear to diverge from that point. The image formed by the mirror in Figure 26.1 is virtual. The image of an object seen in a flat mirror is *always* virtual. Real images can be displayed on a screen (as at a movie), but virtual images cannot be displayed on a screen. We shall see an example of a real image in Section 26.2.

Active Figure 26.2 is an example of a specialized pictorial representation, called a **ray diagram,** that is very useful in studies of mirrors and lenses. In a ray diagram, a small number of the myriad rays leaving a point source are drawn, and the location of the image is found by applying the laws of reflection (and refraction, in the case of refracting surfaces and lenses) to these rays. A carefully drawn ray diagram allows us to build a geometric model so that geometry and trigonometry can be used to solve a problem mathematically.

We can use the simple geometry in Active Figure 26.2 to examine the properties of the images of extended objects formed by flat mirrors. Let us locate the image of the tip of the gray arrow. To find out where the image is formed, it is necessary to follow at least two rays of light as they reflect from the mirror. One of those rays starts at *P*, follows the horizontal path *PQ* to the mirror, and reflects back on itself. The second ray follows the oblique path *PR* and reflects at the same angle according to the law of reflection. We can extend the two reflected rays back to the point from which they appear to diverge, point *P'*. A continuation of this process for points other than *P* on the object would result in an image (drawn as a pink arrow) to the right of the mirror. These rays and the extensions of the rays allow us to build a geometric model for the image formation based on triangles *PQR* and *P'QR*. Because these two triangles are identical, *PQ* = *P'Q*, or *p* = |*q*|. (We use the absolute value notation because, as we shall see shortly, a sign convention is associated with the values of *p* and *q*.) Hence, we conclude that the image formed by an object placed in front of a flat mirror is as far behind the mirror as the object is in front of the mirror.

Our geometric model also shows that the object height *h* equals the image height *h'*. We define the **lateral magnification** (or simply the **magnification**) *M* of an image as follows:

$$M \equiv \frac{\text{image height}}{\text{object height}} = \frac{h'}{h} \qquad \text{26.1} \blacktriangleleft$$

which is a general definition of the magnification for any type of image formed by a mirror or lens. Because *h'* = *h* in this case, *M* = 1 for a flat mirror. We also note that the image is *upright* because the image arrow points in the same direction as the object arrow. An upright image is indicated mathematically by a positive value of the magnification. (Later we discuss situations in which *inverted* images, with negative magnifications, are formed.)

Finally, note that a flat mirror produces an image having an *apparent* left–right reversal. This reversal can be seen by standing in front of a mirror and raising your right hand. The image you see raises its left hand. Likewise, your hair appears to be parted on the opposite side, and a mole on your right cheek appears to be on your left cheek.

This reversal is not *actually* a left–right reversal. Imagine, for example, lying on your left side on the floor, with your body parallel to the mirror surface. Now, your head is on the left and your feet are on the right as you face the mirror. If you shake

---

[2]Your eyes and brain interpret diverging light rays as originating at the point from which the rays diverge. Your eye–brain system can detect the rays only *as they enter your eye* and has no access to information about what experiences the rays underwent before reaching your eyes. Therefore, even though the light rays did not *actually originate* at point *I*, they enter the eye *as if they had,* and *I* is the point at which your brain locates the object.

your feet, the image does not shake its head! If you raise your right hand, however, the image raises its left hand. Therefore, it again appears like a left–right reversal, but in an up–down direction!

The apparent left–right reversal is actually a *front–back* reversal caused by the light rays going forward toward the mirror and then reflecting back from it. Figure 26.3 shows a person's right hand and its image in a flat mirror. Notice that no left–right reversal takes place; rather, the thumbs on both the real hand and the image are on the left side. It is the front–back reversal that makes the image of the right hand appear similar to the real left hand at the left side of the photograph.

An interesting experience with front–back reversal is to stand in front of a mirror while holding an overhead transparency in front of you so that you can read the writing on the transparency. You are also able to read the writing on the image of the transparency. You might have had a similar experience if you have a transparent decal with words on it on the rear window of your car. If the decal is placed so that it can be read from outside the car, you can also read it when looking into your rearview mirror from the front seat.

The thumb is on the left side of both real hands and on the left side of the image. That the thumb is not on the right side of the image indicates that there is no left-to-right reversal.

Figure 26.3 The image in the mirror of a person's right hand is reversed front to back, which makes the image in the mirror appear to be a left hand.

> **QUICK QUIZ 26.1** In the overhead view of Figure 26.4, the image of the stone seen by observer 1 is at *C*. At which of the five points *A, B, C, D,* or *E* does observer 2 see the image?

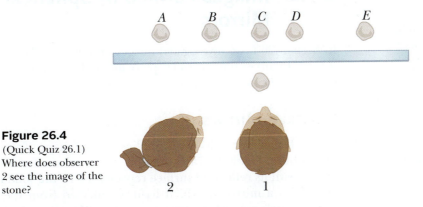

Figure 26.4
(Quick Quiz 26.1)
Where does observer 2 see the image of the stone?

> **QUICK QUIZ 26.2** You are standing approximately 2 m away from a mirror. The mirror has water spots on its surface. True or False: It is possible for you to see the water spots and your image both in focus at the same time.

> **THINKING PHYSICS 26.1**

Most rearview mirrors in cars have a day setting and a night setting. The night setting greatly diminishes the intensity of the image so that lights from trailing vehicles do not temporarily blind the driver. How does such a mirror work?

SOLUTION Figure 26.5 shows a cross-sectional view of a rearview mirror for each setting. The unit consists of a reflective coating on the back of a wedge of glass. In the day setting (Fig. 26.5a), the light from an object behind the car strikes the glass wedge at point 1. Most of the light enters the wedge, refracting as it crosses the front surface, and reflects from the back surface to return to the front surface, where it is refracted again as it re-enters the air as ray *B* (for *bright*). In addition, a small portion of the light is reflected at the front surface of the glass as indicated by ray *D* (for *dim*).

This dim reflected light is responsible for the image observed when the mirror is in the night setting (Fig. 26.5b).

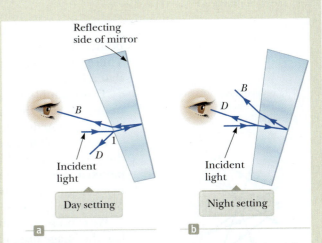

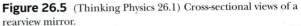

Figure 26.5 (Thinking Physics 26.1) Cross-sectional views of a rearview mirror.

In that case, the wedge is rotated so that the path followed by the bright light (ray *B*) does not lead to the eye. Instead, the dim light reflected from the front surface of the wedge travels to the eye, and the brightness of trailing headlights does not become a hazard. ◄

> ◤ **THINKING PHYSICS 26.2**
>
> Two flat mirrors are perpendicular to each other as in Figure 26.6, and an object is placed at point O. In this situation, multiple images are formed. Locate the positions of these images.
>
> **SOLUTION** The image of the object is at I_1 in mirror 1 (green rays) and at I_2 in mirror 2 (red rays). In addition, a third image is formed at I_3 (blue rays). This third image is the image of I_1 in mirror 2 or, equivalently, the image of I_2 in mirror 1. That is, the image at I_1 (or I_2) serves as the object for I_3. To form this image at I_3, the rays reflect twice after leaving the object at O. ◀

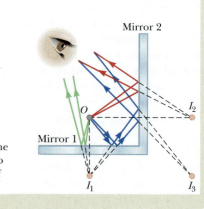

Figure 26.6 (Thinking Physics 26.2) When an object is placed in front of two mutually perpendicular mirrors as shown, three images are formed. Follow the different-colored light rays to understand the formation of each image.

◥ 26.2 | Images Formed by Spherical Mirrors

In Section 26.1, we investigated images formed by a flat reflecting surface. In this section, we will explore images formed by curved mirrors, either from a concave surface of the mirror or a convex surface.

Concave Mirrors

A **spherical mirror,** as its name implies, has the shape of a segment of a sphere. Figure 26.7a shows the cross-section of a spherical mirror with its reflecting surface represented by the solid curved line. Such a mirror in which light is reflected from the inner, concave surface is called a **concave mirror.** The mirror's radius of curvature is R, and its center of curvature is at point C. Point V is the center of the spherical segment, and a line drawn through C and V is called the **principal axis** of the mirror.

Now consider a point source of light placed at point O in Figure 26.7b, on the principal axis and outside point C. Two diverging rays that originate at O are shown. After reflecting from the mirror, these rays converge and meet at I, the image point. They then continue to diverge from I as if a source of light existed there. Therefore, if your eyes detect the rays diverging from point I, you would claim that a light source is located at that point.

This example is the second one we have seen of rays diverging from an image point. Because the light rays pass through the image point in this case, unlike the situation in Active Figure 26.2, the image in Figure 26.7b is a real image.

In what follows, we shall adopt a simplification model that assumes that all rays diverging from an object make small angles with the principal axis. Such rays, called **paraxial rays,** always reflect through the image point as in Figure 26.7b. Rays that make large angles with the principal axis as in Figure 26.8 converge at other points on the principal axis, producing a blurred image.

We can use a geometric model based on the ray diagram in Figure 26.9 to calculate the image distance q if we know the object distance p and radius of curvature R. By convention, these distances are measured from point V. Figure 26.9 shows two of the many light rays leaving the tip of the object. One ray passes through the center of curvature C of the mirror, hitting the mirror perpendicular to the mirror surface and reflecting back on itself. The second ray strikes the mirror at the center point V and reflects as shown, obeying the law of reflection. The image of the tip of the

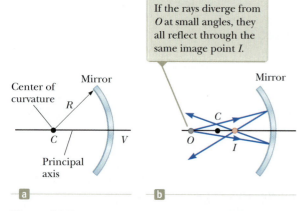

If the rays diverge from O at small angles, they all reflect through the same image point I.

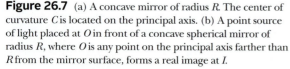

Figure 26.7 (a) A concave mirror of radius R. The center of curvature C is located on the principal axis. (b) A point source of light placed at O in front of a concave spherical mirror of radius R, where O is any point on the principal axis farther than R from the mirror surface, forms a real image at I.

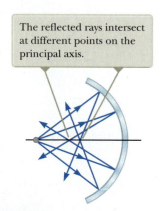

The reflected rays intersect at different points on the principal axis.

Figure 26.8 A spherical concave mirror produces a blurred image when light rays make large angles with the principal axis.

arrow is at the point at which these two reflected rays intersect. Using these rays, we identify the red and yellow model right triangles in Figure 26.9. From the red triangle, we see that $\tan \theta = h/p$, whereas the yellow triangle gives $\tan \theta = -h'/q$. The negative sign signifies that the image is inverted, so h' is a negative number. Therefore, from Equation 26.1 and these results, we find that the magnification of the image is

$$M = \frac{h'}{h} = \frac{-q \tan \theta}{p \tan \theta} = -\frac{q}{p} \qquad \textbf{26.2} \blacktriangleleft$$

We can identify two additional right triangles in the figure (the green one and the smaller red one), with a common point at C and with angle α. These triangles tell us that

$$\tan \alpha = \frac{h}{p - R} \quad \text{and} \quad \tan \alpha = -\frac{h'}{R - q}$$

from which we find that

$$\frac{h'}{h} = -\frac{R - q}{p - R} \qquad \textbf{26.3} \blacktriangleleft$$

If we compare Equations 26.2 and 26.3, we see that

$$\frac{R - q}{p - R} = \frac{q}{p}$$

Algebra reduces this expression to

$$\frac{1}{p} + \frac{1}{q} = \frac{2}{R} \qquad \textbf{26.4} \blacktriangleleft \qquad \blacktriangleright \text{Mirror equation in terms of the radius of curvature}$$

which is called the **mirror equation**. It is applicable only to the paraxial ray simplification model.

If the object is very far from the mirror—that is, if the object distance p is large compared with R, so that p can be said to approach infinity—$1/p \rightarrow 0$, and we see from Equation 26.4 that $q \approx R/2$. In other words, when the object is very far from the mirror, the image point is halfway between the center of curvature and the center of the mirror as in Figure 26.10a. The rays are essentially parallel in this figure because only those few rays traveling parallel to the axis from the distant object encounter the mirror. Rays not parallel to the axis miss the mirror. Figure 26.10b shows an experimental setup of this situation, demonstrating the crossing of the light rays at a single point. The point at which the parallel rays intersect after reflecting from the mirror is called the **focal point** F of the mirror. The focal point is a distance f from the mirror, called the **focal length**. The focal length is a parameter associated with the mirror and is given by

$$f = \frac{R}{2} \qquad \textbf{26.5} \blacktriangleleft$$

The mirror equation can therefore be expressed in terms of the focal length:

$$\boxed{\frac{1}{p} + \frac{1}{q} = \frac{1}{f}} \qquad \textbf{26.6} \blacktriangleleft \qquad \blacktriangleright \text{Mirror equation in terms of focal length}$$

This equation is the commonly used mirror equation, in terms of the focal length of the mirror rather than its radius of curvature, as in Equation 26.4. We shall see how to use this equation in examples that follow shortly.

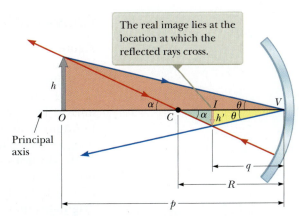

The real image lies at the location at which the reflected rays cross.

Figure 26.9 The image formed by a spherical concave mirror when the object O lies outside the center of curvature C. This geometric construction is used to derive Equation 26.4.

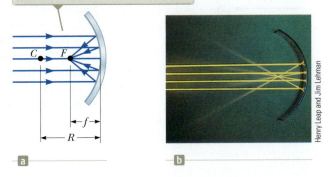

When the object is very far away, the image distance $q \approx R/2 = f$, where f is the focal length of the mirror.

Figure 26.10 (a) Light rays from a distant object ($p \approx \infty$) reflect from a concave mirror through the focal point F. (b) Reflection of parallel rays from a concave mirror.

Henry Leap and Jim Lehman

Convex Mirrors

Figure 26.11 shows the formation of an image by a **convex mirror,** a mirror that is silvered so that light is reflected from the outer, convex surface. Convex mirrors are sometimes called **diverging mirrors** because the rays from any point on an object diverge after reflection as though they were coming from some point behind the mirror. The image in Figure 26.11 is virtual because the reflected rays only appear to originate at the image point as indicated by the dashed lines. Furthermore, the image is always upright and smaller than the object.

We do not derive any equations for convex spherical mirrors because Equations 26.2, 26.4, and 26.6 can be used for either concave or convex mirrors if we adhere to the following procedure. We will refer to the region in which light rays originate and move toward the mirror as the *front side* of the mirror and the other side as the *back side*. For example, in Figures 26.9 and 26.11, the side to the left of the mirrors is the front side and the side to the right of the mirrors is the back side. Figure 26.12 states the sign conventions for object and image distances, and Table 26.1 summarizes the sign conventions for all quantities. One entry in the table, a *virtual object,* is formally introduced in Section 26.4.

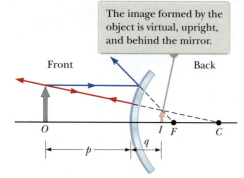

The image formed by the object is virtual, upright, and behind the mirror.

Figure 26.11 Formation of an image by a spherical convex mirror.

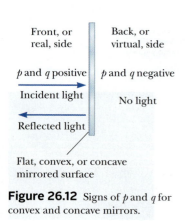

Figure 26.12 Signs of p and q for convex and concave mirrors.

Ray Diagrams for Mirrors

We have been using the specialized pictorial representations called ray diagrams to help us locate images for flat and curved mirrors. Let us now formalize the procedure for drawing accurate ray diagrams. To construct such a diagram, we must know the position of the object and the locations of the focal point and center of curvature of the mirror. We will construct three rays in the examples shown in Active Figure 26.13. Only two rays are necessary to locate the image, but we will include a third as a check. In each part of the figure, the right-hand portion shows a photograph of the situation described by the ray diagram in the left-hand portion. All three rays start from the same object point; in these examples, the top of the object arrow is chosen as the starting point. For the concave mirrors in Active Figure 26.13a and 26.13b, the rays are drawn as follows:

- Ray 1 is drawn from the top of the object parallel to the principal axis and is reflected through the focal point F.
- Ray 2 is drawn from the top of the object through the focal point (or as if coming from the focal point if $p < f$) and is reflected parallel to the principal axis.
- Ray 3 is drawn from the top of the object through the center of curvature C (or as if coming from the center C if $p < f$) and is reflected back on itself.

◄ **TABLE 26.1** | **Sign Conventions for Mirrors**

Quantity	Positive when . . .	Negative when . . .
Object location (p)	object is in front of mirror (real object).	object is in back of mirror (virtual object).
Image location (q)	image is in front of mirror (real image).	image is in back of mirror (virtual image).
Image height (h')	image is upright.	image is inverted.
Focal length (f) and radius (R)	mirror is concave.	mirror is convex.
Magnification (M)	image is upright.	image is inverted.

The image point obtained in this fashion must always agree with the value of q calculated from the mirror equation. With concave mirrors, note what happens as the object is moved closer to the mirror from infinity. The real, inverted image in Active Figure 26.13a moves to the left as the object approaches the mirror. When the object is at the center of curvature, the object and image are at the same distance from the mirror and are the same size. When the object is at the focal point, the image is infinitely far to the left. (Check these last three sentences with the mirror equation!)

Pitfall Prevention | 26.2
Watch Your Signs
Success in working mirror problems (as well as problems involving refracting surfaces and thin lenses) is largely determined by proper sign choices when substituting into the equations. The best way to become adept at these problems is to work a multitude of them on your own.

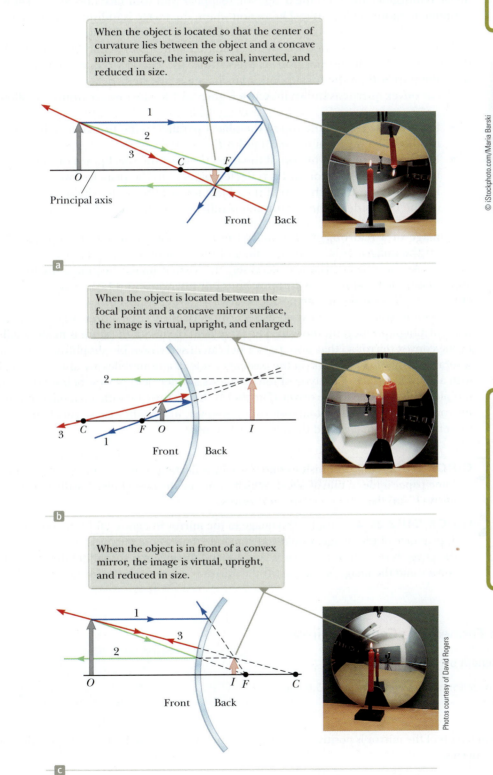

When the object is located so that the center of curvature lies between the object and a concave mirror surface, the image is real, inverted, and reduced in size.

a

When the object is located between the focal point and a concave mirror surface, the image is virtual, upright, and enlarged.

b

When the object is in front of a convex mirror, the image is virtual, upright, and reduced in size.

c

Active Figure 26.13 Ray diagrams for spherical mirrors, along with corresponding photographs of the images of candles.

A satellite-dish antenna is a concave reflector for television signals from a satellite in orbit around the Earth. Because the satellite is so far away, the signals are carried by microwaves that are parallel when they arrive at the dish. These waves reflect from the dish and are focused on the receiver.

Pitfall Prevention | 26.3
The *Focal* Point Is Not The *Focus* Point
The focal point *is usually not* the point at which the light rays focus to form an image. The focal point is determined solely by the curvature of the mirror; it does not depend on the location of the object at all. In general, an image forms at a point different from the focal point of a mirror (or a lens). The *only* exception is when the object is located infinitely far away from the mirror.

When the object lies between the focal point and the mirror surface as in Active Figure 26.13b, the image is virtual, upright, and located on the back side of the mirror. The image is also larger than the object in this case. This situation illustrates the principle behind a shaving mirror or a makeup mirror. Your face is located closer to the concave mirror than the focal point, so you see an enlarged, upright image of your face, to assist you with shaving or applying makeup. If you have such a mirror, look into it and move your face farther from the mirror. Your head will pass through a point at which the image is indistinct and then the image will reappear with your face upside down as you continue to move farther away. The region where the image is indistinct is where your head passes through the focal point and the image is infinitely far away.

Notice that the image of the camera in Active Figures 26.13a and 26.13b is upside down. Regardless of the position of the candle, the camera remains farther away from the mirror than the focal point, so its image is inverted.

For a convex mirror as shown in Active Figure 26.13c, the rays are drawn as follows:

- Ray 1 is drawn from the top of the object parallel to the principal axis and is reflected away from the focal point *F*.
- Ray 2 is drawn from the top of the object toward the focal point on the back side of the mirror and is reflected parallel to the principal axis.
- Ray 3 is drawn from the top of the object toward the center of curvature *C* on the back side of the mirror and is reflected back on itself.

The image of a real object in a convex mirror is always virtual and upright. Notice that the images of both the candle and the camera in Active Figure 26.13c are upright. As the object distance increases, the virtual image becomes smaller and approaches the focal point as *p* approaches infinity. You should construct other diagrams to verify how the image position varies with object position.

Convex mirrors are often used as security devices in large stores, where they are hung at a high position on the wall. The large field of view of the store is made smaller by the convex mirror so that store personnel can observe possible shoplifting activity in several aisles at once. Mirrors on the passenger side of automobiles are also often made with a convex surface. This type of mirror allows a wider field of view behind the automobile to be available to the driver (Fig. 26.14) than is the case with a flat mirror. These mirrors introduce a perceptual distortion, however, in that they cause cars behind the viewer to appear smaller and therefore farther away.

Figure 26.14 An approaching truck is seen in a convex mirror on the right side of an automobile. Notice that the image of the truck is in focus, but the frame of the mirror is not, which demonstrates that the image is not at the same location as the mirror surface.

© Bo Zaunders/Corbis

Figure 26.15 (Quick Quiz 26.4) What type of mirror is shown here?

NASA

> **QUICK QUIZ 26.3** You wish to start a fire by reflecting sunlight from a mirror onto some paper under a pile of wood. Which would be the best choice for the type of mirror? **(a)** flat **(b)** concave **(c)** convex

> **QUICK QUIZ 26.4** Consider the image in the mirror in Figure 26.15. Based on the appearance of this image, would you conclude that **(a)** the mirror is concave and the image is real, **(b)** the mirror is concave and the image is virtual, **(c)** the mirror is convex and the image is real, or **(d)** the mirror is convex and the image is virtual?

Example 26.1 | The Image Formed by a Concave Mirror

A spherical mirror has a focal length of +10.0 cm.

(A) Locate and describe the image for an object distance of 25.0 cm.

SOLUTION

Conceptualize Because the focal length of the mirror is positive, it is a concave mirror (see Table 26.1). We expect the possibilities of both real and virtual images.

Categorize Because the object distance in this part of the problem is larger than the focal length, we expect the image to be real. This situation is analogous to that in Active Figure 26.13a.

26.1 *cont.*

Analyze Find the image distance by using Equation 26.6:

$$\frac{1}{q} = \frac{1}{f} - \frac{1}{p}$$

$$\frac{1}{q} = \frac{1}{10.0 \text{ cm}} - \frac{1}{25.0 \text{ cm}}$$

$$q = \boxed{16.7 \text{ cm}}$$

Find the magnification of the image from Equation 26.2:

$$M = -\frac{q}{p} = -\frac{16.7 \text{ cm}}{25.0 \text{ cm}} = \boxed{-0.667}$$

Finalize The absolute value of *M* is less than unity, so the image is smaller than the object, and the negative sign for *M* tells us that the image is inverted. Because *q* is positive, the image is located on the front side of the mirror and is real. Look into the bowl of a shiny spoon or stand far away from a shaving mirror to see this image.

(B) Locate and describe the image for an object distance of 10.0 cm.

SOLUTION

Categorize Because the object is at the focal point, we expect the image to be infinitely far away.

Analyze Find the image distance by using Equation 26.6:

$$\frac{1}{q} = \frac{1}{f} - \frac{1}{p}$$

$$\frac{1}{q} = \frac{1}{10.0 \text{ cm}} - \frac{1}{10.0 \text{ cm}}$$

$$q = \boxed{\infty}$$

Finalize This result means that rays originating from an object positioned at the focal point of a mirror are reflected so that the image is formed at an infinite distance from the mirror; that is, the rays travel parallel to one another after reflection. Such is the situation in a flashlight or an automobile headlight, where the bulb filament is placed at the focal point of a reflector, producing a parallel beam of light.

(C) Locate and describe the image for an object distance of 5.00 cm.

SOLUTION

Categorize Because the object distance is smaller than the focal length, we expect the image to be virtual. This situation is analogous to that in Active Figure 26.13b.

Analyze Find the image distance by using Equation 26.6:

$$\frac{1}{q} = \frac{1}{f} - \frac{1}{p}$$

$$\frac{1}{q} = \frac{1}{10.0 \text{ cm}} - \frac{1}{5.00 \text{ cm}}$$

$$q = \boxed{-10.0 \text{ cm}}$$

Find the magnification of the image from Equation 26.2:

$$M = -\frac{q}{p} = -\left(\frac{-10.0 \text{ cm}}{5.00 \text{ cm}}\right) = \boxed{+2.00}$$

Finalize The image is twice as large as the object, and the positive sign for *M* indicates that the image is upright (see Active Fig. 26.13b). The negative value of the image distance tells us that the image is virtual, as expected. Put your face close to a shaving mirror to see this type of image.

Example 26.2 | The Image Formed by a Convex Mirror

An automobile rearview mirror as shown in Figure 26.14 shows an image of a truck located 10.0 m from the mirror. The focal length of the mirror is −0.60 m.

(A) Find the position of the image of the truck.

SOLUTION

Conceptualize This situation is depicted in Active Figure 26.13c.

Categorize Because the mirror is convex, we expect it to form an upright, reduced, virtual image for any object position.

Analyze Find the image distance by using Equation 26.6:

$$\frac{1}{q} = \frac{1}{f} - \frac{1}{p}$$

$$\frac{1}{q} = \frac{1}{-0.60 \text{ m}} - \frac{1}{10.0 \text{ m}}$$

$$q = -0.57 \text{ m}$$

(B) Find the magnification of the image.

SOLUTION

Analyze Use Equation 26.2:

$$M = -\frac{q}{p} = -\left(\frac{-0.57 \text{ m}}{10.0 \text{ m}}\right) = +0.057$$

Finalize The negative value of q in part (A) indicates that the image is virtual, or behind the mirror, as shown in Active Figure 26.13c. The magnification in part (B) indicates that the image is much smaller than the truck and is upright because M is positive. The image is reduced in size, so the truck appears to be farther away than it actually is. Because of the image's small size, these mirrors carry the inscription, "Objects in this mirror are closer than they appear." Look into your rearview mirror or the back side of a shiny spoon to see an image of this type.

26.3 | Images Formed by Refraction

In this section, we describe how images are formed by the refraction of rays at the surface of a transparent material. We shall apply the law of refraction and use the simplification model in which we consider only paraxial rays.

Consider two transparent media with indices of refraction n_1 and n_2, where the boundary between the two media is a spherical surface with radius of curvature R (Fig. 26.16). We shall assume that the object at point O is in the medium with index of refraction n_1. As we shall see, all paraxial rays are refracted at the spherical surface and converge to a single point I, the image point.

Let us proceed by considering the geometric construction in Figure 26.17, which shows a single ray leaving point O and passing through point I. Snell's law applied to this refracted ray gives

$$n_1 \sin \theta_1 = n_2 \sin \theta_2$$

Because the angles θ_1 and θ_2 are small for paraxial rays, we can use the approximation $\sin \theta \approx \theta$ (angles in radians). Therefore, Snell's law becomes

$$n_1 \theta_1 = n_2 \theta_2$$

Now we make use of geometric model triangles and recall that an exterior angle of any triangle equals the sum of the two opposite interior angles. Applying this rule to the triangles OPC and PIC in Figure 26.17 gives

$$\theta_1 = \alpha + \beta$$

$$\beta = \theta_2 + \gamma$$

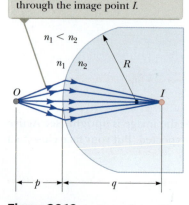

Rays making small angles with the principal axis diverge from a point object at O and are refracted through the image point I.

Figure 26.16 An image formed by refraction at a spherical surface.

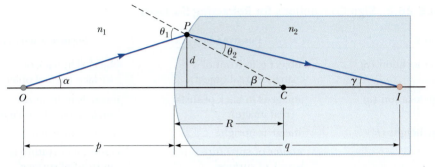

Figure 26.17 Geometry used to derive Equation 26.8, assuming that $n_1 < n_2$. Point C is the center of curvature of the curved refracting surface.

If we combine the last three equations and eliminate θ_1 and θ_2, we find that

$$n_1\alpha + n_2\gamma = (n_2 - n_1)\beta \qquad \textbf{26.7}\blacktriangleleft$$

In the small angle approximation, $\tan\theta \approx \theta$, and so from Figure 26.17 we can write the approximate relations

$$\tan\alpha \approx \alpha \approx \frac{d}{p} \qquad \tan\beta \approx \beta \approx \frac{d}{R} \qquad \tan\gamma \approx \gamma \approx \frac{d}{q}$$

where d is the distance shown in Figure 26.17. We substitute these equations into Equation 26.7 and divide through by d to give

$$\frac{n_1}{p} + \frac{n_2}{q} = \frac{n_2 - n_1}{R} \qquad \textbf{26.8}\blacktriangleleft$$

▶ Relation between object and image distance for a refracting surface

Because this expression does not involve any angles, all paraxial rays leaving an object at distance p from the refracting surface will be focused at the same distance q from the surface on the back side.

By setting up a geometric construction with an object and a refracting surface, we can show that the magnification of an image due to a refracting surface is

$$M = -\frac{n_1 q}{n_2 p} \qquad \textbf{26.9}\blacktriangleleft$$

▶ Magnification of an image formed by a refracting surface

As with mirrors, we must use a sign convention if we are to apply Equations 26.8 and 26.9 to a variety of circumstances. Note that real images are formed on the side of the surface that is *opposite* the side from which the light comes. That is in contrast to mirrors, for which real images are formed on the side where the light originates. Therefore, the sign conventions for spherical refracting surfaces are similar to the conventions for mirrors, recognizing the change in sides of the surface for real and virtual images. For example, in Figure 26.17, p, q, and R are all positive.

The sign conventions for spherical refracting surfaces are summarized in Table 26.2 (page 890). The same conventions will be used for thin lenses discussed in the next section. As with mirrors, we assume that the front of the refracting surface is the side from which the light approaches the surface.

Flat Refracting Surfaces

If the refracting surface is flat, R approaches infinity and Equation 26.8 reduces to

$$\frac{n_1}{p} = -\frac{n_2}{q}$$

or

$$q = -\frac{n_2}{n_1}p \qquad \textbf{26.10}\blacktriangleleft$$

TABLE 26.2 | Sign Conventions for Refracting Surfaces

Quantity	Positive when . . .	Negative when . . .
Object location (p)	object is in front of surface (real object).	object is in back of surface (virtual object).
Image location (q)	image is in back of surface (real image).	image is in front of surface (virtual image).
Image height (h')	image is upright.	image is inverted.
Radius (R)	center of curvature is in back of surface.	center of curvature is in front of surface.

The image is virtual and on the same side of the surface as the object.

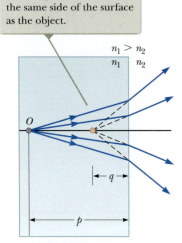

$n_1 > n_2$

n_1 n_2

O

q

p

Active Figure 26.18 The image formed by a flat refracting surface. All rays are assumed to be paraxial.

From Equation 26.10, we see that the sign of q is opposite that of p. Therefore, the image formed by a flat refracting surface is on the same side of the surface as the object. This situation is illustrated in Active Figure 26.18 for the case in which n_1 is greater than n_2, where a virtual image is formed between the object and the surface. Note that the refracted ray bends *away* from the normal in this case because $n_1 > n_2$.

The value of q given by Equation 26.10 is always smaller in magnitude than p when $n_1 > n_2$. This fact indicates that the image of an object located within a material with higher index of refraction than that of the material from which it is viewed is always closer to the flat refracting surface than the object. Therefore, transparent bodies of water such as streams and swimming pools always appear shallower than they are because the image of the bottom of the body of water is closer to the surface than the bottom is in reality.

Example 26.3 | Gaze into the Crystal Ball

A set of coins is embedded in a spherical plastic paperweight having a radius of 3.0 cm. The index of refraction of the plastic is $n_1 = 1.50$. One coin is located 2.0 cm from the edge of the sphere (Fig. 26.19). Find the position of the image of the coin.

Figure 26.19
(Example 26.3) Light rays from a coin embedded in a plastic sphere form a virtual image between the surface of the object and the sphere surface. Because the object is inside the sphere, the front of the refracting surface is the *interior* of the sphere.

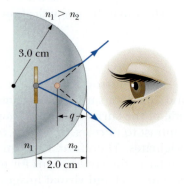

$n_1 > n_2$

3.0 cm

q

n_1 n_2

2.0 cm

SOLUTION

Conceptualize Because $n_1 > n_2$, where $n_2 = 1.00$ is the index of refraction for air, the rays originating from the coin in Figure 26.19 are refracted away from the normal at the surface and diverge outward.

Categorize Because the light rays originate in one material and then pass through a curved surface into another material, this example involves an image formed by refraction.

....

Analyze Apply Equation 26.8, noting from Table 26.2 that R is negative:

$$\frac{n_2}{q} = \frac{n_2 - n_1}{R} - \frac{n_1}{p}$$

$$\frac{1}{q} = \frac{1.00 - 1.50}{-3.0 \text{ cm}} - \frac{1.50}{2.0 \text{ cm}}$$

$$q = \boxed{-1.7 \text{ cm}}$$

....

Finalize The negative sign for q indicates that the image is in front of the surface; in other words, it is in the same medium as the object as shown in Figure 26.19. Therefore, the image must be virtual. (See Table 26.2.) The coin appears to be closer to the paperweight surface than it actually is.

Example 26.4 | **The One That Got Away**

A small fish is swimming at a depth d below the surface of a pond (Fig. 26.20).

(A) What is the apparent depth of the fish as viewed from directly overhead?

SOLUTION

Conceptualize Because $n_1 > n_2$, where $n_2 = 1.00$ is the index of refraction for air, the rays originating from the fish in Figure 26.20a are refracted away from the normal at the surface and diverge outward.

Categorize Because the refracting surface is flat, R is infinite. Hence, we can use Equation 26.10 to determine the location of the image with $p = d$.

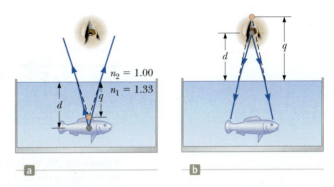

Figure 26.20 (Example 26.4) (a) The apparent depth q of the fish is less than the true depth d. All rays are assumed to be paraxial. (b) Your face appears to the fish to be higher above the surface than it is.

Analyze Use the indices of refraction given in Figure 26.20a in Equation 26.10:

$$q = -\frac{n_2}{n_1}p = -\frac{1.00}{1.33}d = \boxed{-0.752d}$$

Finalize Because q is negative, the image is virtual as indicated by the dashed lines in Figure 26.20a. The apparent depth is approximately three-fourths the actual depth.

(B) If your face is a distance d above the water surface, at what apparent distance above the surface does the fish see your face?

SOLUTION

The light rays from your face are shown in Figure 26.20b.

Conceptualize Because the rays refract toward the normal, your face appears higher above the surface than it actually is.

Categorize Because the refracting surface is flat, R is infinite. Hence, we can use Equation 26.10 to determine the location of the image with $p = d$.

Analyze Use Equation 26.10 to find the image distance:

$$q = -\frac{n_2}{n_1}p = -\frac{1.33}{1.00}d = \boxed{-1.33d}$$

Finalize The negative sign for q indicates that the image is in the medium from which the light originated, which is the air above the water.

26.4 | Images Formed by Thin Lenses

A typical **thin lens** consists of a piece of glass or plastic, ground so that its two surfaces are either segments of spheres or planes. Lenses are commonly used in optical instruments such as cameras, telescopes, and microscopes to form images by refraction.

Figure 26.21 shows cross sections of some representative shapes of lenses. These lenses have been placed in two groups. Those in Figure 26.21a are thicker at the center than at the rim, and those in Figure 26.21b are thinner at the center than at the rim. The lenses in the first group are examples of **converging lenses,** and those in the second group are called **diverging lenses.** The reason for these names will become apparent shortly.

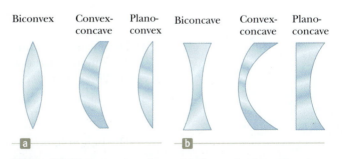

Figure 26.21 Various lens shapes. (a) Converging lenses have a positive focal length and are thickest at the middle. (b) Diverging lenses have a negative focal length and are thickest at the edges.

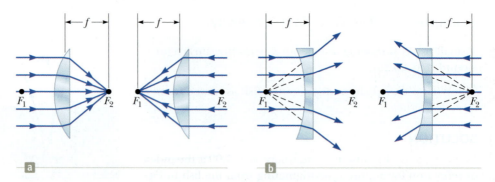

Figure 26.22 Parallel light rays pass through (a) a converging lens and (b) a diverging lens. The focal length is the same for light rays passing through a given lens in either direction. Both focal points F_1 and F_2 are the same distance from the lens.

Pitfall Prevention | 26.4
A Lens Has Two Focal Points but Only One Focal Length
A lens has a focal point on each side, front and back. There is only one focal length, however; each of the two focal points is located the same distance from the lens (Fig. 26.22). As a result, the lens forms an image of an object at the same point if it is turned around. In practice, that might not happen because real lenses are not infinitesimally thin.

As with mirrors, it is convenient to define a point called the **focal point** for a lens. For example, in Figure 26.22a, a group of rays parallel to the principal axis passes through the focal point after being converged by the lens. The distance from the focal point to the lens is again called the **focal length** f. The focal length is the image distance that corresponds to an infinite object distance.

To avoid the complications arising from the thickness of the lens, we adopt a simplification model called the **thin lens approximation,** in which the thickness of the lens is assumed to be negligible. As a result, it makes no difference whether we take the focal length to be the distance from the focal point to the surface of the lens or from the focal point to the center of the lens because the difference in these two lengths is assumed to be negligible. (We will draw lenses in the diagrams with a thickness so that they can be seen.) A thin lens has one focal length and *two* focal points as illustrated in Figure 26.22, corresponding to parallel light rays traveling from the left or right.

Rays parallel to the axis diverge after passing through a lens of the shape shown in Figure 26.22b. In this case, the focal point is defined as the point from which the diverging rays appear to originate, as in Figure 26.22b. Figures 26.22a and 26.22b indicate why the names *converging* and *diverging* are applied to the lenses in Figure 26.21.

Consider now the ray diagram in Figure 26.23 for an object placed a distance p from a converging lens. The red ray from the tip of the object passes through the center of the lens. The blue ray is parallel to the principal axis of the lens (the horizontal axis passing through the center of the lens), and as a result it passes through the focal point F after refraction. The point at which these two rays intersect is the image point at a distance q from the lens.

The tangent of the angle α can be found by using the blue and yellow geometric model triangles in Figure 26.23:

$$\tan \alpha = \frac{h}{p} \quad \text{and} \quad \tan \alpha = -\frac{h'}{q}$$

from which

$$M = \frac{h'}{h} = -\frac{q}{p} \qquad \text{26.11} \blacktriangleleft$$

Therefore, the equation for magnification of an image by a lens is the same as the equation for magnification due to a mirror (Eq. 26.2). We also note from Figure 26.23 that

$$\tan \theta = \frac{d}{f} \quad \text{and} \quad \tan \theta = -\frac{h'}{q - f}$$

The height d, however, is the same as h. Therefore,

$$\frac{h}{f} = -\frac{h'}{q - f} \quad \rightarrow \quad \frac{h'}{h} = -\frac{q - f}{f}$$

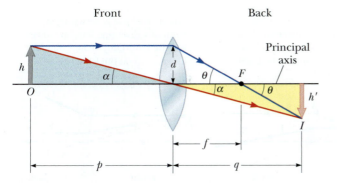

Figure 26.23 A geometric construction for developing the thin lens equation.

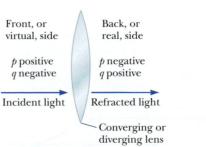

Figure 26.24 A diagram for obtaining the signs of p and q for a thin lens. (This diagram also applies to a refracting surface.)

Using this expression in combination with Equation 26.11 gives us

$$\frac{q}{p} = \frac{q - f}{f}$$

which reduces to

$$\frac{1}{p} + \frac{1}{q} = \frac{1}{f}$$

26.12◀   ▶ Thin lens equation

This equation, called the **thin lens equation** (which is identical to the mirror equation, Eq. 26.6), can be used with either converging or diverging lenses if we adhere to a set of sign conventions. Figure 26.24 is useful for obtaining the signs of p and q. (As with mirrors, we call the side from which the light approaches the *front* of the lens.) The complete sign conventions for lenses are provided in Table 26.3. Note that a converging lens has a positive focal length under this convention and a diverging lens has a negative focal length. Hence, the names *positive* and *negative* are often given to these lenses.

The focal length for a lens in air is related to the curvatures of its surfaces and to the index of refraction n of the lens material by

$$\frac{1}{f} = (n - 1)\left(\frac{1}{R_1} - \frac{1}{R_2}\right)$$

26.13◀   ▶ Lens-makers' equation

where R_1 is the radius of curvature of the front surface and R_2 is the radius of curvature of the back surface. Equation 26.13 enables us to calculate the focal length from the known properties of the lens. It is called the **lens-makers' equation.** Table 26.3 includes the sign conventions for determining the signs of the radii R_1 and R_2.

Ray Diagrams for Thin Lenses

Our specialized pictorial representations called ray diagrams are very convenient for locating the image of a thin lens or system of lenses. They should also help clarify the

TABLE 26.3 | Sign Conventions for Thin Lenses

Quantity	Positive when . . .	Negative when . . .
Object location (p)	object is in front of lens (real object).	object is in back of lens (virtual object).
Image location (q)	image is in back of lens (real image).	image is in front of lens (virtual image).
Image height (h')	image is upright.	image is inverted.
R_1 and R_2	center of curvature is in back of lens.	center of curvature is in front of lens.
Focal length (f)	a converging lens.	a diverging lens.

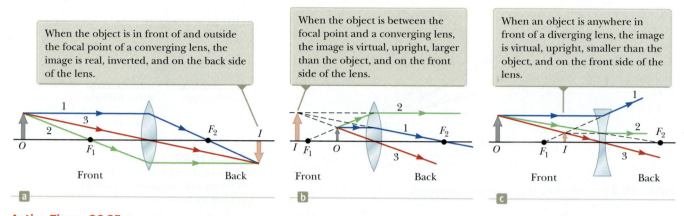

When the object is in front of and outside the focal point of a converging lens, the image is real, inverted, and on the back side of the lens.

When the object is between the focal point and a converging lens, the image is virtual, upright, larger than the object, and on the front side of the lens.

When an object is anywhere in front of a diverging lens, the image is virtual, upright, smaller than the object, and on the front side of the lens.

Active Figure 26.25 Ray diagrams for locating the image formed by a thin lens.

sign conventions we have already discussed. Active Figure 26.25 illustrates this method for three single-lens situations. To locate the image of a converging lens (Active Figs. 26.25a and 26.25b), the following three rays are drawn from the top of the object:

- Ray 1 is drawn parallel to the principal axis. After being refracted by the lens, this ray passes through the focal point on the back side of the lens.
- Ray 2 is drawn through the focal point on the front side of the lens (or as if coming from the focal point if $p < f$) and emerges from the lens parallel to the principal axis.
- Ray 3 is drawn through the center of the lens and continues in a straight line.

To locate the image of a *diverging* lens (Active Fig. 26.25c), the following three rays are drawn from the top of the object:

- Ray 1 is drawn parallel to the principal axis. After being refracted by the lens, this ray emerges directed away from the focal point on the front side of the lens.
- Ray 2 is drawn in the direction toward the focal point on the back side of the lens and emerges from the lens parallel to the principal axis.
- Ray 3 is drawn through the center of the lens and continues in a straight line.

In these ray diagrams, the point of intersection of *any two* of the rays can be used to locate the image. The third ray serves as a check of construction.

For the converging lens in Active Figure 26.25a, where the object is *outside* the front focal point ($p > f$), the image is real and inverted and is located on the back side of the lens. This diagram would be representative of a movie projector, for which the film is the object, the lens is in the projector, and the image is projected on a large screen for the audience to watch. The film is placed in the projector with the scene upside down so that the inverted image is right side up for the audience.

When the object is *inside* the front focal point ($p < f$) as in Active Figure 26.25b, the image is virtual and upright. When used in this way, the lens is acting as a magnifying glass, providing an enlarged upright image for closer study of an object. The object might be a stamp, a fingerprint, or a printed page for someone with failing eyesight.

For the diverging lens of Active Figure 26.25c, the image is virtual and upright for all object locations. A diverging lens is used in a security peephole in a door to give a wide-angle view. Nearsighted individuals use diverging eyeglass lenses or contact lenses. Another use is for a panoramic lens for a camera (although a sophisticated camera "lens" is actually a combination of several lenses). A diverging lens in this application creates a small image of a wide field of view.

QUICK QUIZ 26.5 What is the focal length of a pane of window glass? **(a)** zero **(b)** infinity **(c)** the thickness of the glass **(d)** impossible to determine

QUICK QUIZ 26.6 If you cover the top half of the lens in Active Figure 26.25a with a piece of paper, which of the following happens to the appearance of the image of the object? **(a)** The bottom half disappears. **(b)** The top half disappears. **(c)** The entire image is visible but dimmer. **(d)** There is no change. **(e)** The entire image disappears.

THINKING PHYSICS 26.3   BIO Corrective lenses on diving masks

Diving masks often have lenses built into the glass for divers who do not have perfect vision. This kind of mask allows the individual to dive without the necessity for glasses because the lenses in the faceplate perform the necessary refraction to provide clear vision. Normal glasses have lenses that are curved on both the front and rear surfaces. The lenses in a diving mask faceplate often only have curved surfaces on the *inside* of the glass. Why is this design desirable?

Reasoning The main reason for curving only the inner surface of the lenses in the diving mask faceplate is so that the diver can see clearly when looking at objects straight ahead while underwater *and* in the air. Consider light rays approaching the mask along a normal to the plane of the faceplate. If curved surfaces were

on both the front and the back of the diving lens on the faceplate, refraction would occur at each surface. The lens could be designed so that these two refractions would give clear vision while the diver is in air. When the diver is underwater, however, the refraction between the water and the glass at the first interface is now different because the index of refraction of water is different from that of air. Therefore, the vision would not be clear underwater.

By making the outer surface of the lens flat, light is not refracted at normal incidence to the faceplate at the outer surface *in either air or water;* all the refraction occurs at the inner glass–air surface. Therefore, the same refractive correction exists in water and in air, and the diver can see clearly in both environments. ◀

Example 26.5 | Images Formed by a Converging Lens

A converging lens has a focal length of 10.0 cm.

(A) An object is placed 30.0 cm from the lens. Construct a ray diagram, find the image distance, and describe the image.

SOLUTION

Conceptualize Because the lens is converging, the focal length is positive (see Table 26.3). We expect the possibilities of both real and virtual images.

Categorize Because the object distance is larger than the focal length, we expect the image to be real. The ray diagram for this situation is shown in Figure 26.26a.

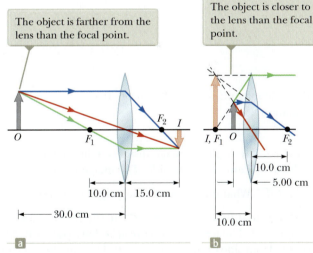

The object is farther from the lens than the focal point.

The object is closer to the lens than the focal point.

Figure 26.26
(Example 26.5) An image is formed by a converging lens.

Analyze Find the image distance by using Equation 26.12:

$$\frac{1}{q} = \frac{1}{f} - \frac{1}{p}$$

$$\frac{1}{q} = \frac{1}{10.0\ \text{cm}} - \frac{1}{30.0\ \text{cm}}$$

$$q = +15.0\ \text{cm}$$

Find the magnification of the image from Equation 26.11:

$$M = -\frac{q}{p} = -\frac{15.0\ \text{cm}}{30.0\ \text{cm}} = -0.500$$

continued

26.5 *cont.*

Finalize The positive sign for the image distance tells us that the image is indeed real and on the back side of the lens. The magnification of the image tells us that the image is reduced in height by one half, and the negative sign for *M* tells us that the image is inverted.

(B) An object is placed 10.0 cm from the lens. Find the image distance and describe the image.

SOLUTION

Categorize Because the object is at the focal point, we expect the image to be infinitely far away.

Analyze Find the image distance by using Equation 26.12:

$$\frac{1}{q} = \frac{1}{f} - \frac{1}{p}$$

$$\frac{1}{q} = \frac{1}{10.0 \text{ cm}} - \frac{1}{10.0 \text{ cm}}$$

$$q = \infty$$

Finalize This result means that rays originating from an object positioned at the focal point of a lens are refracted so that the image is formed at an infinite distance from the lens; that is, the rays travel parallel to one another after refraction.

(C) An object is placed 5.00 cm from the lens. Construct a ray diagram, find the image distance, and describe the image.

SOLUTION

Categorize Because the object distance is smaller than the focal length, we expect the image to be virtual. The ray diagram for this situation is shown in Figure 26.26b.

Analyze Find the image distance by using Equation 26.12:

$$\frac{1}{q} = \frac{1}{f} - \frac{1}{p}$$

$$\frac{1}{q} = \frac{1}{10.0 \text{ cm}} - \frac{1}{5.00 \text{ cm}}$$

$$q = -10.0 \text{ cm}$$

Find the magnification of the image from Equation 26.11:

$$M = -\frac{q}{p} = -\left(\frac{-10.0 \text{ cm}}{5.00 \text{ cm}}\right) = +2.00$$

Finalize The negative image distance tells us that the image is virtual and formed on the side of the lens from which the light is incident, the front side. The image is enlarged, and the positive sign for *M* tells us that the image is upright.

What If? What if the object moves right up to the lens surface so that $p \rightarrow 0$? Where is the image?

Answer In this case, because $p \ll R$, where *R* is either of the radii of the surfaces of the lens, the curvature of the lens can be ignored. The lens should appear to have the same effect as a flat piece of material, which suggests that the image is just on the front side of the lens, at $q = 0$. This conclusion can be verified mathematically by rearranging the thin lens equation:

$$\frac{1}{q} = \frac{1}{f} - \frac{1}{p}$$

If we let $p \rightarrow 0$, the second term on the right becomes very large compared with the first and we can neglect $1/f$. The equation becomes

$$\frac{1}{q} = -\frac{1}{p} \quad \rightarrow \quad q = -p = 0$$

Therefore, *q* is on the front side of the lens (because it has the opposite sign as *p*) and right at the lens surface.

Combinations of Thin Lenses

If two thin lenses are used to form an image, the system can be treated in the following manner. The position of the image of the first lens is calculated as though the second lens were not present. The light then approaches the second lens *as if* it had originally come from the image formed by the first lens. Hence, the image of the first lens is treated as the object of the second lens. The image of the second lens is the final image of the system. If the image of the first lens lies on the back side of the second lens, the image is treated as a *virtual object* for the second lens (i.e., p is negative). The same procedure can be extended to a system of three or more lenses. The overall magnification of a system of thin lenses equals the *product* of the magnifications of the separate lenses.

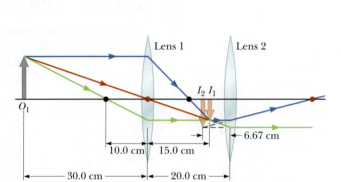

Light from a distant object brought into focus by two converging lenses.

Example 26.6 | Where Is the Final Image?

Two thin converging lenses of focal lengths $f_1 = 10.0$ cm and $f_2 = 20.0$ cm are separated by 20.0 cm as illustrated in Figure 26.27. An object is placed 30.0 cm to the left of lens 1. Find the position and the magnification of the final image.

SOLUTION

Conceptualize Imagine light rays passing through the first lens and forming a real image (because $p > f$) in the absence of a second lens. Figure 26.27 shows these light rays forming the inverted image I_1. Once the light rays converge to the image point, they do not stop. They continue through the image point and interact with the second lens. The rays leaving the image point behave in the same way as the rays leaving an object. Therefore, the image of the first lens serves as the object of the second lens.

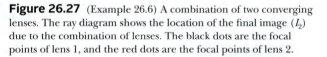

Figure 26.27 (Example 26.6) A combination of two converging lenses. The ray diagram shows the location of the final image (I_2) due to the combination of lenses. The black dots are the focal points of lens 1, and the red dots are the focal points of lens 2.

Categorize We categorize this problem as one in which the thin lens equation is applied in a stepwise fashion to the two lenses.

Analyze Find the location of the image formed by lens 1 from the thin lens equation:

$$\frac{1}{q_1} = \frac{1}{f} - \frac{1}{p_1}$$

$$\frac{1}{q_1} = \frac{1}{10.0 \text{ cm}} - \frac{1}{30.0 \text{ cm}}$$

$$q_1 = +15.0 \text{ cm}$$

Find the magnification of the image from Equation 26.11:

$$M_1 = -\frac{q_1}{p_1} = -\frac{15.0 \text{ cm}}{30.0 \text{ cm}} = -0.500$$

The image formed by this lens acts as the object for the second lens. Therefore, the object distance for the second lens is 20.0 cm − 15.0 cm = 5.00 cm.

Find the location of the image formed by lens 2 from the thin lens equation:

$$\frac{1}{q_2} = \frac{1}{20.0 \text{ cm}} - \frac{1}{5.00 \text{ cm}}$$

$$q_2 = -6.67 \text{ cm}$$

Find the magnification of the image from Equation 26.11:

$$M_2 = -\frac{q_2}{p_2} = -\frac{-6.67 \text{ cm}}{5.00 \text{ cm}} = +1.33$$

The total magnification M of the image due to the two lenses is the product $M_1 M_2$:

$$M = M_1 M_2 = (-0.500)(1.33) = -0.667$$

Finalize The negative sign on the overall magnification indicates that the final image is inverted with respect to the initial object. Because the absolute value of the magnification is less than 1, the final image is smaller than the object. Because q_2 is negative, the final image is on the front, or left, side of lens 2. These conclusions are consistent with the ray diagram in Figure 26.27.

continued

26.6 *cont.*

What If? Suppose you want to create an upright image with this system of two lenses. How must the second lens be moved?

Answer Because the object is farther from the first lens than the focal length of that lens, the first image is inverted. Consequently, the second lens must invert the image once again so that the final image is upright. An inverted image is only formed by a converging lens if the object is outside the focal point. Therefore, the image formed by the first lens must be to the left of the focal point of the second lens in Figure 26.27. To make that happen, you must move the second lens at least as far away from the first lens as the sum $q_1 + f_2 = 15.0\ \text{cm} + 20.0\ \text{cm} = 35.0\ \text{cm}$.

❰ **26.5** | The Eye BIO

Like a camera, a normal eye focuses light and produces a sharp image. The mechanisms by which the eye controls the amount of light admitted and adjusts to produce correctly focused images, however, are far more complex, intricate, and effective than those in even the most sophisticated camera. In all respects, the eye is a physiological wonder.

Figure 26.28 shows the basic parts of the human eye. Light entering the eye passes through a transparent structure called the *cornea* (Fig. 26.29), behind which are a clear liquid (the *aqueous humor*), a variable aperture (the *pupil*, which is an opening in the *iris*), and the *crystalline lens*. Most of the refraction occurs at the outer surface of the eye, where the cornea is covered with a film of tears. Relatively little refraction occurs in the crystalline lens because the aqueous humor in contact with the lens has an average index of refraction close to that of the lens. The iris, which is the colored portion of the eye, is a muscular diaphragm that controls pupil size. The iris regulates the amount of light entering the eye by dilating, or opening, the pupil in low-light conditions and contracting, or closing, the pupil in high-light conditions.

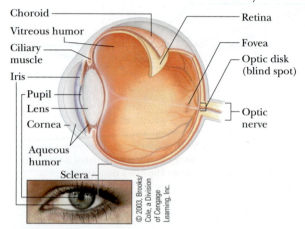

Figure 26.28 Important parts of the eye.

The cornea–lens system focuses light onto the back surface of the eye, the *retina*, which consists of millions of sensitive receptors called *rods* and *cones*. When stimulated by light, these receptors send impulses via the optic nerve to the brain, where an image is perceived. By this process, a distinct image of an object is observed when the image falls on the retina.

The eye focuses on an object by varying the shape of the pliable crystalline lens through a process called **accommodation.** The lens adjustments take place so swiftly that we are not even aware of the change. Accommodation is limited in that objects very close to the eye produce blurred images. The **near point** is the closest distance for which the lens can accommodate to focus light on the retina. This distance usually increases with age and has an average value of 25 cm. At age 10, the near point of the eye is typically approximately 18 cm. It increases to approximately 25 cm at age 20, to 50 cm at age 40, and to 500 cm or greater at age 60. The **far point** of the eye represents the greatest distance for which the lens of the relaxed eye can focus light on the retina. A person with normal vision can see very distant objects and therefore has a far point that can be approximated as infinity.

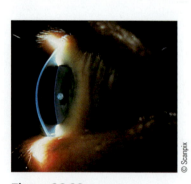

Figure 26.29 Close-up photograph of the cornea of the human eye.

Only three types of color-sensitive cells are present in the retina. They are called red, green, and blue cones because of the peaks of the color ranges to which they respond (Fig. 26.30). If the red and green cones are stimulated simultaneously (as would be the case if yellow light were shining on them), the brain interprets what is seen as yellow. If all three types of cones are stimulated by the separate colors red, blue, and green, white light is seen. If all three types of cones are stimulated by light that contains *all* colors, such as sunlight, again white light is seen.

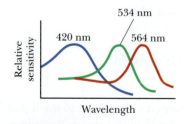

Figure 26.30 Approximate color sensitivity of the three types of cones in the retina.

Televisions and computer monitors take advantage of this visual illusion by having only red, green, and blue dots on the screen. With specific combinations of brightness in these three primary colors, our eyes can be made to see any color in the rainbow. Therefore, the yellow lemon you see in a television commercial is not actually yellow, it is red and green! The paper on which this page is printed is made of tiny, matted, translucent fibers that scatter light in all directions, and the resultant mixture of colors appears white to the eye. Snow, clouds, and white hair are not actually white. In fact, there is no such thing as a white pigment. The appearance of these things is a consequence of the scattering of light containing all colors, which we interpret as white.

Conditions of the Eye

When the eye suffers a mismatch between the focusing range of the lens–cornea system and the length of the eye, with the result that light rays from a near object reach the retina before they converge to form an image as shown in Figure 26.31a, the condition is known as **farsightedness** (or *hyperopia*). A farsighted person can usually see faraway objects clearly but not nearby objects. Although the near point of a normal eye is approximately 25 cm, the near point of a farsighted person is much farther away. The refracting power in the cornea and lens is insufficient to focus the light from all but distant objects satisfactorily. The condition can be corrected by placing a converging lens in front of the eye as shown in Figure 26.31b. The lens refracts the incoming rays more toward the principal axis before entering the eye, allowing them to converge and focus on the retina.

 BIO Hyperopia

A person with **nearsightedness** (or *myopia*), another mismatch condition, can focus on nearby objects but not on faraway objects. The far point of the nearsighted eye is not infinity and may be less than 1 m. The maximum focal length of the nearsighted eye is insufficient to produce a sharp image on the retina, and rays from a distant object converge to a focus in front of the retina. They then continue past that point, diverging before they finally reach the retina and causing blurred vision (Fig. 26.32a). Nearsightedness can be corrected with a diverging lens as shown in Figure 26.32b. The lens refracts the rays away from the principal axis before they enter the eye, allowing them to focus on the retina.

BIO Myopia

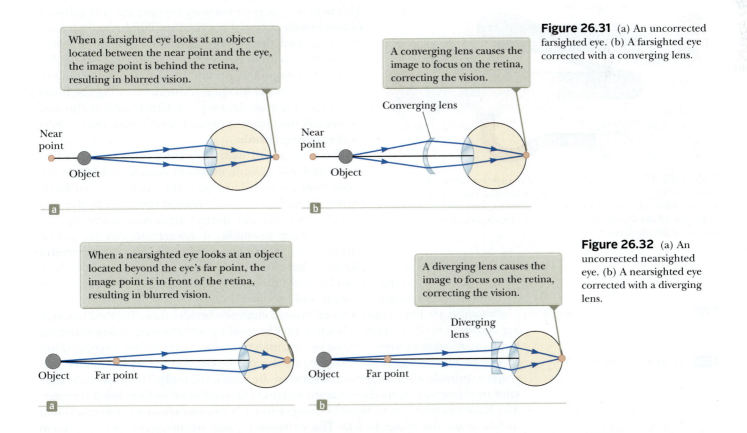

When a farsighted eye looks at an object located between the near point and the eye, the image point is behind the retina, resulting in blurred vision.

A converging lens causes the image to focus on the retina, correcting the vision.

Converging lens

Near point
Object

Near point
Object

a

b

Figure 26.31 (a) An uncorrected farsighted eye. (b) A farsighted eye corrected with a converging lens.

When a nearsighted eye looks at an object located beyond the eye's far point, the image point is in front of the retina, resulting in blurred vision.

A diverging lens causes the image to focus on the retina, correcting the vision.

Diverging lens

Object   Far point

Object   Far point

a

b

Figure 26.32 (a) An uncorrected nearsighted eye. (b) A nearsighted eye corrected with a diverging lens.

BIO Presbyopia

Beginning in middle age, most people lose some of their accommodation ability as their visual muscles weaken and the lens hardens. Unlike farsightedness, which is a mismatch between focusing power and eye length, **presbyopia** (literally, "old-age vision") is due to a reduction in accommodation ability. The cornea and lens do not have sufficient focusing power to bring nearby objects into focus on the retina. The symptoms are the same as those of farsightedness, and the condition can be corrected with converging lenses.

BIO Astigmatism

In eyes having a defect known as **astigmatism,** light from a point source produces a line image on the retina. This condition arises when the cornea, the lens, or both are not perfectly symmetric. Astigmatism can be corrected with lenses that have different curvatures in two mutually perpendicular directions.

Optometrists and ophthalmologists usually prescribe lenses measured in **diopters:** the **power** P of a lens in diopters equals the inverse of the focal length in meters: $P = 1/f$. For example, a converging lens of focal length $+20$ cm has a power of $+5.0$ diopters, and a diverging lens of focal length -40 cm has a power of -2.5 diopters.

> **QUICK QUIZ 26.7** Two campers wish to start a fire during the day. One camper is nearsighted, and one is farsighted. Whose glasses should be used to focus the Sun's rays onto some paper to start the fire? **(a)** either camper **(b)** the nearsighted camper **(c)** the farsighted camper

⟨26.6 | Context Connection: Some Medical Applications BIO

BIO Medical uses of the fiberscope

The first use of optical fibers in medicine appeared with the invention of the *fiberscope* in 1957. Figure 26.33 indicates the construction of a fiberscope, which consists of two bundles of optical fibers. The *illuminating bundle* is an *incoherent* bundle, meaning that no effort is made to match the relative positions of the fibers at the two ends. This matching is not necessary because the sole purpose of this bundle is to deliver light to illuminate the scene. A lens (called the *objective lens*) is used at the internal end of the fiberscope to create a real image of the illuminated scene on the ends of the *viewing bundle* of fibers. The light from the image is transmitted along the fibers to the viewing end. An eyepiece lens is used at this end to magnify the image appearing on the ends of the fibers in the viewing bundle.

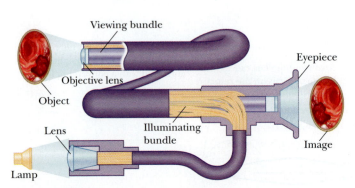

Figure 26.33 The construction of a fiberscope for viewing the interior of the body. The objective lens forms a real image of the scene on the end of a bundle of optical fibers. This image is carried to the other end of the bundle, where an eyepiece lens is used to magnify the image for the physician.

The diameter of such a fiberscope can be as small as 1 mm and still provide excellent optical imaging of the scene to be viewed. Therefore, the fiberscope can be inserted through very small surgical openings in the skin and threaded through narrow areas such as arteries.

As another example, a fiberscope can be passed through the esophagus and into the stomach to enable a physician to look for ulcers. The resulting image can be viewed directly by the physician through the eyepiece lens, but most often it is exhibited on a display monitor and digitized for computer storage and analysis.

BIO Medical uses of the endoscope

Endoscopes are fiberscopes with additional channels besides those for the illuminating and viewing fibers. These channels may be used for withdrawing and introducing fluids, manipulating wires, cutting tissues, injections, and many surgical applications.

BIO The da Vinci Surgical System

The da Vinci Surgical System (see photograph on page 3) makes use of an endoscope to provide a 3-D image of the surgical site within the body. The endoscope contains two lenses to create the separate images to be combined to form the 3-D image. The lenses provide these separate images to each eye of the surgeon to allow him or her to see the image in 3-D. The enhanced image quality provides the surgeon with better visualization of the site, whereas the arms of the robot provide enhanced dexterity and greater precision.

Lasers are used with endoscopes in a variety of medical diagnostic and treatment procedures. As a diagnostic example, the dependence on wavelength of the amount of reflection from a surface allows a fiberscope to be used to make a direct measurement of the blood's oxygen content. Using two laser sources, red light and infrared light are both sent into the blood through optical fibers. Hemoglobin reflects a known fraction of infrared light, regardless of the oxygen carried. Therefore, the measurement of the infrared reflection gives a total hemoglobin count. Red light is reflected much more by hemoglobin carrying oxygen than by hemoglobin that does not. Therefore, the amount of red laser light reflected allows a measurement of the ability of the patient's blood to carry oxygen.

BIO Use of lasers in measuring hemoglobin

Lasers are used to treat medical conditions such as *hydrocephalus,* which occurs in about 0.1% of births. This condition involves an increase in intracranial pressure due to an overproduction of cerebrospinal fluid (CSF), an obstruction of the flow of CSF, or insufficient absorption of CSF. In addition to congenital hydrocephalus, the condition can be acquired later in life due to trauma to the head, brain tumors, or other factors.

BIO Use of lasers in treating hydrocephalus

The older treatment method for obstructive hydrocephalus involved placing a shunt (tube) between ventricular chambers in the brain to allow passage of CSF. A new alternative is *laser-assisted ventriculostomy,* in which a new pathway for CSF is made with an infrared laser beam and an endoscope having a spherical end as shown in Figure 26.34. As the laser beam strikes the spherical end, refraction at the spherical surface causes light waves to spread out in all directions as if the end of the endoscope were a point source of radiation. The result is a rapid decrease in intensity with distance from the sphere, avoiding damage to vital structures in the brain that are close to the area in which a new passageway is to be made. The surface of the spherical end is coated with an infrared radiation-absorbing material, and the absorbed laser energy raises the temperature of the sphere. As the sphere is placed in contact with the location of the desired passageway, the combination of the high temperature and laser radiation leaving the sphere burns a new passageway for the CSF. This treatment requires much less recovery time as well as significantly less postoperative care than that associated with the placement of shunts.

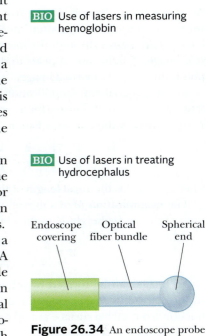

Figure 26.34 An endoscope probe used to open new passageways for cerebrospinal fluid in the treatment of hydrocephalus. Laser light raises the temperature of the sphere and radiates from the sphere to provide energy to tissues for cutting the new passageway.

Lasik (laser-assisted in situ keratomileusis) is a form of refractive eye surgery that uses lasers to correct myopia, hyperopia, and astigmatism. The surgical procedure involves three steps. First, a corneal suction ring is used to immobilize the eye. Once the eye is immobilized, a flap is created on the cornea by using either a metal blade or a laser. The flap is then folded back to reveal the middle section of the cornea called the *stroma.* In the next step of the procedure, the shape of the stroma is modified using a 193-nm wavelength *excimer laser.* The excimer laser vaporizes tissue in a finely controlled manner without damaging adjacent stroma. Finally, once the stroma layer has been reshaped, the flap is repositioned over the treated area and remains in this position by natural adhesion until healing is complete.

BIO Lasik surgery

Tattoos can be removed or modified using a specially designed *Q-switched laser* that penetrates the skin and targets the darker pigments of the tattoo but leaves surrounding tissues unharmed. The Q-switched laser provides very short bursts of energy, measured in nanoseconds, containing a large amount of energy in each burst. Absorption of energy from the laser breaks up the large particles of ink into smaller pieces that can naturally be removed by normal body processes. The short duration of the pulse prevents the energy from spreading into surrounding tissue. It takes several months for the body to eliminate the dissolved tattoo pigments.

BIO Use of lasers in tattoo removal

Patients having an enlarged prostate (benign prostatic hyperplasia) are sometimes treated using laser surgery. In this type of surgery, known by the general name of TURP (transurethral resection of the prostate), the laser removes tissue in the prostate that is blocking the flow of urine. A variety of lasers are used in this procedure, ranging from visible to infrared wavelengths.

BIO Use of lasers in benign prostatic hyperplasia

In Chapter 27, we shall investigate another application of lasers—the technology of *holography*—that has grown tremendously in recent years. In holography, three-dimensional images of objects are recorded on film.

▶ SUMMARY

An **image** of an object is a point from which light either diverges or seems to diverge after interacting with a mirror or lens. If light passes through the image point, the image is a **real image.** If light only appears to diverge from the image point, the image is a **virtual image.**

In the **paraxial ray** simplification model, the object distance p and image distance q for a spherical mirror of radius R are related by the **mirror equation**

$$\frac{1}{p} + \frac{1}{q} = \frac{2}{R} = \frac{1}{f} \qquad \text{26.4, 26.6} \blacktriangleleft$$

where $f = R/2$ is the **focal length** of the mirror.

The **magnification** M of a mirror or lens is defined as the ratio of the image height h' to the object height h:

$$M = \frac{h'}{h} = -\frac{q}{p} \qquad \text{26.2, 26.11} \blacktriangleleft$$

An image can be formed by refraction from a spherical surface of radius R. The object and image distances for refraction from such a surface are related by

$$\frac{n_1}{p} + \frac{n_2}{q} = \frac{n_2 - n_1}{R} \qquad \text{26.8} \blacktriangleleft$$

where the light is incident from the medium of index of refraction n_1 and is refracted in the medium whose index of refraction is n_2.

For a thin lens, and in the paraxial ray approximation, the object and image distances are related by the **thin lens equation:**

$$\frac{1}{p} + \frac{1}{q} = \frac{1}{f} \qquad \text{26.12} \blacktriangleleft$$

The **focal length** f of a thin lens in air is related to the curvature of its surfaces and to the index of refraction n of the lens material by

$$\frac{1}{f} = (n - 1)\left(\frac{1}{R_1} - \frac{1}{R_2}\right) \qquad \text{26.13} \blacktriangleleft$$

Converging lenses have positive focal lengths, and **diverging lenses** have negative focal lengths.

▶ OBJECTIVE QUESTIONS

☐ denotes answer available in *Student Solutions Manual/Study Guide*

1. **BIO** If a woman's eyes are longer than normal, how is her vision affected and how can her vision be corrected? (a) The woman is farsighted (hyperopia), and her vision can be corrected with a diverging lens. (b) The woman is nearsighted (myopia), and her vision can be corrected with a diverging lens. (c) The woman is farsighted, and her vision can be corrected with a converging lens. (d) The woman is nearsighted, and her vision can be corrected with a converging lens. (e) The woman's vision is not correctible.

2. **(i)** When an image of an object is formed by a plane mirror, which of the following statements is *always* true? More than one statement may be correct. (a) The image is virtual. (b) The image is real. (c) The image is upright. (d) The image is inverted. (e) None of those statements is always true. **(ii)** When the image of an object is formed by a concave mirror, which of the preceding statements are *always* true? **(iii)** When the image of an object is formed by a convex mirror, which of the preceding statements are *always* true?

3. An object is located 50.0 cm from a converging lens having a focal length of 15.0 cm. Which of the following statement is true regarding the image formed by the lens? (a) It is virtual, upright, and larger than the object. (b) It is real, inverted, and smaller than the object. (c) It is virtual, inverted, and smaller than the object. (d) It is real, inverted, and larger than the object. (e) It is real, upright, and larger than the object.

4. **(i)** When an image of an object is formed by a converging lens, which of the following statements is *always* true? More than one statement may be correct. (a) The image is virtual. (b) The image is real. (c) The image is upright. (d) The image is inverted. (e) None of those statements is always true. **(ii)** When the image of an object is formed by a diverging lens, which of the statements is *always* true?

5. **BIO** If a man has eyes that are shorter than normal, how is his vision affected and how can it be corrected? (a) The man is farsighted (hyperopia), and his vision can be corrected with a diverging lens. (b) The man is nearsighted (myopia), and his vision can be corrected with a diverging lens. (c) The man is farsighted, and his vision can be corrected with a converging lens. (d) The man is nearsighted, and his vision can be corrected with a converging lens. (e) The man's vision is not correctible.

6. If Josh's face is 30.0 cm in front of a concave shaving mirror creating an upright image 1.50 times as large as the object, what is the mirror's focal length? (a) 12.0 cm (b) 20.0 cm (c) 70.0 cm (d) 90.0 cm (e) none of those answers

7. A converging lens made of crown glass has a focal length of 15.0 cm when used in air. If the lens is immersed in water, what is its focal length? (a) negative (b) less than 15.0 cm (c) equal to 15.0 cm (d) greater than 15.0 cm (e) none of those answers

8. Two thin lenses of focal lengths $f_1 = 15.0$ and $f_2 = 10.0$ cm, respectively, are separated by 35.0 cm along a common axis. The f_1 lens is located to the left of the f_2 lens. An object is now placed 50.0 cm to the left of the f_1 lens, and a final image due to light passing though both lenses forms. By what factor is the final image different in size from the object? (a) 0.600 (b) 1.20 (c) 2.40 (d) 3.60 (e) none of those answers

9. Lulu looks at her image in a makeup mirror. It is enlarged when she is close to the mirror. As she backs away, the image becomes larger, then impossible to identify when she is

30.0 cm from the mirror, then upside down when she is beyond 30.0 cm, and finally small, clear, and upside down when she is much farther from the mirror. **(i)** Is the mirror (a) convex, (b) plane, or (c) concave? **(ii)** Is the magnitude of its focal length (a) 0, (b) 15.0 cm, (c) 30.0 cm, (d) 60.0 cm, or (e) ∞?

10. Model each of the following devices in use as consisting of a single converging lens. Rank the cases according to the ratio of the distance from the object to the lens to the focal length of the lens, from the largest ratio to the smallest. (a) a film-based movie projector showing a movie (b) a magnifying glass being used to examine a postage stamp (c) an astronomical refracting telescope being used to make a sharp image of stars on an electronic detector (d) a searchlight being used to produce a beam of parallel rays from a point source (e) a camera lens being used to photograph a soccer game

11. A converging lens of focal length 8 cm forms a sharp image of an object on a screen. What is the smallest possible

distance between the object and the screen? (a) 0 (b) 4 cm (c) 8 cm (d) 16 cm (e) 32 cm

12. An object, represented by a gray arrow, is placed in front of a plane mirror. Which of the diagrams in Figure OQ26.12 correctly describes the image, represented by the pink arrow?

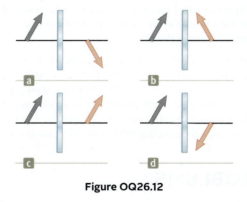

Figure OQ26.12

CONCEPTUAL QUESTIONS

☐ denotes answer available in *Student Solutions Manual/Study Guide*

1. Do the equations $1/p + 1/q = 1/f$ and $M = -q/p$ apply to the image formed by a flat mirror? Explain your answer.

2. **BIO** The optic nerve and the brain invert the image formed on the retina. Why don't we see everything upside down?

3. Consider a spherical concave mirror with the object located to the left of the mirror beyond the focal point. Using ray diagrams, show that the image moves to the left as the object approaches the focal point.

4. In Active Figure 26.25a, assume the gray object arrow is replaced by one that is much taller than the lens. (a) How many rays from the top of the object will strike the lens? (b) How many principal rays can be drawn in a ray diagram?

5. (a) Can a converging lens be made to diverge light if it is placed into a liquid? (b) **What If?** What about a converging mirror?

6. **BIO** Lenses used in eyeglasses, whether converging or diverging, are always designed so that the middle of the lens curves away from the eye like the center lenses of Figures 26.21a and 26.21b. Why?

7. Suppose you want to use a converging lens to project the image of two trees onto a screen. As shown in Figure CQ26.7, one tree is a distance x from the lens and the other is at $2x$. You adjust the screen so that the near tree is in focus. If you now want the far tree to be in focus, do you move the screen toward or away from the lens?

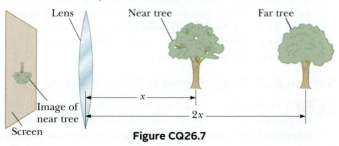

Figure CQ26.7

8. Explain why a fish in a spherical goldfish bowl appears larger than it really is.

9. Why do some emergency vehicles have the symbol ƎƆИA˥UꓭMA written on the front?

10. Explain this statement: "The focal point of a lens is the location of the image of a point object at infinity." (a) Discuss the notion of infinity in real terms as it applies to object distances. (b) Based on this statement, can you think of a simple method for determining the focal length of a converging lens?

11. **BIO** In Figures CQ26.11a and CQ26.11b, which glasses correct nearsightedness and which correct farsightedness?

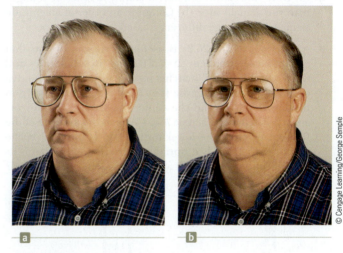

© Cengage Learning/George Semple

Figure CQ26.11 Conceptual Questions 11 and 12.

12. Bethany tries on either her hyperopic grandfather's or her myopic brother's glasses and complains, "Everything looks blurry." Why do the eyes of a person wearing glasses not look blurry? (See Fig. CQ26.11.)

13. **BIO** During LASIK eye surgery (laser-assisted *in situ* keratomileusis), the shape of the cornea is modified by vaporizing some of its material. If the surgery is performed to correct for nearsightedness, how does the cornea need to be reshaped?

14. A solar furnace can be constructed by using a concave mirror to reflect and focus sunlight into a furnace enclosure. What factors in the design of the reflecting mirror would guarantee very high temperatures?

15. Figure CQ26.15 shows a lithograph by M. C. Escher titled *Hand with Reflection Sphere (Self-Portrait in Spherical Mirror)*. Escher said about the work: "The picture shows a spherical mirror, resting on a left hand. But as a print is the reverse of the original drawing on stone, it was my right hand that you see depicted. (Being left-handed, I needed my left hand to make the drawing.) Such a globe reflection collects almost one's whole surroundings in one disk-shaped image. The whole room, four walls, the floor, and the ceiling, everything, albeit distorted, is compressed into that one small circle. Your own head, or more exactly the point between your eyes, is the absolute center. No matter how you turn or twist yourself, you can't get out of that central point. You are immovably the focus, the unshakable core, of your world." Comment on the accuracy of Escher's description.

Figure CQ26.15

▶ PROBLEMS |

WebAssign The problems found in this chapter may be assigned online in Enhanced WebAssign.

   1. denotes straightforward problem; **2.** denotes intermediate problem; **3.** denotes challenging problem

   1. denotes full solution available in the *Student Solutions Manual/Study Guide*

   1. denotes problems most often assigned in Enhanced WebAssign.

BIO denotes biomedical problem

GP denotes guided problem

M denotes Master It tutorial available in Enhanced WebAssign

Q|C denotes asking for quantitative and conceptual reasoning

S denotes symbolic reasoning problem

shaded denotes "paired problems" that develop reasoning with symbols and numerical values

W denotes Watch It video solution available in Enhanced WebAssign

Section 26.1 Images Formed by Flat Mirrors

1. **W** Determine the minimum height of a vertical flat mirror in which a person 178 cm tall can see his or her full image. *Suggestion:* Drawing a ray diagram would be helpful.

2. (a) Does your bathroom mirror show you older or younger than you actually are? (b) Compute an order-of-magnitude estimate for the age difference based on data you specify.

3. A periscope (Fig. P26.3) is useful for viewing objects that cannot be seen directly. It can be used in submarines and when watching golf matches or parades from behind a crowd of people. Suppose the object is a distance p_1 from the upper mirror and the centers of the two flat mirrors are separated by a distance h. (a) What is the distance of the final image from the lower mirror? (b) Is the final image real or virtual? (c) Is it upright or inverted? (d) What is its magnification? (e) Does it appear to be left–right reversed?

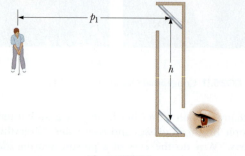

Figure P26.3

4. In a choir practice room, two parallel walls are 5.30 m apart. The singers stand against the north wall. The organist faces the south wall, sitting 0.800 m away from it. To enable her to see the choir, a flat mirror 0.600 m wide is mounted on the south wall, straight in front of her. What width of the north wall can the organist see? *Suggestion:* Draw a top-view diagram to justify your answer.

5. A person walks into a room that has two flat mirrors on opposite walls. The mirrors produce multiple images of the person. Consider *only* the images formed in the mirror on the left. When the person is 2.00 m from the mirror on the left wall and 4.00 m from the mirror on the right wall, find the distance from the person to the first three images seen in the mirror on the left wall.

6. Two flat mirrors have their reflecting surfaces facing each other, with the edge of one mirror in contact with an edge of the other, so that the angle between the mirrors is α. When an object is placed between the mirrors, a number of images are formed. In general, if the angle α is such that $n\alpha = 360°$, where n is an integer, the number of images formed is $n - 1$. Graphically, find all the image positions for the case $n = 6$ when a point object is between the mirrors (but not on the angle bisector).

Section 26.2 Images Formed by Spherical Mirrors

7. **M** A convex spherical mirror has a radius of curvature of magnitude 40.0 cm. Determine the position of the virtual image and the magnification for object distances of 30.0 cm and (b) 60.0 cm. (c) Are the images in parts (a) and (b) upright or inverted?

8. A dentist uses a spherical mirror to examine a tooth. The tooth is 1.00 cm in front of the mirror, and the image is formed 10.0 cm behind the mirror. Determine (a) the mirror's radius of curvature and (b) the magnification of the image.

9. A large hall in a museum has a niche in one wall. On the floor plan, the niche appears as a semicircular indentation of radius 2.50 m. A tourist stands on the centerline of the niche, 2.00 m out from its deepest point, and whispers "Hello." Where is the sound concentrated after reflection from the niche?

10. *Why is the following situation impossible?* At a blind corner in an outdoor shopping mall, a convex mirror is mounted so pedestrians can see around the corner before arriving there and bumping into someone traveling in the perpendicular direction. The installers of the mirror failed to take into account the position of the Sun, and the mirror focuses the Sun's rays on a nearby bush and sets it on fire.

11. **M** A concave spherical mirror has a radius of curvature of magnitude 20.0 cm. (a) Find the location of the image for object distances of **(i)** 40.0 cm, **(ii)** 20.0 cm, and **(iii)** 10.0 cm. For each case, state whether the image is (b) real or virtual and (c) upright or inverted. (d) Find the magnification in each case.

12. **Q|C** **Review.** A ball is dropped at $t = 0$ from rest 3.00 m directly above the vertex of a concave spherical mirror that has a radius of curvature of magnitude 1.00 m and lies in a horizontal plane. (a) Describe the motion of the ball's image in the mirror. (b) At what instant or instants do the ball and its image coincide?

13. **W** (a) A concave spherical mirror forms an inverted image 4.00 times larger than the object. Assuming the distance between object and image is 0.600 m, find the focal length of the mirror. (b) **What If?** Suppose the mirror is convex. The distance between the image and the object is the same as in part (a), but the image is 0.500 the size of the object. Determine the focal length of the mirror.

14. **S** (a) A concave spherical mirror forms an inverted image different in size from the object by a factor $a > 1$. The distance between object and image is d. Find the focal length of the mirror. (b) **What If?** Suppose the mirror is convex, an upright image is formed, and $a < 1$. Determine the focal length of the mirror.

15. **BIO** To fit a contact lens to a patient's eye, a *keratometer* can be used to measure the curvature of the eye's front surface, the cornea. This instrument places an illuminated object of known size at a known distance p from the cornea. The cornea reflects some light from the object, forming an image of the object. The magnification M of the image is measured by using a small viewing telescope that allows comparison of the image formed by the cornea with a second calibrated image projected into the field of view by a prism arrangement. Determine the radius of curvature of the cornea for the case $p = 30.0$ cm and $M = 0.013$ 0.

16. A concave mirror has a radius of curvature of 60.0 cm. Calculate the image position and magnification of an object placed in front of the mirror at distances of (a) 90.0 cm

and (b) 20.0 cm. (c) Draw ray diagrams to obtain the image characteristics in each case.

17. **W** An object 10.0 cm tall is placed at the zero mark of a meterstick. A spherical mirror located at some point on the meterstick creates an image of the object that is upright, 4.00 cm tall, and located at the 42.0-cm mark of the meterstick. Is the mirror convex or concave? (b) Where is the mirror? (c) What is the mirror's focal length?

18. **W** At an intersection of hospital hallways, a convex spherical mirror is mounted high on a wall to help people avoid collisions. The magnitude of the mirror's radius of curvature is 0.550 m. (a) Locate the image of a patient 10.0 m from the mirror. (b) Indicate whether the image is upright or inverted. (c) Determine the magnification of the image.

19. **M** A spherical mirror is to be used to form an image 5.00 times the size of an object on a screen located 5.00 m from the object. (a) Is the mirror required concave or convex? (b) What is the required radius of curvature of the mirror? (c) Where should the mirror be positioned relative to the object?

20. A certain Christmas tree ornament is a silver sphere having a diameter of 8.50 cm. (a) If the size of an image created by reflection in the ornament is three-fourths the reflected object's actual size, determine the object's location. (b) Use a principal-ray diagram to determine whether the image is upright or inverted.

21. **W** A dedicated sports car enthusiast polishes the inside and outside surfaces of a hubcap that is a thin section of a sphere. When she looks into one side of the hubcap, she sees an image of her face 30.0 cm in back of the hubcap. She then flips the hubcap over and sees another image of her face 10.0 cm in back of the hubcap. (a) How far is her face from the hubcap? (b) What is the radius of curvature of the hubcap?

22. You unconsciously estimate the distance to an object from the angle it subtends in your field of view. This angle θ in radians is related to the linear height of the object h and to the distance d by $\theta = h/d$. Assume you are driving a car and another car, 1.50 m high, is 24.0 m behind you. (a) Suppose your car has a flat passenger-side rearview mirror, 1.55 m from your eyes. How far from your eyes is the image of the car following you? (b) What angle does the image subtend in your field of view? (c) **What If?** Now suppose your car has a convex rearview mirror with a radius of curvature of magnitude 2.00 m (as suggested in Fig. 26.14). How far from your eyes is the image of the car behind you? (d) What angle does the image subtend at your eyes? (e) Based on its angular size, how far away does the following car appear to be?

Section 26.3 Images Formed by Refraction

23. One end of a long glass rod ($n = 1.50$) is formed into a convex surface with a radius of curvature of magnitude 6.00 cm. An object is located in air along the axis of the rod. Find the image positions corresponding to object distances of (a) 20.0 cm, (b) 10.0 cm, and (c) 3.00 cm from the convex end of the rod.

24. A cubical block of ice 50.0 cm on a side is placed over a speck of dust on a level floor. Find the location of the image of the speck as viewed from above. The index of refraction of ice is 1.309.

25. A flint glass plate rests on the bottom of an aquarium tank. The plate is 8.00 cm thick (vertical dimension) and is covered with a layer of water 12.0 cm deep. Calculate the apparent thickness of the plate as viewed from straight above the water.

26. **BIO** A simple model of the human eye ignores its lens entirely. Most of what the eye does to light happens at the outer surface of the transparent cornea. Assume that this surface has a radius of curvature of 6.00 mm and that the eyeball contains just one fluid with a refractive index of 1.40. Prove that a very distant object will be imaged on the retina, 21.0 mm behind the cornea. Describe the image.

27. **M** A glass sphere ($n = 1.50$) with a radius of 15.0 cm has a tiny air bubble 5.00 cm above its center. The sphere is viewed looking down along the extended radius containing the bubble. What is the apparent depth of the bubble below the surface of the sphere?

28. **W** A goldfish is swimming at 2.00 cm/s toward the front wall of a rectangular aquarium. What is the apparent speed of the fish measured by an observer looking in from outside the front wall of the tank?

Section 26.4 Images Formed by Thin Lenses

29. **W** A contact lens is made of plastic with an index of refraction of 1.50. The lens has an outer radius of curvature of +2.00 cm and an inner radius of curvature of +2.50 cm. What is the focal length of the lens?

30. **M** An object is located 20.0 cm to the left of a diverging lens having a focal length $f = -32.0$ cm. Determine (a) the location and (b) the magnification of the image. (c) Construct a ray diagram for this arrangement.

31. A thin lens has a focal length of 25.0 cm. Locate and describe the image when the object is placed (a) 26.0 cm and (b) 24.0 cm in front of the lens.

32. A converging lens has a focal length of 20.0 cm. Locate the image for object distances of (a) 40.0 cm, (b) 20.0 cm, and (c) 10.0 cm. For each case, state whether the image is real or virtual and upright or inverted. Find the magnification in each case.

33. The left face of a biconvex lens has a radius of curvature of magnitude 12.0 cm, and the right face has a radius of curvature of magnitude 18.0 cm. The index of refraction of the glass is 1.44. (a) Calculate the focal length of the lens for light incident from the left. (b) **What If?** After the lens is turned around to interchange the radii of curvature of the two faces, calculate the focal length of the lens for light incident from the left.

34. **S** Suppose an object has thickness dp so that it extends from object distance p to $p + dp$. (a) Prove that the thickness dq of its image is given by $(-q^2/p^2)\,dp$. (b) The longitudinal magnification of the object is $M_{long} = dq/dp$. How is the longitudinal magnification related to the lateral magnification M?

35. **W** The projection lens in a certain slide projector is a single thin lens. A slide 24.0 mm high is to be projected so that its image fills a screen 1.80 m high. The slide-to-screen distance is 3.00 m. (a) Determine the focal length of the projection lens. (b) How far from the slide should the lens of the projector be placed so as to form the image on the screen?

36. The use of a lens in a certain situation is described by the equation

$$\frac{1}{p} + \frac{1}{-3.50p} = \frac{1}{7.50 \text{ cm}}$$

Determine (a) the object distance and (b) the image distance. (c) Use a ray diagram to obtain a description of the image. (d) Identify a practical device described by the given equation and write the statement of a problem for which the equation appears in the solution.

37. The nickel's image in Figure P26.37 has twice the diameter of the nickel and is 2.84 cm from the lens. Determine the focal length of the lens.

Figure P26.37

38. **Q|C** In Figure P26.38, a thin converging lens of focal length 14.0 cm forms an image of the square $abcd$, which is $h_c = h_b = 10.0$ cm high and lies between distances of $p_d = 20.0$ cm and $p_a = 30.0$ cm from the lens. Let a', b', c', and d' represent the respective corners

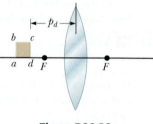

Figure P26.38

of the image. Let q_a represent the image distance for points a' and b', q_d represent the image distance for points c' and d', h'_b represent the distance from point b' to the axis, and h'_c represent the height of c'. (a) Find q_a, q_d, h'_b, and h'_c. (b) Make a sketch of the image. (c) The area of the object is 100 cm². By carrying out the following steps, you will evaluate the area of the image. Let q represent the image distance of any point between a' and d', for which the object distance is p. Let h' represent the distance from the axis to the point at the edge of the image between b' and c' at image distance q. Demonstrate that

$$|h'| = 10.0q\left(\frac{1}{14.0} - \frac{1}{q}\right)$$

where h' and q are in centimeters. (d) Explain why the geometric area of the image is given by

$$\int_{q_a}^{q_d} |h'|\, dq$$

(e) Carry out the integration to find the area of the image.

39. Figure P26.39 diagrams a cross-section of a camera. It has a single lens of focal length 65.0 mm, which is to form an image on the CCD (charge-coupled device) at the back of the camera. Suppose the position of the lens has been adjusted to focus the image of a distant object. How far and in what direction must the lens be moved to form a sharp image of an object that is 2.00 m away?

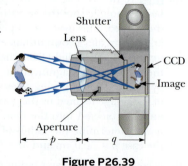

Figure P26.39

40. *Why is the following situation impossible?* An illuminated object is placed a distance $d = 2.00$ m from a screen. By placing a converging lens of focal length $f = 60.0$ cm at two locations between the object and the screen, a sharp, real image of the object can be formed on the screen. In one location of the lens, the image is larger than the object, and in the other, the image is smaller.

41. **W** An antelope is at a distance of 20.0 m from a converging lens of focal length 30.0 cm. The lens forms an image of the animal. (a) If the antelope runs away from the lens at a speed of 5.00 m/s, how fast does the image move? (b) Does the image move toward or away from the lens?

42. An object is at a distance d to the left of a flat screen. A converging lens with focal length $f < d/4$ is placed between object and screen. (a) Show that two lens positions exist that form an image on the screen and determine how far these positions are from the object. (b) How do the two images differ from each other?

43. An object located 32.0 cm in front of a lens forms an image on a screen 8.00 cm behind the lens. (a) Find the focal length of the lens. (b) Determine the magnification. (c) Is the lens converging or diverging?

Section 26.5 The Eye

44. **BIO** A nearsighted person cannot see objects clearly beyond 25.0 cm (her far point). If she has no astigmatism and contact lenses are prescribed for her, what (a) power and (b) type of lens are required to correct her vision?

45. **BIO** The near point of a person's eye is 60.0 cm. To see objects clearly at a distance of 25.0 cm, what should be the (a) focal length and (b) power of the appropriate corrective lens? (Neglect the distance from the lens to the eye.)

46. **BIO** **Q|C** A patient has a near point of 45.0 cm and far point of 85.0 cm. (a) Can a single lens correct the patient's vision? Explain the patient's options. (b) Calculate the power lens needed to correct the near point so that the patient can see objects 25.0 cm away. Neglect the eye–lens distance. (c) Calculate the power lens needed to correct the patient's far point, again neglecting the eye–lens distance.

47. **BIO** **W** The accommodation limits for a nearsighted person's eyes are 18.0 cm and 80.0 cm. When he wears his glasses, he can see faraway objects clearly. At what minimum distance is he able to see objects clearly?

48. **BIO** **Q|C** A person sees clearly wearing eyeglasses that have a power of -4.00 diopters when the lenses are 2.00 cm in front of the eyes. (a) What is the focal length of the lens? (b) Is the person nearsighted or farsighted? (c) If the person wants to switch to contact lenses placed directly on the eyes, what lens power should be prescribed?

49. **BIO** A person is to be fitted with bifocals. She can see clearly when the object is between 30 cm and 1.5 m from the eye. (a) The upper portions of the bifocals (Fig. P26.49) should be designed to enable her to see distant objects

Figure P26.49

clearly. What power should they have? (b) The lower portions of the bifocals should enable her to see objects located 25 cm in front of the eye. What power should they have?

50. **BIO** A certain child's near point is 10.0 cm; her far point (with eyes relaxed) is 125 cm. Each eye lens is 2.00 cm from the retina. (a) Between what limits, measured in diopters, does the power of this lens–cornea combination vary? (b) Calculate the power of the eyeglass lens the child should use for relaxed distance vision. Is the lens converging or diverging?

Section 26.6 Context Connection: Some Medical Applications

51. **BIO** You are designing an endoscope for use inside an air-filled body cavity. A lens at the end of the endoscope will form an image covering the end of a bundle of optical fibers. This image will then be carried by the optical fibers to an eyepiece lens at the outside end of the fiberscope. The radius of the bundle is 1.00 mm. The scene within the body that is to appear within the image fills a circle of radius 6.00 cm. The lens will be located 5.00 cm from the tissues you wish to observe. (a) How far should the lens be located from the end of an optical fiber bundle? (b) What is the focal length of the lens required?

52. **BIO** Consider the endoscope probe used for treating hydrocephalus and shown in Figure 26.34. The spherical end, with refractive index 1.50, is attached to an optical fiber bundle of radius 1.00 mm, which is smaller than the radius of the sphere. The center of the spherical end is on the central axis of the bundle. Consider laser light that travels precisely parallel to the central axis of the bundle and then refracts out from the surface of the sphere into air. (a) In Figure 26.34, does light that refracts out of the sphere and travels toward the upper right come from the top half of the sphere or from the bottom half of the sphere? (b) If laser light that travels along the edge of the optical fiber bundle refracts out of the sphere tangent to the surface of the sphere, what is the radius of the sphere? (c) Find the angle of deviation of the ray considered in part (b), that is, the angle by which its direction changes as it leaves the sphere. (d) Show that the ray considered in part (b) has a greater angle of deviation than any other ray. Show that the light from all parts of the optical fiber bundle does not refract out of the sphere with spherical symmetry, but rather fills a cone around the forward direction. Find the angular diameter of the cone. (e) In reality, however, laser light can diverge from the sphere with approximate spherical symmetry. What considerations that we have not addressed will lead to this approximate spherical symmetry in practice?

Additional Problems

53. The object in Figure P26.53 is midway between the lens and the mirror, which are separated by a distance $d = 25.0$ cm. The magnitude of the mirror's radius of curvature is 20.0 cm, and the lens has a focal length of -16.7 cm. (a) Considering only the light that leaves the object and travels first toward the mirror, locate the final image formed by

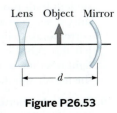

Figure P26.53

this system. (b) Is this image real or virtual? (c) Is it upright or inverted? (d) What is the overall magnification?

54. **GP** **M** In a darkened room, a burning candle is placed 1.50 m from a white wall. A lens is placed between the candle and the wall at a location that causes a larger, inverted image to form on the wall. When the lens is in this position, the object distance is p_1. When the lens is moved 90.0 cm toward the wall, another image of the candle is formed on the wall. From this information, we wish to find p_1 and the focal length of the lens. (a) From the lens equation for the first position of the lens, write an equation relating the focal length f of the lens to the object distance p_1, with no other variables in the equation. (b) From the lens equation for the second position of the lens, write another equation relating the focal length f of the lens to the object distance p_1. (c) Solve the equations in parts (a) and (b) simultaneously to find p_1. (d) Use the value in part (c) to find the focal length f of the lens.

55. The distance between an object and its upright image is 20.0 cm. If the magnification is 0.500, what is the focal length of the lens being used to form the image?

56. **S** The distance between an object and its upright image is d. If the magnification is M, what is the focal length of the lens being used to form the image?

57. The lens and mirror in Figure P26.57 are separated by $d = 1.00$ m and have focal lengths of $+80.0$ cm and -50.0 cm, respectively. An object is placed $p = 1.00$ m to the left of the lens as shown. (a) Locate the final image, formed by light that has gone through the lens twice. (b) Determine the overall magnification of the image and (c) state whether the image is upright or inverted.

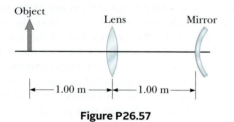

Figure P26.57

58. *Why is the following situation impossible?* Consider the lens–mirror combination shown in Figure P26.58. The lens has a focal length of $f_L = 0.200$ m, and the mirror has a focal length of $f_M = 0.500$ m. The lens and mirror are placed a distance $d = 1.30$ m apart, and an object is placed at $p = 0.300$ m from the lens. By moving a screen to various positions to the left of the lens, a student finds two different positions of the screen that produce a sharp image of the object. One of these positions corresponds to light leaving the object and traveling to the left through the lens. The other position corresponds to light traveling to the right from the object, reflecting from the mirror and then passing through the lens.

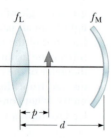

Figure P26.58

59. **Review.** A spherical lightbulb of diameter 3.20 cm radiates light equally in all directions, with power 4.50 W. (a) Find the light intensity at the surface of the lightbulb. (b) Find

the light intensity 7.20 m away from the center of the lightbulb. (c) At this 7.20-m distance, a lens is set up with its axis pointing toward the lightbulb. The lens has a circular face with a diameter of 15.0 cm and has a focal length of 35.0 cm. Find the diameter of the lightbulb's image. (d) Find the light intensity at the image.

60. **S** Derive the lens-makers' equation as follows. Consider an object in vacuum at $p_1 = \infty$ from a first refracting surface of radius of curvature R_1. Locate its image. Use this image as the object for the second refracting surface, which has nearly the same location as the first because the lens is thin. Locate the final image, proving it is at the image distance q_2 given by

$$\frac{1}{q_2} = (n - 1)\left(\frac{1}{R_1} - \frac{1}{R_2}\right)$$

61. **M** An object is placed 12.0 cm to the left of a diverging lens of focal length -6.00 cm. A converging lens of focal length 12.0 cm is placed a distance d to the right of the diverging lens. Find the distance d so that the final image is infinitely far away to the right.

62. **S** An object is placed a distance p to the left of a diverging lens of focal length f_1. A converging lens of focal length f_2 is placed a distance d to the right of the diverging lens. Find the distance d so that the final image is infinitely far away to the right.

63. The lens-makers' equation applies to a lens immersed in a liquid if n in the equation is replaced by n_2/n_1. Here n_2 refers to the index of refraction of the lens material and n_1 is that of the medium surrounding the lens. (a) A certain lens has focal length 79.0 cm in air and index of refraction 1.55. Find its focal length in water. (b) A certain mirror has focal length 79.0 cm in air. Find its focal length in water.

64. **Q|C** In many applications, it is necessary to expand or decrease the diameter of a beam of parallel rays of light, which can be accomplished by using a converging lens and a diverging lens in combination. Suppose you have a converging lens of focal length 21.0 cm and a diverging lens of focal length -12.0 cm. (a) How can you arrange these lenses to increase the diameter of a beam of parallel rays? (b) By what factor will the diameter increase?

65. A real object is located at the zero end of a meterstick. A large concave spherical mirror at the 100-cm end of the meterstick forms an image of the object at the 70.0-cm position. A small convex spherical mirror placed at the 20.0-cm position forms a final image at the 10.0-cm point. What is the radius of curvature of the convex mirror?

66. A *zoom lens* system is a combination of lenses that produces a variable magnification of a fixed object as it maintains a fixed image position. The magnification is varied by moving one or more lenses along the axis. Multiple lenses are used in practice, but the effect of zooming in on an object can be demonstrated with a simple two-lens system. An object, two converging lenses, and a screen are mounted on an optical bench. Lens 1, which is to the right of the object, has a focal length of $f_1 = 5.00$ cm, and lens 2, which is to the right of the first lens, has a focal length of $f_2 = 10.0$ cm. The screen is to the right of lens 2. Initially, an object is situated at a distance of 7.50 cm to the left of lens 1, and the image formed

on the screen has a magnification of +1.00. (a) Find the distance between the object and the screen. (b) Both lenses are now moved along their common axis while the object and the screen maintain fixed positions until the image formed on the screen has a magnification of +3.00. Find the displacement of each lens from its initial position in part (a). (c) Can the lenses be displaced in more than one way?

67. The disk of the Sun subtends an angle of 0.533° at the Earth. What are (a) the position and (b) the diameter of the solar image formed by a concave spherical mirror with a radius of curvature of magnitude 3.00 m?

68. **Q|C** A floating strawberry illusion is achieved with two parabolic mirrors, each having a focal length 7.50 cm, facing each other as shown in Figure P26.68. If a strawberry is placed on the lower mirror, an image of the strawberry is formed at the small opening at the center of the top mirror, 7.50 cm above the lowest point of the bottom mirror. The position of the eye in Figure P26.68a corresponds to the view of the apparatus in Figure P26.68b. Consider the light path marked *A*. Notice that this light path is blocked by the upper mirror so that the strawberry itself is not directly observable. The light path marked *B* corresponds to the eye viewing the image of the strawberry that is formed at the opening at the top of the apparatus. (a) Show that the final image is formed at that location and describe its characteristics. (b) A very startling effect is to shine a flashlight beam on this image. Even at a glancing angle, the incoming light beam is seemingly reflected from the image! Explain.

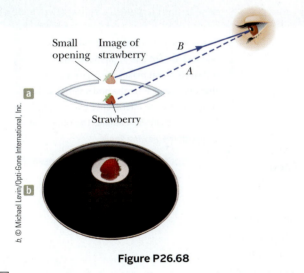

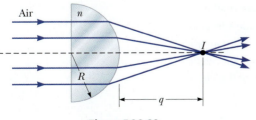

Figure P26.68

69. A parallel beam of light enters a glass hemisphere perpendicular to the flat face as shown in Figure P26.69. The magnitude of the radius of the hemisphere is $R = 6.00$ cm, and its index of refraction is $n = 1.560$. Assuming paraxial rays, determine the point at which the beam is focused.

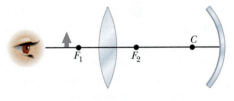

Figure P26.69

70. An object 2.00 cm high is placed 40.0 cm to the left of a converging lens having a focal length of 30.0 cm. A diverging lens with a focal length of −20.0 cm is placed 110 cm to the right of the converging lens. Determine (a) the position and (b) the magnification of the final image. (c) Is the image upright or inverted? (d) **What If?** Repeat parts (a) through (c) for the case in which the second lens is a converging lens having a focal length of 20.0 cm.

71. An observer to the right of the mirror–lens combination shown in Figure P26.71 (not to scale) sees two real images that are the same size and in the same location. One image is upright, and the other is inverted. Both images are 1.50 times larger than the object. The lens has a focal length of 10.0 cm. The lens and mirror are separated by 40.0 cm. Determine the focal length of the mirror.

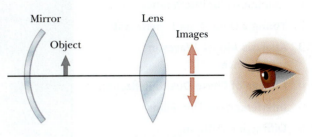

Figure P26.71

72. Figure P26.72 shows a thin converging lens for which the radii of curvature of its surfaces have magnitudes of 9.00 cm and 11.0 cm. The lens is in front of a concave spherical mirror with the radius of curvature $R = 8.00$ cm. Assume the focal points F_1 and F_2 of the lens are 5.00 cm from the center of the lens. (a) Determine the index of refraction of the lens material. The lens and mirror are 20.0 cm apart, and an object is placed 8.00 cm to the left of the lens. Determine (b) the position of the final image and (c) its magnification as seen by the eye in the figure. (d) Is the final image inverted or upright? Explain.

Figure P26.72

73. Assume the intensity of sunlight is 1.00 kW/m² at a particular location. A highly reflecting concave mirror is to be pointed toward the Sun to produce a power of at least 350 W at the image point. (a) Assuming the disk of the Sun subtends an angle of 0.533° at the Earth, find the required radius R_a of the circular face area of the mirror. (b) Now suppose the light intensity is to be at least 120 kW/m² at the image. Find the required relationship between R_a and the radius of curvature R of the mirror.

Chapter 27

Wave Optics

Chapter Outline

The colors in many of a hummingbird's feathers are not due to pigment. The iridescence that makes the brilliant colors that often appear on the bird's throat and belly is due to an interference effect caused by structures in the feathers. The colors will vary with the viewing angle.

Dec Hogan/Shutterstock.com

In Chapters 25 and 26, we used the ray approximation to examine what happens when light reflects from a surface or refracts into a new medium. We used the general term *geometric optics* for these discussions. This chapter is concerned with **wave optics**, a subject that addresses the optical phenomena of interference and diffraction. These phenomena cannot be adequately explained with the ray approximation. We must address the wave nature of light to be able to understand these phenomena.

We introduced the concept of wave interference in Chapter 14 for one-dimensional waves. This phenomenon depends on the principle of superposition, which tells us that when two or more traveling mechanical waves combine at a given point, the resultant displacement of the elements of the medium at that point is the sum of the displacements due to the individual waves.

We shall see the full richness of the waves in interference model in this chapter as we apply it to light. We used one-dimensional waves on strings to introduce interference in Active Figures 14.1 and 14.2. As we discuss the interference of light waves, two major changes from this previous discussion must be noted. First, we shall no longer focus on one-dimensional waves, so we must build geometric models to analyze the situation in two or three dimensions. Second, we shall study electromagnetic waves rather than mechanical waves. Therefore, the principle of superposition needs to be cast in terms of addition of field vectors rather than displacements of the elements of the medium.

27.1 | Conditions for Interference

In our discussion of wave interference for mechanical waves in Chapter 14, we found that two waves can add together constructively or destructively. In constructive interference between waves, the amplitude of the resultant wave is greater than that of either individual wave, whereas in destructive interference, the resultant amplitude is less than that of either individual wave. Electromagnetic waves also undergo interference. Fundamentally, all interference associated with electromagnetic waves arises as a result of combining the electric and magnetic fields that constitute the individual waves.

In Figure 14.4, we described a device that allows interference to be observed for sound waves. Interference effects in visible electromagnetic waves are not easy to observe because of their short wavelengths (from about 4×10^{-7} to 7×10^{-7} m). Two sources producing two waves of identical wavelengths are needed to create interference. To produce a stable interference pattern, however, the individual waves must maintain a constant phase relationship with one another; they must be **coherent.** As an example, the sound waves emitted by two side-by-side loudspeakers driven by a single amplifier can produce interference because the two loudspeakers respond to the amplifier in the same way at the same time.

If two separate light sources are placed side by side, no interference effects are observed because the light waves from one source are emitted independently of the other source; hence, the emissions from the two sources do not maintain a constant phase relationship with each other over the time of observation. An ordinary light source undergoes random changes in time intervals less than a nanosecond. Therefore, the conditions for constructive interference, destructive interference, or some intermediate state are maintained only for such short time intervals. The result is that no interference effects are observed because the eye cannot follow such rapid changes. Such light sources are said to be **incoherent.**

27.2 | Young's Double-Slit Experiment

A common method for producing two coherent light sources is to use a monochromatic source to illuminate a barrier containing two small openings (usually in the shape of slits). The light emerging from the two slits is coherent because a single source produces the original light beam and the two slits serve only to separate the original beam into two parts (which, after all, is what was done to the sound signal from the side-by-side loudspeakers in the preceding section). Any random change in the light emitted by the source occurs in both beams at the same time, and, as a result, interference effects can be observed when the light from the two slits arrives at a viewing screen.

If the light traveled only in its original direction after passing through the slits as shown in Figure 27.1a, the waves would not overlap and no interference pattern would be seen. Instead, as we have discussed in our treatment of Huygens's principle (Section 25.6), the waves spread out from the slits as shown in Figure 27.1b. In other words, the light deviates from a straight-line path and enters the region that would otherwise be shadowed. As noted in Section 25.2, this divergence of light from its initial line of travel is called **diffraction.**

Interference in light waves from two sources was first demonstrated by Thomas Young in 1801. A schematic diagram of the apparatus that Young used is shown in Active Figure 27.2a (page 912). Plane light waves arrive at a barrier that contains two parallel slits S_1 and S_2. These two slits serve as a pair of coherent light sources because waves emerging from them originate from the same wave front and therefore maintain a constant phase relationship. The light from S_1 and S_2 produces on a viewing screen a visible pattern of bright and dark parallel bands

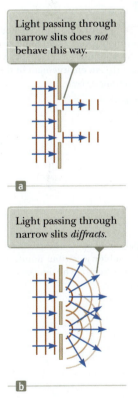

Light passing through narrow slits does *not* behave this way.

a

Light passing through narrow slits *diffracts*.

b

Figure 27.1 (a) If light waves did not spread out after passing through the slits, no interference would occur. (b) The light waves from the two slits overlap as they spread out, filling what we expect to be shadowed regions with light and producing interference fringes on a screen placed to the right of the slits.

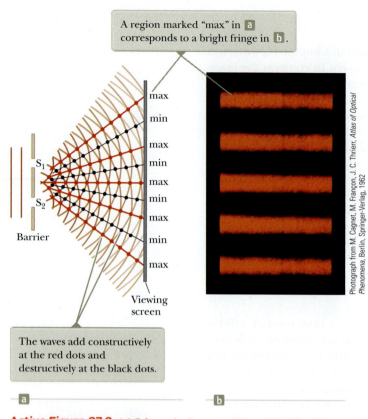

A region marked "max" in **a** corresponds to a bright fringe in **b**.

max
min
max
min
max
min
max
min
max

S_1

S_2

Barrier

Viewing screen

The waves add constructively at the red dots and destructively at the black dots.

Photograph from M. Cagnet, M. Françon, J. C. Thrier, *Atlas of Optical Phenomena*, Berlin, Springer-Verlag, 1962

a   **b**

Active Figure 27.2 (a) Schematic diagram of Young's double-slit experiment. Slits S_1 and S_2 behave as coherent sources of light waves that produce an interference pattern on the viewing screen (drawing not to scale). (b) An enlargement of the center of a fringe pattern formed on the viewing screen.

called **fringes** (Active Fig. 27.2b). When the light from S_1 and that from S_2 both arrive at a point on the screen such that constructive interference occurs at that location, a bright fringe appears. When the light from the two slits combines destructively at any location on the screen, a dark fringe results.

Figure 27.3 is a schematic diagram that allows us to generate a mathematical representation by modeling the interference as if waves combine at the viewing screen.[1] In Figure 27.3a, two waves leave the two slits in phase and strike the screen at the central point O. Because these waves travel equal distances, they arrive in phase at O. As a result, constructive interference occurs at this location and a bright fringe is observed. In Figure 27.3b, the two light waves again start in phase, but the lower wave has to travel one wavelength farther to reach point P on the screen. Because the lower wave falls behind the upper one by exactly one wavelength, they still arrive in phase at P. Hence, a second bright fringe appears at this location. Now consider point R located between O and P in Figure 27.3c. At this location, the lower wave has fallen half a wavelength behind the upper wave when they arrive at the screen. Hence, the trough from the lower wave overlaps the crest from the upper wave, giving rise to destructive interference at R. For this reason, one observes a dark fringe at this location.

Young's double-slit experiment is the prototype for many interference effects. Interference of waves occurs relatively commonly in technological applications, so this phenomenon represents an important analysis model to understand. In the next section, we develop the mathematical representation for interference of light.

Figure 27.3 Waves leave the slits and combine at various points on the viewing screen. (The figures are not to scale.)

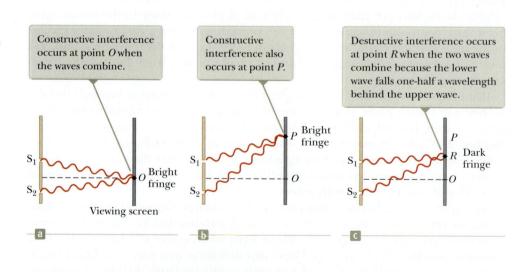

Constructive interference occurs at point O when the waves combine.

Constructive interference also occurs at point P.

Destructive interference occurs at point R when the two waves combine because the lower wave falls one-half a wavelength behind the upper wave.

S_1

S_2

O Bright fringe

Viewing screen

P Bright fringe

O

S_1

S_2

P

R Dark fringe

O

S_1

S_2

a   **b**   **c**

[1]The interference occurs everywhere between the slits and the screen, not only at the screen. See Thinking Physics 27.1. The model we have proposed will give us a valid result.

⟨27.3⟩ Analysis Model: Waves in Interference

We can obtain a quantitative description of Young's experiment with the help of a geometric model constructed from Figure 27.4a. The viewing screen is located a perpendicular distance L from the slits S_1 and S_2, which are separated by a distance d. Consider point P on the screen. Angle θ is measured from a line perpendicular to the screen from the midpoint between the slits and a line from the midpoint to point P. We identify r_1 and r_2 as the distances the waves travel from slit to screen. Let us assume that the source is monochromatic. Under these conditions, the waves emerging from S_1 and S_2 have the same wavelength and amplitude and are in phase. The light intensity on the screen at P is the result of the superposition of the light coming from both slits. Note from the geometric model triangle in yellow in Figure 27.4a that a wave from the lower slit travels farther than a wave from the upper slit by an amount δ (Greek letter delta). This distance is called the **path difference.**

If L is much greater than d, the two paths are very close to being parallel. We shall adopt a simplification model in which the two paths are exactly parallel. In this case, from Figure 27.4b, we see that

$$\delta = r_2 - r_1 = d \sin \theta \qquad \textbf{27.1} \blacktriangleleft \qquad \blacktriangleright \text{ Path difference}$$

In Figure 27.4a, the condition $L \gg d$ is not satisfied because the figure is not to scale; in Figure 27.4b, the rays leave the slits as if the condition is satisfied. As noted earlier, the value of this path difference determines whether the two waves are in phase or out of phase when they arrive at P. If the path difference is either zero or some integral multiple of the wavelength, the two waves are in phase at P and **constructive interference** results. The condition for bright fringes at P is therefore

$$\delta = d \sin \theta_{\text{bright}} = m\lambda \qquad m = 0, \pm 1, \pm 2, \ldots \qquad \textbf{27.2} \blacktriangleleft \qquad \blacktriangleright \text{ Conditions for constructive interference}$$

The number m is an integer called the **order number.** The central bright fringe at $\theta_{\text{bright}} = 0$ is associated with the order number $m = 0$ and is called the **zeroth-order maximum.** The first maximum on either side, for which $m = \pm 1$, is called the **first-order maximum,** and so forth.

Similarly, when the path difference is an odd multiple of $\lambda/2$, the two waves arriving at P are 180° out of phase and give rise to **destructive interference.** Therefore, the condition for dark fringes at P is

$$\delta = d \sin \theta_{\text{dark}} = (m + \tfrac{1}{2})\lambda \qquad m = 0, \pm 1, \pm 2, \ldots \qquad \textbf{27.3} \blacktriangleleft \qquad \blacktriangleright \text{ Conditions for destructive interference}$$

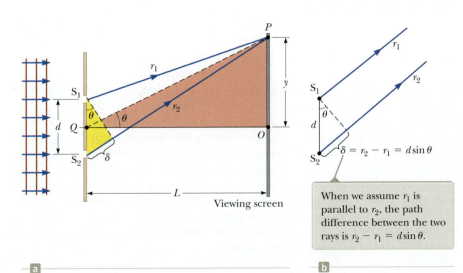

Figure 27.4 (a) Geometric construction for describing Young's double-slit experiment (not to scale). (b) The slits are represented as sources, and the outgoing light rays are assumed to be parallel as they travel to P. To achieve that in practice, it is essential that $L \gg d$.

When we assume r_1 is parallel to r_2, the path difference between the two rays is $r_2 - r_1 = d \sin \theta$.

Viewing screen

These equations provide the *angular* positions of the fringes. It is also useful to obtain expressions for the *linear* positions measured along the screen from O to P. From the geometric model triangle OPQ in Figure 27.4a, we see that

$$\tan \theta = \frac{y}{L} \qquad \text{27.4} \blacktriangleleft$$

Using this result, we can see that the linear positions of bright and dark fringes are given by

$$y_{\text{bright}} = L \tan \theta_{\text{bright}} \qquad \text{27.5} \blacktriangleleft$$

$$y_{\text{dark}} = L \tan \theta_{\text{dark}} \qquad \text{27.6} \blacktriangleleft$$

where θ_{bright} and θ_{dark} are given by Equations 27.2 and 27.3.

When the angles to the fringes are small, the fringes are equally spaced near the center of the pattern. To verify this statement, note that, for small angles, $\tan \theta \approx \sin \theta$ and Equation 27.5 gives the positions of the bright fringes as $y_{\text{bright}} = L \sin \theta_{\text{bright}}$. Incorporating Equation 27.2, we find that

$$y_{\text{bright}} = L \left(\frac{m\lambda}{d} \right) \qquad \text{(small angles)} \qquad \text{27.7} \blacktriangleleft$$

and we see that y_{bright} is linear in the order number m, so the fringes are equally spaced. Similarly, for dark fringes,

$$y_{\text{dark}} = L \frac{(m + \frac{1}{2})\lambda}{d} \qquad \text{(small angles)} \qquad \text{27.8} \blacktriangleleft$$

As demonstrated in Example 27.1, Young's double-slit experiment provides a method for measuring the wavelength of light. In fact, Young used this technique to do precisely that. In addition, his experiment gave the wave model of light a great deal of credibility. It was inconceivable that particles of light coming through the slits could cancel one another in a way that would explain the dark fringes.

The principles discussed in this section are the basis of the **waves in interference** analysis model. This model was applied to mechanical waves in one dimension in Chapter 14. Here we see the details of applying this model in three dimensions to light.

> **QUICK QUIZ 27.1** Which of the following causes the fringes in a two-slit interference pattern to move farther apart? (**a**) decreasing the wavelength of the light (**b**) decreasing the screen distance L (**c**) decreasing the slit spacing d (**d**) immersing the entire apparatus in water

> **THINKING PHYSICS 27.1**
>
> Consider a double-slit experiment in which a laser beam is passed through a pair of very closely spaced slits and a clear interference pattern is displayed on a distant screen. Now suppose you place smoke particles between the double slit and the screen. With the presence of the smoke particles, will you see the effects of the interference in the space between the slits and the screen, or will you only see the effects on the screen?
>
> **Reasoning** You see the effects in the area filled with smoke. Bright beams of light are directed toward the bright areas on the screen, and dark regions are directed toward the dark areas on the screen. The geometrical construction shown in Figure 27.4a is important for developing the mathematical description of interference. It is subject to misinterpretation, however, because it might suggest that the interference can only occur at the position of the screen. A better diagram for this situation is Active Figure 27.2a, which shows *paths* of destructive and constructive interference all the way from the slits to the screen. These paths are made visible by the smoke. ◄

Example 27.1 | Measuring the Wavelength of a Light Source

A viewing screen is separated from a double slit by 4.80 m. The distance between the two slits is 0.030 0 mm. Monochromatic light is directed toward the double slit and forms an interference pattern on the screen. The first dark fringe is 4.50 cm from the center line on the screen.

(A) Determine the wavelength of the light.

SOLUTION

Conceptualize Study Figure 27.4 to be sure you understand the phenomenon of interference of light waves. The distance of 4.50 cm is y in Figure 27.4.

Categorize We determine results using equations developed in this section, so we categorize this example as a substitution problem. Because $L \gg y$, the angles for the fringes are small.

Solve Equation 27.8 for the wavelength and substitute numerical values, taking $m = 0$ for the first dark fringe:

$$\lambda = \frac{y_{dark}d}{(m + \frac{1}{2})L} = \frac{(4.50 \times 10^{-2}\,\text{m})(3.00 \times 10^{-5}\,\text{m})}{(0 + \frac{1}{2})(4.80\,\text{m})}$$

$$= 5.62 \times 10^{-7}\,\text{m} = \boxed{562\,\text{nm}}$$

(B) Calculate the distance between adjacent bright fringes.

SOLUTION

Find the distance between adjacent bright fringes from Equation 27.7 and the results of part (A):

$$y_{m+1} - y_m = L\frac{(m+1)\lambda}{d} - L\frac{m\lambda}{d}$$

$$= L\frac{\lambda}{d} = 4.80\,\text{m}\left(\frac{5.62 \times 10^{-7}\,\text{m}}{3.00 \times 10^{-5}\,\text{m}}\right)$$

$$= 9.00 \times 10^{-2}\,\text{m} = \boxed{9.00\,\text{cm}}$$

For practice, find the wavelength of the sound in Example 14.1 using the procedure in part (A) of this example.

Intensity Distribution of the Double-Slit Interference Pattern

We shall now discuss briefly the distribution of light intensity I (the energy delivered by the light per unit area per unit time) associated with the double-slit interference pattern. Again, suppose the two slits represent coherent sources of sinusoidal waves. In this case, the two waves have the same angular frequency ω and a constant phase difference ϕ. Although the waves have equal phase at the slits, their phase difference ϕ at P depends on the path difference $\delta = r_2 - r_1 = d \sin \theta$. Because a path difference of λ corresponds to a phase difference of 2π rad, we can establish the equality of the ratios:

$$\frac{\delta}{\phi} = \frac{\lambda}{2\pi}$$

$$\phi = \frac{2\pi}{\lambda}\delta = \frac{2\pi}{\lambda}d \sin \theta \qquad \textbf{27.9} \blacktriangleleft$$

This equation tells us how the phase difference ϕ depends on the angle θ.

Although we shall not prove it here, a careful analysis of the electric fields arriving at the screen from the two very narrow slits shows that the **time-averaged light intensity** at a given angle θ is

$$I = I_{max} \cos^2\left(\frac{\pi d \sin \theta}{\lambda}\right) \qquad \textbf{27.10} \blacktriangleleft$$

where I_{max} is the intensity at point O in Figure 27.4a, directly behind the midpoint between the slits. Intensity versus $d \sin \theta$ is plotted in Figure 27.5.

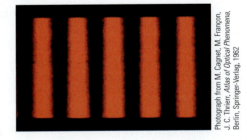

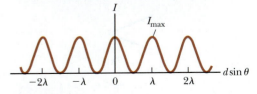

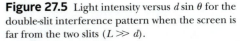

Figure 27.5 Light intensity versus $d \sin \theta$ for the double-slit interference pattern when the screen is far from the two slits ($L \gg d$).

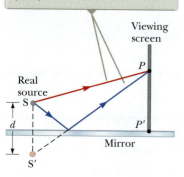

An interference pattern is produced on the screen as a result of the combination of the direct ray (red) and the reflected ray (blue).

Figure 27.6 Lloyd's mirror. The reflected ray undergoes a phase change of 180°.

27.4 | Change of Phase Due to Reflection

Young's method of producing two coherent light sources involves illuminating a pair of slits with a single source. Another simple arrangement for producing an interference pattern with a single light source is known as *Lloyd's mirror*. A point light source S is placed close to a mirror as illustrated in Figure 27.6. Waves can reach the point *P* either by the direct path from S to *P* or by the indirect path involving reflection from the mirror. The reflected ray strikes the screen as if it originated from a source S′ located below the mirror.

At points far from the source, one would expect an interference pattern due to waves from S and S′, just as is observed for two real coherent sources at these points. An interference pattern is indeed observed. The positions of the dark and bright fringes, however, are *reversed* relative to the pattern of two real coherent sources (Young's experiment) because the coherent sources S and S′ differ in phase by 180°. This 180° phase change is produced on reflection. In general, an electromagnetic wave undergoes a phase change of 180° on reflection from a medium of higher index of refraction than the one in which it is traveling.

It is useful to draw an analogy between reflected light waves and the reflections of a transverse wave on a stretched string when the wave meets a boundary (Section 13.4) as in Figure 27.7. The reflected pulse on a string undergoes a phase change of 180° when it is reflected from a rigid end, and no phase change when it is reflected from a free end, as illustrated in Active Figures 13.12 and 13.13. If the boundary is between two strings, the transmitted wave exhibits no phase change. Similarly, an electromagnetic wave undergoes a 180° phase change when reflected from the boundary of a medium of higher index of refraction than the one in which it is traveling. There is no phase change for the reflected ray when the wave is incident on a boundary leading to a medium of lower index of refraction. In either case, the transmitted wave exhibits no phase change.

27.5 | Interference in Thin Films

Interference effects can be observed in many situations in which one beam of light is split and then recombined. A common occurrence is the appearance of colored bands in a film of oil on water or in a soap bubble illuminated with white light. The colors in these situations result from the interference of waves reflected from the opposite surfaces of the film.

Consider a film of uniform thickness t and index of refraction n as in Figure 27.8. We adopt a simplification model in which the light ray is incident on the film from above and nearly normal to the surface of the film. Two rays are reflected from the film, one from the upper surface and one from the lower surface after the refracted

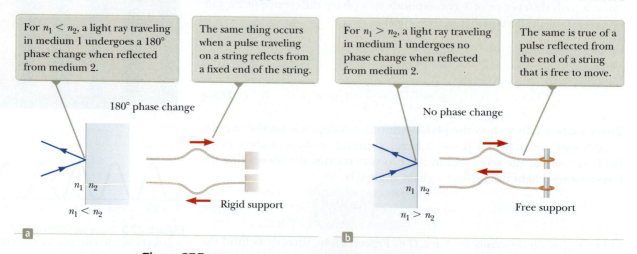

For $n_1 < n_2$, a light ray traveling in medium 1 undergoes a 180° phase change when reflected from medium 2.

The same thing occurs when a pulse traveling on a string reflects from a fixed end of the string.

For $n_1 > n_2$, a light ray traveling in medium 1 undergoes no phase change when reflected from medium 2.

The same is true of a pulse reflected from the end of a string that is free to move.

180° phase change

No phase change

n_1 n_2

$n_1 < n_2$

Rigid support

n_1 n_2

$n_1 > n_2$

Free support

Figure 27.7 Comparisons of reflections of light waves and waves on strings.

ray has traveled through the film. Because the film is thin and has parallel sides, the reflected rays are parallel. Hence, rays reflected from the top surface can interfere with rays reflected from the bottom surface. To determine whether the reflected rays interfere constructively or destructively, we first note the following facts:

- An electromagnetic wave traveling from a medium of index of refraction n_1 toward a medium of index of refraction n_2 undergoes a 180° phase change on reflection when $n_2 > n_1$. No phase change occurs in the reflected wave if $n_2 < n_1$.
- The wavelength λ_n of light in a medium whose index of refraction n is

$$\lambda_n = \frac{\lambda}{n} \qquad \text{27.11} \blacktriangleleft$$

where λ is the wavelength of light in free space.

Let us apply these rules to the film of Figure 27.8. According to the first rule, ray 1, which is reflected from the upper surface (A), undergoes a phase change of 180° with respect to the incident wave. Ray 2, which is reflected from the lower surface (B), undergoes no phase change with respect to the incident wave. Therefore, ignoring the path difference for now, outgoing ray 1 is 180° out of phase with respect to ray 2, a phase difference that is equivalent to a path difference of $\lambda_n/2$. We must also consider, however, that ray 2 travels an extra distance approximately equal to $2t$ before the waves recombine. The *total* phase difference arises from a combination of the path difference and the 180° phase change on reflection. For example, if $2t = \lambda_n/2$, rays 1 and 2 will recombine in phase and constructive interference will result. In general, the condition for constructive interference is

$$2t = (m + \tfrac{1}{2})\lambda_n \qquad m = 0, 1, 2, \ldots \qquad \text{27.12} \blacktriangleleft$$

This condition takes into account two factors: (a) the difference in optical path length for the two rays (the term $m\lambda_n$) and (b) the 180° phase change on reflection (the term $\lambda_n/2$). Because $\lambda_n = \lambda/n$, we can write Equation 27.12 in the form

$$2nt = (m + \tfrac{1}{2})\lambda \qquad m = 0, 1, 2, \ldots \qquad \text{27.13} \blacktriangleleft$$

▶ Condition for constructive interference in thin films

If the extra distance $2t$ traveled by ray 2 corresponds to an integer multiple of λ_n, the two waves will combine out of phase and destructive interference results. The general equation for destructive interference is

$$2nt = m\lambda \qquad m = 0, 1, 2, \ldots \qquad \text{27.14} \blacktriangleleft$$

▶ Condition for destructive interference in thin films

The preceding conditions for constructive and destructive interference are valid when the medium above the top surface of the film is the same as the medium below the bottom surface. The surrounding medium may have a refractive index less than or greater than that of the film. In either case, the rays reflected from the two surfaces will be out of phase by 180°. The conditions are also valid if different media are above and below the film and if both have n less than or larger than that of the film.

If the film is placed between two different media, one with $n < n_{\text{film}}$ and the other with $n > n_{\text{film}}$, the conditions for constructive and destructive interference are reversed. In this case, either a phase change of 180° takes place for both ray 1 reflecting from surface A and ray 2 reflecting from surface B, or no phase change occurs for either ray; hence, the net change in relative phase due to the reflections is zero.

Rays 3 and 4 in Figure 27.8 lead to interference effects in the light transmitted through the thin film. The analysis of these effects is similar to that of the reflected light.

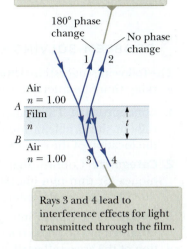

Interference in light reflected from a thin film is due to a combination of rays 1 and 2 reflected from the upper and lower surfaces of the film.

180° phase change

No phase change

Air
$n = 1.00$

A Film
n

B Air
$n = 1.00$

t

Rays 3 and 4 lead to interference effects for light transmitted through the film.

Figure 27.8 Light paths through a thin film.

QUICK QUIZ 27.2 In a laboratory accident, you spill two liquids onto water, neither of which mixes with the water. They both form thin films on the water surface. When the films become very thin as they spread, you observe that one film becomes bright and the other dark in reflected light. The film that appears dark **(a)** has an index of refraction higher than that of water, **(b)** has an index of refraction lower than that of water, **(c)** has an index of refraction equal to that of water, or **(d)** has an index of refraction lower than that of the bright film.

> **QUICK QUIZ 27.3** One microscope slide is placed on top of another with their left edges in contact and a human hair under the right edge of the upper slide. As a result, a wedge of air exists between the slides. An interference pattern results when monochromatic light is reflected from the wedge. What is at the left edge of the slides? **(a)** a dark fringe **(b)** a bright fringe **(c)** impossible to determine

---

> **PROBLEM-SOLVING STRATEGY: Thin-Film Interference**

The following suggestions should be kept in mind while working thin-film interference problems:

1. **Conceptualize** Think about what is going on physically in the problem. Identify the light source and the location of the observer.

2. **Categorize** Confirm that you should use the techniques for thin film interference by identifying the thin film causing the interference.

3. **Analyze** The type of interference that occurs is determined by the phase relationship between the portion of the wave reflected at the upper surface of the film and the portion reflected at the lower surface. Phase differences between the two portions of the wave have two causes: (a) differences in the distances traveled by the two portions and (b) phase changes occurring on reflection. *Both* causes must be considered when determining which type of interference occurs. If the media above and below the film both have index of refraction larger than that of the film or if both indices are smaller, use Equation 27.13 for constructive interference and Equation 27.14 for destructive interference. If the film is located between two different media, one with $n < n_{film}$ and the other with $n > n_{film}$, reverse these two equations for constructive and destructive interference.

4. **Finalize** Inspect your final results to see if they make sense physically and are of an appropriate size.

> ### Example 27.2 | Interference in a Soap Film

Calculate the minimum thickness of a soap-bubble film that results in constructive interference in the reflected light if the film is illuminated with light whose wavelength in free space is $\lambda = 600$ nm. The index of refraction of the soap film is 1.33.

SOLUTION

Conceptualize Imagine that the film in Figure 27.8 is soap, with air on both sides.

Categorize We determine the result using an equation from this section, so we categorize this example as a substitution problem.

The minimum film thickness for constructive interference in the reflected light corresponds to $m = 0$ in Equation 27.13. Solve this equation for t and substitute numerical values:

$$t = \frac{(0 + \frac{1}{2})\lambda}{2n} = \frac{\lambda}{4n} = \frac{(600 \text{ nm})}{4(1.33)} = \boxed{113 \text{ nm}}$$

What If? What if the film is twice as thick? Does this situation produce constructive interference?

Answer Using Equation 27.13, we can solve for the thicknesses at which constructive interference occurs:

$$t = (m + \frac{1}{2})\frac{\lambda}{2n} = (2m + 1)\frac{\lambda}{4n} \qquad m = 0, 1, 2, \ldots$$

The allowed values of m show that constructive interference occurs for *odd* multiples of the thickness corresponding to $m = 0$, $t = 113$ nm. Therefore, constructive interference does *not* occur for a film that is twice as thick.

> ### Example 27.3 | Nonreflective Coatings for Solar Cells

Solar cells—devices that generate electricity when exposed to sunlight—are often coated with a transparent, thin film of silicon monoxide (SiO, $n = 1.45$) to minimize reflective losses from the surface. Suppose a silicon solar cell ($n = 3.5$) is coated with a thin film of silicon monoxide for this purpose (Fig. 27.9a). Determine the minimum film thickness that produces the least reflection at a wavelength of 550 nm, near the center of the visible spectrum.

27.3 *cont.*

SOLUTION

Conceptualize Figure 27.9a helps us visualize the path of the rays in the SiO film that result in interference in the reflected light.

Categorize Based on the geometry of the SiO layer, we categorize this example as a thin-film interference problem.

Analyze The reflected light is a minimum when rays 1 and 2 in Figure 27.9a meet the condition of destructive interference. In this situation, *both* rays undergo a 180° phase change upon reflection: ray 1 from the upper SiO surface and ray 2 from the lower SiO surface. The net change in phase due to reflection is therefore zero, and the condition for a reflection minimum requires a path difference of $\lambda_n/2$, where λ_n is the wavelength of the light in SiO. Hence, $2nt = \lambda/2$, where λ is the wavelength in air and n is the index of refraction of SiO.

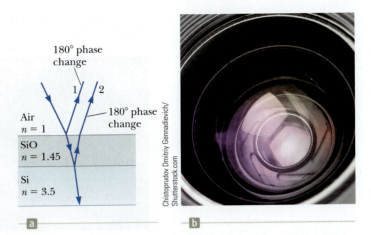

Figure 27.9 (Example 27.3) (a) Reflective losses from a silicon solar cell are minimized by coating the surface of the cell with a thin film of silicon monoxide. (b) The reflected light from a coated camera lens often has a reddish-violet appearance.

Solve the equation $2nt = \lambda/2$ for t and substitute numerical values:

$$t = \frac{\lambda}{4n} = \frac{550 \text{ nm}}{4(1.45)} = \boxed{94.8 \text{ nm}}$$

Finalize A typical uncoated solar cell has reflective losses as high as 30%, but a coating of SiO can reduce this value to about 10%. This significant decrease in reflective losses increases the cell's efficiency because less reflection means that more sunlight enters the silicon to create charge carriers in the cell. No coating can ever be made perfectly nonreflecting because the required thickness is wavelength-dependent and the incident light covers a wide range of wavelengths.

Glass lenses used in cameras and other optical instruments are usually coated with a transparent thin film to reduce or eliminate unwanted reflection and to enhance the transmission of light through the lenses. The camera lens in Figure 27.9b has several coatings (of different thicknesses) to minimize reflection of light waves having wavelengths near the center of the visible spectrum. As a result, the small amount of light that is reflected by the lens has a greater proportion of the far ends of the spectrum and often appears reddish violet.

27.6 | Diffraction Patterns

In Sections 25.2 and 27.2, we discussed briefly the phenomenon of **diffraction,** and now we shall investigate this phenomenon more fully for light waves. In general, diffraction occurs when waves pass through small openings, around obstacles, or by sharp edges.

We might expect that the light passing through one such small opening would simply result in a broad region of light on a screen due to the spreading of the light as it passes through the opening. We find something more interesting, however. A **diffraction pattern** consisting of light and dark areas is observed, somewhat similar to the interference patterns discussed earlier. For example, when a narrow slit is placed between a distant light source (or a laser beam) and a screen, the light produces a diffraction pattern like that in Figure 27.10. The pattern consists of a broad, intense central band (called the **central maximum**), flanked by a series of narrower, less intense additional bands (called **side maxima**) and a series of dark bands (or **minima**).

Figure 27.11 (page 920) shows the shadow of a penny, which displays bright and dark rings of a diffraction pattern. The bright spot at the center (called the *Arago bright spot* after its discoverer, Dominique Arago) can be explained using the wave theory of light. Waves that diffract from all points on the edge of the penny travel the same distance to the midpoint on the screen. Therefore, the midpoint is a region of constructive interference and a bright spot appears. In contrast, from the viewpoint of geometric optics, the center of the pattern would be completely screened by the penny, and so an approach that does not include the wave nature of light would not predict a central bright spot.

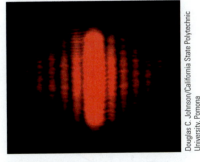

Figure 27.10 The diffraction pattern that appears on a screen when light passes through a narrow vertical slit. The pattern consists of a broad central band and a series of less intense and narrower side bands.

Notice the bright spot at the center.

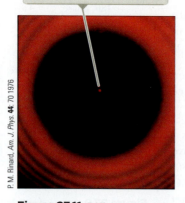

Figure 27.11 Diffraction pattern created by the illumination of a penny, with the penny positioned midway between the screen and light source.

The pattern consists of a central bright fringe flanked by much weaker maxima alternating with dark fringes.

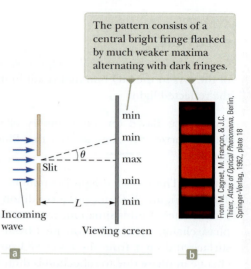

Active Figure 27.12
(a) Geometry for analyzing the Fraunhofer diffraction pattern of a single slit. (Drawing not to scale.) (b) Photograph of a single-slit Fraunhofer diffraction pattern.

Pitfall Prevention | 27.1
Diffraction versus Diffraction Pattern
Diffraction refers to the general behavior of waves spreading out as they pass through a slit. We used diffraction in explaining the existence of an interference pattern. A *diffraction pattern* is actually a misnomer, but it is deeply entrenched in the language of physics. The diffraction pattern seen on a screen when a single slit is illuminated is actually another interference pattern. The interference is between parts of the incident light illuminating different regions of the slit.

Each portion of the slit acts as a point source of light waves.

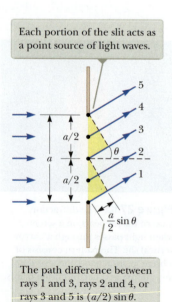

The path difference between rays 1 and 3, rays 2 and 4, or rays 3 and 5 is $(a/2) \sin \theta$.

Figure 27.13 Paths of light rays that encounter a narrow slit of width a and diffract toward a screen in the direction described by angle θ.

Let us consider a common situation, that of light passing through a narrow opening modeled as a slit and projected onto a screen. As a simplification model, we assume that the observing screen is far from the slit so that the rays reaching the screen are approximately parallel. This situation can also be achieved experimentally by using a converging lens to focus the parallel rays on a nearby screen. In this model, the pattern on the screen is called a **Fraunhofer diffraction pattern.**[2]

Active Figure 27.12a shows light entering a single slit from the left and diffracting as it propagates toward a screen. Active Figure 27.12b is a photograph of a single-slit Fraunhofer diffraction pattern. A bright fringe is observed along the axis at $\theta = 0$, with alternating dark and bright fringes on each side of the central bright fringe.

Until now, we assumed that slits act as point sources of light. In this section, we shall determine how their finite widths are the basis for understanding the nature of the Fraunhofer diffraction pattern produced by a single slit. We can deduce some important features of this problem by examining waves coming from various portions of the slit as shown in the geometric model of Figure 27.13. According to Huygens's principle, each portion of the slit acts as a source of waves. Hence, light from one portion of the slit can interfere with light from another portion, and the resultant intensity on the screen depends on the direction θ.

To analyze the diffraction pattern, it is convenient to divide the slit into two halves as in Figure 27.13. All the waves that originate at the slit are in phase. Consider waves 1 and 3, which originate at the bottom and center of the slit, respectively. To reach the same point on the viewing screen, wave 1 travels farther than wave 3 by an amount equal to the path difference $(a/2) \sin \theta$, where a is the width of the slit. Similarly, the path difference between waves 3 and 5 is also $(a/2) \sin \theta$. If the path difference between two waves is exactly one-half of a wavelength (corresponding to a phase difference of 180°), the two waves cancel each other and destructive interference results. That is true, in fact, for any two waves that originate at points separated by half the slit width because the phase difference between two such points is 180°.

---

[2]If the screen were brought close to the slit (and no lens is used), the pattern is a *Fresnel* diffraction pattern. The Fresnel pattern is more difficult to analyze, so we shall restrict our discussion to Fraunhofer diffraction.

Therefore, waves from the upper half of the slit interfere *destructively* with waves from the lower half of the slit when

$$\frac{a}{2} \sin \theta = \pm \frac{\lambda}{2}$$

or when

$$\sin \theta = \pm \frac{\lambda}{a}$$

If we divide the slit into four parts rather than two and use similar reasoning, we find that the screen is also dark when

$$\sin \theta = \pm \frac{2\lambda}{a}$$

Likewise, we can divide the slit into six parts and show that darkness occurs on the screen when

$$\sin \theta = \pm \frac{3\lambda}{a}$$

Therefore, the general condition for destructive interference is

$$\sin \theta_{dark} = m\frac{\lambda}{a} \qquad m = \pm 1, \pm 2, \pm 3, \ldots \qquad \text{27.15} \blacktriangleleft$$

▶ Condition for destructive interference in a diffraction pattern

Equation 27.15 gives the values of θ for which the diffraction pattern has zero intensity, that is, a dark fringe is formed. Equation 27.15, however, tells us nothing about the variation in intensity along the screen. The general features of the intensity distribution are shown in Active Figure 27.14: a broad central bright fringe flanked by much weaker, alternating bright fringes. The various dark fringes (points of zero intensity) occur at the values of θ that satisfy Equation 27.15. The position of the points of constructive interference lie approximately halfway between the dark fringes. Note that the central bright fringe is twice as wide as the weaker maxima.

QUICK QUIZ 27.4 Suppose the slit width in Active Figure 27.14 is made half as wide. Does the central bright fringe (a) become wider, (b) remain the same, or (c) become narrower?

Pitfall Prevention | 27.2
Similar Equations
Equation 27.15 has exactly the same form as Equation 27.2, with d, the slit separation, used in Equation 27.2 and a, the slit width, in Equation 27.15. Keep in mind, however, that Equation 27.2 describes the *bright* regions in a two-slit interference pattern, whereas Equation 27.15 describes the *dark* regions in a single-slit diffraction pattern. Furthermore, $m = 0$ does not represent a dark fringe in the diffraction pattern.

> **THINKING PHYSICS 27.2**

If a classroom door is open slightly, you can hear sounds coming from the hallway. Yet you cannot see what is happening in the hallway. What accounts for the difference?

Reasoning The space between the slightly open door and the wall is acting as a single slit for waves. Sound waves have wavelengths larger than the slit width, so sound is effectively diffracted by the opening and spread throughout the room. The sound is then reflected from walls, floor, and ceiling, further distributing the sound throughout the room. Light wavelengths are much smaller than the slit width, so virtually no diffraction for the light occurs. You must have a direct line of sight to detect the light waves. ◀

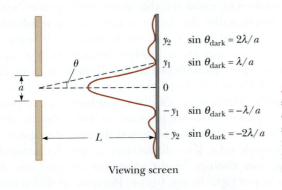

Viewing screen

Active Figure 27.14 Light intensity distribution for the Fraunhofer diffraction pattern from a single slit of width a. The positions of two minima on each side of the central maximum are labeled. (Drawing not to scale.)

Example 27.4 | Where Are the Dark Fringes?

Light of wavelength 580 nm is incident on a slit having a width of 0.300 mm. The viewing screen is 2.00 m from the slit. Find the width of the central bright fringe.

SOLUTION

Conceptualize Based on the problem statement, we imagine a single-slit diffraction pattern similar to that in Active Figure 27.14.

Categorize We categorize this example as a straightforward application of our discussion of single-slit diffraction patterns.

Analyze Evaluate Equation 27.15 for the two dark fringes that flank the central bright fringe, which correspond to $m = \pm 1$:

$$\sin \theta_{dark} = \pm \frac{\lambda}{a}$$

Let y represent the vertical position along the viewing screen in Active Figure 27.14, measured from the point on the screen directly behind the slit. Then, $\tan \theta_{dark} = y_1/L$, where the subscript 1 refers to the first dark fringe. Because θ_{dark} is very small, we can use the approximation $\sin \theta_{dark} \approx \tan \theta_{dark}$; therefore, $y_1 = L \sin \theta_{dark}$.

The width of the central bright fringe is twice the absolute value of y_1:

$$2|y_1| = 2|L \sin \theta_{dark}| = 2\left|\pm L\frac{\lambda}{a}\right| = 2L\frac{\lambda}{a} = 2(2.00 \text{ m})\frac{580 \times 10^{-9} \text{ m}}{0.300 \times 10^{-3} \text{ m}}$$

$$= 7.73 \times 10^{-3} \text{ m} = \boxed{7.73 \text{ mm}}$$

Finalize Notice that this value is much greater than the width of the slit. Let's explore below what happens if we change the slit width.

What If? What if the slit width is increased by an order of magnitude to 3.00 mm? What happens to the diffraction pattern?

Answer Based on Equation 27.15, we expect that the angles at which the dark bands appear will decrease as a increases. Therefore, the diffraction pattern narrows.

Repeat the calculation with the larger slit width:

$$2|y_1| = 2L\frac{\lambda}{a} = 2(2.00 \text{ m})\frac{580 \times 10^{-9} \text{ m}}{3.00 \times 10^{-3} \text{ m}} = 7.73 \times 10^{-4} \text{ m} = \boxed{0.773 \text{ mm}}$$

Notice that this result is *smaller* than the width of the slit. In general, for large values of a, the various maxima and minima are so closely spaced that only a large, central bright area resembling the geometric image of the slit is observed. This concept is very important in the performance of optical instruments such as telescopes.

❮27.7 | Resolution of Single-Slit and Circular Apertures

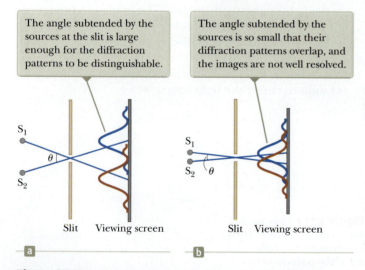

The angle subtended by the sources at the slit is large enough for the diffraction patterns to be distinguishable.

The angle subtended by the sources is so small that their diffraction patterns overlap, and the images are not well resolved.

S_1

θ

S_2

Slit   Viewing screen

S_1

S_2 θ

Slit   Viewing screen

a

b

Figure 27.15 Two point sources far from a narrow slit each produce a diffraction pattern. (a) The sources are separated by a large angle. (b) The sources are separated by a small angle. (Notice that the angles are greatly exaggerated. The drawing is not to scale.)

Imagine you are driving in the middle of a dark desert at night, along a road that is perfectly straight and flat for many kilometers. You see another vehicle coming toward you from a distance. When the vehicle is far away, you might be unable to determine whether it is an automobile with two headlights or a motorcycle with one. As it approaches you, at some point you will be able to distinguish the two headlights and determine that it is an automobile. Once you are able to see two separate headlights, you describe the light sources as being **resolved.**

The ability of optical systems to distinguish between closely spaced objects is limited because of the wave nature of light. To understand this limitation, consider Figure 27.15, which shows two light sources far from a narrow slit. The sources can be considered as two point sources S_1 and S_2 that are incoherent. For example, they could be two distant stars observed through the aperture of a telescope tube. If no diffraction occurred, one would observe two distinct bright spots (or images) on the screen at the right in the figure. Because of diffraction,

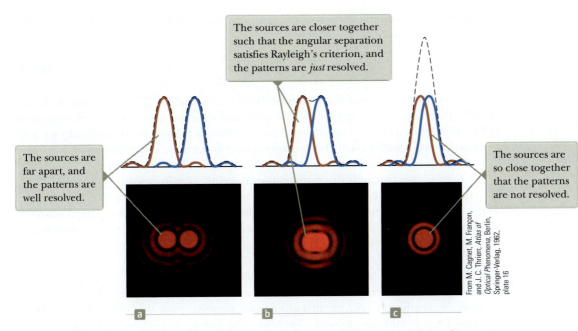

The sources are closer together such that the angular separation satisfies Rayleigh's criterion, and the patterns are *just* resolved.

The sources are far apart, and the patterns are well resolved.

The sources are so close together that the patterns are not resolved.

From M. Cagnet, M. Françon, and J. C. Thrierr, *Atlas of Optical Phenomena*, Berlin, Springer-Verlag, 1962, plate 16

Figure 27.16 Individual diffraction patterns of two point sources (solid curves) and the resultant patterns (dashed curves) for various angular separations of the sources as the light passes through a circular aperture. In each case, the dashed curve is the sum of the two solid curves.

however, each source is imaged as a bright central region flanked by weaker bright and dark bands. What is observed on the screen is the sum of two diffraction patterns: one from S_1 and the other from S_2.

If the two sources are far enough apart to ensure that their central maxima do not overlap as in Figure 27.15a, their images can be distinguished and are said to be resolved. If the sources are close together, however, as in Figure 27.15b, the two central maxima may overlap and the sources are not resolved. To decide when two sources are resolved, the following condition is often used:

When the central maximum of one image falls on the first minimum of another image, the images are said to be just resolved. This limiting condition of resolution is known as **Rayleigh's criterion.**

▶ Rayleigh's criterion

Figure 27.16 shows the diffraction patterns from circular apertures for three situations. When the objects are far apart, they are well resolved (Fig. 27.16a). They are just resolved when their angular separation satisfies Rayleigh's criterion (Fig. 27.16b). Finally, the sources are not resolved in Figure 27.16c.

From Rayleigh's criterion, we can determine the minimum angular separation θ_{min} subtended by the sources at a slit such that the sources are just resolved. In Section 27.6, we found that the first minimum in a single-slit diffraction pattern occurs at the angle that satisfies the relationship

$$\sin \theta = \frac{\lambda}{a}$$

where a is the width of the slit. According to Rayleigh's criterion, this expression gives the smallest angular separation for which the two sources are resolved. Because $\lambda \ll a$ in most situations, $\sin \theta$ is small and we can use the approximation $\sin \theta \approx \theta$. Therefore, the limiting angle of resolution for a slit of width a is

$$\theta_{min} = \frac{\lambda}{a}$$

27.16 ◀   ▶ Limiting angle of resolution for a slit

where θ_{min} is expressed in radians. Hence, the angle subtended by the two sources at the slit must be *greater* than λ / a if the sources are to be resolved.

Many optical systems use circular apertures rather than slits. The diffraction pattern of a circular aperture, as seen in Figure 27.16, consists of a central circular bright disk surrounded by progressively fainter rings. Analysis shows that the limiting angle of resolution of the circular aperture is

▶ Limiting angle of resolution for a circular aperture

$$\theta_{min} = 1.22 \frac{\lambda}{D}$$

27.17 ◀

where D is the diameter of the aperture. Note that Equation 27.17 is similar to Equation 27.16 except for the factor of 1.22, which arises from a mathematical analysis of diffraction from a circular aperture. This equation is related to the difficulty we had seeing the two headlights at the beginning of this section. When observing with the eye, D in Equation 27.17 is the diameter of the pupil. The diffraction pattern formed when light passes through the pupil causes the difficulty in resolving the headlights.

Another example of the effect of diffraction on resolution for circular apertures is the astronomical telescope. The end of the tube through which the light passes is circular, so the ability of the telescope to resolve light from closely spaced stars is limited by the diameter of this opening.

QUICK QUIZ 27.5 Suppose you are observing a binary star with a telescope and are having difficulty resolving the two stars. You decide to use a colored filter to maximize the resolution. (A filter of a given color transmits only that color of light.) What color filter should you choose? **(a)** blue **(b)** green **(c)** yellow **(d)** red

---

▶ **THINKING PHYSICS 27.3**

Cats' eyes have pupils that can be modeled as vertical slits. At night, are cats more successful in resolving headlights on a distant car or vertically separated lights on the mast of a distant boat?

Reasoning The effective slit width in the vertical direction of the cat's eye is larger than that in the horizontal direction. Therefore, the eye has more resolving power for lights separated in the vertical direction and would be more effective at resolving the mast lights on the boat. ◀

Example 27.5 | **Resolution of the Eye**   BIO

Light of wavelength 500 nm, near the center of the visible spectrum, enters a human eye. Although pupil diameter varies from person to person, let's estimate a daytime diameter of 2 mm.

(A) Estimate the limiting angle of resolution for this eye, assuming its resolution is limited only by diffraction.

SOLUTION

Conceptualize In Figure 27.16, identify the aperture through which the light travels to be the pupil of the eye. Light passing through this small aperture causes diffraction patterns to occur on the retina.

Categorize We determine the result using equations developed in this section, so we categorize this example as a substitution problem.

Use Equation 27.17, taking $\lambda = 500$ nm and $D = 2$ mm:

$$\theta_{min} = 1.22 \frac{\lambda}{D} = 1.22 \left(\frac{5.00 \times 10^{-7}\,\text{m}}{2 \times 10^{-3}\,\text{m}} \right)$$

$$= 3 \times 10^{-4}\,\text{rad} \approx 1\,\text{min of arc}$$

(B) Determine the minimum separation distance d between two point sources that the eye can distinguish if the point sources are a distance $L = 25$ cm from the observer (Fig. 27.17).

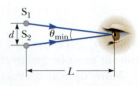

Figure 27.17 (Example 27.5) Two point sources separated by a distance d as observed by the eye.

27.5 cont.

SOLUTION

Noting that θ_{min} is small, find d:

$$\sin \theta_{min} \approx \theta_{min} \approx \frac{d}{L} \quad \rightarrow \quad d = L\theta_{min}$$

Substitute numerical values:

$$d = (25 \text{ cm})(3 \times 10^{-4} \text{ rad}) = \boxed{8 \times 10^{-3} \text{ cm}}$$

This result is approximately equal to the thickness of a human hair.

27.8 | The Diffraction Grating

The **diffraction grating,** a useful device for analyzing light sources, consists of a large number of equally spaced parallel slits. A grating can be made by cutting parallel, equally spaced grooves on a glass or metal plate with a precision ruling machine. In a *transmission grating,* the spaces between lines are transparent to the light and hence act as separate slits. In a *reflection grating,* the spaces between lines are highly reflective. Gratings with many lines very close to one another can have very small slit spacings. For example, a grating ruled with 5 000 lines/cm has a slit spacing of $d = (1/5\,000)$ cm $= 2 \times 10^{-4}$ cm.

Figure 27.18 shows a pictorial representation of a section of a flat diffraction grating. A plane wave is incident from the left, normal to the plane of the grating. The pattern observed on the screen at the right in Figure 27.18 is the result of the combined effects of interference and diffraction. Each slit produces diffraction, and the diffracted beams interfere with one another to produce the final pattern. Each slit acts as a source of waves, and all waves start at the slits in phase. For some arbitrary direction θ measured from the horizontal, however, the waves must travel different path lengths before reaching a particular point on the screen. From Figure 27.18, note that the path difference between waves from any two adjacent slits is equal to $d \sin \theta$. (We assume once again that the distance L to the screen is much larger than d.) If this path difference equals one wavelength or some integral multiple of a wavelength, waves from all slits will be in phase at the screen and a bright line will be observed. When the light is incident normally on the plane of the grating, the condition for *maxima* in the interference pattern at the angle θ is therefore[3]

$$d \sin \theta_{bright} = m\lambda \quad m = 0, \pm 1, \pm 2, \pm 3, \ldots \qquad \text{27.18} \blacktriangleleft$$

Pitfall Prevention | 27.3
A Diffraction Grating Is an Interference Grating
As with diffraction pattern, diffraction grating is a misnomer, but it is deeply entrenched in the language of physics. The diffraction grating depends on diffraction in the same way as the double slit, spreading the light so that light from different slits can interfere. It would be more correct to call it an interference grating, but diffraction grating is the name in use.

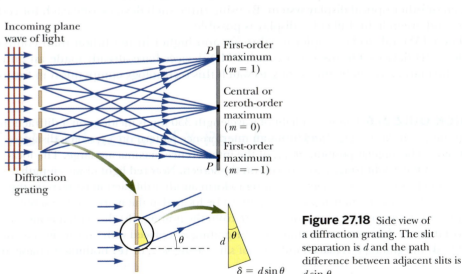

Incoming plane wave of light

P — First-order maximum ($m = 1$)

Central or zeroth-order maximum ($m = 0$)

First-order maximum ($m = -1$)

P

Diffraction grating

θ

d

θ

$\delta = d \sin \theta$

Figure 27.18 Side view of a diffraction grating. The slit separation is d and the path difference between adjacent slits is $d \sin \theta$.

---

[3]Notice that this equation is identical to Equation 27.2. This equation can be used for a number of slits from two to any number N. The intensity distribution will change with the number of slits, but the locations of the maxima are the same.

NASA, ESA, and M. Livio and the Hubble 20th Anniversary Team (STScI)

Figure 3 This image of a portion of the Carina Nebula, 7 500 light-years from the Earth, was released to celebrate the 20th anniversary of the launching and deployment of the Hubble Space Telescope (April 24, 1990). Significant new star development is causing this display. Several jets of material, such as those at the top of the image, are caused by gravitational attraction of material to the new stars' surfaces. The colors in the image are due to light emitted from individual atoms of oxygen, nitrogen, and hydrogen. The Carina Nebula is a member of our own galaxy. Across the entire sky, it is estimated that the Hubble Space Telescope can detect 100 billion galaxies. It is also estimated that this is a very small fraction of all the galaxies in the visible part of the Universe. To develop a theory of the origin of this tremendously large system, we need to understand quarks, the most fundamental particles in the current theory of particle physics.

smaller scales allow us to advance our understanding of the larger and larger systems that are familiar to us. This connection between the small and the large is the theme of this Context.

Let us consider some examples of macroscopic systems and their connection to the behavior of microscopic particles. Consider your experiences with common electronic devices that are used today to view information on a liquid crystal display: handheld calculators, smartphones, tablet computers, and flat panel televisions. The events you observe—the appearance of numbers, to-do lists, or photographs on an LCD display—are macroscopic, but what controls these macroscopic events? They are controlled by a microprocessor within the electronic device. The operation of the microprocessor depends on the behavior of electrons within the integrated circuit chip. The design and manufacture of the macroscopic electronic device are not possible without an understanding of the behavior of the electrons.

As a second example, a supernova explosion is clearly a macroscopic event; it is a star with a radius on the order of billions of meters undergoing a violent event. We have been able to advance our understanding of such events by studying the atomic nucleus, which is on the order of 10^{-15} m in size.

If we imagine an even larger system than a star—the entire Universe—we can advance our understanding of its origin by thinking about particles even smaller than the nucleus. Consider the constituents of protons and neutrons, called *quarks*. Models based on quarks provide further understanding of a theory of the origin of the Universe called the *Big Bang*. In this Context, we shall study both quarks and the Big Bang.

It seems that the larger the system we wish to investigate, the smaller are the particles whose behavior we must understand! We shall explore this relationship and study the principles of quantum physics as we respond to our central question:

How can we connect the physics of microscopic particles to the physics of the Universe?

Quantum Physics

Chapter Outline

© Robert Harding Picture Library Ltd./Alamy

A scanning electron microscope photograph shows significant detail of a cheese mite, *Tyrolichus casei*. The mite is so small, with a maximum length of 0.70 mm, that ordinary microscopes do not reveal minute anatomical details. The operation of the electron microscope is based on the wave nature of electrons, a central feature in quantum physics.

In the earlier chapters of this book, we focused on the physics of particles. The particle model was a simplification model that allowed us to ignore the unnecessary details of an object when studying its behavior. We later combined particles into additional simplification models of systems and rigid objects. In Chapter 13, we introduced the wave as yet another simplification model and found that we could understand the motion of vibrating strings and the intricacies of sound by studying simple waves. In Chapters 24 to 27, we found that the wave model for light helped us understand many phenomena associated with optics.

It is hoped that you now have confidence in your abilities to analyze problems in the very different worlds of particles and waves. Your confidence may have been shaken somewhat by the discussion at the beginning of Chapter 25 in which we indicated that light has both wave-like and particle-like behaviors.

In this chapter, we return to this dual nature of light and study it in more detail. This study leads to a new simplification model, the quantum particle, and a new analysis model, the quantum particle under boundary conditions. A careful analysis of these two models shows that particles and waves are not as unrelated as you might expect.

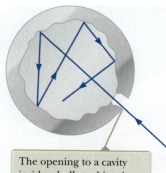

The opening to a cavity inside a hollow object is a good approximation of a black body: the hole acts as a perfect absorber.

Figure 28.1 A physical model of a black body.

28.1 | Blackbody Radiation and Planck's Theory

As we discussed in Chapter 17, an object at any temperature emits energy referred to as **thermal radiation.** The characteristics of this radiation depend on the temperature and properties of the surface of the object. If the surface is at room temperature, the wavelengths of the thermal radiation are primarily in the infrared region and hence are not observed by the eye. As the temperature of the surface increases, the object eventually begins to glow red. At sufficiently high temperatures, the object appears to be white, as in the glow of the hot tungsten filament of a lightbulb. A careful study of thermal radiation shows that it consists of a continuous distribution of wavelengths from all portions of the electromagnetic spectrum.

From a classical viewpoint, thermal radiation originates from accelerated charged particles near the surface of the object. The thermally agitated charges can have a distribution of accelerations, which accounts for the continuous spectrum of radiation emitted by the object. By the end of the 19th century, it had become apparent that this classical explanation of thermal radiation was inadequate. The basic problem was in understanding the observed distribution of wavelengths in the radiation emitted by an ideal object called a black body. As mentioned in Chapter 24, a **black body** is an ideal system that absorbs all radiation incident on it. A good approximation of a black body is a small hole leading to the inside of a hollow object as shown in Figure 28.1. The nature of the radiation emitted from the hole depends only on the temperature of the cavity walls.

The wavelength distribution of radiation from cavities was studied extensively in the late 19th century. Experimental data for the distribution of energy in **blackbody radiation** at three temperatures are shown in Active Figure 28.2. The distribution of radiated energy varies with wavelength and temperature. Two regular features of the distribution were noted in these experiments:

1. **The total power of emitted radiation increases with temperature.** We discussed this feature briefly in Chapter 17, where we introduced **Stefan's law,** Equation 17.36, for the power emitted from a surface of area A and temperature T:

▶ Stefan's law

$$P = \sigma A e T^4$$ **28.1** ◀

For a black body, the emissivity is $e = 1$ exactly.

2. **The peak of the wavelength distribution shifts to shorter wavelengths as the temperature increases.** This shift was found experimentally to obey the following relationship, called **Wien's displacement law:**

▶ Wien's displacement law

$$\lambda_{max} T = 2.898 \times 10^{-3} \text{ m} \cdot \text{K}$$ **28.2** ◀

The glow emanating from the spaces between these hot charcoal briquettes is, to a close approximation, blackbody radiation. The color of the light depends on the temperature of the briquettes.

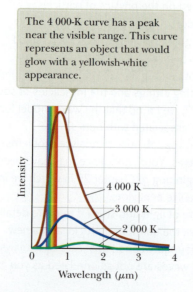

The 4 000-K curve has a peak near the visible range. This curve represents an object that would glow with a yellowish-white appearance.

Active Figure 28.2 Intensity of blackbody radiation versus wavelength at three temperatures. The visible range of wavelengths is between 0.4 μm and 0.7 μm. At approximately 6 000 K, the peak is in the center of the visible wavelengths and the object appears white.

where λ_{max} is the wavelength at which the curve in Active Figure 28.2 peaks and T is the absolute temperature of the surface emitting the radiation.

A successful theoretical model for blackbody radiation must predict the shape of the curve in Active Figure 28.2, the dependence of power on temperature expressed in Stefan's law, and the shift of the peak with temperature described by Wien's displacement law. Early attempts to use classical ideas to explain the shapes of the curves in Active Figure 28.2 failed.

Let's consider one of these early attempts. To describe the distribution of energy from a black body, we define $I(\lambda,T)\,d\lambda$ to be the intensity, or power per unit area, emitted in the wavelength interval $d\lambda$. The result of a calculation based on a classical theory of blackbody radiation known as the **Rayleigh–Jeans law** is

$$I(\lambda,T) = \frac{2\pi c k_B T}{\lambda^4} \qquad \textbf{28.3} \blacktriangleleft$$

▶ Rayleigh–Jeans law

where k_B is Boltzmann's constant. The black body is modeled as the hole leading into a cavity (Fig. 28.1), resulting in many modes of oscillation of the electromagnetic field caused by accelerated charges in the cavity walls and the emission of electromagnetic waves at all wavelengths. In the classical theory used to derive Equation 28.3, the average energy for each wavelength of the standing-wave modes is assumed to be proportional to $k_B T$, based on the theorem of equipartition of energy discussed in Section 16.5.

An experimental plot of the blackbody radiation spectrum, together with the theoretical prediction of the Rayleigh–Jeans law, is shown in Figure 28.3. At long wavelengths, the Rayleigh–Jeans law is in reasonable agreement with experimental data, but at short wavelengths, major disagreement is apparent.

As λ approaches zero, the function $I(\lambda,T)$ given by Equation 28.3 approaches infinity. Hence, according to classical theory, not only should short wavelengths predominate in a blackbody spectrum, but also the energy emitted by any black body should become infinite in the limit of zero wavelength. In contrast to this prediction, the experimental data plotted in Figure 28.3 show that as λ approaches zero, $I(\lambda,T)$ also approaches zero. This mismatch of theory and experiment was so disconcerting that scientists called it the *ultraviolet catastrophe*. (This "catastrophe"—infinite energy—occurs as the wavelength approaches zero; the word *ultraviolet* was applied because ultraviolet wavelengths are short.)

In 1900, Max Planck developed a structural model for blackbody radiation that leads to a theoretical equation for the wavelength distribution that is in complete agreement with experimental results at all wavelengths. His model represents the dawn of **quantum physics.**

Using the components of structural models introduced in Section 11.2, we can describe Planck's model as follows.

1. *A description of the physical components of the system:* Planck identified blackbody radiation as arising from *oscillators,* related to the charged particles within the molecules of the black body. The components of the system are the oscillators and the radiation emitted from them.
2. *A description of where the components are located relative to one another and how they interact:* The oscillators emitting observable blackbody radiation are located at the surface of the black body. The energy of the oscillator is quantized; that is, it can have only certain *discrete* amounts of energy E_n given by

$$E_n = nhf \qquad \textbf{28.4} \blacktriangleleft$$

where n is a positive integer called a **quantum number,**[1] f is the oscillator frequency, and h is **Planck's constant,** first introduced in Chapter 11. Because the energy of each oscillator can have only discrete values given by Equation 28.4,

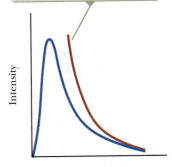

The classical theory (red-brown curve) shows intensity growing without bound for short wavelengths, unlike the experimental data (blue curve).

Figure 28.3 Comparison of experimental results and the curve predicted by the Rayleigh–Jeans law for the distribution of blackbody radiation.

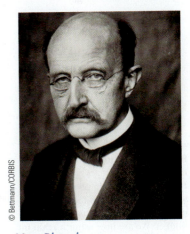

Max Planck
German Physicist (1858–1947)
Planck introduced the concept of a "quantum of action" (Planck's constant, *h*) in an attempt to explain the spectral distribution of blackbody radiation, which laid the foundations for quantum theory. In 1918, he was awarded the Nobel Prize in Physics for this discovery of the quantized nature of energy.

---

[1] We first introduced the notion of a quantum number for microscopic systems in Section 11.5, in which we incorporated it into the Bohr model of the hydrogen atom. We put it in bold again here because it is an important notion for the remaining chapters in this book.

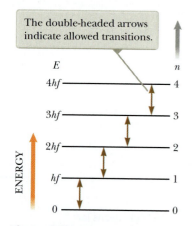

The double-headed arrows indicate allowed transitions.

Figure 28.4 Allowed energy levels for an oscillator with a natural frequency f.

we say that the energy is **quantized.** Each discrete energy value corresponds to a different **quantum state,** represented by the quantum number n. When the oscillator is in the $n = 1$ quantum state, its energy is hf; when it is in the $n = 2$ quantum state, its energy is $2hf$; and so on. An oscillator radiates or absorbs energy only when it changes quantum states. If it remains in one quantum state, no energy is absorbed or emitted.

3. *A description of the time evolution of the system:* As noted at the end of structural model component 2, radiation is emitted or absorbed only when the oscillator makes a transition from one energy state to a different state. The oscillators emit or absorb energy in discrete units, similar to the transitions discussed in the Bohr model in Chapter 11. The entire energy difference between the initial and final states in the transition is emitted as a single quantum of radiation. If the transition is from one state to an adjacent state—say, from the $n = 3$ state to the $n = 2$ state—Equation 28.4 shows that the amount of energy radiated by the oscillator is

$$E = hf \qquad \textbf{28.5} \blacktriangleleft$$

Figure 28.4 shows the quantized energy levels and allowed transitions proposed by Planck.

4. *A description of the agreement between predictions of the model and actual observations and, possibly, predictions of new effects that have not yet been observed:* The test of Planck's model will be this: Does the model predict a wavelength distribution curve that matches the experimental results in Active Figure 28.2 better than the expression in Equation 28.3 does?

These ideas may not sound bold to you because we have seen them in the Bohr model of the hydrogen atom in Chapter 11. It is important to keep in mind, however, that the Bohr model was not introduced until 1913, whereas Planck made his assumptions in 1900. The key point in Planck's theory is the radical assumption of quantized energy states.

Using this approach, Planck generated a theoretical expression for the wavelength distribution that agreed remarkably well with the experimental curves in Active Figure 28.2:

▶ Planck's wavelength distribution function

$$I(\lambda, T) = \frac{2\pi hc^2}{\lambda^5 (e^{hc/\lambda k_B T} - 1)} \qquad \textbf{28.6} \blacktriangleleft$$

This function includes the parameter h, which Planck adjusted so that his curve matched the experimental data at all wavelengths. The value of this parameter is found to be independent of the material of which the black body is made and independent of the temperature; it is a fundamental constant of nature. The value of h, Planck's constant, which was first introduced in Chapter 11, is

▶ Planck's constant

$$h = 6.626 \times 10^{-34} \, \text{J} \cdot \text{s} \qquad \textbf{28.7} \blacktriangleleft$$

At long wavelengths, Equation 28.6 reduces to the Rayleigh–Jeans expression, Equation 28.3, and at short wavelengths, it predicts an exponential decrease in $I(\lambda, T)$ with decreasing wavelength, in agreement with experimental results.

When Planck presented his theory, most scientists (including Planck!) did not consider the quantum concept realistic. It was believed to be a mathematical trick that happened to predict the correct results. Hence, Planck and others continued to search for what they believed to be a more rational explanation of blackbody radiation. Subsequent developments, however, showed that a theory based on the quantum concept (rather than on classical concepts) was required to explain a number of other phenomena at the atomic level.

We don't see quantum effects on an everyday basis because the energy change in a macroscopic system due to a transition between adjacent states is such a small fraction of the total energy of the system that we could never expect to detect the change. (See Example 28.2 for a numerical example.) Therefore, even though changes in the energy of a macroscopic system are indeed quantized and proceed

Pitfall Prevention | 28.2
n **Is Again an Integer**
In the preceding chapters on optics, we used the symbol n for the index of refraction, which was not an integer. We are now using n again in the manner in which it was used in Chapter 11 to indicate the quantum number of a Bohr orbit and in Chapter 14 to indicate the standing wave mode on a string or in an air column. In quantum physics, n is often used as an integer quantum number to identify a particular quantum state of a system.

by small quantum jumps, our senses perceive the decrease as continuous. Quantum effects become important and measurable only on the submicroscopic level of atoms and molecules. Furthermore, quantum results must blend smoothly with classical results when the quantum number becomes large. This statement is known as the **correspondence principle.**

You may have had your body temperature measured at the doctor's office by an *ear thermometer,* which can read your temperature in a matter of seconds (Fig. 28.5). This type of thermometer measures the amount of infrared radiation emitted by the eardrum in a fraction of a second. It then converts the amount of radiation into a temperature reading. This thermometer is very sensitive because temperature is raised to the fourth power in Stefan's law. Problem 3 at the end of the chapter allows you to explore the sensitivity of this device.

BIO The ear thermometer

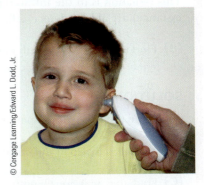

Figure 28.5 An ear thermometer measures a patient's temperature by detecting the intensity of infrared radiation leaving the eardrum.

QUICK QUIZ 28.1 Figure 28.6 shows two stars in the constellation Orion. Betelgeuse appears to glow red, whereas Rigel looks blue in color. Which star has a higher surface temperature? (a) Betelgeuse (b) Rigel (c) both have the same (d) impossible to determine

Figure 28.6 (Quick Quiz 28.1) Which star is hotter, Betelgeuse or Rigel?

THINKING PHYSICS 28.1

You are observing a yellow candle flame, and your laboratory partner claims that the light from the flame is atomic in origin. You disagree, claiming that the candle flame is hot, so the radiation must be thermal in origin. Before this disagreement leads to fisticuffs, how could you determine who is correct?

Reasoning A simple determination could be made by observing the light from the candle flame through a diffraction grating spectrometer, which was discussed in Section 27.8. If the spectrum of the light is continuous, it is thermal in origin. If the spectrum shows discrete lines, it is atomic in origin. The results of the experiment show that the light is primarily thermal in origin and originates in the hot particles of soot in the candle flame. ◀

Example 28.1 | **Thermal Radiation from Different Objects**   **BIO**

(A) Find the peak wavelength of the blackbody radiation emitted by the human body when the skin temperature is 35°C.

SOLUTION

Conceptualize Thermal radiation is emitted from the surface of any object. The peak wavelength is related to the surface temperature through Wien's displacement law (Eq. 28.2).

Categorize We evaluate results using an equation developed in this section, so we categorize this example as a substitution problem.

continued

28.1 *cont.*

Solve Equation 28.2 for λ_{max}:

$$(1) \quad \lambda_{max} = \frac{2.898 \times 10^{-3} \text{ m} \cdot \text{K}}{T}$$

Substitute the surface temperature:

$$\lambda_{max} = \frac{2.898 \times 10^{-3} \text{ m} \cdot \text{K}}{308 \text{ K}} = \boxed{9.41 \ \mu\text{m}}$$

This radiation is in the infrared region of the spectrum and is invisible to the human eye. Some animals (pit vipers, for instance) are able to detect radiation of this wavelength and therefore can locate warm-blooded prey even in the dark.

(B) Find the peak wavelength of the blackbody radiation emitted by the tungsten filament of a lightbulb, which operates at 2 000 K.

SOLUTION

Substitute the filament temperature into Equation (1):

$$\lambda_{max} = \frac{2.898 \times 10^{-3} \text{ m} \cdot \text{K}}{2 \ 000 \text{ K}} = \boxed{1.45 \ \mu\text{m}}$$

This radiation is also in the infrared, meaning that most of the energy emitted by a lightbulb is not visible to us.

(C) Find the peak wavelength of the blackbody radiation emitted by the Sun, which has a surface temperature of approximately 5 800 K.

SOLUTION

Substitute the surface temperature into Equation (1):

$$\lambda_{max} = \frac{2.898 \times 10^{-3} \text{ m} \cdot \text{K}}{5 \ 800 \text{ K}} = \boxed{0.500 \ \mu\text{m}}$$

This radiation is near the center of the visible spectrum, near the color of a yellow-green tennis ball. Because it is the most prevalent color in sunlight, our eyes have evolved to be most sensitive to light of approximately this wavelength.

(D) What total power is emitted by your skin, assuming that it emits like a black body?

SOLUTION

We need to make an estimate of the surface area of your skin. Model your body as a rectangular box of height 2 m, width 0.3 m, and depth 0.2 m, and find its total surface area:

$$A = 2(2 \text{ m})(0.3 \text{ m}) + 2(2 \text{ m})(0.2 \text{ m}) + 2(0.2 \text{ m})(0.3 \text{ m})$$
$$\approx 2 \text{ m}^2$$

Use Stefan's law, Equation 28.1, to find the power of the emitted radiation:

$$P = \sigma A e T^4 \approx (5.7 \times 10^{-8} \text{ W/m}^2 \cdot \text{K}^4)(2 \text{ m}^2)(1)(308 \text{ K})^4$$
$$\approx \boxed{10^3 \text{ W}}$$

(E) Based on your answer to part (D), why don't you glow as brightly as several lightbulbs?

SOLUTION

The answer to part (D) indicates that your skin is radiating energy at approximately the same rate as that which enters ten 100-W lightbulbs by electrical transmission. You are not visibly glowing, however, because most of this radiation is in the infrared range, as we found in part (A), and our eyes are not sensitive to infrared radiation.

Example **28.2** | **The Quantized Oscillator**

A 2.00-kg block is attached to a massless spring that has a force constant of $k = 25.0$ N/m. The spring is stretched 0.400 m from its equilibrium position and released from rest.

(A) Find the total energy of the system and the frequency of oscillation according to classical calculations.

SOLUTION

Conceptualize We understand the details of the block's motion from our study of simple harmonic motion in Chapter 12. Review that material if you need to.

28.2 *cont.*

Categorize The phrase "according to classical calculations" tells us to categorize this part of the problem as a classical analysis of the oscillator. We model the block as a particle in simple harmonic motion.

Analyze Based on the way the block is set into motion, its amplitude is 0.400 m.

Evaluate the total energy of the block–spring system using Equation 12.21:

$$E = \tfrac{1}{2}kA^2 = \tfrac{1}{2}(25.0 \text{ N/m})(0.400 \text{ m})^2 = \boxed{2.00 \text{ J}}$$

Evaluate the frequency of oscillation from Equation 12.14:

$$f = \frac{1}{2\pi}\sqrt{\frac{k}{m}} = \frac{1}{2\pi}\sqrt{\frac{25.0 \text{ N/m}}{2.00 \text{ kg}}} = \boxed{0.563 \text{ Hz}}$$

(B) Assuming the energy of the oscillator is quantized, find the quantum number n for the system oscillating with this amplitude.

SOLUTION

Categorize This part of the problem is categorized as a quantum analysis of the oscillator. We model the block–spring system as a Planck oscillator.

Analyze Solve Equation 28.4 for the quantum number n:

$$n = \frac{E_n}{hf}$$

Substitute numerical values:

$$n = \frac{2.00 \text{ J}}{(6.626 \times 10^{-34} \text{ J} \cdot \text{s})(0.563 \text{ Hz})} = \boxed{5.36 \times 10^{33}}$$

Finalize Notice that 5.36×10^{33} is a very large quantum number, which is typical for macroscopic systems. Changes between quantum states for the oscillator are explored next.

What If? Suppose the oscillator makes a transition from the $n = 5.36 \times 10^{33}$ state to the state corresponding to $n = 5.36 \times 10^{33} - 1$. By how much does the energy of the oscillator change in this one-quantum change?

Answer From Equation 28.5 and the result to part (A), the energy carried away due to the transition between states differing in n by 1 is

$$E = hf = (6.626 \times 10^{-34} \text{ J} \cdot \text{s})(0.563 \text{ Hz}) = 3.73 \times 10^{-34} \text{ J}$$

This energy change due to a one-quantum change is fractionally equal to 3.73×10^{-34} J/2.00 J, or on the order of one part in 10^{34}! It is such a small fraction of the total energy of the oscillator that it cannot be detected. Therefore, even though the energy of a macroscopic block–spring system is quantized and does indeed decrease by small quantum jumps, our senses perceive the decrease as continuous. Quantum effects become important and detectable only on the submicroscopic level of atoms and molecules.

28.2 | The Photoelectric Effect

Blackbody radiation was historically the first phenomenon to be explained with a quantum model. In the latter part of the 19th century, at the same time as data were being taken on thermal radiation, experiments showed that light incident on certain metallic surfaces causes electrons to be emitted from the surfaces. As mentioned in Section 25.1, this phenomenon, first discovered by Hertz, is known as the **photoelectric effect**. The emitted electrons are called **photoelectrons.**[2]

Active Figure 28.7 (page 952) is a schematic diagram of a photoelectric effect apparatus. An evacuated glass or quartz tube contains a metal plate E connected to the negative terminal of a battery. Another metal plate C is maintained at a positive potential by the battery. When the tube is kept in the dark, the ammeter reads zero, indicating that there is no current in the circuit. When light of the appropriate

---

[2]Photoelectrons are not different from other electrons. They are given this name solely because of their ejection from the metal by photons in the photoelectric effect.

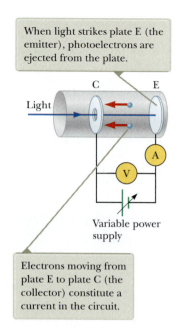

When light strikes plate E (the emitter), photoelectrons are ejected from the plate.

C E

Light

A

V

Variable power supply

Electrons moving from plate E to plate C (the collector) constitute a current in the circuit.

Active Figure 28.7 A circuit diagram for studying the photoelectric effect.

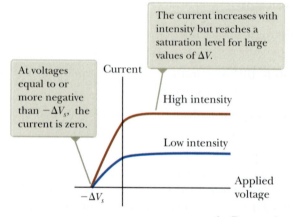

The current increases with intensity but reaches a saturation level for large values of ΔV.

At voltages equal to or more negative than $-\Delta V_s$, the current is zero.

Current

High intensity

Low intensity

Applied voltage

$-\Delta V_s$

Active Figure 28.8 Photoelectric current versus applied potential difference for two light intensities.

wavelength shines on plate E, however, a current is detected by the ammeter, indicating a flow of charges across the gap between E and C. This current arises from electrons emitted from the negative plate E (the emitter) and collected at the positive plate C (the collector).

Active Figure 28.8, a graphical representation of the results of a photoelectric experiment, plots the photoelectric current versus the potential difference ΔV between E and C for two light intensities. For large positive values of ΔV, the current reaches a maximum value. In addition, the current increases as the incident light intensity increases, as you might expect. Finally, when ΔV is negative—that is, when the battery polarity is reversed to make E positive and C negative—the current drops because many of the photoelectrons emitted from E are repelled by the negative collecting plate C. Only those electrons ejected from the metal with a kinetic energy greater than $e|\Delta V|$ will reach C, where e is the magnitude of the charge on the electron. When the magnitude of ΔV is equal to ΔV_s, the **stopping potential**, no electrons reach C and the current is zero.

Let us consider the combination of the electric field between the plates and an electron ejected from plate E with the maximum kinetic energy to be an isolated system. Suppose this electron stops just as it reaches plate C. Applying the energy version of the isolated system model, Equation 7.2 becomes:

$$\Delta K + \Delta U = 0 \quad \rightarrow \quad K_f + U_f = K_i + U_i$$

where the initial configuration of the system refers to the instant that the electron leaves the metal with the maximum possible kinetic energy K_{max} and the final configuration is when the electron stops just before touching plate C. If we define the electric potential energy of the system in the initial configuration to be zero, the energy equation above can be written

$$0 + (-e)(-\Delta V_s) = K_{max} + 0$$

$$K_{max} = e\,\Delta V_s \qquad \qquad \textbf{28.8} \blacktriangleleft$$

This equation allows us to measure K_{max} experimentally by measuring the voltage at which the current drops to zero.

The following are several features of the photoelectric effect in which the predictions made by a structural model based on a classical approach, using the wave model for light, are compared with the experimental results. Notice the strong contrast between the predictions and the results.

1. Dependence of photoelectron kinetic energy on light intensity

   Classical prediction: Electrons should absorb energy continuously from the electromagnetic waves. As the light intensity incident on a metal is increased, energy should be transferred into the metal at a higher rate and the electrons should be ejected with more kinetic energy.

   Experimental result: The maximum kinetic energy of photoelectrons is *independent* of light intensity as shown in Active Figure 28.8 with both curves falling to zero at the *same* negative voltage. (According to Equation 28.8, the maximum kinetic energy is proportional to the stopping potential.)

2. Time interval between incidence of light and ejection of photoelectrons

   Classical prediction: At low light intensities, a measurable time interval should pass between the instant the light is turned on and the time an electron is ejected from the metal. This time interval is required for the electron to absorb the incident radiation before it acquires enough energy to escape from the metal.

   Experimental result: Electrons are emitted from the surface of the metal almost *instantaneously* (less than 10^{-9} s after the surface is illuminated), even at very low light intensities.

3. Dependence of ejection of electrons on light frequency

   Classical prediction: Electrons should be ejected from the metal at any incident light frequency, as long as the light intensity is high enough, because energy is transferred to the metal regardless of the incident light frequency.

Experimental result: No electrons are emitted if the incident light frequency falls below some **cutoff frequency** f_c, whose value is characteristic of the material being illuminated. No electrons are ejected below this cutoff frequency *regardless* of the light intensity.

4. Dependence of photoelectron kinetic energy on light frequency

Classical prediction: There should be *no* relationship between the frequency of the light and the electron kinetic energy. The kinetic energy should be related to the intensity of the light.

Experimental result: The maximum kinetic energy of the photoelectrons increases with increasing light frequency.

Notice that *all four* predictions of the classical model are incorrect. A successful explanation of the photoelectric effect was given by Einstein in 1905, the same year he published his special theory of relativity. As part of a general paper on electromagnetic radiation, for which he received the Nobel Prize in Physics in 1921, Einstein extended Planck's concept of quantization to electromagnetic waves. He assumed that light (or any other electromagnetic wave) of frequency f can be considered to be a stream of quanta, regardless of the source of the radiation. Today we call these quanta **photons.** Each photon has an energy E given by Equation 28.5, $E = hf$, and moves in a vacuum at the speed of light c, where is $c = 3.00 \times 10^8$ m/s.

> **QUICK QUIZ 28.2** While standing outdoors one evening, you are exposed to the following four types of electromagnetic radiation: yellow light from a sodium street lamp, radio waves from an AM radio station, radio waves from an FM radio station, and microwaves from an antenna of a communications system. Rank these types of waves in terms of photon energy from highest to lowest.

Let us organize Einstein's model for the photoelectric effect using the components of structural models:

1. *A description of the physical components of the system:* We imagine the system to consist of two physical components: (1) an electron that is to be ejected by an incoming photon and (2) the remainder of the metal.
2. *A description of where the components are located relative to one another and how they interact:* In Einstein's model, a photon of the incident light gives *all* its energy hf to a *single* electron in the metal. Therefore, the absorption of energy by the electrons is not a continuous process as envisioned in the wave model, but rather a discontinuous process in which energy is delivered to the electrons in bundles. The energy transfer is accomplished via a one-photon/one-electron event.
3. *A description of the time evolution of the system:* We can describe the time evolution of the system by applying the nonisolated system model for energy over a time interval that includes the absorption of one photon and the ejection of the corresponding electron. Energy is transferred into the system by electromagnetic radiation, the photon. The system has two types of energy: the potential energy of the metal–electron system and the kinetic energy of the ejected electron. Therefore, we can write the conservation of energy equation (Eq. 7.2) as

$$\Delta K + \Delta U = T_{ER} \qquad \text{28.9} \blacktriangleleft$$

The energy transfer into the system is that of the photon, $T_{ER} = hf$. During the process, the kinetic energy of the electron increases from zero to its final value, which we assume to be the maximum possible value K_{max}. The potential energy of the system increases because the electron is pulled away from the metal to which it is attracted. We define the potential energy of the system when the electron is outside the metal as zero. The potential energy of the system when the electron is in the metal is $U = -\phi$, where ϕ is called the **work function** of the metal. The work function represents the minimum energy with which an electron is bound in the metal and is on the order of a few electron volts. Table 28.1 lists selected values. The increase in potential energy of the system

> **TABLE 28.1** |
> **Work Functions of Selected Metals**
>
> | Metal | ϕ (eV) |
> |-------|-------------|
> | Na | 2.46 |
> | Al | 4.08 |
> | Cu | 4.70 |
> | Zn | 4.31 |
> | Ag | 4.73 |
> | Pt | 6.35 |
> | Pb | 4.14 |
> | Fe | 4.50 |
>
> *Note:* Values are typical for metals listed. Actual values may vary depending on whether the metal is a single crystal or polycrystalline. Values may also depend on the face from which electrons are ejected from crystalline metals. Furthermore, different experimental procedures may produce differing values.

when the electron is removed from the metal is the work function ϕ. Substituting these energies into Equation 28.9, we have

$$(K_{max} - 0) + [0 - (-\phi)] = hf$$

$$K_{max} + \phi = hf \qquad \text{28.10} \blacktriangleleft$$

If the electron makes collisions with other electrons or metal ions as it is being ejected, some of the incoming energy is transferred to the metal and the electron is ejected with less kinetic energy than K_{max}.

4. *A description of the agreement between predictions of the model and actual observations and, possibly, predictions of new effects that have not yet been observed:* The prediction made by Einstein is an equation for the maximum kinetic energy of an ejected electron as a function of frequency of the illuminating radiation. This equation can be found by rearranging Equation 28.10:

▶ Photoelectric effect equation

$$K_{max} = hf - \phi \qquad \text{28.11} \blacktriangleleft$$

With Einstein's structural model, one can explain the observed features of the photoelectric effect that cannot be understood using classical concepts:

1. Dependence of photoelectron kinetic energy on light intensity

   Equation 28.11 shows that K_{max} is independent of the light intensity. The maximum kinetic energy of any one electron, which equals $hf - \phi$, depends only on the light frequency and the work function. If the light intensity is doubled, the number of photons arriving per unit time is doubled, which doubles the rate at which photoelectrons are emitted. The maximum kinetic energy of any one photoelectron, however, is unchanged.

2. Time interval between incidence of light and ejection of photoelectrons

   Near-instantaneous emission of electrons is consistent with the photon model of light. The incident energy appears in small packets, and there is a one-to-one interaction between photons and electrons. If the incident light has very low intensity, there are very few photons arriving per unit time interval; each photon, however, can have sufficient energy to eject an electron immediately.

3. Dependence of ejection of electrons on light frequency

   Because the photon must have energy greater than the work function ϕ to eject an electron, the photoelectric effect cannot be observed below a certain cutoff frequency. If the energy of an incoming photon does not satisfy this requirement, an electron cannot be ejected from the surface, even though many photons per unit time are incident on the metal in a very intense light beam.

4. Dependence of photoelectron kinetic energy on light frequency

   A photon of higher frequency carries more energy and therefore ejects a photoelectron with more kinetic energy than does a photon of lower frequency.

   Einstein's theoretical result (Eq. 28.11) predicts a linear relationship between the maximum electron kinetic energy K_{max} and the light frequency f. Experimental observation of such a linear relationship would be a final confirmation of Einstein's theory. Indeed, such a linear relationship is observed as sketched in Active Figure 28.9. The slope of the curves for all metals is Planck's constant h. The absolute value of the intercept on the vertical axis is the work function ϕ, which varies from one metal to another. The intercept on the horizontal axis is the cutoff frequency, which is related to the work function through the relation $f_c = \phi/h$. This cutoff frequency corresponds to a **cutoff wavelength** of

▶ Cutoff wavelength

$$\lambda_c = \frac{c}{f_c} = \frac{c}{\phi/h} = \frac{hc}{\phi} \qquad \text{28.12} \blacktriangleleft$$

where c is the speed of light. Light with wavelength *greater* than λ_c incident on a material with a work function of ϕ does not result in the emission of photoelectrons.

The combination hc occurs often when relating the energy of a photon to its wavelength. A common shortcut to use in solving problems is to express this combination in useful units according to the numerical value

$$hc = 1\ 240\ \text{eV} \cdot \text{nm}$$

One of the first practical uses of the photoelectric effect was as a detector in a light meter of a camera. Light reflected from the object to be photographed strikes a photoelectric surface in the meter, causing it to emit photoelectrons that then pass through a sensitive ammeter. The magnitude of the current in the ammeter depends on the light intensity.

The phototube, another early application of the photoelectric effect, acts much like a switch in an electric circuit. It produces a current in the circuit when light of sufficiently high frequency falls on a metal plate in the phototube, but produces no current in the dark. Phototubes were used in burglar alarms and in the detection of the soundtrack on motion picture film. Modern semiconductor devices have now replaced older devices based on the photoelectric effect.

The photoelectric effect is used today in the operation of photomultiplier tubes. Figure 28.10 shows the structure of such a device. A photon striking the photocathode ejects an electron by means of the photoelectric effect. This electron is accelerated across the potential difference between the photocathode and the first *dynode*, shown as being at +200 V relative to the photocathode in Figure 28.10. This high-energy electron strikes the dynode and ejects several more electrons. This process is repeated through a series of dynodes at ever higher potentials until an electrical pulse is produced as millions of electrons strike the last dynode. Therefore, the tube is called a *multiplier* because one photon at the input has resulted in millions of electrons at the output.

The photomultiplier tube is used in nuclear detectors to detect the presence of gamma rays emitted from radioactive nuclei, which we will study in Chapter 30. It is also used in astronomy in a technique called *photoelectric photometry*. In this technique, the light collected by a telescope from a single star is allowed to fall on a photomultiplier tube for a time interval. The tube measures the total light energy during the time interval, which can then be converted to a luminosity of the star.

The photomultiplier tube is being replaced in many astronomical observations with a *charge-coupled device* (CCD), which is the same device that is used in a digital camera. In this device, an array of pixels are formed on the silicon surface of an integrated circuit. When the surface is exposed to light from an astronomical scene through a telescope or a terrestrial scene through a digital camera, electrons generated by the photoelectric effect are caught in "traps" beneath the surface. The number of electrons is related to the intensity of the light striking the surface. A signal processor measures the number of electrons associated with each pixel and converts this information into a digital code that a computer can use to reconstruct and display the scene.

The *electron bombardment CCD camera* allows higher sensitivity than a conventional CCD. In this device, electrons ejected from a photocathode by the photoelectric effect are accelerated through a high voltage before striking a CCD array. The higher energy of the electrons results in a very sensitive detector of low-intensity radiation.

The explanation of the photoelectric effect with a quantum model, combined with Planck's quantum model for blackbody radiation, laid a strong foundation for further investigation into quantum physics. In the next section, we present a

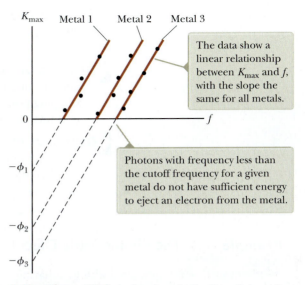

Active Figure 28.9 A plot of results for K_{max} of photoelectrons versus frequency of incident light in a typical photoelectric effect experiment.

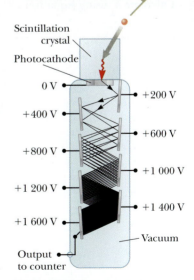

Figure 28.10 The multiplication of electrons in a photomultiplier tube.

third experimental result that provides further strong evidence of the quantum nature of light.

QUICK QUIZ 28.3 Consider one of the curves in Active Figure 28.8. Suppose the intensity of the incident light is held fixed but its frequency is increased. Does the stopping potential in Active Figure 28.8 (**a**) remain fixed, (**b**) move to the right, or (**c**) move to the left?

QUICK QUIZ 28.4 Suppose classical physicists had the idea of plotting K_{max} versus f as in Active Figure 28.9. Draw a graph of what the expected plot would look like, based on the wave model for light.

Example 28.3 | **The Photoelectric Effect for Sodium**

A sodium surface is illuminated with light having a wavelength of 300 nm. As indicated in Table 28.1, the work function for sodium metal is 2.46 eV.

(A) Find the maximum kinetic energy of the ejected photoelectrons.

SOLUTION

Conceptualize Imagine a photon striking the metal surface and ejecting an electron. The electron with the maximum energy is one near the surface that experiences no interactions with other particles in the metal that would reduce its energy on its way out of the metal.

Categorize We evaluate the results using equations developed in this section, so we categorize this example as a substitution problem.

Find the energy of each photon in the illuminating light beam from Equation 28.5:

$$E = hf = \frac{hc}{\lambda}$$

From Equation 28.11, find the maximum kinetic energy of an electron:

$$K_{max} = \frac{hc}{\lambda} - \phi = \frac{1\ 240\ \text{eV} \cdot \text{nm}}{300\ \text{nm}} - 2.46\ \text{eV} = \boxed{1.67\ \text{eV}}$$

(B) Find the cutoff wavelength λ_c for sodium.

SOLUTION

Calculate λ_c using Equation 28.12:

$$\lambda_c = \frac{hc}{\phi} = \frac{1\ 240\ \text{eV} \cdot \text{nm}}{2.46\ \text{eV}} = \boxed{504\ \text{nm}}$$

28.3 | The Compton Effect

In 1919, Einstein proposed that a photon of energy E carries a momentum equal to $E/c = hf/c$. In 1923, Arthur Holly Compton carried Einstein's idea of photon momentum further with the **Compton effect.**

Prior to 1922, Compton and his coworkers had accumulated evidence that showed that the classical wave theory of light failed to explain the scattering of x-rays from electrons. According to classical theory, incident electromagnetic waves of frequency f_0 should have two effects: (1) the electrons should accelerate in the direction of propagation of the x-ray by radiation pressure (see Section 24.5), and (2) the oscillating electric field should set the electrons into oscillation at the apparent frequency of the radiation as detected by the moving electron. The apparent frequency detected by the electron differs from f_0 due to the Doppler effect (see Section 24.3) because the electron absorbs as a moving particle. The electron then re-radiates as a moving particle, exhibiting another Doppler shift in the frequency of emitted radiation.

Because different electrons move at different speeds, depending on the amount of energy absorbed from the electromagnetic waves, the scattered wave frequency at a given angle should show a distribution of Doppler-shifted values. Contrary to this prediction, Compton's experiment showed that, at a given angle, only *one* frequency of radiation was observed that was different from that of the incident radiation. Compton and his coworkers realized that the scattering of x-ray photons from electrons could be explained by treating photons as point-like particles with energy hf and momentum hf/c and by assuming that the energy and momentum of the isolated system of the photon and the electron are conserved in a two-dimensional collision. By doing so, Compton was adopting a particle model for something that was well known as a wave, as had Einstein in his explanation of the photoelectric effect. Figure 28.11 shows the quantum picture of the exchange of momentum and energy between an individual x-ray photon and an electron. In the classical model, the electron is pushed along the direction of propagation of the incident x-ray by radiation pressure. In the quantum model in Figure 28.11, the electron is scattered through an angle ϕ with respect to this direction as if it were a billiard-ball type collision.

Figure 28.12 is a schematic diagram of the apparatus used by Compton. The x-rays, scattered from a carbon target, were diffracted by a rotating crystal spectrometer, and the intensity was measured with an ionization chamber that generated a current proportional to the intensity. The incident beam consisted of monochromatic x-rays of wavelength $\lambda_0 = 0.071$ nm. The experimental intensity-versus-wavelength plots observed by Compton for four scattering angles (corresponding to θ in Fig. 28.11) are shown in Figure 28.13 (page 958). The graphs for the three nonzero angles show two peaks, one at λ_0 and one at $\lambda' > \lambda_0$. The shifted peak at λ' is caused by the scattering of x-rays from free electrons, which was predicted by Compton to depend on scattering angle as

$$\lambda' - \lambda_0 = \frac{h}{m_e c}(1 - \cos\theta)$$ **28.13** ◄   ▶ Compton shift equation

In this expression, known as the **Compton shift equation,** m_e is the mass of the electron; $h/m_e c$ is called the **Compton wavelength** λ_C for the electron and has the value

$$\lambda_C = \frac{h}{m_e c} = 0.002\ 43 \text{ nm}$$ **28.14** ◄   ▶ Compton wavelength

Compton's measurements were in excellent agreement with the predictions of Equation 28.13. They were the first experimental results to convince most physicists of the fundamental validity of the quantum theory!

Arthur Holly Compton
American Physicist (1892–1962)
Compton was born in Wooster, Ohio, and attended Wooster College and Princeton University. He became the director of the laboratory at the University of Chicago, where experimental work concerned with sustained nuclear chain reactions was conducted. This work was of central importance to the construction of the first nuclear weapon. His discovery of the Compton effect led to his sharing of the 1927 Nobel Prize in Physics with Charles Wilson.

© Mary Evans Picture Library/Alamy

The electron recoils just as if struck by a classical particle, revealing the particle-like nature of the photon.

Figure 28.11 The quantum model for x-ray scattering from an electron.

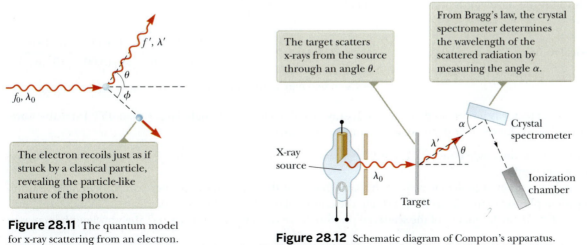

The target scatters x-rays from the source through an angle θ.

From Bragg's law, the crystal spectrometer determines the wavelength of the scattered radiation by measuring the angle α.

X-ray source

Target

Crystal spectrometer

Ionization chamber

Figure 28.12 Schematic diagram of Compton's apparatus.

The Compton effect and x-ray technicians

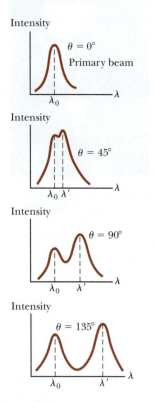

Figure 28.13 Scattered x-ray intensity versus wavelength for Compton scattering at $\theta = 0°$, 45°, 90°, and 135°.

The Compton effect should be kept in mind by x-ray technicians working in hospitals and radiology laboratories. X-rays directed into the patient's body are Compton scattered by electrons in the body in all directions. Equation 28.13 shows that the scattered wavelength is still well within the x-ray region so that these scattered x-rays can damage human tissue. In general, technicians operate the x-ray machine from behind an absorbing wall to avoid exposure to the scattered x-rays. Furthermore, when dental x-rays are taken, a lead apron is placed over the patient to reduce the absorption of scattered x-rays by other parts of the patient's body.

> **THINKING PHYSICS 28.2**
>
> The Compton effect involves a change in wavelength as photons are scattered through different angles. Suppose we illuminate a piece of material with a beam of light and then view the material from different angles relative to the beam of light. Will we see a *color change* corresponding to the change in wavelength of the scattered light?
>
> **Reasoning** Visible light scattered by the material undergoes a change in wavelength, but the change is far too small to detect as a color change. The largest possible change in wavelength, at 180° scattering, is twice the Compton wavelength, about 0.005 nm, which represents a change of less than 0.001% of the wavelength of red light. The Compton effect is only detectable for wavelengths that are very short to begin with, so the Compton wavelength is an appreciable fraction of the incident wavelength. As a result, the usual radiation for observing the Compton effect is in the x-ray range of the electromagnetic spectrum. ◀

> *Example* 28.4 | **Compton Scattering at 45°**

X-rays of wavelength $\lambda_0 = 0.200\,000$ nm are scattered from a block of material. The scattered x-rays are observed at an angle of 45.0° to the incident beam. Calculate their wavelength.

SOLUTION

Conceptualize Imagine the process in Figure 28.11, with the photon scattered at 45° to its original direction.

Categorize We evaluate the result using an equation developed in this section, so we categorize this example as a substitution problem.

Solve Equation 28.13 for the wavelength of the scattered x-ray:

$$(1) \quad \lambda' = \lambda_0 + \frac{h(1 - \cos\theta)}{m_e c}$$

Substitute numerical values:

$$\lambda' = 0.200\,000 \times 10^{-9}\,\text{m} + \frac{(6.626 \times 10^{-34}\,\text{J} \cdot \text{s})(1 - \cos 45.0°)}{(9.11 \times 10^{-31}\,\text{kg})(3.00 \times 10^8\,\text{m/s})}$$

$$= 0.200\,000 \times 10^{-9}\,\text{m} + 7.10 \times 10^{-13}\,\text{m} = \boxed{0.200\,710\,\text{nm}}$$

What If? What if the detector is moved so that scattered x-rays are detected at an angle larger than 45°? Does the wavelength of the scattered x-rays increase or decrease as the angle θ increases?

Answer In Equation (1), if the angle θ increases, $\cos\theta$ decreases. Consequently, the factor $(1 - \cos\theta)$ increases. Therefore, the scattered wavelength increases.

We could also apply an energy argument to achieve this same result. As the scattering angle increases, more energy is transferred from the incident photon to the electron. As a result, the energy of the scattered photon decreases with increasing scattering angle. Because $E = hf$, the frequency of the scattered photon decreases, and because $\lambda = c/f$, the wavelength increases.

28.4 | Photons and Electromagnetic Waves

The agreement between experimental measurements and theoretical predictions based on quantum models for phenomena such as the photoelectric effect and the Compton effect offers clear evidence that when light and matter interact, the light behaves as if it were composed of particles with energy hf and momentum hf/c. An obvious question at this point is, "How can light be considered a photon when it exhibits wave-like properties?" We describe light in terms of photons having energy and momentum, which are parameters of the particle model. Remember, however, that light and other electromagnetic waves exhibit interference and diffraction effects, which are consistent only with the wave model.

Which model is correct? Is light a wave or a particle? The answer depends on the phenomenon being observed. Some experiments can be explained better, or solely, with the photon model, whereas others are best described, or can only be described, with a wave model. The end result is that we must accept both models and admit that the true nature of light is not describable in terms of any single classical picture. Hence, light has a dual nature in that it exhibits both wave and particle characteristics. You should recognize, however, that the same beam of light that can eject photoelectrons from a metal can also be diffracted by a grating. In other words, the particle model and the wave model of light complement each other.

The success of the particle model of light in explaining the photoelectric effect and the Compton effect raises many other questions. Because a photon is a particle, what is the meaning of its "frequency" and "wavelength," and which determines its energy and momentum? Is light in some sense simultaneously a wave and a particle? Although photons have no rest energy, can some simple expression describe the effective mass of a "moving" photon? If a "moving" photon has mass, do photons experience gravitational attraction? What is the spatial extent of a photon, and how does an electron absorb or scatter one photon? Some of these questions can be answered, but others demand a view of atomic processes that is too pictorial and literal. Furthermore, many of these questions stem from classical analogies such as colliding billiard balls and water waves breaking on a shore. Quantum mechanics gives light a more fluid and flexible nature by treating the particle model and wave model of light as both necessary and complementary. Neither model can be used exclusively to describe all properties of light. A complete understanding of the observed behavior of light can be attained only if the two models are combined in a complementary manner. Before discussing this combination in more detail, we now turn our attention from electromagnetic waves to the behavior of entities that we have called particles.

28.5 | The Wave Properties of Particles

We feel quite comfortable in adopting a particle model for matter because we have studied such concepts as conservation of energy and momentum for particles as well as extended objects. It might therefore be even more difficult than it was for light to accept that *matter* also has a dual nature!

In 1923, in his doctoral dissertation, Louis Victor de Broglie postulated that because photons have wave and particle characteristics, perhaps all forms of matter have wave as well as particle properties. This postulate was a highly revolutionary idea with no experimental confirmation at that time. According to de Broglie, an electron in motion exhibits both wave and particle characteristics. De Broglie explained the source of this assertion in his 1929 Nobel Prize acceptance speech:

> On the one hand the quantum theory of light cannot be considered satisfactory since it defines the energy of a light corpuscle by the equation $E = hf$ containing the frequency f. Now a purely corpuscular theory contains nothing that enables us to define a frequency; for this reason alone, therefore, we are compelled, in the case of light, to introduce the idea of a corpuscle and that of periodicity simultaneously. On the other hand, determination of the stable motion of electrons in the atom introduces integers, and up to this point the

SSPL/Getty Images

Louis de Broglie
French Physicist (1892–1987)
De Broglie was born in Dieppe, France. At the Sorbonne in Paris, he studied history in preparation for what he hoped would be a career in the diplomatic service. The world of science is lucky he changed his career path to become a theoretical physicist. De Broglie was awarded the Nobel Prize in Physics in 1929 for his prediction of the wave nature of electrons.

only phenomena involving integers in physics were those of interference and of normal modes of vibration. This fact suggested to me the idea that electrons too could not be considered simply as corpuscles, but that periodicity must be assigned to them also.

In Chapter 9, we found that the relationship between energy and momentum for a photon is $p = E/c$. We also know from Equation 28.5 that the energy of a photon is $E = hf = hc/\lambda$. Therefore, the momentum of a photon can be expressed as

$$p = \frac{E}{c} = \frac{hf}{c} = \frac{hc}{c\lambda} = \frac{h}{\lambda}$$

From this equation, we see that the photon wavelength can be specified by its momentum: $\lambda = h/p$. De Broglie suggested that material particles of momentum p should also have wave properties and a corresponding wavelength given by the same expression. Because the magnitude of the momentum of a nonrelativistic particle of mass m and speed u is $p = mu$, the **de Broglie wavelength** of that particle is[3]

▶ De Broglie wavelength of a particle

$$\lambda = \frac{h}{p} = \frac{h}{mu}$$

28.15 ◀

Furthermore, in analogy with photons, de Broglie postulated that particles obey the Einstein relation $E = hf$, so the frequency of a particle is

▶ Frequency of a particle

$$f = \frac{E}{h}$$

28.16 ◀

The dual nature of matter is apparent in these last two equations because each contains both particle concepts (p and E) and wave concepts (λ and f). These relationships are established experimentally for photons. Is there experimental verification for the wave nature of a particle, such as the electron? Let's find out.

The Davisson–Germer Experiment

De Broglie's proposal that any kind of particle exhibits both wave and particle properties was first regarded as pure speculation. If particles such as electrons had wave-like properties, under the correct conditions they should exhibit diffraction effects. In 1927, three years after de Broglie published his work, C. J. Davisson and L. H. Germer of the United States succeeded in observing these diffraction effects and measuring the wavelength of electrons. Their important discovery provided the first experimental confirmation of the wave nature of particles proposed by de Broglie.

Interestingly, the intent of the initial Davisson–Germer experiment was not to confirm the de Broglie hypothesis. In fact, the discovery was made by accident (as is often the case). The experiment involved the scattering of low-energy electrons (≈ 54 eV) projected toward a nickel target in a vacuum. During one experiment, the nickel surface was badly oxidized because of an accidental break in the vacuum system. After the nickel target was heated in a flowing stream of hydrogen to remove the oxide coating, electrons scattered by it exhibited intensity maxima and minima at specific angles. The experimenters finally realized that the nickel had formed large crystal regions on heating and that the regularly spaced planes of atoms in the crystalline regions served as a diffraction grating (Section 27.8) for electrons.

Shortly thereafter, Davisson and Germer performed more extensive diffraction measurements on electrons scattered from single-crystal targets. Their results showed conclusively the wave nature of electrons and confirmed the de Broglie relation $p = h/\lambda$. A year later in 1928, G. P. Thomson of Scotland observed electron diffraction patterns by passing electrons through very thin gold foils. Diffraction patterns have since been observed for helium atoms, hydrogen atoms, and neutrons. Hence, the wave nature of particles has been established in a variety of ways.

Pitfall Prevention | **28.3**
What's Waving?
If particles have wave properties, what's waving? You are familiar with waves on strings, which are very concrete. Sound waves are more abstract, but you are likely comfortable with them. Electromagnetic waves are even more abstract, but at least they can be described in terms of physical variables and electric and magnetic fields. In contrast, waves associated with particles are completely abstract and cannot be associated with a physical variable. Later in this chapter, we describe the wave associated with a particle in terms of probability.

---

[3]The de Broglie wavelength for a particle moving at any speed u, including relativistic speeds, is $\lambda = h/\gamma mu$, where $\gamma = (1 - u^2/c^2)^{-1/2}$. Recall that in Chapter 9 we used u for particle speed to distinguish it from v, the speed of a reference frame.

QUICK QUIZ 28.5 An electron and a proton both moving at nonrelativistic speeds have the same de Broglie wavelength. Which of the following quantities are also the same for the two particles? (a) speed (b) kinetic energy (c) momentum (d) frequency

QUICK QUIZ 28.6 We have discussed two wavelengths associated with the electron, the Compton wavelength and the de Broglie wavelength. Which is an actual *physical* wavelength associated with the electron? (a) the Compton wavelength (b) the de Broglie wavelength (c) both wavelengths (d) neither wavelength

---

Example 28.5 | Wavelengths for Microscopic and Macroscopic Objects

(A) Calculate the de Broglie wavelength for an electron ($m_e = 9.11 \times 10^{-31}$ kg) moving at 1.00×10^7 m/s.

SOLUTION

Conceptualize Imagine the electron moving through space. From a classical viewpoint, it is a particle under constant velocity. From the quantum viewpoint, the electron has a wavelength associated with it.

Categorize We evaluate the result using an equation developed in this section, so we categorize this example as a substitution problem.

Evaluate the de Broglie wavelength using Equation 28.15:

$$\lambda = \frac{h}{m_e u} = \frac{6.63 \times 10^{-34}\, \text{J} \cdot \text{s}}{(9.11 \times 10^{-31}\, \text{kg})(1.00 \times 10^7\, \text{m/s})} = \boxed{7.27 \times 10^{-11}\, \text{m}}$$

The wave nature of this electron could be detected by diffraction techniques such as those in the Davisson–Germer experiment.

(B) A rock of mass 50 g is thrown with a speed of 40 m/s. What is its de Broglie wavelength?

SOLUTION

Evaluate the de Broglie wavelength using Equation 28.15:

$$\lambda = \frac{h}{mu} = \frac{6.63 \times 10^{-34}\, \text{J} \cdot \text{s}}{(50 \times 10^{-3}\, \text{kg})(40\, \text{m/s})} = \boxed{3.3 \times 10^{-34}\, \text{m}}$$

This wavelength is much smaller than any aperture through which the rock could possibly pass. Hence, we could not observe diffraction effects, and as a result, the wave properties of large-scale objects cannot be observed.

Example 28.6 | An Accelerated Charge

A particle of charge q and mass m is accelerated from rest through an electric potential difference ΔV. Assuming that the particle moves with a nonrelativistic speed, find its de Broglie wavelength.

SOLUTION

Conceptualize Imagine the motion of the particle. It starts from rest and then accelerates due to the force from the electric field. As the speed of the particle increases, its de Broglie wavelength decreases.

Categorize We identify the system as the particle and the electric field and apply the energy version of the isolated system model. The initial configuration of the system occurs at the instant the particle starts to move and the final configuration is when the particle reaches its final speed after accelerating through the potential difference ΔV. We define the electric potential energy of the system in the initial configuration to be zero.

Analyze

Write the conservation of energy equation (Eq. 7.2) for the isolated system:

$$\Delta K + \Delta U = 0$$

Substitute the initial and final energies, recognizing that a positive charge accelerates in the direction of *decreasing* electric potential:

$$\left(\tfrac{1}{2}mu^2 - 0\right) + (-q\Delta V - 0) = 0$$

continued

28.6 *cont.*

Solve for the final speed u:

$$u = \sqrt{\frac{2q\Delta V}{m}}$$

Substitute into Equation 28.15:

$$\lambda = \frac{h}{mu} = \frac{h}{m}\sqrt{\frac{m}{2q\Delta V}} = \frac{h}{\sqrt{2mq\Delta V}}$$

Finalize Notice that increasing the charge of the particle or the potential difference will decrease the wavelength. This result occurs because either of these changes will cause the particle to move with a higher speed. Increasing the mass of the particle would decrease its speed if everything else remains the same, so it might be surprising to see that increasing the mass would also decrease the de Broglie wavelength. Keep in mind, however, that the de Broglie wavelength depends on the *momentum* of the particle. The speed decreases according to the square root of the inverse of the mass in this situation, but the general expression for momentum shows a direct proportionality to the mass. As a result, for this situation, the momentum is proportional to the square root of the mass.

The Electron Microscope BIO

A practical device that relies on the wave characteristics of electrons is the **electron microscope.** A *transmission* electron microscope, used for viewing flat, thin samples, is shown in Figure 28.14. In many respects, it is similar to an optical microscope, but the electron microscope has a much greater resolving power because it can accelerate electrons to very high kinetic energies, giving them very short wavelengths. No microscope can resolve details that are significantly smaller than the wavelength of the waves used to illuminate the object. Typically, the wavelengths of electrons are about 100 times shorter than those of the visible light used in optical microscopes. As a result, an electron microscope with ideal lenses would be able to distinguish details about 100 times smaller than those distinguished by an optical microscope.

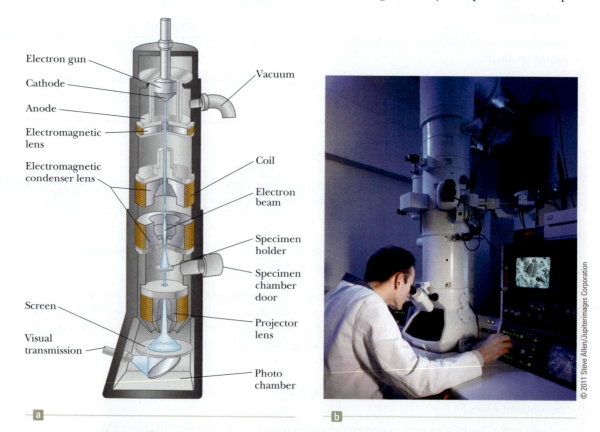

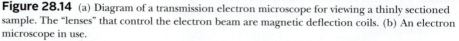

© 2011 Steve Allen/Jupiterimages Corporation

Figure 28.14 (a) Diagram of a transmission electron microscope for viewing a thinly sectioned sample. The "lenses" that control the electron beam are magnetic deflection coils. (b) An electron microscope in use.

(Electromagnetic radiation of the same wavelength as the electrons in an electron microscope is in the x-ray region of the spectrum.)

The electron beam in an electron microscope is controlled by electrostatic or magnetic deflection, which acts on the electrons to focus the beam and form an image. Rather than examining the image through an eyepiece as in an optical microscope, the viewer looks at an image formed on a monitor or other type of display screen. The photograph at the beginning of this chapter shows the amazing detail available with an electron microscope.

28.6 | A New Model: The Quantum Particle

The discussions presented in previous sections may be quite disturbing because we considered the particle and wave models to be distinct in earlier chapters. The notion that both light and material particles have both particle and wave properties does not fit with this distinction. We have experimental evidence, however, that this dual nature is just what we must accept. This acceptance leads to a new simplification model, the **quantum particle model.** We add the quantum particle to our other simplification models from which we build analysis models: the particle, the system, the rigid object, and the wave. In this model, entities have both particle and wave characteristics, and we must choose one appropriate behavior—particle or wave—to understand a particular phenomenon.

In this section, we shall investigate this model, which might bring you more comfort with this idea. As we shall demonstrate, we can construct from waves an entity that exhibits properties of a particle.

Let us first review the characteristics of ideal particles and waves. In the particle model, an ideal particle has zero size. As mentioned in Section 13.2, in the wave model, an ideal wave has a single frequency and is infinitely long. Therefore, an essential identifying feature of a particle that differentiates it from a wave is that it is *localized* in space. Let us show that we can build a localized entity from infinitely long waves. Imagine drawing one wave along the x axis, with one of its crests located at $x = 0$, as in Figure 28.15a. Now, draw a second wave, of the same amplitude but a different frequency, with one of its crests also at $x = 0$. The result of the superposition of these two waves is a *beat* because the waves are alternately in phase and out of phase. (Beats were discussed in Section 14.5.) Figure 28.15b shows the results of superposing these two waves.

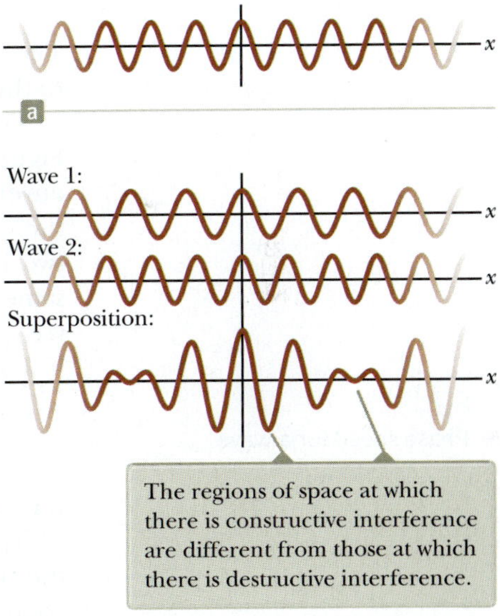

Notice that we have already introduced some localization by doing so. A single wave has the same amplitude everywhere in space; no point in space is any different from any other point. By adding a second wave, however, something is different between the in-phase and the out-of-phase points in space.

Now imagine that more and more waves are added to our original two, each new wave having a new frequency. Each new wave is added so that one of its crests is at $x = 0$. The result at $x = 0$ is that all the waves add constructively. When we consider a large number of waves, the probability of a positive value of a wave function at any point x is equal to the probability of a negative value and destructive interference occurs *everywhere* except near $x = 0$, where we superposed all the crests. The result is shown in Active Figure 28.16. The small region of constructive interference is called a **wave packet.** This wave packet is a localized region of space that is different from all other regions, because the result of the superposition of the waves

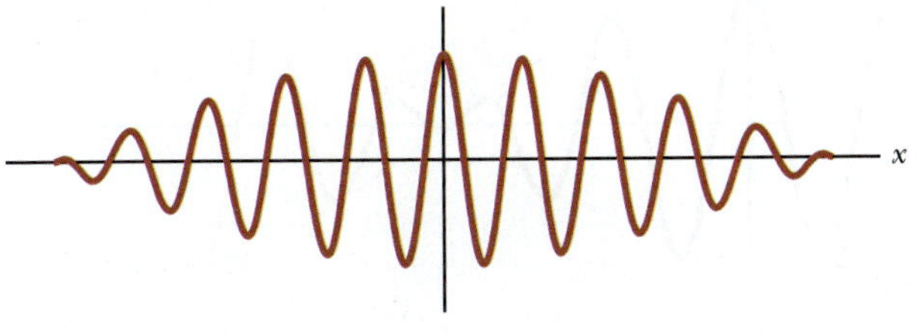

a

Wave 1:

Wave 2:

Superposition:

> The regions of space at which there is constructive interference are different from those at which there is destructive interference.

b

Figure 28.15 (a) An idealized wave of an exact single frequency is the same throughout space and time. (b) If two ideal waves with slightly different frequencies are combined, beats result (Section 14.5).

Active Figure 28.16 If a large number of waves are combined, the result is a wave packet, which represents a particle.

everywhere else is zero. We can identify the wave packet as a particle because it has the localized nature of what we have come to recognize as a particle!

The localized nature of this entity is the *only* characteristic of a particle that was generated with this process. We have not addressed how the wave packet might achieve such particle characteristics as mass, electric charge, spin, and so on. Therefore, you may not be completely convinced that we have built a particle. As further evidence that the wave packet can represent the particle, let us show that the wave packet has another characteristic of a particle.

Let us return to our combination of only two waves so as to make the mathematical representation simple. Consider two waves with equal amplitudes but different frequencies f_1 and f_2. We can represent the waves mathematically as

$$y_1 = A \cos(k_1 x - \omega_1 t) \qquad \text{and} \qquad y_2 = A \cos(k_2 x - \omega_2 t)$$

where, as in Chapter 13, $\omega = 2\pi f$ and $k = 2\pi/\lambda$. Using the superposition principle, we add the waves:

$$y = y_1 + y_2 = A \cos(k_1 x - \omega_1 t) + A \cos(k_2 x - \omega_2 t)$$

It is convenient to write this expression in a form that uses the trigonometric identity

$$\cos a + \cos b = 2 \cos\left(\frac{a - b}{2}\right) \cos\left(\frac{a + b}{2}\right)$$

Letting $a = k_1 x - \omega_1 t$ and $b = k_2 x - \omega_2 t$, we find that

$$y = 2A \cos\left[\frac{(k_1 x - \omega_1 t) - (k_2 x - \omega_2 t)}{2}\right] \cos\left[\frac{(k_1 x - \omega_1 t) + (k_2 x - \omega_2 t)}{2}\right]$$

$$= \left[2A \cos\left(\frac{\Delta k}{2} x - \frac{\Delta \omega}{2} t\right)\right] \cos\left(\frac{k_1 + k_2}{2} x - \frac{\omega_1 + \omega_2}{2} t\right) \qquad \textbf{28.17} \blacktriangleleft$$

The second cosine factor represents a wave with a wave number and frequency equal to the averages of the values for the individual waves.

The factor in brackets represents the envelope of the wave as shown in Active Figure 28.17. Notice that this factor also has the mathematical form of a wave. This envelope of the combination can travel through space with a different speed than the individual waves. As an extreme example of this possibility, imagine combining two identical waves moving in opposite directions. The two waves move with the same speed, but the envelope has a speed of *zero* because we have built a standing wave, which we studied in Section 14.2.

For an individual wave, the speed is given by Equation 13.11:

▶ Phase speed for a wave

$$v_{\text{phase}} = \frac{\omega}{k}$$

It is called the **phase speed** because it is the rate of advance of a crest on a single wave, which is a point of fixed phase. This equation can be interpreted as the following: the phase speed of a wave is the ratio of the coefficient of the time variable t to the coefficient of the space variable x in the equation for the wave, $y = A \cos(kx - \omega t)$.

Active Figure 28.17 The beat pattern of Figure 28.15b, with an envelope function (black dashed line) superimposed.

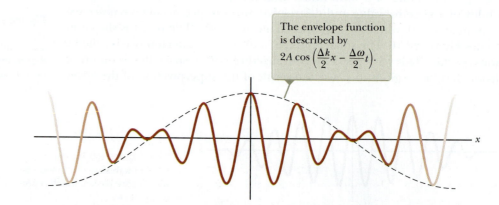

The envelope function is described by $2A \cos\left(\frac{\Delta k}{2} x - \frac{\Delta \omega}{2} t\right)$.

The factor in brackets in Equation 28.17 is of the form of a wave, so it moves with a speed given by this same ratio:

$$v_g = \frac{\text{coefficient of time variable } t}{\text{coefficient of space variable } x} = \frac{(\Delta\omega/2)}{(\Delta k/2)} = \frac{\Delta\omega}{\Delta k}$$

The subscript g on the speed indicates that it is commonly called the **group speed**, or the speed of the wave packet (the *group* of waves) that we have built. We have generated this expression for a simple addition of two waves. For a superposition of a very large number of waves to form a wave packet, this ratio becomes a derivative:

$$v_g = \frac{d\omega}{dk} \qquad \text{28.18} \blacktriangleleft \qquad \blacktriangleright \text{ Group speed for a wave packet}$$

Let us multiply the numerator and the denominator by $\hbar$, where $\hbar = h/2\pi$:

$$v_g = \frac{\hbar \, d\omega}{\hbar \, dk} = \frac{d(\hbar\omega)}{d(\hbar k)} \qquad \text{28.19} \blacktriangleleft$$

We look at the terms in the parentheses in the numerator and denominator in this equation separately. For the numerator

$$\hbar\omega = \frac{h}{2\pi}(2\pi f) = hf = E$$

For the denominator,

$$\hbar k = \frac{h}{2\pi}\left(\frac{2\pi}{\lambda}\right) = \frac{h}{\lambda} = p$$

Therefore, Equation 28.19 can be written as

$$v_g = \frac{d(\hbar\omega)}{d(\hbar k)} = \frac{dE}{dp} \qquad \text{28.20} \blacktriangleleft$$

Because we are exploring the possibility that the envelope of the combined waves represents the particle, consider a free particle moving with a speed u that is small compared with that of light. The energy of the particle is its kinetic energy:

$$E = \tfrac{1}{2}mu^2 = \frac{p^2}{2m}$$

Differentiating this equation with respect to p, where $p = mu$, gives

$$v_g = \frac{dE}{dp} = \frac{d}{dp}\left(\frac{p^2}{2m}\right) = \frac{1}{2m}(2p) = u \qquad \text{28.21} \blacktriangleleft$$

That is, the group speed of the wave packet is identical to the speed of the particle that it is modeled to represent! Therefore, we have further confidence that the wave packet is a reasonable way to model a particle.

QUICK QUIZ 28.7 As an analogy to wave packets, consider an "automobile packet" that occurs near the scene of an accident on a freeway. The phase speed is analogous to the speed of individual automobiles as they move through the backup caused by the accident. The group speed can be identified as the speed of the leading edge of the packet of cars. For the automobile packet, is the group speed (**a**) the same as the phase speed, (**b**) less than the phase speed, or (**c**) greater than the phase speed?

28.7 | The Double-Slit Experiment Revisited

One way to crystallize our ideas about the electron's wave–particle duality is to consider a hypothetical experiment in which electrons are fired at a double slit. Consider a parallel beam of monoenergetic electrons that is incident on a double slit

Figure 28.18 Electron interference. The slit separation d is much greater than the individual slit widths and much less than the distance between the slit and the detector screen.

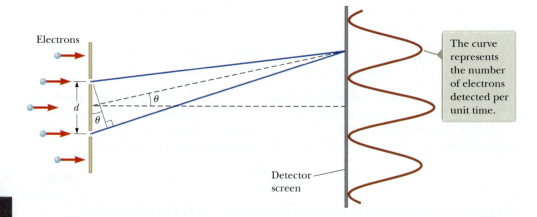

Electrons

θ

d

θ

Detector screen

The curve represents the number of electrons detected per unit time.

After just 28 electrons, no regular pattern appears

a

After 1 000 electrons, a pattern of fringes begins to appear.

b

After 10 000 electrons, the pattern looks very much like the experimental results shown in d.

c

Two-slit electron pattern (experimental results)

d

Active Figure 28.19 (a)–(c) Computer-simulated interference patterns for a beam of electrons incident on a double slit. (d) Photograph of a double-slit interference pattern produced by electrons.

as in Figure 28.18. We shall assume that the slit widths are small compared with the electron wavelength, so we need not worry about diffraction maxima and minima as discussed for light in Section 27.6. An electron detector screen is positioned far from the slits at a distance much greater than the separation distance d of the slits. If the detector screen collects electrons for a long enough time interval, one finds a typical wave interference pattern for the counts per minute, or probability of arrival of electrons. Such an interference pattern would not be expected if the electrons behaved as classical particles. It is clear that electrons are interfering, which is a distinct wave-like behavior.

If we measure the angles θ at which the maximum intensity of electrons arrives at the detector screen in Figure 28.18, we find they are described by exactly the same equation as that for light, $d \sin \theta = m\lambda$ (Eq. 27.2), where m is the order number and λ is the electron wavelength. Therefore, the dual nature of the electron is clearly shown in this experiment: the electrons are detected as particles at a localized spot on the detector screen at some instant of time, but the probability of arrival at that spot is determined by finding the intensity of two interfering waves.

Now imagine that we lower the beam intensity so that one electron at a time arrives at the double slit. It is tempting to assume the electron goes through either slit 1 or slit 2. You might argue that there are no interference effects because there is not a second electron going through the other slit to interfere with the first. This assumption places too much emphasis on the particle model of the electron, however. The interference pattern is still observed if the time interval for the measurement is sufficiently long for many electrons to arrive at the detector screen! This situation is illustrated by the computer-simulated patterns in Active Figure 28.19 where the interference pattern becomes clearer as the number of electrons reaching the detector screen increases. Hence, our assumption that the electron is localized and goes through only one slit when both slits are open must be wrong (a painful conclusion!).

To interpret these results, we are forced to conclude that an electron interacts with both slits *simultaneously*. If you try to determine experimentally which slit the electron goes through, the act of measuring destroys the interference pattern. It is impossible to determine which slit the electron goes through. In effect, we can say only that the electron passes through *both* slits! The same arguments apply to photons.

If we restrict ourselves to a pure particle model, it is an uncomfortable notion that the electron can be present at both slits at once. From the quantum particle model, however, the particle can be considered to be built from waves that exist throughout space. Therefore, the wave components of the electron are present at both slits at the same time, and this model leads to a more comfortable interpretation of this experiment.

28.8 | The Uncertainty Principle

Whenever one measures the position or velocity of a particle at any instant, experimental uncertainties are built into the measurements. According to classical mechanics, there is no fundamental barrier to an ultimate refinement of the apparatus or experimental procedures. In other words, it is possible, in principle, to make such measurements with arbitrarily small uncertainty. Quantum theory predicts, however, that it is fundamentally impossible to make simultaneous measurements of a particle's position and momentum with infinite accuracy.

In 1927, Werner Heisenberg introduced this notion, which is now known as the **Heisenberg uncertainty principle:**

If a measurement of the position of a particle is made with uncertainty Δx and a simultaneous measurement of its momentum is made with uncertainty Δp_x, the product of the two uncertainties can never be smaller than $\hbar/2$:

$$\Delta x \, \Delta p_x \geq \frac{\hbar}{2}$$ 28.22 ◀

◀ Uncertainty principle for momentum and position

That is, it is physically impossible to simultaneously measure the exact position and exact momentum of a particle. Heisenberg was careful to point out that the inescapable uncertainties Δx and Δp_x do not arise from imperfections in practical measuring instruments. Furthermore, they do not arise due to any perturbation of the system that we might cause in the measuring process. Rather, the uncertainties arise from the quantum structure of matter.

To understand the uncertainty principle, consider a particle for which we know the wavelength *exactly*. According to the de Broglie relation $\lambda = h/p$, we would know the momentum to infinite accuracy, so $\Delta p_x = 0$.

In reality, as we have mentioned, a single-wavelength wave would exist throughout space. Any region along this wave is the same as any other region (see Fig. 28.15a). If we were to ask, "Where is the particle that this wave represents?" no special location in space along the wave could be identified with the particle because all points along the wave are the same. Therefore, we have *infinite* uncertainty in the position of the particle and we know nothing about where it is. Perfect knowledge of the momentum has cost us all information about the position.

In comparison, now consider a particle with some uncertainty in momentum so that a range of values of momentum are possible. According to the de Broglie relation, the result is a range of wavelengths. Therefore, the particle is not represented by a single wavelength, but a combination of wavelengths within this range. This combination forms a wave packet as we discussed in Section 28.6 and illustrated in Active Figure 28.16. Now, if we are asked to determine the location of the particle, we can only say that it is somewhere in the region defined by the wave packet because a distinct difference exists between this region and the rest of space. Therefore, by losing some information about the momentum of the particle, we have gained information about its position.

If we were to lose all information about the momentum, we would be adding together waves of all possible wavelengths. The result would be a wave packet of zero length. Therefore, if we know nothing about the momentum, we know exactly where the particle is.

The mathematical form of the uncertainty principle argues that the product of the uncertainties in position and momentum will always be larger than some minimum value. This value can be calculated from the types of arguments discussed earlier, which result in the value of $\hbar/2$ in Equation 28.22.

Another form of the uncertainty principle can be generated by reconsidering Active Figure 28.16. Imagine that the horizontal axis is time rather than spatial position x. We can then make the same arguments that we made about knowledge of wavelength and position in the time domain. The corresponding variables would be frequency and time. Because frequency is related to the energy of the particle by $E = hf$, the uncertainty principle in this form is

$$\Delta E \, \Delta t \geq \frac{\hbar}{2}$$ 28.23 ◀

Werner Heisenberg
German Theoretical Physicist
(1901–1976)
Heisenberg obtained his Ph.D. in 1923 at the University of Munich. While other physicists tried to develop physical models of quantum phenomena, Heisenberg developed an abstract mathematical model called *matrix mechanics*. The more widely accepted physical models were shown to be equivalent to matrix mechanics. Heisenberg made many other significant contributions to physics, including his famous uncertainty principle for which he received a Nobel Prize in Physics in 1932, the prediction of two forms of molecular hydrogen, and theoretical models of the nucleus.

Pitfall Prevention | 28.4
The Uncertainty Principle
Some students incorrectly interpret the uncertainty principle as meaning that a measurement interferes with the system. For example, if an electron is observed in a hypothetical experiment using an optical microscope, the photon used to see the electron collides with it and makes it move, giving it an uncertainty in momentum. This scenario does *not* represent the basis of the uncertainty principle. The uncertainty principle is independent of the measurement process and is based on the wave nature of matter.

◀ Uncertainty principle for energy and time

This form of the uncertainty principle suggests that energy conservation can appear to be violated by an amount ΔE as long as it is only for a short time interval Δt consistent with Equation 28.23. We shall use this notion to estimate the rest energies of particles in Chapter 31.

Example 28.7 | **Locating an Electron**

The speed of an electron is measured to be 5.00×10^3 m/s to an accuracy of 0.003 00%. Find the minimum uncertainty in determining the position of this electron.

SOLUTION

Conceptualize The fractional value given for the accuracy of the electron's speed can be interpreted as the fractional uncertainty in its momentum. This uncertainty corresponds to a minimum uncertainty in the electron's position through the uncertainty principle.

Categorize We evaluate the result using concepts developed in this section, so we categorize this example as a substitution problem.

Assume the electron is moving along the x axis and find the uncertainty in p_x, letting f represent the accuracy of the measurement of its speed:

$$\Delta p_x = m\,\Delta u_x = mfu_x$$

Solve Equation 28.22 for the uncertainty in the electron's position and substitute numerical values:

$$\Delta x \geq \frac{\hbar}{2\,\Delta p_x} = \frac{\hbar}{2mfu_x} = \frac{1.055 \times 10^{-34}\,\text{J}\cdot\text{s}}{2(9.11 \times 10^{-31}\,\text{kg})(0.000\,030\,0)(5.00 \times 10^3\,\text{m/s})}$$

$$= 3.86 \times 10^{-4}\,\text{m} = \boxed{0.386\,\text{mm}}$$

28.9 | An Interpretation of Quantum Mechanics

We have been introduced to some new and strange ideas so far in this chapter. In an effort to understand the concepts of quantum physics better, let us investigate another bridge between particles and waves. We first think about electromagnetic radiation from the particle point of view. For a particular situation in which electromagnetic radiation exists, the probability per unit volume of finding a photon in a given region of space at an instant of time is proportional to the number of photons per unit volume at that time:

$$\frac{\text{Probability}}{V} \propto \frac{N}{V}$$

The number of photons per unit volume is proportional to the intensity of the radiation:

$$\frac{N}{V} \propto I$$

Now, we form the bridge to the wave model by recalling that the intensity of electromagnetic radiation is proportional to the square of the electric field amplitude for the electromagnetic wave (Section 24.4):

$$I \propto E^2$$

Equating the beginning and the end of this string of proportionalities, we have

$$\frac{\text{Probability}}{V} \propto E^2 \qquad\qquad \textbf{28.24} \blacktriangleleft$$

Therefore, for electromagnetic radiation, the probability per unit volume of finding a particle associated with this radiation (the photon) is proportional to the square of the amplitude of the wave associated with the particle.

Recognizing the wave–particle duality of both electromagnetic radiation and matter, we should suspect a parallel proportionality for a material particle. That is, the probability per unit volume of finding the particle is proportional to the square of the amplitude of a wave representing the particle. In Section 28.5 we learned that there is a de Broglie wave associated with every particle. The amplitude of the de Broglie wave associated with a particle is not a measurable quantity (because the wave function representing a particle is generally a complex function, as we discuss below). In contrast, the electric field is a measurable quantity for an electromagnetic wave. The matter analog to Equation 28.24 relates the square of the wave's amplitude to the probability per unit volume of finding the particle. As a result, we call the amplitude of the wave associated with the particle the **probability amplitude,** or the **wave function,** and give it the symbol Ψ. In general, the complete wave function Ψ for a system depends on the positions of all the particles in the system and on time; therefore, it can be written $\Psi(\vec{r}_1, \vec{r}_2, \vec{r}_3, \ldots, \vec{r}_j, \ldots, t)$, where $\vec{r}_j$ is the position vector of the jth particle in the system. For many systems of interest, including all those in this text, the wave function Ψ is mathematically separable in space and time and can be written as a product of a space function ψ for one particle of the system and a complex time function:[4]

$$\Psi(\vec{r}_1, \vec{r}_2, \vec{r}_3, \ldots, \vec{r}_j, \ldots, t) = \psi(\vec{r}_j) e^{-i\omega t} \qquad \textbf{28.25} \blacktriangleleft$$

▶ Space- and time-dependent wave function Ψ

where $\omega \ (= 2\pi f)$ is the angular frequency of the wave function and $i = \sqrt{-1}$.

For any system in which the potential energy is time-independent and depends only on the positions of particles within the system, the important information about the system is contained within the space part of the wave function. The time part is simply the factor $e^{-i\omega t}$. Therefore, the understanding of ψ is the critical aspect of a given problem.

The wave function ψ is often complex-valued. The quantity $|\psi|^2 = \psi^*\psi$, where ψ^* is the complex conjugate[5] of ψ, is always real and positive, and is proportional to the probability per unit volume of finding a particle at a given point at some instant. The wave function contains within it all the information that can be known about the particle.

This probability interpretation of the wave function was first suggested by Max Born (1882–1970) in 1928. In 1926, Erwin Schrödinger (1887–1961) proposed a wave equation that describes the manner in which the wave function changes in space and time. The *Schrödinger wave equation*, which we shall examine in Section 28.12, represents a key element in the theory of quantum mechanics.

In Section 28.5, we found that the de Broglie equation relates the momentum of a particle to its wavelength through the relation $p = h/\lambda$. If an ideal free particle has a precisely known momentum p_x, its wave function is a sinusoidal wave of wavelength $\lambda = h/p_x$ and the particle has equal probability of being at any point along the x axis. The wave function for such a free particle moving along the x axis can be written as

$$\psi(x) = Ae^{ikx} \qquad \textbf{28.26} \blacktriangleleft$$

▶ Wave function for a free particle

where $k = 2\pi/\lambda$ is the angular wave number and A is a constant amplitude.[6]

Although we cannot measure ψ, we can measure the quantity $|\psi|^2$, the absolute square of ψ, which can be interpreted as follows. If ψ represents a single particle, $|\psi|^2$—called the **probability density**—is the relative probability per unit volume that

Pitfall Prevention | 28.5
The Wave Function Belongs to a System
The common language in quantum mechanics is to associate a wave function with a particle. The wave function, however, is determined by the particle *and* its interaction with its environment, so it more rightfully belongs to a system. In many cases, the particle is the only part of the system that experiences a change, which is why the common language has developed. You will see examples in the future in which it is more proper to think of the system wave function rather than the particle wave function.

---

[4]The standard form of a complex number is $a + ib$. The notation $e^{i\theta}$ is equivalent to the standard form as follows:

$$e^{i\theta} = \cos\theta + i\sin\theta$$

Therefore, the notation $e^{-i\omega t}$ in Equation 28.25 is equivalent to $\cos(-\omega t) + i\sin(-\omega t) = \cos\omega t - i\sin\omega t$.

[5]For a complex number $z = a + ib$, the complex conjugate is found by changing i to $-i$: $z^* = a - ib$. The product of a complex number and its complex conjugate is always real and positive:

$$z^*z = (a - ib)(a + ib) = a^2 - (ib)^2 = a^2 - (i)^2 b^2 = a^2 + b^2$$

[6]For the free particle, the full wave function, based on Equation 28.25 is

$$\Psi(x, t) = Ae^{ikx}e^{-i\omega t} = Ae^{i(kx - \omega t)} = A[\cos(kx - \omega t) + i\sin(kx - \omega t)]$$

The real part of this wave function has the same form as the waves that we added together to form wave packets in Section 28.6.

the particle will be found at any given point in the volume. This interpretation can also be stated in the following manner. If dV is a small volume element surrounding some point, the probability of finding the particle in that volume element is $|\psi|^2\,dV$. In this section, we deal only with one-dimensional systems, where the particle must be located along the x axis, and we therefore replace dV with dx. In this case, the probability $P(x)\,dx$ that the particle will be found in the infinitesimal interval dx around the point x is

$$P(x)\,dx = |\psi|^2\,dx \qquad \textbf{28.27} \blacktriangleleft$$

Because the particle must be somewhere along the x axis, the sum of the probabilities over all values of x must be 1:

▶ Normalization condition on ψ

$$\int_{-\infty}^{\infty} |\psi|^2\,dx = 1 \qquad \textbf{28.28} \blacktriangleleft$$

Any wave function satisfying Equation 28.28 is said to be **normalized.** Normalization is simply a statement that the particle exists at some point at all times.

Although it is not possible to specify the position of a particle with complete certainty, it is possible through $|\psi|^2$ to specify the probability of observing it in a small region surrounding a given point. The probability of finding the particle in the arbitrarily sized interval $a \le x \le b$ is

The probability of a particle being in the interval $a \le x \le b$ is the area under the probability density curve from a to b.

$$P_{ab} = \int_{a}^{b} |\psi|^2\,dx \qquad \textbf{28.29} \blacktriangleleft$$

The probability P_{ab} is the area under the curve of $|\psi|^2$ versus x between the points $x = a$ and $x = b$ as in Figure 28.20.

Experimentally, the probability is finite of finding a particle in an interval near some point at some instant. The value of that probability must lie between the limits 0 and 1. For example, if the probability is 0.3, there is a 30% chance of finding the particle in the interval.

The wave function ψ satisfies a wave equation, just as the electric field associated with an electromagnetic wave satisfies a wave equation that follows from Maxwell's equations. As mentioned earlier, the wave equation satisfied by ψ is the Schrödinger equation (Section 28.12), and ψ can be computed from it. Although ψ is not a measurable quantity, all the measurable quantities of a particle, such as its energy and momentum, can be derived from a knowledge of ψ. For example, once the wave function for a particle is known, it is possible to calculate the average position at which you would find the particle after many measurements. This average position is called the **expectation value** of x and is defined by the equation

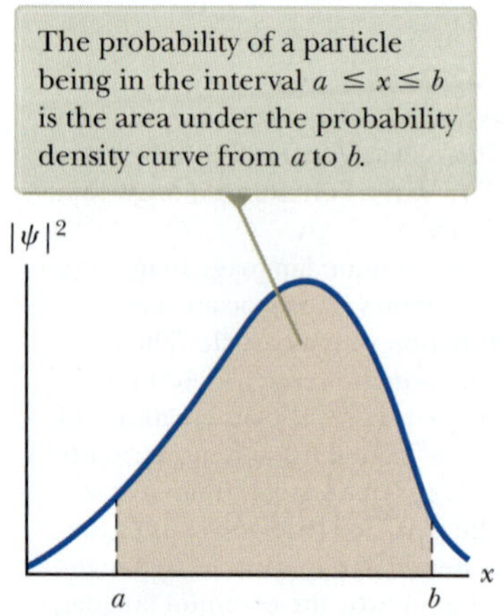

Figure 28.20 An arbitrary probability density curve for a particle.

▶ Expectation value for position x

$$\langle x \rangle \equiv \int_{-\infty}^{\infty} \psi^* x \psi\,dx \qquad \textbf{28.30} \blacktriangleleft$$

where brackets $\langle\ \rangle$ are used to denote expectation values. Furthermore, one can find the expectation value of any function $f(x)$ associated with the particle by using the following equation:

▶ Expectation value for a function $f(x)$

$$\langle f(x) \rangle \equiv \int_{-\infty}^{\infty} \psi^* f(x) \psi\,dx \qquad \textbf{28.31} \blacktriangleleft$$

◀ 28.10 | A Particle in a Box

In this section, we shall apply some of the ideas we have developed to a sample problem. Let us choose a simple problem: a particle confined to a one-dimensional region of space, called the *particle in a box* (even though the "box" is one-dimensional!). From a classical viewpoint, if a particle is confined to bouncing back and forth along the x axis between two impenetrable walls as in the pictorial representation in Figure 28.21a, its motion is easy to describe. If the speed of the particle is u, the magnitude of its momentum mu remains constant, as does its kinetic energy. Classical physics places no restrictions on the values of a particle's momentum and energy. The quantum mechanics approach to this problem is quite different and requires that we find the appropriate wave function consistent with the conditions of the situation.[7]

Because the walls are impenetrable, the probability of finding the particle outside the box is zero, so the wave function $\psi(x)$ must be zero for $x < 0$ and for $x > L$, where L is the distance between the two walls. A mathematical condition for any wave function is that it must be continuous in space.[8] Therefore, if ψ is zero outside the walls, it must also be zero *at* the walls; that is, $\psi(0) = 0$ and $\psi(L) = 0$. Only those wave functions that satisfy this condition are allowed.

Figure 28.21b shows a graphical representation of the particle in a box problem, which graphs the potential energy of the particle–environment system as a function of the position of the particle. When the particle is inside the box, the potential energy of the system does not depend on the particle's location and we can choose its value to be zero. Outside the box, we have to ensure that the wave function is zero. We can do so by defining the potential energy of the system as infinitely large if the particle were outside the box. Because kinetic energy is necessarily positive, the only way a particle could be outside the box is if the system has an infinite amount of energy.

The wave function for a particle in the box can be expressed as a real sinusoidal function:[9]

$$\psi(x) = A \sin\left(\frac{2\pi x}{\lambda}\right) \qquad \textbf{28.32} \blacktriangleleft$$

This wave function must satisfy the boundary conditions at the walls. The boundary condition at $x = 0$ is satisfied already because the sine function is zero when $x = 0$. For the boundary condition at $x = L$, we have

$$\psi(L) = 0 = A \sin\left(\frac{2\pi L}{\lambda}\right)$$

which can only be true if

$$\frac{2\pi L}{\lambda} = n\pi \quad \rightarrow \quad \lambda = \frac{2L}{n} \qquad \textbf{28.33} \blacktriangleleft$$

where $n = 1, 2, 3, \ldots$. Therefore, only certain wavelengths for the particle are allowed! Each of the allowed wavelengths corresponds to a quantum state for the system, and n is the quantum number. Expressing the wave function in terms of the quantum number n, we have

$$\psi_n(x) = A \sin\left(\frac{2\pi x}{\lambda}\right) = A \sin\left(\frac{2\pi x}{2L/n}\right) = A \sin\left(\frac{n\pi x}{L}\right) \qquad \textbf{28.34} \blacktriangleleft$$

▶ Wave functions for a particle in a box

Active Figures 28.22a and 28.22b (page 972) are graphical representations of ψ_n versus x and $|\psi_n|^2$ versus x for $n = 1, 2$, and 3 for the particle in a box. Note that although ψ_n can be positive or negative, $|\psi_n|^2$ is always positive. Because $|\psi_n|^2$ represents a probability density, a negative value for $|\psi_n|^2$ is meaningless.

Further inspection of Active Figure 28.22b shows that $|\psi_n|^2$ is zero at the boundaries, satisfying our boundary condition. In addition, $|\psi_n|^2$ is zero at other points, depending on the value of n. For $n = 2$, $|\psi_n|^2 = 0$ at $x = L/2$; for $n = 3$, $|\psi_n|^2 = 0$ at

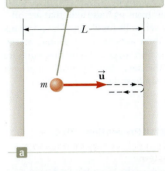

This figure is a *pictorial representation* showing a particle of mass m and speed u bouncing between two impenetrable walls separated by a distance L.

a

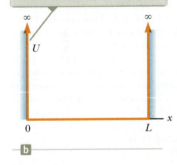

This figure is a *graphical representation* showing the potential energy of the particle–box system. The blue areas are classically forbidden.

b

Figure 28.21 (a) The particle in a box. (b) The potential energy function for the system.

---

[7]Before continuing, you might want to review Sections 14.2 and 14.3 on standing mechanical waves.

[8]If the wave function is not continuous at a point, the derivative of the wave function at that point is infinite. This issue leads to problems in the Schrödinger equation, for which the wave function is a solution and which is discussed in Section 28.12.

[9]We show that this function is the correct one explicitly in Section 28.12.

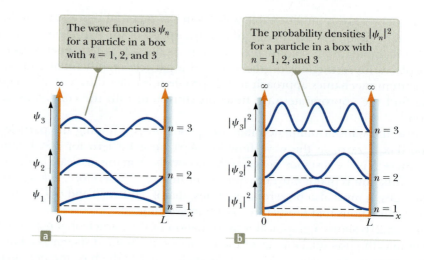

The wave functions ψ_n for a particle in a box with $n = 1$, 2, and 3

The probability densities $|\psi_n|^2$ for a particle in a box with $n = 1$, 2, and 3

Pitfall Prevention | 28.6
**Reminder: Energy Belongs to a
System**
We describe Equation 28.35 as
representing the energy of the
particle; it is commonly used
language for the particle in a box
problem. In reality, we are analyzing
the energy of the *system* of the
particle and whatever environment is
establishing the impenetrable walls.
In the case of a particle in a box,
the only nonzero type of energy is
kinetic and it belongs to the particle.
In general, energies that we calculate
using quantum physics are associated
with a system of interacting particles,
such as the electron and proton
in the hydrogen atom studied in
Chapter 11.

▶ Quantized energies for a particle
in a box

$x = L/3$ and $x = 2L/3$. The number of zero points increases by one each time the
quantum number increases by one.

Because the wavelengths of the particle are restricted by the condition $\lambda = 2L/n$,
the magnitude of the momentum of the particle is also restricted to specific values
that we can find from the expression for the de Broglie wavelength, Equation 28.15:

$$p = \frac{h}{\lambda} = \frac{h}{2L/n} = \frac{nh}{2L}$$

From this expression, we find that the allowed values of the energy, which is simply
the kinetic energy of the particle, are

$$E_n = \tfrac{1}{2}mu^2 = \frac{p^2}{2m} = \frac{(nh/2L)^2}{2m}$$

$$E_n = \left(\frac{h^2}{8mL^2}\right)n^2 \qquad n = 1, 2, 3, \ldots \qquad \textbf{28.35} \blacktriangleleft$$

As we see from this expression, the energy of the particle is quantized, similar to our
quantization of energy in the hydrogen atom in Chapter 11. The lowest allowed energy corresponds to $n = 1$, for which $E_1 = h^2/8mL^2$. Because $E_n = n^2E_1$, the excited
states corresponding to $n = 2, 3, 4, \ldots$ have energies given by $4E_1, 9E_1, 16E_1, \ldots$.

Active Figure 28.23 is an energy level diagram[10] describing the energy values of
the allowed states. Note that the state $n = 0$, for which E would be equal to zero, is
not allowed. Therefore, according to quantum mechanics, the particle can never be
at rest. The least energy it can have, corresponding to $n = 1$, is called the **zero-point
energy.** This result is clearly contradictory to the classical viewpoint, in which $E = 0$
is an acceptable state, as are all positive values of E.

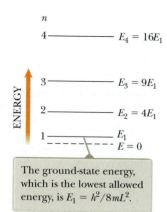

The ground-state energy,
which is the lowest allowed
energy, is $E_1 = h^2/8mL^2$.

Active Figure 28.23 Energy level
diagram for a particle confined to a
one-dimensional box of length L.

QUICK QUIZ 28.8 Consider an electron, a proton, and an alpha particle (a helium
nucleus), each trapped separately in identical boxes. **(i)** Which particle corresponds
to the highest ground-state energy? (a) the electron (b) the proton (c) the alpha
particle (d) The ground-state energy is the same in all three cases. **(ii)** Which particle
has the longest wavelength when the system is in the ground state? (a) the electron
(b) the proton (c) the alpha particle (d) All three particles have the same wavelength.

QUICK QUIZ 28.9 A particle is in a box of length L. Suddenly, the length of the box
is increased to $2L$. What happens to the energy levels shown in Active Figure 28.23?
(a) Nothing; they are unaffected. **(b)** They move farther apart. **(c)** They move closer
together.

---

[10]We introduced the energy level diagram as a specialized semigraphical representation in Chapter 11.

Example 28.8 | Microscopic and Macroscopic Particles in Boxes

(A) An electron is confined between two impenetrable walls 0.200 nm apart. Determine the energy levels for the states $n = 1, 2,$ and 3.

SOLUTION

Conceptualize In Figure 28.21a, imagine that the particle is an electron and the walls are very close together.

Categorize We evaluate the energy levels using an equation developed in this section, so we categorize this example as a substitution problem.

Use Equation 28.35 for the $n = 1$ state:

$$E_1 = \frac{h^2}{8m_e L^2}(1)^2 = \frac{(6.63 \times 10^{-34}\,\text{J} \cdot \text{s})^2}{8(9.11 \times 10^{-31}\,\text{kg})(2.00 \times 10^{-10}\,\text{m})^2}$$

$$= 1.51 \times 10^{-18}\,\text{J} = 9.42\,\text{eV}$$

Using $E_n = n^2 E_1$, find the energies of the $n = 2$ and $n = 3$ states:

$$E_2 = (2)^2 E_1 = 4(9.42\,\text{eV}) = 37.7\,\text{eV}$$

$$E_3 = (3)^2 E_1 = 9(9.42\,\text{eV}) = 84.8\,\text{eV}$$

(B) Find the speed of the electron in the $n = 1$ state.

SOLUTION

Solve the classical expression for kinetic energy for the particle speed:

$$K = \tfrac{1}{2}m_e u^2 \quad \rightarrow \quad u = \sqrt{\frac{2K}{m_e}}$$

Recognize that the kinetic energy of the particle is equal to the system energy and substitute E_n for K:

$$(1) \quad u = \sqrt{\frac{2E_n}{m_e}}$$

Substitute numerical values from part (A):

$$u = \sqrt{\frac{2(1.51 \times 10^{-18}\,\text{J})}{9.11 \times 10^{-31}\,\text{kg}}} = 1.82 \times 10^6\,\text{m/s}$$

Simply placing the electron in the box results in a *minimum* speed of the electron equal to 0.6% of the speed of light!

(C) A 0.500-kg baseball is confined between two rigid walls of a stadium that can be modeled as a box of length 100 m. Calculate the minimum speed of the baseball.

SOLUTION

Conceptualize In Figure 28.21a, imagine that the particle is a baseball and the walls are those of the stadium.

Categorize This part of the example is a substitution problem in which we apply a quantum approach to a macroscopic object.

Use Equation 28.35 for the $n = 1$ state:

$$E_1 = \frac{h^2}{8mL^2}(1)^2 = \frac{(6.63 \times 10^{-34}\,\text{J} \cdot \text{s})^2}{8(0.500\,\text{kg})(100\,\text{m})^2} = 1.10 \times 10^{-71}\,\text{J}$$

Use Equation (1) to find the speed:

$$u = \sqrt{\frac{2(1.10 \times 10^{-71}\,\text{J})}{0.500\,\text{kg}}} = 6.63 \times 10^{-36}\,\text{m/s}$$

This speed is so small that the object can be considered to be at rest, which is what one would expect for the minimum speed of a macroscopic object.

What If? What if a sharp line drive is hit so that the baseball is moving with a speed of 150 m/s? What is the quantum number of the state in which the baseball now resides?

Answer We expect the quantum number to be very large because the baseball is a macroscopic object.

Evaluate the kinetic energy of the baseball:

$$\tfrac{1}{2}mu^2 = \tfrac{1}{2}(0.500\,\text{kg})(150\,\text{m/s})^2 = 5.62 \times 10^3\,\text{J}$$

From Equation 28.35, calculate the quantum number n:

$$n = \sqrt{\frac{8mL^2 E_n}{h^2}} = \sqrt{\frac{8(0.500\,\text{kg})(100\,\text{m})^2(5.62 \times 10^3\,\text{J})}{(6.63 \times 10^{-34}\,\text{J} \cdot \text{s})^2}} = 2.26 \times 10^{37}$$

continued

28.8 *cont.*

This result is a tremendously large quantum number. As the baseball pushes air out of the way, hits the ground, and rolls to a stop, it moves through more than 10^{37} quantum states. These states are so close together in energy that we cannot observe the transitions from one state to the next. Rather, we see what appears to be a smooth variation in the speed of the ball. The quantum nature of the universe is simply not evident in the motion of macroscopic objects.

28.11 | Analysis Model: Quantum Particle Under Boundary Conditions

The discussion of the particle in a box is very similar to the discussion in Chapter 14 of standing waves on strings:

- Because the ends of the string must be nodes, the wave functions for allowed waves must be zero at the boundaries of the string. Because the particle in a box cannot exist outside the box, the allowed wave functions for the particle must be zero at the boundaries.
- The boundary conditions on the string waves lead to quantized wavelengths and frequencies of the waves. The boundary conditions on the wave function for the particle in a box lead to quantized wavelengths and frequencies of the particle.

In quantum mechanics, it is very common for particles to be subject to boundary conditions. We therefore introduce a new analysis model, the **quantum particle under boundary conditions.** In many ways, this model is similar to the waves under boundary conditions model studied in Section 14.3. In fact, the allowed wavelengths for the wave function of a particle in a box (Eq. 28.33) are identical in form to the allowed wavelengths for mechanical waves on a string fixed at both ends (Eq. 14.5).

The quantum particle under boundary conditions model *differs* in some ways from the waves under boundary conditions model:

- In most cases of quantum particles, the wave function is *not* a simple sinusoidal function like the wave function for waves on strings. Furthermore, the wave function for a quantum particle may be a complex function.
- For a quantum particle, frequency is related to energy through $E = hf$, so the quantized frequencies lead to quantized energies.
- There may be no stationary "nodes" associated with the wave function of a quantum particle under boundary conditions. Systems more complicated than the particle in a box have more complicated wave functions, and some boundary conditions may not lead to zeroes of the wave function at fixed points.

In general,

---

an interaction of a quantum particle with its environment represents one or more boundary conditions, and, if the interaction restricts the particle to a finite region of space, results in quantization of the energy of the system.

Boundary conditions on quantum wave functions are related to the coordinates describing the problem. For the particle in a box, the wave function must be zero at two values of *x*. In the case of a three-dimensional system such as the hydrogen atom we shall discuss in Chapter 29, the problem is best presented in *spherical coordinates*. These coordinates, an extension of the plane polar coordinates introduced in Section 1.6, consist of a radial coordinate *r* and two angular coordinates. The generation of the wave function and application of the boundary conditions for the hydrogen atom are beyond the scope of this book. We shall, however, examine the behavior of some of the hydrogen-atom wave functions in Chapter 29.

Boundary conditions on wave functions that exist for all values of *x* require that the wave function approach zero as $x \rightarrow \infty$ (so that the wave function can be normalized) and remain finite as $x \rightarrow 0$. One boundary condition on any angular parts of wave functions is that adding 2π radians to the angle must return the wave function to the same value because an addition of 2π results in the same angular position.

28.12 | The Schrödinger Equation

In Section 24.3, we discussed a wave equation for electromagnetic radiation. The waves associated with particles also satisfy a wave equation. We might guess that the wave equation for material particles is different from that associated with photons because material particles have a nonzero rest energy. The appropriate wave equation was developed by Schrödinger in 1926. In applying the quantum particle under boundary conditions model to a quantum system, the approach is to determine a solution to this equation and then apply the appropriate boundary conditions to the solution. The solution yields the allowed wave functions and energy levels of the system under consideration. Proper manipulation of the wave function then enables one to calculate all measurable features of the system.

The Schrödinger equation as it applies to a particle of mass m confined to moving along the x axis and interacting with its environment through a potential energy function $U(x)$ is

$$-\frac{\hbar^2}{2m}\frac{d^2\psi}{dx^2} + U\psi = E\psi$$

28.36 ◄   ► Time-independent Schrödinger equation

where E is the total energy of the system (particle and environment). Because this equation is independent of time, it is commonly referred to as the **time-independent Schrödinger equation.** (We shall not discuss the time-dependent Schrödinger equation, whose solution is Ψ, Eq. 28.25, in this text.)

This equation is consistent with the energy version of the isolated system model. The system is the particle and its environment. Problem 54 shows, both for a free particle and a particle in a box, that the first term in the Schrödinger equation reduces to the kinetic energy of the particle multiplied by the wave function. Therefore, Equation 28.36 tells us that the total energy is the sum of the kinetic energy and the potential energy and that the total energy is a constant: $K + U = E = $ constant.

In principle, if the potential energy $U(x)$ for the system is known, one can solve Equation 28.36 and obtain the wave functions and energies for the allowed states of the system. Because U may vary discontinuously with position, it may be necessary to solve the equation separately for various regions. In the process, the wave functions for the different regions must join smoothly at the boundaries and we require that $\psi(x)$ be *continuous*. Furthermore, so that $\psi(x)$ obeys the normalization condition, we require that $\psi(x)$ approach zero as x approaches $\pm\infty$. Finally, $\psi(x)$ must be *single-valued* and $d\psi/dx$ must also be continuous[11] for finite values of $U(x)$.

The task of solving the Schrödinger equation may be very difficult, depending on the form of the potential energy function. As it turns out, the Schrödinger equation has been extremely successful in explaining the behavior of atomic and nuclear systems, whereas classical physics has failed to do so. Furthermore, when quantum mechanics is applied to macroscopic objects, the results agree with classical physics, as required by the correspondence principle.

The Particle in a Box via the Schrödinger Equation

To see how the Schrödinger equation is applied to a problem, let us return to our particle in a one-dimensional box of length L (see Fig. 28.21) and analyze it with the Schrödinger equation. In association with Figure 28.21b, we discussed the potential energy diagram that describes the problem. A potential energy diagram such as this one is a useful representation for understanding and solving problems with the Schrödinger equation.

© INTERFOTO/Alamy

Erwin Schrödinger
Austrian Theoretical Physicist (1887–1961)
Schrödinger is best known as one of the creators of quantum mechanics. His approach to quantum mechanics was demonstrated to be mathematically equivalent to the more abstract matrix mechanics developed by Heisenberg. Schrödinger also produced important papers in the fields of statistical mechanics, color vision, and general relativity.

---

[11]If $d\psi/dx$ were not continuous, we would not be able to evaluate $d^2\psi/dx^2$ in Equation 28.36 at the point of discontinuity.

Because of the shape of the curve in Figure 28.21b, the particle in a box is some-times said to be in a **square well,**[12] where a **well** is an upward-facing region of the curve in a potential energy diagram. (A downward-facing region is called a *barrier,* which we shall investigate in Section 28.13.)

In the region $0 < x < L$, where $U = 0$, we can express the Schrödinger equation in the form

$$\frac{d^2\psi}{dx^2} = -\frac{2mE}{\hbar^2}\psi = -k^2\psi \qquad \qquad \text{28.37} \blacktriangleleft$$

where

$$k = \frac{\sqrt{2mE}}{\hbar} \qquad \qquad \text{28.38} \blacktriangleleft$$

The solution to Equation 28.37 is a function whose second derivative is the negative of the same function multiplied by a constant k^2. We recognize both the sine and cosine functions as satisfying this requirement. Therefore, the most general solution to the equation is a linear combination of both solutions:

$$\psi(x) = A \sin kx + B \cos kx$$

where A and B are constants determined by the boundary conditions.

Our first boundary condition is that $\psi(0) = 0$:

$$\psi(0) = A \sin 0 + B \cos 0 = 0 + B = 0 \quad \rightarrow \quad B = 0$$

Therefore, our solution reduces to

$$\psi(x) = A \sin kx$$

The second boundary condition, $\psi(L) = 0$, when applied to the reduced solution, gives

$$\psi(L) = A \sin kL = 0$$

which is satisfied only if kL is an integral multiple of π, that is, if $kL = n\pi$, where n is an integer. Because $k = \sqrt{2mE}/\hbar$, we have

$$kL = \frac{\sqrt{2mE}}{\hbar}L = n\pi$$

For each integer choice for n, this equation determines a quantized energy E_n. Solving for the allowed energies E_n gives

$$E_n = \left(\frac{h^2}{8mL^2}\right)n^2 \qquad \qquad \text{28.39} \blacktriangleleft$$

which are identical to the allowed energies in Equation 28.35.

Substituting the values of k in the wave function, the allowed wave functions $\psi_n(x)$ are given by

$$\psi_n(x) = A \sin\left(\frac{n\pi x}{L}\right) \qquad \qquad \text{28.40} \blacktriangleleft$$

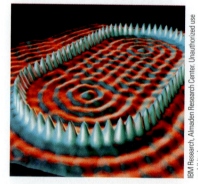

Figure 28.24 This is an image of a quantum corral consisting of a ring of 48 iron atoms located on a copper surface. The diameter of the ring is 143 nm, and the image was obtained using a low-temperature scanning tunneling microscope (STM) as mentioned in Section 28.13. Corrals and other structures are able to confine surface electron waves. The study of such structures will play an important role in determining the future of small electronic devices.

This wave function agrees with Equation 28.34.

Normalizing this relationship shows that $A = \sqrt{(2/L)}$. (See Problem 56 at the end of this chapter.) Therefore, the normalized wave function is

$$\psi_n(x) = \sqrt{\frac{2}{L}}\sin\left(\frac{n\pi x}{L}\right) \qquad \qquad \text{28.41} \blacktriangleleft$$

The notion of trapping particles in potential wells is used in the burgeoning field of **nanotechnology,** which refers to the design and application of devices having di-mensions ranging from 1 to 100 nm. The fabrication of these devices often involves manipulating single atoms or small groups of atoms to form structures such as the quantum corral in Figure 28.24.

One area of nanotechnology of interest to researchers is the **quantum dot.** The quantum dot, a small region that is grown in a silicon crystal, acts as a potential well.

---

[12]It is called a square well even if it has a rectangular shape in a potential energy diagram.

This region can trap electrons into states with quantized energies. The wave functions for a particle in a quantum dot look similar to those in Active Figure 28.22a if L is on the order of nanometers. The storage of binary information using quantum dots is an active field of research. A simple binary scheme would involve associating a one with a quantum dot containing an electron and a zero with an empty dot. Other schemes involve cells of multiple dots such that arrangements of electrons among the dots correspond to ones and zeros. Several research laboratories are studying the properties and potential applications of quantum dots. Information should be forthcoming from these laboratories at a steady rate in the next few years.

> ### *Example* 28.9 | The Expectation Values for the Particle in a Box
>
> A particle of mass m is confined to a one-dimensional box between $x = 0$ and $x = L$. Find the expectation value of the position x of the particle in the state characterized by quantum number n.

SOLUTION

Conceptualize Active Figure 28.22b shows that the probability for the particle to be at a given location varies with position within the box. Can you predict what the expectation value of x will be from the symmetry of the wave functions?

Categorize The statement of the example categorizes the problem for us: we focus on a quantum particle in a box and on the calculation of its expectation value of x.

Analyze In Equation 28.30, the integration from $-\infty$ to ∞ reduces to the limits 0 to L because $\psi = 0$ everywhere except in the box.

Substitute Equation 28.41 into Equation 28.30 to find the expectation value for x:

$$\langle x \rangle = \int_{-\infty}^{\infty} \psi_n^* x \psi_n \, dx = \int_0^L x \left[\sqrt{\frac{2}{L}} \sin \left(\frac{n\pi x}{L} \right) \right]^2 dx$$

$$= \frac{2}{L} \int_0^L x \sin^2 \left(\frac{n\pi x}{L} \right) dx$$

Evaluate the integral by consulting an integral table or by mathematical integration:[13]

$$\langle x \rangle = \frac{2}{L} \left[\frac{x^2}{4} - \frac{x \sin \left(2 \frac{n\pi x}{L} \right)}{4 \frac{n\pi}{L}} - \frac{\cos \left(2 \frac{n\pi x}{L} \right)}{8 \left(\frac{n\pi}{L} \right)^2} \right]_0^L$$

$$= \frac{2}{L} \left[\frac{L^2}{4} \right] = \frac{L}{2}$$

Finalize This result shows that the expectation value of x is at the center of the box for all values of n, which you would expect from the symmetry of the square of the wave functions (the probability density) about the center (Active Fig. 28.22b).

The $n = 2$ wave function in Active Figure 28.22b has a value of zero at the midpoint of the box. Can the expectation value of the particle be at a position at which the particle has zero probability of existing? Remember that the expectation value is the *average* position. Therefore, the particle is as likely to be found to the right of the midpoint as to the left, so its average position is at the midpoint even though its probability of being there is zero. As an analogy, consider a group of students for whom the average final examination score is 50%. There is no requirement that some student achieve a score of exactly 50% for the average of all students to be 50%.

28.13 | Tunneling Through a Potential Energy Barrier

Consider the potential energy function shown in Figure 28.25 (page 978), in which the potential energy of the system is zero everywhere except for a region of width L where the potential energy has a constant value of U. This type of potential energy function is called a **square barrier,** and U is called the **barrier height.** A very

---

[13]To integrate this function, first replace $\sin^2 (n\pi x/L)$ with $\frac{1}{2}(1 - \cos 2n\pi x/L)$ (refer to Table B.3 in Appendix B), which allows $\langle x \rangle$ to be expressed as two integrals. The second integral can then be evaluated by partial integration (Section B.7 in Appendix B).

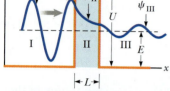

Figure 28.25 Wave function ψ for a particle incident from the left on a barrier of height U and width L. The wave function is plotted vertically from an axis positioned at the energy of the particle.

Pitfall Prevention | 28.8
"Height" on an Energy Diagram
The word *height* (as in *barrier height*) refers to an energy in discussions of barriers in potential energy diagrams. For example, we might say the height of the barrier is 10 eV. On the other hand, the barrier *width* refers to the traditional usage of such a word and is an actual physical length measurement between the two locations of the vertical sides of the barrier.

interesting and peculiar phenomenon occurs when a moving particle encounters such a barrier of finite height and width. Consider a particle of energy $E < U$ that is incident on the barrier from the left (see Fig. 28.25). Classically, the particle is reflected by the barrier. If the particle were to exist in region II, its kinetic energy would be negative, which is not allowed classically. Therefore, region II, and in turn region III, are both classically *forbidden* to the particle incident from the left. According to quantum mechanics, however, all regions are accessible to the particle, regardless of its energy. (Although all regions are accessible, the probability of the particle being in a region that is classically forbidden is very low.) According to the uncertainty principle, the particle can be within the barrier as long as the time interval during which it is in the barrier is short and consistent with Equation 28.23. If the barrier is relatively narrow, this short time interval can allow the particle to move across the barrier. Therefore, it is possible for us to understand the passing of the particle through the barrier with the help of the uncertainty principle.

Let us approach this situation using a mathematical representation. The Schrödinger equation has valid solutions in all three regions I, II, and III. The solutions in regions I and III are sinusoidal as in Equation 28.26. In region II, the solution is exponential. Applying the boundary conditions that the wave functions in the three regions must join smoothly at the boundaries, we find that a full solution can be found such as that represented by the curve in Figure 28.25. Therefore, Schrödinger's equation and the boundary conditions are satisfied, which tells us mathematically that such a process can theoretically occur according to the quantum theory.

Because the probability of locating the particle is proportional to $|\psi|^2$, we conclude that the chance of finding the particle beyond the barrier in region III is nonzero. This result is in complete disagreement with classical physics. The movement of the particle to the far side of the barrier is called **tunneling** or **barrier penetration.**

The probability of tunneling can be described with a **transmission coefficient** T and a **reflection coefficient** R. The transmission coefficient represents the probability that the particle penetrates to the other side of the barrier, and the reflection coefficient is the probability that the particle is reflected by the barrier. Because the incident particle is either reflected or transmitted, we require that $T + R = 1$. An approximate expression for the transmission coefficient, obtained when $T \ll 1$ (a very wide barrier or a very high barrier, that is, $U \gg E$), is

$$T \approx e^{-2CL} \qquad \textbf{28.42} \blacktriangleleft$$

where

$$C = \frac{\sqrt{2m(U - E)}}{\hbar} \qquad \textbf{28.43} \blacktriangleleft$$

According to quantum physics, Equation 28.42 tells us that T can be nonzero, which is in contrast to the classical point of view that requires that $T = 0$. That we experimentally observe the phenomenon of tunneling provides further confidence in the principles of quantum physics.

Figure 28.25 shows the wave function of a particle tunneling through a barrier in one dimension. A similar wave function having spherical symmetry describes the barrier penetration of a particle leaving a radioactive nucleus, which we will study in Chapter 30. The wave function exists both inside and outside the nucleus, and its amplitude is constant in time. In this way, the wave function correctly describes the small but constant probability that the nucleus will decay. The moment of decay cannot be predicted. In general, quantum mechanics implies that the future is indeterminate. (This feature is in contrast to classical mechanics, from which the trajectory of an object can be calculated to arbitrarily high precision from precise knowledge of its initial position and velocity and of the forces exerted on it.) We must conclude that the fundamental laws of nature are probabilistic.

A radiation detector can be used to show that a nucleus decays by radiating a particle at a particular moment and in a particular direction. To point out the contrast between this experimental result and the wave function describing it, Schrödinger imagined a box containing a cat, a radioactive sample, a radiation counter, and a vial of poison. When a nucleus in the sample decays, the counter triggers the administration of lethal poison to the cat. Quantum mechanics correctly predicts the probability of finding the cat dead when the box is opened. Before the box is opened, does the animal have a wave function describing it as a fractionally dead cat, with some chance of being alive?

This question is currently under investigation, never with actual cats, but sometimes with interference experiments building upon the experiment described in Section 28.7. Does the act of measurement change the system from a probabilistic to a definite state? When a particle emitted by a radioactive nucleus is detected at one particular location, does the wave function describing the particle drop instantaneously to zero everywhere else in the Universe? (Einstein called such a state change a "spooky action at a distance.") Is there a fundamental difference between a quantum system and a macroscopic system? The answers to these questions are basically unknown.

> **QUICK QUIZ 28.10** Which of the following changes would increase the probability of transmission of a particle through a potential barrier? (You may choose more than one answer.) (a) decreasing the width of the barrier (b) increasing the width of the barrier (c) decreasing the height of the barrier (d) increasing the height of the barrier (e) decreasing the kinetic energy of the incident particle (f) increasing the kinetic energy of the incident particle

Example 28.10 | Transmission Coefficient for an Electron

A 30-eV electron is incident on a square barrier of height 40 eV.

(A) What is the probability that the electron tunnels through the barrier if its width is 1.0 nm?

SOLUTION

Conceptualize Because the particle energy is smaller than the height of the potential barrier, we expect the electron to reflect from the barrier with a probability of 100% according to classical physics. Because of the tunneling phenomenon, however, there is a finite probability that the particle can appear on the other side of the barrier.

Categorize We evaluate the probability using an equation developed in this section, so we categorize this example as a substitution problem.

Evaluate the quantity $U - E$ that appears in Equation 28.43:

$$U - E = 40 \text{ eV} - 30 \text{ eV} = 10 \text{ eV} \left(\frac{1.6 \times 10^{-19} \text{ J}}{1 \text{ eV}} \right) = 1.6 \times 10^{-18} \text{ J}$$

Evaluate the quantity $2CL$ using Equation 28.43:

$$(1) \quad 2CL = 2 \frac{\sqrt{2(9.11 \times 10^{-31} \text{ kg})(1.6 \times 10^{-18} \text{ J})}}{1.055 \times 10^{-34} \text{ J} \cdot \text{s}} (1.0 \times 10^{-9} \text{ m}) = 32.4$$

From Equation 28.42, find the probability of tunneling through the barrier:

$$T \approx e^{-2CL} = e^{-32.4} = \boxed{8.5 \times 10^{-15}}$$

(B) What is the probability that the electron tunnels through the barrier if its width is 0.10 nm?

SOLUTION

In this case, the width L in Equation (1) is one-tenth as large, so evaluate the new value of $2CL$:

$$2CL = (0.1)(32.4) = 3.24$$

From Equation 28.42, find the new probability of tunneling through the barrier:

$$T \approx e^{-2CL} = e^{-3.24} = \boxed{0.039}$$

In part (A), the electron has approximately 1 chance in 10^{14} of tunneling through the barrier. In part (B), however, the electron has a much higher probability (3.9%) of penetrating the barrier. Therefore, reducing the width of the barrier by only one order of magnitude increases the probability of tunneling by about 12 orders of magnitude!

Applications of Tunneling

As we have seen, tunneling is a quantum phenomenon, a result of the wave nature of matter. Many applications may be understood only on the basis of tunneling.

- **Alpha decay.** One form of radioactive decay is the emission of alpha particles (the nuclei of helium atoms) by unstable, heavy nuclei (Chapter 30). For an alpha particle to escape from the nucleus, it must penetrate a barrier whose height is several times larger than the energy of the nucleus–alpha particle system. The barrier is due to a combination of the attractive nuclear force (discussed in Chapter 30) and the Coulomb repulsion (discussed in detail in Chapter 19) between the alpha particle and the rest of the nucleus. Occasionally, an alpha particle tunnels through the barrier, which explains the basic mechanism for this type of decay and the large variations in the mean lifetimes of various radioactive nuclei.

- **Nuclear fusion.** The basic reaction that powers the Sun and, indirectly, almost everything else in the solar system is fusion, which we will study in Chapter 30. In one step of the process that occurs at the core of the Sun, protons must approach each other to within such a small distance that they fuse to form a deuterium nucleus. According to classical physics, these protons cannot overcome and penetrate the barrier caused by their mutual electrical repulsion. Quantum-mechanically, however, the protons are able to tunnel through the barrier and fuse together.

- **Scanning tunneling microscope.** The scanning tunneling microscope, or STM, is a remarkable device that uses tunneling to create images of surfaces with resolution comparable to the size of a single atom. A small probe with a very fine tip is made to scan very close to the surface of a specimen. A tunneling current is maintained between the probe and specimen; the current (which is related to the probability of tunneling) is very sensitive to the barrier height (which is related to the separation between the tip and specimen) as seen in Example 28.10. Maintaining a constant tunneling current produces a feedback signal that is used to raise and lower the probe as the surface is scanned. Because the vertical motion of the probe follows the contour of the specimen's surface, an image of the surface is obtained. The image of the quantum corral shown in Figure 28.24 is made with a scanning tunneling microscope.

◣ 28.14 | Context Connection: The Cosmic Temperature

Now that we have introduced the concepts of quantum physics for microscopic particles and systems, let us see how we can connect these concepts to processes occurring on a cosmic scale. For our first such connection, consider the Universe as a system. It is widely believed that the Universe began with a cataclysmic explosion called the **Big Bang,** first mentioned in Chapter 5. Because of this explosion, all the material in the Universe is moving apart. This expansion causes a Doppler shift in radiation left over from the Big Bang such that the wavelength of the radiation lengthens. In the 1940s, Ralph Alpher, George Gamow, and Robert Hermann developed a structural model of the Universe in which they predicted that the thermal radiation from the Big Bang should still be present and that it should now have a wavelength distribution consistent with a black body with a temperature of a few kelvins.

In 1965, Arno Penzias and Robert Wilson of Bell Telephone Laboratories were measuring radiation from the Milky Way galaxy using a special 20-ft antenna as a radio telescope. They noticed a consistent background "noise" of radiation in the signals from the antenna. Despite their great efforts to test alternative hypotheses for the origin of the noise in terms of interference from the Sun, an unknown source in the Milky Way, structural problems in the antenna, and even the presence of pigeon droppings in the antenna, none of the hypotheses was sufficient to explain the noise.

What Penzias and Wilson were detecting was the thermal radiation from the Big Bang. That it was detected by their system regardless of the direction of the antenna was consistent with the radiation being spread throughout the Universe, as the Big Bang model predicts. A measurement of the intensity of this radiation suggested that the temperature associated with the radiation was about 3 K, consistent with Alpher, Gamow, and Hermann's prediction from the 1940s. Although the measured intensity was consistent with their prediction, the measurement was taken at only a single wavelength. Full agreement with the model of the Universe as a black body would come only if measurements at many wavelengths demonstrated a distribution in wavelengths consistent with Active Figure 28.2.

In the years following Penzias and Wilson's discovery, other researchers made measurements at different wavelengths. In 1989, the COBE (*COsmic Background Explorer*) satellite was launched by NASA and added critical measurements at wavelengths below 0.1 cm. The results of these measurements led to a Nobel Prize in Physics for the principal investigators in 2006. Several data points from COBE are shown in Figure 28.26. The Wilkinson Microwave Anisotropy Probe, launched in June 2001, exhibits data that allow observation of temperature differences in the cosmos in the microkelvin range. Ongoing observations are also being made from Earth-based facilities, associated with projects such as QUaD, Qubic, and the South Pole Telescope. In addition, the Planck satellite was launched in May 2009 by the European Space Agency. This space-based observatory should measure the cosmic background radiation with higher sensitivity than the Wilkinson probe. The series of measurements taken since 1965 are consistent with thermal radiation associated with a temperature of 2.7 K. The whole story of the cosmic temperature is a remarkable example of science at work: building a model, making a prediction, taking measurements, and testing the measurements against the predictions.

The first chapter of our *Cosmic Connection* Context describes the first example of this connection. By studying the thermal radiation from microscopic vibrating objects, we learn something about the origin of our Universe. In Chapter 29, we shall see more examples of this fascinating connection.

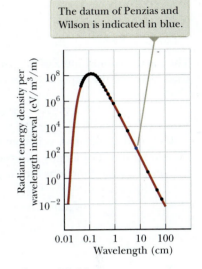

The datum of Penzias and Wilson is indicated in blue.

Figure 28.26 Theoretical blackbody (brown curve) and measured radiation spectra (black points) of the Big Bang. Most of the data were collected from the Cosmic Background Explorer, or COBE, satellite.

❯ SUMMARY |

The characteristics of **blackbody radiation** cannot be explained by classical concepts. Planck introduced the first model of **quantum physics** when he argued that the atomic oscillators responsible for this radiation exist only in discrete **quantum states.**

In the **photoelectric effect,** electrons are ejected from a metallic surface when light is incident on that surface. Einstein provided a successful explanation of this effect by extending Planck's quantum theory to the electromagnetic field. In this model, light is viewed as a stream of particles called **photons,** each with energy $E = hf$, where f is the frequency and h is **Planck's constant.** The maximum kinetic energy of the ejected photoelectron is given by

$$K_{max} = hf - \phi \qquad \text{28.11} ◀$$

where ϕ is the **work function** of the metal.

X-rays striking a target are scattered at various angles by electrons in the target. A shift in wavelength is observed for the scattered x-rays, and the phenomenon is known as the **Compton effect.** Classical physics does not correctly explain the experimental results of this effect. If the x-ray is treated as a photon, conservation of energy and momentum applied

to the isolated system of the photon and the electron yields for the Compton shift the expression

$$\lambda' - \lambda_0 = \frac{h}{m_e c}(1 - \cos\theta) \qquad \text{28.13} ◀$$

where m_e is the mass of the electron, c is the speed of light, and θ is the scattering angle for the photon.

Every object of mass m and momentum p has wave-like properties, with a **de Broglie wavelength** given by the relation

$$\lambda = \frac{h}{p} = \frac{h}{mu} \qquad \text{28.15} ◀$$

The wave–particle duality is the basis of the **quantum particle model.** It can be interpreted by imagining a particle to be made up of a combination of a large number of waves. These waves interfere constructively in a small region of space called a **wave packet.**

The **uncertainty principle** states that if a measurement of position is made with uncertainty Δx and a *simultaneous* measurement of momentum is made with uncertainty Δp_x, the product of the two uncertainties can never be less than $\hbar/2$:

$$\Delta x\, \Delta p_x \geq \frac{\hbar}{2} \qquad \text{28.22} ◀$$

The uncertainty principle is a natural outgrowth of the wave packet model.

Particles are represented by a **wave function** $\psi(x, y, z)$. The **probability density** that a particle will be found at a point is $|\psi|^2$. If the particle is confined to moving along the x axis, the probability that it will be located in an interval dx is given by $|\psi|^2\, dx$. Furthermore, the wave function must be **normalized**:

$$\int_{-\infty}^{\infty} |\psi|^2\, dx = 1 \qquad \text{28.28} \blacktriangleleft$$

The measured position x of the particle, averaged over many trials, is called the **expectation value** of x and is defined by

$$\langle x \rangle \equiv \int_{-\infty}^{\infty} \psi^* x \psi\, dx \qquad \text{28.30} \blacktriangleleft$$

If a particle of mass m is confined to moving in a one-dimensional box of length L whose walls are perfectly rigid, the allowed wave functions for the particle are

$$\psi_n(x) = A \sin\left(\frac{n\pi x}{L}\right) \qquad \text{28.34} \blacktriangleleft$$

where n is an integer quantum number starting at 1. The particle has a well-defined wavelength λ whose values are such that the length L of the box is equal to an integral number of half wavelengths, that is, $L = n\lambda/2$. The energies of a particle in a box are quantized and are given by

$$E_n = \left(\frac{h^2}{8mL^2}\right) n^2 \qquad n = 1, 2, 3, \ldots \qquad \text{28.35} \blacktriangleleft$$

The wave function must satisfy the Schrödinger equation. The **time-independent Schrödinger equation** for a particle confined to moving along the x axis is

$$-\frac{\hbar^2}{2m}\frac{d^2\psi}{dx^2} + U\psi = E\psi \qquad \text{28.36} \blacktriangleleft$$

where E is the total energy of the system and U is the potential energy of the system.

When a particle of energy E meets a barrier of height U, where $E < U$, the particle has a finite probability of penetrating the barrier. This process, called **tunneling,** is the basic mechanism that explains the operation of the scanning tunneling microscope and the phenomenon of alpha decay in some radioactive nuclei.

Analysis Model for Problem Solving

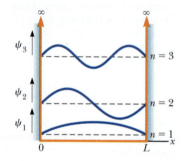

Quantum Particle Under Boundary Conditions. An interaction of a quantum particle with its environment represents one or more boundary conditions. If the interaction restricts the particle to a finite region of space, the energy of the system is quantized. All wave functions must satisfy the following four boundary conditions: (1) $\psi(x)$ must remain finite as x approaches 0, (2) $\psi(x)$ must approach zero as x approaches $\pm\infty$, (3) $\psi(x)$ must be continuous for all values of x, and (4) $d\psi/dx$ must be continuous for all finite values of $U(x)$. If the solution to Equation 28.36 is piecewise, conditions (3) and (4) must be applied at the boundaries between regions of x in which Equation 28.36 has been solved.

OBJECTIVE QUESTIONS

☐ denotes answer available in *Student Solutions Manual/Study Guide*

1. Is each one of the following statements (a) through (e) true or false for an electron? (a) It is a quantum particle, behaving in some experiments like a classical particle and in some experiments like a classical wave. (b) Its rest energy is zero. (c) It carries energy in its motion. (d) It carries momentum in its motion. (e) Its motion is described by a wave function that has a wavelength and satisfies a wave equation.

2. Which of the following phenomena most clearly demonstrates the particle nature of light? (a) diffraction (b) the photoelectric effect (c) polarization (d) interference (e) refraction

3. In a Compton scattering experiment, a photon of energy E is scattered from an electron at rest. After the scattering event occurs, which of the following statements is true? (a) The frequency of the photon is greater than E/h. (b) The energy of the photon is less than E. (c) The wavelength of the photon is less than hc/E. (d) The momentum of the photon increases. (e) None of those statements is true.

4. The probability of finding a certain quantum particle in the section of the x axis between $x = 4$ nm and $x = 7$ nm is 48%. The particle's wave function $\psi(x)$ is constant over this range. What numerical value can be attributed to $\psi(x)$, in units of $\text{nm}^{-1/2}$? (a) 0.48 (b) 0.16 (c) 0.12 (d) 0.69 (e) 0.40

5. A proton, an electron, and a helium nucleus all move at speed v. Rank their de Broglie wavelengths from largest to smallest.

6. Consider (a) an electron (b) a photon, and (c) a proton, all moving in vacuum. Choose all correct answers for each question. (i) Which of the three possess rest energy? (ii) Which have charge? (iii) Which carry energy? (iv) Which carry momentum? (v) Which move at the speed of light? (vi) Which have a wavelength characterizing their motion?

7. In a certain experiment, a filament in an evacuated light-bulb carries a current I_1 and you measure the spectrum of

light emitted by the filament, which behaves as a black body at temperature T_1. The wavelength emitted with highest intensity (symbolized by λ_{max}) has the value λ_1. You then increase the potential difference across the filament by a factor of 8, and the current increases by a factor of 2. **(i)** After this change, what is the new value of the temperature of the filament? (a) $16T_1$ (b) $8T_1$ (c) $4T_1$ (d) $2T_1$ (e) still T_1 **(ii)** What is the new value of the wavelength emitted with highest intensity? (a) $4\lambda_1$ (b) $2\lambda_1$ (c) λ_1 (d) $\frac{1}{2}\lambda_1$ (e) $\frac{1}{4}\lambda_1$

8. What is the de Broglie wavelength of an electron accelerated from rest through a potential difference of 50.0 V? (a) 0.100 nm (b) 0.139 nm (c) 0.174 nm (d) 0.834 nm (e) none of those answers

9. A quantum particle of mass m_1 is in a square well with infinitely high walls and length 3 nm. Rank the situations (a) through (e) according to the particle's energy from highest to lowest, noting any cases of equality. (a) The particle of mass m_1 is in the ground state of the well. (b) The same particle is in the $n = 2$ excited state of the same well. (c) A particle with mass $2m_1$ is in the ground state of the same well. (d) A particle of mass m_1 in the ground state of the same well, and the uncertainty principle has become inoperative; that is, Planck's constant has been reduced to zero. (e) A particle of mass m_1 is in the ground state of a well of length 6 nm.

10. A particle in a rigid box of length L is in the first excited state, for which $n = 2$ (Fig. OQ28.10). Where is the particle most likely to be found? (a) At the center of the box. (b) At either end of the box. (c) All points in the box are equally likely. (d) One-fourth of the way from either end of the box. (e) None of those answers is correct.

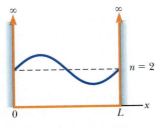

Figure OQ28.10

11. **BIO** Which of the following is most likely to cause sunburn by delivering more energy to individual molecules in skin cells? (a) infrared light (b) visible light (c) ultraviolet light (d) microwaves (e) Choices (a) through (d) are equally likely.

12. An x-ray photon is scattered by an originally stationary electron. Relative to the frequency of the incident photon, is the frequency of the scattered photon (a) lower, (b) higher, or (c) unchanged?

13. A beam of quantum particles with kinetic energy 2.00 eV is reflected from a potential barrier of small width and

original height 3.00 eV. How does the fraction of the particles that are reflected change as the barrier height is reduced to 2.01 eV? (a) It increases. (b) It decreases. (c) It stays constant at zero. (d) It stays constant at 1. (e) It stays constant with some other value.

14. Suppose a tunneling current in an electronic device goes through a potential-energy barrier. The tunneling current is small because the width of the barrier is large and the barrier is high. To increase the current most effectively, what should you do? (a) Reduce the width of the barrier. (b) Reduce the height of the barrier. (c) Either choice (a) or choice (b) is equally effective. (d) Neither choice (a) nor choice (b) increases the current.

15. Figure OQ28.15 represents the wave function for a hypothetical quantum particle in a given region. From the choices *a* through *e*, at what value of *x* is the particle most likely to be found?

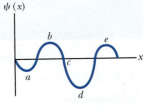

Figure OQ28.15

16. Which of the following statements are true according to the uncertainty principle? More than one statement may be correct. (a) It is impossible to simultaneously determine both the position and the momentum of a particle along the same axis with arbitrary accuracy. (b) It is impossible to simultaneously determine both the energy and momentum of a particle with arbitrary accuracy. (c) It is impossible to determine a particle's energy with arbitrary accuracy in a finite amount of time. (d) It is impossible to measure the position of a particle with arbitrary accuracy in a finite amount of time. (e) It is impossible to simultaneously measure both the energy and position of a particle with arbitrary accuracy.

17. Rank the wavelengths of the following quantum particles from the largest to the smallest. If any have equal wavelengths, display the equality in your ranking. (a) a photon with energy 3 eV (b) an electron with kinetic energy 3 eV (c) a proton with kinetic energy 3 eV (d) a photon with energy 0.3 eV (e) an electron with momentum 3 eV/c

18. Both an electron and a proton are accelerated to the same speed, and the experimental uncertainty in the speed is the same for the two particles. The positions of the two particles are also measured. Is the minimum possible uncertainty in the electron's position (a) less than the minimum possible uncertainty in the proton's position, (b) the same as that for the proton, (c) more than that for the proton, or (d) impossible to tell from the given information?

☐ denotes answer available in *Student Solutions Manual/Study Guide*

► CONCEPTUAL QUESTIONS

1. The classical model of blackbody radiation given by the Rayleigh–Jeans law has two major flaws. (a) Identify the flaws and (b) explain how Planck's law deals with them.

2. All objects radiate energy. Why, then, are we not able to see all objects in a dark room?

3. **BIO** *Iridescence* is the phenomenon that gives shining colors to the feathers of peacocks, hummingbirds (see page 910),

resplendent quetzals, and even ducks and grackles. Without pigments, it colors Morpho butterflies (Fig. CQ28.3, page 984), Urania moths, some beetles and flies, rainbow trout, and mother-of-pearl in abalone shells. Iridescent colors change as you turn an object. They are produced by a wide variety of intricate structures in different species. Problem 68 in Chapter 27 describes the structures that produce iridescence in a peacock feather. These structures were all unknown until the invention of the electron microscope. Explain why light microscopes cannot reveal them.

Figure CQ28.3

4. What is the significance of the wave function ψ?

5. Discuss the relationship between ground-state energy and the uncertainty principle.

6. For a quantum particle in a box, the probability density at certain points is zero as seen in Figure CQ28.6. Does this value imply that the particle cannot move across these points? Explain.

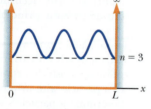

Figure CQ28.6

7. If the photoelectric effect is observed for one metal, can you conclude that the effect will also be observed for another metal under the same conditions? Explain.

8. In the photoelectric effect, explain why the stopping potential depends on the frequency of light but not on the intensity.

9. Why does the existence of a cutoff frequency in the photoelectric effect favor a particle theory for light over a wave theory?

10. In quantum mechanics, it is possible for the energy E of a particle to be less than the potential energy, but classically this condition is not possible. Explain.

11. Consider the wave functions in Figure CQ28.11. Which of them are not physically significant in the interval shown? For those that are not, state why they fail to qualify.

12. Why are the following wave functions not physically possible for all values of x? (a) $\psi(x) = Ae^x$ (b) $\psi(x) = A \tan x$

13. Which has more energy, a photon of ultraviolet radiation or a photon of yellow light? Explain.

14. How does the Compton effect differ from the photoelectric effect?

15. Is an electron a wave or a particle? Support your answer by citing some experimental results.

16. How is the Schrödinger equation useful in describing quantum phenomena?

17. Suppose a photograph were made of a person's face using only a few photons. Would the result be simply a very faint image of the face? Explain your answer.

18. Is light a wave or a particle? Support your answer by citing specific experimental evidence.

19. If matter has a wave nature, why is this wave-like characteristic not observable in our daily experiences?

20. In describing the passage of electrons through a slit and arriving at a screen, physicist Richard Feynman said that "electrons arrive in lumps, like particles, but the probability of arrival of these lumps is determined as the intensity of the waves would be. It is in this sense that the electron behaves sometimes like a particle and sometimes like a wave." Elaborate on this point in your own words. For further discussion, see R. Feynman, *The Character of Physical Law* (Cambridge, MA: MIT Press, 1980), chap. 6.

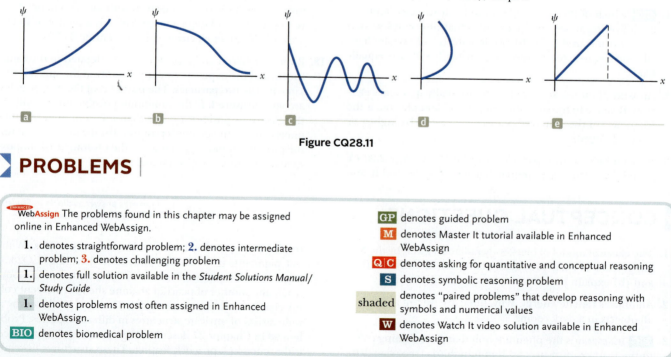

Figure CQ28.11

PROBLEMS

Section 28.1 Blackbody Radiation and Planck's Hypothesis

1. **BIO** **M** The average threshold of dark-adapted (scotopic) vision is 4.00×10^{-11} W/m² at a central wavelength of 500 nm. If light with this intensity and wavelength enters the eye and the pupil is open to its maximum diameter of 8.50 mm, how many photons per second enter the eye?

2. **BIO** **Q|C** Figure P28.2 shows the spectrum of light emitted by a firefly. (a) Determine the temperature of a black body that would emit radiation peaked at the same wavelength. (b) Based on your result, explain whether firefly radiation is blackbody radiation.

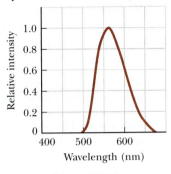

Figure P28.2

3. **BIO** With young children and the elderly, use of a traditional fever thermometer has risks of bacterial contamination and tissue perforation. The radiation thermometer shown in Figure 28.5 works fast and avoids most risks. The instrument measures the power of infrared radiation from the ear canal. This cavity is accurately described as a black body and is close to the hypothalamus, the body's temperature control center. Take normal body temperature as 37.0°C. If the body temperature of a feverish patient is 38.3°C, what is the percentage increase in radiated power from his ear canal?

4. **M** (i) Calculate the energy, in electron volts, of a photon whose frequency is (a) 620 THz, (b) 3.10 GHz, and (c) 46.0 MHz. (ii) Determine the corresponding wavelengths for the photons listed in part (i) and (iii) state the classification of each on the electromagnetic spectrum.

5. **Review.** This problem is about how strongly matter is coupled to radiation, the subject with which quantum mechanics began. For a simple model, consider a solid iron sphere 2.00 cm in radius. Assume its temperature is always uniform throughout its volume. (a) Find the mass of the sphere. (b) Assume the sphere is at 20.0°C and has emissivity 0.860. Find the power with which it radiates electromagnetic waves. (c) If it were alone in the Universe, at what rate would the sphere's temperature be changing? (d) Assume Wien's law describes the sphere. Find the wavelength λ_{max} of electromagnetic radiation it emits most strongly. Although it emits a spectrum of waves having all different wavelengths, assume its power output is carried by photons of wavelength λ_{max}. Find (e) the energy of one photon and (f) the number of photons it emits each second.

6. **W** The radius of our Sun is 6.96×10^8 m, and its total power output is 3.85×10^{26} W. (a) Assuming the Sun's surface emits as a black body, calculate its surface temperature. (b) Using the result of part (a), find λ_{max} for the Sun.

7. A simple pendulum has a length of 1.00 m and a mass of 1.00 kg. The maximum horizontal displacement of the pendulum bob from equilibrium is 3.00 cm. Calculate the quantum number n for the pendulum.

8. **M** An FM radio transmitter has a power output of 150 kW and operates at a frequency of 99.7 MHz. How many photons per second does the transmitter emit?

9. **BIO** **W** The human eye is most sensitive to 560-nm (green) light. What is the temperature of a black body that would radiate most intensely at this wavelength?

Section 28.2 The Photoelectric Effect

10. **Q|C** From the scattering of sunlight, J. J. Thomson calculated the classical radius of the electron as having the value 2.82×10^{-15} m. Sunlight with an intensity of 500 W/m² falls on a disk with this radius. Assume light is a classical wave and the light striking the disk is completely absorbed. (a) Calculate the time interval required to accumulate 1.00 eV of energy. (b) Explain how your result for part (a) compares with the observation that photoelectrons are emitted promptly (within 10^{-9} s).

11. Molybdenum has a work function of 4.20 eV. (a) Find the cutoff wavelength and cutoff frequency for the photoelectric effect. (b) What is the stopping potential if the incident light has a wavelength of 180 nm?

12. **GP** **Q|C** The work function for platinum is 6.35 eV. Ultraviolet light of wavelength 150 nm is incident on the clean surface of a platinum sample. We wish to predict the stopping voltage we will need for electrons ejected from the surface. (a) What is the photon energy of the ultraviolet light? (b) How do you know that these photons will eject electrons from platinum? (c) What is the maximum kinetic energy of the ejected photoelectrons? (d) What stopping voltage would be required to arrest the current of photoelectrons?

13. Electrons are ejected from a metallic surface with speeds of up to 4.60×10^5 m/s when light with a wavelength of 625 nm is used. (a) What is the work function of the surface? (b) What is the cutoff frequency for this surface?

14. **Review.** An isolated copper sphere of radius 5.00 cm, initially uncharged, is illuminated by ultraviolet light of wavelength 200 nm. The work function for copper is 4.70 eV. What charge does the photoelectric effect induce on the sphere?

15. Two light sources are used in a photoelectric experiment to determine the work function for a particular metal surface. When green light from a mercury lamp ($\lambda = 546.1$ nm) is used, a stopping potential of 0.376 V reduces the photocurrent to zero. (a) Based on this measurement, what is the work function for this metal? (b) What stopping potential would be observed when using the yellow light from a helium discharge tube ($\lambda = 587.5$ nm)?

Section 28.3 The Compton Effect

16. Calculate the energy and momentum of a photon of wavelength 700 nm.

17. **W** A photon having energy $E_0 = 0.880$ MeV is scattered by a free electron initially at rest such that the scattering angle of the scattered electron is equal to that of the scattered photon as shown in Figure P28.17 (page 986). (a) Determine the scattering angle of the photon and the electron. (b) Determine the energy and momentum of the scattered photon. (c) Determine the kinetic energy and momentum of the scattered electron.

18. **S** A photon having energy E_0 is scattered by a free electron initially at rest such that the scattering angle of the scattered

▸ 29.1 | Early Structural Models of the Atom

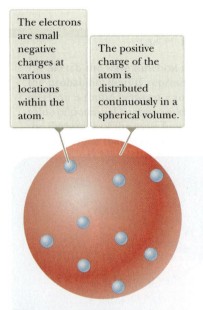

The electrons are small negative charges at various locations within the atom.

The positive charge of the atom is distributed continuously in a spherical volume.

Figure 29.1 Thomson's model of the atom.

The structural model of the atom in Newton's day described the atom as a tiny, hard, indestructible sphere, a particle model that ignored any internal structure of the atom. Although this model was a good basis for the kinetic theory of gases (Chapter 16), new structural models had to be devised when later experiments revealed the electrical nature of atoms. J. J. Thomson suggested a structural model that describes the atom as a continuous volume of positive charge with electrons embedded throughout it (Fig. 29.1).

In 1911, Ernest Rutherford and his students Hans Geiger and Ernst Marsden performed a critical experiment whose results were inconsistent with those predicted by Thomson's model. In this experiment, a beam of positively charged alpha particles was projected into a thin metal foil as in Figure 29.2a. Most of the particles passed through the foil as if it were empty space, which is consistent with the Thomson model. Some of the results of the experiment, however, were astounding. Many alpha particles were deflected from their original direction of travel through large angles. Some particles were even deflected backward, reversing their direction of travel. When Geiger informed Rutherford of these results, Rutherford wrote, "It was quite the most incredible event that has ever happened to me in my life. It was almost as incredible as if you fired a 15-inch [artillery] shell at a piece of tissue paper and it came back and hit you."

Such large deflections were not expected on the basis of Thomson's model. According to this model, a positively charged alpha particle would never come close enough to a sufficiently large concentration of positive charge to cause any large-angle deflections. Furthermore, the electrons in Thomson's model have far too little mass to cause such a large deflection of the massive alpha particles. Rutherford explained his astounding results with a new structural model, as introduced in Section 11.5: he assumed that the positive charge was concentrated in a region that was small relative to the size of the atom. He called this concentration of positive charge the **nucleus** of the atom. Any electrons belonging to the atom were assumed to be outside the nucleus. To explain why these electrons were not pulled into the nucleus by the attractive electric force, Rutherford imagined that the electrons move in orbits about the nucleus in the same manner as the planets orbit the Sun, as in Figure 29.2b.

There are two basic difficulties with Rutherford's planetary structural model. As we saw in Chapter 11, an atom emits discrete characteristic frequencies of electromagnetic radiation and no others; the Rutherford model is unable to explain this phenomenon. A second difficulty is that Rutherford's electrons experience a centripetal acceleration. According to Maxwell's equations in electromagnetism, charges orbiting with frequency f experience centripetal acceleration and therefore should radiate electromagnetic waves of frequency f. Unfortunately, this classical model leads to disaster when applied to the atom. As the electron radiates energy from the electron–proton system, the radius of the orbit of the electron steadily decreases and its frequency of revolution increases. Energy is continuously transferred out of the system by electromagnetic radiation. As a result, the energy of the system decreases, resulting in the decay of the orbit of the electron. This decrease in total

Figure 29.2 (a) Rutherford's technique for observing the scattering of alpha particles from a thin foil target. The source is a naturally occurring radioactive substance, such as radium. (b) Rutherford's planetary model of the atom.

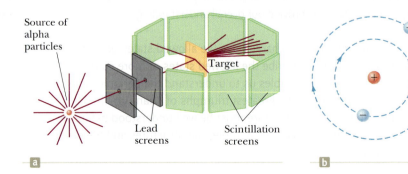

Source of alpha particles

Target

Lead screens

Scintillation screens

a

b

energy leads to an increase in the kinetic energy of the electron,[1] an ever-increasing frequency of emitted radiation, and a rapid collapse of the atom as the electron plunges into the nucleus (Fig. 29.3).

The stage was set for Bohr and his model that we discussed in Chapter 11. To circumvent the unsuccessful predictions of the Rutherford model—electrons falling into the nucleus and a continuous emission spectrum from elements—Bohr postulated that classical radiation theory does not hold for atomic-sized systems. He overcame the problem of an atom that continuously loses energy by applying Planck's ideas of quantized energy levels to orbiting atomic electrons. Therefore, as described in Section 11.5, Bohr postulated that electrons in atoms are generally confined to stable, nonradiating orbits called stationary states. Furthermore, he applied Einstein's concept of the photon to arrive at an expression for the frequency of radiation emitted when the atom makes a transition from one stationary state to another.

While the Bohr model was more successful than Rutherford's model in terms of predicting stable atoms and wavelengths of emitted radiation, it failed to predict more subtle spectral details, as mentioned in Pitfall Prevention 11.4. One of the first indications that the Bohr theory needed modification arose when improved spectroscopic techniques were used to examine the spectral lines of hydrogen. It was found that many of the lines in the Balmer and other series were not single lines at all. Instead, each was a group of closely spaced lines. An additional difficulty arose when it was observed that, in some situations, some single spectral lines were split into three closely spaced lines when the atoms were placed in a strong magnetic field. The Bohr model cannot explain this phenomenon.

Efforts to explain these difficulties with the Bohr model led to improvements in the structural model of the atom. One of the changes introduced was the concept that the electron has an intrinsic angular momentum called *spin*, which we introduced in Chapter 22 in terms of the contribution of spin to the magnetic properties of materials. We shall discuss spin in more detail in this chapter.

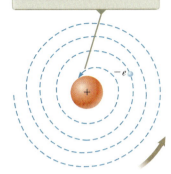

Because the accelerating electron radiates energy, the size of the orbit decreases until the electron falls into the nucleus.

Figure 29.3 The classical model of the nuclear atom predicts that the atom decays.

29.2 | The Hydrogen Atom Revisited

A quantum treatment of the hydrogen atom requires a solution to the Schrödinger equation (Eq. 28.36), with U being the electric potential energy of the electron–proton system. The full mathematical solution of the Schrödinger equation as applied to the hydrogen atom gives a complete and beautiful description of the atom's properties. The mathematical procedures that make up the solution are beyond the scope of this text, however, and so the details shall be omitted. The solutions for some states of hydrogen will be discussed, together with the quantum numbers used to characterize allowed stationary states. We also discuss the physical significance of the quantum numbers.

Let us outline the components of a quantum structural model for the hydrogen atom:

1. *A description of the physical components of the system:* As in the earlier hydrogen atom models, the physical components are the electron and a positive charge, which we model as being concentrated in a very small nucleus. We also model the electron as a localized concentration of charge, but we will find that our results suggest that this assumption must be relaxed.
2. *A description of where the components are located relative to one another and how they interact:* Because of the small size of the atom, we assume that the electron and the nucleus are close together. We also assume that they interact via the electric force. We do *not* assume any sort of orbit for the electron.
3. *A description of the time evolution of the system:* We wish to understand the details of the stable hydrogen atom. We also wish to understand the process that occurs

---

[1]As an orbital system that interacts via an inverse square force law loses energy, the kinetic energy of the orbiting object increases but the potential energy of the system decreases by a larger amount; therefore, the change in the total energy of the system is negative.

⟨ **TABLE 29.4** | Quantum Numbers for the $n = 2$ State of Hydrogen

n	ℓ	m_ℓ	m_s	Subshell	Shell	Number of States in Subshell
2	0	0	$\frac{1}{2}$			
2	0	0	$-\frac{1}{2}$	$2s$	L	2
2	1	1	$\frac{1}{2}$			
2	1	1	$-\frac{1}{2}$			
2	1	0	$\frac{1}{2}$			
2	1	0	$-\frac{1}{2}$	$2p$	L	6
2	1	-1	$\frac{1}{2}$			
2	1	-1	$-\frac{1}{2}$			

to $m_s = \pm\frac{1}{2}$. Electrons with spin $+\frac{1}{2}$ are deflected in one direction by the nonuniform magnetic field, and those with spin $-\frac{1}{2}$ are deflected in the opposite direction.

The Stern–Gerlach experiment provided two important results. First, it verified the concept of space quantization. Second, it showed that spin angular momentum exists even though this property was not recognized until long after the experiments were performed.

As mentioned earlier, there are eight quantum states corresponding to $n = 2$ in the hydrogen atom, not four as found in Example 29.1. Each of the four states in Example 29.1 is actually two states because of the two possible values of m_s. Table 29.4 shows the quantum numbers corresponding to these eight states.

⟩ **THINKING PHYSICS 29.1**

Does the Stern–Gerlach experiment differentiate between orbital angular momentum and spin angular momentum?

Reasoning A magnetic force on the magnetic moment arises from both orbital angular momentum and spin angular momentum. In this sense, the experiment does not differentiate between the two. The number of components on the screen does tell us something, however, because orbital angular momenta are described by an integral quantum number ℓ, whereas spin angular momentum depends on a half-integral quantum number s. If an odd number of components occur on the screen, three possibilities arise: the atom

has (1) orbital angular momentum only, (2) an even number of electrons with spin angular momentum, or (3) a combination of orbital angular momentum and an even number of electrons with spin angular momentum. If an even number of components occurs on the screen, at least one unpaired spin angular momentum exists, possibly in combination with orbital angular momentum. The only numbers of components for which we can specify the type of angular momentum are one component (no orbital, no spin) and two components (spin of one electron). Once we see more than two components, multiple possibilities arise because of various combinations of $\vec{L}$ and $\vec{S}$. ◄

⟨ **29.5** | **The Exclusion Principle and the Periodic Table**

The quantum model for hydrogen generated from the Schrödinger equation, including the notion of electron spin, is based on a system consisting of one electron and one proton. As soon as we consider the next atom, helium, we introduce complicating factors. The two electrons in helium both interact with the nucleus, so we can define a potential energy function for those interactions. They also interact with each other, however. The line of action of the electron–nucleus interaction is along a line between the electron and the nucleus. The line of action of the electron–electron interaction is along the line between the two electrons, which is different from that of the electron–nucleus interaction. Therefore, the Schrödinger equation is extremely difficult to solve. As we consider atoms with more and more electrons, the possibility of an algebraic solution of the Schrödinger equation becomes hopeless.

We find, however, that despite our inability to solve the Schrödinger equation, we can use the same four quantum numbers developed for hydrogen for the electrons in heavier atoms. We are not able to calculate the quantized energy levels easily, but we can gain information about the levels from theoretical models and experimental measurements.

Because a quantum state in any atom is specified by four quantum numbers, n, ℓ, m_ℓ, and m_s, an obvious and important question is, "How many electrons in an atom can have a particular set of quantum numbers?" Pauli provided an answer in 1925 in a powerful statement known as the **exclusion principle:**

No two electrons in an atom can ever be in the same quantum state; that is, no two electrons in the same atom can have the same set of quantum numbers.

It is interesting that if this principle were not valid, every atom would radiate energy by means of photons and end up with all electrons in the lowest energy state. The chemical behavior of the elements would be grossly modified because this behavior depends on the electronic structure of atoms. Nature as we know it would not exist! In reality, we can view the electronic structure of complex atoms as a succession of filled levels increasing in energy, where the outermost electrons are primarily responsible for the chemical properties of the element.

Imagine building an atom by forming the nucleus and then filling in the available quantum states with electrons until the atom is neutral. We shall use the common language here that "electrons go into available states." Keep in mind, however, that the states are those of the *system* of the atom. As a general rule, the order of filling of an atom's subshells with electrons is as follows. Once one subshell is filled, the next electron goes into the vacant subshell that is lowest in energy.

Before we discuss the electronic configurations of some elements, it is convenient to define an **orbital** as the state of an electron characterized by the quantum numbers n, ℓ, and m_ℓ. From the exclusion principle, we see that at most two electrons can be in any orbital. One of these electrons has $m_s = +\frac{1}{2}$ and the other has $m_s = -\frac{1}{2}$. Because each orbital is limited to two electrons, the numbers of electrons that can occupy the shells are also limited.

Table 29.5 shows the allowed quantum states for an atom up to $n = 3$. Each square in the bottom row of the table represents one orbital, with the ↑ arrows representing $m_s = +\frac{1}{2}$ and the ↓ arrows representing $m_s = -\frac{1}{2}$. The $n = 1$ shell can accommodate only two electrons because only one orbital is allowed with $m_\ell = 0$. The $n = 2$ shell has two subshells, with $\ell = 0$ and $\ell = 1$. The $\ell = 0$ subshell is limited to only two electrons because $m_\ell = 0$. The $\ell = 1$ subshell has three allowed orbitals, corresponding to $m_\ell = 1$, 0, and -1. Because each orbital can accommodate two electrons, the $\ell = 1$ subshell can hold six electrons (and the $n = 2$ shell can hold eight). The $n = 3$ shell has three subshells and nine orbitals and can accommodate up to 18 electrons. In general, each shell can accommodate up to $2n^2$ electrons.

The results of the exclusion principle can be illustrated by an examination of the electronic arrangement in a few of the lighter atoms. For example, **hydrogen** has only one electron, which, in its ground state, can be described by either of two sets of quantum numbers: $1, 0, 0, +\frac{1}{2}$ or $1, 0, 0, -\frac{1}{2}$. The electronic configuration of this atom is often designated as $1s^1$. The notation $1s$ refers to a state for which $n = 1$ and $\ell = 0$, and the superscript indicates that one electron is present in the s subshell.

Wolfgang Pauli
Austrian Theoretical Physicist
(1900–1958)
An extremely talented Austrian theoretical physicist, Pauli made important contributions in many areas of modern physics. Pauli gained public recognition at the age of 21 with a masterful review article on relativity, which is still considered one of the finest and most comprehensive introductions to the subject. Other major contributions were the discovery of the exclusion principle, the explanation of the connection between particle spin and statistics, and theories of relativistic quantum electrodynamics, the neutrino hypothesis, and the hypothesis of nuclear spin.

Pitfall Prevention | 29.3
The Exclusion Principle Is More General
The exclusion principle stated here is a limited form of the more general exclusion principle, which states that no two *fermions*, which are *all* particles with half-integral spin $\frac{1}{2}, \frac{3}{2}, \frac{5}{2}, \ldots$ can be in the same quantum state. The present form is satisfactory for our discussions of atomic physics, and we will discuss the general form further in Chapter 31.

TABLE 29.5 | Allowed Quantum States for an Atom Up to $n = 3$

n	1	2			3								
ℓ	0	0	1		0	1			2				
m_ℓ	0	0	1	0	−1	0	1	0	−1	2	1	0	−1	−2
m_s	↑↓	↑↓	↑↓	↑↓	↑↓	↑↓	↑↓	↑↓	↑↓	↑↓	↑↓	↑↓	↑↓	↑↓

Atom	1s	2s		2p		Electronic configuration
Li	↑↓	↑				$1s^2 2s^1$
Be	↑↓	↑↓				$1s^2 2s^2$
B	↑↓	↑↓	↑			$1s^2 2s^2 2p^1$
C	↑↓	↑↓	↑	↑		$1s^2 2s^2 2p^2$
N	↑↓	↑↓	↑	↑	↑	$1s^2 2s^2 2p^3$
O	↑↓	↑↓	↑↓	↑	↑	$1s^2 2s^2 2p^4$
F	↑↓	↑↓	↑↓	↑↓	↑	$1s^2 2s^2 2p^5$
Ne	↑↓	↑↓	↑↓	↑↓	↑↓	$1s^2 2s^2 2p^6$

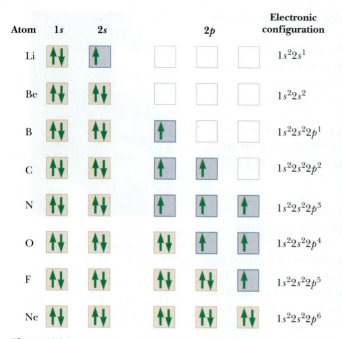

Figure 29.11 The filling of electronic states must obey both the exclusion principle and Hund's rules.

Neutral **helium** has two electrons. In the ground state, the quantum numbers for these two electrons are 1, 0, 0, $+\frac{1}{2}$ and 1, 0, 0, $-\frac{1}{2}$. No other combinations of quantum numbers are possible for this level, and we say that the K shell is filled. The electronic configuration of helium is designated as $1s^2$.

The electronic configurations of some successive elements are given in Figure 29.11. Neutral **lithium** has three electrons. In the ground state, two of them are in the $1s$ subshell and the third is in the $2s$ subshell because this subshell is lower in energy than the $2p$ subshell. (In addition to the simple dependence of E on n in Eq. 29.2, there is an additional dependence on ℓ, which will be addressed in Section 29.6.) Hence, the electronic configuration for lithium is $1s^2 2s^1$.

Note that the electronic configuration of **beryllium,** with its four electrons, is $1s^2 2s^2$, and **boron** has a configuration of $1s^2 2s^2 2p^1$. The $2p$ electron in boron may be described by one of six sets of quantum numbers, corresponding to six states of equal energy.

Carbon has six electrons, and a question arises concerning how to assign the two $2p$ electrons. Do they go into the same orbital with paired spins (↑↓), or do they occupy different orbitals with unpaired spins (↑↑ or ↓↓)? Experimental data show that the lowest energy configuration is the latter, where the spins are unpaired. Hence, the two $2p$ electrons in carbon and the three $2p$ electrons in nitrogen have unpaired spins in the ground state (see Fig. 29.11). The general rules that govern such situations throughout the periodic table are called **Hund's rules.** The rule appropriate for elements like carbon is that when an atom has orbitals of equal energy, the order in which they are filled by electrons is such that a maximum number of electrons will have unpaired spins. Some exceptions to this rule occur in elements having subshells close to being filled or half-filled.

An early attempt to find some order among the elements was made by a Russian chemist, Dmitri Mendeleev (1834–1907), in 1871. He developed a tabular representation of the elements, which has become one of the most important, as well as well-recognized, tools of science. He arranged the atoms in a table similar to that shown in Figure 29.12 according to their atomic masses and chemical similarities. The first table Mendeleev proposed contained many blank spaces, and he boldly stated that the gaps were there only because the elements had not yet been discovered. By noting the columns in which these missing elements should be located, he was able to make rough predictions about their chemical properties. Within 20 years of Mendeleev's announcement, the missing elements were indeed discovered. The predictions made possible by this table represent an excellent example of the power of presenting information in an alternative representation.

The elements in the **periodic table** (Fig. 29.12) are arranged so that all those in a vertical column have similar chemical properties. For example, consider the elements in the last column: He (helium), Ne (neon), Ar (argon), Kr (krypton), Xe (xenon), and Rn (radon). The outstanding characteristic of all these elements is that they do not normally take part in chemical reactions; that is, they do not readily join with other atoms to form molecules. They are therefore called *inert gases.*

We can partially understand this behavior by looking at the electronic configurations in Figure 29.12. The element helium is one in which the electronic configuration is $1s^2$; in other words, one shell is filled. Additionally, it is found that the energy associated with this filled shell is considerably lower than the energy of the next available level, the $2s$ level. Next, look at the electronic configuration for neon, $1s^2 2s^2 2p^6$. Again, the outermost shell is filled, and a gap in energy occurs between the $2p$ level and the $3s$ level. Argon has the configuration $1s^2 2s^2 2p^6 3s^2 3p^6$. Here, the $3p$ subshell is filled, and a gap in energy arises between the $3p$ subshell and the $3d$ subshell. We could continue this procedure through all the inert gases; the pattern remains the

Group I	Group II	Transition elements										Group III	Group IV	Group V	Group VI	Group VII	Group 0
H 1 $1s^1$																H 1 $1s^1$	He 2 $1s^2$
Li 3 $2s^1$	Be 4 $2s^2$											B 5 $2p^1$	C 6 $2p^2$	N 7 $2p^3$	O 8 $2p^4$	F 9 $2p^5$	Ne 10 $2p^6$
Na 11 $3s^1$	Mg 12 $3s^2$											Al 13 $3p^1$	Si 14 $3p^2$	P 15 $3p^3$	S 16 $3p^4$	Cl 17 $3p^5$	Ar 18 $3p^6$
K 19 $4s^1$	Ca 20 $4s^2$	Sc 21 $3d^14s^2$	Ti 22 $3d^24s^2$	V 23 $3d^34s^2$	Cr 24 $3d^54s^1$	Mn 25 $3d^54s^2$	Fe 26 $3d^64s^2$	Co 27 $3d^74s^2$	Ni 28 $3d^84s^2$	Cu 29 $3d^{10}4s^1$	Zn 30 $3d^{10}4s^2$	Ga 31 $4p^1$	Ge 32 $4p^2$	As 33 $4p^3$	Se 34 $4p^4$	Br 35 $4p^5$	Kr 36 $4p^6$
Rb 37 $5s^1$	Sr 38 $5s^2$	Y 39 $4d^15s^2$	Zr 40 $4d^25s^2$	Nb 41 $4d^45s^1$	Mo 42 $4d^55s^1$	Tc 43 $4d^55s^2$	Ru 44 $4d^75s^1$	Rh 45 $4d^85s^1$	Pd 46 $4d^{10}$	Ag 47 $4d^{10}5s^1$	Cd 48 $4d^{10}5s^2$	In 49 $5p^1$	Sn 50 $5p^2$	Sb 51 $5p^3$	Te 52 $5p^4$	I 53 $5p^5$	Xe 54 $5p^6$
Cs 55 $6s^1$	Ba 56 $6s^2$	57–71*	Hf 72 $5d^26s^2$	Ta 73 $5d^36s^2$	W 74 $5d^46s^2$	Re 75 $5d^56s^2$	Os 76 $5d^66s^2$	Ir 77 $5d^76s^2$	Pt 78 $5d^96s^1$	Au 79 $5d^{10}6s^1$	Hg 80 $5d^{10}6s^2$	Tl 81 $6p^1$	Pb 82 $6p^2$	Bi 83 $6p^3$	Po 84 $6p^4$	At 85 $6p^5$	Rn 86 $6p^6$
Fr 87 $7s^1$	Ra 88 $7s^2$	89–103**	Rf 104 $6d^27s^2$	Db 105 $6d^37s^2$	Sg 106 $6d^47s^2$	Bh 107 $6d^57s^2$	Hs 108 $6d^67s^2$	Mt 109 $6d^77s^2$	Ds 110 $6d^97s^1$	Rg 111	Cn 112		114		116		

*Lanthanide series	La 57 $5d^16s^2$	Ce 58 $5d^14f^16s^2$	Pr 59 $4f^36s^2$	Nd 60 $4f^46s^2$	Pm 61 $4f^56s^2$	Sm 62 $4f^66s^2$	Eu 63 $4f^76s^2$	Gd 64 $5d^14f^76s^2$	Tb 65 $5d^14f^86s^2$	Dy 66 $4f^{10}6s^2$	Ho 67 $4f^{11}6s^2$	Er 68 $4f^{12}6s^2$	Tm 69 $4f^{13}6s^2$	Yb 70 $4f^{14}6s^2$	Lu 71 $5d^14f^{14}6s^2$
**Actinide series	Ac 89 $6d^17s^2$	Th 90 $6d^27s^2$	Pa 91 $5f^26d^17s^2$	U 92 $5f^36d^17s^2$	Np 93 $5f^46d^17s^2$	Pu 94 $5f^67s^2$	Am 95 $5f^77s^2$	Cm 96 $5f^76d^17s^2$	Bk 97 $5f^86d^17s^2$	Cf 98 $5f^{10}7s^2$	Es 99 $5f^{11}7s^2$	Fm 100 $5f^{12}7s^2$	Md 101 $5f^{13}7s^2$	No 102 $5f^{14}7s^2$	Lr 103 $5f^{14}6d^17s^2$

Figure 29.12 The periodic table of the elements is an organized tabular representation of the elements that shows their periodic chemical behavior. Elements in a given column have similar chemical behavior. This table shows the chemical symbol for the element, the atomic number, and the electronic configuration. A more complete periodic table is available in Appendix C.

same. An inert gas is formed when either a shell or a subshell is filled and a gap in energy occurs before the next possible level is encountered.

If we consider the column to the left of the inert gases in the periodic table, we find a group of elements called the *halogens*: fluorine, chlorine, bromine, iodine, and astatine. At room temperature, fluorine and chlorine are gases, bromine is a liquid, and iodine and astatine are solids. In each of these atoms, the outer subshell is one electron short of being filled. As a result, the halogens are chemically very active, readily accepting an electron from another atom to form a closed shell. The halogens tend to from strong ionic bonds with atoms at the other side of the periodic table. In a halogen lightbulb, bromine or iodine atoms combine with tungsten atoms evaporated from the filament and return them to the filament, resulting in a longer-lasting bulb. In addition, the filament can be operated at a higher temperature than in ordinary lightbulbs, giving a brighter and whiter light.

At the left side of the periodic table, the Group I elements consist of hydrogen and the *alkali metals*, lithium, sodium, potassium, rubidium, cesium, and francium. Each of these atoms contains one electron in a subshell outside of a closed subshell. Therefore, these elements easily form positive ions because the lone electron is bound with a relatively low energy and is easily removed. Therefore, the alkali metal atoms are chemically active and form very strong bonds with halogen atoms. For example, table salt, NaCl, is a combination of an alkali metal and a halogen. Because the outer electron is weakly bound, pure alkali metals tend to be good electrical conductors, although, because of their high chemical activity, pure alkali metals are not generally found in nature.

It is interesting to plot ionization energy versus the atomic number Z as in Figure 29.13 (page 1008). Note the pattern of differences in atomic numbers between the peaks in the graph: 8, 8, 18, 18, 32. This pattern follows from the Pauli exclusion

Figure 29.13 Ionization energy of the elements versus atomic number.

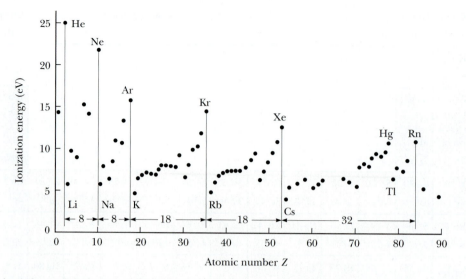

principle and helps explain why the elements repeat their chemical properties in groups. For example, the peaks at $Z = 2$, 10, 18, and 36 correspond to the elements He, Ne, Ar, and Kr, which have filled shells. These elements have similar chemical behavior.

QUICK QUIZ 29.4 Rank the energy necessary to remove the outermost electron from the following three elements, smallest to largest: lithium, potassium, cesium.

BIO Treatment of cancers with proton therapy

A variety of treatment plans are available for battling cancerous tumors. Some of these options involving atomic and nuclear phenomena will be discussed in this chapter and the next. One of these treatment procedures is called *proton therapy*. In this procedure, a beam of protons is used to irradiate the cancerous tissue. Protons are one form of *ionizing radiation*, that is, radiation that will ionize atoms of diseased tissue with the goal of destroying the tissue. A major advantage of using protons is that the dose delivered to the tissue, that is, the ionizing energy deposited in the tissue, is a maximum over the last few millimeters of the particle's range. As a result, relatively little ionization occurs along the first part of the protons' path, leaving healthy tissue unharmed. By adjusting the incoming energy of the protons, up to 250 MeV, the depth at which the majority of the energy is delivered can be tuned to coincide with the location of the tumor. Special nozzles at the end of the proton beam shape the beam according to the three-dimensional shape of the tumor, allowing the entire tumor to receive uniform irradiation. As a result, the cancerous tissue is damaged while surrounding healthy tissue experiences much less damage.

Proton therapy has been used for prostate cancer, sarcomas, medically inoperable lung cancer, acoustic neuromas, and a variety of ocular tumors. Proton therapy procedures have been performed since the early 1950s using particle accelerators built for physics research. Beginning in 1990, dedicated hospital-based proton therapy centers were built. At the time of this printing, there are ten such centers in the United States and 37 worldwide.

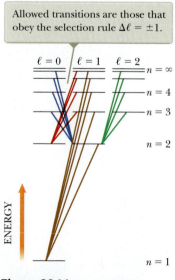

Figure 29.14 Some allowed electronic transitions for hydrogen, represented by the colored lines.

29.6 | More on Atomic Spectra: Visible and X-Ray

In Chapter 11, we briefly discussed the origin of the spectral lines for hydrogen and hydrogen-like ions. Recall that an atom in an excited state will emit electromagnetic radiation if it makes a transition to a lower energy state.

The energy level diagram for hydrogen is shown in Figure 29.14. This semigraphical representation is different from Active Figure 11.18 in that the individual states corresponding to different values of ℓ within a given value of n are spread out horizontally. Figure 29.14 shows only those states up to $\ell = 2$; the shells from $n = 4$ upward would have more sets of states to the right, which are not shown.

The diagonal lines in Figure 29.14 represent allowed transitions between stationary states. Whenever an atom makes a transition from a higher energy state to a lower one, a photon of light is emitted. The frequency of this photon is $f = \Delta E/h$, where ΔE is the energy difference between the two states and h is Planck's constant. The **selection rules** for the allowed transitions are

$$\Delta \ell = \pm 1 \quad \text{and} \quad \Delta m_\ell = 0 \text{ or } \pm 1 \qquad \text{29.16} \blacktriangleleft$$

▶ Selection rules for allowed atomic transitions

Transitions that do not obey the above selection rules are said to be **forbidden.** (Such transitions can occur, but their probability is negligible relative to the probability of the allowed transitions.) For example, any transition represented by a vertical line in Figure 29.14 is forbidden because the quantum number ℓ does not change.

Because the orbital angular momentum of an atom changes when a photon is emitted or absorbed (i.e., as a result of a transition) and because angular momentum of the isolated system of the atom and the photon must be conserved, we conclude that the photon involved in the process must carry angular momentum. In fact, the photon has an intrinsic angular momentum equivalent to that of a particle with a spin of $s = 1$, compared with the electron with $s = \frac{1}{2}$. Hence, a photon possesses energy, linear momentum, and angular momentum. This example is the first one we have seen of a single particle with *integral* spin.

Equation 29.2 gives the energies of the allowed quantum states for hydrogen. We can also apply the Schrödinger equation to other one-electron systems, such as the He^+ and Li^{++} ions. The primary difference between these ions and the hydrogen atom is the different number of protons Z in the nucleus. The result is a generalization of Equation 29.2 for these other one-electron systems:

$$E_n = -\frac{(13.6 \text{ eV}) Z^2}{n^2} \qquad \text{29.17} \blacktriangleleft$$

For outer electrons in multielectron atoms, the nuclear charge Ze is largely canceled or shielded by the negative charge of the inner-core electrons. Hence, the outer electrons interact with a net charge that is reduced below the actual charge of the nucleus. (According to Gauss's law, the electric field at the position of an outer electron depends on the net charge of the nucleus and the electrons closer to the nucleus.) The expression for the allowed energies for multielectron atoms has the same form as Equation 29.17, with Z replaced by an effective atomic number Z_{eff}. That is,

$$E_n \approx -\frac{(13.6 \text{ eV}) Z_{\text{eff}}^2}{n^2} \qquad \text{29.18} \blacktriangleleft$$

where Z_{eff} depends on n and ℓ.

▶ THINKING PHYSICS 29.2

A physics student is watching a meteor shower in the early morning hours. She notices that the streaks of light from the meteoroids entering the very high regions of the atmosphere last for up to 2 or 3 s before fading.

She also notices a lightning storm off in the distance. The streaks of light from the lightning fade away almost immediately after the flash, certainly in much less than 1 s. Both lightning and meteors cause the air to turn into a plasma because of the very high temperatures generated. The light is emitted from both sources when the stripped electrons in the plasma recombine with the ionized molecules. Why would this light last longer for meteors than for lightning?

Reasoning The answer lies in the subtle phrase in the description of the meteoroids "entering the very high regions of the atmosphere." In the very high regions of the atmosphere, the pressure of the air is very low. The

density of the air is therefore very low, so molecules of the air are relatively far apart. Therefore, after the air is ionized by the passing meteoroid, the probability per unit time interval of freed electrons encountering an ionized molecule with which to recombine is relatively low. As a result, the recombination process for all freed electrons occurs over a relatively long time interval, measured in seconds.

On the other hand, lightning occurs in the lower regions of the atmosphere (the troposphere) where the pressure and density are relatively high. After the ionization by the lightning flash, the freed electrons and ionized molecules are much closer together than in the upper atmosphere. The probability per unit time interval of a recombination is much higher, and the time interval for the recombination of all the electrons and ions to occur is much shorter. ◀

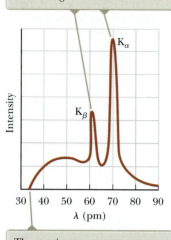

Figure 29.15 The x-ray spectrum of a metal target. The data shown were obtained when 37-keV electrons bombarded a molybdenum target.

BIO Medical imaging with x-rays

Figure 29.16 Bremsstrahlung is created by this machine and used to treat cancer in a patient.

X-Ray Spectra

X-rays are emitted when high-energy electrons or any other charged particles bombard a metal target. The x-ray spectrum typically consists of a broad continuous band containing a series of sharp lines as shown in Figure 29.15. In Section 24.6, we mentioned that an accelerated electric charge emits electromagnetic radiation. The x-rays in Figure 29.15 are the result of the slowing down of high-energy electrons as they strike the target. It may take several interactions with the atoms of the target before the electron loses all its kinetic energy. The amount of kinetic energy lost in any given interaction can vary from zero up to the entire kinetic energy of the electron. Therefore, the wavelength of radiation from these interactions lies in a continuous range from some minimum value up to infinity. It is this general slowing down of the electrons that provides the continuous curve in Figure 29.15, which shows the cutoff of x-rays below a minimum wavelength value that depends on the kinetic energy of the incoming electrons. X-ray radiation with its origin in the slowing down of electrons is called **bremsstrahlung,** the German word for "braking radiation."

Extremely high-energy bremsstrahlung can be used for the treatment of cancerous tissues in a process known as *external beam radiotherapy*. Figure 29.16 shows a machine that uses a linear accelerator to accelerate electrons up to 18 MeV and smash them into a tungsten target. The result is a beam of photons, up to a maximum energy of 18 MeV, which is actually in the gamma-ray range in Figure 24.11. This radiation is directed at the tumor in the patient.

In the previous section, we discussed treating cancerous tissues with energetic protons. The advantage of that technique is that most of the energy of the protons is delivered within the cancerous tissue, leaving healthy tissue relatively unharmed. The disadvantage is the size and cost of the cyclotron or synchrotron necessary to accelerate the protons up to therapeutic energies. External beam radiotherapy carries a higher probability of damage to healthy tissue, but the equipment necessary to accelerate electrons to therapeutic energies is much smaller and more inexpensive than that for proton therapy.

X-rays have been used for medical imaging since the early 20th century. As x-rays pass through the body, tissues of varying density and composition absorb different amounts of energy. By allowing the x-rays to expose photographic film after passing through the body, the film shows a shadowy image of the internal structure of the body. In an advance called *fluoroscopy*, the photographic film is replaced with a fluorescent screen (first decade of the 20th century) or a detector screen and video monitor (1950s). This procedure allows real-time evaluation of the x-ray image. In the 1970s, a major advance was made with the development of *computed tomography* or *CT scans*. A CT scan is created by a combination of an x-ray source and detector that rotate around the body and a computer to analyze the data. The result is a series of images representing cross-sectional slices through the body. CT scans, including newer versions providing three-dimensional images, are widely used in medical diagnostic procedures. While the use of CT scans has some advantages over the use of MRI (magnetic resonance imaging) scans, such as in the imaging of tumors in the thoracic region, it has a distinct disadvantage in exposing the patient to x-rays, which can cause damage to healthy tissue. MRI scans, on the other hand, expose the patient only to a strong magnetic field and harmless radio waves.

The discrete lines in Figure 29.15, called **characteristic x-rays** and discovered in 1908, have a different origin from that of bremsstrahlung. Their origin remained unexplained until the details of atomic structure were understood. The first step in the production of characteristic x-rays occurs when a bombarding electron collides with a target atom. The incoming electron must have sufficient energy to remove an inner-shell electron from the atom. The vacancy created in the shell is filled when an electron in a higher shell drops down into the shell containing the vacancy. The time interval required for this to happen is very short, less than 10^{-9} s. As usual, this transition is accompanied by the emission of a photon whose energy equals the difference in energy between the two shells. Typically, the energy of such transitions is greater than 1 000 eV, and the emitted x-ray photons have wavelengths in the range of 0.01 to 1 nm.

Let us assume that the incoming electron has dislodged an atomic electron from the innermost shell, the K shell. If the vacancy is filled by an electron dropping from the next higher shell, the L shell, the photon emitted in the process has an energy corresponding to the K_α line on the curve of Figure 29.15. If the vacancy is filled by an electron dropping from the M shell, the line produced is called the K_β line. In this notation, the letter K represents the final shell into which the electron drops and the subscript provides a Greek letter corresponding to the number of the shell above the final shell in which the electron originates. Therefore, K_α indicates that the final shell is the K shell, whereas the initial shell is the first shell above K (because α is the first letter in the Greek alphabet), which is the L shell.

Other characteristic x-ray lines are formed when electrons drop from upper shells to vacancies in shells other than the K shell. For example, L lines are produced when vacancies in the L shell are filled by electrons dropping from higher shells. An L_α line is produced as an electron drops from the M shell to the L shell, and an L_β line is produced by a transition from the N shell to the L shell.

Although multielectron atoms cannot be analyzed exactly using either the Bohr model or the Schrödinger equation, we can apply our knowledge of Gauss's law from Chapter 19 to make some surprisingly accurate estimates of expected x-ray energies and wavelengths. Consider an atom of atomic number Z in which one of the two electrons in the K shell has been ejected. Imagine that we draw a gaussian sphere just inside the most probable radius of the L electrons. The electric field at the position of the L electrons is a combination of that due to the nucleus, the single K electron, the other L electrons, and the outer electrons. The wave functions of the outer electrons are such that they have a very high probability of being farther from the nucleus than the L electrons are. Therefore, they are much more likely to be outside the gaussian surface than inside and, on the average, do not contribute significantly to the electric field at the position of the L electrons. The effective charge inside the gaussian surface is the positive nuclear charge and one negative charge due to the single K electron. If we ignore the interactions between L electrons, a single L electron behaves as if it experiences an electric field due to a charge enclosed by the gaussian surface of $(Z - 1)e$. The nuclear charge is in effect shielded by the electron in the K shell such that Z_{eff} in Equation 29.18 is $Z - 1$. For higher-level shells, the nuclear charge is shielded by electrons in all the inner shells.

We can now use Equation 29.18 to estimate the energy associated with an electron in the L shell:

$$E_{\text{L}} \approx -(Z - 1)^2 \frac{13.6 \text{ eV}}{2^2}$$

After the atom makes the transition, there are two electrons in the K shell. We can approximate the energy associated with one of these electrons as that of a one-electron atom. (In reality, the nuclear charge is reduced somewhat by the negative charge of the other electron, but let's ignore this effect.) Therefore,

$$E_{\text{K}} \approx -Z^2(13.6 \text{ eV}) \qquad \text{29.19} \blacktriangleleft$$

As we show in Example 29.5, the energy of the atom with an electron in an M shell can be estimated in a similar fashion. Taking the energy difference between the initial and final levels, the energy and wavelength of the emitted photon can then be calculated.

In 1914, Henry G. J. Moseley (1887–1915) plotted $\sqrt{1/\lambda}$ versus the Z values for a number of elements, where λ is the wavelength of the K_α line of each element. He found that the curve is a straight line as in Figure 29.17. This finding is consistent with rough calculations of the energy levels given by Equation 29.19. From this plot, Moseley was able to determine the Z values of some missing elements, which provided a periodic table in excellent agreement with the known chemical properties of the elements.

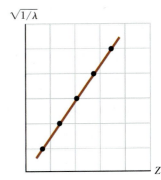

Figure 29.17 A Moseley plot of $\sqrt{1/\lambda}$ versus Z, where λ is the wavelength of the K_α x-ray line of the element with atomic number Z.

> **QUICK QUIZ 29.5** True or False: It is possible for an x-ray spectrum to show the continuous spectrum of x-rays without the presence of the characteristic x-rays.

> **QUICK QUIZ 29.6** In an x-ray tube, as you increase the energy of the electrons striking the metal target, do the wavelengths of the characteristic x-rays (**a**) increase, (**b**) decrease, or (**c**) remain constant?

Example 29.5 | Estimating the Energy of an X-Ray

Estimate the energy of the characteristic x-ray emitted from a tungsten target when an electron drops from an M shell ($n = 3$ state) to a vacancy in the K shell ($n = 1$ state). The atomic number for tungsten is $Z = 74$.

SOLUTION

Conceptualize Imagine an accelerated electron striking a tungsten atom and ejecting an electron from the K shell. Subsequently, an electron in the M shell drops down to fill the vacancy and the energy difference between the states is emitted as an x-ray photon.

Categorize We estimate the results using equations developed in this section, so we categorize this example as a substitution problem.

Use Equation 29.19 and $Z = 74$ for tungsten to estimate the energy associated with the electron in the K shell:

$$E_K \approx -(74)^2 (13.6 \text{ eV}) = -7.4 \times 10^4 \text{ eV}$$

Use Equation 29.18 and that nine electrons shield the nuclear charge (eight electrons in the $n = 2$ state and one electron in the $n = 1$ state) to estimate the energy of the M shell:

$$E_M \approx -\frac{(13.6 \text{ eV})(74 - 9)^2}{(3)^2} \approx -6.4 \times 10^3 \text{ eV}$$

Find the energy of the emitted x-ray photon:

$$hf = E_M - E_K \approx -6.4 \times 10^3 \text{ eV} - (-7.4 \times 10^4 \text{ eV})$$
$$\approx 6.8 \times 10^4 \text{ eV} = \boxed{68 \text{ keV}}$$

Consultation of x-ray tables shows that the M–K transition energies in tungsten vary from 66.9 keV to 67.7 keV, where the range of energies is due to slightly different energy values for states of different ℓ. Therefore, our estimate differs from the midpoint of this experimentally measured range by approximately 1%.

29.7 | Context Connection: Atoms in Space

We have spent quite a bit of time on the hydrogen atom in this chapter. Let us now consider hydrogen atoms located in space. Because hydrogen is the most abundant element in the Universe, its role in astronomy and cosmology is very important.

Let us begin by considering pictures of some nebulae you might have seen in an astronomy text, such as Figure 29.18. Time-exposure photographs of these objects show a variety of colors. What causes the colors in these clouds of gas and grains of dust? Let us imagine a cloud of hydrogen atoms in space near a very hot star. The high-energy photons from the star can interact with the hydrogen atoms, either raising them to a high-energy state or ionizing them. As the atoms fall back to the lower states, many atoms emit the Balmer series of wavelengths. Therefore, these atoms provide red, green, blue, and violet colors to the nebula, corresponding to the colors seen in the hydrogen spectrum in Chapter 11.

In practice, nebulae are classified into three groups depending on the transitions occurring in the hydrogen atoms. **Emission nebulae** (Fig. 29.18a) are near a hot star, so hydrogen atoms are excited by light from the star as described above. Therefore, the light from an emission nebula is dominated by discrete emission spectral lines and contains colors. **Reflection nebulae** (Fig. 29.18b) are near a cool star. In these cases, most of the light from the nebula is the starlight reflected from larger grains of material in the nebula rather than emitted by excited atoms. Therefore, the spectrum of the light from the nebula is the same as that from the star: an absorption

a, C. R. O'Dell (Rice University) and NASA;
b, © Science Photo Library / Alamy;
c, A. Caulet (ST-ECF, ESA) and NASA) 1007 29.16

Figure 29.18 Types of astronomical nebulae. (a) The central part of the Orion Nebula represents an emission nebula, from which colored light is emitted from atoms. (b) The Pleiades. The clouds of light surrounding the stars represent a reflection nebula, from which starlight is reflected by dust particles. (c) The Lagoon Nebula shows the effects of a dark nebula, in which clouds of dust block starlight and appear as a dark silhouette against the light from stars farther away.

spectrum with dark lines corresponding to atoms and ions in the outer regions of the star. The light from these nebulae tends to appear white. Finally, **dark nebulae** (Fig. 29.18c) are not close to a star. Therefore, little radiation is available to excite atoms or reflect from grains of dust. As a result, the material in these nebulae screens out light from stars beyond them, and they appear as black patches against the brightness of the more distant stars.

In addition to hydrogen, some other atoms and ions in space are raised to higher energy states by radiation from stars and proceed to emit various colors. Some of the more prominent colors are violet (373 nm) from the O^+ ion and green (496 nm and 501 nm) from the O^{++} ion. Helium and nitrogen also provide strong colors.

In our discussion of the quantum numbers for the hydrogen atom, we claimed that two states are possible in the $1s$ shell, corresponding to up or down spin, and that these two states are equivalent in energy in the absence of a magnetic field. If we modify our structural model to include the spin of the proton, however, we find that the two atomic states corresponding to the electron spin are not the same in energy. The state in which the electron and proton spins are parallel is slightly higher in energy than the state in which they are antiparallel. The energy difference is only 5.9×10^{-6} eV. Because these two states differ in energy, it is possible for the atom to make a transition between the states. If the transition is from the parallel state to the antiparallel state, a photon is emitted, with energy equal to the difference in energy between the states. The wavelength of this photon is

$$\lambda = \frac{c}{f} = \frac{hc}{hf} = \frac{hc}{E} = \frac{1\,240 \text{ eV} \cdot \text{nm}}{5.9 \times 10^{-6} \text{ eV}} \left(\frac{10^{-9} \text{ m}}{1 \text{ nm}} \right)$$

$$= 0.21 \text{ m} = 21 \text{ cm}$$

This radiation is called, for obvious reasons, **21-cm radiation.** It is radiation with a wavelength that is identifiable with the hydrogen atom. Therefore, by looking for this radiation in space, we can detect hydrogen atoms. Furthermore, if the wavelength of the observed radiation is not equal to 21 cm, we can infer that it has been Doppler shifted due to relative motion between the Earth and the source. This Doppler shift can then be used to measure the relative speed of the source toward or away from the Earth. This technique has been extensively used to study the hydrogen distribution in the Milky Way galaxy and to detect the presence of spiral arms in our galaxy, similar to the spiral arms in other galaxies.

Our study of atomic physics allows us to understand an important connection between the microscopic world of quantum physics and the macroscopic Universe. Atoms throughout the Universe act as transmitters of information to us about the local conditions. In Chapter 30, which deals with nuclear physics, we shall see how our understanding of microscopic processes helps us understand the local conditions at the center of a star.

SUMMARY

The methods of quantum mechanics can be applied to the hydrogen atom using the appropriate potential energy function $U(r) = -k_e e^2/r$ in the Schrödinger equation. The solution to this equation yields the wave functions for the allowed states and the allowed energies, given by

$$E_n = -\left(\frac{k_e e^2}{2a_0}\right)\frac{1}{n^2} = -\frac{13.606 \text{ eV}}{n^2} \qquad n = 1, 2, 3, \ldots \textbf{29.2} \blacktriangleleft$$

which is precisely the result obtained in the Bohr theory. The allowed energy depends only on the **principal quantum number** n. The allowed wave functions depend on three quantum numbers, n, ℓ, and m_ℓ, where ℓ is the **orbital quantum number** and m_ℓ is the **orbital magnetic quantum number.** The restrictions on the quantum numbers are as follows:

$$n = 1, 2, 3, \ldots$$
$$\ell = 0, 1, 2, \ldots, n - 1$$
$$m_\ell = -\ell, -\ell + 1, \ldots, \ell - 1, \ell$$

All states with the same principal quantum number n form a **shell,** identified by the letters K, L, M, . . . (corresponding to $n = 1, 2, 3, \ldots$). All states with the same values of both n and ℓ form a **subshell,** designated by the letters s, p, d, f, . . . (corresponding to $\ell = 0, 1, 2, 3, \ldots$).

An atom in a state characterized by a specific n can have the following values of **orbital angular momentum** L:

$$L = \sqrt{\ell(\ell + 1)}\hbar \qquad \ell = 0, 1, 2, \ldots, n - 1 \qquad \textbf{29.9} \blacktriangleleft$$

The allowed values of the projection of the angular momentum vector $\vec{\mathbf{L}}$ along the z axis are given by

$$L_z = m_\ell \hbar \qquad \textbf{29.10} \blacktriangleleft$$

where m_ℓ is restricted to integer values lying between $-\ell$ and ℓ. Only discrete values of L_z are allowed, and they are determined by the restrictions on m_ℓ. This quantization of L_z is referred to as **space quantization.**

To describe a quantum state of the hydrogen atom completely, it is necessary to include a fourth quantum number m_s, called the **spin magnetic quantum number.** This quantum number can have only two values, $\pm\frac{1}{2}$. In effect, this additional quantum number doubles the number of allowed states specified by the quantum numbers n, ℓ, and m_ℓ.

The electron has an intrinsic angular momentum called **spin angular momentum.** That is, the total angular momentum of an atom can have two contributions: one arising from the spin of the electron ($\vec{\mathbf{S}}$) and one arising from the orbital motion of the electron ($\vec{\mathbf{L}}$).

Electronic spin can be described by a quantum number $s = \frac{1}{2}$. The **magnitude of the spin angular momentum** is

$$S = \frac{\sqrt{3}}{2}\hbar \qquad \textbf{29.12} \blacktriangleleft$$

and the z component of $\vec{\mathbf{S}}$ is

$$S_z = m_s\hbar = \pm\tfrac{1}{2}\hbar \qquad \textbf{29.13} \blacktriangleleft$$

The magnetic moment $\vec{\boldsymbol{\mu}}_{\text{spin}}$ associated with the spin angular momentum of an electron is

$$\vec{\boldsymbol{\mu}}_{\text{spin}} = -\frac{e}{m_e}\vec{\mathbf{S}} \qquad \textbf{29.14} \blacktriangleleft$$

The z component of $\vec{\boldsymbol{\mu}}_{\text{spin}}$ can have the values

$$\mu_{\text{spin},z} = \pm\frac{e\hbar}{2m_e} \qquad \textbf{29.15} \blacktriangleleft$$

The quantity $e\hbar/2m_e$ is called the **Bohr magneton** μ_B and has the numerical value 9.274×10^{-24} J/T.

The **exclusion principle** states that no two electrons in an atom can have the same set of quantum numbers n, ℓ, m_ℓ, and m_s. Using this principle, one can determine the electronic configuration of the elements. This procedure serves as a basis for understanding atomic structure and the chemical properties of the elements.

The allowed electronic transitions between any two states in an atom are governed by the **selection rules**

$$\Delta\ell = \pm 1 \qquad \text{and} \qquad \Delta m_\ell = 0 \text{ or } \pm 1 \qquad \textbf{29.16} \blacktriangleleft$$

The **x-ray spectrum** of a metal target consists of a set of sharp characteristic lines superimposed on a broad, continuous spectrum. **Bremsstrahlung** is x-radiation with its origin in the slowing down of high-energy electrons as they encounter the target. **Characteristic x-rays** are emitted by atoms when an electron undergoes a transition from an outer shell into an electron vacancy in one of the inner shells.

OBJECTIVE QUESTIONS

☐ denotes answer available in *Student Solutions Manual/Study Guide*

1. Consider the $n = 3$ energy level in a hydrogen atom. How many electrons can be placed in this level? (a) 1 (b) 2 (c) 8 (d) 9 (e) 18

2. When an electron collides with an atom, it can transfer all or some of its energy to the atom. A hydrogen atom is in its ground state. Incident on the atom are several electrons, each having a kinetic energy of 10.5 eV. What is the result? (a) The atom can be excited to a higher allowed state. (b) The atom is ionized. (c) The electrons pass by the atom without interaction.

3. When an atom emits a photon, what happens? (a) One of its electrons leaves the atom. (b) The atom moves to a state of higher energy. (c) The atom moves to a state of lower energy. (d) One of its electrons collides with another particle. (e) None of those events occur.

4. The periodic table is based on which of the following principles? (a) The uncertainty principle. (b) All electrons in an atom must have the same set of quantum numbers. (c) Energy is conserved in all interactions. (d) All electrons in an atom are in orbitals having the same energy.

(e) No two electrons in an atom can have the same set of quantum numbers.

5. If an electron in an atom has the quantum numbers $n = 3$, $\ell = 2$, $m_\ell = 1$, and $m_s = \frac{1}{2}$, what state is it in? (a) $3s$ (b) $3p$ (c) $3d$ (d) $4d$ (e) $3f$

6. What can be concluded about a hydrogen atom with its electron in the d state? (a) The atom is ionized. (b) The orbital quantum number is $\ell = 1$. (c) The principal quantum number is $n = 2$. (d) The atom is in its ground state. (e) The orbital angular momentum of the atom is not zero.

7. Which of the following electronic configurations are *not* allowed for an atom? Choose all correct answers. (a) $2s^2 2p^6$ (b) $3s^2 3p^7$ (c) $3d^7 4s^2$ (d) $3d^{10} 4s^2 4p^6$ (e) $1s^2 2s^2 2d^1$

8. (a) In the hydrogen atom, can the quantum number n increase without limit? (b) Can the frequency of possible discrete lines in the spectrum of hydrogen increase without limit? (c) Can the wavelength of possible discrete lines in the spectrum of hydrogen increase without limit?

9. Consider the quantum numbers (a) n, (b) ℓ, (c) m_ℓ, and (d) m_s. (i) Which of these quantum numbers are fractional as opposed to being integers? (ii) Which can sometimes attain negative values? (iii) Which can be zero?

10. (i) What is the principal quantum number of the initial state of an atom as it emits an M_β line in an x-ray spectrum? (a) 1 (b) 2 (c) 3 (d) 4 (e) 5 (ii) What is the principal quantum number of the final state for this transition? Choose from the same possibilities as in part (i).

☐ denotes answer available in *Student Solutions Manual/Study Guide*

CONCEPTUAL QUESTIONS

1. Suppose the electron in the hydrogen atom obeyed classical mechanics rather than quantum mechanics. Why should a gas of such hypothetical atoms emit a continuous spectrum rather than the observed line spectrum?

2. Why are three quantum numbers needed to describe the state of a one-electron atom (ignoring spin)?

3. Compare the Bohr theory and the Schrödinger treatment of the hydrogen atom, specifically commenting on their treatment of total energy and orbital angular momentum of the atom.

4. (a) According to Bohr's model of the hydrogen atom, what is the uncertainty in the radial coordinate of the electron? (b) What is the uncertainty in the radial component of the velocity of the electron? (c) In what way does the model violate the uncertainty principle?

5. Could the Stern–Gerlach experiment be performed with ions rather than neutral atoms? Explain.

6. An energy of about 21 eV is required to excite an electron in a helium atom from the $1s$ state to the $2s$ state. The same transition for the He$^+$ ion requires approximately twice as much energy. Explain.

7. Why do lithium, potassium, and sodium exhibit similar chemical properties?

8. It is easy to understand how two electrons (one spin up, one spin down) fill the $n = 1$ or K shell for a helium atom. How is it possible that eight more electrons are allowed in the $n = 2$ shell, filling the K and L shells for a neon atom?

9. Why is a *nonuniform* magnetic field used in the Stern–Gerlach experiment?

10. Discuss some consequences of the exclusion principle.

PROBLEMS

ENHANCED WebAssign The problems found in this chapter may be assigned online in Enhanced WebAssign.

1. denotes straightforward problem; 2. denotes intermediate problem; 3. denotes challenging problem

1. denotes full solution available in the *Student Solutions Manual/Study Guide*

1. denotes problems most often assigned in Enhanced WebAssign.

BIO denotes biomedical problem

GP denotes guided problem

M denotes Master It tutorial available in Enhanced WebAssign

QC denotes asking for quantitative and conceptual reasoning

S denotes symbolic reasoning problem

shaded denotes "paired problems" that develop reasoning with symbols and numerical values

W denotes Watch It video solution available in Enhanced WebAssign

Section 29.1 Early Structural Models of the Atom

1. **Review.** In the Rutherford scattering experiment, 4.00-MeV alpha particles scatter off gold nuclei (containing 79 protons and 118 neutrons). Assume a particular alpha particle moves directly toward the gold nucleus and scatters backward at $180°$, and that the gold nucleus remains fixed throughout the entire process. Determine (a) the distance of closest approach of the alpha particle to the gold nucleus and (b) the maximum force exerted on the alpha particle.

2. According to classical physics, a charge e moving with an acceleration a radiates energy at a rate

$$\frac{dE}{dt} = -\frac{1}{6\pi\epsilon_0} \frac{e^2 a^2}{c^3}$$

(a) Show that an electron in a classical hydrogen atom (see Fig. 29.3) spirals into the nucleus at a rate

$$\frac{dr}{dt} = -\frac{e^4}{12\pi^2 \epsilon_0^2 m_e^2 c^3} \left(\frac{1}{r^2}\right)$$

0 to 4.00 in steps of 0.250. (c) Find the value of r for which the probability of finding the electron outside a sphere of radius r is equal to the probability of finding the electron inside this sphere. You must solve a transcendental equation numerically, and your graph is a good starting point.

53. The force on a magnetic moment μ_z in a nonuniform magnetic field B_z is given by $F_z = \mu_z(dB_z/dz)$. If a beam of silver atoms travels a horizontal distance of 1.00 m through such a field and each atom has a speed of 100 m/s, how strong must be the field gradient dB_z/dz to deflect the beam 1.00 mm?

54. [S] Assume three identical uncharged particles of mass m and spin $\frac{1}{2}$ are contained in a one-dimensional box of length L. What is the ground-state energy of this system?

55. Review. (a) Is the mass of a hydrogen atom in its ground state larger or smaller than the sum of the masses of a proton and an electron? (b) What is the mass difference? (c) How large is the difference as a percentage of the total mass? (d) Is it large enough to affect the value of the atomic mass listed to six decimal places in Table A.3 in Appendix A?

56. [GP] [S] We wish to show that the most probable radial position for an electron in the 2s state of hydrogen is $r = 5.236a_0$. (a) Use Equations 29.6 and 29.8 to find the radial probability density for the 2s state of hydrogen. (b) Calculate the derivative of the radial probability density with respect to r. (c) Set the derivative in part (b) equal to zero and identify three values of r that represent minima in the function. (d) Find two values of r that represent maxima in the function. (e) Identify which of the values in part (c) represents the highest probability.

57. For hydrogen in the 1s state, what is the probability of finding the electron farther than $2.50a_0$ from the nucleus?

58. [S] For hydrogen in the 1s state, what is the probability of finding the electron farther than βa_0 from the nucleus, where β is an arbitrary number?

59. [W] **Review.** (a) How much energy is required to cause an electron in hydrogen to move from the $n = 1$ state to the $n = 2$ state? (b) Suppose the atom gains this energy through collisions among hydrogen atoms at a high temperature. At what temperature would the average atomic kinetic energy $\frac{3}{2}k_BT$ be great enough to excite the electron? Here k_B is Boltzmann's constant.

60. [Q][C] All atoms have the same size, to an order of magnitude. (a) To demonstrate this fact, estimate the atomic diameters for aluminum (with molar mass 27.0 g/mol and density 2.70 g/cm^3) and uranium (molar mass 238 g/mol and density 18.9 g/cm^3). (b) What do the results of part (a) imply about the wave functions for inner-shell electrons as we progress to higher and higher atomic mass atoms?

61. [M] In the technique known as electron spin resonance (ESR), a sample containing unpaired electrons is placed in a magnetic field. Consider a situation in which a single electron (*not* contained in an atom) is immersed in a magnetic field. In this simple situation, only two energy states are possible, corresponding to $m_s = \pm\frac{1}{2}$. In ESR, the absorption of a photon causes the electron's spin magnetic moment to flip from the lower energy state to the higher energy state. According to the result of Problem 24 in Chapter 22, the change in energy is $2\mu_BB$. (The lower energy state corresponds to the case in which the z component of the magnetic moment $\vec{\mu}_{spin}$ is aligned with the magnetic field, and the higher energy state corresponds to the case in which the z component of $\vec{\mu}_{spin}$ is aligned opposite to the field.) What is the photon frequency required to excite an ESR transition in a 0.350-T magnetic field?

62. [S] Show that the wave function for a hydrogen atom in the 2s state

$$\psi_{2s}(r) = \frac{1}{4\sqrt{2\pi}}\left(\frac{1}{a_0}\right)^{3/2}\left(2 - \frac{r}{a_0}\right)e^{-r/2a_0}$$

satisfies the spherically symmetric Schrödinger equation given in Problem 16.

63. Find the average (expectation) value of $1/r$ in the 1s state of hydrogen. Note that the general expression is given by

$$\langle 1/r \rangle = \int_{\text{all space}} |\psi|^2 (1/r)\, dV = \int_0^\infty P(r)\,(1/r)\, dr$$

Is the result equal to the inverse of the average value of r?

64. *Why is the following situation impossible?* An experiment is performed on an atom. Measurements of the atom when it is in a particular excited state show five possible values of the z component of orbital angular momentum, ranging between 3.16×10^{-34} kg · m^2/s and -3.16×10^{-34} kg · m^2/s.

65. An electron in chromium moves from the $n = 2$ state to the $n = 1$ state without emitting a photon. Instead, the excess energy is transferred to an outer electron (one in the $n = 4$ state), which is then ejected by the atom. In this Auger (pronounced "ohjay") process, the ejected electron is referred to as an Auger electron. Use the Bohr theory to find the kinetic energy of the Auger electron.

66. Review. Steven Chu, Claude Cohen-Tannoudji, and William Phillips received the 1997 Nobel Prize in Physics for "the development of methods to cool and trap atoms with laser light." One part of their work was with a beam of atoms (mass $\sim 10^{-25}$ kg) that move at a speed on the order of 1 km/s, similar to the speed of molecules in air at room temperature. An intense laser light beam tuned to a visible atomic transition (assume 500 nm) is directed straight into the atomic beam; that is, the atomic beam and the light beam are traveling in opposite directions. An atom in the ground state immediately absorbs a photon. Total system momentum is conserved in the absorption process. After a lifetime on the order of 10^{-8} s, the excited atom radiates by spontaneous emission. It has an equal probability of emitting a photon in any direction. Therefore, the average "recoil" of the atom is zero over many absorption and emission cycles. (a) Estimate the average deceleration of the atomic beam. (b) What is the order of magnitude of the distance over which the atoms in the beam are brought to a halt?

Nuclear Physics

© Vienna Report Agency/Sygma/Corbis

In 1896, the year that marked the birth of nuclear physics, Antoine-Henri Becquerel (1852–1908) introduced the world of science to radioactivity in uranium compounds by accidentally discovering that uranyl potassium sulfate crystals emit an invisible radiation that can darken a photographic plate when the plate is covered to exclude light. After a series of experiments, he concluded that the radiation emitted by the crystals was of a new type, one that requires no external stimulation and is so penetrating that it can darken protected photographic plates and ionize gases.

A great deal of research followed as scientists attempted to understand the radiation emitted by radioactive nuclei. Pioneering work by Rutherford showed that the radiation was of three types, which he called alpha, beta, and gamma rays. Later experiments showed that alpha rays are helium nuclei, beta rays are electrons or related particles called positrons, and gamma rays are high-energy photons.

As we saw in Section 29.1, the 1911 experiments of Rutherford established that the nucleus of an atom has a very small volume and that most of the atomic mass is contained in the nucleus. Furthermore, such studies demonstrated a new type of force, the nuclear force, first introduced in Section 5.5, that is predominant at distances on the order of 10^{-15} m and essentially zero at distances larger than that.

In this chapter, we discuss the structure of the atomic nucleus. We shall describe the basic properties of nuclei, nuclear forces, nuclear binding energy, the phenomenon of radioactivity, and nuclear reactions.

Ötzi the Iceman, a Copper Age man, was discovered by German tourists in the Italian Alps in 1991 when a glacier melted enough to expose his remains. Analysis of his corpse has exposed his last meal, illnesses he suffered, and places he lived. Radioactivity was used to determine that he lived in about 3300 BC. Ötzi can be seen today in the Südtiroler Archäologiemuseum (South Tyrol Museum of Archaeology) in Bolzano, Italy.

nuclei. This process of **fusion** is made difficult because the nuclei must overcome a very strong Coulomb repulsion before they become close enough together to fuse. One way to assist the nuclei in overcoming this repulsion is to cause them to move with very high kinetic energy by raising the system of nuclei to a very high temperature. If the density of nuclei is high also, the probability of nuclei colliding is high and fusion can occur. The technological problem of creating very high temperatures and densities is a major challenge in the area of Earth-based controlled fusion research.

At some natural locations (e.g., the cores of stars), the necessary high temperatures and densities exist. Consider a collection of gas and dust somewhere in the Universe to be an isolated system. What happens as this system collapses under its own gravitational attraction? Energy of the system is conserved, and the gravitational potential energy associated with the separated particles decreases while the kinetic energy of the particles increases. As the falling particles collide with the particles that have already fallen into the central region of collapse, their kinetic energy is distributed to the other particles by collisions and randomized; it becomes internal energy, which is related to the temperature of the collection of particles.

If the temperature and density of the system's core rise to the point where fusion can occur, the system becomes a star. The primary constituent of the Universe is hydrogen, so the fusion reaction at the center of a star combines hydrogen nuclei—protons—into helium nuclei. A common reaction process for stars with relatively cool cores ($T < 15 \times 10^6$ K) is the **proton–proton cycle.** In the first step of the process, two protons combine to form deuterium:

$$\,^1_1\text{H} + \,^1_1\text{H} \;\rightarrow\; \,^2_1\text{H} + e^+ + \nu$$

Notice the implicit ^2_2He nucleus that is formed but that does not appear in the reaction equation. This nucleus is highly unstable and decays very rapidly by beta-plus decay to the deuterium nucleus, a positron, and a neutrino.

In the next step, the deuterium nucleus undergoes fusion with another proton to form a helium-3 nucleus:

$$\,^1_1\text{H} + \,^2_1\text{H} \;\rightarrow\; \,^3_2\text{He} + \gamma$$

Finally, two helium-3 nuclei formed in such reactions can fuse to form helium-4 and two protons:

$$\,^3_2\text{He} + \,^3_2\text{He} \;\rightarrow\; \,^4_2\text{He} + \,^1_1\text{H} + \,^1_1\text{H}$$

The net result of this cycle has been the joining of four protons to form a helium-4 nucleus, with the release of energy that eventually leaves the star as electromagnetic radiation from its surface. In addition, notice that the reaction releases neutrinos, which serve as a signal for beta decay occurring within the star. The observation of increased neutrino flow from a supernova is an important tool in analyzing the event.

For stars with hotter cores ($T > 15 \times 10^6$ K), another process, called the **carbon cycle,** dominates. At such high temperatures, hydrogen nuclei can fuse into nuclei heavier than helium such as carbon. In the first of six steps in the cycle, a carbon nucleus fuses with a proton to form nitrogen:

$$\,^1_1\text{H} + \,^{12}_6\text{C} \;\rightarrow\; \,^{13}_7\text{N}$$

The nitrogen nucleus is proton-rich and undergoes beta-plus decay:

$$\,^{13}_7\text{N} \;\rightarrow\; \,^{13}_6\text{C} + e^+ + \nu$$

The resulting carbon-13 nucleus fuses with another proton, with the emission of a gamma ray:

$$\,^1_1\text{H} + \,^{13}_6\text{C} \;\rightarrow\; \,^{14}_7\text{N} + \gamma$$

The nitrogen-14 fuses with another proton, with more gamma emission:

$$\,^1_1\text{H} + \,^{14}_7\text{N} \;\rightarrow\; \,^{15}_8\text{O} + \gamma$$

The oxygen nucleus undergoes beta-plus decay:

$$\,^{15}_8\text{O} \;\rightarrow\; \,^{15}_7\text{N} + e^+ + \nu$$

Finally, the nitrogen-15 fuses with another proton:

$$_1^1\text{H} + _7^{15}\text{N} \rightarrow _6^{12}\text{C} + _2^4\text{He}$$

Notice that the net effect of this process is to combine four protons into a helium nucleus, just like the proton–proton cycle. The carbon-12 with which we began the process is returned at the end, so it acts only as a catalyst to the process and is not consumed.

Depending on its mass, a star transforms energy in its core at a rate between 10^{23} and 10^{33} W. The energy transformed from the rest energy of the nuclei in the core is transferred outward through the surrounding layers by matter transfer in two forms. First, neutrinos carry energy directly through these layers to space because these particles interact only weakly with matter. Second, energy carried by photons from the core is absorbed by the gases in layers outside the core and slowly works its way to the surface by convection. This energy is eventually radiated from the surface of the star by electromagnetic radiation, mostly in the infrared, visible, and ultraviolet regions of the electromagnetic spectrum. The weight of the layers outside the core keeps the core from exploding. The whole system of a star is stable as long as the supply of hydrogen in the core lasts.

In the previous chapters, we presented examples of the applications of quantum physics and atomic physics to processes in space. In this chapter, we have seen that nuclear processes also have an important role in the cosmos. The formation of stars is a critical process in the development of the Universe. The energy provided by stars is crucial to life on planets such as the Earth. In our next, and final, chapter, we shall discuss the processes that occur on an even smaller scale, the scale of *elementary particles*. We shall find again that looking at a smaller scale allows us to advance our understanding of the largest scale system, the Universe.

SUMMARY

A nuclide can be represented by $_Z^A\text{X}$, where A is the **mass number,** the total number of nucleons, and Z is the **atomic number,** the total number of protons. The total number of neutrons in a nucleus is the **neutron number** N, where $A = N + Z$. Elements with the same Z but different A and N values are called **isotopes.**

Assuming that a nucleus is spherical, its radius is

$$r = aA^{1/3} \qquad \textbf{30.1} \blacktriangleleft$$

where $a = 1.2$ fm.

Nuclei are stable because of the **nuclear force** between nucleons. This short-range force dominates the Coulomb repulsive force at distances of less than about 2 fm and is independent of charge.

Light nuclei are most stable when the number of protons equals the number of neutrons. Heavy nuclei are most stable when the number of neutrons exceeds the number of protons. In addition, many stable nuclei have Z and N values that are both even. Nuclei with unusually high stability have Z or N values of 2, 8, 20, 28, 50, 82, and 126, called **magic numbers.**

Nuclei have an intrinsic angular momentum (spin) of magnitude $\sqrt{I(I + 1)}\,\hbar$, where I is the **nuclear spin quantum number.** The magnetic moment of a nucleus is measured in terms of the **nuclear magneton** μ_n, where

$$\mu_n \equiv \frac{e\hbar}{2m_p} = 5.05 \times 10^{-27}\,\text{J/T} \qquad \textbf{30.3} \blacktriangleleft$$

The difference in mass between the separate nucleons and the nucleus containing these nucleons, when multiplied by c^2, gives the **binding energy** E_b of the nucleus. We can calculate the binding energy of any nucleus $_Z^A\text{X}$ using the expression

$$E_b(\text{MeV}) = [ZM(\text{H}) + Nm_n - M(_Z^A\text{X})] \times 931.494\,\text{MeV/u} \qquad \textbf{30.4} \blacktriangleleft$$

Radioactive processes include alpha decay, beta decay, and gamma decay. An alpha particle is a ^4He nucleus, a beta particle is either an electron (e^-) or a positron (e^+), and a gamma particle is a high-energy photon.

If a radioactive material contains N_0 radioactive nuclei at $t = 0$, the number N of nuclei remaining at time t is

$$N = N_0 e^{-\lambda t} \qquad \textbf{30.6} \blacktriangleleft$$

where λ is the **decay constant,** or **disintegration constant.** The **decay rate,** or **activity,** of a radioactive substance is given by

$$R = \left|\frac{dN}{dt}\right| = N_0\lambda e^{-\lambda t} = R_0 e^{-\lambda t} \qquad \textbf{30.7} \blacktriangleleft$$

where $R_0 = N_0\lambda$ is the activity at $t = 0$. The **half-life** $T_{1/2}$ is defined as the time interval required for half of a given number of radioactive nuclei to decay, where

$$T_{1/2} = \frac{\ln 2}{\lambda} = \frac{0.693}{\lambda} \qquad \textbf{30.8} \blacktriangleleft$$

Alpha decay can occur because according to quantum mechanics some nuclei have barriers that can be penetrated by the alpha particles (the tunneling process). This process is energetically more favorable for those nuclei having large excesses of neutrons. A nucleus can undergo **beta decay** in two ways. It can emit either an electron (e^-) and an antineutrino ($\bar{\nu}$) or a positron (e^+) and a neutrino (ν). In the **electron capture** process, the nucleus of an atom absorbs one of its own electrons (usually from the K shell) and emits a neutrino. In **gamma decay,** a nucleus in an excited state decays to its ground state and emits a gamma ray.

Nuclear reactions can occur when a target nucleus X is bombarded by a particle a, resulting in a nucleus Y and an outgoing particle b:

$$a + X \rightarrow Y + b \quad \text{or} \quad X(a, b)Y \qquad \textbf{30.24}\blacktriangleleft$$

The rest energy transformed to kinetic energy in such a reaction, called the **reaction energy** Q, is

$$Q = (M_a + M_X - M_Y - M_b)c^2 \qquad \textbf{30.25}\blacktriangleleft$$

A reaction for which Q is positive is called **exothermic.** A reaction for which Q is negative is called **endothermic.** The minimum kinetic energy of the incoming particle necessary for such a reaction to occur is called the **threshold energy.**

OBJECTIVE QUESTIONS

1. When $^{32}_{15}$P decays to $^{32}_{16}$S, which of the following particles is emitted? (a) a proton (b) an alpha particle (c) an electron (d) a gamma ray (e) an antineutrino

2. Two samples of the same radioactive nuclide are prepared. Sample G has twice the initial activity of sample H. **(i)** How does the half-life of G compare with the half-life of H? (a) It is two times larger. (b) It is the same. (c) It is half as large. **(ii)** After each has passed through five half-lives, how do their activities compare? (a) G has more than twice the activity of H. (b) G has twice the activity of H. (c) G and H have the same activity. (d) G has lower activity than H.

3. If a radioactive nuclide ^A_ZX decays by emitting a gamma ray, what happens? (a) The resulting nuclide has a different Z value. (b) The resulting nuclide has the same A and Z values. (c) The resulting nuclide has a different A value. (d) Both A and Z decrease by one. (e) None of those statements is correct.

4. Does a nucleus designated as $^{40}_{18}$X contain (a) 20 neutrons and 20 protons, (b) 22 protons and 18 neutrons, (c) 18 protons and 22 neutrons, (d) 18 protons and 40 neutrons, or (e) 40 protons and 18 neutrons?

5. In the decay $^{234}_{90}$Th $\rightarrow$ ^A_ZRa $+$ $^4_2$He, identify the mass number and the atomic number of the Ra nucleus: (a) $A = 230$, $Z = 92$ (b) $A = 238$, $Z = 88$ (c) $A = 230$, $Z = 88$ (d) $A = 234$, $Z = 88$ (e) $A = 238$, $Z = 86$

6. When $^{144}_{60}$Nd decays to $^{140}_{58}$Ce, identify the particle that is released. (a) a proton (b) an alpha particle (c) an electron (d) a neutron (e) a neutrino

7. When the $^{95}_{36}$Kr nucleus undergoes beta decay by emitting an electron and an antineutrino, does the daughter nucleus (Rb) contain (a) 58 neutrons and 37 protons, (b) 58 protons and 37 neutrons, (c) 54 neutrons and 41 protons, or (d) 55 neutrons and 40 protons?

8. What is the Q value for the reaction ^{9}Be $+ \alpha \rightarrow$ ^{12}C $+$ n? (a) 8.4 MeV (b) 7.3 MeV (c) 6.2 MeV (d) 5.7 MeV (e) 4.2 MeV

9. The half-life of radium-224 is about 3.6 days. What approximate fraction of a sample remains undecayed after two weeks? (a) $\frac{1}{2}$ (b) $\frac{1}{4}$ (c) $\frac{1}{8}$ (d) $\frac{1}{16}$ (e) $\frac{1}{32}$

10. A free neutron has a half-life of 614 s. It undergoes beta decay by emitting an electron. Can a free proton undergo a similar decay? (a) yes, the same decay (b) yes, but by emitting a positron (c) yes, but with a very different half-life (d) no

11. Which of the following quantities represents the reaction energy of a nuclear reaction? (a) (final mass $-$ initial mass)$/c^2$ (b) (initial mass $-$ final mass)$/c^2$ (c) (final mass $-$ initial mass)c^2 (d) (initial mass $-$ final mass)c^2 (e) none of those quantities

12. In the first nuclear weapon test carried out in New Mexico, the energy released was equivalent to approximately 17 kilotons of TNT. Estimate the mass decrease in the nuclear fuel representing the energy converted from rest energy into other forms in this event. *Note:* One ton of TNT has the energy equivalent of 4.2×10^9 J. (a) 1 μg (b) 1 mg (c) 1 g (d) 1 kg (e) 20 kg

CONCEPTUAL QUESTIONS

1. A student claims that a heavy form of hydrogen decays by alpha emission. How do you respond?

2. In beta decay, the energy of the electron or positron emitted from the nucleus lies somewhere in a relatively large range of possibilities. In alpha decay, however, the alpha-particle energy can only have discrete values. Explain this difference.

3. In Rutherford's experiment, assume an alpha particle is headed directly toward the nucleus of an atom. Why doesn't the alpha particle make physical contact with the nucleus?

4. Explain why nuclei that are well off the line of stability in Figure 30.4 tend to be unstable.

5. Compare and contrast the properties of a photon and a neutrino.

6. Why do nearly all the naturally occurring isotopes lie above the $N = Z$ line in Figure 30.4?

7. Why are very heavy nuclei unstable?

8. "If no more people were to be born, the law of population growth would strongly resemble the radioactive decay law." Discuss this statement.

9. Can carbon-14 dating be used to measure the age of a rock? Explain.

10. In positron decay, a proton in the nucleus becomes a neutron and its positive charge is carried away by the positron. A neutron, though, has a larger rest energy than a proton. How is that possible?

11. Consider two heavy nuclei X and Y having similar mass numbers. If X has the higher binding energy, which nucleus tends to be more unstable? Explain your answer.

12. What fraction of a radioactive sample has decayed after two half-lives have elapsed?

13. Figure CQ30.13 shows a watch from the early 20th century. The numbers and the hands of the watch are painted with a paint that contains a small amount of natural radium $^{226}_{88}\text{Ra}$ mixed with a phosphorescent material. The decay of the radium causes the phosphorescent material to glow continuously. The radioactive nuclide $^{226}_{88}\text{Ra}$ has a half-life of approximately 1.60×10^3 years. Being that the solar system is approximately 5 billion years old, why was this isotope still available in the 20th century for use on this watch?

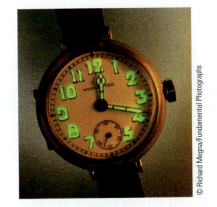

Figure CQ30.13

14. Suppose it could be shown that the cosmic-ray intensity at the Earth's surface was much greater 10 000 years ago. How would this difference affect what we accept as valid carbon-dated values of the age of ancient samples of once-living matter? Explain your answer.

15. (a) How many values of I_z are possible for $I = \frac{5}{2}$? (b) For $I = 3$?

16. Can a nucleus emit alpha particles that have different energies? Explain.

17. If a nucleus such as ^{226}Ra initially at rest undergoes alpha decay, which has more kinetic energy after the decay, the alpha particle or the daughter nucleus? Explain your answer.

▶ PROBLEMS

WebAssign The problems found in this chapter may be assigned online in Enhanced WebAssign.

1. denotes straightforward problem; 2. denotes intermediate problem; 3. denotes challenging problem

1. denotes full solution available in the *Student Solutions Manual/ Study Guide*

1. denotes problems most often assigned in Enhanced WebAssign.

BIO denotes biomedical problem

GP denotes guided problem

M denotes Master It tutorial available in Enhanced WebAssign

Q|C denotes asking for quantitative and conceptual reasoning

S denotes symbolic reasoning problem

shaded denotes "paired problems" that develop reasoning with symbols and numerical values

W denotes Watch It video solution available in Enhanced WebAssign

Note: Atomic masses are listed in Table A.3 in Appendix A.

Section 30.1 Some Properties of Nuclei

1. **M** (a) Use energy methods to calculate the distance of closest approach for a head-on collision between an alpha particle having an initial energy of 0.500 MeV and a gold nucleus (^{197}Au) at rest. Assume the gold nucleus remains at rest during the collision. (b) What minimum initial speed must the alpha particle have to approach as close as 300 fm to the gold nucleus?

2. (a) What is the order of magnitude of the number of protons in your body? (b) Of the number of neutrons? (c) Of the number of electrons?

3. **Review.** Singly ionized carbon is accelerated through 1 000 V and passed into a mass spectrometer to determine the isotopes present (see Chapter 22). The magnitude of the magnetic field in the spectrometer is 0.200 T. The orbit radius for a ^{12}C isotope as it passes through the field is $r = 7.89$ cm. Find the radius of the orbit of a ^{13}C isotope.

4. **S Review.** Singly ionized carbon is accelerated through a potential difference ΔV and passed into a mass spectrometer to determine the isotopes present (see Chapter 22). The magnitude of the magnetic field in the spectrometer is B. The orbit radius for an isotope of mass m_1 as it passes through the field is r_1. Find the radius of the orbit of an isotope of mass m_2.

5. Find the radius of (a) a nucleus of $^4_2$He and (b) a nucleus of $^{238}_{92}$U.

6. **QC** In a Rutherford scattering experiment, alpha particles having kinetic energy of 7.70 MeV are fired toward a gold nucleus that remains at rest during the collision. The alpha particles come as close as 29.5 fm to the gold nucleus before turning around. (a) Calculate the de Broglie wavelength for the 7.70-MeV alpha particle and compare it with the distance of closest approach, 29.5 fm. (b) Based on this comparison, why is it proper to treat the alpha particle as a particle and not as a wave in the Rutherford scattering experiment?

7. A star ending its life with a mass of four to eight times the Sun's mass is expected to collapse and then undergo a supernova event. In the remnant that is not carried away by the supernova explosion, protons and electrons combine to form a neutron star with approximately twice the mass of the Sun. Such a star can be thought of as a gigantic atomic nucleus. Assume $r = aA^{1/3}$ (Eq. 30.1). If a star of mass 3.98×10^{30} kg is composed entirely of neutrons ($m_n = 1.67 \times 10^{-27}$ kg), what would its radius be?

8. **Review.** Two golf balls each have a 4.30-cm diameter and are 1.00 m apart. What would be the gravitational force exerted by each ball on the other if the balls were made of nuclear matter?

9. The radio frequency at which a nucleus having a magnetic moment of magnitude μ displays resonance absorption between spin states is called the Larmor frequency and is given by

$$f = \frac{\Delta E}{h} = \frac{2\mu B}{h}$$

Calculate the Larmor frequency for (a) free neutrons in a magnetic field of 1.00 T, (b) free protons in a magnetic field of 1.00 T, and (c) free protons in the Earth's magnetic field at a location where the magnitude of the field is 50.0 μT.

Section 30.2 Nuclear Binding Energy

10. Using the graph in Figure 30.6, estimate how much energy is released when a nucleus of mass number 200 fissions into two nuclei each of mass number 100.

11. Calculate the binding energy per nucleon for (a) ^{2}H, (b) ^{4}He, (c) ^{56}Fe, and (d) ^{238}U.

12. A pair of nuclei for which $Z_1 = N_2$ and $Z_2 = N_1$ are called *mirror isobars* (the atomic and neutron numbers are interchanged). Binding-energy measurements on these nuclei can be used to obtain evidence of the charge independence of nuclear forces (that is, proton–proton, proton–neutron, and neutron–neutron nuclear forces are equal). Calculate the difference in binding energy for the two mirror isobars $^{15}_8$O and $^{15}_7$N. The electric repulsion among eight protons rather than seven accounts for the difference.

13. Nuclei having the same mass numbers are called *isobars*. The isotope $^{139}_{57}$La is stable. A radioactive isobar, $^{139}_{59}$Pr, is located below the line of stable nuclei as shown in Figure P30.13 and decays by e^+ emission. Another radioactive isobar of $^{139}_{57}$La, $^{139}_{55}$Cs, decays by e^- emission and is located above the line of stable nuclei in Figure P30.13. (a) Which of these three isobars has the highest neutron-to-proton ratio? (b) Which

has the greatest binding energy per nucleon? (c) Which do you expect to be heavier, $^{139}_{59}$Pr, or $^{139}_{55}$Cs?

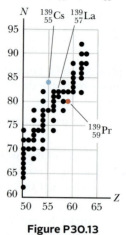

Figure P30.13

Section 30.3 Radioactivity

14. The half-life of ^{131}I is 8.04 days. On a certain day, the activity of an iodine-131 sample is 6.40 mCi. What is its activity 40.2 days later?

15. **M** The radioactive isotope ^{198}Au has a half-life of 64.8 h. A sample containing this isotope has an initial activity ($t = 0$) of 40.0 μCi. Calculate the number of nuclei that decay in the time interval between $t_1 = 10.0$ h and $t_2 = 12.0$ h.

16. **S** A radioactive nucleus has half-life $T_{1/2}$. A sample containing these nuclei has initial activity R_0 at $t = 0$. Calculate the number of nuclei that decay during the interval between the later times t_1 and t_2.

17. A sample of radioactive material contains 1.00×10^{15} atoms and has an activity of 6.00×10^{11} Bq. What is its half-life?

18. **M** A freshly prepared sample of a certain radioactive isotope has an activity of 10.0 mCi. After 4.00 h, its activity is 8.00 mCi. Find (a) the decay constant and (b) the half-life. (c) How many atoms of the isotope were contained in the freshly prepared sample? (d) What is the sample's activity 30.0 h after it is prepared?

19. What time interval elapses while 90.0% of the radioactivity of a sample of $^{72}_{33}$As disappears as measured by its activity? The half-life of $^{72}_{33}$As is 26 h.

20. **QC** (a) The daughter nucleus formed in radioactive decay is often radioactive. Let N_{10} represent the number of parent nuclei at time $t = 0$, $N_1(t)$ the number of parent nuclei at time t, and λ_1 the decay constant of the parent. Suppose the number of daughter nuclei at time $t = 0$ is zero. Let $N_2(t)$ be the number of daughter nuclei at time t, and let λ_2 be the decay constant of the daughter. Show that $N_2(t)$ satisfies the differential equation

$$\frac{dN_2}{dt} = \lambda_1 N_1 - \lambda_2 N_2$$

(b) Verify by substitution that this differential equation has the solution

$$N_2(t) = \frac{N_{10}\lambda_1}{\lambda_1 - \lambda_2}(e^{-\lambda_2 t} - e^{-\lambda_1 t})$$

This equation is the law of successive radioactive decays. (c) ^{218}Po decays into ^{214}Pb with a half-life of 3.10 min, and

^{214}Pb decays into ^{214}Bi with a half-life of 26.8 min. On the same axes, plot graphs of $N_1(t)$ for ^{218}Po and $N_2(t)$ for ^{214}Pb. Let $N_{10} = 1\,000$ nuclei, and choose values of t from 0 to 36 min in 2-min intervals. (d) The curve for ^{214}Pb obtained in part (c) at first rises to a maximum and then starts to decay. At what instant t_m is the number of ^{214}Pb nuclei a maximum? (e) By applying the condition for a maximum $dN_2/dt = 0$, derive a symbolic equation for t_m in terms of λ_1 and λ_2. (f) Explain whether the value obtained in part (c) agrees with this equation.

21. **BIO** In an experiment on the transport of nutrients in a plant's root structure, two radioactive nuclides X and Y are used. Initially, 2.50 times more nuclei of type X are present than of type Y. At a time 3.00 d later, there are 4.20 times more nuclei of type X than of type Y. Isotope Y has a half-life of 1.60 d. What is the half-life of isotope X?

22. **BIO** **S** In an experiment on the transport of nutrients in a plant's root structure, two radioactive nuclides X and Y are used. Initially, the ratio of the number of nuclei of type X present to that of type Y is r_1. After a time interval Δt, the ratio of the number of nuclei of type X present to that of type Y is r_2. Isotope Y has a half-life of T_Y. What is the half-life of isotope X?

Section 30.4 The Radioactive Decay Processes

23. Find the energy released in the alpha decay

$$^{238}_{92}U \rightarrow {}^{234}_{90}Th + {}^{4}_{2}He$$

24. **BIO** **GP** **W** A living specimen in equilibrium with the atmosphere contains one atom of ^{14}C (half-life = 5 730 yr) for every 7.70×10^{11} stable carbon atoms. An archeological sample of wood (cellulose, $C_{19}H_{22}O_{11}$) contains 21.0 mg of carbon. When the sample is placed inside a shielded beta counter with 88.0% counting efficiency, 837 counts are accumulated in one week. We wish to find the age of the sample. (a) Find the number of carbon atoms in the sample. (b) Find the number of carbon-14 atoms in the sample. (c) Find the decay constant for carbon-14 in inverse seconds. (d) Find the initial number of decays per week just after the specimen died. (e) Find the corrected number of decays per week from the current sample. (f) From the answers to parts (d) and (e), find the time interval in years since the specimen died.

25. The nucleus $^{15}_{8}O$ decays by electron capture. The nuclear reaction is written

$$^{15}_{8}O + e^- \rightarrow {}^{15}_{7}N + \nu$$

(a) Write the process going on for a single particle within the nucleus. (b) Disregarding the daughter's recoil, determine the energy of the neutrino.

26. Identify the unknown nuclide or particle (X).
(a) $X \rightarrow {}^{65}_{28}Ni + \gamma$
(b) $^{215}_{84}Po \rightarrow X + \alpha$
(c) $X \rightarrow {}^{55}_{26}Fe + e^+ + \nu$

27. Enter the correct nuclide symbol in each open tan rectangle in Figure P30.27, which shows the sequences of decays in the natural radioactive series starting with the long-lived isotope uranium-235 and ending with the stable nucleus lead-207.

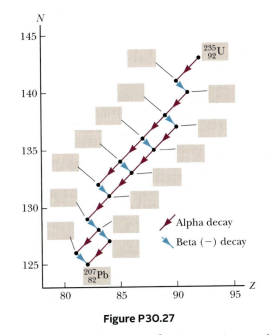

Figure P30.27

28. A ^{3}H nucleus beta decays into ^{3}He by creating an electron and an antineutrino according to the reaction

$$^{3}_{1}H \rightarrow {}^{3}_{2}He + e^- + \bar{\nu}$$

Determine the total energy released in this decay.

29. **BIO** *Indoor air pollution.* Uranium is naturally present in rock and soil. At one step in its series of radioactive decays, ^{238}U produces the chemically inert gas radon-222, with a half-life of 3.82 days. The radon seeps out of the ground to mix into the atmosphere, typically making open air radioactive with activity 0.3 pCi/L. In homes, ^{222}Rn can be a serious pollutant, accumulating to reach much higher activities in enclosed spaces. If the radon activity exceeds 4 pCi/L, the Environmental Protection Agency suggests taking action to reduce it, such as by reducing infiltration of air from the ground. (a) Convert the activity 4 pCi/L to units of becquerel per cubic meter. (b) How many ^{222}Rn atoms are in one cubic meter of air displaying this activity? (c) What fraction of the mass of the air does the radon constitute?

Section 30.5 Nuclear Reactions

Note: Problem 73 in Chapter 20 can be assigned with this section.

30. The following fission reaction is typical of those occurring in a nuclear electric generating station:

$$^{1}_{0}n + {}^{235}_{92}U \rightarrow {}^{141}_{56}Ba + {}^{92}_{36}Kr + 3({}^{1}_{0}n)$$

(a) Find the energy released in the reaction. The masses of the products are 140.914 411 u for $^{141}_{56}Ba$ and 91.926 156 u for $^{92}_{36}Kr$. (b) What fraction of the initial rest energy of the system is transformed to other forms?

31. Identify the unknown nuclides and particles X and X' in the nuclear reactions
(a) $X + {}^{4}_{2}He \rightarrow {}^{24}_{12}Mg + {}^{1}_{0}n$
(b) $^{235}_{92}U + {}^{1}_{0}n \rightarrow {}^{90}_{38}Sr + X + 2({}^{1}_{0}n)$
(c) $2({}^{1}_{1}H) \rightarrow {}^{2}_{1}H + X + X'$

32. Natural gold has only one isotope, $^{197}_{79}Au$. If natural gold is irradiated by a flux of slow neutrons, electrons are emitted.

(a) Write the reaction equation. (b) Calculate the maximum energy of the emitted electrons.

33. **W** A beam of 6.61-MeV protons is incident on a target of $^{27}_{13}$Al. Those that collide produce the reaction

$$p + {}^{27}_{13}\text{Al} \rightarrow {}^{27}_{14}\text{Si} + n$$

Ignoring any recoil of the product nucleus, determine the kinetic energy of the emerging neutrons.

34. Of all the hydrogen in the oceans, 0.030 0% of the mass is deuterium. The oceans have a volume of 317 million mi³. (a) If nuclear fusion were controlled and all the deuterium in the oceans were fused to $^4_2$He, how many joules of energy would be released? (b) **What If?** World power consumption is approximately 1.50×10^{13} W. If consumption were 100 times greater, how many years would the energy calculated in part (a) last?

35. **M** **Review.** Suppose seawater exerts an average frictional drag force of 1.00×10^5 N on a nuclear-powered ship. The fuel consists of enriched uranium containing 3.40% of the fissionable isotope $^{235}_{92}$U, and the ship's reactor has an efficiency of 20.0%. Assuming 200 MeV is released per fission event, how far can the ship travel per kilogram of fuel?

Section 30.6 Context Connection: The Engine of the Stars

36. The Sun radiates energy at the rate of 3.85×10^{26} W. Suppose the net reaction $4(^1_1\text{H}) + 2(_{-1}^0\text{e}) \rightarrow {}^4_2\text{He} + 2\nu + \gamma$ accounts for all the energy released. Calculate the number of protons fused per second.

37. Carbon detonations are powerful nuclear reactions that temporarily tear apart the cores inside massive stars late in their lives. These blasts are produced by carbon fusion, which requires a temperature of approximately 6×10^8 K to overcome the strong Coulomb repulsion between carbon nuclei. (a) Estimate the repulsive energy barrier to fusion, using the temperature required for carbon fusion. (In other words, what is the average kinetic energy of a carbon nucleus at 6×10^8 K?) (b) Calculate the energy (in MeV) released in each of these "carbon-burning" reactions:

$$^{12}\text{C} + {}^{12}\text{C} \rightarrow {}^{20}\text{Ne} + {}^4\text{He}$$
$$^{12}\text{C} + {}^{12}\text{C} \rightarrow {}^{24}\text{Mg} + \gamma$$

(c) Calculate the energy in kilowatt-hours given off when 2.00 kg of carbon completely fuse according to the first reaction.

38. **Q|C** After determining that the Sun has existed for hundreds of millions of years, but before the discovery of nuclear physics, scientists could not explain why the Sun has continued to burn for such a long time interval. For example, if it were a coal fire, it would have burned up in about 3 000 yr. Assume the Sun, whose mass is equal to 1.99×10^{30} kg, originally consisted entirely of hydrogen and its total power output is 3.85×10^{26} W. (a) Assuming the energy-generating mechanism of the Sun is the fusion of hydrogen into helium via the net reaction

$$4(^1_1\text{H}) + 2(\text{e}^-) \rightarrow {}^4_2\text{He} + 2\nu + \gamma$$

calculate the energy (in joules) given off by this reaction. (b) Take the mass of one hydrogen atom to be equal to

1.67×10^{-27} kg. Determine how many hydrogen atoms constitute the Sun. (c) If the total power output remains constant, after what time interval will all the hydrogen be converted into helium, making the Sun die? (d) How does your answer to part (c) compare with current estimates of the expected life of the Sun, which are 4 billion to 7 billion years?

39. A theory of nuclear astrophysics proposes that all the elements heavier than iron are formed in supernova explosions ending the lives of massive stars. Assume equal amounts of ^{235}U and ^{238}U were created at the time of the explosion and the present ^{235}U/^{238}U ratio on the Earth is 0.007 25. The half-lives of ^{235}U and ^{238}U are 0.704×10^9 yr and 4.47×10^9 yr, respectively. How long ago did the star(s) explode that released the elements that formed the Earth?

40. Consider the two nuclear reactions

$$A + B \rightarrow C + E \quad \text{(I)}$$
$$C + D \rightarrow F + G \quad \text{(II)}$$

(a) Show that the net disintegration energy for these two reactions ($Q_{net} = Q_I + Q_{II}$) is identical to the disintegration energy for the net reaction

$$A + B + D \rightarrow E + F + G$$

(b) One chain of reactions in the proton–proton cycle in the Sun's core is

$$^1_1\text{H} + {}^1_1\text{H} \rightarrow {}^2_1\text{H} + {}^0_1\text{e} + \nu$$
$$^0_1\text{e} + {}_{-1}^0\text{e} \rightarrow 2\gamma$$
$$^1_1\text{H} + {}^2_1\text{H} \rightarrow {}^3_2\text{He} + \gamma$$
$$^1_1\text{H} + {}^3_2\text{He} \rightarrow {}^4_1\text{H} + {}^0_1\text{e} + \nu$$
$$^0_1\text{e} + {}_{-1}^0\text{e} \rightarrow 2\gamma$$

Based on part (a), what is Q_{net} for this sequence?

41. In addition to the proton–proton cycle, the carbon cycle, first proposed by Hans Bethe (1906–2005) in 1939, is another cycle by which energy is released in stars as hydrogen is converted to helium. The carbon cycle requires higher temperatures than the proton–proton cycle. The series of reactions is

$$^{12}\text{C} + {}^1\text{H} \rightarrow {}^{13}\text{N} + \gamma$$
$$^{13}\text{N} \rightarrow {}^{13}\text{C} + e^+ + \nu$$
$$e^+ + e^- \rightarrow 2\gamma$$
$$^{13}\text{C} + {}^1\text{H} \rightarrow {}^{14}\text{N} + \gamma$$
$$^{14}\text{N} + {}^1\text{H} \rightarrow {}^{15}\text{O} + \gamma$$
$$^{15}\text{O} \rightarrow {}^{15}\text{N} + e^+ + \nu$$
$$e^+ + e^- \rightarrow 2\gamma$$
$$^{15}\text{N} + {}^1\text{H} \rightarrow {}^{12}\text{C} + {}^4\text{He}$$

(a) Assuming that the proton–proton cycle requires a temperature of 1.5×10^7 K, estimate by proportion the temperature required for the carbon cycle. (b) Calculate the Q value for each step in the carbon cycle and the overall energy released. (c) Do you think that the energy carried off by the neutrinos is deposited in the star? Explain.

Additional Problems

42. **S** As part of his discovery of the neutron in 1932, James Chadwick determined the mass of the newly identified particle by firing a beam of fast neutrons, all having the same speed, at two different targets and measuring the maximum recoil speeds of the target nuclei. The maximum speeds arise when an elastic head-on collision occurs between a neutron and a stationary target nucleus. (a) Represent the masses and final speeds of the two target nuclei as m_1, v_1, m_2, and v_2 and assume Newtonian mechanics applies. Show that the neutron mass can be calculated from the equation

$$m_n = \frac{m_1 v_1 - m_2 v_2}{v_2 - v_1}$$

(b) Chadwick directed a beam of neutrons (produced from a nuclear reaction) on paraffin, which contains hydrogen. The maximum speed of the protons ejected was found to be 3.30×10^7 m/s. Because the velocity of the neutrons could not be determined directly, a second experiment was performed using neutrons from the same source and nitrogen nuclei as the target. The maximum recoil speed of the nitrogen nuclei was found to be 4.70×10^6 m/s. The masses of a proton and a nitrogen nucleus were taken as 1.00 u and 14.0 u, respectively. What was Chadwick's value for the neutron mass?

43. (a) Find the radius of the $^{12}_{6}$C nucleus. (b) Find the force of repulsion between a proton at the surface of an $^{12}_{6}$C nucleus and the remaining five protons. (c) How much work (in MeV) has to be done to overcome this electric repulsion in transporting the last proton from a large distance up to the surface of the nucleus? (d) Repeat parts (a), (b), and (c) for $^{238}_{92}$U.

44. When the nuclear reaction represented by Equation 30.24 is endothermic, the reaction energy Q is negative. For the reaction to proceed, the incoming particle must have a minimum energy called the threshold energy, E_{th}. Some fraction of the energy of the incident particle is transferred to the compound nucleus to conserve momentum. Therefore, E_{th} must be greater than Q. (a) Show that

$$E_{th} = -Q\left(1 + \frac{M_a}{M_X}\right)$$

(b) Calculate the threshold energy of the incident alpha particle in the reaction

$$^4_2\text{He} + ^{14}_7\text{N} \rightarrow ^{17}_8\text{O} + ^1_1\text{H}$$

45. (a) One method of producing neutrons for experimental use is bombardment of light nuclei with alpha particles. In the method used by James Chadwick in 1932, alpha particles emitted by polonium are incident on beryllium nuclei:

$$^4_2\text{He} + ^9_4\text{Be} \rightarrow ^{12}_6\text{C} + ^1_0\text{n}$$

What is the Q value of this reaction? (b) Neutrons are also often produced by small-particle accelerators. In one design, deuterons accelerated in a Van de Graaff generator bombard other deuterium nuclei and cause the reaction

$$^2_1\text{H} + ^2_1\text{H} \rightarrow ^3_2\text{He} + ^1_0\text{n}$$

Calculate the Q value of the reaction. (c) Is the reaction in part (b) exothermic or endothermic?

46. **Q|C** (a) Why is the beta decay p → n + e⁺ + ν forbidden for a free proton? (b) **What If?** Why is the same reaction possible if the proton is bound in a nucleus? For example, the following reaction occurs:

$$^{13}_7\text{N} \rightarrow ^{13}_6\text{C} + e^+ + \nu$$

(c) How much energy is released in the reaction given in part (b)?

47. **BIO** After the sudden release of radioactivity from the Chernobyl nuclear reactor accident in 1986, the radioactivity of milk in Poland rose to 2 000 Bq/L due to iodine-131 present in the grass eaten by dairy cattle. Radioactive iodine, with half-life 8.04 days, is particularly hazardous because the thyroid gland concentrates iodine. The Chernobyl accident caused a measurable increase in thyroid cancers among children in Poland and many other Eastern European countries. (a) For comparison, find the activity of milk due to potassium. Assume 1.00 liter of milk contains 2.00 g of potassium, of which 0.011 7% is the isotope ^{40}K with half-life 1.28×10^9 yr. (b) After what elapsed time would the activity due to iodine fall below that due to potassium?

48. The activity of a radioactive sample was measured over 12 h, with the net count rates shown in the accompanying table. (a) Plot the logarithm of the counting rate as a function of time. (b) Determine the decay constant and half-life of the radioactive nuclei in the sample. (c) What counting rate would you expect for the sample at $t = 0$? (d) Assuming the efficiency of the counting instrument is 10.0%, calculate the number of radioactive atoms in the sample at $t = 0$.

Time (h)	Counting Rate (counts/min)
1.00	3 100
2.00	2 450
4.00	1 480
6.00	910
8.00	545
10.0	330
12.0	200

49. On July 4, 1054, a brilliant light appeared in the constellation Taurus the Bull. The supernova, which could be seen in daylight for some days, was recorded by Arab and Chinese astronomers. As it faded, it remained visible for years, dimming for a time with the 77.1-day half-life of the radioactive cobalt-56 that had been created in the explosion. (a) The remains of the star now form the Crab nebula (see Fig. 10.24). In it, the cobalt-56 has now decreased to what fraction of its original activity? (b) Suppose that an American, of the people called the Anasazi, made a charcoal drawing of the supernova. The carbon-14 in the charcoal has now decayed to what fraction of its original activity?

50. When, after a reaction or disturbance of any kind, a nucleus is left in an excited state, it can return to its normal (ground) state by emission of a gamma-ray photon (or several photons). This process is illustrated by Equation 30.21. The emitting nucleus must recoil to conserve both energy and momentum. (a) Show that the recoil energy of the nucleus is

$$E_r = \frac{(\Delta E)^2}{2Mc^2}$$

where ΔE is the difference in energy between the excited and ground states of a nucleus of mass M. (b) Calculate the recoil energy of the ^{57}Fe nucleus when it decays by gamma emission from the 14.4-keV excited state. For this calculation, take the mass to be 57 u. *Suggestion:* Assume $hf \ll Mc^2$.

51. **BIO** **W** A small building has become accidentally contaminated with radioactivity. The longest-lived material in the building is strontium-90. ($^{90}_{38}$Sr has an atomic mass 89.907 7 u, and its half-life is 29.1 yr. It is particularly dangerous because it substitutes for calcium in bones.) Assume the building initially contained 5.00 kg of this substance uniformly distributed throughout the building and the safe level is defined as less than 10.0 decays/min (which is small compared with background radiation). How long will the building be unsafe?

52. **Q|C** In a piece of rock from the Moon, the ^{87}Rb content is assayed to be 1.82×10^{10} atoms per gram of material and the ^{87}Sr content is found to be 1.07×10^9 atoms per gram. The relevant decay relating these nuclides is ^{87}Rb $\rightarrow$ ^{87}Sr + e$^-$ + $\bar{\nu}$. The half-life of the decay is 4.75×10^{10} yr. (a) Calculate the age of the rock. (b) **What If?** Could the material in the rock actually be much older? What assumption is implicit in using the radioactive dating method?

53. When gamma rays are incident on matter, the intensity of the gamma rays passing through the material varies with depth x as $I(x) = I_0 e^{-\mu x}$, where I_0 is the intensity of the radiation at the surface of the material (at $x = 0$) and μ is the linear absorption coefficient. For 0.400-MeV gamma rays in lead, the linear absorption coefficient is 1.59 cm^{-1}. (a) Determine the "half-thickness" for lead, that is, the thickness of lead that would absorb half the incident gamma rays. (b) What thickness reduces the radiation by a factor of 10^4?

54. **S** When gamma rays are incident on matter, the intensity of the gamma rays passing through the material varies with depth x as $I(x) = I_0 e^{-\mu x}$, where I_0 is the intensity of the radiation at the surface of the material (at $x = 0$) and μ is the linear absorption coefficient. (a) Determine the "half thickness" for a material with linear absorption coefficient μ, that is, the thickness of the material that would absorb half the incident gamma rays. (b) What thickness changes the radiation by a factor of f?

55. **M** When gamma rays are incident on matter, the intensity of the gamma rays passing through the material varies with depth x as $I(x) = I_0 e^{-\mu x}$, where I_0 is the intensity of the radiation at the surface of the material (at $x = 0$) and μ is the linear absorption coefficient. For low-energy gamma rays in steel, take the absorption coefficient to be 0.720 mm^{-1}. (a) Determine the "half-thickness" for steel, that is, the thickness of steel that would absorb half the incident gamma rays. (b) In a steel mill, the thickness of sheet steel passing into a roller is measured by monitoring the intensity of gamma radiation reaching a detector below the rapidly moving metal from a small source immediately above the metal. If the thickness of the sheet changes from 0.800 mm to 0.700 mm, by what percentage does the gamma-ray intensity change?

56. *Why is the following situation impossible?* In an effort to study positronium, a scientist places ^{57}Co and ^{14}C in proximity. The ^{57}Co nuclei decay by e$^+$ emission, and the ^{14}C nuclei decay by e$^-$ emission. Some of the positrons and electrons from these decays combine to form sufficient amounts of positronium for the scientist to gather data.

57. **W** The alpha-emitter plutonium-238 ($^{238}_{94}$Pu, atomic mass 238.049 560 u, half-life 87.7 yr) was used in a nuclear energy source on the Apollo Lunar Surface Experiments Package (Fig. P30.57). The energy source, called the Radioisotope Thermoelectric Generator, is the small gray object to the left of the gold-shrouded Central Station in the photograph. Assume the source contains 3.80 kg of ^{238}Pu and the efficiency for conversion of radioactive decay energy to energy transferred by electrical transmission is 3.20%. Determine the initial power output of the source.

Figure P30.57

58. **Q|C** (a) Calculate the energy (in kilowatt-hours) released if 1.00 kg of ^{239}Pu undergoes complete fission and the energy released per fission event is 200 MeV. (b) Calculate the energy (in electron volts) released in the deuterium–tritium fusion reaction

$$^2_1\text{H} + {}^3_1\text{H} \rightarrow {}^4_2\text{He} + {}^1_0\text{n}$$

(c) Calculate the energy (in kilowatt-hours) released if 1.00 kg of deuterium undergoes fusion according to this reaction. (d) **What If?** Calculate the energy (in kilowatt-hours) released by the combustion of 1.00 kg of carbon in coal if each $C + O_2 \rightarrow CO_2$ reaction yields 4.20 eV. (e) List advantages and disadvantages of each of these methods of energy generation.

59. The radioactive isotope ^{137}Ba has a relatively short half-life and can be easily extracted from a solution containing its parent ^{137}Cs. This barium isotope is commonly used in an undergraduate laboratory exercise for demonstrating the radioactive decay law. Undergraduate students using modest experimental equipment took the data presented in Figure P30.59. Determine the half-life for the decay of ^{137}Ba using their data.

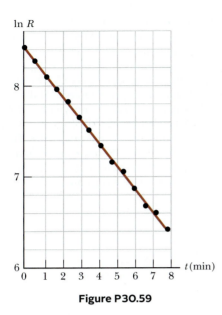

Figure P30.59

60. **S** A method called *neutron activation analysis* can be used for chemical analysis at the level of isotopes. When a sample is irradiated by neutrons, radioactive atoms are produced continuously and then decay according to their character- istic half-lives. (a) Assume one species of radioactive nuclei is produced at a constant rate R and its decay is described by the conventional radioactive decay law. Assuming irradia- tion begins at time $t = 0$, show that the number of radioac- tive atoms accumulated at time t is

$$N = \frac{R}{\lambda}(1 - e^{-\lambda t})$$

(b) What is the maximum number of radioactive atoms that can be produced?

61. During the manufacture of a steel engine component, ra- dioactive iron (^{59}Fe) with a half-life of 45.1 d is included in the total mass of 0.200 kg. The component is placed in a test engine when the activity due to this isotope is 20.0 μCi. After a 1 000-h test period, some of the lubricat- ing oil is removed from the engine and found to contain enough ^{59}Fe to produce 800 disintegrations/min/L of oil. The total volume of oil in the engine is 6.50 L. Calculate the total mass worn from the engine component per hour of operation.

62. **Q|C** Approximately 1 of every 3 300 water molecules con- tains one deuterium atom. (a) If all the deuterium nuclei in 1 L of water are fused in pairs according to the D–D fu- sion reaction ^{2}H + ^{2}H → ^{3}He + n + 3.27 MeV, how much energy in joules is liberated? (b) **What If?** Burning gasoline produces approximately 3.40×10^7 J/L. State how the en- ergy obtainable from the fusion of the deuterium in 1 L of water compares with the energy liberated from the burning of 1 L of gasoline.

63. **Review.** A nuclear power plant operates by using the energy released in nuclear fission to convert 20°C water into 400°C steam. How much water could theoretically be converted to steam by the complete fissioning of 1.00 g of ^{235}U at 200 MeV/fission?

64. **S** **Review.** A nuclear power plant operates by using the en- ergy released in nuclear fission to convert liquid water at T_c into steam at T_h. How much water could theoretically be converted to steam by the complete fissioning of a mass m of ^{235}U if the energy released per fission event is E?

65. **BIO** To destroy a cancerous tumor, a dose of gamma ra- diation with a total energy of 2.12 J is to be delivered in 30.0 days from implanted sealed capsules containing palladium-103. Assume this isotope has a half-life of 17.0 d and emits gamma rays of energy 21.0 keV, which are entirely absorbed within the tumor. (a) Find the initial activity of the set of capsules. (b) Find the total mass of radioactive palla- dium these "seeds" should contain.

66. **Q|C** Natural uranium must be processed to produce ura- nium enriched in ^{235}U for weapons and power plants. The processing yields a large quantity of nearly pure ^{238}U as a by-product, called "depleted uranium." Because of its high mass density, ^{238}U is used in armor-piercing artillery shells. (a) Find the edge dimension of a 70.0-kg cube of ^{238}U ($\rho = 19.1 \times 10^3$ kg/m^3). (b) The isotope ^{238}U has a long half-life of 4.47×10^9 yr. As soon as one nucleus decays, a relatively rapid series of 14 steps begins that together constitute the net reaction

$$^{235}_{92}\text{U} \rightarrow 8(^4_2\text{He}) + 6(^0_{-1}\text{e}) + ^{206}_{82}\text{Pb} + 6\bar{\nu} + Q_{\text{net}}$$

Find the net decay energy. (Refer to Table A.3.) (c) Argue that a radioactive sample with decay rate R and decay en- ergy Q has power output $P = QR$. (d) Consider an artillery shell with a jacket of 70.0 kg of ^{238}U. Find its power output due to the radioactivity of the uranium and its daughters. Assume the shell is old enough that the daughters have reached steady-state amounts. Express the power in joules per year. (e) **What If?** A 17-year-old soldier of mass 70.0 kg works in an arsenal where many such artillery shells are stored. Assume his radiation exposure is limited to absorb- ing 45.5 mJ per year per kilogram of body mass. Find the net rate at which he can absorb energy of radiation, in joules per year.

67. **M** Free neutrons have a characteristic half-life of 10.4 min. What fraction of a group of free neutrons with kinetic energy 0.040 0 eV decays before traveling a distance of 10.0 km?

68. *Why is the following situation impossible?* A ^{10}B nucleus is struck by an incoming alpha particle. As a result, a proton and a ^{12}C nucleus leave the site after the reaction.

69. On August 6, 1945, the United States dropped on Hiroshima a nuclear bomb that released 5×10^{13} J of en- ergy, equivalent to that from 12 000 tons of TNT. The fis- sion of one $^{235}_{92}$U nucleus releases an average of 208 MeV. Estimate (a) the number of nuclei fissioned and (b) the mass of this $^{235}_{92}$U.

70. **Review.** The first nuclear bomb was a fissioning mass of plutonium-239 that exploded in the Trinity test before dawn on July 16, 1945, at Alamogordo, New Mexico. Enrico Fermi was 14 km away, lying on the ground fac- ing away from the bomb. After the whole sky had flashed with unbelievable brightness, Fermi stood up and began dropping bits of paper to the ground. They first fell at his feet in the calm and silent air. As the shock wave passed,

about 40 s after the explosion, the paper then in flight jumped approximately 2.5 m away from ground zero. (a) Equation 13.27 describes the relationship between the pressure amplitude ΔP_{max} of a sinusoidal air compression wave and its displacement amplitude s_{max}. The compression pulse produced by the bomb explosion was not a sinusoidal wave, but let's use the same equation to compute an estimate for the pressure amplitude, taking $\omega \sim 1\ s^{-1}$ as an estimate for the angular frequency at which the pulse ramps up and down. (b) Find the change in volume ΔV of a sphere of radius 14 km when its radius increases by 2.5 m. (c) The energy carried by the blast wave is the work done by one layer of air on the next as the wave crest passes. An extension of the logic used to derive Equation 17.7 shows that this work is given by $(\Delta P_{max})(\Delta V)$. Compute an estimate for this energy. (d) Assume the blast wave carried on the order of one-tenth of the explosion's energy. Make an order-of-magnitude estimate of the bomb yield. (e) One ton of exploding TNT releases 4.2 GJ of energy. What was the order of magnitude of the energy of the Trinity test in equivalent tons of TNT? Fermi's immediate knowledge of the bomb yield agreed with that determined days later by analysis of elaborate measurements.

Particle Physics

Chapter Outline

Courtesy of Brookhaven National Laboratory

A shower of particle tracks from a head-on collision of gold nuclei, each moving with energy 100 GeV. This collision occurred at the Relativistic Heavy Ion Collider (RHIC) at Brookhaven National Laboratory and was recorded with the STAR (Solenoidal Tracker at RHIC) detector. The tracks represent many fundamental particles arising from the energy of the collision.

In the early chapters of this book, we discussed the particle model, which treats an object as a particle of zero size with no structure. Some behaviors of objects, such as thermal expansion, can be understood by modeling the object as a collection of particles: atoms. In these models, any internal structure of the atom is ignored. We could not ignore the internal structure of the atom to understand such phenomena as atomic spectra, however. Modeling the hydrogen atom as a system of an electron in orbit about a particle-like nucleus helped in this regard (Section 11.5). In Chapter 30, however, we could not model the nucleus as a particle and ignore its structure to understand behavior such as nuclear stability and radioactive decay. We had to model the nucleus as a collection of smaller particles, nucleons. What about these nuclear constituents, the protons and neutrons? Can we apply the particle model to these entities? As we shall see, even protons and neutrons have structure, which leads to a puzzling question. As we continue to investigate the structure of smaller and smaller "particles," will we ever reach a level at which the building blocks are truly and completely described by the particle model?

In this concluding chapter, we explore this question by examining the properties and classifications of the various known subatomic particles and the fundamental interactions that govern their behavior. We also discuss the current model of elementary particles, in which all matter is believed to be constructed from only two families of particles: quarks and leptons.

The word *atom* is from the Greek *atomos*, which means "indivisible." At one time, atoms were thought to be the indivisible constituents of matter; that is, they were regarded as elementary particles. After 1932, physicists viewed all matter as consisting of only three constituent particles: electrons, protons, and neutrons. (The neutron was observed and identified in 1932.) With the exception of the free neutron (as opposed to a neutron within a nucleus), these particles are very stable. Beginning in 1945, many new particles were discovered in experiments involving high-energy collisions between known particles. These new particles are characteristically very unstable and have very short half-lives, ranging between 10^{-6} s and 10^{-23} s. So far, more than 300 of these unstable, temporary particles have been catalogued.

Since the 1930s, many powerful particle accelerators have been constructed throughout the world, making it possible to observe collisions of highly energetic particles under controlled laboratory conditions so as to reveal the subatomic world in finer detail. Until the 1960s, physicists were bewildered by the large number and variety of subatomic particles being discovered. They wondered if the particles had no systematic relationship connecting them, or whether a pattern was emerging that would provide a better understanding of the elaborate structure in the subnuclear world. Since that time, physicists have advanced our knowledge of the structure of matter tremendously by developing a structural model in which most of the ever-growing number of particles are made of smaller particles called *quarks*. Therefore, protons and neutrons, for example, are not truly elementary but are systems of tightly bound quarks.

Pitfall Prevention | 31.1
The Nuclear Force and the Strong Force
The nuclear force discussed in Chapter 30 was historically called the strong force. Once the quark theory (Section 31.9) was established, however, the phrase *strong force* was reserved for the force between quarks. We shall follow this convention: the strong force is between quarks or particles built from quarks, and the nuclear force is between nucleons in a nucleus. The nuclear force is a secondary result of the strong force as discussed in Section 31.10. It is sometimes called the residual strong force. Because of this historical development of the names for these forces, other books sometimes refer to the nuclear force as the strong force.

▶ Field particles

31.1 | The Fundamental Forces in Nature

As we learned in Chapter 5, all natural phenomena can be described by four fundamental forces between particles. In order of decreasing strength, they are the **strong** force, the **electromagnetic** force, the **weak** force, and the **gravitational** force. In current models, the electromagnetic and weak forces are considered to be two manifestations of a single interaction, the **electroweak force,** as discussed in Section 31.11.

The **nuclear force,** as we mentioned in Chapter 30, holds nucleons together. It is very short range and is negligible for separations greater than about 2 fm (about the size of the nucleus). The electromagnetic force, which binds atoms and molecules together to form ordinary matter, has about 10^{-2} times the strength of the nuclear force. It is a long-range force that decreases in strength as the inverse square of the separation between interacting particles. The weak force is a short-range force that accounts for radioactive decay processes such as beta decay, and its strength is only about 10^{-5} times that of the nuclear force. Finally, the gravitational force is a long-range force that has a strength of only about 10^{-39} times that of the nuclear force. Although this familiar interaction is the force that holds the planets, stars, and galaxies together, its effect on elementary particles is negligible.

In modern physics, interactions between particles are often described in terms of a structural model that involves the exchange of **field particles,** or **exchange particles.** Field particles are also called **gauge bosons.**[1] (In general, all particles with integral spin are called *bosons*.) In the case of the familiar electromagnetic interaction, for instance, the field particles are photons. In the language of modern physics, we say that the electromagnetic force is *mediated* by photons and that photons are the quanta of the electromagnetic field. Likewise, the nuclear force is mediated by field particles called **gluons,** the weak force is mediated by particles called the **W** and **Z** **bosons,** and the gravitational force is mediated by quanta of the gravitational field called **gravitons.** These forces, their ranges, and their relative strengths are summarized in Table 31.1.

---

[1]The word *gauge* comes from *gauge theory*, which is a sophisticated mathematical analysis that is beyond the scope of this book.

TABLE 31.1 | Fundamental Forces

Force	Relative Strength	Range of Force	Mediating Field Particle	Mass of Field Particle (GeV/c^2)
Nuclear/Strong	1	Short (~1 fm)	Gluon	0
Electromagnetic	10^{-2}	∞	Photon	0
Weak	10^{-5}	Short (~10^{-3} fm)	W$^\pm$, Z^0 bosons	80.4, 80.4, 91.2
Gravitational	10^{-39}	∞	Graviton	0

31.2 | Positrons and Other Antiparticles

In the 1920s, English theoretical physicist Paul Adrien Maurice Dirac developed a version of quantum mechanics that incorporated special relativity. Dirac's theory explained the origin of electron spin and its magnetic moment. It also presented a major difficulty, however. Dirac's relativistic wave equation required solutions corresponding to negative energy states even for free electrons. If negative energy states existed, however, one would expect an electron in a state of positive energy to make a rapid transition to one of these states, emitting a photon in the process. Dirac avoided this difficulty by postulating a structural model in which all negative energy states are filled. The electrons occupying these negative energy states are collectively called the *Dirac sea*. Electrons in the Dirac sea are not directly observable because the Pauli exclusion principle does not allow them to react to external forces; there are no states available to which an electron can make a transition in response to an external force. Therefore, an electron in such a state acts as an isolated system unless an interaction with the environment is strong enough to excite the electron to a positive energy state. Such an excitation causes one of the negative energy states to be vacant, as in Figure 31.1, leaving a hole in the sea of filled states. (Notice that positive energy states exist only for $E > m_e c^2$, representing the rest energy of the electron. Similarly, negative energy states exist only for $E < -m_e c^2$.) *The hole can react to external forces and is observable.* The hole reacts in a way similar to that of the electron, except that it has a positive charge. It is the **antiparticle** to the electron.

The profound implication of this model is that *every particle has a corresponding antiparticle.* The antiparticle has the same mass as the particle, but the opposite charge. For example, the electron's antiparticle, called a **positron,** has a mass of 0.511 MeV/c^2 and a positive charge of 1.60×10^{-19} C.

Carl Anderson (1905–1991) observed and identified the positron in 1932, and in 1936 he was awarded the Nobel Prize in Physics for that achievement. Anderson discovered the positron while examining tracks in a cloud chamber created by electron-like particles of positive charge. (A cloud chamber contains a gas that has been supercooled to just below its usual condensation point. An energetic radioactive particle passing through ionizes the gas and leaves a visible track. These early experiments used cosmic rays—mostly energetic protons passing through interstellar space—to initiate high-energy reactions in the upper atmosphere, which resulted in the production of positrons at ground level.) To discriminate between positive and negative charges, Anderson placed the cloud chamber in a magnetic field, causing moving charged particles to follow curved paths as discussed in Section 22.3. He noted that some of the electron-like tracks deflected in a direction corresponding to a positively charged particle.

Since Anderson's discovery, the positron has been observed in a number of experiments. A common process for producing positrons is **pair production.** In this process, a gamma-ray photon with sufficiently high energy interacts with a nucleus and an electron–positron pair is created. In the Dirac sea model, an electron in a negative energy state is excited to a positive energy state, resulting in a new observable electron and a hole, which is the positron. Because the total rest energy of the electron–positron pair is $2m_e c^2 = 1.022$ MeV, the photon must have at least this much energy to create an electron–positron pair. Therefore, energy in the form of a

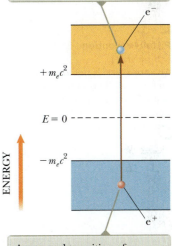

An electron can make a transition out of the Dirac sea only if it is provided with energy equal to or larger than $2m_e c^2$.

An upward transition of an electron leaves a vacancy in the Dirac sea, which can behave as a particle identical to the electron except for its positive charge.

Figure 31.1 Dirac's model for the existence of antielectrons (positrons).

Paul Adrien Maurice Dirac
British Physicist (1902–1984)
Dirac was instrumental in the understanding of antimatter and the unification of quantum mechanics and relativity. He made many contributions to the development of quantum physics and cosmology. In 1933, Dirac won a Nobel Prize in Physics.

Figure 31.2 (a) Bubble-chamber tracks of electron–positron pairs produced by 300-MeV gamma rays striking a lead sheet from the left. (b) The pertinent pair-production events. The positrons deflect upward and the electrons deflect downward in an applied magnetic field.

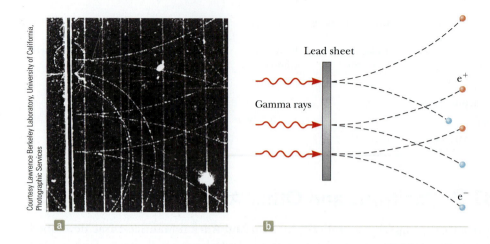

Courtesy Lawrence Berkeley Laboratory, University of California, Photographic Services

Lead sheet

Gamma rays

e⁺

e⁻

gamma-ray photon is converted to rest energy in accordance with Einstein's relationship $E_R = mc^2$. We can use the isolated system model to describe this process. The energy of the system of the photon and the nucleus is conserved and transformed to rest energy of the electron and positron, kinetic energy of these particles, and some small amount of kinetic energy associated with the nucleus. Figure 31.2a shows tracks of electron–positron pairs created by 300-MeV gamma rays striking a lead plate.

QUICK QUIZ 31.1 Given the identification of the particles in Figure 31.2b, is the direction of the external magnetic field in Figure 31.2a (a) into the page, (b) out of the page, (c) impossible to determine?

The reverse process can also occur. Under the proper conditions, an electron and positron can annihilate each other to produce two gamma-ray photons (see Thinking Physics 31.1) that have a combined energy of at least 1.022 MeV:

$$e^- + e^+ \rightarrow 2\gamma$$

BIO Positron-emission tomography (PET)

Electron–positron annihilation is used in the medical diagnostic technique called *positron-emission tomography* (PET). The patient is injected with a glucose solution containing a radioactive substance that decays by positron emission (often ¹⁸F), and the material is carried by the blood throughout the body. A positron emitted during a decay event in one of the radioactive nuclei in the glucose solution annihilates with an electron in the immediately surrounding tissue, resulting in two gamma-ray photons emitted in opposite directions. A gamma detector surrounding the patient pinpoints the source of the photons and, with the assistance of a computer, displays an image of the sites at which the glucose accumulates. (Glucose is metabolized rapidly in cancerous tumors and accumulates in these sites, providing a strong signal for a PET detector system.) The images from a PET scan can indicate a wide variety of disorders in the brain, including Alzheimer's disease (Fig. 31.3). In addition, because glucose metabolizes more rapidly in active areas of the brain, a PET scan can indicate which areas of the brain are involved when the patient is engaging in such activities as language use, music, or vision.

Prior to 1955, on the basis of the Dirac theory, it was expected that every particle has a corresponding antiparticle, but antiparticles such as the antiproton and antineutron had not been detected experimentally. Because the relativistic Dirac theory had some failures (it predicted the wrong-size magnetic moment for the photon) as well as many successes, it was important to determine whether the antiproton really existed. In 1955, a team led by Emilio Segrè (1905–1989) and Owen Chamberlain (1920–2006) used the Bevatron particle accelerator at the University of California–Berkeley to produce antiprotons and antineutrons. They therefore established

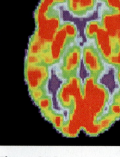

National Institutes of Health

Figure 31.3 PET scans of the brain of a healthy older person (*left*) and that of a patient suffering from Alzheimer's disease (*right*). Lighter regions contain higher concentrations of radioactive glucose, indicating higher metabolism rates and therefore increased brain activity.

with certainty the existence of antiparticles. For this work, Segrè and Chamberlain received the Nobel Prize in Physics in 1959. It is now established that every particle has a corresponding antiparticle with equal mass and spin, and with charge, magnetic moment, and strangeness of equal magnitude but opposite sign. (The property of strangeness is explained in Section 31.6.) The only exception to these rules for particles and antiparticles are the neutral photon, pion, and eta, each of which is its own antiparticle.

An intriguing aspect of the existence of antiparticles is that if we replace every proton, neutron, and electron in an atom with its antiparticle, we can create a stable antiatom; combinations of antiatoms should form antimolecules and eventually antiworlds. As far as we know, everything would behave in the same way in an antiworld as in our world. In principle, it is possible that some distant antimatter galaxies exist, separated from normal-matter galaxies by millions of light-years. Unfortunately, because the photon is its own antiparticle, the light emitted from an antimatter galaxy is no different from that from a normal-matter galaxy, so astronomical observations cannot determine if the galaxy is composed of matter or antimatter. Although no evidence of antimatter galaxies exists at present, it is awe-inspiring to imagine the cosmic spectacle that would result if matter and antimatter galaxies were to collide: a gigantic eruption of jets of annihilation radiation, transforming the entire galactic mass into energetic particles fleeing the collision point.

> ### THINKING PHYSICS 31.1
>
> When an electron and a positron meet at low speed in free space, why are two 0.511-MeV gamma rays produced rather than one gamma ray with an energy of 1.022 MeV?
>
> **Reasoning** Gamma rays are photons, and photons carry momentum. We apply the momentum version of the isolated system model to the system, which consists initially of the electron and positron. If the system, assumed to be at rest, transformed to only one photon, momentum would not be conserved because the initial momentum of the electron–positron system is zero, whereas the final system consists of a single photon of energy 1.022 MeV and nonzero momentum. On the other hand, the two gamma-ray photons travel in *opposite* directions, so the total momentum of the final system—two photons—is zero, and momentum is conserved. ◄

❰ 31.3 | Mesons and the Beginning of Particle Physics

In the mid-1930s, physicists had a fairly simple view of the structure of matter. The building blocks were the proton, the electron, and the neutron. Three other particles were known or had been postulated at the time: the photon, the neutrino, and the positron. These six particles were considered the fundamental constituents of matter. With this marvelously simple picture of the world, however, no one was able to answer an important question. Because many protons in proximity in any nucleus should strongly repel one another due to their positive charges, what is the nature of the force that holds the nucleus together? Scientists recognized that this mysterious force, which we now call the nuclear force, must be much stronger than anything encountered in nature up to that time.

In 1935, Japanese physicist Hideki Yukawa proposed the first theory to successfully explain the nature of the nuclear force, an effort that later earned him the Nobel Prize in Physics. To understand Yukawa's theory, it is useful to first recall that in the modern structural model of electromagnetic interactions, charged particles interact by exchanging photons. Yukawa used this idea to explain the nuclear force by proposing a new particle whose exchange between nucleons in the nucleus produces the nuclear force. Furthermore, he established that the range of the force is inversely proportional to the mass of this particle and predicted that the mass would be about 200 times the mass of the electron. Because the new particle would have a

Hideki Yukawa
Japanese Physicist (1907–1981)
Yukawa was awarded the Nobel Prize in Physics in 1949 for predicting the existence of mesons. This photograph of him at work was taken in 1950 in his office at Columbia University. Yukawa came to Columbia in 1949 after spending the early part of his career in Japan.

mass between that of the electron and that of the proton, it was called a meson (from the Greek *meso*, meaning "middle").

In an effort to substantiate Yukawa's predictions, physicists began an experimental search for the meson by studying cosmic rays entering the Earth's atmosphere. In 1937, Anderson and his collaborators discovered a particle of mass 106 MeV/c^2, about 207 times the mass of the electron. Subsequent experiments showed that the particle interacted very weakly with matter, however, and hence could not be the carrier of the nuclear force. The puzzling situation inspired several theoreticians to propose that two mesons existed with slightly different masses. This idea was confirmed by the discovery in 1947 of the **pi (π) meson**, or simply **pion**, by Cecil Frank Powell (1903–1969) and Giuseppe P. S. Occhialini (1907–1993). The particle discovered by Anderson in 1937, the one thought to be Yukawa's meson, is not really a meson. (We shall discuss the requirements for a particle to be a meson in Section 31.4.) Instead, it takes part in the weak and electromagnetic interactions only and is now called the **muon** (μ). We first discussed the muon in Section 9.4, with regard to time dilation.

The pion, Yukawa's carrier of the nuclear force, comes in three varieties corresponding to three charge states: π^+, π^-, and π^0. The π^+ and π^- particles have masses of 139.6 MeV/c^2, and the π^0 particle has a mass of 135.0 MeV/c^2. Pions and muons are very unstable particles. For example, the π^-, which has a mean lifetime of 2.6×10^{-8} s, first decays to a muon and an antineutrino. The muon, which has a mean lifetime of 2.2 μs, then decays into an electron, a neutrino, and an antineutrino:

$$\pi^- \rightarrow \mu^- + \bar{\nu} \qquad \qquad \textbf{31.1} \blacktriangleleft$$
$$\mu^- \rightarrow e^- + \nu + \bar{\nu}$$

Note that for chargeless particles (as well as some charged particles such as the proton), a bar over the symbol indicates an antiparticle.

Figure 31.4 Feynman diagram representing a photon mediating the electromagnetic force between two electrons.

The interaction between two particles can be represented in a simple qualitative graphical representation called a **Feynman diagram,** developed by American physicist Richard P. Feynman. Figure 31.4 is such a diagram for the electromagnetic interaction between two electrons approaching each other. A Feynman diagram is a qualitative graph of time in the vertical direction versus space in the horizontal direction. It is qualitative in the sense that the actual values of time and space are not important, but the overall appearance of the graph provides a representation of the process. The time evolution of the process can be approximated by starting at the bottom of the diagram and moving your eyes upward.

In the simple case of the electron–electron interaction in Figure 31.4, a photon is the field particle that mediates the electromagnetic force between the electrons. Notice that the entire interaction is represented in such a diagram as if it occurs at a single point in time. Therefore, the paths of the electrons appear to undergo a discontinuous change in direction at the moment of interaction. This representation is correct on a microscopic level over a time interval that includes the exchange of one photon. It is different from the paths produced over the much longer interval during which we watch the interaction from a macroscopic point of view. In this case, the paths would be curved (as in Fig. 31.2) due to the continuous exchange of large numbers of field particles, illustrating another aspect of the qualitative nature of Feynman diagrams.

In the electron–electron interaction, the photon, which transfers energy and momentum from one electron to the other, is called a *virtual photon* because it vanishes during the interaction without having been detected. In Chapter 28, we discussed that a photon has energy $E = hf$, where f is its frequency. Consequently, for a system of two electrons initially at rest, the system has energy $2m_e c^2$ before a virtual photon is released and energy $2m_e c^2 + hf$ after the virtual photon is released (plus any kinetic energy of the electron resulting from the emission of the photon). Is that a violation of the law of conservation of energy for an isolated system? No; this process does *not* violate the law of conservation of energy because the virtual photon has a very short lifetime Δt that makes the uncertainty in the energy $\Delta E \approx \hbar / 2\ \Delta t$ of the system consisting of two electrons and the photon greater than the photon energy. Therefore,

Richard Feynman
American Physicist (1918–1988)
Inspired by Dirac, Feynman developed quantum electrodynamics, the theory of the interaction of light and matter on a relativistic and quantum basis. In 1965, Feynman won the Nobel Prize in Physics. The prize was shared by Feynman, Julian Schwinger, and Sin Itiro Tomonaga. Early in Feynman's career, he was a leading member of the team developing the first nuclear weapon in the Manhattan Project. Toward the end of his career, he worked on the commission investigating the 1986 *Challenger* tragedy and demonstrated the effects of cold temperatures on the rubber O-rings used in the space shuttle.

within the constraints of the uncertainty principle, the energy of the system is conserved.

Now consider a pion exchange between a proton and a neutron according to Yukawa's model (Fig. 31.5a). The energy ΔE_R needed to create a pion of mass m_π is given by Einstein's equation $\Delta E_R = m_\pi c^2$. As with the photon in Figure 31.4, the very existence of the pion would appear to violate the law of conservation of energy if the particle existed for a time greater than $\Delta t \approx \hbar/2\,\Delta E_R$ (from the uncertainty principle), where Δt is the time interval required for the pion to transfer from one nucleon to the other. Therefore,

$$\Delta t \approx \frac{\hbar}{2\,\Delta E_R} = \frac{\hbar}{2m_\pi c^2}$$

and the rest energy of the pion is

$$m_\pi c^2 = \frac{\hbar}{2\,\Delta t} \qquad \textbf{31.2} \blacktriangleleft$$

Because the pion cannot travel faster than the speed of light, the maximum distance d it can travel in a time interval Δt is $c\,\Delta t$. Therefore, using Equation 31.2 and $d = c\,\Delta t$, we find

$$m_\pi c^2 = \frac{\hbar c}{2d} \qquad \textbf{31.3} \blacktriangleleft$$

From Chapter 30, we know that the range of the nuclear force is on the order of 10^{-15} fm. Using this value for d in Equation 31.3, we estimate the rest energy of the pion to be

$$m_\pi c^2 \approx \frac{(1.055 \times 10^{-34}\,\text{J} \cdot \text{s})(3.00 \times 10^8\,\text{m/s})}{2(1 \times 10^{-15}\,\text{m})}$$

$$= 1.6 \times 10^{-11}\,\text{J} \approx 100\,\text{MeV}$$

which corresponds to a mass of $100\,\text{MeV}/c^2$ (approximately 200 times the mass of the electron). This value is in reasonable agreement with the observed pion mass.

The concept we have just described is quite revolutionary. In effect, it says that a system of two nucleons can change into two nucleons plus a pion as long as it returns to its original state in a very short time interval. (Remember that this model is the older, historical one, which assumes that the pion is the field particle for the nuclear force.) Physicists often say that a nucleon undergoes *fluctuations* as it emits and absorbs pions. As we have seen, these fluctuations are a consequence of a combination of quantum mechanics (through the uncertainty principle) and special relativity (through Einstein's mass–energy relationship $E_R = mc^2$).

This section has dealt with the particles that mediate the nuclear force, pions, and the mediators of the electromagnetic force, photons. Current ideas indicate that the nuclear force is more fundamentally described as an average or residual effect of the force between quarks, as will be explained in Section 31.10. The graviton, which is the mediator of the gravitational force, has yet to be observed. The $\text{W}^\pm$ and Z^0 particles that mediate the weak force were discovered in 1983 by Italian physicist Carlo Rubbia (b. 1934) and his associates using a proton–antiproton collider. Rubbia and Simon van der Meer (1925–2011), both at CERN (European Organization for Nuclear Research), shared the 1984 Nobel Prize in Physics for the detection and identification of the $\text{W}^\pm$ and Z^0 particles and the development of the proton–antiproton collider. In this accelerator, protons and antiprotons undergo head-on collisions with each other. In some of the collisions, $\text{W}^\pm$ and Z^0 particles are produced, which in turn are identified by their decay products. Figure 31.5b shows a Feynman diagram for a weak interaction mediated by a Z^0 boson.

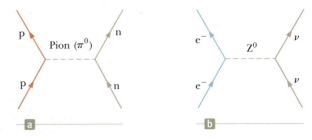

Figure 31.5 (a) Feynman diagram representing a proton and a neutron interacting via the nuclear force with a neutral pion mediating the force. (This model is *not* the most fundamental model for nucleon interaction.) (b) Feynman diagram for an electron and a neutrino interacting via the weak force with a Z^0 boson mediating the force.

31.4 | Classification of Particles

All particles other than field particles can be classified into two broad categories, *hadrons* and *leptons*. The criterion for separating these particles into categories is whether or not they interact via the strong force. This force increases with separation distance, similar to the force exerted by a stretched spring. The nuclear force between nucleons in a nucleus is a particular manifestation of the strong force, but, as mentioned in Pitfall Prevention 31.1, we will use the term *strong force* in general to refer to any interaction between particles made up of more elementary units called quarks. Table 31.2 provides a summary of the properties of some of these particles.

Hadrons

Particles that interact through the strong force are called **hadrons.** The two classes of hadrons—*mesons* and *baryons*—are distinguished by their masses and spins.

TABLE 31.2 | Some Particles and Their Properties

Category	Particle Name	Symbol	Anti-particle	Mass (MeV/c^2)	B	L_e	L_μ	L_τ	S	Lifetime(s)	Spin
Leptons	Electron	e^-	e^+	0.511	0	+1	0	0	0	Stable	$\frac{1}{2}$
	Electron–neutrino	ν_e	$\bar{\nu}_e$	$< 2\text{eV}/c^2$	0	+1	0	0	0	Stable	$\frac{1}{2}$
	Muon	μ^-	μ^+	105.7	0	0	+1	0	0	2.20×10^{-6}	$\frac{1}{2}$
	Muon–neutrino	ν_μ	$\bar{\nu}_\mu$	< 0.17	0	0	+1	0	0	Stable	$\frac{1}{2}$
	Tau	τ^-	τ^+	1 784	0	0	0	+1	0	$< 4 \times 10^{-13}$	$\frac{1}{2}$
	Tau–neutrino	ν_τ	$\bar{\nu}_\tau$	< 18	0	0	0	+1	0	Stable	$\frac{1}{2}$
Hadrons											
Mesons	Pion	π^+	π^-	139.6	0	0	0	0	0	2.60×10^{-8}	0
		π^0	Self	135.0	0	0	0	0	0	0.83×10^{-16}	0
	Kaon	K^+	K^-	493.7	0	0	0	0	+1	1.24×10^{-8}	0
		K_S^0	$\bar{K}_S^0$	497.7	0	0	0	0	+1	0.89×10^{-10}	0
		K_L^0	$\bar{K}_L^0$	497.7	0	0	0	0	+1	5.2×10^{-8}	0
	Eta	η	Self	548.8	0	0	0	0	0	$< 10^{-18}$	0
		η'	Self	958	0	0	0	0	0	2.2×10^{-21}	0
Baryons	Proton	p	$\bar{p}$	938.3	+1	0	0	0	0	Stable	$\frac{1}{2}$
	Neutron	n	$\bar{n}$	939.6	+1	0	0	0	0	614	$\frac{1}{2}$
	Lambda	Λ^0	$\bar{\Lambda}^0$	1 115.6	+1	0	0	0	−1	2.6×10^{-10}	$\frac{1}{2}$
	Sigma	Σ^+	$\bar{\Sigma}^-$	1 189.4	+1	0	0	0	−1	0.80×10^{-10}	$\frac{1}{2}$
		Σ^0	$\bar{\Sigma}^0$	1 192.5	+1	0	0	0	−1	6×10^{-20}	$\frac{1}{2}$
		Σ^-	$\bar{\Sigma}^+$	1 197.3	+1	0	0	0	−1	1.5×10^{-10}	$\frac{1}{2}$
	Delta	Δ^{++}	$\bar{\Delta}^{--}$	1 230	+1	0	0	0	0	6×10^{-24}	$\frac{3}{2}$
		Δ^+	$\bar{\Delta}^-$	1 231	+1	0	0	0	0	6×10^{-24}	$\frac{3}{2}$
		Δ^0	$\bar{\Delta}^0$	1 232	+1	0	0	0	0	6×10^{-24}	$\frac{3}{2}$
		Δ^-	$\bar{\Delta}^+$	1 234	+1	0	0	0	0	6×10^{-24}	$\frac{3}{2}$
	Xi	Ξ^0	$\bar{\Xi}^0$	1 315	+1	0	0	0	−2	2.9×10^{-10}	$\frac{1}{2}$
		Ξ^-	Ξ^+	1 321	+1	0	0	0	−2	1.64×10^{-10}	$\frac{1}{2}$
	Omega	Ω^-	Ω^+	1 672	+1	0	0	0	−3	0.82×10^{-10}	$\frac{3}{2}$

Mesons all have zero or integer spin (0 or 1).[2] As indicated in Section 31.3, the origin of the name comes from the expectation that Yukawa's proposed meson mass would lie between the mass of the electron and the mass of the proton. Several meson masses do lie in this range, although there are heavier mesons that have masses larger than that of the proton.

All mesons are known to decay into final products including electrons, positrons, neutrinos, and photons. The pions are the lightest of the known mesons; they have masses of about 140 MeV/c^2 and a spin of 0. Another is the K meson, with a mass of approximately 500 MeV/c^2 and a spin of 0.

Baryons, the second class of hadrons, have masses equal to or greater than the proton mass (*baryon* means "heavy" in Greek), and their spins are always an odd half-integer value ($\frac{1}{2}$ or $\frac{3}{2}$). Protons and neutrons are baryons, as are many other particles. With the exception of the proton, all baryons decay in such a way that the end products include a proton. For example, the baryon called the Ξ hyperon decays to the Λ^0 baryon in about 10^{-10} s. The Λ^0 baryon then decays to a proton and a π^- in approximately 3×10^{-10} s.

Today it is believed that hadrons are not elementary particles, but rather are composed of more elementary units called quarks. We shall discuss quarks in Section 31.9.

Leptons

Leptons (from the Greek *leptos,* meaning "small" or "light") are a group of particles that participate in the electromagnetic (if charged) and weak interactions. All leptons have spins of $\frac{1}{2}$. Unlike hadrons, which have size and structure, leptons appear to be truly elementary particles with no structure.

Quite unlike hadrons, the number of known leptons is small. Currently, scientists believe that only six leptons exist: the electron, the muon, and the tau, e^-, μ^-, τ^-, and a neutrino associated with each, ν_e, ν_μ, ν_τ. The tau lepton, discovered in 1975, has a mass equal to about twice that of the proton. Direct experimental evidence for the neutrino associated with the tau was announced by the Fermi National Accelerator Laboratory (Fermilab) in July 2000. Each of these six leptons has an antiparticle.

Current studies indicate that neutrinos may have a small but nonzero mass. If they do have mass, they cannot travel at the speed of light. Also, so many neutrinos exist that their combined mass may be sufficient to cause all the matter in the Universe to eventually collapse to infinite density and then explode and create a completely new Universe! We shall discuss this concept in more detail in Section 31.12.

31.5 | Conservation Laws

We have seen the importance of conservation laws for isolated systems many times in earlier chapters and have solved problems using conservation of energy, linear momentum, angular momentum, and electric charge. Conservation laws are important in understanding why certain decays and reactions occur but others do not. In general, our familiar conservation laws provide us with a set of rules that all processes must follow.

Certain new conservation laws have been identified through experimentation and are important in the study of elementary particles. The members of the isolated system change identity during a decay or reaction. The initial particles before the decay or reaction are different from the final particles afterward.

---

[2]Therefore, the particle discovered by Anderson in 1937, the muon, is not a meson; the muon has spin $\frac{1}{2}$. It belongs in the *lepton* classification described shortly.

Baryon Number

Experimental results tell us that whenever a baryon is created in a nuclear reaction or decay, an antibaryon is also created. This scheme can be quantified by assigning a baryon number $B = +1$ for all baryons, $B = -1$ for all antibaryons, and $B = 0$ for all other particles. Therefore, the **law of conservation of baryon number** states that

▶ Conservation of baryon number

whenever a reaction or decay occurs, the sum of the baryon numbers of the system before the process must equal the sum of the baryon numbers after the process.

An equivalent statement is that the net number of baryons remains constant in any process.

If baryon number is absolutely conserved, the proton must be absolutely stable. For example, a decay of the proton to a positron and a neutral pion would satisfy conservation of energy, momentum, and electric charge. Such a decay has never been observed, however. At present, we can say only that the proton has a half-life of at least 10^{33} years (the estimated age of the Universe is only 10^{10} years). Therefore, it is extremely unlikely that one would see a given proton undergo a decay process. If we collect a huge number of protons, however, perhaps we might see *some* proton in the collection undergo a decay, as addressed in Example 31.2.

QUICK QUIZ 31.2 Consider the following decays: **(i)** $n \rightarrow \pi^+ + \pi^- + \mu^+ + \mu^-$ and **(ii)** $n \rightarrow p + \pi^-$. From the following choices, which conservation laws are violated by each decay? **(a)** energy **(b)** electric charge **(c)** baryon number **(d)** angular momentum **(e)** no conservation laws

---

Example **31.1** | **Checking Baryon Numbers**

Use the law of conservation of baryon number to determine whether each of the following reactions can occur:

(A) $p + n \rightarrow p + p + n + \bar{p}$

SOLUTION

Conceptualize The mass on the right is larger than the mass on the left. Therefore, one might be tempted to claim that the reaction violates energy conservation. The reaction can indeed occur, however, if the initial particles have sufficient kinetic energy to allow for the increase in rest energy of the system.

Categorize We use a conservation law developed in this section, so we categorize this example as a substitution problem.

Evaluate the total baryon number for the left side of the reaction: $1 + 1 = 2$

Evaluate the total baryon number for the right side of the reaction: $1 + 1 + 1 + (-1) = 2$

Therefore, baryon number is conserved and the reaction can occur.

(B) $p + n \rightarrow p + p + \bar{p}$

SOLUTION

Evaluate the total baryon number for the left side of the reaction: $1 + 1 = 2$

Evaluate the total baryon number for the right side of the reaction: $1 + 1 + (-1) = 1$

Because baryon number is not conserved, the reaction cannot occur.

Example 31.2 | **Detecting Proton Decay**

Measurements taken at the Super Kamiokande neutrino detection facility (Fig. 31.6) indicate that the half-life of protons is at least 10^{33} yr.

(A) Estimate how long we would have to watch, on average, to see a proton in a glass of water decay.

SOLUTION

Conceptualize Imagine the number of protons in a glass of water. Although this number is huge, the probability of a single proton undergoing decay is small, so we would expect to wait for a long time interval before observing a decay.

Categorize Because a half-life is provided in the problem, we categorize this problem as one in which we can apply our statistical analysis techniques from Section 30.3.

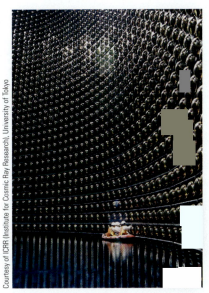

Figure 31.6 (Example 31.2) This detector at the Super Kamiokande neutrino facility in Japan is used to study photons and neutrinos. It holds 50 000 metric tons of highly purified water and 13 000 photomultipliers. The photograph was taken while the detector was being filled. Technicians in a raft clean the photodetectors before they are submerged.

Courtesy of ICRR (Institute for Cosmic Ray Research), University of Tokyo

Analyze Let's estimate that a drinking glass contains a mass $m = 250$ g of water, with a molar mass $M = 18$ g/mol.

Find the number of molecules of water in the glass:

$$N_{\text{molecules}} = nN_A = \frac{m}{M}N_A$$

Each water molecule contains one proton in each of its two hydrogen atoms plus eight protons in its oxygen atom, for a total of ten protons. Therefore, there are $N = 10N_{\text{molecules}}$ protons in the glass of water.

Find the activity of the protons from Equations 30.5, 30.7, and 30.8:

$$(1) \quad R = \lambda N = \frac{\ln 2}{T_{1/2}}\left(10\frac{m}{M}N_A\right) = \frac{\ln 2}{10^{33}\text{ yr}}(10)\left(\frac{250\text{ g}}{18\text{ g/mol}}\right)(6.02 \times 10^{23}\text{ mol}^{-1})$$

$$= 5.8 \times 10^{-8}\text{ yr}^{-1}$$

Finalize The decay constant represents the probability that *one* proton decays in one year. The probability that *any* proton in our glass of water decays in the one-year interval is given by Equation (1). Therefore, we must watch our glass of water for $1/R \approx$ **17 million years!** That indeed is a long time interval, as expected.

(B) The Super Kamiokande neutrino facility contains 50 000 metric tons of water. Estimate the average time interval between detected proton decays in this much water if the half-life of a proton is 10^{33} yr.

SOLUTION

Analyze The proton decay rate R in a sample of water is proportional to the number N of protons. Set up a ratio of the decay rate in the Super Kamiokande facility to that in a glass of water:

$$\frac{R_{\text{Kamiokande}}}{R_{\text{glass}}} = \frac{N_{\text{Kamiokande}}}{N_{\text{glass}}} \quad \rightarrow \quad R_{\text{Kamiokande}} = \frac{N_{\text{Kamiokande}}}{N_{\text{glass}}}R_{\text{glass}}$$

The number of protons is proportional to the mass of the sample, so express the decay rate in terms of mass:

$$R_{\text{Kamiokande}} = \frac{m_{\text{Kamiokande}}}{m_{\text{glass}}}R_{\text{glass}}$$

Substitute numerical values:

$$R_{\text{Kamiokande}} = \left(\frac{50\,000\text{ metric tons}}{0.250\text{ kg}}\right)\left(\frac{1\,000\text{ kg}}{1\text{ metric ton}}\right)(5.8 \times 10^{-8}\text{ yr}^{-1}) \approx 12\text{ yr}^{-1}$$

Finalize The average time interval between decays is about one-twelfth of a year, or approximately **one month.** That is much shorter than the time interval in part (A) due to the tremendous amount of water in the detector facility. Despite this rosy prediction of one proton decay per month, a proton decay has never been observed. This suggests that the half-life of the proton may be larger than 10^{33} years or that proton decay simply does not occur.

Lepton Number

From observations of commonly occurring decays of the electron, muon, and tau, we arrive at three conservation laws involving lepton numbers, one for each variety of lepton. The **law of conservation of electron lepton number** states that

▶ Conservation of electron lepton number

> the sum of the electron lepton numbers of the system before a reaction or decay must equal the sum of the electron lepton numbers after the reaction or decay.

The electron and the electron neutrino are assigned a positive electron lepton number $L_e = +1$, the antileptons e^+ and $\bar{\nu}_e$ are assigned a negative electron lepton number $L_e = -1$; all others have $L_e = 0$. For example, consider the decay of the neutron

$$n \rightarrow p + e^- + \bar{\nu}_e$$

Before the decay, the electron lepton number is $L_e = 0$; after the decay, it is $0 + 1 + (-1) = 0$. Therefore, the electron lepton number is conserved. It is important to recognize that the baryon number must also be conserved; which can easily be checked by noting that before the decay $B = +1$ and after the decay B is $+1 + 0 + 0 = +1$.

Similarly, when a decay involves muons, the muon lepton number L_μ is conserved. The μ^- and the ν_μ are assigned positive numbers, $L_\mu = +1$, the antimuons μ^+ and $\bar{\nu}_\mu$ are assigned negative numbers, $L_\mu = -1$; all others have $L_\mu = 0$. Finally, the tau lepton number L_τ is conserved, and similar assignments can be made for the tau lepton and its neutrino.

> **QUICK QUIZ 31.3** Consider the following decay: $\pi^0 \rightarrow \mu^- + e^+ + \nu_\mu$. What conservation laws are violated by this decay? **(a)** energy **(b)** angular momentum **(c)** electric charge **(d)** baryon number **(e)** electron lepton number **(f)** muon lepton number **(g)** tau lepton number **(h)** no conservation laws

> **QUICK QUIZ 31.4** Suppose a claim is made that the decay of the neutron is given by $n \rightarrow p + e^-$. What conservation laws are violated by this decay? **(a)** energy **(b)** angular momentum **(c)** electric charge **(d)** baryon number **(e)** electron lepton number **(f)** muon lepton number **(g)** tau lepton number **(h)** no conservation laws

---

Example 31.3 | Checking Lepton Numbers

Use the law of conservation of lepton numbers to determine whether each of the following decay schemes (A) and (B) can occur:

(A) $\mu^- \rightarrow e^- + \bar{\nu}_e + \nu_\mu$

SOLUTION

Conceptualize Because this decay involves a muon and an electron, L_μ and L_e must each be conserved separately if the decay is to occur.

Categorize We use a conservation law developed in this section, so we categorize this example as a substitution problem.

Evaluate the lepton numbers before the decay: $L_\mu = +1$ $L_e = 0$

Evaluate the total lepton numbers after the decay: $L_\mu = 0 + 0 + 1 = +1$ $L_e = +1 + (-1) + 0 = 0$

Therefore, both numbers are conserved and on this basis the decay is possible.

(B) $\pi^+ \rightarrow \mu^+ + \nu_\mu + \nu_e$

31.3 *cont.*

SOLUTION

Evaluate the lepton numbers before the decay: $L_\mu = 0$ $L_e = 0$

Evaluate the total lepton numbers after the decay: $L_\mu = -1 + 1 + 0 = 0$ $L_e = 0 + 0 + 1 = 1$

Therefore, the decay is not possible because electron lepton number is not conserved.

31.6 | Strange Particles and Strangeness

Many particles discovered in the 1950s were produced by the nuclear interaction of pions with protons and neutrons in the atmosphere. A group of these particles—the kaon (K), lambda (Λ), and sigma (Σ) particles—exhibited unusual properties in production and decay and hence were called *strange particles*.

One unusual property is that these particles are always produced in pairs. For example, when a pion collides with a proton, two neutral strange particles are produced with high probability:

$$\pi^- + p \rightarrow \Lambda^0 + K^0$$

On the other hand, the reaction $\pi^- + p \rightarrow n^0 + K^0$ in which only one of the final particles is strange never occurs, even though no conservation laws known in the 1950s are violated and the energy of the pion is sufficient to initiate the reaction.

The second peculiar feature of strange particles is that, although they are produced by the strong force at a high rate, they do not decay at a very high rate into particles that interact via the strong force. Instead, they decay very slowly, which is characteristic of the weak interaction. Their half-lives are in the range 10^{-10} s to 10^{-8} s; most other particles that interact via the strong force have very short lifetimes, on the order of 10^{-20} s or less.

Such observations indicate the necessity to make modifications in our model. To explain these unusual properties of strange particles, a new quantum number S, called **strangeness,** was introduced into our model of elementary particles, together with a new conservation law. The strangeness numbers for some particles are given in Table 31.2. The production of strange particles in pairs is handled by assigning $S = +1$ to one of the particles and $S = -1$ to the other. All nonstrange particles are assigned strangeness $S = 0$. The **law of conservation of strangeness** states that

---

whenever a reaction or decay occurs via the strong force, the sum of the strangeness numbers of the system before the process must equal the sum of the strangeness numbers after the process. In processes that occur via the weak interaction, strangeness may not be conserved.

▶ Conservation of strangeness

The low decay rate of strange particles can be explained by assuming that the nuclear and electromagnetic interactions obey the law of conservation of strangeness, but the weak interaction does not. Because the decay reaction involves the loss of one strange particle, it violates strangeness conservation and hence proceeds slowly via the weak interaction.

Example 31.4 | Is Strangeness Conserved?

(A) Use the law of strangeness conservation to determine whether the reaction $\pi^0 + n \rightarrow K^+ + \Sigma^-$ occurs.

SOLUTION

Conceptualize We recognize that there are strange particles appearing in this reaction, so we see that we will need to investigate conservation of strangeness.

Categorize We use a conservation law developed in this section, so we categorize this example as a substitution problem.

continued

31.4 *cont.*

Evaluate the strangeness for the left side of the reaction using Table 31.2: $S = 0 + 0 = 0$

Evaluate the strangeness for the right side of the reaction: $S = +1 - 1 = 0$

Therefore, strangeness is conserved and the reaction is allowed.

(B) Show that the reaction $\pi^- + p \rightarrow \pi^- + \Sigma^+$ does not conserve strangeness.

SOLUTION

Evaluate the strangeness for the left side of the reaction: $S = 0 + 0 = 0$

Evaluate the strangeness for the right side of the reaction: $S = 0 + (-1) = -1$

Therefore, strangeness is not conserved.

31.7 | Measuring Particle Lifetimes

The bewildering array of entries in Table 31.2 leaves one yearning for firm ground. In fact, it is natural to wonder about an entry, for example, that shows a particle (Σ^0) that exists for 10^{-20} s and has a mass of 1192.5 MeV/c^2. How is it possible to detect a particle that exists for only 10^{-20} s?

Most particles are unstable and are created in nature only rarely, in cosmic ray showers. In the laboratory, however, large numbers of these particles are created in controlled collisions between high-energy particles and a suitable target. The incident particles must have very high energy, and it takes a considerable time interval for electromagnetic fields to accelerate particles to high energies. Therefore, stable charged particles such as electrons or protons generally make up the incident beam. Similarly, targets must be simple and stable, and the simplest target, hydrogen, serves nicely as both target (the proton) and detector.

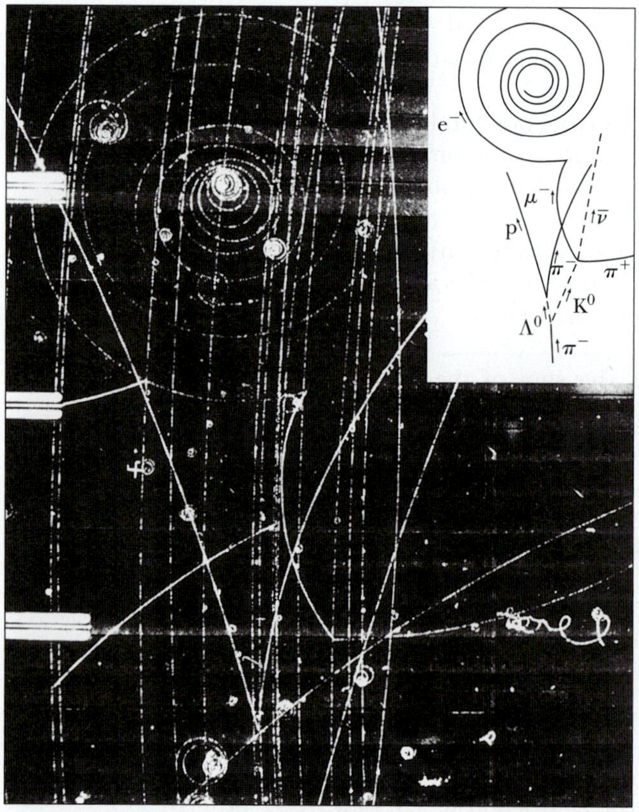

Courtesy Lawrence Berkeley Laboratory, University of California, Photographic Services

Figure 31.7 This bubble-chamber photograph shows many events, and the inset is a drawing of identified tracks. The strange particles Λ^0 and K^0 are formed at the bottom as the π^- interacts with a proton according to $\pi^- + p \rightarrow \Lambda^0 + K^0$. (Notice that the neutral particles leave no tracks, as indicated by the dashed lines in the insert.) The Λ^0 then decays in the reaction $\Lambda^0 \rightarrow \pi^- + p$ and the K^0 in the reaction $K^0 \rightarrow \pi^0 + \mu^- + \bar{\nu}_\mu$.

Figure 31.7 shows a typical event in which hydrogen in a bubble chamber served as both target source and detector. (A bubble chamber is a device in which the tracks of charged particles are made visible in liquid hydrogen that is maintained near its boiling point.) Many parallel tracks of negative pions are visible entering the photograph from the bottom. As the labels in the inset drawing show, one of the pions has hit a stationary proton in the hydrogen, producing two strange particles, Λ^0 and K^0, according to the reaction

$$\pi^- + p \rightarrow \Lambda^0 + K^0$$

Neither neutral strange particle leaves a track, but their subsequent decays into charged particles can be clearly seen as indicated in Figure 31.7. A magnetic field directed into the plane of the photograph causes the track of each charged particle to curve, and from the measured curvature one can determine the particle's charge and linear momentum. If the mass and momentum of the incident particle are known, we can then usually calculate the product particle mass, kinetic energy, and speed from conservation of momentum and energy. Finally, by combining a product particle's speed with a measurable decay track length, we can calculate the product particle's lifetime. Figure 31.7 shows that sometimes one can use this lifetime technique even for a neutral particle, which leaves no track. As long as the beginning and end points of the missing track are known as well as the particle speed, one can infer the missing track length and find the lifetime of the neutral particle.

Resonance Particles

With clever experimental technique and much effort, decay track lengths as short as 10^{-6} m can be measured. Therefore, lifetimes as short as 10^{-16} s can be measured for high-energy particles traveling at about the speed of light. We arrive at this result by assuming that a decaying particle travels 1 μm at a speed of $0.99c$ in the reference frame of the laboratory, yielding a lifetime of $\Delta t_{lab} = 1 \times 10^{-6}$ m$/0.99c \approx 3.4 \times 10^{-15}$ s. This result is not our final one, however, because we must account for the relativistic effects of time dilation. Because the proper lifetime Δt_p as measured in the decaying particle's reference frame is shorter than the laboratory frame value Δt_{lab} by a factor of $\sqrt{1 - (v^2/c^2)}$ (see Eq. 9.6), we can calculate the proper lifetime:

$$\Delta t_p = \Delta t_{lab} \sqrt{1 - \frac{v^2}{c^2}} = (3.4 \times 10^{-15} \text{ s}) \sqrt{1 - \frac{(0.99c)^2}{c^2}} = 4.8 \times 10^{-16} \text{ s}$$

Unfortunately, even with Einstein's help, the best answer we can obtain with the track length method is several orders of magnitude away from lifetimes of 10^{-20} s. How then can we detect the presence of particles that exist for time intervals like 10^{-20} s? For such short-lived particles, known as **resonance particles,** all we can do is infer their masses, their lifetimes, and, indeed, their very existence from data on their decay products.

31.8 | Finding Patterns in the Particles

A tool scientists use to help understand nature is the detection of patterns in data. One of the best examples of the use of this tool is the development of the periodic table, which provides fundamental understanding of the chemical behavior of the elements. The periodic table explains how more than a hundred elements can be formed from three particles: the electron, proton, and neutron. The number of observed particles and resonances observed by particle physicists is even larger than the number of elements. Is it possible that a small number of entities could exist from which all these particles could be built? Motivated by the success of the periodic table, let us explore the historical search for patterns among the particles.

Many classification schemes have been proposed for grouping particles into families. Consider, for instance, the baryons listed in Table 31.2 that have spins of $\frac{1}{2}$: p, n, $\Lambda^0, \Sigma^+, \Sigma^0, \Sigma^-, \Xi^0$, and Ξ^-. If we plot strangeness versus charge for these baryons using a sloping coordinate system, as in Figure 31.8a, we observe a fascinating pattern. Six of the baryons form a hexagon, and the remaining two are at the hexagon's center.[3]

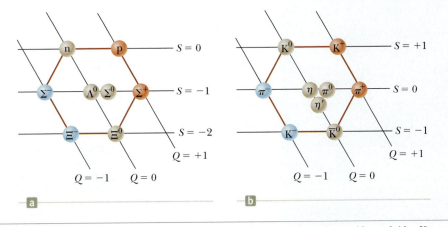

Figure 31.8 (a) The hexagonal eightfold-way pattern for the eight spin-$\frac{1}{2}$ baryons. This strangeness-versus-charge plot uses a sloping axis for charge number Q and a horizontal axis for strangeness S. (b) The eightfold-way pattern for the nine spin-zero mesons.

---

[3]The reason for the sloping coordinate system is so that a *regular* hexagon is formed, one with equal sides. If a normal orthogonal coordinate system is used, the pattern still appears, but the hexagonal shape does not have equal sides. Try it!

Figure 31.9 The pattern for the higher-mass, spin-$\frac{3}{2}$ baryons known at the time the pattern was proposed.

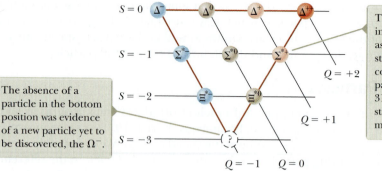

The particles indicated with an asterisk are excited states of the corresponding particles in Figure 31.8. These excited states have higher mass and spin $\frac{3}{2}$.

The absence of a particle in the bottom position was evidence of a new particle yet to be discovered, the Ω^-.

Murray Gell-Mann
American Physicist (b. 1929)
In 1969, Murray Gell-Mann was awarded the Nobel Prize in Physics for his theoretical studies dealing with subatomic particles.

As a second example, consider the following nine spin-zero mesons listed in Table 31.2: π^+, π^0, π^-, K^+, K^0, K^-, η, η', and the antiparticle $\overline{K}^0$. Figure 31.8b is a plot of strangeness versus charge for this family. Again, a hexagonal pattern emerges. In this case, each particle on the perimeter of the hexagon lies opposite its antiparticle, and the remaining three (which form their own antiparticles) are at its center. These and related symmetric patterns were developed independently in 1961 by Murray Gell-Mann and Yuval Ne'eman (1925–2006). Gell-Mann called the patterns the **eightfold way,** after the eightfold path to nirvana in Buddhism.

Groups of baryons and mesons can be displayed in many other symmetric patterns within the framework of the eightfold way. For example, the family of spin-$\frac{3}{2}$ baryons known in 1961 contains nine particles arranged in a pattern like that of the pins in a bowling alley as in Figure 31.9. [The particles Σ^{*+}, Σ^{*0}, Σ^{*-}, Ξ^{*0}, and Ξ^{*-} are excited states of the particles Σ^+, Σ^0, Σ^-, Ξ^0, and Ξ^-. In these higher-energy states, the spins of the three quarks (see Section 31.9) making up the particle are aligned so that the total spin of the particle is $\frac{3}{2}$.] When this pattern was proposed, an empty spot occurred in it (at the bottom position), corresponding to a particle that had never been observed. Gell-Mann predicted that the missing particle, which he called the omega minus (Ω^-), should have spin $\frac{3}{2}$, charge -1, strangeness -3, and rest energy of approximately 1 680 MeV. Shortly thereafter, in 1964, scientists at the Brookhaven National Laboratory found the missing particle through careful analyses of bubble-chamber photographs (Fig. 31.10) and confirmed all its predicted properties.

The prediction of the missing particle from the eightfold way has much in common with the prediction of missing elements in the periodic table. Whenever a vacancy occurs in an organized pattern of information, experimentalists have a guide for their investigations.

Figure 31.10 Discovery of the Ω^- particle. The photograph on the left shows the original bubble-chamber tracks. The drawing on the right isolates the tracks of the important events.

The K^- particle at the bottom collides with a proton to produce the first detected Ω^- particle plus a K^0 and a K^+.

❰ 31.9 | Quarks

As we have noted, leptons appear to be truly elementary particles because they occur in a small number of types, have no measurable size or internal structure, and do not seem to break down to smaller units. Hadrons, on the other hand, are complex particles having size and structure. The existence of the eightfold-way patterns suggests that hadrons have a more elemental substructure. Furthermore, we know that hundreds of types of hadrons exist and that many of them decay into other hadrons. These facts strongly suggest that hadrons cannot be truly elementary. In this section, we show that the complexity of hadrons can be explained by a simple substructure.

The Original Quark Model: A Structural Model for Hadrons

In 1963, Gell-Mann and George Zweig (b. 1937) independently proposed that hadrons have a more elemental substructure. According to their structural model, all hadrons are composite systems of two or three fundamental constituents called **quarks** (pronounced to rhyme with *forks*). (Gell-Mann borrowed the word *quark* from the passage "Three quarks for Muster Mark" in James Joyce's *Finnegan's Wake*.) The model proposes that three types of quarks exist, designated by the symbols u, d, and s. They are given the arbitrary names **up, down,** and **strange.** The various types of quarks are called **flavors.** Baryons consist of three quarks, and mesons consist of a quark and an antiquark. Active Figure 31.11 is a pictorial representation of the quark composition of several hadrons.

An unusual property of quarks is that they carry a fractional electronic charge. The u, d, and s quarks have charges of $+\frac{2}{3}e$, $-\frac{1}{3}e$, and $-\frac{1}{3}e$, respectively, where e is the elementary charge 1.6×10^{-19} C. These and other properties of quarks and antiquarks are given in Table 31.3. Notice that quarks have spin $\frac{1}{2}$, which means that all quarks are *fermions*, defined as any particle having half-integral spin. As Table 31.3

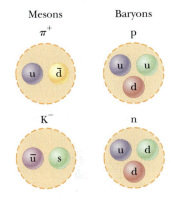

Active Figure 31.11 Quark compositions of two mesons and two baryons.

❰ TABLE 31.3 | Properties of Quarks and Antiquarks

Quarks

Name	Symbol	Spin	Charge	Baryon Number	Strangeness	Charm	Bottomness	Topness
Up	u	$\frac{1}{2}$	$+\frac{2}{3}e$	$\frac{1}{3}$	0	0	0	0
Down	d	$\frac{1}{2}$	$-\frac{1}{3}e$	$\frac{1}{3}$	0	0	0	0
Strange	s	$\frac{1}{2}$	$-\frac{1}{3}e$	$\frac{1}{3}$	−1	0	0	0
Charmed	c	$\frac{1}{2}$	$+\frac{2}{3}e$	$\frac{1}{3}$	0	+1	0	0
Bottom	b	$\frac{1}{2}$	$-\frac{1}{3}e$	$\frac{1}{3}$	0	0	+1	0
Top	t	$\frac{1}{2}$	$+\frac{2}{3}e$	$\frac{1}{3}$	0	0	0	+1

Antiquarks

Name	Symbol	Spin	Charge	Baryon Number	Strangeness	Charm	Bottomness	Topness
Anti-up	$\overline{u}$	$\frac{1}{2}$	$-\frac{2}{3}e$	$-\frac{1}{3}$	0	0	0	0
Anti-down	$\overline{d}$	$\frac{1}{2}$	$+\frac{1}{3}e$	$-\frac{1}{3}$	0	0	0	0
Anti-strange	$\overline{s}$	$\frac{1}{2}$	$+\frac{1}{3}e$	$-\frac{1}{3}$	+1	0	0	0
Anti-charmed	$\overline{c}$	$\frac{1}{2}$	$-\frac{2}{3}e$	$-\frac{1}{3}$	0	−1	0	0
Anti-bottom	$\overline{b}$	$\frac{1}{2}$	$+\frac{1}{3}e$	$-\frac{1}{3}$	0	0	−1	0
Anti-top	$\overline{t}$	$\frac{1}{2}$	$-\frac{2}{3}e$	$-\frac{1}{3}$	0	0	0	−1

shows, associated with each quark is an antiquark of opposite charge, baryon number, and strangeness.

The composition of all hadrons known when Gell-Mann and Zweig presented their models can be completely specified by three simple rules:

- A meson consists of one quark and one antiquark, giving it a baryon number of 0, as required.
- A baryon consists of three quarks.
- An antibaryon consists of three antiquarks.

The theory put forth by Gell-Mann and Zweig is referred to as the *original quark model*.

QUICK QUIZ 31.5 Using a coordinate system like that in Figure 31.8, draw an eightfold-way diagram for the three quarks in the original quark model.

Charm and Other Developments

Although the original quark model was highly successful in classifying particles into families, some discrepancies were evident between predictions of the model and certain experimental decay rates. It became clear that the structural model needed to be modified to remove these discrepancies. Consequently, several physicists proposed a fourth quark in 1967. They argued that if four leptons exist (as was thought at the time: the electron, the muon, and a neutrino associated with each), four quarks should also exist because of an underlying symmetry in nature. The fourth quark, designated by c, was given a property called **charm.** A **charmed quark** has charge $+\frac{2}{3}e$, but its charm distinguishes it from the other three quarks. This addition introduces a new quantum number C, representing charm. The new quark has charm $C = +1$, its antiquark has charm $C = -1$, and all other quarks have $C = 0$ as indicated in Table 31.3. Charm, like strangeness, is conserved in strong and electromagnetic interactions, but not in weak interactions.

Evidence that the charmed quark exists began to accumulate in 1974 when a new heavy particle called the J/Ψ particle (or simply Ψ) was discovered independently by two groups, one led by Burton Richter (b. 1931) at the Stanford Linear Accelerator (SLAC), and the other led by Samuel Ting (b. 1936) at the Brookhaven National Laboratory. Richter and Ting were awarded the Nobel Prize in Physics in 1976 for this work. The J/Ψ particle does not fit into the three-quark structural model; instead, it has properties of a combination of the proposed charmed quark and its antiquark ($c\bar{c}$). It is much more massive than the other known mesons ($\sim$3 100 MeV/c^2), and its lifetime is much longer than the lifetimes of particles that decay via the strong force. Soon, related mesons were discovered, corresponding to such quark combinations as $\bar{c}d$ and $c\bar{d}$, which all have large masses and long lifetimes. The existence of these new mesons provided firm evidence for the fourth quark flavor.

In 1975, researchers at Stanford University reported strong evidence for the tau (τ) lepton with a mass of 1 784 MeV/c^2. It is the fifth type of lepton to be discovered, which led physicists to propose that more flavors of quarks may exist, based on symmetry arguments similar to those leading to the proposal of the charmed quark. These proposals led to more elaborate quark models and the prediction of two new quarks: **top** (t) and **bottom** (b). To distinguish these quarks from the original four, quantum numbers called *topness* and *bottomness* (with allowed values +1, 0, −1) are assigned to all quarks and antiquarks (Table 31.3). In 1977, researchers at the Fermi National Laboratory, under the direction of Leon Lederman (b. 1922), reported the discovery of a very massive new meson Υ whose composition is considered to be $b\bar{b}$, providing evidence for the bottom quark. In March 1995, researchers at Fermilab announced the discovery of the top quark (supposedly the last of the quarks to be found), with a mass of 173 GeV/c^2.

TABLE 31.4 | Quark Composition of Mesons

		Antiquarks				
		$\bar{b}$	$\bar{c}$	$\bar{s}$	$\bar{d}$	$\bar{u}$
Quarks	**b**	Υ ($\bar{b}b$)	B_c^- ($\bar{c}b$)	$\bar{B}_s^{\,0}$ ($\bar{s}b$)	$\bar{B}_d^{\,0}$ ($\bar{d}b$)	B^- ($\bar{u}b$)
	c	B_c^+ ($\bar{b}c$)	J/Ψ ($\bar{c}c$)	D_s^+ ($\bar{s}c$)	D^+ ($\bar{d}c$)	D^0 ($\bar{u}c$)
	s	B_s^0 ($\bar{b}s$)	D_s^- ($\bar{c}s$)	η, η' ($\bar{s}s$)	$\bar{K}^{\,0}$ ($\bar{d}s$)	K^- ($\bar{u}s$)
	d	B_d^0 ($\bar{b}d$)	D^- ($\bar{c}d$)	K^0 ($\bar{s}d$)	π^0, η, η' ($\bar{d}d$)	π^- ($\bar{u}d$)
	u	B^+ ($\bar{b}u$)	$\bar{D}^{\,0}$ ($\bar{c}u$)	K^+ ($\bar{s}u$)	π^+ ($\bar{d}u$)	π^0, η, η' ($\bar{u}u$)

Note: The top quark does not form mesons because it decays too quickly.

Table 31.4 lists the quark compositions of mesons formed from the up, down, strange, charmed, and bottom quarks. Table 31.5 shows the quark combinations for the baryons listed in Table 31.2. Note that only two flavors of quarks, u and d, are contained in all hadrons encountered in ordinary matter (protons and neutrons).

You are probably wondering if such discoveries will ever end. How many "building blocks" of matter really exist? At present, physicists believe that the fundamental particles in nature are six quarks and six leptons (together with their antiparticles) listed in Table 31.6 and the field particles listed in Table 31.1. Table 31.6 lists the rest energies and charges of the quarks and leptons.

Despite extensive experimental effort, no isolated quark has ever been observed. Physicists now believe that quarks are permanently confined inside hadrons because of the strong force, which prevents them from escaping. Current efforts are under way to form a **quark-gluon plasma,** a state of matter in which the quarks are freed from neutrons and protons. In 2000, scientists at CERN announced evidence for a quark-gluon plasma formed by colliding lead nuclei. In 2005, scientists at the Relativistic Heavy Ion Collider (RHIC) at Brookhaven reported evidence from four experimental studies of a new state of matter that may be a quark-gluon plasma. Neither the CERN nor RHIC results are entirely conclusive and have not been independently verified. Three experimental detectors at the new Large Hadron Collider (LHC) at CERN will look for evidence of the creation of a quark-gluon plasma.

TABLE 31.5 |
Quark Composition of Several Baryons

Particle	Quark Composition
p	uud
n	udd
Λ^0	uds
Σ^+	uus
Σ^0	uds
Σ^-	dds
Δ^{++}	uuu
Δ^+	uud
Δ^0	udd
Δ^-	ddd
Ξ^0	uss
Ξ^-	dss
Ω^-	sss

Note: Some baryons have the same quark composition, such as the p and the Δ^+ and the n and the Δ^0. In these cases, the Δ particles are considered to be excited states of the proton and neutron.

TABLE 31.6 | The Elementary Particles and Their Rest Energies and Charges

Particle	Approximate Rest Energy	Charge
Quarks		
u	2.4 MeV	$+\frac{2}{3}e$
d	4.8 MeV	$-\frac{1}{3}e$
s	104 MeV	$-\frac{1}{3}e$
c	1.27 GeV	$+\frac{2}{3}e$
b	4.2 GeV	$-\frac{1}{3}e$
t	173 GeV	$+\frac{2}{3}e$
Leptons		
e^-	511 keV	$-e$
μ^-	105.7 MeV	$-e$
τ^-	1.78 GeV	$-e$
ν_e	< 2 eV	0
ν_μ	< 0.17 MeV	0
ν_τ	< 18 MeV	0

> **THINKING PHYSICS 31.2**

We have seen a law of conservation of *lepton number* and a law of conservation of *baryon number*. Why isn't there a law of conservation of *meson number*?

Reasoning We can argue from the point of view of creating particle–antiparticle pairs from available energy. (Review pair production in Section 31.2.) If energy is converted to rest energy of a lepton–antilepton pair, no net change occurs in lepton number because the lepton has a lepton number of +1 and the antilepton −1. Energy can also be transformed into rest energy

of a baryon–antibaryon pair. The baryon has baryon number +1, the antibaryon −1, and no net change in baryon number occurs.

Now, however, suppose energy is transformed into rest energy of a quark–antiquark pair. By definition in quark theory, a quark–antiquark pair *is a meson*. Therefore, we have created a meson from energy because no meson existed before, now one does. Therefore, meson number is not conserved. With more energy, we can create more mesons, with no restriction from a conservation law other than that of energy. ◄

❰31.10 | Multicolored Quarks

Shortly after the concept of quarks was proposed, scientists recognized that certain particles had quark compositions that violated the Pauli exclusion principle. As noted in Pitfall Prevention 29.3 in Chapter 29, all fermions obey the exclusion principle. Because all quarks are fermions with spin $\frac{1}{2}$, they are expected to follow the exclusion principle. One example of a particle that appears to violate the exclusion principle is the Ω^- (sss) baryon that contains three s quarks having parallel spins, giving it a total spin of $\frac{3}{2}$. Other examples of baryons that have identical quarks with parallel spins are the Δ^{++} (uuu) and the Δ^- (ddd). To resolve this problem, in 1965 Moo-Young Han (b. 1934) and Yoichiro Nambu (b. 1921) suggested a modification of the structural model of quarks in which quarks possess a new property called **color**. This property is similar in many respects to electric charge except that it occurs in three varieties called **red, green,** and **blue**. The antiquarks have the colors **antired, antigreen,** and **antiblue**. To satisfy the exclusion principle, all three quarks in a baryon must have different colors. Just as a combination of actual colors of light can produce the neutral color white, a combination of three quarks with different colors is also described as white, or colorless. A meson consists of a quark of one color and an antiquark of the corresponding anticolor. The result is that baryons and mesons are always colorless (or white).

Although the concept of color in the quark model was originally conceived to satisfy the exclusion principle, it also provided a better theory for explaining certain experimental results. For example, the modified theory correctly predicts the lifetime of the π^0 meson. The theory of how quarks interact with one another is called **quantum chromodynamics,** or QCD, to parallel quantum electrodynamics (the theory of interaction between electric charges). In QCD, the quark is said to carry a **color charge,** in analogy to electric charge. The strong force between quarks is often called the **color force.**

The color force between quarks is analogous to the electric force between charges; like colors repel and opposite colors attract. Therefore, two green quarks repel each other, but a green quark is attracted to an antigreen quark. The attraction between quarks of opposite color to form a meson ($q\overline{q}$) is indicated in Figure 31.12a. Differently colored quarks also attract one another, but with less strength than opposite colors of quark and antiquark. For example, a cluster of red, blue, and green quarks all attract one another to form a baryon as indicated in Figure 31.12b. Therefore, every baryon contains three quarks of three different colors.

As stated earlier, the strong force between quarks is carried by massless particles that travel at the speed of light called **gluons.** According to QCD, there are eight gluons, all carrying two color charges, a color and an anticolor such as a "blue–antired" gluon. When a quark emits or absorbs a gluon, its color changes. For example, a blue quark that emits a blue–antired gluon becomes a red quark, and a red quark that absorbs this gluon becomes a blue quark.

> **Pitfall Prevention | 31.3**
> **Color Charge Is Not Really Color**
> The description of color for a quark has nothing to do with visual sensation from light. It is simply a convenient name for a property that is analogous to electric charge.

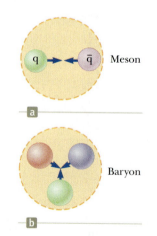

Figure 31.12 (a) A green quark is attracted to an antigreen quark, forming a meson whose quark structure is ($q\overline{q}$). (b) Three quarks of different colors attract one another to form a baryon.

Figure 31.13a shows a Feynman diagram representing the interaction between a neutron and a proton by means of Yukawa's pion, in this case a π^-. In Figure 31.13a, the charged pion carries charge from one nucleon to the other, so the nucleons change identities and the proton becomes a neutron and the neutron becomes a proton. (This process differs from Fig. 31.5a, in which the field particle is a π^0, resulting in no transfer of charge from one nucleon to the other.)

Let us look at the same interaction from the viewpoint of the quark model shown in Figure 31.13b. In this Feynman diagram, the proton and neutron are represented by their quark constituents. Each quark in the neutron and proton is continuously emitting and absorbing gluons. The energy of a gluon can result in the creation of quark–antiquark pairs. This is similar to the creation of electron–positron pairs in pair production, which we investigated in Section 31.2. When the neutron and proton approach to within 1 to 2 fm of each other, these gluons and quarks can be exchanged between the two nucleons, and such exchanges produce the strong force. Figure 31.13b depicts one possibility for the process shown in Figure 31.13a. A down quark in the neutron on the right emits a gluon. The energy of the gluon is then transformed to create a $u\bar{u}$ pair. The u quark stays within the nucleon (which has now changed to a proton), and the recoiling d quark and the $\bar{u}$ antiquark are transmitted to the proton on the left side of the diagram. Here the $\bar{u}$ annihilates a u quark within the proton and the d is captured. Therefore, the net effect is to change a u quark to a d quark, and the proton has changed to a neutron.

As the d quark and $\bar{u}$ antiquark in Figure 31.13b transfer between the nucleons, the d and $\bar{u}$ exchange gluons with each other and can be considered to be bound to each other by means of the strong force. If we look back at Table 31.4, we see that this combination is a π^-, which is Yukawa's field particle! Therefore, the quark model of interactions between nucleons is consistent with the pion-exchange model.

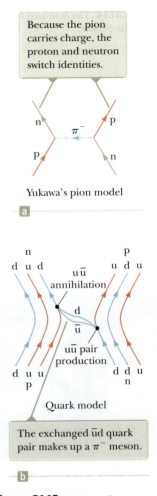

Figure 31.13 (a) A nuclear interaction between a proton and a neutron explained in terms of Yukawa's pion-exchange model. (b) The same interaction, explained in terms of quarks and gluons.

31.11 | The Standard Model

Scientists now believe that there are three classifications of truly elementary particles: leptons, quarks, and field particles. These three particles are further classified as either fermions or bosons. Quarks and leptons have spin $\frac{1}{2}$ and hence are fermions, whereas the field particles have integral spin of 1 or higher and are bosons.

Recall from Section 31.1 that the weak force is believed to be mediated by the W^+, W^-, and Z^0 bosons. These particles are said to have *weak charge* just as quarks have color charge. Therefore, each elementary particle can have mass, electric charge, color charge, and weak charge. Of course, one or more of these could be zero.

In 1979, Sheldon Glashow (b. 1932), Abdus Salam (1926–1996), and Steven Weinberg (b. 1933) won the Nobel Prize in Physics for developing a theory that unified the electromagnetic and weak interactions. This **electroweak theory** postulates that the weak and electromagnetic interactions have the same strength at very high particle energies. The two interactions are viewed as two different manifestations of a single unifying electroweak interaction. The photon and the three massive bosons ($W^\pm$ and Z^0) play a key role in the electroweak theory. The theory makes many concrete predictions, but perhaps the most spectacular is the prediction of the masses of the W and Z particles at about 82 GeV/c^2 and 93 GeV/c^2, respectively. As mentioned earlier, the 1984 Nobel Prize in Physics was awarded to Carlo Rubbia and Simon van der Meer for their work leading to the discovery of these particles at these energies at the CERN Laboratory in Geneva, Switzerland.

The combination of the electroweak theory and QCD for the strong interaction form what is referred to in high-energy physics as the **Standard Model.** Although the details of the Standard Model are complex, its essential ingredients can be summarized with the help of Figure 31.14 on page 1074. (The Standard Model does not include the gravitational force at present; we include gravity in Fig. 31.14, however, because physicists hope to eventually incorporate this force into a unified theory.)

31.6 *cont.*

Analyze Write an expression for the total mechanical energy of the system and set it equal to zero, representing the galaxy moving at the escape speed:

$$E_{\text{total}} = K + U = \tfrac{1}{2}mv^2 - \frac{GmM}{R} = 0$$

Substitute for the mass M contained within the sphere the product of the critical density and the volume of the sphere:

$$\tfrac{1}{2}mv^2 = \frac{Gm(\tfrac{4}{3}\pi R^3 \rho_c)}{R}$$

Solve for the critical density:

$$\rho_c = \frac{3v^2}{8\pi G R^2}$$

From Hubble's law, substitute for the ratio $v/R = H$:

$$(1) \quad \rho_c = \frac{3}{8\pi G}\left(\frac{v}{R}\right)^2 = \frac{3H^2}{8\pi G}$$

(B) Estimate a numerical value for the critical density in grams per cubic centimeter.

SOLUTION

In Equation (1), substitute numerical values for H and G:

$$\rho_c = \frac{3H^2}{8\pi G} = \frac{3[22 \times 10^{-3}\,\text{m}/(\text{s} \cdot \text{ly})]^2}{8\pi(6.67 \times 10^{-11}\,\text{N} \cdot \text{m}^2/\text{kg}^2)} = 8.7 \times 10^5\,\text{kg/m} \cdot (\text{ly})^2$$

Reconcile the units by converting light-years to meters:

$$\rho_c = 8.7 \times 10^5\,\text{kg/m} \cdot (\text{ly})^2 \left(\frac{1\,\text{ly}}{9.46 \times 10^{15}\,\text{m}}\right)^2$$

$$= 9.7 \times 10^{-27}\,\text{kg/m}^3 = \boxed{9.7 \times 10^{-30}\,\text{g/cm}^3}$$

Finalize Because the mass of a hydrogen atom is 1.67×10^{-24} g, this value of ρ_c corresponds to 6×10^{-6} hydrogen atoms per cubic centimeter or 6 atoms per cubic meter.

Missing Mass in the Universe?

The luminous matter in galaxies averages out to a Universe density of about 5×10^{-33} g/cm³. The radiation in the Universe has a mass equivalent of approximately 2% of the visible matter. The total mass of all nonluminous matter (such as interstellar gas and black holes) may be estimated from the speeds of galaxies orbiting one another in a cluster. The higher the galaxy speeds, the more mass in the cluster. Measurements on the Coma cluster of galaxies indicate that the amount of nonluminous matter is 20 to 30 times the amount of luminous matter present in stars and luminous gas clouds. Yet even this large invisible component of dark matter, if extrapolated to the Universe as a whole, leaves the observed mass density a factor of 10 less than ρ_c. The deficit, called *missing mass*, has been the subject of intense theoretical and experimental work. Exotic particles such as axions, photinos, and superstring particles have been suggested as candidates for the missing mass. More mundane proposals argue that the missing mass is present in certain galaxies as neutrinos. In fact, neutrinos are so abundant that a tiny neutrino rest energy on the order of only 20 eV would furnish the missing mass and "close" the Universe. Therefore, current experiments designed to measure the rest energy of the neutrino will affect predictions for the future of the Universe, showing a clear connection between one of the smallest pieces of the Universe and the Universe as a whole!

Mysterious Energy in the Universe?

A surprising twist in the story of the Universe arose in 1998 with the observation of a class of supernovae that have a fixed absolute brightness. By combining the apparent brightness and the redshift of light from these explosions, their distance and speed of recession of the Earth can be determined. These observations led to the conclusion that the expansion of the Universe is not slowing down but rather is accelerating! Observations by other groups also led to the same interpretation.

To explain this acceleration, physicists have proposed *dark energy,* which is energy possessed by the vacuum of space. In the early life of the Universe, gravity dominated over the dark energy. As the Universe expanded and the gravitational force between galaxies became smaller because of the great distances between them, the dark energy became more important. The dark energy results in an effective repulsive force that causes the expansion rate to increase.[4]

Although we have some degree of certainty about the beginning of the Universe, we are uncertain about how the story will end. Will the Universe keep on expanding forever, or will it someday collapse and then expand again, perhaps in an endless series of oscillations? Results and answers to these questions remain inconclusive, and the exciting controversy continues.

▶ SUMMARY

There are four fundamental forces in nature: **strong, electromagnetic, weak,** and **gravitational.** The strong force is the force between quarks. A residual effect of the strong force is the **nuclear force** between nucleons that keeps the nucleus together. The weak force is responsible for beta decay. The electromagnetic and weak forces are now considered to be manifestations of a single force called the **electroweak force.** Every fundamental interaction is mediated by the exchange of **field particles.** The electromagnetic interaction is mediated by the photon; the weak interaction is mediated by the $W^\pm$ and Z^0 **bosons;** the gravitational interaction is mediated by **gravitons;** the strong interaction is mediated by **gluons.**

An **antiparticle** and a particle have the same mass, but opposite charge, and other properties may have opposite values such as lepton number and baryon number. It is possible to produce particle–antiparticle pairs in nuclear reactions if the available energy is greater than $2mc^2$, where m is the mass of the particle (or antiparticle).

Particles other than field particles are classified as hadrons or leptons. **Hadrons** interact through the strong force. They have size and structure and are not elementary particles. Hadrons are of two types, baryons and mesons. **Mesons** have baryon number zero and have either zero or integral spin. **Baryons,** which generally are the most massive particles, have nonzero baryon number and a spin of $\frac{1}{2}$ or $\frac{3}{2}$. The neutron and proton are examples of baryons.

Leptons have no structure or size and are considered truly elementary. They interact through the weak and electromagnetic forces. The six leptons are the electron e^-, the muon μ^-, the tau τ^-; and their neutrinos ν_e, ν_μ, and ν_τ.

In all reactions and decays, quantities such as energy, linear momentum, angular momentum, electric charge, baryon number, and lepton number are strictly conserved. Certain particles have properties called **strangeness** and **charm.** These unusual properties are conserved only in those reactions and decays that occur via the strong force.

Theories in elementary particle physics have postulated that all hadrons are composed of smaller units known as **quarks.** Quarks have fractional electric charge and come in six "flavors": **up** (u), **down** (d), **strange** (s), **charmed** (c), **top** (t), and **bottom** (b). Each baryon contains three quarks, and each meson contains one quark and one antiquark.

According to the theory of **quantum chromodynamics,** quarks have a property called **color charge,** and the strong force between quarks is referred to as the **color force.**

▶ OBJECTIVE QUESTIONS

☐ denotes answer available in *Student Solutions Manual/Study Guide*

1. An isolated stationary muon decays into an electron, an electron antineutrino, and a muon neutrino. Is the total kinetic energy of these three particles (a) zero, (b) small, or (c) large compared to their rest energies, or (d) none of those choices are possible?

2. Define the average density of the solar system ρ_{SS} as the total mass of the Sun, planets, satellites, rings, asteroids, icy outliers, and comets, divided by the volume of a sphere around the Sun large enough to contain all these objects. The sphere extends about halfway to the nearest star, with a radius of approximately 2×10^{16} m, about two light-years.

   How does this average density of the solar system compare with the critical density ρ_c required for the Universe to stop its Hubble's-law expansion? (a) ρ_{SS} is much greater than ρ_c. (b) ρ_{SS} is approximately or precisely equal to ρ_c. (c) ρ_{SS} is much less than ρ_c. (d) It is impossible to determine.

3. What interactions affect protons in an atomic nucleus? More than one answer may be correct. (a) the nuclear interaction (b) the weak interaction (c) the electromagnetic interaction (d) the gravitational interaction

4. Place the following events into the correct sequence from the earliest in the history of the Universe to the latest.

---

[4]For a discussion of dark energy, see S. Perlmutter, "Supernovae, Dark Energy, and the Accelerating Universe," *Physics Today,* 56(4): 53–60, April 2003.

(a) Neutral atoms form. (b) Protons and neutrons are no longer annihilated as fast as they form. (c) The Universe is a quark–gluon soup. (d) The Universe is like the core of a normal star today, forming helium by nuclear fusion. (e) The Universe is like the surface of a hot star today, consisting of a plasma of ionized atoms. (f) Polyatomic molecules form. (g) Solid materials form.

5. When an electron and a positron meet at low speed in empty space, they annihilate each other to produce two 0.511-MeV gamma rays. What law would be violated if they produced one gamma ray with an energy of 1.02 MeV? (a) conservation of energy (b) conservation of momentum (c) conservation of charge (d) conservation of baryon number (e) conservation of electron lepton number

6. Which of the following field particles mediates the strong force? (a) photon (b) gluon (c) graviton (d) W^+ and Z bosons (e) none of those field particles

7. The Ω^- particle is a baryon with spin $\frac{3}{2}$. Does the Ω^- particle have (a) three possible spin states in a magnetic field, (b) four possible spin states, (c) three times the charge of a spin $-\frac{1}{2}$ particle, or (d) three times the mass of a spin $-\frac{1}{2}$ particle, or (e) are none of those choices correct?

8. In one experiment, two balls of clay of the same mass travel with the same speed v toward each other. They collide head-on and come to rest. In a second experiment, two clay balls of the same mass are again used. One ball hangs at rest, suspended from the ceiling by a thread. The second ball is fired toward the first at speed v, to collide, stick to the first ball, and continue to move forward. Is the kinetic energy that is transformed into internal energy in the first experiment (a) one-fourth as much as in the second experiment, (b) one-half as much as in the second experiment, (c) the same as in the second experiment, (d) twice as much as in the second experiment, or (e) four times as much as in the second experiment?

☐ denotes answer available in *Student Solutions Manual/Study Guide*

CONCEPTUAL QUESTIONS

1. The Ξ^0 particle decays by the weak interaction according to the decay mode $\Xi^0 \rightarrow \Lambda^0 + \pi^0$. Would you expect this decay to be fast or slow? Explain.

2. Are the laws of conservation of baryon number, lepton number, and strangeness based on fundamental properties of nature (as are the laws of conservation of momentum and energy, for example)? Explain.

3. An antibaryon interacts with a meson. Can a baryon be produced in such an interaction? Explain.

4. Describe the essential features of the Standard Model of particle physics.

5. Name the four fundamental interactions and the field particle that mediates each.

6. What are the differences between hadrons and leptons?

7. Kaons all decay into final states that contain no protons or neutrons. What is the baryon number for kaons?

8. Describe the properties of baryons and mesons and the important differences between them.

9. How many quarks are in each of the following: (a) a baryon, (b) an antibaryon, (c) a meson, (d) an antimeson? (e) How do you explain that baryons have half-integral spins, whereas mesons have spins of 0 or 1?

10. In the theory of quantum chromodynamics, quarks come in three colors. How would you justify the statement that "all baryons and mesons are colorless"?

11. The W and Z bosons were first produced at CERN in 1983 by causing a beam of protons and a beam of antiprotons to meet at high energy. Why was this discovery important?

12. How did Edwin Hubble determine in 1928 that the Universe is expanding?

13. Neutral atoms did not exist until hundreds of thousands of years after the Big Bang. Why?

PROBLEMS

WebAssign The problems found in this chapter may be assigned online in Enhanced WebAssign.

1. denotes straightforward problem; 2. denotes intermediate problem; 3. denotes challenging problem

1. denotes full solution available in the *Student Solutions Manual/Study Guide*

1. denotes problems most often assigned in Enhanced WebAssign.

BIO denotes biomedical problem

GP denotes guided problem

M denotes Master It tutorial available in Enhanced WebAssign

Q|C denotes asking for quantitative and conceptual reasoning

S denotes symbolic reasoning problem

shaded denotes "paired problems" that develop reasoning with symbols and numerical values

W denotes Watch It video solution available in Enhanced WebAssign

Section 31.1 The Fundamental Forces in Nature

Section 31.2 Positrons and Other Antiparticles

1. A photon produces a proton–antiproton pair according to the reaction $\gamma \rightarrow p + \bar{p}$. (a) What is the minimum possible frequency of the photon? (b) What is its wavelength?

2. **BIO** At some time in your life, you may find yourself in a hospital to have a PET, or positron-emission tomography, scan. In the procedure, a radioactive element that undergoes e^+ decay is introduced into your body. The equipment detects the gamma rays that result from pair annihilation when the emitted positron encounters an electron in your

body's tissue. During such a scan, suppose you receive an injection of glucose containing on the order of 10^{10} atoms of ^{14}O, with half-life 70.6 s. Assume the oxygen remaining after 5 min is uniformly distributed through 2 L of blood. What is then the order of magnitude of the oxygen atoms' activity in 1 cm^3 of the blood?

3. Model a penny as 3.10 g of pure copper. Consider an anti-penny minted from 3.10 g of copper anti-atoms, each with 29 positrons in orbit around a nucleus comprising 29 anti-protons and 34 or 36 antineutrons. (a) Find the energy released if the two coins collide. (b) Find the value of this energy at the unit price of $0.11/kWh, a representative retail rate for energy from the electric company.

4. Two photons are produced when a proton and an antiproton annihilate each other. In the reference frame in which the center of mass of the proton–antiproton system is stationary, what are (a) the minimum frequency and (b) the corresponding wavelength of each photon?

5. **M** A photon with an energy $E_\gamma = 2.09$ GeV creates a proton–antiproton pair in which the proton has a kinetic energy of 95.0 MeV. What is the kinetic energy of the antiproton? *Note:* $m_p c^2 = 938.3$ MeV.

Section 31.3 Mesons and the Beginning of Particle Physics

6. When a high-energy proton or pion traveling near the speed of light collides with a nucleus, it travels an average distance of 3×10^{-15} m before interacting. From this information, find the order of magnitude of the time interval required for the strong interaction to occur.

7. A neutral pion at rest decays into two photons according to $\pi^0 \rightarrow \gamma + \gamma$. Find the (a) energy, (b) momentum, and (c) frequency of each photon.

8. Occasionally, high-energy muons collide with electrons and produce two neutrinos according to the reaction $\mu^+ + e^- \rightarrow 2\nu$. What kind of neutrinos are they?

9. One mediator of the weak interaction is the Z^0 boson, with mass 91 GeV/c^2. Use this information to find the order of magnitude of the range of the weak interaction.

10. **Q|C** (a) Prove that the exchange of a virtual particle of mass m can be associated with a force with a range given by

$$d \approx \frac{1\,240}{4\pi mc^2} = \frac{98.7}{mc^2}$$

where d is in nanometers and mc^2 is in electron volts. (b) State the pattern of dependence of the range on the mass. (c) What is the range of the force that might be produced by the virtual exchange of a proton?

11. **Q|C** **W** A free neutron beta decays by creating a proton, an electron, and an antineutrino according to the reaction $n \rightarrow p + e^- + \bar\nu$. Imagine that a free neutron were to decay by creating a proton and electron according to the reaction

$$n \rightarrow p + e^-$$

and assume that the neutron is initially at rest in the laboratory. (a) Determine the energy released in this reaction. (b) Determine the speeds of the proton and electron after the reaction. (Energy and momentum are conserved in the reaction.) (c) Is either of these particles moving at a relativistic speed? Explain.

Section 31.4 Classification of Particles

12. Identify the unknown particle on the left side of the following reaction

$$? + p \rightarrow n + \mu^+$$

13. Name one possible decay mode for Ω^+, $\overline{K}_S^0$, $\overline{\Lambda}^0$, and $\bar n$.

Section 31.5 Conservation Laws

14. **Q|C** (a) Show that baryon number and charge are conserved in the following reactions of a pion with a proton.

$$(1)\ \pi^+ + p \rightarrow K^+ + \Sigma^+$$
$$(2)\ \pi^+ + p \rightarrow \pi^+ + \Sigma^+$$

(b) The first reaction is observed, but the second never occurs. Explain.

15. Each of the following reactions is forbidden. Determine what conservation laws are violated for each reaction.

(a) $p + \bar p \rightarrow \mu^+ + e^-$ (b) $\pi^- + p \rightarrow p + \pi^+$
(c) $p + p \rightarrow p + p + n$ (d) $\gamma + p \rightarrow n + \pi^0$
(e) $\nu_e + p \rightarrow n + e^+$

16. The first of the following two reactions can occur, but the second cannot. Explain.

$K_S^0 \rightarrow \pi^+ + \pi^-$ (can occur)
$\Lambda^0 \rightarrow \pi^+ + \pi^-$ (cannot occur)

17. The following reactions or decays involve one or more neutrinos. In each case, supply the missing neutrino (ν_e, ν_μ, or ν_τ) or antineutrino.

(a) $\pi^- \rightarrow \mu^- + ?$ (b) $K^+ \rightarrow \mu^+ + ?$
(c) $? + p \rightarrow n + e^+$ (d) $? + n \rightarrow p + e^-$
(e) $? + n \rightarrow p + \mu^-$ (f) $\mu^- \rightarrow e^- + ? + ?$

18. Determine the type of neutrino or antineutrino involved in each of the following processes.

(a) $\pi^+ \rightarrow \pi^0 + e^+ + ?$ (b) $? + p \rightarrow \mu^- + p + \pi^+$
(c) $\Lambda^0 \rightarrow p + \mu^- + ?$ (d) $\tau^+ \rightarrow \mu^+ + ? + ?$

19. A $\overline{K}_S^0$ particle at rest decays into a π^+ and a π^-. The mass of the $\overline{K}_S^0$ is 497.7 MeV/c^2, and the mass of each π meson is 139.6 MeV/c^2. What is the speed of each pion?

20. (a) Show that the proton-decay $p \rightarrow e^+ + \gamma$ cannot occur because it violates the conservation of baryon number. (b) **What If?** Imagine that this reaction does occur and the proton is initially at rest. Determine the energies and magnitudes of the momentum of the positron and photon after the reaction. (c) Determine the speed of the positron after the reaction.

21. Determine which of the following reactions can occur. For those that cannot occur, determine the conservation law (or laws) violated.

(a) $p \rightarrow \pi^+ + \pi^0$ (b) $p + p \rightarrow p + p + \pi^0$
(c) $p + p \rightarrow p + \pi^+$ (d) $\pi^+ \rightarrow \mu^+ + \nu_\mu$
(e) $n \rightarrow p + e^- + \bar\nu_e$ (f) $\pi^+ \rightarrow \mu^+ + n$

Section 31.6 Strange Particles and Strangeness

22. **QC** The neutral meson ρ^0 decays by the strong interaction into two pions:

$$\rho^0 \rightarrow \pi^+ + \pi^- \qquad (T_{1/2} \sim 10^{-23} \text{ s})$$

The neutral kaon also decays into two pions:

$$K_S^0 \rightarrow \pi^+ + \pi^- \qquad (T_{1/2} \sim 10^{-10} \text{ s})$$

How do you explain the difference in half-lives?

23. Determine whether or not strangeness is conserved in the following decays and reactions.

(a) $\Lambda^0 \rightarrow p + \pi^-$
(b) $\pi^- + p \rightarrow \Lambda^0 + K^0$
(c) $\bar{p} + p \rightarrow \bar{\Lambda}^0 + \Lambda^0$
(d) $\pi^- + p \rightarrow \pi^- + \Sigma^+$
(e) $\Xi^- \rightarrow \Lambda^0 + \pi^-$
(f) $\Xi^0 \rightarrow p + \pi^-$

24. For each of the following forbidden decays, determine what conservation law is violated.

(a) $\mu^- \rightarrow e^- + \gamma$
(b) $n \rightarrow p + e^- + \nu_e$
(c) $\Lambda^0 \rightarrow p + \pi^0$
(d) $p \rightarrow e^+ + \pi^0$
(e) $\Xi^0 \rightarrow n + \pi^0$

25. Which of the following processes are allowed by the strong interaction, the electromagnetic interaction, the weak interaction, or no interaction at all?

(a) $\pi^- + p \rightarrow 2\eta$
(b) $K^- + n \rightarrow \Lambda^0 + \pi^-$
(c) $K^- \rightarrow \pi^- + \pi^0$
(d) $\Omega^- \rightarrow \Xi^- + \pi^0$
(e) $\eta \rightarrow 2\gamma$

26. Identify the conserved quantities in the following processes.

(a) $\Xi^- \rightarrow \Lambda^0 + \mu^- + \nu_\mu$
(b) $K_S^0 \rightarrow 2\pi^0$
(c) $K^- + p \rightarrow \Sigma^0 + n$
(d) $\Sigma^0 \rightarrow \Lambda^0 + \gamma$
(e) $e^+ + e^- \rightarrow \mu^+ + \mu^-$
(f) $\bar{p} + n \rightarrow \bar{\Lambda}^0 + \Sigma^-$

27. Fill in the missing particle. Assume reaction (a) occurs via the strong interaction and reactions (b) and (c) involve the weak interaction. Assume also the total strangeness changes by one unit if strangeness is not conserved.

(a) $K^+ + p \rightarrow ? + p$
(b) $\Omega^- \rightarrow ? + \pi^-$
(c) $K^+ \rightarrow ? + \mu^+ + \nu_\mu$

Section 31.7 Measuring Particle Lifetimes

28. **GP** The particle decay $\Sigma^+ \rightarrow \pi^+ + n$ is observed in a bubble chamber. Figure P31.28 represents the curved tracks of the particles Σ^+ and π^+ and the invisible track of the neutron in the presence of a uniform magnetic field of 1.15 T directed out of the page. The measured radii of curvature are 1.99 m for the Σ^+ particle and 0.580 m for the π^+ particle. From this information, we wish to determine the mass of the Σ^+ particle. (a) Find the magnitudes of the momenta of the Σ^+ and the π^+ particles in units of MeV/c. (b) The angle between the momenta of the Σ^+ and the π^+ particles at the moment of decay is $\theta = 64.5°$. Find the magnitude of the momentum of the neutron. (c) Calculate the total energy of the π^+ particle and of

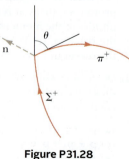

Figure P31.28

the neutron from their known masses ($m_\pi = 139.6$ MeV/c^2, $m_n = 939.6$ MeV/c^2) and the relativistic energy–momentum relation. (d) What is the total energy of the Σ^+ particle? (e) Calculate the mass of the Σ^+ particle. (f) Compare the mass with the value in Table 31.2.

29. **M** If a K_S^0 meson at rest decays in 0.900×10^{-10} s, how far does a K_S^0 meson travel if it is moving at $0.960c$?

30. A particle of mass m_1 is fired at a stationary particle of mass m_2, and a reaction takes place in which new particles are created out of the incident kinetic energy. Taken together, the product particles have total mass m_3. The minimum kinetic energy that the bombarding particle must have to induce the reaction is called the threshold energy. At this energy, the kinetic energy of the products is a minimum, so the fraction of the incident kinetic energy that is available to create new particles is a maximum. This situation occurs when all the product particles have the same velocity; then the particles have no kinetic energy of motion relative to one another. (a) By using conservation of relativistic energy and momentum, and the relativistic energy-momentum relation, show that the threshold energy is given by

$$K_{min} = \frac{[m_3^2 - (m_1 + m_2)^2]c^2}{2m_2}$$

Calculate the threshold energy for each of the following reactions:

(b) $p + p \rightarrow p + p + p + \bar{p}$

(One of the initial protons is at rest. Antiprotons are produced.)

(c) $\pi^- + p \rightarrow K^0 + \Lambda^0$

(The proton is at rest. Strange particles are produced.)

(d) $p + p \rightarrow p + p + \pi^0$

(One of the initial protons is at rest. Pions are produced.)

(e) $p + \bar{p} \rightarrow Z^0$

[One of the initial particles is at rest. Z^0 particles (mass 91.2 GeV/c^2) are produced.]

Section 31.8 Finding Patterns in the Particles

Section 31.9 Quarks

Section 31.10 Multicolored Quarks

Section 31.11 The Standard Model

31. The quark composition of the proton is uud, whereas that of the neutron is udd. Show that the charge, baryon number, and strangeness of these particles equal the sums of these numbers for their quark constituents.

32. (a) Find the number of electrons and the number of each species of quarks in 1 L of water. (b) Make an order-of-magnitude estimate of the number of each kind of fundamental matter particle in your body. State your assumptions and the quantities you take as data.

33. **What If?** Imagine that binding energies could be ignored. Find the masses of the u and d quarks from the masses of the proton and neutron.

34. The quark compositions of the K^0 and Λ^0 particles are $d\bar{s}$ and uds, respectively. Show that the charge, baryon number, and strangeness of these particles equal the sums of these numbers for the quark constituents.

35. A Σ^0 particle traveling through matter strikes a proton; then a Σ^+ and a gamma ray as well as a third particle emerge. Use the quark model of each to determine the identity of the third particle.

36. The reaction $\pi^- + p \rightarrow K^0 + \Lambda^0$ occurs with high probability, whereas the reaction $\pi^- + p \rightarrow K^0 + n$ never occurs. Analyze these reactions at the quark level. Show that the first reaction conserves the total number of each type of quark and the second reaction does not.

37. Analyze each of the following reactions in terms of constituent quarks and show that each type of quark is conserved. (a) $\pi^+ + p \rightarrow K^+ + \Sigma^+$ (b) $K^- + p \rightarrow K^+ + K^0 + \Omega^-$ (c) Determine the quarks in the final particle for this reaction: $p + p \rightarrow K^0 + p + \pi^+ + ?$ (d) In the reaction in part (c), identify the mystery particle.

38. Identify the particles corresponding to the quark states (a) suu, (b) $\bar{u}d$, (c) $\bar{s}d$, and (d) ssd.

39. What is the electrical charge of the baryons with the quark compositions (a) $\bar{u}\,\bar{u}\,d$ and (b) $\bar{u}\,d\,d$? (c) What are these baryons called?

Section 31.12 Context Connection: Investigating the Smallest System to Understand the Largest

Note: Problem 14 in Chapter 24, Problems 42 and 43 in Chapter 29, and Problems 61 and 63 in Chapter 28 can be assigned with this section.

40. Hubble's law can be stated in vector form as $\vec{v} = H\vec{R}$. Outside the local group of galaxies, all objects are moving away from us with velocities proportional to their positions relative to us. In this form, it sounds as if our location in the Universe is specially privileged. Prove that Hubble's law is equally true for an observer elsewhere in the Universe. Proceed as follows. Assume we are at the origin of coordinates, one galaxy cluster is at location $\vec{R}_1$ and has velocity $\vec{v}_1 = H\vec{R}_1$ relative to us, and another galaxy cluster has position vector $\vec{R}_2$ and velocity $\vec{v}_2 = H\vec{R}_2$. Suppose the speeds are nonrelativistic. Consider the frame of reference of an observer in the first of these galaxy clusters. (a) Show that our velocity relative to her, together with the position vector of our galaxy cluster from hers, satisfies Hubble's law. (b) Show that the position and velocity of cluster 2 relative to cluster 1 satisfy Hubble's law.

41. **W** Using Hubble's law, find the wavelength of the 590-nm sodium line emitted from galaxies (a) 2.00×10^6 ly, (b) 2.00×10^8 ly, and (c) 2.00×10^9 ly away from the Earth.

42. **Q|C** Assume dark matter exists throughout space with a uniform density of 6.00×10^{-28} kg/m³. (a) Find the amount of such dark matter inside a sphere centered on the Sun, having the Earth's orbit as its equator. (b) Explain whether the gravitational field of this dark matter would have a measurable effect on the Earth's revolution.

43. Scientists have proposed one possibility for the origin of dark matter: WIMPs, or *weakly interacting massive particles*. Another proposal is that dark matter consists of large planet-sized objects, called MACHOs, or *massive astrophysical compact halo objects*, that drift through interstellar space and are not bound to a solar system. Whether WIMPs or MACHOs, suppose astronomers perform theoretical calculations and determine the average density of the observable Universe to be $1.20\rho_c$. If this value were correct, how many times larger will the Universe become before it begins to collapse? That is, by what factor will the distance between remote galaxies increase in the future?

44. If the average density of the Universe is small compared with the critical density, the expansion of the Universe described by Hubble's law proceeds with speeds that are nearly constant over time. (a) Prove that in this case the age of the Universe is given by the inverse of the Hubble constant. (b) Calculate $1/H$ and express it in years.

45. The early Universe was dense with gamma-ray photons of energy $\sim k_B T$ and at such a high temperature that protons and antiprotons were created by the process $\gamma \rightarrow p + \bar{p}$ as rapidly as they annihilated each other. As the Universe cooled in adiabatic expansion, its temperature fell below a certain value and proton pair production became rare. At that time, slightly more protons than antiprotons existed, and essentially all the protons in the Universe today date from that time. (a) Estimate the order of magnitude of the temperature of the Universe when protons condensed out. (b) Estimate the order of magnitude of the temperature of the Universe when electrons condensed out.

46. Assume that the average density of the Universe is equal to the critical density. (a) Prove that the age of the Universe is given by $2/3H$. (b) Calculate $2/3H$ and express it in years.

47. Gravitation and other forces prevent Hubble's-law expansion from taking place except in systems larger than clusters of galaxies. **What If?** Imagine that these forces could be ignored and all distances expanded at a rate described by the Hubble constant of 22×10^{-3} m/s · ly. (a) At what rate would the 1.85-m height of a basketball player be increasing? (b) At what rate would the distance between the Earth and the Moon be increasing?

Additional Problems

48. *Why is the following situation impossible?* A gamma-ray photon with energy 1.05 MeV strikes a stationary electron, causing the following reaction to occur:

$$\gamma^- + e^- \rightarrow e^- + e^- + e^+$$

Assume all three final particles move with the same speed in the same direction after the reaction.

49. **Review.** Supernova Shelton 1987A, located approximately 170 000 ly from the Earth, is estimated to have emitted a burst of neutrinos carrying energy $\sim 10^{46}$ J (Fig. P31.49 on page 1084). Suppose the average neutrino energy was 6 MeV and your mother's body presented cross-sectional area 5 000 cm². To an order of magnitude, how many of these neutrinos passed through her?

Figure P31.49 Problems 49 and 50. The giant star cataloged as Sanduleak −69° 202 in the "before" picture (*top*) became Supernova Shelton 1987A in the "after" picture (*bottom*).

50. The most recent naked-eye supernova was Supernova Shelton 1987A (Fig. P31.49). It was 170 000 ly away in the next galaxy to ours, the Large Magellanic Cloud. About 3 h before its optical brightening was noticed, two continuously running neutrino detection experiments simultaneously registered the first neutrinos from an identified source other than the Sun. The Irvine–Michigan–Brookhaven experiment in a salt mine in Ohio registered eight neutrinos over a 6-s period, and the Kamiokande II experiment in a zinc mine in Japan counted eleven neutrinos in 13 s. (Because the supernova is far south in the sky, these neutrinos entered the detectors from below. They passed through the Earth before they were by chance absorbed by nuclei in the detectors.) The neutrino energies were between about 8 MeV and 40 MeV. If neutrinos have no mass, neutrinos of all energies should travel together at the speed of light, and the data are consistent with this possibility. The arrival times could show scatter simply because neutrinos were created at different moments as the core of the star collapsed into a neutron star. If neutrinos have nonzero mass, lower-energy neutrinos should move comparatively slowly. The data are consistent with a 10 MeV neutrino requiring at most about 10 s more than a photon would require to travel from the supernova to us. Find the upper limit that this observation sets on the mass of a neutrino. (Other evidence sets an even tighter limit.)

51. An unstable particle, initially at rest, decays into a proton (rest energy 938.3 MeV) and a negative pion (rest energy 139.6 MeV). A uniform magnetic field of 0.250 T exists perpendicular to the velocities of the created particles. The radius of curvature of each track is found to be 1.33 m. What is the mass of the original unstable particle?

52. **S** An unstable particle, initially at rest, decays into a positively charged particle of charge $+e$ and rest energy E_+ and a negatively charged particle of charge $+e$ and rest energy E_-. A uniform magnetic field of magnitude B exists perpendicular to the velocities of the created particles. The radius of curvature of each track is r. What is the mass of the original unstable particle?

53. For each of the following decays or reactions, name at least one conservation law that prevents it from occurring.

(a) $\pi^- + p \rightarrow \Sigma^+ + \pi^0$

(b) $\mu^- \rightarrow \pi^- + \nu_e$

(c) $p \rightarrow \pi^+ + \pi^+ + \pi^-$

54. **M** The energy flux carried by neutrinos from the Sun is estimated to be on the order of 0.400 W/m² at the Earth's surface. Estimate the fractional mass loss of the Sun over 10^9 yr due to the emission of neutrinos. The mass of the Sun is 1.989×10^{30} kg. The Earth–Sun distance is equal to 1.496×10^{11} m.

55. Two protons approach each other head-on, each with 70.4 MeV of kinetic energy, and engage in a reaction in which a proton and positive pion emerge at rest. What third particle, obviously uncharged and therefore difficult to detect, must have been created?

56. A rocket engine for space travel using photon drive and matter–antimatter annihilation has been suggested. Suppose the fuel for a short-duration burn consists of N protons and N antiprotons, each with mass m. (a) Assume that all the fuel is annihilated to produce photons. When the photons are ejected from the rocket, what momentum can be imparted to it? (b) If half of the protons and antiprotons annihilate each other and the energy released is used to eject the remaining particles, what momentum could be given to the rocket? Which scheme results in the greatest change in speed for the rocket?

57. Two protons approach each other with velocities of equal magnitude in opposite directions. What is the minimum kinetic energy of each proton if the two are to produce a π^+ meson at rest in the reaction $p + p \rightarrow p + n + \pi^+$?

58. Identify the mediators for the two interactions described in the Feynman diagrams shown in Figure P31.58.

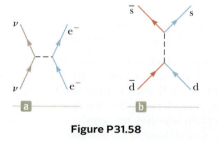

Figure P31.58

59. (a) What processes are described by the Feynman diagrams in Figure P31.59? (b) What is the exchanged particle in each process?

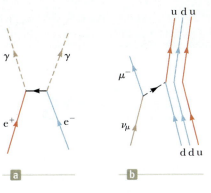

Figure P31.59

60. The cosmic rays of highest energy are mostly protons, accelerated by unknown sources. Their spectrum shows a cutoff at an energy on the order of 10^{20} eV. Above that energy, a proton will interact with a photon of cosmic microwave background radiation to produce mesons, for example, according to

$$p + \gamma \rightarrow p + \pi^0$$

Demonstrate this fact by taking the following steps. (a) Find the minimum photon energy required to produce this reaction in the reference frame where the total momentum of the photon–proton system is zero. The reaction was observed experimentally in the 1950s with photons of a few hundred MeV. (b) Use Wien's displacement law to find the wavelength of a photon at the peak of the blackbody spectrum of the primordial microwave background radiation, with a temperature of 2.73 K. (c) Find the energy of this photon. (d) Consider the reaction in part (a) in a moving reference frame so that the photon is the same as that in part (c). Calculate the energy of the proton in this frame, which represents the Earth reference frame.

61. Assume that the half-life of free neutrons is 614 s. What fraction of a group of free thermal neutrons with kinetic energy 0.040 0 eV will decay before traveling a distance of 10.0 km?

62. A π^- meson at rest decays according to $\pi^- \rightarrow \mu^- + \bar{\nu}_\mu$. Assume the antineutrino has no mass and moves off with the speed of light. Take $m_\pi c^2 = 139.6$ MeV and $m_\mu c^2 = 105.7$ MeV. What is the energy carried off by the neutrino?

63. **Review.** Use the Boltzmann distribution function $e^{-E/k_B T}$ to calculate the temperature at which 1.00% of a population of photons will have energy greater than 1.00 eV. The energy required to excite an atom is on the order of 1 eV. Therefore, as the temperature of the Universe fell below the value you calculate, neutral atoms could form from plasma and the Universe became transparent. The cosmic background radiation represents our vastly red-shifted view of the opaque fireball of the Big Bang as it was at this time and temperature. The fireball surrounds us; we are embers.

64. A Σ^0 particle at rest decays according to $\Sigma^0 \rightarrow \Lambda^0 + \gamma$. Find the gamma-ray energy.

65. Determine the kinetic energies of the proton and pion resulting from the decay of a Λ^0 at rest:

$$\Lambda^0 \rightarrow p + \pi^-$$

Problems and Perspectives

We have now investigated the principles of quantum physics and have seen many connections to our central question for the *Cosmic Connection* Context:

> **How can we connect the physics of microscopic particles to the physics of the Universe?**

While particle physicists have been exploring the realm of the very small, cosmologists have been exploring cosmic history back to within the first second after the Big Bang. Observation of events that occur when two particles collide in an accelerator is essential in reconstructing the early moments in cosmic history. The key to understanding the early Universe is first to understand the world of elementary particles. Cosmologists and physicists now find that they have many common goals and are joining hands to attempt to understand the physical world at its most fundamental level.

Problems

We have made great progress in understanding the Universe and its underlying structure, but a multitude of questions remain unanswered. Why does so little antimatter exist in the Universe? Do neutrinos have a small rest energy, and if so, how do they contribute to the "dark matter" of the Universe? Is there "dark energy" in the Universe? Is it possible to unify the strong and electroweak forces in a logical and consistent manner? Can gravity be unified with the other forces? Why do quarks and leptons form three similar but distinct families? Are muons the same as electrons (apart from their difference in mass), or do they have other subtle differences that have not been detected? Why are some particles charged and others neutral? Why do quarks carry a fractional charge? What determines the masses of the fundamental constituents? Can isolated quarks exist? Do leptons and quarks have a substructure?

String Theory: A New Perspective

Let us briefly discuss one current effort at answering some of these questions by proposing a new perspective on particles. As you read this book, you may recall starting off with the particle model and doing quite a bit of physics with it. In the *Earthquakes* Context, we introduced the wave model, and more physics was used to investigate the properties of waves. We used a wave model for light in the *Lasers* Context. Early in this Context, however, we saw the need to return to the particle model for light. Furthermore, we found that material particles had wave-like characteristics. The quantum particle model of Chapter 28 allowed us to build particles out of waves, suggesting that a wave is the fundamental entity. In Chapter 31, however, we discussed the elementary particles as the fundamental entities. It seems as if we cannot make up our mind! In some sense, that is true because the wave–particle duality is still an area of active research. In this final Context Conclusion, we shall discuss a current research effort to build particles out of waves and vibrations.

String theory is an effort to unify the four fundamental forces by modeling all particles as various quantized vibrational modes of a single entity, an incredibly small string. The typical length of such a string is on the order of 10^{-35} m, called the **Planck length.** We have seen quantized modes before with the frequencies of vibrating guitar strings in Chapter 14 and the quantized energy levels of atoms in Chapter 29. In string theory, each quantized mode of vibration of the string corresponds to a different elementary particle in the Standard Model.

One complicating factor in string theory is that it requires space–time to have ten dimensions. Despite the theoretical and conceptual difficulties in dealing with ten dimensions, string theory holds promise in incorporating gravity with the other forces. Four of the ten dimensions are visible to us—three space dimensions and one time dimension—and the other six are *compactified*. In other words, the six dimensions are curled up so tightly that they are not visible in the macroscopic world.

As an analogy, consider a soda straw. We can build a soda straw by cutting a rectangular piece of paper (Fig. 1a), which clearly has two dimensions, and rolling it up into a small tube (Fig. 1b). From far away, the soda straw looks like a one-dimensional straight line. The second dimension has been curled up and is not visible. String theory claims that six space–time dimensions are curled up in an analogous way, with the curling on the size of the Planck length and impossible to see from our viewpoint.

Another complicating factor with string theory is that it is difficult for string theorists to guide experimentalists in how and what to look for in an experiment. The Planck length is so incredibly small that direct experimentation on strings is impossible. Until the theory has been further developed, string theorists are restricted to applying the theory to known results and testing for consistency.

One of the predictions of string theory is called **supersymmetry** (SUSY), which suggests that every elementary particle has a superpartner that has not yet been observed. It is believed that supersymmetry is a broken symmetry (like the broken electroweak symmetry at low energies) and that the masses of the superpartners are above our current capabilities of detection by accelerators. Some theorists claim that the mass of superpartners is the missing mass discussed in the Context Conclusion of Chapter 31. Keeping with the whimsical trend in naming particles and their properties that we saw in Chapter 31, superpartners are given names such as the *squark* (the superpartner to a quark), the *selectron* (electron), and the *gluinos* (gluon).

Other theorists are working on **M-theory,** which is an 11-dimensional theory based on membranes rather than strings. In a way reminiscent of the correspondence principle, M-theory is claimed to reduce to string theory if one compactifies from 11 dimensions to 10.

The questions that we listed at the beginning of this Context Conclusion go on and on. Because of the rapid advances and new discoveries in the field of particle physics, by the time you read this book some of these questions may be resolved and other new questions may emerge.

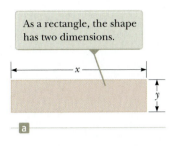

As a rectangle, the shape has two dimensions.

The curled-up second dimension is not visible when viewed from a distance that is large compared with the diameter of the straw.

Figure 1 (a) A piece of paper is cut into a rectangular shape. (b) The paper is rolled up into a soda straw.

Question

1. **Review question.** A girl and her grandmother grind corn while the woman tells the girl stories about what is most important. A boy keeps crows away from ripening corn while his grandfather sits in the shade and explains to him the Universe and his place in it. What the children do not understand this year they will better understand next year. Now you must take the part of the adults. State the most general, most fundamental, most universal truths that you know. If you need to repeat someone else's ideas, get the best version of those ideas you can and state your source. If there is something you do not understand, make a plan to understand it better within the next year.

Problem

1. Classical general relativity views the structure of space–time as deterministic and well defined down to arbitrarily small distances. On the other hand, quantum general relativity forbids distances smaller than the Planck length given

by $L = (\hbar G/c^3)^{1/2}$. (a) Calculate the value of the Planck length. The quantum limitation suggests that after the Big Bang, when all the presently observable section of the Universe was contained within a point-like singularity, nothing could be observed until that singularity grew larger than the Planck length. Because the size of the singularity grew at the speed of light, we can infer that no observations were possible during the time interval required for light to travel the Planck length. (b) Calculate this time interval, known as the Planck time T, and compare it with the ultrahot epoch mentioned in the text. (c) Does this answer suggest that we may never know what happened between the time $t = 0$ and the time $t = T$?

The Meaning of Success

To earn the respect of intelligent people and to win the affection of children;
To appreciate the beauty in nature and all that surrounds us;
To seek out and nurture the best in others;
To give the gift of yourself to others without the slightest thought of return, for it is in giving that we receive;
To have accomplished a task, whether it be saving a lost soul, healing a sick child, writing a book, or risking your life for a friend;
To have celebrated and laughed with great joy and enthusiasm and sung with exultation;
To have hope even in times of despair, for as long as you have hope, you have life;
To love and be loved;
To be understood and to understand;
To know that even one life has breathed easier because you have lived;
This is the meaning of success.

Ralph Waldo Emerson and modified by Ray Serway

Appendix A

Tables

TABLE A.1 | Conversion Factors

Length

	m	cm	km	in.	ft	mi
1 meter	1	10^2	10^{-3}	39.37	3.281	6.214×10^{-4}
1 centimeter	10^{-2}	1	10^{-5}	0.393 7	3.281×10^{-2}	6.214×10^{-6}
1 kilometer	10^3	10^5	1	3.937×10^4	3.281×10^3	0.621 4
1 inch	2.540×10^{-2}	2.540	2.540×10^{-5}	1	8.333×10^{-2}	1.578×10^{-5}
1 foot	0.304 8	30.48	3.048×10^{-4}	12	1	1.894×10^{-4}
1 mile	1 609	1.609×10^5	1.609	6.336×10^4	5 280	1

Mass

	kg	g	slug	u
1 kilogram	1	10^3	6.852×10^{-2}	6.024×10^{26}
1 gram	10^{-3}	1	6.852×10^{-5}	6.024×10^{23}
1 slug	14.59	1.459×10^4	1	8.789×10^{27}
1 atomic mass unit	1.660×10^{-27}	1.660×10^{-24}	1.137×10^{-28}	1

Note: 1 metric ton = 1 000 kg.

Time

	s	min	h	day	yr
1 second	1	1.667×10^{-2}	2.778×10^{-4}	1.157×10^{-5}	3.169×10^{-8}
1 minute	60	1	1.667×10^{-2}	6.994×10^{-4}	1.901×10^{-6}
1 hour	3 600	60	1	4.167×10^{-2}	1.141×10^{-4}
1 day	8.640×10^4	1 440	24	1	2.738×10^{-5}
1 year	3.156×10^7	5.259×10^5	8.766×10^3	365.2	1

Speed

	m/s	cm/s	ft/s	mi/h
1 meter per second	1	10^2	3.281	2.237
1 centimeter per second	10^{-2}	1	3.281×10^{-2}	2.237×10^{-2}
1 foot per second	0.304 8	30.48	1	0.681 8
1 mile per hour	0.447 0	44.70	1.467	1

Note: 1 mi/min = 60 mi/h = 88 ft/s.

Force

	N	lb
1 newton	1	0.224 8
1 pound	4.448	1

(Continued)

⟨ **TABLE A.1** | Conversion Factors (*continued*)

Energy, Energy Transfer

	J	ft · lb	eV
1 joule	1	0.737 6	6.242×10^{18}
1 foot-pound	1.356	1	8.464×10^{18}
1 electron volt	1.602×10^{-19}	1.182×10^{-19}	1
1 calorie	4.186	3.087	2.613×10^{19}
1 British thermal unit	1.055×10^3	7.779×10^2	6.585×10^{21}
1 kilowatt-hour	3.600×10^6	2.655×10^6	2.247×10^{25}

	cal	Btu	kWh
1 joule	0.238 9	9.481×10^{-4}	2.778×10^{-7}
1 foot-pound	0.323 9	1.285×10^{-3}	3.766×10^{-7}
1 electron volt	3.827×10^{-20}	1.519×10^{-22}	4.450×10^{-26}
1 calorie	1	3.968×10^{-3}	1.163×10^{-6}
1 British thermal unit	2.520×10^2	1	2.930×10^{-4}
1 kilowatt-hour	8.601×10^5	3.413×10^2	1

Pressure

	Pa	atm
1 pascal	1	9.869×10^{-6}
1 atmosphere	1.013×10^5	1
1 centimeter mercury[a]	1.333×10^3	1.316×10^{-2}
1 pound per square inch	6.895×10^3	6.805×10^{-2}
1 pound per square foot	47.88	4.725×10^{-4}

	cm Hg	lb/in.^{2}	lb/ft^2
1 pascal	7.501×10^{-4}	1.450×10^{-4}	2.089×10^{-2}
1 atmosphere	76	14.70	2.116×10^3
1 centimeter mercury[a]	1	0.194 3	27.85
1 pound per square inch	5.171	1	144
1 pound per square foot	3.591×10^{-2}	6.944×10^{-3}	1

[a]At 0°C and at a location where the free-fall acceleration has its "standard" value, 9.806 65 m/s².

⟨ **TABLE A.2** | Symbols, Dimensions, and Units of Physical Quantities

Quantity	Common Symbol	Unit[a]	Dimensions[b]	Unit in Terms of Base SI Units
Acceleration	$\vec{a}$	m/s²	L/T²	m/s²
Amount of substance	n	MOLE		mol
Angle	θ, ϕ	radian (rad)	1	
Angular acceleration	$\vec{\alpha}$	rad/s²	T^{-2}	s^{-2}
Angular frequency	ω	rad/s	T^{-1}	s^{-1}
Angular momentum	$\vec{L}$	kg · m²/s	ML²/T	kg · m²/s
Angular velocity	$\vec{\omega}$	rad/s	T^{-1}	s^{-1}
Area	A	m²	L²	m²
Atomic number	Z			
Capacitance	C	farad (F)	Q²T²/ML²	A² · s⁴/kg · m²
Charge	q, Q, e	coulomb (C)	Q	A · s

(*Continued*)

TABLE A.2 | Symbols, Dimensions, and Units of Physical Quantities (*continued*)

Quantity	Common Symbol	Unit[a]	Dimensions[b]	Unit in Terms of Base SI Units
Charge density				
Line	λ	C/m	Q/L	$A \cdot s/m$
Surface	σ	C/m^2	Q/L^2	$A \cdot s/m^2$
Volume	ρ	C/m^3	Q/L^3	$A \cdot s/m^3$
Conductivity	σ	$1/\Omega \cdot m$	Q^2T/ML^3	$A^2 \cdot s^3/kg \cdot m^3$
Current	I	AMPERE	Q/T	A
Current density	J	A/m^2	Q/TL^2	A/m^2
Density	ρ	kg/m^3	M/L^3	kg/m^3
Dielectric constant	κ			
Electric dipole moment	$\vec{p}$	$C \cdot m$	QL	$A \cdot s \cdot m$
Electric field	$\vec{E}$	V/m	ML/QT^2	$kg \cdot m/A \cdot s^3$
Electric flux	Φ_E	$V \cdot m$	ML^3/QT^2	$kg \cdot m^3/A \cdot s^3$
Electromotive force	$\mathcal{E}$	volt (V)	ML^2/QT^2	$kg \cdot m^2/A \cdot s^3$
Energy	E, U, K	joule (J)	ML^2/T^2	$kg \cdot m^2/s^2$
Entropy	S	J/K	ML^2/T^2K	$kg \cdot m^2/s^2 \cdot K$
Force	$\vec{F}$	newton (N)	ML/T^2	$kg \cdot m/s^2$
Frequency	f	hertz (Hz)	T^{-1}	s^{-1}
Heat	Q	joule (J)	ML^2/T^2	$kg \cdot m^2/s^2$
Inductance	L	henry (H)	ML^2/Q^2	$kg \cdot m^2/A^2 \cdot s^2$
Length	ℓ, L	METER	L	m
Displacement	$\Delta x, \Delta\vec{r}$			
Distance	d, h			
Position	$x, y, z, \vec{r}$			
Magnetic dipole moment	$\vec{\mu}$	$N \cdot m/T$	QL^2/T	$A \cdot m^2$
Magnetic field	$\vec{B}$	tesla (T) $(= Wb/m^2)$	M/QT	$kg/A \cdot s^2$
Magnetic flux	Φ_B	weber (Wb)	ML^2/QT	$kg \cdot m^2/A \cdot s^2$
Mass	m, M	KILOGRAM	M	kg
Molar specific heat	C	$J/mol \cdot K$		$kg \cdot m^2/s^2 \cdot mol \cdot K$
Moment of inertia	I	$kg \cdot m^2$	ML^2	$kg \cdot m^2$
Momentum	$\vec{p}$	$kg \cdot m/s$	ML/T	$kg \cdot m/s$
Period	T	s	T	s
Permeability of free space	μ_0	$N/A^2 \ (= H/m)$	ML/Q^2	$kg \cdot m/A^2 \cdot s^2$
Permittivity of free space	ϵ_0	$C^2/N \cdot m^2 \ (= F/m)$	Q^2T^2/ML^3	$A^2 \cdot s^4/kg \cdot m^3$
Potential	V	volt (V) $(= J/C)$	ML^2/QT^2	$kg \cdot m^2/A \cdot s^3$
Power	P	watt (W) $(= J/s)$	ML^2/T^3	$kg \cdot m^2/s^3$
Pressure	P	pascal (Pa) $(= N/m^2)$	M/LT^2	$kg/m \cdot s^2$
Resistance	R	ohm $(\Omega) \ (= V/A)$	ML^2/Q^2T	$kg \cdot m^2/A^2 \cdot s^3$
Specific heat	c	$J/kg \cdot K$	L^2/T^2K	$m^2/s^2 \cdot K$
Speed	v	m/s	L/T	m/s
Temperature	T	KELVIN	K	K
Time	t	SECOND	T	s
Torque	$\vec{\tau}$	$N \cdot m$	ML^2/T^2	$kg \cdot m^2/s^2$
Velocity	$\vec{v}$	m/s	L/T	m/s
Volume	V	m^3	L^3	m^3
Wavelength	λ	m	L	m
Work	W	joule (J) $(= N \cdot m)$	ML^2/T^2	$kg \cdot m^2/s^2$

[a]The base SI units are given in uppercase letters.

[b]The symbols M, L, T, K, and Q denote mass, length, time, temperature, and charge, respectively.

Appendix B

Mathematics Review

This appendix in mathematics is intended as a brief review of operations and methods. Early in this course, you should be totally familiar with basic algebraic techniques, analytic geometry, and trigonometry. The sections on differential and integral calculus are more detailed and are intended for students who have difficulty applying calculus concepts to physical situations.

◖B.1 | Scientific Notation

Many quantities used by scientists often have very large or very small values. The speed of light, for example, is about 300 000 000 m/s, and the ink required to make the dot over an i in this textbook has a mass of about 0.000 000 001 kg. Obviously, it is very cumbersome to read, write, and keep track of such numbers. We avoid this problem by using a method incorporating powers of the number 10:

$$10^0 = 1$$
$$10^1 = 10$$
$$10^2 = 10 \times 10 = 100$$
$$10^3 = 10 \times 10 \times 10 = 1\ 000$$
$$10^4 = 10 \times 10 \times 10 \times 10 = 10\ 000$$
$$10^5 = 10 \times 10 \times 10 \times 10 \times 10 = 100\ 000$$

and so on. The number of zeros corresponds to the power to which ten is raised, called the **exponent** of ten. For example, the speed of light, 300 000 000 m/s, can be expressed as 3.00×10^8 m/s.

In this method, some representative numbers smaller than unity are the following:

$$10^{-1} = \frac{1}{10} = 0.1$$

$$10^{-2} = \frac{1}{10 \times 10} = 0.01$$

$$10^{-3} = \frac{1}{10 \times 10 \times 10} = 0.001$$

$$10^{-4} = \frac{1}{10 \times 10 \times 10 \times 10} = 0.000\ 1$$

$$10^{-5} = \frac{1}{10 \times 10 \times 10 \times 10 \times 10} = 0.000\ 01$$

In these cases, the number of places the decimal point is to the left of the digit 1 equals the value of the (negative) exponent. Numbers expressed as some power of ten multiplied by another number between one and ten are said to be in **scientific notation.** For example, the scientific notation for 5 943 000 000 is 5.943×10^9 and that for 0.000 083 2 is 8.32×10^{-5}.

When numbers expressed in scientific notation are being multiplied, the following general rule is very useful:

$$10^n \times 10^m = 10^{n+m}$$

B.1 ◀

where n and m can be *any* numbers (not necessarily integers). For example, $10^2 \times 10^5 = 10^7$. The rule also applies if one of the exponents is negative: $10^3 \times 10^{-8} = 10^{-5}$.

When dividing numbers expressed in scientific notation, note that

$$\frac{10^n}{10^m} = 10^n \times 10^{-m} = 10^{n-m}$$

B.2 ◀

Exercises

With help from the preceding rules, verify the answers to the following equations:

1. $86\,400 = 8.64 \times 10^4$
2. $9\,816\,762.5 = 9.816\,762\,5 \times 10^6$
3. $0.000\,000\,039\,8 = 3.98 \times 10^{-8}$
4. $(4.0 \times 10^8)(9.0 \times 10^9) = 3.6 \times 10^{18}$
5. $(3.0 \times 10^7)(6.0 \times 10^{-12}) = 1.8 \times 10^{-4}$
6. $\dfrac{75 \times 10^{-11}}{5.0 \times 10^{-3}} = 1.5 \times 10^{-7}$
7. $\dfrac{(3 \times 10^6)(8 \times 10^{-2})}{(2 \times 10^{17})(6 \times 10^5)} = 2 \times 10^{-18}$

◖B.2 | Algebra

Some Basic Rules

When algebraic operations are performed, the laws of arithmetic apply. Symbols such as x, y, and z are usually used to represent unspecified quantities, called the **unknowns.**

First, consider the equation

$$8x = 32$$

If we wish to solve for x, we can divide (or multiply) each side of the equation by the same factor without destroying the equality. In this case, if we divide both sides by 8, we have

$$\frac{8x}{8} = \frac{32}{8}$$

$$x = 4$$

Next consider the equation

$$x + 2 = 8$$

In this type of expression, we can add or subtract the same quantity from each side. If we subtract 2 from each side, we have

$$x + 2 - 2 = 8 - 2$$

$$x = 6$$

In general, if $x + a = b$, then $x = b - a$.

Now consider the equation

$$\frac{x}{5} = 9$$

If we multiply each side by 5, we are left with x on the left by itself and 45 on the right:

$$\left(\frac{x}{5}\right)(5) = 9 \times 5$$

$$x = 45$$

In all cases, *whatever operation is performed on the left side of the equality must also be performed on the right side.*

The following rules for multiplying, dividing, adding, and subtracting fractions should be recalled, where a, b, c, and d are four numbers:

	Rule	Example
Multiplying	$\left(\dfrac{a}{b}\right)\left(\dfrac{c}{d}\right) = \dfrac{ac}{bd}$	$\left(\dfrac{2}{3}\right)\left(\dfrac{4}{5}\right) = \dfrac{8}{15}$
Dividing	$\dfrac{(a/b)}{(c/d)} = \dfrac{ad}{bc}$	$\dfrac{2/3}{4/5} = \dfrac{(2)(5)}{(4)(3)} = \dfrac{10}{12}$
Adding	$\dfrac{a}{b} \pm \dfrac{c}{d} = \dfrac{ad \pm bc}{bd}$	$\dfrac{2}{3} - \dfrac{4}{5} = \dfrac{(2)(5) - (4)(3)}{(3)(5)} = -\dfrac{2}{15}$

Exercises

In the following exercises, solve for x.

Answers

1. $a = \dfrac{1}{1 + x}$ $x = \dfrac{1 - a}{a}$

2. $3x - 5 = 13$ $x = 6$

3. $ax - 5 = bx + 2$ $x = \dfrac{7}{a - b}$

4. $\dfrac{5}{2x + 6} = \dfrac{3}{4x + 8}$ $x = -\dfrac{11}{7}$

Powers

When powers of a given quantity x are multiplied, the following rule applies:

$$x^n x^m = x^{n+m} \qquad \text{B.3} \blacktriangleleft$$

For example, $x^2 x^4 = x^{2+4} = x^6$.

When dividing the powers of a given quantity, the rule is

$$\frac{x^n}{x^m} = x^{n-m} \qquad \text{B.4} \blacktriangleleft$$

For example, $x^8/x^2 = x^{8-2} = x^6$.

A power that is a fraction, such as $\frac{1}{3}$, corresponds to a root as follows:

$$x^{1/n} = \sqrt[n]{x} \qquad \text{B.5} \blacktriangleleft$$

For example, $4^{1/3} = \sqrt[3]{4} = 1.587\,4$. (A scientific calculator is useful for such calculations.)

Finally, any quantity x^n raised to the mth power is

$$(x^n)^m = x^{nm} \qquad \text{B.6} \blacktriangleleft$$

◀ **TABLE B.1** | **Rules of Exponents**

$x^0 = 1$
$x^1 = x$
$x^n x^m = x^{n+m}$
$x^n/x^m = x^{n-m}$
$x^{1/n} = \sqrt[n]{x}$
$(x^n)^m = x^{nm}$

Table B.1 summarizes the rules of exponents.

Exercises

Verify the following equations:

1. $3^2 \times 3^3 = 243$
2. $x^5 x^{-8} = x^{-3}$
3. $x^{10}/x^{-5} = x^{15}$
4. $5^{1/3} = 1.709\,976$ (Use your calculator.)
5. $60^{1/4} = 2.783\,158$ (Use your calculator.)
6. $(x^4)^3 = x^{12}$

Factoring

Some useful formulas for factoring an equation are the following:

$$ax + ay + az = a(x + y + z) \quad \text{common factor}$$

$$a^2 + 2ab + b^2 = (a + b)^2 \quad \text{perfect square}$$

$$a^2 - b^2 = (a + b)(a - b) \quad \text{differences of squares}$$

Quadratic Equations

The general form of a quadratic equation is

$$ax^2 + bx + c = 0 \qquad \text{B.7} \blacktriangleleft$$

where x is the unknown quantity and a, b, and c are numerical factors referred to as **coefficients** of the equation. This equation has two roots, given by

$$x = \frac{-b \pm \sqrt{b^2 - 4ac}}{2a} \qquad \text{B.8} \blacktriangleleft$$

If $b^2 \geq 4ac$, the roots are real.

Example B.1

The equation $x^2 + 5x + 4 = 0$ has the following roots corresponding to the two signs of the square-root term:

$$x = \frac{-5 \pm \sqrt{5^2 - (4)(1)(4)}}{2(1)} = \frac{-5 \pm \sqrt{9}}{2} = \frac{-5 \pm 3}{2}$$

$$x_+ = \frac{-5 + 3}{2} = -1 \quad x_- = \frac{-5 - 3}{2} = -4$$

where x_+ refers to the root corresponding to the positive sign and x_- refers to the root corresponding to the negative sign.

Exercises

Solve the following quadratic equations:

Answers

1. $x^2 + 2x - 3 = 0$ $x_+ = 1$ $x_- = -3$
2. $2x^2 - 5x + 2 = 0$ $x_+ = 2$ $x_- = \frac{1}{2}$
3. $2x^2 - 4x - 9 = 0$ $x_+ = 1 + \sqrt{22}/2$ $x_- = 1 - \sqrt{22}/2$

Linear Equations

A linear equation has the general form

$$y = mx + b \qquad \text{B.9} \blacktriangleleft$$

where m and b are constants. This equation is referred to as linear because the graph of y versus x is a straight line as shown in Figure B.1. The constant b, called the **y-intercept,** represents the value of y at which the straight line intersects the y axis. The constant m is equal to the **slope** of the straight line. If any two points on the straight line are specified by the coordinates (x_1, y_1) and (x_2, y_2) as in Figure B.1, the slope of the straight line can be expressed as

$$\text{Slope} = \frac{y_2 - y_1}{x_2 - x_1} = \frac{\Delta y}{\Delta x} \qquad \text{B.10} \blacktriangleleft$$

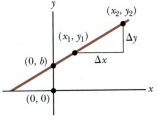

Figure B.1 A straight line graphed on an xy coordinate system. The slope of the line is the ratio of Δy to Δx.

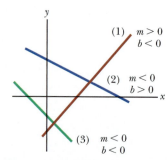

Figure B.2 The brown line has a positive slope and a negative y-intercept. The blue line has a negative slope and a positive y-intercept. The green line has a negative slope and a negative y-intercept.

Note that m and b can have either positive or negative values. If $m > 0$, the straight line has a *positive* slope as in Figure B.1. If $m < 0$, the straight line has a *negative* slope. In Figure B.1, both m and b are positive. Three other possible situations are shown in Figure B.2.

Exercises

1. Draw graphs of the following straight lines: (a) $y = 5x + 3$ (b) $y = -2x + 4$
   (c) $y = -3x - 6$
2. Find the slopes of the straight lines described in Exercise 1.

Answers   (a) 5 (b) -2 (c) -3

3. Find the slopes of the straight lines that pass through the following sets of
   points: (a) $(0, -4)$ and $(4, 2)$ (b) $(0, 0)$ and $(2, -5)$ (c) $(-5, 2)$ and $(4, -2)$

Answers   (a) $\frac{3}{2}$ (b) $-\frac{5}{2}$ (c) $-\frac{4}{9}$

Solving Simultaneous Linear Equations

Consider the equation $3x + 5y = 15$, which has two unknowns, x and y. Such an equation does not have a unique solution. For example, $(x = 0, y = 3)$, $(x = 5, y = 0)$, and $(x = 2, y = \frac{9}{5})$ are all solutions to this equation.

If a problem has two unknowns, a unique solution is possible only if we have *two* pieces of information. In most common cases, those two pieces of information are equations. In general, if a problem has n unknowns, its solution requires n equations. To solve two simultaneous equations involving two unknowns, x and y, we solve one of the equations for x in terms of y and substitute this expression into the other equation.

In some cases, the two pieces of information may be (1) one equation and (2) a condition on the solutions. For example, suppose we have the equation $m = 3n$ and the condition that m and n must be the smallest positive nonzero integers possible. Then, the single equation does not allow a unique solution, but the addition of the condition gives us that $n = 1$ and $m = 3$.

Example B.2

Solve the two simultaneous equations.

$$(1)\ \ 5x + y = -8$$

$$(2)\ \ 2x - 2y = 4$$

SOLUTION

From Equation (2), $x = y + 2$. Substitution of this equation into Equation (1) gives

$$5(y + 2) + y = -8$$

$$6y = -18$$

$$y = \boxed{-3}$$

$$x = y + 2 = \boxed{-1}$$

Alternative Solution Multiply each term in Equation (1) by the factor 2 and add the result to Equation (2):

$$10x + 2y = -16$$

$$\underline{2x - 2y = 4}$$

$$12x \qquad\quad = -12$$

$$x = \boxed{-1}$$

$$y = x - 2 = \boxed{-3}$$

Two linear equations containing two unknowns can also be solved by a graphical method. If the straight lines corresponding to the two equations are plotted in a conventional coordinate system, the intersection of the two lines represents the solution. For example, consider the two equations

$$x - y = 2$$

$$x - 2y = -1$$

These equations are plotted in Figure B.3. The intersection of the two lines has the coordinates $x = 5$ and $y = 3$, which represents the solution to the equations. You should check this solution by the analytical technique discussed earlier.

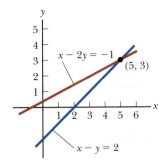

Figure B.3 A graphical solution for two linear equations.

Exercises

Solve the following pairs of simultaneous equations involving two unknowns:

	Answers
1. $x + y = 8$ $x - y = 2$	$x = 5, y = 3$
2. $98 - T = 10a$ $T - 49 = 5a$	$T = 65, a = 3.27$
3. $6x + 2y = 6$ $8x - 4y = 28$	$x = 2, y = -3$

Logarithms

Suppose a quantity x is expressed as a power of some quantity a:

$$x = a^y \qquad \text{B.11} \blacktriangleleft$$

The number a is called the **base** number. The **logarithm** of x with respect to the base a is equal to the exponent to which the base must be raised to satisfy the expression $x = a^y$:

$$y = \log_a x \qquad \text{B.12} \blacktriangleleft$$

Conversely, the **antilogarithm** of y is the number x:

$$x = \text{antilog}_a y \qquad \text{B.13} \blacktriangleleft$$

In practice, the two bases most often used are base 10, called the *common* logarithm base, and base $e = 2.718\ 282$, called Euler's constant or the *natural* logarithm base. When common logarithms are used,

$$y = \log_{10} x \quad (\text{or } x = 10^y) \qquad \text{B.14} \blacktriangleleft$$

When natural logarithms are used,

$$y = \ln x \quad (\text{or } x = e^y) \qquad \text{B.15} \blacktriangleleft$$

For example, $\log_{10} 52 = 1.716$, so antilog$_{10}$ 1.716 $= 10^{1.716} = 52$. Likewise, ln 52 $= 3.951$, so antiln 3.951 $= e^{3.951} = 52$.

In general, note you can convert between base 10 and base e with the equality

$$\ln x = (2.302\ 585) \log_{10} x \qquad \text{B.16} \blacktriangleleft$$

Finally, some useful properties of logarithms are the following:

$$\left. \begin{aligned} \log(ab) &= \log a + \log b \\ \log(a/b) &= \log a - \log b \\ \log(a^n) &= n \log a \end{aligned} \right\} \quad \text{any base}$$

$$\ln e = 1$$

$$\ln e^a = a$$

$$\ln \left(\frac{1}{a} \right) = -\ln a$$

TABLE B.4 | Derivative for Several Functions

$$\frac{d}{dx}(a) = 0$$

$$\frac{d}{dx}(ax^n) = nax^{n-1}$$

$$\frac{d}{dx}(e^{ax}) = ae^{ax}$$

$$\frac{d}{dx}(\sin ax) = a \cos ax$$

$$\frac{d}{dx}(\cos ax) = -a \sin ax$$

$$\frac{d}{dx}(\tan ax) = a \sec^2 ax$$

$$\frac{d}{dx}(\cot ax) = -a \csc^2 ax$$

$$\frac{d}{dx}(\sec x) = \tan x \sec x$$

$$\frac{d}{dx}(\csc x) = -\cot x \csc x$$

$$\frac{d}{dx}(\ln ax) = \frac{1}{x}$$

$$\frac{d}{dx}(\sin^{-1} ax) = \frac{a}{\sqrt{1 - a^2x^2}}$$

$$\frac{d}{dx}(\cos^{-1} ax) = \frac{-a}{\sqrt{1 - a^2x^2}}$$

$$\frac{d}{dx}(\tan^{-1} ax) = \frac{a}{\sqrt{1 + a^2x^2}}$$

Note: The symbols a and n represent constants.

If $y(x)$ is a polynomial or algebraic function of x, we apply Equation B.29 to *each* term in the polynomial and take $d[\text{constant}]/dx = 0$. In Examples B.4 through B.7, we evaluate the derivatives of several functions.

Special Properties of the Derivative

A. Derivative of the product of two functions   If a function $f(x)$ is given by the product of two functions—say, $g(x)$ and $h(x)$—the derivative of $f(x)$ is defined as

$$\frac{d}{dx}f(x) = \frac{d}{dx}[g(x)h(x)] = g\frac{dh}{dx} + h\frac{dg}{dx} \qquad \text{B.30} \blacktriangleleft$$

B. Derivative of the sum of two functions   If a function $f(x)$ is equal to the sum of two functions, the derivative of the sum is equal to the sum of the derivatives:

$$\frac{d}{dx}f(x) = \frac{d}{dx}[g(x) + h(x)] = \frac{dg}{dx} + \frac{dh}{dx} \qquad \text{B.31} \blacktriangleleft$$

C. Chain rule of differential calculus   If $y = f(x)$ and $x = g(z)$, then dy/dz can be written as the product of two derivatives:

$$\frac{dy}{dz} = \frac{dy}{dx}\frac{dx}{dz} \qquad \text{B.32} \blacktriangleleft$$

D. The second derivative   The second derivative of y with respect to x is defined as the derivative of the function dy/dx (the derivative of the derivative). It is usually written as

$$\frac{d^2y}{dx^2} = \frac{d}{dx}\left(\frac{dy}{dx}\right) \qquad \text{B.33} \blacktriangleleft$$

Some of the more commonly used derivatives of functions are listed in Table B.4.

Example **B.4** |

Suppose $y(x)$ (that is, y as a function of x) is given by

$$y(x) = ax^3 + bx + c$$

where a and b are constants. It follows that

$$y(x + \Delta x) = a(x + \Delta x)^3 + b(x + \Delta x) + c$$

$$= a(x^3 + 3x^2 \, \Delta x + 3x \, \Delta x^2 + \Delta x^3) + b(x + \Delta x) + c$$

so

$$\Delta y = y(x + \Delta x) - y(x) = a(3x^2 \, \Delta x + 3x \, \Delta x^2 + \Delta x^3) + b \, \Delta x$$

Substituting this into Equation B.28 gives

$$\frac{dy}{dx} = \lim_{\Delta x \to 0} \frac{\Delta y}{\Delta x} = \lim_{\Delta x \to 0} [3ax^2 + 3ax \, \Delta x + a \, \Delta x^2] + b$$

$$\frac{dy}{dx} = \boxed{3ax^2 + b}$$

Example B.5

Find the derivative of

$$y(x) = 8x^5 + 4x^3 + 2x + 7$$

SOLUTION

Applying Equation B.29 to each term independently and remembering that d/dx (constant) $= 0$, we have

$$\frac{dy}{dx} = 8(5)x^4 + 4(3)x^2 + 2(1)x^0 + 0$$

$$\frac{dy}{dx} = \boxed{40x^4 + 12x^2 + 2}$$

Example B.6

Find the derivative of $y(x) = x^3/(x+1)^2$ with respect to x.

SOLUTION

We can rewrite this function as $y(x) = x^3(x+1)^{-2}$ and apply Equation B.30:

$$\frac{dy}{dx} = (x+1)^{-2}\frac{d}{dx}(x^3) + x^3\frac{d}{dx}(x+1)^{-2}$$

$$= (x+1)^{-2}\,3x^2 + x^3\,(-2)(x+1)^{-3}$$

$$\frac{dy}{dx} = \boxed{\frac{3x^2}{(x+1)^2}} - \boxed{\frac{2x^3}{(x+1)^3}} = \boxed{\frac{x^2(x+3)}{(x+1)^3}}$$

Example B.7

A useful formula that follows from Equation B.30 is the derivative of the quotient of two functions. Show that

$$\frac{d}{dx}\left[\frac{g(x)}{h(x)}\right] = \frac{h\dfrac{dg}{dx} - g\dfrac{dh}{dx}}{h^2}$$

SOLUTION

We can write the quotient as gh^{-1} and then apply Equations B.29 and B.30:

$$\frac{d}{dx}\left(\frac{g}{h}\right) = \frac{d}{dx}(gh^{-1}) = g\frac{d}{dx}(h^{-1}) + h^{-1}\frac{d}{dx}(g)$$

$$= -gh^{-2}\frac{dh}{dx} + h^{-1}\frac{dg}{dx}$$

$$= \frac{h\dfrac{dg}{dx} - g\dfrac{dh}{dx}}{h^2}$$

◄B.7 | Integral Calculus

We think of integration as the inverse of differentiation. As an example, consider the expression

$$f(x) = \frac{dy}{dx} = 3ax^2 + b \qquad \text{B.34} ◄$$

which was the result of differentiating the function

$$y(x) = ax^3 + bx + c$$

in Example B.4. We can write Equation B.34 as $dy = f(x)\,dx = (3ax^2 + b)\,dx$ and obtain $y(x)$ by "summing" over all values of x. Mathematically, we write this inverse operation as

$$y(x) = \int f(x)\,dx$$

For the function $f(x)$ given by Equation B.34, we have

$$y(x) = \int (3ax^2 + b)\,dx = ax^3 + bx + c$$

where c is a constant of the integration. This type of integral is called an *indefinite integral* because its value depends on the choice of c.

A general **indefinite integral** $I(x)$ is defined as

$$I(x) = \int f(x)\,dx \qquad \text{B.35} ◄$$

where $f(x)$ is called the *integrand* and $f(x) = dI(x)/dx$.

For a *general continuous* function $f(x)$, the integral can be described as the area under the curve bounded by $f(x)$ and the x axis, between two specified values of x, say, x_1 and x_2, as in Figure B.15.

The area of the blue element in Figure B.15 is approximately $f(x_i)\,\Delta x_i$. If we sum all these area elements between x_1 and x_2 and take the limit of this sum as $\Delta x_i \to 0$, we obtain the *true* area under the curve bounded by $f(x)$ and the x axis, between the limits x_1 and x_2:

$$\text{Area} = \lim_{\Delta x_i \to 0} \sum_i f(x_i)\,\Delta x_i = \int_{x_1}^{x_2} f(x)\,dx \qquad \text{B.36} ◄$$

Integrals of the type defined by Equation B.36 are called **definite integrals.**

One common integral that arises in practical situations has the form

$$\int x^n\,dx = \frac{x^{n+1}}{n+1} + c \quad (n \neq -1) \qquad \text{B.37} ◄$$

This result is obvious, being that differentiation of the right-hand side with respect to x gives $f(x) = x^n$ directly. If the limits of the integration are known, this integral becomes a *definite integral* and is written

$$\int_{x_1}^{x_2} x^n\,dx = \frac{x^{n+1}}{n+1}\bigg|_{x_1}^{x_2} = \frac{x_2^{\,n+1} - x_1^{\,n+1}}{n+1} \quad (n \neq -1) \qquad \text{B.38} ◄$$

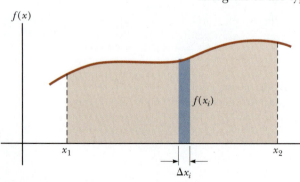

$f(x)$

$f(x_i)$

x_1 x_2

Δx_i

Figure B.15 The definite integral of a function is the area under the curve of the function between the limits x_1 and x_2.

Examples |

1. $\displaystyle\int_0^a x^2 \, dx = \frac{x^3}{3}\bigg|_0^a = \frac{a^3}{3}$

3. $\displaystyle\int_3^5 x \, dx = \frac{x^2}{2}\bigg|_3^5 = \frac{5^2 - 3^2}{2} = 8$

2. $\displaystyle\int_0^b x^{3/2} \, dx = \frac{x^{5/2}}{5/2}\bigg|_0^b = \frac{2}{5} b^{5/2}$

Partial Integration

Sometimes it is useful to apply the method of *partial integration* (also called "integrating by parts") to evaluate certain integrals. This method uses the property

$$\int u \, dv = uv - \int v \, du \qquad\qquad \text{B.39} \blacktriangleleft$$

where u and v are *carefully* chosen so as to reduce a complex integral to a simpler one. In many cases, several reductions have to be made. Consider the function

$$I(x) = \int x^2 \, e^x \, dx$$

which can be evaluated by integrating by parts twice. First, if we choose $u = x^2$, $v = e^x$, we obtain

$$\int x^2 \, e^x \, dx = \int x^2 \, d(e^x) = x^2 \, e^x - 2\int e^x \, x \, dx + c_1$$

Now, in the second term, choose $u = x$, $v = e^x$, which gives

$$\int x^2 \, e^x \, dx = x^2 \, e^x - 2x \, e^x + 2\int e^x \, dx + c_1$$

or

$$\int x^2 \, e^x \, dx = x^2 \, e^x - 2xe^x + 2e^x + c_2$$

The Perfect Differential

Another useful method to remember is that of the *perfect differential,* in which we look for a change of variable such that the differential of the function is the differential of the independent variable appearing in the integrand. For example, consider the integral

$$I(x) = \int \cos^2 x \sin x \, dx$$

This integral becomes easy to evaluate if we rewrite the differential as $d(\cos x) = -\sin x \, dx$. The integral then becomes

$$\int \cos^2 x \sin x \, dx = -\int \cos^2 x \, d(\cos x)$$

If we now change variables, letting $y = \cos x$, we obtain

$$\int \cos^2 x \sin x \, dx = -\int y^2 \, dy = -\frac{y^3}{3} + c = -\frac{\cos^3 x}{3} + c$$

Table B.5 lists some useful indefinite integrals. Table B.6 gives Gauss's probability integral and other definite integrals. A more complete list can be found in various handbooks, such as *The Handbook of Chemistry and Physics* (Boca Raton, FL: CRC Press, published annually).

TABLE B.5 | Some Indefinite Integrals (An arbitrary constant should be added to each of these integrals.)

$$\int x^n \, dx = \frac{x^{n+1}}{n+1} \text{ (provided } n \neq 1)$$

$$\int \ln ax \, dx = (x \ln ax) - x$$

$$\int \frac{dx}{x} = \int x^{-1} \, dx = \ln x$$

$$\int xe^{ax} \, dx = \frac{e^{ax}}{a^2} (ax - 1)$$

$$\int \frac{dx}{a + bx} = \frac{1}{b} \ln (a + bx)$$

$$\int \frac{dx}{a + be^{cx}} = \frac{x}{a} - \frac{1}{ac} \ln (a + be^{cx})$$

$$\int \frac{x \, dx}{a + bx} = \frac{x}{b} - \frac{a}{b^2} \ln (a + bx)$$

$$\int \sin ax \, dx = -\frac{1}{a} \cos ax$$

$$\int \frac{dx}{x(x + a)} = -\frac{1}{a} \ln \frac{x + a}{x}$$

$$\int \cos ax \, dx = \frac{1}{a} \sin ax$$

$$\int \frac{dx}{(a + bx)^2} = -\frac{1}{b(a + bx)}$$

$$\int \tan ax \, dx = -\frac{1}{a} \ln (\cos ax) = \frac{1}{a} \ln (\sec ax)$$

$$\int \frac{dx}{a^2 + x^2} = \frac{1}{a} \tan^{-1} \frac{x}{a}$$

$$\int \cot ax \, dx = \frac{1}{a} \ln (\sin ax)$$

$$\int \frac{dx}{a^2 - x^2} = \frac{1}{2a} \ln \frac{a + x}{a - x} \left(a^2 - x^2 > 0 \right)$$

$$\int \sec ax \, dx = \frac{1}{a} \ln (\sec ax + \tan ax) = \frac{1}{a} \ln \left[\tan \left(\frac{ax}{2} + \frac{\pi}{4} \right) \right]$$

$$\int \frac{dx}{x^2 - a^2} = \frac{1}{2a} \ln \frac{x - a}{x + a} \, (x^2 - a^2 > 0)$$

$$\int \csc ax \, dx = \frac{1}{a} \ln (\csc ax - \cot ax) = \frac{1}{a} \ln \left(\tan \frac{ax}{2} \right)$$

$$\int \frac{x \, dx}{a^2 \pm x^2} = \pm \frac{1}{2} \ln (a^2 \pm x^2)$$

$$\int \sin^2 ax \, dx = \frac{x}{2} - \frac{\sin 2ax}{4a}$$

$$\int \frac{dx}{\sqrt{a^2 - x^2}} = \sin^{-1} \frac{x}{a} = -\cos^{-1} \frac{x}{a} \, (a^2 - x^2 > 0)$$

$$\int \cos^2 ax \, dx = \frac{x}{2} + \frac{\sin 2ax}{4a}$$

$$\int \frac{dx}{\sqrt{x^2 \pm a^2}} = \ln (x + \sqrt{x^2 \pm a^2})$$

$$\int \frac{dx}{\sin^2 ax} = -\frac{1}{a} \cot ax$$

$$\int \frac{x \, dx}{\sqrt{a^2 - x^2}} = -\sqrt{a^2 - x^2}$$

$$\int \frac{dx}{\cos^2 ax} = \frac{1}{a} \tan ax$$

$$\int \frac{x \, dx}{\sqrt{x^2 \pm a^2}} = \sqrt{x^2 \pm a^2}$$

$$\int \tan^2 ax \, dx = \frac{1}{a} (\tan ax) - x$$

(Continued)

TABLE B.5 | Some Indefinite Integrals (continued)

$$\int \sqrt{a^2 - x^2}\, dx = \tfrac{1}{2}\left(x\sqrt{a^2 - x^2} + a^2 \sin^{-1} \frac{x}{|a|}\right)$$

$$\int \cot^2 ax\, dx = -\frac{1}{a}(\cot ax) - x$$

$$\int x\sqrt{a^2 - x^2}\, dx = -\tfrac{1}{3}(a^2 - x^2)^{3/2}$$

$$\int \sin^{-1} ax\, dx = x(\sin^{-1} ax) + \frac{\sqrt{1 - a^2 x^2}}{a}$$

$$\int \sqrt{x^2 \pm a^2}\, dx = \tfrac{1}{2}\left[x\sqrt{x^2 \pm a^2} \pm a^2 \ln(x + \sqrt{x^2 \pm a^2})\right]$$

$$\int \cos^{-1} ax\, dx = x(\cos^{-1} ax) - \frac{\sqrt{1 - a^2 x^2}}{a}$$

$$\int x(\sqrt{x^2 \pm a^2})\, dx = \tfrac{1}{3}(x^2 \pm a^2)^{3/2}$$

$$\int \frac{dx}{(x^2 + a^2)^{3/2}} = \frac{x}{a^2 \sqrt{x^2 + a^2}}$$

$$\int e^{ax}\, dx = \frac{1}{a} e^{ax}$$

$$\int \frac{x\, dx}{(x^2 + a^2)^{3/2}} = -\frac{1}{\sqrt{x^2 + a^2}}$$

TABLE B.6 | Gauss's Probability Integral and Other Definite Integrals

$$\int_0^\infty x^n\, e^{-ax}\, dx = \frac{n!}{a^{n+1}}$$

$$I_0 = \int_0^\infty e^{-ax^2}\, dx = \frac{1}{2}\sqrt{\frac{\pi}{a}} \quad \text{(Gauss's probability integral)}$$

$$I_1 = \int_0^\infty x e^{-ax^2}\, dx = \frac{1}{2a}$$

$$I_2 = \int_0^\infty x^2 e^{-ax^2}\, dx = -\frac{dI_0}{da} = \frac{1}{4}\sqrt{\frac{\pi}{a^3}}$$

$$I_3 = \int_0^\infty x^3 e^{-ax^2}\, dx = -\frac{dI_1}{da} = \frac{1}{2a^2}$$

$$I_4 = \int_0^\infty x^4 e^{-ax^2}\, dx = \frac{d^2 I_0}{da^2} = \frac{3}{8}\sqrt{\frac{\pi}{a^5}}$$

$$I_5 = \int_0^\infty x^5 e^{-ax^2}\, dx = \frac{d^2 I_1}{da^2} = \frac{1}{a^3}$$

$$\vdots$$

$$I_{2n} = (-1)^n \frac{d^n}{da^n} I_0$$

$$I_{2n+1} = (-1)^n \frac{d^n}{da^n} I_1$$

5. Path A is isovolumetric, path B is adiabatic, path C is isothermal, and path D is isobaric.

6. (i) (a) (ii) (c)

7. (b)

Answers to Odd-Numbered Problems

1. 0.281°C
3. (a) 2.26×10^6 J (b) 2.80×10^4 steps (c) 6.99×10^3 steps
5. 87.0°C
7. 1.78×10^4 kg
9. 23.6°C
11. 29.6°C
13. 1.22×10^5 J
15. (a) 0°C (b) 114 g
17. 0.294 g
19. 0.415 kg
21. -1.18 MJ
23. -466 J
25. (a) -12.0 MJ (b) $+12.0$ MJ
27. 720 J
29.

Process	Q	W	ΔE_{int}
BC	$-$	0	$-$
CA	$-$	$+$	$-$
AB	$+$	$-$	$+$

31. (a) 7.50 kJ (b) 900 K
33. (a) $-0.048\,6$ J (b) 16.2 kJ (c) 16.2 kJ
35. (a) -9.08 kJ (b) 9.08 kJ
37. (a) $0.041\,0$ m³ (b) $+5.48$ kJ (c) -5.48 kJ
39. (a) 3.46 kJ (b) 2.45 kJ (c) -1.01 kJ
41. 74.8 J
43. (a) 719 J/kg · °C (b) 0.811 kg (c) 233 kJ (d) 327 kJ
45. (a) 0 (b) 209 J (c) 317 K
47. (a)

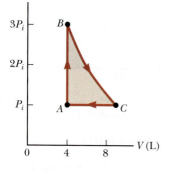

(b) 8.77 L (c) 900 K (d) 300 K (e) -336 J
49. (a) a factor of 0.118 (b) a factor of 2.35 (c) 0 (d) 135 J (e) 135 J
51. 227 K
53. (a) 2.45×10^{-4} m³ (b) 9.97×10^{-3} mol (c) 9.01×10^5 Pa (d) 5.15×10^{-5} m³ (e) 560 K (f) 53.9 J (g) 6.79×10^{-6} m³ (h) 53.3 g (i) 2.24 K
55. 2.32×10^{-21} J
57. (a) 41.6 J/K (b) 58.2 J/K (c) 58.2 J/K, 74.8 J/K
59. 51.2°C

61. 74.8 kJ
63. 3.85×10^{26} W
65. 364 K
67. (a) 13.0°C (b) -0.532°C/s
69. (a) 17.2 L (b) 0.351 L/s
71. 1.90×10^3 J/kg · °C
73. (a) 9.31×10^{10} J (b) -8.47×10^{12} J (c) 8.38×10^{12} J
75. 2.27×10^3 m
77. (5.87×10^4)°C
79. (a) 2 000 W (b) 4.46°C
81. 38.6 m³/day
83. (i) (a) 100 kPa (b) 66.5 L (c) 400 K (d) $+5.82$ kJ (e) $+7.48$ kJ (f) -1.66 kJ (ii) (a) 133 kPa (b) 49.9 L (c) 400 K (d) $+5.82$ kJ (e) $+5.82$ kJ (f) 0 (iii) (a) 120 kPa (b) 41.6 L (c) 300 K (d) 0 (e) -909 J (f) $+909$ J (iv) (a) 120 kPa (b) 43.3 L (c) 312 K (d) $+722$ J (e) 0 (f) $+722$ J
85. (a) 300 K (b) 1.00 atm
87. 0.480°C
89. 3.76 m/s

CHAPTER 18

Answers to Quick Quizzes

1. (i) (c) (ii) (b)
2. C, B, A
3. (d)
4. (a) one (b) six
5. (b)
6. (a)
7. false

Answers to Odd-Numbered Problems

1. (a) 6.94% (b) 335 J
3. 55.4%
5. (a) 10.7 kJ (b) 0.533 s
7. (a) 67.2% (b) 58.8 kW
9. (a) 8.70×10^8 J (b) 3.30×10^8 J
11. (a) 741 J (b) 459 J
13. 33.0%
15. (a) 564 K (b) 212 kW (c) 47.5%
17. (a) 24.0 J (b) 144 J
19. 1.86
21. 9.00
23. 11.8
25. 1.17 J
27. (a) isobaric (b) 402 kJ (c) 1.20 kJ/K
29. 195 J/K
31. (a)

Macrostate	Microstates	Number of Ways to Draw
All R	RRR	1
2 R, 1G	GRR, RGR, RRG	3
1 R, 2G	GGR, GRG, RGG	3
All G	GGG	1

(b)

Macrostate	Microstates	Number of Ways to Draw
All R	RRRR	1
4R, 1G	GRRRR, RGRRR, RRGRR, RRRGR, RRRRG	5
3R, 2G	GGRRR, GRGRR, GRRGR, GRRRG, RGGRR, RGRGR, RGRRG, RRGGR, RRGRG, RRRGG	10
2R, 3G	RRGGG, RGRGG, RGGRG, RGGGR, GRRGG, GRGRG, GRGGR, GGRRG, GGRGR, GGGRR	10
1R, 4G	RGGGG, GRGGG, GGRGG, GGGRG, GGGGR	5
All G	GGGGG	1

33. (a) -3.45 J/K (b) $+8.06$ J/K (c) $+4.62$ J/K
35. (a) one (b) six
37. 0.507 J/K
39. 1.02 kJ/K
41. (a) 5.76 J/K (b) no change in temperature
43. (a) 5.2×10^{17} J (b) 1.8×10^{3} s
45. (a) 5.00 kW (b) 763 W
47. (a) $3nRT_i$ (b) $3nRT_i \ln 2$ (c) $-3nRT_i$ (d) $-nRT_i \ln 2$ (e) $3nRT_i (1 + \ln 2)$ (f) $2nrT_i \ln 2$ (g) 0.273
49. 32.9 kJ
51. (a) 39.4 J (b) 65.4 rad/s = 625 rev/min (c) 293 rad/s = 2.79×10^{3} rev/min
53. 5.97×10^{4} kg/s
55. (a) 8.48 kW (b) 1.52 kW (c) 1.09×10^{4} J/K (d) drop by 20.0%
57. 23.1 mW
59. 1.18 J/K
61. (a) $P_A = 25.0$ atm, $V_A = 1.97 \times 10^{-3}$ m³; $P_B = 4.13$ atm, $V_B = 1.19 \times 10^{-2}$ m³; $P_C = 1.00$ atm, $V_C = 3.28 \times 10^{-2}$ m³; $P_D = 6.05$ atm, $V_D = 5.43 \times 10^{-3}$ m³ (b) 2.99×10^{3} J
63. (a) 4.10×10^{3} J (b) 1.42×10^{4} J (c) 1.01×10^{4} J (d) 28.8% (e) Because $e_C = 80.0\%$, the efficiency of the cycle is much lower than that of a Carnot engine operating between the same temperature extremes.
65. 8.36×10^{6} J/K · s

Context 5 Conclusion

1. 298 K
2. 60 km
3. (c) 336 K (d) The troposphere and stratosphere are too thick to be accurately modeled as having uniform temperatures. (e) 227 K (f) 107 (g) The multilayer model should be better for Venus than for the Earth. There are many layers, so the temperature of each can reasonably be modeled as uniform.

CHAPTER 19

Answers to Quick Quizzes

1. (a), (c), (e)
2. (e)
3. (b)
4. (a)
5. A, B, C
6. (b) and (d)
7. (i) (c) (ii) (d)

Answers to Odd-Numbered Problems

1. (a) $+1.60 \times 10^{-19}$ C, 1.67×10^{-27} kg
(b) $+1.60 \times 10^{-19}$ C, 3.82×10^{-26} kg
(c) -1.60×10^{-19} C, 5.89×10^{-26} kg
(d) $+3.20 \times 10^{-19}$ C, 6.65×10^{-26} kg
(e) -4.80×10^{-19} C, 2.33×10^{-26} kg
(f) $+6.40 \times 10^{-19}$ C, 2.33×10^{-26} kg
(g) $+1.12 \times 10^{-18}$ C, 2.33×10^{-26} kg
(h) -1.60×10^{-19} C, 2.99×10^{-26} kg
3. $\sim 10^{26}$ N
5. (a) 2.16×10^{-5} N toward the other (b) 8.99×10^{-7} N away from the other
7. (a) 0.951 m (b) yes, if the third bead has positive charge
9. 0.872 N at 330°
11. (a) 8.24×10^{-8} N (b) 2.19×10^{6} m/s
13. 2.25×10^{-9} N/m
15. (a) 6.64×10^{6} N/C (b) 2.41×10^{7} N/C (c) 6.39×10^{6} N/C (d) 6.64×10^{5} N/C
17. 1.82 m to the left of the -2.50-μC charge
19. (a) $(-0.599\hat{\mathbf{i}} - 2.70\hat{\mathbf{j}})$ kN/C (b) $(-3.00\hat{\mathbf{i}} - 13.5\hat{\mathbf{j}})$ μN
21. (a) 2.16×10^{7} N/C (b) to the left
23. (a) 1.59×10^{6} N/C (b) toward the rod
25. (a) 2.00×10^{-10} C (b) 1.41×10^{-10} C (c) 5.89×10^{-11} C
27. (a) $\dfrac{k_e q}{a^2}(3.06\hat{\mathbf{i}} + 5.06\hat{\mathbf{j}})$ (b) $\dfrac{k_e q^2}{a^2}(3.06\hat{\mathbf{i}} + 5.06\hat{\mathbf{j}})$
29.

31. (a) $-5.76 \times 10^{13}\hat{\mathbf{i}}$ m/s² (b) $\vec{\mathbf{v}}_i = 2.84 \times 10^{6}\hat{\mathbf{i}}$ m/s (c) 4.93×10^{-8} s
33. (a) 6.13×10^{10} m/s² (b) 1.96×10^{-5} s (c) 11.7 m (d) 1.20×10^{-15} J
35. (a) 111 ns (b) 5.68 mm (c) $(450\hat{\mathbf{i}} + 102\hat{\mathbf{j}})$ km/s
37. 4.14 MN/C
39. (a) -6.89 MN · m²/C (b) less than
41. -18.8 kN · m²/C
43. (a) -55.7 nC (b) negative, spherically symmetric
45. (a) 0 (b) 3.65×10^{5} N/C (c) 1.46×10^{6} N/C (d) 6.49×10^{5} N/C
47. 3.50 kN
49. -2.48 μC/m²
51. 508 kN/C up
53. (a) 0 (b) 7.19 MN/C away from the center
55. (a) 0 (b) 5.39×10^{3} N/C outward (c) 539 N/C outward

57. (a) 0 (b) 79.9 MN/C radially outward (c) 0
(d) 7.34 MN/C radially outward
59. (a) 80.0 nC/m² (b) $9.04\hat{\mathbf{k}}$ kN/C (c) $-9.04\hat{\mathbf{k}}$ kN/C
61. (a) 708 nC/m², positive on the face whose normal is in
the same direction as the electric field, negative on the
other face (b) 177 nC, positive on the face whose normal is in the same direction as the electric field, negative on the other face
63. 1.77×10^{-12} C/m³; positive
65. (a) 40.8 N (b) 263°
67. 5.25 μC
69. 1.67×10^{-5} C
71. $(-1.36\hat{\mathbf{i}} + 1.96\hat{\mathbf{j}})$ kN/C
73. (a) 0 (b) $\dfrac{\sigma}{\epsilon_0}$ to the right (c) 0
(d) (1) $\dfrac{\sigma}{\epsilon_0}$ to the left (2) 0 (3) $\dfrac{\sigma}{\epsilon_0}$ to the right
75. (a) $Q\left(\dfrac{r}{a}\right)^3$ (b) $k_e\left(\dfrac{Qr}{a^3}\right)$ (c) Q (d) $k_e\left(\dfrac{Q}{r^2}\right)$ (e) $E = 0$ (f) $-Q$
(g) $+Q$ (h) inner surface of radius b
77. 0.205 μC

CHAPTER 20

Answers to Quick Quizzes

1. (i) (b) (ii) (a)
2. Ⓑ to Ⓒ, Ⓒ to Ⓓ, Ⓐ to Ⓑ, Ⓓ to Ⓔ
3. (i) (b) (ii) (c)
4. (i) (c) (ii) (a)
5. (i) (a) (ii) (a)
6. (d)
7. (a)
8. (b)
9. (a)

Answers to Odd-Numbered Problems

1. +260 V
3. (a) 1.52×10^5 m/s (b) 6.49×10^6 m/s
5. (a) −38.9 V (b) the origin
7. (a) 1.44×10^{-7} V (b) -7.19×10^{-8} V
(c) -1.44×10^{-7} V, $+7.19 \times 10^{-8}$ V
9. (a) 0 (b) 0 (c) 44.9 kV
11. -1.10×10^7 V
13. 8.94 J
15. (a) 10.8 m/s and 1.55 m/s (b) They would be greater.
The conducting spheres will polarize each other, with
most of the positive charge of one and the negative
charge of the other on their inside faces. Immediately
before the spheres collide, their centers of charge will
be closer than their geometric centers, so they will have
less electric potential energy and more kinetic energy.
17. (a) 32.2 kV (b) −0.096 5 J
19. (a) −45.0 μV (b) 34.6 km/s
21. (a) −4.83 m (b) 0.667 m and −2.00 m
23. (a) $\vec{\mathbf{E}} = (-5 + 6xy)\hat{\mathbf{i}} + (3x^2 - 2z^2)\hat{\mathbf{j}} - 4yz\hat{\mathbf{k}}$ (b) 7.07 N/C
25. $-0.553 k_e \dfrac{Q}{R}$

27. $\dfrac{1}{2}k_e\alpha L \ln\left(\dfrac{\sqrt{4b^2 + L^2} + L}{\sqrt{4b^2 + L^2} - L}\right)$
29. −1.51 MV
31. (a) 1.35×10^5 V (b) larger sphere: 2.25×10^6 V/m
(away from the center); smaller sphere: 6.74×10^6 V/m
(away from the center)
33. (a) 48.0 μC (b) 6.00 μC
35. (a) 1.33 μC/m² (b) 13.4 pF
37. (a) 11.1 kV/m toward the negative plate (b) 98.4 nC/m²
(c) 3.74 pF (d) 74.8 pC
39. (a) 2.69 nF (b) 3.02 kV
41. (a) 11.1 nF (b) 26.6 C
43. (a) 3.53 μF (b) 6.35 V on 5.00 μF, 2.65 V on 12.0 μF
(c) 31.8 μC on each capacitor
45. (a) 5.96 μF
(b) 89.5 μC on 20 μF, 63.2 μC on 6 μF, and 26.3 μC on
15 μF and 3 μF
47. (a) in series (b) 398 μF (b) in parallel; 2.20 μF
49. ten
51. 12.9 μF
53. (a) 1.50 μC (b) 1.83 kV
55. (a) 2.50×10^{-2} J (b) 66.7 V (c) 3.33×10^{-2} J (d) Positive
work is done by the agent pulling the plates apart.
57. (a)

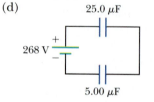

(b) 0.150 J (c) 268 V
(d)

59. (a) 216 μJ (b) 54.0 μJ
61. (a) 2.51×10^{-3} m³ = 2.51 L
63. (a) 81.3 pF (b) 2.40 kV
65. (a) 13.3 nC (b) 272 nC
67. $-9.43 \times 10^{-2}\hat{\mathbf{i}}$ N
69. (a) $6.00\hat{\mathbf{i}}$ m/s (b) 3.64 m (c) $-9.00\hat{\mathbf{i}}$ m/s (d) $12.0\hat{\mathbf{i}}$ m/s
71. (a) volume 9.09×10^{-16} m³, area 4.54×10^{-10} m²
(b) 2.01×10^{-13} F (c) 2.01×10^{-14} C; 1.26×10^5 electronic charges
73. 253 MeV
75. 579 V
77. 702 J
79. (a) 40.0 μJ (b) 500 V
81. 0.188 m²
83. 23.3 V across the 5.00-μF capacitor, 26.7 V across the
10.0-μF capacitor
85. (a) $\dfrac{\epsilon_0\ell}{d}[\ell + x(\kappa - 1)]$ (b) $\dfrac{Q^2 d}{2\epsilon_0\ell[\ell + x(\kappa - 1)]}$
(c) $\dfrac{Q^2 d(\kappa - 1)}{2\epsilon_0\ell[\ell + x(\kappa - 1)]^2}\hat{\mathbf{i}}$ (d) $205\hat{\mathbf{i}}$ μN
87. $\dfrac{4}{3}C$

CHAPTER 21

Answers to Quick Quizzes

1. (a) > (b) = (c) > (d)
2. (b)
3. (a)
4. $I_a = I_b > I_c = I_d > I_e = I_f$
5. (b)
6. (a)
7. (i) (b) (ii) (a) (iii) (a) (iv) (b)
8. (i) (c) (ii) (d)

Answers to Odd-Numbered Problems

1. 7.50×10^{15} electrons
3. (a) 17.0 A (b) 85.0 kA/m^2
5. (a) 2.55 A/m^2 (b) 5.30×10^{10} m^{-3} (c) 1.21×10^{10} s
7. 0.129 mm/s
9. (a) 31.5 n$\Omega \cdot$ m (b) 6.35 MA/m^2 (c) 49.9 mA
   (d) 658 μm/s (e) 0.400 V
11. (a) 1.82 m (b) 280 μm
13. 1.98 A
15. (a) no change (b) doubles (c) doubles (d) no change
17. 0.18 V/m
19. 448 A
21. (a) 184 W (b) 461°C
23. $0.319
25. 36.1%
27. $0.494/day
29. 5.00 A, 24.0 Ω
31. (a) 667 A (b) 50.0 km
33. (a) 6.73 Ω (b) 1.97 Ω
35. (a) 12.4 V (b) 9.65 V
37. (a) 17.1 Ω
    (b) 1.99 A for 4 Ω and 9 Ω, 1.17 A for 7 Ω, 0.818 A for 10 Ω
39. (a) 227 mA (b) 5.68 V
41. (a) 75.0 V (b) 25.0 W, 6.25 W, and 6.25 W (c) 37.5 W
43. 14.2 W to 2.00 Ω, 28.4 W to 4.00 Ω, 1.33 W to 3.00 Ω, 4.00 W to 1.00 Ω
45. (a) 0.714 A (b) 1.29 A (c) 12.6 V
47. (a) 0.846 A down in the 8.00-Ω resistor, 0.462 A down in the middle branch, 1.31 A up in the right-hand branch (b) -222 J by the 4.00-V battery, 1.88 kJ by the 12.0-V battery
    (c) 687 J to 8.00 Ω, 128 J to 5.00 Ω, 25.6 J to the 1.00-Ω resistor in the center branch, 616 J to 3.00 Ω, 205 J to the 1.00-Ω resistor in the right branch
    (d) Chemical energy in the 12.0-V battery is transformed into internal energy in the resistors. The 4.00-V battery is being charged, so its chemical potential energy is increasing at the expense of some of the chemical potential energy in the 12.0-V battery.
    (e) 1.66 kJ
49. 50.0 mA from a to e
51. 1.00 A upward in 200 Ω; 4.00 A upward in 70.0 Ω; 3.00 A upward in 80.0 Ω; 8.00 A downward in 20.0 Ω (b) 200 V
53. (a) 5.00 s (b) 150 μC (c) 4.06 μA
55. (a) -61.6 mA (b) 0.235 μC (c) 1.96 A
57. (a) 1.50 s (b) 1.00 s (c) $I = 200 + 100e^{-t}$, where I is in microamperes and t is in seconds
59. (a) 6.00 V (b) 8.29 μs
61. $6.00 \times 10^{-15}/\Omega \cdot$ m
63. (a) 8.00 V/m in the positive x direction (b) 0.637 Ω
    (c) 6.28 A in the positive x direction (d) 200 MA/m^2
65. 2.22 h
67. (a) 9.93 μC (b) 33.7 nA (c) 335 nW (d) 337 nW
69. (a) 222 μC (b) 444 μC
71. (a) $R_x = R_2 - \frac{1}{4}R_1$ (b) No; $R_x = 2.75$ Ω, so the station is inadequately grounded.
73. (a) $R \to \infty$ (b) $R \to 0$ (c) $R = r$
75. (a) For the heater, 12.5 A; for the toaster 6.25 A; for the grill, 8.33A (b) The current draw is greater than 25.0 amps, so this circuit will trip the circuit breaker.
77. (a) 0.991 (b) 0.648 (c) The energy flows are precisely analogous to the currents in parts (a) and (b). The ceiling has the smallest thermal resistance of the thermal resistors in parallel, so increasing its thermal resistance will produce the biggest reduction in the total energy flow.

Context 6 Conclusion

1. (a) 87.0 s (b) 261 s (c) $t \to \infty$
2. (a) 0.01 s (b) 7×10^6
3. (a) 3×10^6 (b) 9×10^6

CHAPTER 22

Answers to Quick Quizzes

1. (e)
2. (i) (b) (ii) (a)
3. (c)
4. $B > C > A$
5. (a)
6. $c > a > d > b$
7. $a = c = d > b = 0$
8. (c)

Answers to Odd-Numbered Problems

1. (a) 8.67×10^{-14} N (b) 5.19×10^{13} m/s^2
3. (a) 7.91×10^{-12} N (b) zero
5. (a) west (b) zero deflection (c) up (d) down
7. $-20.9\hat{j}$ mT
9. (a) 5.00 cm (b) 8.79×10^6 m/s
11. 115 keV
13. 244 kV/m
15. 0.278 m
17. 70.0 mT
19. $-2.88\hat{j}$ N
21. (a) east (b) 0.245 T
23. (a) 4.73 N (b) 5.46 N (c) 4.73 N
25. (a) 9.98 N $\cdot$ m (b) clockwise as seen looking down from a position on the positive y axis
27. (a) 5.41 mA $\cdot$ m^2 (b) 4.33 mN $\cdot$ m
29. 1.60×10^{-6} T
31. 5.52 μT into the page
33. (a) at $y = -0.420$ m (b) $-3.47 \times 10^{-2}\hat{j}$ N
    (c) $-1.73 \times 10^4\hat{j}$ N/C
35. 262 nT into the page

37. (a) 53.3 μT toward the bottom of the page (b) 20.0 μT toward the bottom of the page (c) zero

39. (a) 21.5 mA (b) 4.51 V (c) 96.7 mW

41. (a) 28.3 μT into the page (b) 24.7 μT into the page

43. $-27.0\hat{\mathbf{i}}$ μN

45. (a) 10 μT (b) 80 μN toward the other wire (c) 16 μT (d) 80 μN toward the other wire

47. (a) 200 μT toward the top of the page (b) 133 μT toward the bottom of the page

49. (a) 3.60 T (b) 1.94 T

51. (a) 4.00 m (b) 7.50 nT (c) 1.26 m (d) zero

53. 5.40 cm

55. (a) 226 μN away from the center of the loop (b) zero

57. 31.8 mA

59. (a) 8.63×10^{45} electrons (b) 4.01×10^{20} kg

61. (a) 4.0×10^{-3} N $\cdot$ m (b) -6.9×10^{-3} J

63. (a) $(3.52\hat{\mathbf{i}} - 1.60\hat{\mathbf{j}}) \times 10^{-18}$ N (b) 24.4°

65. 2.75 Mrad/s

67. 39.2 mT

69. (a) 2.46 N upward (b) Equation 22.23 is the expression for the magnetic field produced a distance x above the center of the loop. The magnetic field at the center of the loop or on its axis is much weaker than the magnetic field just outside the wire. The wire has negligible curvature on the scale of 1 mm, so we model the lower loop as a long straight wire to find the field it creates at the location of the upper wire (c) 107 m/s² upward

71. (a) 1.79×10^{-8} s (b) 35.1 eV

73. (a) 1.33 m/s (b) Positive ions carried by the blood flow experience an upward force resulting in the upper wall of the blood vessel at electrode A becoming positively charged and the lower wall of the blood vessel at electrode B becoming negatively charged (c) No. Negative ions moving in the direction of v would be deflected toward point B, giving A a higher potential than B. Positive ions moving in the direction of v would be deflected toward A, again giving A a higher potential than B. Therefore, the sign of the potential difference does not depend on whether the ions in the blood are positively or negatively charged.

75. (a) $\alpha = 8.90°$ (b) -8.00×10^{-21} kg $\cdot$ m/s

77. 143 pT

79. (a) $B \sim 10^{-1}$ T (b) $\tau \sim 10^{-1}$ N $\cdot$ m (c) $I \sim 1$ A $= 10^0$ A (d) $A \sim 10^{-3}$ m² (e) $N \sim 10^3$

81. $\dfrac{\mu_0 I}{2\pi w} \ln\left(1 + \dfrac{w}{b}\right)\hat{\mathbf{k}}$

CHAPTER 23

Answers to Quick Quizzes

1. (c)

2. $c, d = e, b, a$

3. (b)

4. (c)

5. (b)

6. (d)

7. (b)

8. (b)

9. (a), (d)

Answers to Odd-Numbered Problems

1. 61.8 mV

3. 0.800 mA

5. (a) 1.60 A counterclockwise (b) 20.1 μT (c) left

7. (a) $\dfrac{\mu_0 IL}{2\pi} \ln\left(1 + \dfrac{w}{h}\right)$ (b) 4.80 μV (c) counterclockwise

9. +9.82 mV

11. 160 A

13. $\varepsilon = -(1.42 \times 10^{-2}) \cos(120t)$, where t is in seconds and ε is in mV

15. (a) 3.00 N to the right (b) 6.00 W

17. (a) 0.500 A (b) 2.00 W (c) 2.00 W

19. (a) 233 Hz (b) 1.98 mV

21. 24.1 V with the outer contact positive

23. $\varepsilon = 28.6 \sin 4.00\pi t$, where ε is in millivolts and t is in seconds

25. 360 T

27. (a) 7.54 kV (b) The plane of the loop is parallel to $\vec{\mathbf{B}}$.

29. (a) 8.01×10^{-21} N (b) clockwise (c) $t = 0$ or $t = 1.33$ s

31. 19.5 mV

33. $\varepsilon = -18.8 \cos 120\pi t$, where ε is in volts and t is in seconds

35. (a) 360 mV (b) 180 mV (c) 3.00 s

37. (a) 0.139 s (b) 0.461 s

39. (a) 2.00 ms (b) 0.176 A (c) 1.50 A (d) 3.22 ms

41. (a) 20.0% (b) 4.00%

43. (a) $I_L = 0.500(1 - e^{-10.0t})$, where I_L is in amperes and t is in seconds (b) $I_S = 1.50 - 0.250e^{-10.0t}$, where I_S is in amperes and t is in seconds

45. 92.8 V

47. (a) 5.66 ms (b) 1.22 A (c) 58.1 ms

49. 2.44 μJ

51. (a) 44.3 nJ/m³ (b) 995 μJ/m³

53. (a) 113 V (b) 300 V/m

55. $\varepsilon = -7.22 \cos(1\,046\pi t)$, where ε is in millivolts and t is in seconds

57. (a) 43.8 A (b) 38.3 W

59. $\sim 10^{-4}$ V, by reversing a 20-turn coil of diameter 3 cm in 0.1 s in a field of 10^{-3} T

61. 6.00 A

63. 10.2 μV

65. $\varepsilon = -87.1 \cos(200\pi t + \phi)$, where ε is in millivolts and t is in seconds

67. 91.2 μH

69. (a) $\frac{1}{2}\mu_0\pi N^2 R$ (b) 10^{-7} H (c) 10^{-9} s

71. (a) 6.25×10^{10} J (b) 2.00×10^3 N/m

73. (a) 50.0 mT (b) 20.0 mT (c) 2.29 MJ (d) 318 Pa

75. 2.29 μC

Context 7 Conclusion

1. (a) 29.2 MHz (b) 42.6 MHz (c) 2.13 kHz

2. (a) 2.47×10^3 A (b) 0.986 T $\cdot$ m² (c) 0.197 V (d) 64.0 kg

CHAPTER 24

Answers to Quick Quizzes

1. (i) (b) (ii) (c)

2. (c)

3. (d)

4. (b), (c)
5. (c)
6. (a)
7. (b)

Answers to Odd-Numbered Problems

1. (a) out of the page (b) 1.85×10^{-18} T
3. (a) 11.3 GV $\cdot$ m/s (b) 0.100 A
5. $(-2.87\hat{\mathbf{j}} + 5.75\hat{\mathbf{k}}) \times 10^9$ m/s^2
7. (a) 503 Hz (b) 12.0 μC (c) 37.9 mA (d) 72.0 μJ
9. 2.25×10^8 m/s
11. 74.9 MHz
13. (a) 6.00 MHz (b) $-73.4\hat{\mathbf{k}}$ nT
    (c) $\vec{\mathbf{B}} = -73.4\hat{\mathbf{k}} \cos(0.126x - 3.77 \times 10^7 t)$
15. 2.9×10^8 m/s $\pm 5\%$
17. $0.220c$
19. 1.13×10^4 Hz
21. (a) 13.4 m/s toward the station and 13.4 m/s away from the station (b) 0.056 7 rad/s
23. 307 μW/m^2
25. 3.34 μJ/m^3
27. 49.5 mV
29. 3.33×10^3 m^2
31. (a) 332 kW/m^2 radially inward
    (b) 1.88 kV/m and 222 μT
33. (a) 1.90 kN/C (b) 50.0 pJ (c) 1.67×10^{-19} kg $\cdot$ m/s
35. 545 THz
37. (a) 6.00 pm (b) 7.49 cm
39. 56.2 m
41. (a) 4.16 m to 4.54 m (b) 3.41 m to 3.66 m
    (c) 1.61 m to 1.67 m
43. (a) 0.690 wavelengths (b) 58.9 wavelengths
45. (a) 54.7° (b) 63.4° (c) 71.6°
47. 0.375
49. (a) six (b) 7.50°
51. (a) 2.33 mT (b) 650 MW/m^2 (c) 511 W
53. (a) 4.24×10^{15} W/m^2 (b) 1.20×10^{-12} J
55. (a) 28.3 THz (b) 10.6 μm (c) infrared
57. 3.49×10^{16} photons
59. (a) 3.85×10^{26} W (b) 1.02 kV/m and 3.39 μT
61. $\sim 10^6$ J
63. (a) 6.67×10^{-16} T (b) 5.31×10^{-17} W/m^2
    (c) 1.67×10^{-14} W (d) 5.56×10^{-23} N
65. 378 nm
67. (a) 625 kW/m^2 (b) 21.7 kV/m (c) 72.4 μT (d) 17.8 min
69. (a) 0.161 m (b) 0.163 m^2 (c) 76.8 W (d) 470 W/m^2
    (e) 595 V/m (f) 1.98 mT (g) 119 W
71. (a) 388 K (b) 363 K
73. (a) 584 nT (b) 419 m^{-1} (c) 1.26×10^{11} s^{-1}
    (d) $\vec{\mathbf{B}}$ vibrates in the xz plane. (e) $40.6\hat{\mathbf{i}}$ W/m^2
    (f) 271 nPa (g) $407\hat{\mathbf{i}}$ nm/s^2
75. (a) 22.6 h (b) 30.6 s

CHAPTER 25

Answers to Quick Quizzes

1. (d)
2. Beams ② and ④ are reflected; beams ③ and ⑤ are refracted.
3. (c)

4. (a)
5. False
6. (i) (b) (ii) (b)
7. (c)

Answers to Odd-Numbered Problems

1. 86.8°
3. (a) 1.94 m (b) 50.0° above the horizontal
5. six times from the mirror on the left and five times from the mirror on the right
7. (a) 4.74×10^{14} Hz (b) 422 nm (c) 2.00×10^8 m/s
9. 25.7°
11. 22.5°
13. (a) 2.0×10^8 m/s (b) 4.74×10^{14} Hz (c) 4.2×10^{-7} m
15. (a) 1.81×10^8 m/s (b) 2.25×10^8 m/s
    (c) 1.36×10^8 m/s
17. 3.39 m
19. $\tan^{-1} n$
21. 6.30 cm
23. 4.61°
25. 30.0° and 19.5° at entry, 40.5° and 77.1° at exit
27. (a) 27.0° (b) 37.1° (c) 49.8°
29. 27.9°
31. 1.000 07
33. 62.5°
35. 67.1°
37. 2.27 m
39. 23.1°
41. (a) 334 μs (b) 0.014 6%
43. 3.79 m
45. (a) 1.20 (b) 3.40 ns
47. 62.2%
49. (a) 0.042 6 or 4.26% (b) no difference
51. 70.6%
53. 27.5°
55. 1.93
57. 36.5°
59. (a) $\left(\dfrac{4x^2 + L^2}{L}\right)\omega$ (b) 0 (c) $L\omega$ (d) $2L\omega$ (e) $\dfrac{\pi}{8\omega}$
61. $\sin^{-1}\left[\dfrac{L}{R^2}\left(\sqrt{n^2R^2 - L^2} - \sqrt{R^2 - L^2}\right)\right]$ or
    $\sin^{-1}\left[n\sin\left(\sin^{-1}\dfrac{L}{R} - \sin^{-1}\dfrac{L}{nR}\right)\right]$

CHAPTER 26

Answers to Quick Quizzes

1. C
2. false
3. (b)
4. (b)
5. (b)
6. (c)
7. (c)

Answers to Odd-Numbered Problems

1. 89.0 cm
3. (a) $p_1 + h$, behind the lower mirror (b) virtual
   (c) upright (d) 1.00 (e) no

15. 797

17. $3\hbar$

19. (a) $\sqrt{6}\hbar = 2.58 \times 10^{-34}\,\text{J}\cdot\text{s}$
    (b) $2\sqrt{3}\hbar = 3.65 \times 10^{-34}\,\text{J}\cdot\text{s}$

21. $\ell = 4$

23. $\sqrt{6}\hbar = 2.58 \times 10^{-34}\,\text{J}\cdot\text{s}$

25. (a) 2 (b) 8 (c) 18 (d) 32 (e) 50

27. aluminum

29. (a) 30 (b) 36

31. $1s^2 2s^2 2p^6 3s^2 3p^6 3d^{10} 4s^2 4p^6 4d^{10} 4f^{14} 5s^2 5p^6 5d^{10} 5f^{14} 6s^2 6p^6 6d^8 7s^2$

33. 18.4 T

35. 0.068 nm

37. manganese

39. (a) If $\ell = 2$, then $m_\ell = 2, 1, 0, -1, -2$; if $\ell = 1$, then $m_\ell = 1, 0, -1$; if $\ell = 0$, then $m_\ell = 0$. (b) -6.05 eV

41. (a) parallel spins, with antiparallel magnetic moments
    (b) 5.87×10^{-6} eV (c) 10^{-30} eV

43. (a) $0.160c$ (b) 2.18×10^9 ly

45. (a) -8.16 eV, -2.04 eV, -0.902 eV, -0.508 eV, -0.325 eV
    (b) 1 090 nm, 811 nm, 724 nm, and 609 nm (c) 122 nm, 103 nm, 97.3 nm, 95.0 nm, 91.2 nm (d) The spectrum could be that of hydrogen, Doppler-shifted by motion away from us at speed $0.471c$.

47. (a) 486 nm (b) 0.815 m/s

49. (a) 609 μeV (b) 6.9 μeV (c) 147 GHz (d) 2.04 mm

51. (a) $1.57 \times 10^{14}\,\text{m}^{-3/2}$ (b) $2.47 \times 10^{28}\,\text{m}^{-3}$
    (c) $8.69 \times 10^8\,\text{m}^{-1}$

53. 0.386 T/m

55. (a) smaller (b) 1.46×10^{-8} u (c) 1.45×10^{-6} % (d) no

57. 0.125

59. (a) 1.63×10^{-18} J (b) 7.88×10^4 K

61. 9.80 GHz

63. $\dfrac{1}{a_0}$, no

65. 6.03 keV

CHAPTER 30

Answers to Quick Quizzes

1. (i) (b) (ii) (a) (iii) (c)
2. (c)
3. (e)
4. (e)
5. (b)
6. (c)

Answers to Odd-Numbered Problems

1. (a) 455 fm (b) 6.05×10^6 m/s
3. 8.21 cm
5. (a) 1.90 fm (b) 7.44 fm
7. 16 km
9. (a) 29.2 MHz (b) 42.6 MHz (c) 2.13 kHz
11. (a) 1.11 MeV (b) 7.07 MeV (c) 8.79 MeV
    (d) 7.57 MeV
13. (a) $^{139}_{55}\text{Cs}$ (b) $^{139}_{57}\text{La}$ (c) $^{139}_{55}\text{Cs}$
15. 9.47×10^9 nuclei
17. 1.16×10^3 s

19. 86.4 h

21. 2.66 d

23. 4.27 MeV

25. (a) $e^- + p \rightarrow n + \nu$ (b) 2.75 MeV

27.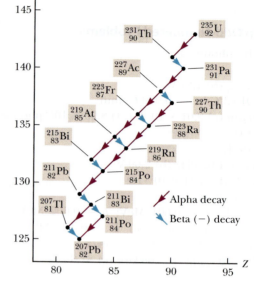

29. (a) 148 Bq/m^3 (b) 7.05×10^7 atoms/m^3
    (c) 2.17×10^{-17}

31. (a) $^{21}_{10}\text{Ne}$ (b) $^{144}_{54}\text{Xe}$ (c) $e^+ + \nu$

33. 1.02 MeV

35. 5.58×10^6 m

37. (a) 8×10^4 eV (b) 4.62 MeV and 13.9 MeV
    (c) 1.03×10^7 kWh

39. 5.94 Gyr

41. (a) 5×10^7 K (b) 1.94 MeV, 1.20 MeV, 1.02 MeV, 7.55 MeV, 7.30 MeV, 1.73 MeV, 1.02 MeV, 4.97 MeV, 26.7 MeV (c) Most of the neutrinos leave the star directly after their creation, without interacting with any other particles.

43. (a) 2.7 fm (b) 1.5×10^2 N (c) 2.6 MeV (d) $r = 7.4$ fm; $F = 3.8 \times 10^2$ N; $W = 18$ MeV

45. (a) 5.70 MeV (b) 3.27 MeV (c) exothermic

47. (a) 61.8 Bq (b) 40.3 d

49. (a) $\sim 10^{-1362}$ (b) 0.891

51. 1.66×10^3 yr

53. (a) 0.436 cm (b) 5.79 cm

55. (a) 0.963 mm (b) It increases by 7.47%.

57. 69.0 W

59. 2.7 min

61. 4.44×10^{-8} kg/h

63. 2.57×10^4 kg

65. (a) 421 MBq (b) 153 ng

67. 0.401%

69. (a) 1.5×10^{24} nuclei (b) 0.6 kg

CHAPTER 31

Answers to Quick Quizzes

1. (a)
2. (i) (c), (d) (ii) (a)
3. (b), (e), (f)
4. (b), (e)

5.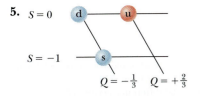

Answers to Odd-Numbered Problems

1. (a) 4.54×10^{23} Hz (b) 6.61×10^{-16} m
3. (a) 5.57×10^{14} J (b) \$$1.70 \times 10^7$
5. 118 MeV
7. (a) 67.5 MeV (b) 67.5 MeV/c (c) 1.63×10^{22} Hz
9. $\sim 10^{-18}$ m
11. (a) 0.782 MeV (b) $v_e = 0.919c$, $v_p = 380$ km/s
   (c) The electron is relativistic; the proton is not.
13. $\Omega^+ \rightarrow \overline{\Lambda}{}^0 + K^+$, $\overline{K}{}_S^0 \rightarrow \pi^+ + \pi^-$, $\overline{\Lambda}{}^0 \rightarrow \overline{p} + \pi^+$,
   $\overline{n} \rightarrow \overline{p} + e^+ + \nu_e$
15. (a) muon lepton number and electron lepton number
   (b) charge (c) baryon number (d) charge
   (e) electron lepton number
17. (a) $\overline{\nu}_\mu$ (b) ν_μ (c) $\overline{\nu}_e$ (d) ν_e (e) ν_μ (f) $\overline{\nu}_e + \nu_\mu$
19. $0.828c$
21. (a) It cannot occur because it violates baryon number conservation. (b) It can occur. (c) It cannot occur because it violates baryon number conservation. (d) It can occur. (e) It can occur. (f) It cannot occur because it violates baryon number conservation, muon lepton number conservation, and energy conservation.
23. (a) Strangeness is not conserved. (b) Strangeness is conserved. (c) Strangeness is conserved. (d) Strangeness is not conserved. (e) Strangeness is not conserved. (f) Strangeness is not conserved.

25. (a) It is not allowed because baryon number is not conserved. (b) strong interaction (c) weak interaction (d) weak interaction (e) electromagnetic interaction
27. (a) K^+ (scattering event) (b) Ξ^0 (c) π^0
29. 9.25 cm
33. $m_u = 312$ MeV/c^2; $m_d = 314$ MeV/c^2
35. The unknown particle is a neutron, udd.
37. (a) The reaction has a net of 3u, 0d, and 0s before and after (b) the reaction has a net of 1u, 1d, and 1s before and after (c) (uds) before and after (d) Λ^0 or Σ^0
39. (a) $-e$ (b) 0 (c) antiproton; antineutron
41. (a) 590.09 nm (b) 599 nm (c) 684 nm
43. 6.00
45. (a) $\sim 10^{13}$ K (b) $\sim 10^{10}$ K
47. (a) 4.30×10^{-18} m/s (b) 0.892 nm/s
49. $\sim 10^{14}$
51. 1.12 GeV/c^2
53. (a) Charge is not conserved. (b) Energy, muon lepton number, and electron lepton number are not conserved. (c) Baryon number is not conserved.
55. neutron
57. 70.4 MeV
59. (a) electron–positron annihilation; e^- (b) A neutrino collides with a neutron, producing a proton and a muon; W^+.
61. 0.407%
63. 2.52×10^3 K
65. 5.35 MeV and 32.3 MeV

Context 9 Conclusion

1. (a) 1.61×10^{-35} m (b) 5.38×10^{-44} s (c) yes

Standard Abbreviations and Symbols for Units

Symbol	Unit	Symbol	Unit
A	ampere	K	kelvin
u	atomic mass unit	kg	kilogram
atm	atmosphere	kmol	kilomole
Btu	British thermal unit	L	liter
C	coulomb	lb	pound
°C	degree Celsius	ly	light-year
cal	calorie	m	meter
d	day	min	minute
eV	electron volt	mol	mole
°F	degree Fahrenheit	N	newton
F	farad	Pa	pascal
ft	foot	rad	radian
G	gauss	rev	revolution
g	gram	s	second
H	henry	T	tesla
h	hour	V	volt
hp	horsepower	W	watt
Hz	hertz	Wb	weber
in.	inch	yr	year
J	joule	Ω	ohm

Mathematical Symbols Used in the Text and Their Meaning

Symbol	Meaning
$=$	is equal to
$\equiv$	is defined as
$\neq$	is not equal to
$\propto$	is proportional to
$\sim$	is on the order of
$>$	is greater than
$<$	is less than
$\gg (\ll)$	is much greater (less) than
$\approx$	is approximately equal to
Δx	the change in x
$\sum\limits_{i=1}^{N} x_i$	the sum of all quantities x_i from $i = 1$ to $i = N$
$|x|$	the absolute value of x (always a nonnegative quantity)
$\Delta x \to 0$	Δx approaches zero
$\dfrac{dx}{dt}$	the derivative of x with respect to t
$\dfrac{\partial x}{\partial t}$	the partial derivative of x with respect to t
$\int$	integral